Spring Cloud Alibaba 微服务架构实战派

（上册）

胡弦◎著

電子工業出版社
Publishing House of Electronics Industry
北京•BEIJING

内 容 简 介

本书覆盖了微服务架构的主要技术点，包括分布式服务治理、分布式配置管理、分布式流量防护、分布式事务处理、分布式消息处理、分布式网关、分布式链路追踪、分布式Job、分库分表、读写分离、分布式缓存、服务注册/订阅路由、全链路蓝绿发布和灰度发布。这些技术点采用“是什么→怎么用→什么原理（源码解析）”的主线来讲解。

为了方便读者在企业中落地Spring Cloud Alibaba项目，本书还包括几个相对完整的项目实战：全链路日志平台、中台架构、数据迁移平台、业务链路告警平台。

本书的目标是：①让读者在动手中学习，而不是“看书时好像全明白了，一动手却发现什么都不会”；②读者可以掌握微服务全栈技术，而不仅仅是Spring Cloud Alibaba框架，对于相关的技术（Seata、RocketMQ），基本都是从零讲起，这样避免了读者为了学会微服务技术，得找Spring Cloud Alibaba的书、Seata的书、RocketMQ的书……本书是一站式解决方案。

本书适合对微服务架构感兴趣的开发人员。无论读者是否接触过微服务开发，只要具备一定的Java开发基础，都能通过本书的学习快速掌握微服务开发技能，快速搭建出可以在企业中应用的微服务架构。

图书在版编目（CIP）数据

Spring Cloud Alibaba 微服务架构实战派. 上下册 / 胡弦著. —北京：电子工业出版社，2022.1
ISBN 978-7-121-42313-0

Ⅰ. ①S… Ⅱ. ①胡… Ⅲ. ①互联网络－网络服务器 Ⅳ. ①TP368.5

中国版本图书馆 CIP 数据核字（2021）第 226217 号

责任编辑：吴宏伟
印　　刷：三河市良远印务有限公司
装　　订：三河市良远印务有限公司
出版发行：电子工业出版社
　　　　　北京市海淀区万寿路 173 信箱　　邮编：100036
开　　本：787×980　1/16　印张：60.25　字数：1446 千字
版　　次：2022 年 1 月第 1 版
印　　次：2024 年 5 月第 5 次印刷
定　　价：236.00 元（上下册）

凡所购买电子工业出版社图书有缺损问题，请向购买书店调换。若书店售缺，请与本社发行部联系，联系及邮购电话：（010）88254888，88258888。

质量投诉请发邮件至 zlts@phei.com.cn，盗版侵权举报请发邮件至 dbqq@phei.com.cn。

本书咨询联系方式：010-51260888-819，faq@phei.com.cn。

前言

在写这本书之前，我先后在两家杭州的“独角兽”公司担任技术负责人，并负责推进公司核心业务的“中台化”改造。在落地业务中台和技术中台的过程中，我督促并指导开发人员统一使用 Spring Cloud Alibaba 作为中台服务最底层的基础框架。为了快速推进业务服务 Spring Cloud Alibaba 化的进度，我冲在业务的第一线，收集和整理开发人员在使用 Spring Cloud Alibaba 过程中反馈的技术问题，并提供有效的技术解决方案，直至项目落地。

我每周都会做技术复盘，通过分析大量的问题总结出一个结论：开发人员反馈的问题大部分都是由于 Spring Cloud Alibaba 使用不合理所造成的。也就是说，很多开发人员并不了解 Spring Cloud Alibaba 的原理及如何落地实践。于是，我就产生了把我这几年落地 Spring Cloud Alibaba 的经验通过图书的方式输出的想法。

1. 本书特色

本书聚焦于 Spring Cloud Alibaba 微服务架构实战，全面分析了基于 Spring Cloud Alibaba 的微服务架构全栈技术原理。本书有如下特色：

（1）技术新。

Spring Cloud Alibaba 是一个将 Spring Cloud“阿里巴巴化”的微服务架构框架，它具备 Spring Cloud 所有的能力，并添加了 Nacos、Dubbo、RocketMQ 等 Spring Cloud 不具备的微服务架构能力。简单来说就是：搭建微服务架构，使用 Spring Cloud Alibaba 比使用 Spring Cloud 更高效，更简单，开发的技术成本更低。

本书中所有代码采用 Spring Cloud Alibaba 目前的最新版本（2.2. 5.RELEASE）来编写，与 Spring Cloud Alibaba 相关的微服务技术（Seata、RocketMQ 等）也采用的是目前最新的稳定版本。

（2）精心设计的主线：零基础入门，循序渐进，直至项目实战。

本书精心研究了程序类、架构类知识的认知规律，全书总共分为五个部分：入门篇、基础篇、中级篇、高级篇及项目实战篇，设计了一条相对科学的主线“它是什么→怎么搭建基础开发环境→怎么进行单项技术开发→怎么完成一个完整的项目”，让读者快速从“菜鸟”向微服务架构实战高手迈进。

（3）不只介绍 Spring Cloud Alibaba 框架本身，而是微服务架构全栈技术。

有的同类书只介绍 Spring Cloud Alibaba 框架本身，假定读者对支撑 Spring Cloud Alibaba 微服务架构相关技术（比如 Seata、Skywalking、Redis、RocketMQ 等）是了解的。这样就会存在一个问题——为了学会微服务技术，得找 Spring Cloud Alibaba 的书、Seata 的书、RocketMQ 的书……而这恰恰是难点所在——怎样将它们平滑地衔接起来学习。

本书是以“实现完整的 Spring Cloud Alibaba 微服务架构”为目标，为了这个目标，除介绍 Spring Cloud Alibaba 这个“主角”外，对于支撑 Spring Cloud Alibaba 微服务架构的技术（比如 Seata、Skywalking、Redis、RocketMQ 等）也基本都是从零讲起，保证读者能够平滑地学习。本书是“一站到底”的解决方案：读者只需从这里上车，中途无需转乘，读者需要什么，本书就提供什么，直达终点。

（4）绘制了大量的图，便于理解原理、架构、流程。

一图胜千文，书中在涉及原理、架构、流程的地方都尽量配有插图，以便读者有直观的理解。

（5）实战性强。

本书介绍了大量的实战案例，能让读者“动起来”，在实践中体会功能，而不只是一种概念上的理解。

在讲解每一个知识模块时，我们都在思考：在这个知识模块中，哪些是读者必须实现的“标准动作”（实例）；哪些“标准动作”是可以先完成的，以求读者能快速有一个感知；哪些“标准动作”是有一定难度，需要放到后面完成的。读者在跟随书中实例一个个实践之后，再去理解那些抽象的概念和原理就是水道渠成了。

本书的目标之一是，让读者在动手中学习，而不是“看书时好像全明白了，一动手却发现什都不会”。本书相信“知行合一”理念，不是“只知，而无行”，避免眼高手低。

（6）深入剖析原理。

本书以系统思维的方式，从业务功能视角剖析微服务架构中技术的底层原理，使读者具备快速阅读新框架源码的能力。读者只有具备了这种功能，才能举一反三，实现更复杂的功能，应对更复杂的应用场景。

（7）采用真实项目，实现“从树木到森林”的突破。

本书的“项目实战篇”是从架构、代码和业务的视角，在真实项目中验证“Spring Cloud Alibaba 微服务架构”的架构方法论及核心技术原理，让读者有身临生产级场景的感觉。

（8）衔接运维，一键部署。

本书中所有的技术框架都有详细的“搭建技术框架运维环境的步骤”，读者只需要按照本书的安装步骤，就可以快速搭建运维环境，从而在本地环境中快速运行本书的实例。

（9）干货丰富，知识的“巨无霸”。

本书共分为上、下两册，总计 18 章，近 1000 页，内容非常丰富，算得上是相关领域图书中的“巨无霸”。

2. 阅读本书，你能学到什么

- 掌握 Spring Cloud Alibaba 的核心原理及微服务架构项目实战经验；
- 掌握 Nacos 注册中心和配置中心的核心原理及微服务架构项目实战经验；
- 掌握 Sentinel 的核心原理及微服务架构项目实战经验；
- 掌握 Seata 的核心原理及微服务架构项目实战经验；
- 掌握 RocketMQ 的核心原理及微服务架构项目实战经验；
- 掌握 Skywalking 的核心原理及微服务架构项目实战经验；
- 掌握 Elastic Job 的核心原理及微服务架构项目实战经验；
- 掌握 ShardingSphere 的核心原理及微服务架构项目实战经验；
- 掌握 Spring Cloud Gateway 的核心原理及微服务架构项目实战经验；
- 掌握分布式缓存 Redis 的集群管理和分布式锁的原理及微服务架构项目实战经验；
- 掌握 Discovery 的核心原理及微服务架构项目实战经验；
- 掌握在业务中台和技术中台中落地“基于 Spring Cloud Alibaba 微服务架构”的项目实战经验；
- 掌握在微服务架构中“基于 DataX 的异构数据迁移”的项目实战经验；
- 掌握在微服务架构中“基于 Skywalking 的链路告警平台”的项目实战经验；
- 掌握在微服务架构中“基于 ELK 和 Sywalking 的全链路日志平台”的项目实战经验。

3. 读者对象

本书读者对象如下：

◎ 初学 Java 的自学者；
◎ 软件开发工程师；
◎ Java 语言中高级开发人员；
◎ 编程爱好者；
◎ 中间件爱好者；
◎ 技术总监；
◎ 培训机构的老师和学员；
◎ 高等院校计算机相关专业学生；
◎ Spring Cloud Alibaba 初学者；
◎ DevOps 运维人员；
◎ 技术经理；
◎ 其他对 Spring Cloud Alibaba 感兴趣的 IT 人员。

4. 致谢

Spring Cloud Alibaba 是我深度使用的微服务框架之一，也是我实现职业生涯飞跃的敲门砖。虽然我在微服务架构领域中有很多技术沉淀，但是作为一个技术人员，将自己懂的技术通过文字输出给读者，还是需要很强的技术布道能力及文字编排能力的。

感谢我的家人，特别是我太太陈益超和我三岁的儿子胡辰昱，在我写书期间对我的支持。同时也要感谢电子工业出版社的编辑吴宏伟老师，将我带进“通过文字进行技术知识输出”的大门。

胡弦

2021.09.17

读者服务

微信扫码回复：42313

- 获取本书配套代码
- 加入本书读者交流群，与作者互动
- 获取【百场业界大咖直播合集】（持续更新），仅需 1 元

目录

入门篇

基础篇

中级篇

入门篇

第 1 章 进入 Spring Cloud Alibaba 的世界

Spring Cloud Alibaba 致力于提供微服务开发的一站式解决方案，它是 Spring Cloud 组件被植入 Alibaba 元素之后的产物。利用 Spring Cloud Alibaba，可以快速搭建微服务架构并完成技术升级。中小企业如果需要快速落地业务中台和技术中台，并向数字化业务转型，那 Spring Cloud Alibaba 绝对是一个“神器”。

本章带大家熟悉一些 IT 领域的基础架构理论，算是一盘“开胃菜”。

1.1 了解微服务架构

应用架构经历了从单体架构、SOA 架构，到微服务架构、服务网格架构、Serverless 架构的演变。

单体架构和微服务架构是应用架构阶段中比较成熟的架构模式，下面会重点阐述单体架构和微服务架构的关系。

1.1.1 单体架构与微服务架构的区别

单体架构与微服务架构到底有什么区别，其实可以从文字上推敲一下：

- 单体架构重点强调“单”。所有的单体架构应用程序都采用一样的架构，业务功能强耦合在一个工程中，整体部署。（采用单体架构的应用程序被称为“单体应用程序”。）

- 微服务架构重点强调“微”。“微”这个词在微服务架构中的含义是“简单”。微服务架构就是将复杂的“巨无霸”服务拆分为简单的服务，并分布式部署，输出完整的业务能力。（采用微服务架构的应用程序被称为“微服务应用程序”。）

1. 单体架构

采用单体架构的应用程序会耦合很多子模块，子模块相互配合才能完成特定领域的业务功能。

在单体应用程序中，所有服务 API 都被打包在一个 War 包中，War 包在启动后会作为一个进程运行。用户界面（比如浏览器）、数据访问层（比如 Tomcat 集群）和数据存储层（比如 MySQL 数据库）是紧密耦合的。通常小型团队在软件开发的早期阶段会采用单体架构。一般的单体架构如图 1-1 所示。

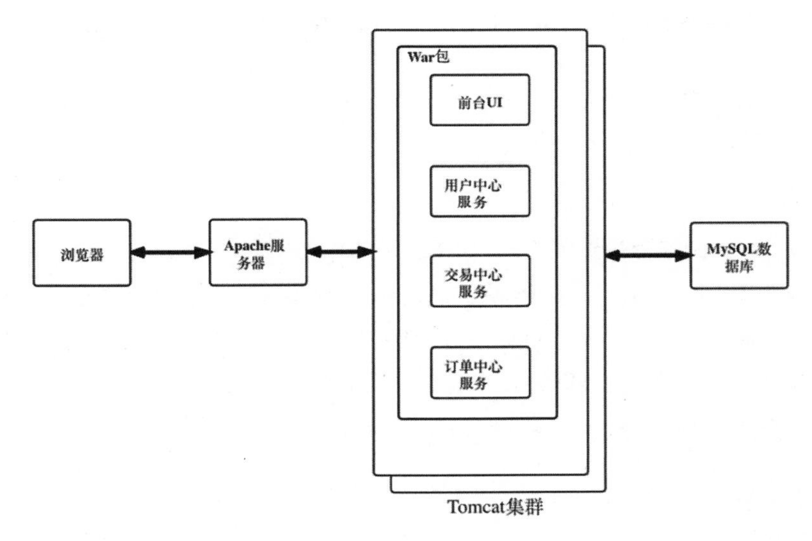

图 1-1

通常，用户通过浏览器访问业务功能；用 Apache 服务器作为前端的负载均衡器，整体后端业务被通过一个 War 包部署到 Tomcat 集群；业务功能可以整体扩展，DB 共用一个数据库（例如 MySQL 数据库）。

从以上的单体架构中可以看到：前台 UI、用户中心服务、交易中心服务及订单中心服务被部署在一个进程中（如果前后端没有分离，则前台 UI 也会和后端一起部署）；服务都是本地调用，不存在网络抖动的风险，在一定程度上提升了部分业务接口的性能，并且开发流程比较简单。

随着团队规模变大、业务拓新和裂变，团队会不断引入新技术，以提升业务功能的交付效率。业务拓新和裂变会带来代码量的急增，单体架构也就变得很难维护，技术就成了业务发展的绊脚石。

但是单体架构也是有它的可取之处。

单体架构的优点：

- 开发单体架构应用程序通常简单。
- 所有组件都被打包成一个 War 包，非常易于部署。
- 易于扩展整个应用程序。

单体架构的缺点：

- 代码库庞大，从而增加了代码开发和维护的复杂度。
- CI/DI 集成和部署复杂度高，需要专门的团队来构建和部署。
- 开发人员利用 IDE 工具开发服务的成本剧增，庞大的代码库使 IDE 变慢，从而使得构建时间增加。
- 由于单体架构不满足“高内聚，低耦合”的架构设计原则，所以导致技术架构升级和代码重构成本较高。
- 应用启动耗时多。
- 故障影响范围大。

2. 微服务架构

在软件领域，对于微服务的概念并没有严格的定义。

我们可以将微服务定义为：一堆松散耦合的组件，它们一起工作以执行任务；这些轻量级的组件可以通过各种语言（Java、PHP 或 Python）完成开发，并且它们可以使用各种协议（HTTP、HTTPS 或 JMS）进行通信；大多数微服务通过 REST API 暴露服务，完成跨平台服务调用。

微服务架构遵循“高内聚，低耦合”的架构设计原则。

微服务的优点：

- 每个微服务都很小，都专注于特定领域的功能和业务需求。所以，微服务通常会和领域模型关联。
- 微服务可以由小型开发人员团队（通常为 2~5 个开发人员）独立开发。
- 微服务是松散耦合的，这意味着服务的开发和部署是独立的。
- 可以使用不同的编程语言来开发微服务（团队可以统一技术栈，便于技术管理）。
- 微服务提供了一种轻松灵活的方式来集成自动部署工具与持续集成工具（例如：Jenkins、Hudson、Bamboo 等）。
- 微服务架构可以快速引进最新技术（框架、编程语言、编程实践等）。
- 微服务架构易于根据需求进行扩展。
- 微服务架构流程编排非常容易。

- 一个组件出现故障，不会导致整个系统停机。
- 在开发整体解决方案时，可以多个小型团队并行开发不同的微服务任务。因此，它有助于减少开发时间。
- CI / CD 非常容易。

典型的微服务架构如图 1-2 所示。

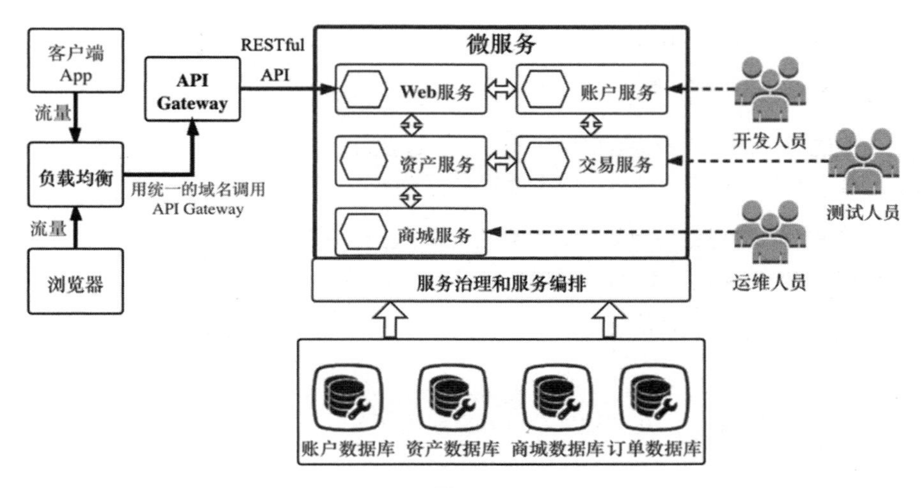

图 1-2

微服务架构通常包含如下内容。

- 客户端 App 和浏览器：微服务的服务对象，即功能体验者。
- 负载均衡器：通常指 Nginx 和 SLB 等。
- API Gateway：通常指业务网关。常规的技术栈包括 Spring Cloud Gateway 和 Zuul 等。
- Web 服务：通常指 RESTful API 服务。它通过后台管理系统统一配置 API 的路由信息，并实时地将路由信息推送到由 API Gateway，API Gateway 将 RESTful API 通过统一的域名暴露给 App 或者浏览器。

Web 服务与微服务通过 RPC 通信，当然也可以通过 HTTP 通信。API Gateway 也可以直接与微服务通信，比如 Spring Cloud Gateway 可同时支持 Dubbo 和 RESTful API 通信 。

在电商系统中有账户服务、资产服务、商城服务和订单服务等，这些服务对于微服务架构来说，都只是某一个领域里的能力输出点。微服务面向的技术人员，主要包括开发人员、测试人员和运维人员。在微服务架构中，所有的服务会有独立的数据库，数据都会被物理隔离。

微服务架构的缺点：

- 代码维护成本比单体架构较高。

- 在跨进程通信时，需要引入分布式技术栈以保证系统之间的高可用、高性能和高并发性，技术复杂度成几何数递增。
- 在采用微服务架构后，业务调用链路比单体架构更长，网络性能开销成倍增长，开发人员需要具备更强的业务理解能力和技术拓新能力。

1.1.2 分布式架构与微服务架构的区别

下面分析下分布式架构与微服务架构的区别。

1. 分布式架构

分布式系统的组件位于不同的互联网计算机上，这些计算机通过相互传递消息进行通信和协调它们的行动指令。这些组件相互交互，以实现共同的目标。

分布式系统中的网络是由使用分布中间件连接的自治计算机组成的，这有助于分享不同的资源和能力，为用户提供单一和综合连贯的网络。

分布式系统是由自治的组件或计算机组成的。从任何用户或应用程序看来，分布式系统都是一个单一的系统。

构建分布式系统的关键之一是“如何在这些自治组件之间建立通信和协调”，这个也是分布式架构面临的最大挑战。

分布式架构通常要考虑如下因素。

（1）服务调度和编排管理。

为了应对不断增长服务实例和容器的规模带来的技术挑战，服务调度和编排管理成为分布式系统中关键的组成部分。服务调度和编排管理领域比较成熟的产品有 Docker Swarm、Kubernetes、Mesos、Marathon 等。

（2）服务注册与发现。

从几个服务到几百个，甚至上千个微服务，服务之间的调用关系越来越复杂。Consul、ZooKeeper、Etcd、Confd、Nacos 和 Eureka 等是这个领域的一些成熟产品。这些产品大多支持跨服务实例的 RPC 流量的负载均衡。

（3）系统状态管理和集群管理。

随着集群规模的增长，需要管理集群中机器节点的运行状态，比如每个服务的 SRV、容量规划、负载流量等。Docker Swarm Agent、Kubernetes、Mesos、Containership 等是这个领域的一些成熟产品。Skywalking、Cat 等分布式链路追踪工具，能监控集群中机器节点的运行状态。

（4）数据存储。

容器存储是临时的。这意味着，需要存储在容器生命周期之外的数据都必须存储在外部存储设备中，比如使用 Docker Volume Plugin、Flocker 和 Kubernetes Persistent volume 等。

（5）网络。

网络是分布式架构中服务性能的关键指标。服务之间通过 RPC 通信，通信肯定是要联网的。通常开发人员和运维人员在沟通线上服务器如何部署时，总是最先确认服务之间网络的可用性、网络的带宽是多少。

网络也是分布式架构中影响服务调用可用性的重要因素。通常开发人员做性能优化，最新想到的是优化网络的 I/O 开销。

（6）容器中运行着不同的服务进程，需要管理这些进程及进程之间的访问权限。

如果多个容器在同一主机上运行，则在共享网络资源时可能需要创建安全组以进行容器隔离。同样，如果希望容器能发现跨主机托管的服务，则需要一个简单的模型来访问这些服务。Flannel、Weaveworks 和 Calico 等是这个领域的一些产品（容器隔离技术）。

（7）监控、审核和日志。

如果有成千上万个容器在运行，则监视/审核/记录所有容器是一个棘手的问题，因为需要从每个容器中提取数据/日志进行分析。Loggly、Fluentd、日志条目、Datadog 和 ELK 堆栈等是这个领域的一些产品。

2. 如何区分微服务架构与分布式架构

从概念理解的角度来看，分布式服务架构强调的是“服务化”及“服务的分散化”，而微服务架构则更强调“服务的专业化”和“精细分工”。

从实践的角度来看，微服务架构通常是分布式服务架构，反之则未必成立。

所以，选择微服务架构通常意味着需要解决分布式架构的各种难题。

微服务架构是目前技术团队面对互联网产品爆发式增长的最优选择，要解决的是快速迭代、高可靠和高可用等问题。

把复杂度很高的产品拆分成一些较小的模块（每一个模块用 5～9 个小团队来维护），并遵循康威定律，这样可以减少沟通成本，提高协作效率，更好地实现快速迭代和弹性扩展。

1.2 如何构建微服务架构

1.2.1 构建微服务架构的目标

微服务可以对企业产生积极的影响，一般来说，构建微服务架构有以下四个目标。

- 降低设计、实现和维护 IT 服务的总体成本。
- 提高从服务构建、打包到部署的速度。
- 提高系统的弹性。
- 使得在微服务运行过程中，服务的状态是可度量和可观测的。

1.2.2 构建微服务架构的关键点

通常在构建微服务架构时要考虑如下关键点。

1. 可伸缩性

可伸缩性是反映软件系统计算处理能力的指标，它代表了一种弹性：在系统扩展过程中，软件能够保证旺盛的生命力，通过很少的改动（甚至只是硬件设备的添置），就能实现整个系统处理能力的线性增长，实现高吞吐量、低延迟和高性能。

提升可伸缩性和纯粹的性能调优有着本质区别：

- 提升可伸缩性，是综合考量和平衡高性能、低成本和可维护性等诸多因素，讲究平滑且线性的性能提升，侧重于系统的水平伸缩，通过廉价的服务器实现分布式计算。
- 纯粹的性能调优，优化的只是单台机器的性能。

提升可伸缩性和纯粹的性能调优的共同点是：根据应用系统的特点，在吞吐量和延迟之间进行一个侧重选择。当然，水平伸缩分区会受到 CAP 定理约束。

软件的可扩展性设计非常重要，但又比较难以掌握。业界试图采用云计算或高并发语言等方式来节省开发者的精力。但无论采用什么技术，如果应用系统内部是“铁板”一块（例如严重依赖数据库），则在系统达到一定访问量时，负载都会集中到一两台数据库服务器上，这时进行分区扩展和伸缩就比较困难。正如 Hibernate 框架创建人 Gavin King 所说：关系数据库是最不可扩展的。

2. 可用性

业界通常用多少个“9”来衡量系统的可用性，用 99.99%表示一年中系统有 1 小时左右的不可用时间。

任何一个服务的可用性都不会是 100%，在服务运行时间里很有可能发生故障。把功能集中的单体架构拆分成多个相互独立的微服务架构，虽然可以降低一损俱损的全局故障风险，但由于微服务之间存在大量的依赖关系，随着微服务个数的增多，依赖关系也会变得越来越复杂。而且，每个微服务架构都有可能发生故障，如果不能进行故障隔离，避免故障的连锁反应，则结果可能比单体架构更糟糕。

$$\text{服务可用性}=\frac{\text{服务周期总分钟数-服务不可用分钟数}}{\text{服务周期总分数}}\times 100\%$$

- 服务周期：一个服务周期为一个自然月。
- 服务周期总分钟数：按“服务周期内的总天数×24（h）×60（min）”计算。
- 服务不可用分钟数：如果在某一分钟内，试图与指定微服务注册中心建立连接的连续尝试均失败，则视为该分钟内该微服务不可用。在一个服务周期内，“微服务不可用分钟数之和”即“服务不可用分钟数”。

假设有 100 个微服务，并且每个微服务只发生 1 种故障，那么总共会有 100 种不同的故障（其实每个微服务可能有不止 1 种故障）。

确保微服务架构的高可用性所面临的巨大挑战如下：

- 如果服务 A 发生故障，如何确保依赖服务 A 的服务 B 的可用性。
- 如果系统出现故障，如何将故障服务自动地隔离，如何实现故障服务的调用方的优雅降级。

3. 弹性能力

开发人员越来越依赖用微服务架构将应用程序构建为一组细粒度、重点狭窄且独立的服务，每个服务均被独立开发和部署。

尽管微服务方法促进了敏捷性，但它也带来了新的挑战：这些服务必须通过网络调用在相互之间进行交互，以及与其他系统（例如 Web API 和数据库）进行交互，但由于网络始终有不可靠的因素，所以此类交互随时可能出现故障。

因此，基于微服务的应用程序的弹性（即从某些类型的故障中恢复并保持功能的能力），在很大程度上取决于该应用程序处理不可靠网络上的服务间通信的能力。因此，基于微服务的应用程序的弹性，在很大程度上取决于您所实现的微服务通信的弹性。

4. 可扩展性

可扩展性是一个系统或应用程序的属性，表示系统或应用程序可以处理更多的工作，或者很容易地进行扩展，以应对网络访问、数据处理、数据库访问和日益增长的文件系统访问需求。

可扩展性通常分为水平扩展性和垂直扩展性。

（1）水平扩展性：在系统进行扩展时，通过添加与现有节点功能相同的新节点，在所有节点之间重新分配负载（可以横向扩展或向上扩展）。

对于 SOA 系统和 Web 服务器，可以通过向负载平衡网络中添加更多的服务器来扩展，以将传入的请求分布在所有服务器中。

（2）垂直扩展性：向系统中的机器节点添加处理器、内存、存储设备或网络接口来扩展，以提升系统的 TPS 处理能力，系统会垂直或向上扩展。

虚拟主机通过增加处理器的数量或主内存的数量来扩大规模，以便在同一个硬件中承载更多的虚拟服务器。

5. 高内聚低耦合性

“高内聚，低耦合”是软件工程中的概念，是判断设计好坏的标准。“高内聚，低耦合”要求开发人员要面向对象进行系统设计。

6. 服务治理能力

当应用从单体架构变为微服务架构后，我们需要管理更多的应用服务，服务数量可能会成倍增加。面对这样的挑战，我们必须具备服务治理能力。

7. 故障隔离

微服务架构通过定义明确的服务边界来有效地隔离故障。和其他分布式系统一样，微服务架构在网络、硬件和应用层都会存在很多问题。由于服务间是互相依赖的，因此任何服务出错都会导致用户不可用。为了尽可能减少故障的影响范围，我们需要构建容错能力强的服务，通过容错来隔离故障。

微服务架构将应用逻辑拆分成若干个服务，服务之间通过网络交互。服务是通过网络被调用的，而不是在进程中被调用的，这给需要在多个物理和逻辑组件间进行协作的系统带来了潜在的问题和复杂性。分布式系统越复杂，则网络特定故障发生的可能性越大。

相比于传统应用庞大的结构，微服务架构最大的一个优点是：团队能独立地设计、开发和部署各自的服务。团队能掌控各自服务的整个生命周期。这也意味着，服务之间的调用链路关系会非常长，并且会存在强依赖的调用关系。

在微服务架构中，通常服务的调用链路都是跨团队的。服务的发布和配置等会导致被调用的服务暂时不可用。因此，调用方要确保自己的服务具备隔离故障的能力。

8. 通过 DevOps 持续交付

DevOps 是指，开发团队和运维团队不再是分开的两个团队，而是“你中有我，我中有你”的一个团队。

如果从字面上来理解，DevOps 只是“Dev（开发人员）+Ops（运维人员）”。但实际上，它是一组过程、方法与系统的统称。

DevOps 概念在 2009 年首次被提出，发展到现在其内容也在不断丰富，有理论也有实践，包括组织文化、自动化、精益、反馈和分享等不同方面：

- 组织架构、企业文化与理念等，需要自上而下设计，用于促进开发部门、运维部门和质量保障部门之间的沟通、协作与整合。简单而言，其组织形式类似于系统分层设计。
- 自动化是指，所有的操作都不需要人工参与，全部依赖系统自动完成。比如上述的持续交付过程必须实现自动化才有可能实现快速迭代。
- DevOps 的出现是因为软件行业日益清晰地认识到：为了按时交付软件产品和服务，开发部门和运维部门必须紧密合作。

总之，DevOps 强调的是团队之间可以通过自动化工具来进行协作和沟通，以完成软件的生命周期管理，从而更快、更频繁地交付更稳定的软件。

1.3 认识 Spring Cloud Alibaba

2018 年 10 月 31 日，Spring Cloud Alibaba 正式开源，并提交了第一个正式稳定版本 0.1.0.RELEASE。Spring Cloud Alibaba 用了不到两年的时间，完成了从 0 到 1，成为 GitHub 上的明星项目。

笔者认为，除社区孜孜不倦地持续迭代功能外，最关键还是 Spring Cloud Alibaba 的架构设计理念非常契合目前微服务架构生态的演进路径。微服务架构本质上就是分布式架构，Spring Cloud Alibaba 为开发分布式应用程序提供了一站式解决方案，它包含开发分布式应用程序所需的所有组件。

1. 功能特性

开发者只需要添加一些注释和少量配置，即可将 Spring Cloud 应用程序连接到 Spring Cloud Alibaba 的分布式解决方案，并使用 Spring Cloud Alibaba 的中间件构建分布式应用程序。

Spring Cloud Alibaba 的主要功能：流量控制和服务降级，服务注册与发现，分布式配置，事件驱动，消息总线，分布式事务，Dubbo RPC。

2. 组件能力

Spring Cloud Alibaba 的组件功能既有免费版本，也有收费版本。组件如下。

- Sentinel：以“流”为切入点，在流量控制、并发性、容错降级和负载保护等方面提供解决方案，以保护服务的稳定性。
- Nacos：一个具备动态服务发现和分布式配置等功能的管理平台，主要用于构建云原生应用程序。
- RocketMQ：一个高性能、高可用、高吞吐量的金融级消息中间件。Spring Cloud Alibaba 将 RocketMQ 定制化封装，开发人员可“开箱即用”。
- Dubbo：一个基于 Java 的高性能开源 RPC 框架。
- Seata：一个高性能且易于使用的分布式事务解决方案，可用于微服务架构。
- 阿里云 OSS（阿里云对象存储服务）：一种加密的安全云存储服务，可以存储、处理和访问来自世界任何地方的大量数据。
- 阿里云 SchedulerX：一款分布式任务调度产品，支持定期任务和在指定时间点触发任务。
- 阿里云 SMS：一种覆盖全球的消息服务，提供便捷、高效和智能的通信功能，可帮助企业快速联系其客户。

1.4 学习 Spring Cloud Alibaba 的建议

在学习 Spring Cloud Alibaba 之前，需要先熟悉 Spring Boot 和 Spring Cloud 的基础原理（这里默认读者是具备 Spring Framework 项目开发能力的）。由于本书主要是针对 Spring Cloud Alibaba 的，所以下面只简单地讲解一下 Spring Boot 和 Spring Cloud，如果需要深入研究，可以自行查找相关资料。

1.4.1 熟悉 Spring Boot

很多 Spring 的初学者经常会因为其烦琐的配置文件而却步，特别是对于一些还没有被拆分的老项目。很难想象在增加一个新的需求，或在引入一个新的技术组件时，业务开发人员需要重复地增加多少冗余的配置文件。

在 Spring 中，如果需要开启一个新的项目，则一般是利用复制功能快速搭建出项目骨架。但是每个项目往往又存在很多差异，所以还需要手动地一个个去修改，这就增加了项目的技术复杂度。业务开发人员不仅要关注项目的业务逻辑，还要花时间熟悉一些比较底层的技术细节。

很多技术专家和架构师会去尝试封装一些基础组件，但是这些组件很多都不能做到“开箱即用”。

Spring Boot 是一款“开箱即用”的基础服务框架，它通过 Factory 机制完成普通 POJO 对象的初始化，并将这些对象注入 IOC 容器中；通过条件注解控制 Bean 对象之间的依赖关系，业务开发人员通过注解即可完成指定功能的注入。

1.4.2　熟悉 Spring Cloud

Spring Cloud 是一系列框架的有序集合。它利用 Spring Boot 的开发便利性，巧妙地简化了分布式系统基础设施的开发流程。例如，服务发现/注册、配置中心、消息总线、负载均衡、断路器、数据监控等功能，都可以用 Spring Boot 的开发风格做到一键部署和启动。

Spring Cloud 并没有“重复制造轮子”，它只是将各家公司开发的比较成熟、经得起实际考验的服务框架组合起来，通过 Spring Boot 风格进行再封装，屏蔽了复杂的配置和实现原理，最终给开发者提供了一套简单易懂、易部署和易维护的分布式系统开发工具包。

Spring Cloud 并不是 Spring Framework 团队全新研发的框架，它只是把一些“能解决微服务架构中常见问题的优秀开源框架”基于 Spring Boot 和 Spring Cloud 规范进行了整合，通过 Spring Boot 框架进行二次封装，屏蔽了负载的 XML 文件配置，给业务开发者提供了“开箱即用”的良好开发体验。

可以总结出：Spring Cloud 是一套标准的开发规范，是微服务架构的“一篮子”解决方案。

1. 功能特性

Spring Cloud 具有以下特性：分布式/版本化配置、服务注册与发现、服务路由、服务负载均衡、用于服务降级的断路器、确保数据一致性的全局锁、分布式集群选举算法、分布式消息总线。

2. 组件

Spring Cloud 的子项目大致可分成：

（1）对现有成熟框架 Spring Boot 的封装和抽象——这也是数量最多的子项目。

（2）对一部分分布式基础能力的实现，如 Spring Cloud Stream 通过消息总线封装了 Kafka 和 ActiveMQ。

对于想快速实践微服务的开发者来说，第（1）类子项目就已经足够使用了。Spring Cloud 常用组件见表 1-1。

表 1-1　Spring Cloud 的组件

Spring Cloud 组件	功能描述
spring-cloud-sleuth	分布式链路追踪和性能监控
spring-cloud-config	分布式配置服务
spring-cloud-aws	支持 AWS 云
spring-cloud-consul	支持微服务治理框架 Consul
spring-cloud-gateway	支持高性能业务网关 Gateway
spring-cloud-security	用来实现微服务架构中的安全认证
spring-cloud-openfeign	支持 HTTP 客户端 Openfeign
spring-cloud-netflix	支持 Netflix 微服务治理“全家桶”，比如 Euraka
spring-cloud-stream-binder-kafka	支持分布式消息中间件 Kafka
spring-cloud-dataflow	支持数据流服务
spring-cloud-commons	Spring Cloud 服务治理的公共包
spring-cloud-circuitbreaker	熔断和降级
spring-cloud-zookeeper	支持分布式组件 ZooKeeper
spring-cloud-bus	分布式消息总线
spring-cloud-kubernetes	支持基于 K8s 部署的应用集群管理
spring-cloud-task	支持分布式定时任务
spring-cloud-function	支持函数式编程

1.4.3　Spring Cloud Alibaba 的版本演进

Spring Cloud Alibaba 的版本演进如下。

- Spring Cloud Alibaba 的第一个正式开源版本是 0.1.0.RELEASE。
- 2.1.0.RELEASE 版本，引入了 Sentinel、Nacos Config、RocketMQ 及 Seata。
- 2.1.1.RELEASE 版本，在 2.1.0.RELEASE 的基础上新增了 Spring Cloud Alibaba Sidecar。
- 2.2.0.RELEASE 版本，支持 Sentinel 的 1.7.1 版本、Dubbo 的 2.7.4.1 版本。
- 2.2.1.RELEASE 版本，支持 Nacos Client 的 1.2.1 版本、Seata 的 1.1.0 版本、Dubbo 的 2.7.6 版本。
- 2.2.2.RELEASE 版本，支持 Nacos Client 的 1.3.2 版本、Seata 的 1.3.0 版本、Sentinel 的 1.8.0 版本、Dubbo 的 2.7.8 版本。
- 2.2.3.RELEASE 版本，支持 Nacos Client 的 1.3.3 版本。

Spring Cloud Alibaba 目前（2021 年 3 月）最新版本为 2.2.5.RELEASE。

1.5　Spring Cloud Alibaba 与 Spring Cloud 的关系

Spring Cloud Alibaba 与 Spring Cloud 的关系如图 1-3 所示。Spring Cloud Alibaba 是 Spring Cloud 的超集。

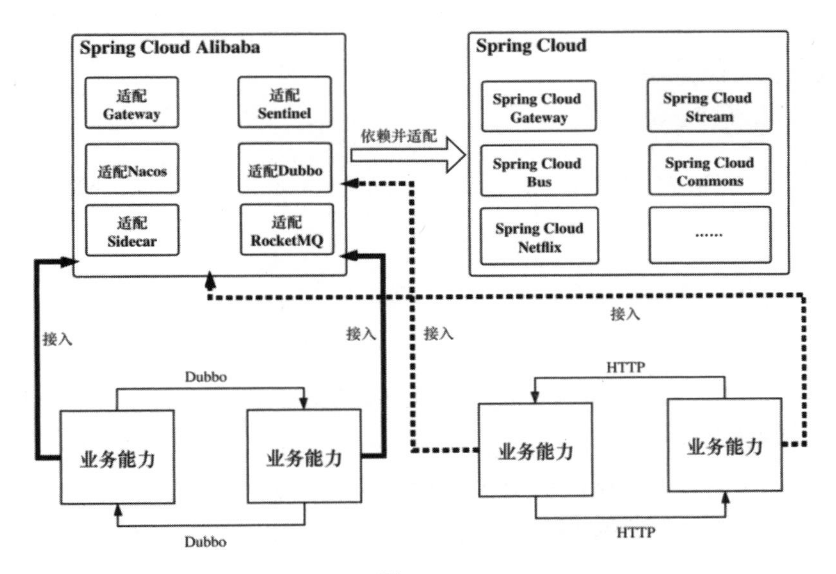

图 1-3

在图 1-3 中，业务能力代表应用的服务，服务之间通过 HTTP 或者 Dubbo 进行通信。

在没有 Spring Cloud Alibaba 之前，如果应用使用 Spring Cloud 作为基础框架，并使用 Spring Cloud 生态的服务治理能力（只支持 HTTP 或者 DNS），则其是不能兼容 Dubbo 服务的。在 Spring Cloud Alibaba 出现后，服务可以快速地完成 HTTP 和 Dubbo 通信的适配。

Spring Cloud Alibaba 依赖 Spring Cloud，并通过二次开发让开发者使用 Spring Cloud 更加简单和灵活。开发者可以直接使用 Spring Cloud Alibaba 来替换 Spring Cloud。在 Spring Cloud Alibaba 出现后，开发者可以在应用中采用 Spring Cloud 的协议和 Dubbo 的协议混合通信。

如果业务原先采用“Spring Cloud + Eureka/Consul”微服务体系，则可以快速切换到 Spring Cloud Alibaba 微服务体系。

如果业务还停留在单体架构，则可以几乎零成本地切换到 Spring Cloud Alibaba 微服务体系，风险很小。

如果业务应用要上阿里云，则可以直接采用 Spring Cloud Alibaba 的商业版来实现，成本非常小。

1.6 搭建基础环境

开发人员要使用 Spring Cloud Alibaba 进行开发，则必须先搭建基础环境，包括 Maven 和 Git。

1.6.1 安装 Maven

Maven 是一个软件项目管理工具，它基于项目对象模型（POM）的概念。

开发团队几乎不用花多少时间就能完成工程的基础构建配置，因为 Maven 使用了一个标准的目录结构和一个默认的构建生命周期。

如果有多个开发团队，则 Maven 能够在很短的时间内使得每项工作都按照标准进行。因为大部分的工程配置操作都非常简单并且可复用，所以在创建报告、检查、构建和测试自动配置时，Maven 可以让开发者的工作变得更简单。

Maven 能够帮助开发者完成以下工作：构建代码 、生成代码文档、度量代码、添加依赖、添加 SCM（软件配置管理）、发布代码、分发代码、添加邮件列表。

总的来说，Maven 简化了工程的构建过程，并对其进行了标准化。它可以无缝衔接代码编译、发布、文档生成、团队合作和其他任务。Maven 提高了代码功能的重用性，并且能够高效地构建生产环境的程序包。

1. 安装 Maven

安装 Maven 很简单：只需提取存档，并将带有 mvn 命令的 bin 文件夹添加到 PATH 中即可。

详细步骤如下：

（1）安装 JDK，并配置环境变量。

（2）下载 Maven 的安装包，并解压缩。

（3）配置 Maven 的环境变量：

```
export M2_HOME=/Users/huxian/Downloads/apache-maven-3.5.0
export PATH=${PATH}:${M2_HOME}/bin
```

（4）输入“mvn -version”命令，如果出现如图 1-4 所示的界面，则表示 Maven 配置成功。

```
huxians-MacBook-Air:~ huxian$ mvn -version
Apache Maven 3.5.0 (ff8f5e7444045639af65f6095c62210b5713f426; 2017-04-04T03:39:0
6+08:00)
Maven home: /Users/huxian/Downloads/apache-maven-3.5.0
Java version: 1.8.0_131, vendor: Oracle Corporation
Java home: /Library/Java/JavaVirtualMachines/jdk1.8.0_131.jdk/Contents/Home/jre
Default locale: en_US, platform encoding: UTF-8
OS name: "mac os x", version: "10.12.6", arch: "x86_64", family: "mac"
```

图 1-4

2. Maven 的专业术语

（1）POM。

POM 代表工程对象模型。它是使用 Maven 打包代码的一个基本组件，也是一个 XML 文件。它被放在工程的根目录下，文件名为 pom.xml。

POM 包含工程和各种配置细节的信息，Maven 使用这些信息构建工程。POM 也包含目标和插件。在执行一个任务或者目标时，Maven 会查找当前目录下的 POM，从中读取所需的配置信息，然后执行。

（2）构建生命周期。

构建生命周期是一组“阶段”的序列（sequence of phases），每个“阶段”定义了 POM 依赖执行的顺序。这里的“阶段”是生命周期的一部分。

（3）构建配置文件。

构建配置文件是一组配置的集合，用来设置或者覆盖 Maven 构建的默认配置。可以使用构建配置文件为不同的环境（例如 Producation 和 Development 环境）定制构建过程。

在 pom.xml 中，使用“activeProfiles/profiles”元素来指定 Profile，并且可以用很多方式触发它。可以在构建 Profile 时修改 POM，并为变量设置不同的目标环境（例如，在开发、测试和产品环境中的数据库服务器的路径）。

（4）仓库。

在 Maven 中，仓库是一个位置，例如目录，它可以存储所有的工程 Jar 文件、Library Jar 文件、插件，以及其他由工程指定的文件。Maven 的仓库有 3 种类型：本地（local）、中央（central）和远程（remote）。

（5）插件。

Maven 实际上是一个依赖插件的框架，所有任务实际上是由插件完成的。Maven 插件通常被用来创建 Jar 文件、创建 War 文件、编译代码文件、测试代码单元、创建工程文档和创建工程报告。

（6）快照与版本。

快照是一个特殊的版本（当前开发文件的一个副本）。与常规版本不同，Maven 在每一次构建时都会从远程仓库中“克隆”一份新的快照。

开发人员将每次更新的快照（例如 data-service:1.0-SNAPSHOT）发布到仓库中，以替换旧的快照。

- 对于版本：Maven 一旦下载了指定的版本（例如 data-service:1.0），就将不会再从仓库下载一个新的相同版本（这里指不会下载 1.0 版版本）。如果要下载新的代码，则数据服务版本需要被升级到 1.1 版本。
- 对于快照：每次在用户接口团队构建他们的项目时，Maven 将自动获取最新的快照（data-service:1.0-SNAPSHOT）。

1.6.2 熟悉 Git

Git 是一个免费的开源分布式版本控制系统，旨在快速、高效地处理从小型项目到大型项目的所有内容。它易于学习，占用空间小，具有闪电般的快速性能。

相比 Subversion、CVS、Perforce 和 ClearCase 之类的 SCM 工具，Git 具有廉价的本地分支、方便的暂存区域和多个工作流等功能。

如果读者已经安装了 Git，则可以输入“git version”命令来验证 Git 是否可用，如图 1-5 所示。

```
[huxian@huxians-MacBook-Pro target % git version
git version 2.24.3 (Apple Git-128)
huxian@huxians-MacBook-Pro target % 
```

图 1-5

第 2 章 熟用开发工具

熟练使用开发工具，是开发人员必备的技能。使用开发工具，开发人员可以提炼出开发规范，并将这些规范应用到项目的开发流程中。

在开发的过程中，经常会碰到一些因为开发工具使用不规范而造成的沟通阻塞问题，具体如下：

- 团队没有使用统一的开发工具，导致项目代码的风格和规范不一致，降低了代码的可维护性。
- 团队开发人员不熟悉开发工具，没有形成一些标准的规范和通用的沟通语言，导致在开发过程中不能进行对称和有效的沟通。
- 高效率地阅读代码是开发人员的基础技能。如果开发工具使用不当，则会直接影响开发人员阅读代码的效率。

能够提效开发人员工作效率的开发工具有很多，本章将介绍一些常用的开发工具。

2.1 安装开发工具 IntelliJ IDEA

开发人员要先安装 IntelliJ IDEA，才能够体验它的功能。

1. 安装环境要求

在安装 IntelliJ IDEA 前，要确保要达到硬件要求和软件要求。

（1）硬件要求。

- 至少需要 2 GB 的 RAM，推荐使用 4 GB 的 RAM。

- 至少需要 1.5 GB 硬盘空间 + 1 GB 的缓存。
- 屏幕分辨率最低 1024 px × 768 px。

（2）软件要求。

IntelliJ IDEA 是与 JRE 1.8 发行版捆绑在一起的，开发人员不需要在电脑上安装 JRE 就可以运行 IntelliJ IDEA，IntelliJ IDEA 支持的操作系统见表 2-1。

表 2-1　IntelliJ IDEA 支持的操作系统

操作系统	系统要求
Windows	微软的 Windows 10/8/7（sp1）或 Vista（sp2），32 位或 64 位版本
Mac	MacOS 10.5 或更高版本（仅支持 64 位操作系统）
Linux	Linux (建议使用 64 位系统)

2. 下载 IDEA

IntelliJ IDEA 有两个版本：Ultimate 和 Community。

- Community 版本：免费开源版本，功能较少。
- Ultimate 版本：商业化版本，提供了一整套优秀的工具和功能。

在 IntelliJ IDEA 的官网找到下载页，单击 Download 按钮下载相应版本，如图 2-1 所示。

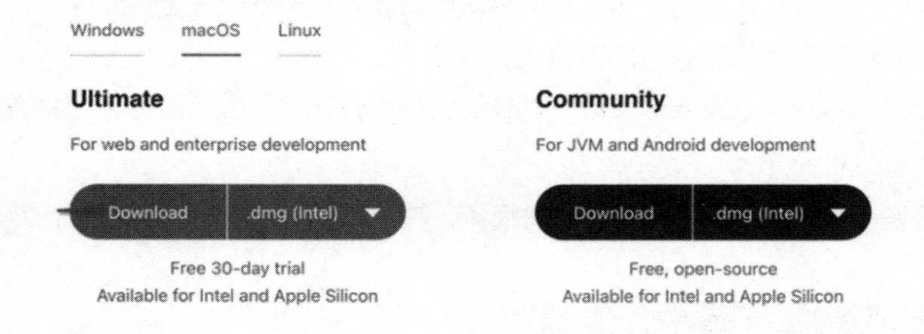

图 2-1

2.2 【实例】用 Spring Cloud Alibaba 开发一个 RESTful API 服务

本实例的源码在本书配套资源的“/spring-cloud-alibaba-practice/chaptertwo/restful-practice/”目录下。

下面带读者快速开发一个简单的 RESTful API 服务。

1. 搭建 Spring Cloud Alibaba 环境

（1）下载 Nacos 的安装包，本程序使用的是 Nacos 的 1.3.1 版本。

（2）将安装包复制到电脑的安装目录下（安装目录要具备“写”权限），解压缩安装包 nacos-server-1.3.1.tar.gz。

（3）切换到 bin 目录，使用脚本“sh startup.sh -m standalone”以单机模式启动 Nacos。

（4）使用“http://127.0.0.1:8848/nacos/”访问 Nacos 的控制台，并创建命名空间“spring cloud alibaba practice”，如图 2-2 所示。

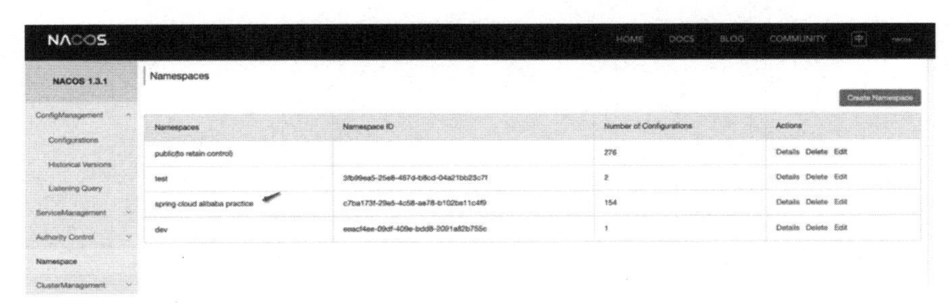

图 2-2

2. 创建项目

下面通过 IDEA 创建新项目。

（1）新建项目，具体操作如图 2-3 所示。

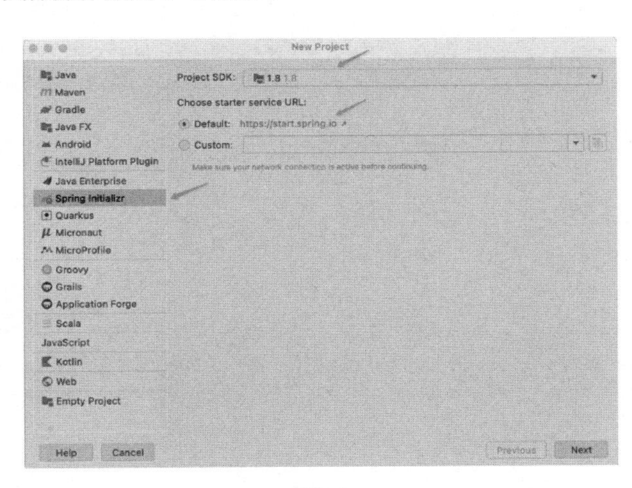

图 2-3

- 选择 Spring Initializr。
- 配置 Project SDK，本书所有源码默认的 JDK 版本是 1.8 版本。

（2）配置项目元数据。

项目元数据的配置如图 2-4 所示。开发人员只需结合所在项目组的项目规范，输入指定项目的元数据即可，工具会自动生成代码。

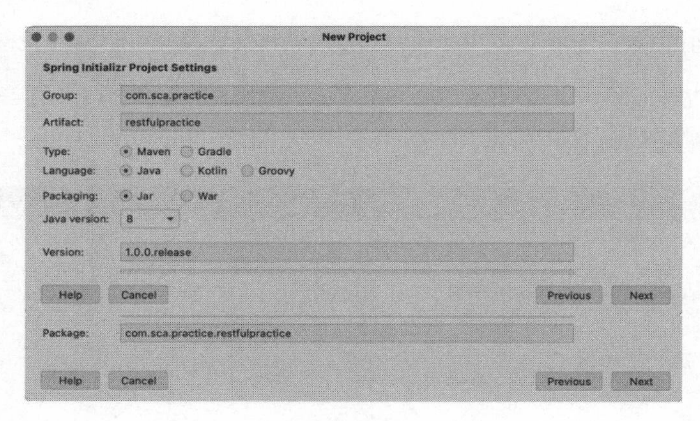

图 2-4

（3）添加组件的依赖关系。

如图 2-5 所示，在左边选中 Web，在右边的列表菜单中选中 Spring Web。同理可以添加 Nacos Configuration 和 Nacos Service Discovery 组件的依赖关系。

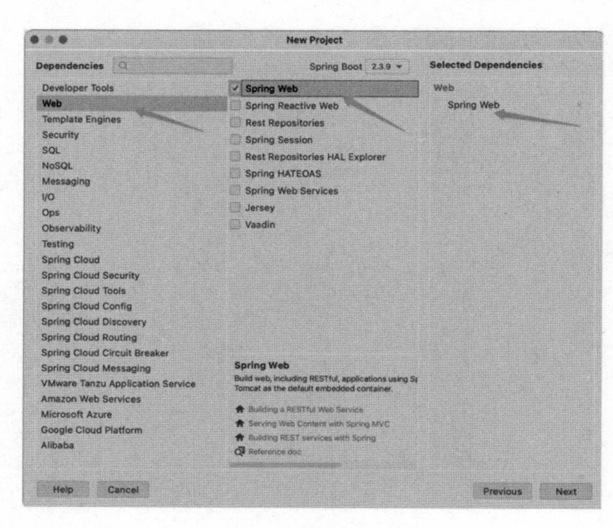

图 2-5

（4）配置项目环境 resources/application.yaml：

```
  # Spring Cloud 配置
spring:
  application:
    name: restful-practice
  jackson:
    default-property-inclusion: non_null
  cloud:
    # Nacos 注册中心的配置
    nacos:
      discovery:
            # 对应本机的 IP 地址
        server-addr: 127.0.0.1:8848
            # 对应在 Nacos 中创建的命名空间
        namespace: b9dd148f-1916-4346-9f44-770f1becc503
server:
  # Tomcat 的端口号，也可以随机选择
  port: 8747

# Dubbo 配置项，对应 DubboConfigurationProperties 类
dubbo:
  # Dubbo 服务注册中心的配置，对应 RegistryConfig 类
  registry:
    # 指定 Dubbo 服务注册中心的地址
    address: spring-cloud://localhost
  # Spring Cloud Alibaba Dubbo 的专属配置项，对应 DubboCloudProperties 类
  cloud:
    # 设置订阅的应用列表，多个用 （，）号分隔，默认为（ * ）表示订阅所有应用
     subscribed-services: "*"
```

（5）添加 RESTful API：

```
  @RestController
public class RestfulPracticeController {

    @GetMapping(value = "/testRestfulApi")
    public String testRestfulApi(){
        return "test";
    }
}
```

（6）启动服务，完成 RESTful API 开发，并将服务注册到 Nacos 中，如图 2-6 所示。

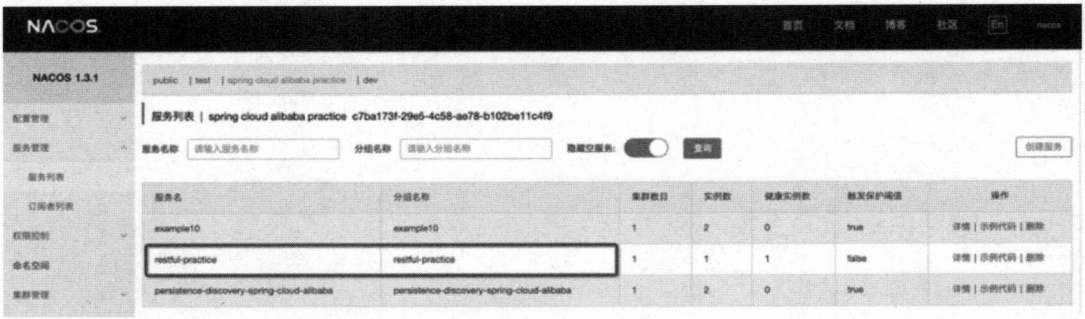

图 2-6

2.3　了解 Spring Framework 官方开发工具 STS

Spring Tools（简称 STS），目前最新的版本是 4.10.0。STS 是由 Spring Framework 团队打造的 Spring Framework 开发工具。它融合了优秀的中间件及开发者工具。它以独立的进程运行。它在构建之初就考虑了性能问题，并且采用了最新的 Spring Framwork 技术，为开发基于 Spring Framework 的企业应用提供了技术支持。

最新版本的 STS 可在各种编码环境中使用，例如 Eclipse、Visual Studio Code 和 Theia。

从 Spring Framework 官网下载对应操作系统的安装包，如图 2-7 所示。

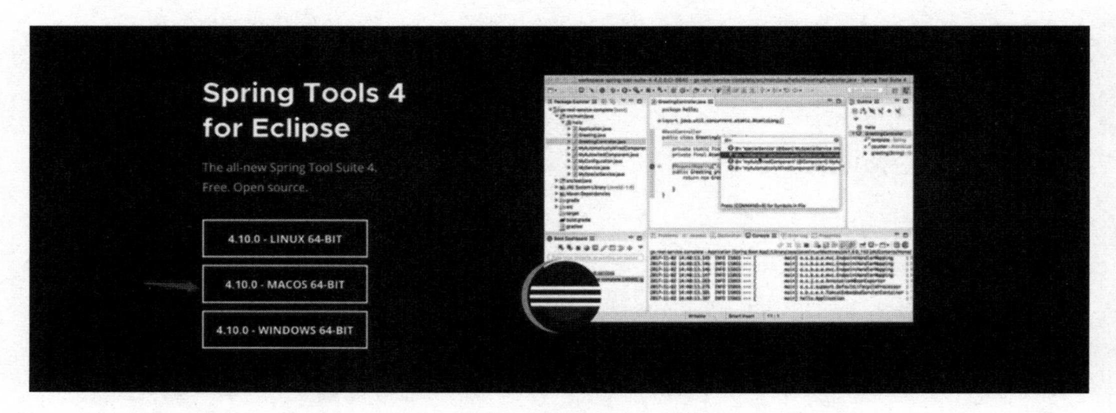

图 2-7

下载完成后，可以按照步骤安装 STS。如图 2-8 所示，将图中标注的菜单拖入 Applications 图标中即可安装完成。

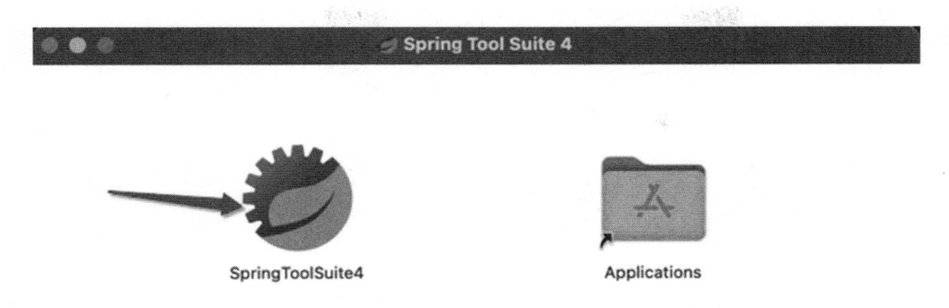

图 2-8

2.4　了解 Spring Framework 官方脚手架工具

Spring Framework 官方也提供了脚手架工具。在官网查看到的脚手架说明如图 2-9 所示。

图 2-9

该脚手架工具有以下功能特性：

- 代码编译工具可以选择 Maven 或者 Gadle。
- 代码语言可以选择 Java、Kotlin 或 Groovy。
- 可以选择 Spring Boot 的版本。
- 可以添加组件的依赖关系。
- 可以配置项目元数据，包括：Group、Artifact、Name、Description 和 Package name。

基础篇

第 3 章 Spring Cloud Alibaba 基础实战

本章通过基础实战带领读者感受 Spring Cloud Alibaba 的魅力。

3.1 Spring Cloud Alibaba“牛刀小试”

开发人员在用 Spring Cloud Albaba 作为基础框架开发应用程序时，必须要考虑开发过程中的一些技术痛点，包括乐观锁、动态切换多数据源、SQL 语句中表名的动态替换等。

3.1.1 【实例】实现乐观锁

本实例的源码在本书配套资源的“/spring-cloud-alibaba-practice/chapterthree/optimistic-locking-spring-cloud-alibaba/”目录下。

比如，线程 A 正在更新数据库中的某一条记录，如果这时线程 B 也更新该数据库中的同一条记录，并且线程 A 和线程 B 是跨实例的。在这种业务场景下，本地事务也无法控制数据的一致性。

下面采用 Spring Cloud Alibaba 演示如何通过 MyBatis-Plus 实现乐观锁，保证在高并发业务下面场景下的数据一致性。

MyBatis-Plus（简称 MP）是 MyBatis 的增强工具，它在 MyBatis 的基础上做了增强，为简化开发、提高效率而生。

1. 设计思路

设计思路如下：

（1）取出当前需要更新的记录，获取当前 version；

（2）在更新时带上这个 version；

（3）在执行更新时，在 SQL 中加入如下逻辑：

```
set version = newVersion where version = oldVersion
```

（4）如果 version 和 oldVersion 不相等，则更新失败。

2. 代码实现

代码实现如下：

（1）在实例的数据库表 example2_product 中添加 version 字段，默认值为 0。

（2）添加 Mybatis-Plus 的乐观锁的全局配置，具体代码如下：

```
@Configuration
public class MybatisPlusOptLockerConfig {
    @Bean
public MybatisPlusInterceptor mybatisPlusInterceptor() {
        MybatisPlusInterceptor interceptor = new MybatisPlusInterceptor();
        interceptor.addInnerInterceptor(new
OptimisticLockerInnerInterceptor());
        return interceptor;
    }
}
```

（3）在数据库实体 Example2ProductEntity 类中添加 Mybatis-Plus 的版本号注解 @Version，具体代码如下：

```
@Data
@Accessors(chain = true)
@TableName(value = "example2_product")
public class Example2ProductEntity implements Serializable {
    static final long serialVersionUID = -232434345545442L;
    @TableId(value = "id", type = IdType.AUTO)
    private Long id;
    private Long goodId;
    private String goodName;
private Long num;
    //用注解@Version 来开启乐观锁
    @Version
```

```
    private Integer version;
}
```

（4）模拟两个线程同时更新商品库存的业务场景，具体代码如下：

```
//同时更新商品库存：原始库存为 100，依次增加库存 50 和扣减库存 30，理论上最终库存是 120
public void updateByEntitySucc() {
    QueryWrapper<Example2ProductEntity> ew = new QueryWrapper<>();
    ew.eq("id",3477374334L);
    Long id;
    Example2ProductEntity goodA=example2ProductMapper.selectOne(ew);
    Example2ProductEntity goodB=example2ProductMapper.selectOne(ew);
    id=goodB.getId();
    goodA.setNum(goodA.getNum()+50);
    example2ProductMapper.updateById(goodA);
    goodB.setNum(goodB.getNum()-30);
    int result=example2ProductMapper.updateById(goodB);
    if(result==0){
        Example2ProductEntity
goodC=example2ProductMapper.selectById(goodB.getId());
        goodC.setNum(goodC.getNum()-30);
        example2ProductMapper.updateById(goodC);
    }
    Example2ProductEntity newEntity=example2ProductMapper.selectById(id);
    System.out.println("当前库存为："+newEntity.getNum());
}
```

在执行库存扣减之前，数据库表 example2_product 中的原始库存为 100，如图 3-1 所示。

	id	good_id	num	good_name	version
1	3467374334	7878	983848034626479	手机	0
2	3477374334	7878	100	苹果笔记本电脑	0

图 3-1

- 如果没有加乐观锁，则最终商品库存为 70，实际的 SQL 语句执行日志如下：

①执行增加库存 50，SQL 语句执行日志如图 3-2 所示。

```
Creating a new SqlSession
SqlSession [org.apache.ibatis.session.defaults.DefaultSqlSession@7144fd26] was not registered for synchronization because synchronization is not active
JDBC Connection [HikariProxyConnection@1478075595 wrapping com.mysql.cj.jdbc.ConnectionImpl@7afa281b] will not be managed by Spring
==>  Preparing: UPDATE example2_product SET good_id=?, good_name=?, num=?, version=? WHERE id=?
==> Parameters: 7878(Long), 苹果笔记本电脑(String), 150(Long), 0(Integer), 3477374334(Long)
<==    Updates: 1
Closing non transactional SqlSession [org.apache.ibatis.session.defaults.DefaultSqlSession@7144fd26]
```

图 3-2

②执行扣减库存 30，SQL 语句执行日志如图 3-3 所示。

```
Creating a new SqlSession
SqlSession [org.apache.ibatis.session.defaults.DefaultSqlSession@319d28e5] was not registered for synchronization because synchronization is not active
JDBC Connection [HikariProxyConnection@1118057081 wrapping com.mysql.cj.jdbc.ConnectionImpl@7afa281b] will not be managed by Spring
==>  Preparing: UPDATE example2_product SET good_id=?, good_name=?, num=?, version=? WHERE id=?
==> Parameters: 7878(Long), 苹果笔记本电脑(String), 70(Long), 0(Integer), 3477374334(Long)
<==    Updates: 1
Closing non transactional SqlSession [org.apache.ibatis.session.defaults.DefaultSqlSession@319d28e5]
```

图 3-3

③在更新商品库存完成后，查询更新结果的 SQL 语执行日志，如图 3-4 所示。

```
Creating a new SqlSession
SqlSession [org.apache.ibatis.session.defaults.DefaultSqlSession@4cd737c7] was not registered for synchronization because synchronization is not active
JDBC Connection [HikariProxyConnection@1471395282 wrapping com.mysql.cj.jdbc.ConnectionImpl@7afa281b] will not be managed by Spring
==>  Preparing: SELECT id,good_id,good_name,num,version FROM example2_product WHERE id=?
==> Parameters: 3477374334(Long)
<==    Columns: id, good_id, good_name, num, version
<==        Row: 3477374334, 7878, 苹果笔记本电脑, 70, 0
<==      Total: 1
Closing non transactional SqlSession [org.apache.ibatis.session.defaults.DefaultSqlSession@4cd737c7]
当前库存为: 70
```

图 3-4

- 在添加乐观锁配置后，最终库存为 120，库存更新逻辑正确。实际的 SQL 语句执行日志如下：

①执行增加库存 50，SQL 语句执行日志如图 3-5 所示，执行设置库存为 150 的 SQL 语句成功，但是执行设置库存为 70 的 SQL 语句失败，确保库存扣减的数据一致性。

```
Creating a new SqlSession
SqlSession [org.apache.ibatis.session.defaults.DefaultSqlSession@d0bfd2f] was not registered for synchronization because synchronization is not active
JDBC Connection [HikariProxyConnection@448288658 wrapping com.mysql.cj.jdbc.ConnectionImpl@71bb2250] will not be managed by Spring
==>  Preparing: UPDATE example2_product SET good_id=?, good_name=?, num=?, version=? WHERE id=? AND version=?
==> Parameters: 7878(Long), 苹果笔记本电脑(String), 150(Long), 1(Integer), 3477374334(Long), 0(Integer)
<==    Updates: 1
Closing non transactional SqlSession [org.apache.ibatis.session.defaults.DefaultSqlSession@d0bfd2f]
Creating a new SqlSession
SqlSession [org.apache.ibatis.session.defaults.DefaultSqlSession@77d85aab] was not registered for synchronization because synchronization is not active
JDBC Connection [HikariProxyConnection@317940623 wrapping com.mysql.cj.jdbc.ConnectionImpl@71bb2250] will not be managed by Spring
==>  Preparing: UPDATE example2_product SET good_id=?, good_name=?, num=?, version=? WHERE id=? AND version=?
==> Parameters: 7878(Long), 苹果笔记本电脑(String), 70(Long), 1(Integer), 3477374334(Long), 0(Integer)
<==    Updates: 0
Closing non transactional SqlSession [org.apache.ibatis.session.defaults.DefaultSqlSession@77d85aab]
```

图 3-5

②执行扣减库存 30，SQL 语句执行日志如图 3-6 所示。

```
Creating a new SqlSession
SqlSession [org.apache.ibatis.session.defaults.DefaultSqlSession@6223424] was not registered for synchronization because synchronization is not active
JDBC Connection [HikariProxyConnection@1660776075 wrapping com.mysql.cj.jdbc.ConnectionImpl@71bb2250] will not be managed by Spring
==>  Preparing: UPDATE example2_product SET good_id=?, good_name=?, num=?, version=? WHERE id=? AND version=?
==> Parameters: 7878(Long), 苹果笔记本电脑(String), 120(Long), 2(Integer), 3477374334(Long), 1(Integer)
<==    Updates: 1
Closing non transactional SqlSession [org.apache.ibatis.session.defaults.DefaultSqlSession@6223424]
Creating a new SqlSession
```

图 3-6

③更新商品库存完成后，查询更新结果的 SQL 语句执行日志，如图 3-7 所示。

```
Creating a new SqlSession
SqlSession [org.apache.ibatis.session.defaults.DefaultSqlSession@1be9f5f1] was not registered for synchronization because synchronization is not active
JDBC Connection [HikariProxyConnection@2126577093 wrapping com.mysql.cj.jdbc.ConnectionImpl@4083cc87] will not be managed by Spring
==>  Preparing: SELECT id,good_id,good_name,num,version FROM example2_product WHERE id=?
==> Parameters: 3477374334(Long)
<==    Columns: id, good_id, good_name, num, version
<==        Row: 3477374334, 7878, 苹果笔记本电脑, 120, 2
<==      Total: 1
Closing non transactional SqlSession [org.apache.ibatis.session.defaults.DefaultSqlSession@1be9f5f1]
当前库存为: 120
```

图 3-7

④查询数据库，可以发现乐观锁的版本号已经从 0 变更为 2，说明乐观锁已经成功地控制了 SQL 语句更新的次数为 2，如图 3-8 所示，符合预期的控制并发的效果。

Example2UserManager.java　Example2UserInit.java　Example2ProductMapper.java　example2_product　bootstrap.yaml

pom.xml (optimistic-locking-spring-cloud-alibaba)

2 rows　Tx: Auto　Comma-...d (CSV)　example2_product

<Filter Criteria>

	id	good_id	num	good_name	version
1	3467374334	7878	983848034626479	手机	0
2	3477374334	7878	120	苹果笔记本电脑	2

图 3-8

3.1.2 【实例】实现多数据源

本实例的源码在本书配套资源的“/spring-cloud-alibaba-practice/chapterthree/multidata-source-spring-cloud-alibaba/”目录下。

在微服务开发过程中会存在使用多个数据源的业务场景，比如主从分离。本实例将演示如何实现多数据源。

1. 准备环境

在 MySQL 中新建 3 个数据源 master、slave_1 和 slave_2，如图 3-9 所示。

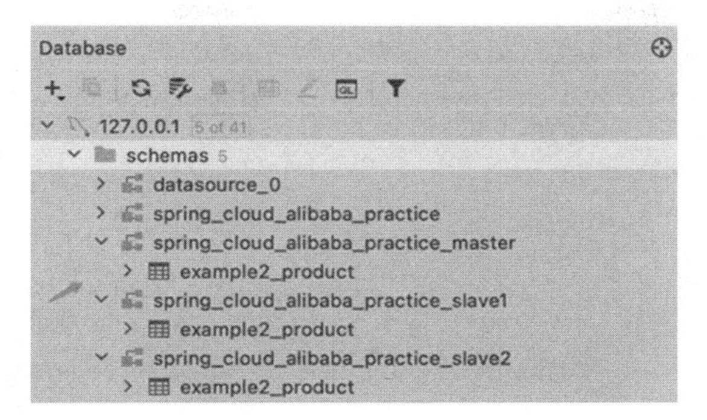

图 3-9

其中，master 数据源的环境配置信息如下所示，slave_1 和 slave_2 数据源的环境配置信息请参考本实例的配套代码中的文件“resources/bootstrap.yaml”。

```
spring:
  datasource:
    dynamic:
      primary: master
      lazy: false
      strict: false
      datasource:
        master:
            url:
jdbc:mysql://127.0.0.1:3306/spring_cloud_alibaba_practice_master?characterEn
coding=utf8&connectTimeout=1000&socketTimeout=3000&autoReconnect=true&useUni
code=true&useSSL=false&serverTimezone=UTC
            username: root
            password: 123456huxian
            maximumPoolSize: 10
            minimumIdle: 2
            idleTimeout: 600000
            connectionTimeout: 30000
            maxLifetime: 1800000
  ...
```

2. 代码实现

（1）引入 dynamic-datasource-spring-boot-starter：

```
<dependency>
    <groupId>com.baomidou</groupId>
    <artifactId>dynamic-datasource-spring-boot-starter</artifactId>
    <version>3.2.0</version>
</dependency>
```

（2）使用多个数据源，具体代码如下：

```
    @Service
    @DS("master")
    public class Example4UserServiceImpl extends
ServiceImpl<Example2ProductMapper,Example2ProductEntity> implements
Example4UserService{

        @Resource
        private Example2ProductMapper example2ProductMapper;

        @DS("master")
        @Override
        public List<Example2ProductEntity> selectFromMaster(Example2ProductBo
example2ProductBo) {
            return
example2ProductMapper.queryGoodInfoByGoodId(example2ProductBo);
        }

        @DS("slave_1")
        @Override
        public List<Example2ProductEntity> selectFromSlave1(Example2ProductBo
example2ProductBo) {
            return
example2ProductMapper.queryGoodInfoByGoodId(example2ProductBo);
        }

        @DS("slave_2")
        @Override
        public List<Example2ProductEntity> selectFromSlave2(Example2ProductBo
example2ProductBo) {
            return
example2ProductMapper.queryGoodInfoByGoodId(example2ProductBo);
        }
    }
```

3. 运行结果

启动实例，如果看到如下日志，则说明多数据源已经配置成功。

```
    ...
    2021-03-14 22:16:21.300  INFO 28412 --- [           main]
c.b.d.d.DynamicRoutingDataSource         : dynamic-datasource - load a
datasource named [slave_2] success
    2021-03-14 22:16:21.300  INFO 28412 --- [           main]
c.b.d.d.DynamicRoutingDataSource         : dynamic-datasource - load a
datasource named [master] success
```

```
    2021-03-14 22:16:21.300  INFO 28412 --- [           main]
c.b.d.d.DynamicRoutingDataSource         : dynamic-datasource - load a
datasource named [slave_1] success
```

3.1.3 【实例】实现 SQL 语句中表名的动态替换

本实例的源码在本书配套资源的“/spring-cloud-alibaba-practice/chapterthree/replace-table-spring-cloud-alibaba/”目录下。

在多租户的业务场景中，通常会用到物理隔离和逻辑隔离。物理隔离可以通过分表来实现。如果采用分表来解决数据的物理隔离，则需要实现 SQL 语句中表名的动态替换。

1. 数据库的建表语句

数据的库建表语句可以参考本实例的配套代码中的文件“resources/init.sql”。

2. 执行过程分析

（1）添加表名动态替换的全局配置，具体代码如下：

```
    @Configuration
    @MapperScan("com.alibaba.cloud.youxia.mapper")
    public class MybatisPlusConfig {
        @Resource
        private UserService userService;
        @Bean
        public MybatisPlusInterceptor mybatisPlusInterceptor() {
            MybatisPlusInterceptor interceptor = new MybatisPlusInterceptor();
            DynamicTableNameInnerInterceptor dynamicTableNameInnerInterceptor =
new DynamicTableNameInnerInterceptor();
            HashMap<String, TableNameHandler> map = new HashMap<String,
TableNameHandler>(2) {{
                put("example5_order", (sql, tableName) -> {
                    //按照租户 ID 实现表名的动态替换
                    UserDTO userDTO=userService.getLoginUser();
                    return tableName + "_"+userDTO.getTenantId();
                });
            }};
            dynamicTableNameInnerInterceptor.setTableNameHandlerMap(map);
            interceptor.addInnerInterceptor(dynamicTableNameInnerInterceptor);
            return interceptor;
        }
    }
```

具体逻辑如下：

①通过用户服务获取当前登录的用户信息，然后从用户信息中获取租户 ID。

②采用组合规则“example5_order_*租户 ID*”重新生成表名，它对应数据库中实际分表之后的物理表名称。

③通过动态表名拦截器类 DynamicTableNameInnerInterceptor，进行 SQL 语句中表名的动态替换。

（2）配置 Mybatis-Plus 对应的 SQL 映射语句，具体代码如下：

```
    <?xml version="1.0" encoding="UTF-8" ?>
    <!DOCTYPE mapper PUBLIC "-//mybatis.org//DTD Mapper 3.0//EN"
"http://mybatis.org/dtd/mybatis-3-mapper.dtd" >
    <mapper namespace="com.alibaba.cloud.youxia.mapper.Example5OrderMapper">
        <select id="selectOrder"
resultType="com.alibaba.cloud.youxia.entity.Example5OrderEntity"
              parameterType="com.alibaba.cloud.youxia.bo.Example5OrderBo">
            //如果在数据库中不存在表名 example5_order，则需要动态替换
            select * from example5_order
            where order_id>0
        </select>
    </mapper>
```

3. 执行过程

运行本实例，具体 SQL 语句的执行过程如下：

```
    JDBC Connection [HikariProxyConnection@1290940607 wrapping
com.mysql.cj.jdbc.ConnectionImpl@3dc55719] will not be managed by Spring
    ==>  Preparing: select * from example5_order_6767 where order_id>0
    ==> Parameters:
    <==    Columns: id, order_name, order_id
    <==        Row: 2323233, 租户 6767 的订单, 2388923923
    <==      Total: 1
```

可以看到：在 SQL 语句的执行过程中，数据库表名 example5_order 被替换为对应租户的数据库表名 example5_order_6767。

在接下的 3.2 节和 3.3 节中，将实现两个常用但又相对复杂一点的实例。

3.2 【实例】用 Maven 和 Spring Cloud Alibaba 实现多环境部署

本实例的源码在本书配套资源的“/spring-cloud-alibaba-practice/chapterthree/multi-environment-deployment/”目录下。

在开发过程中，经常要碰到多环境部署的情况。比如一套代码要打包并部署在开发环境、测试环境、预发布环境及线上环境中。

本实例演示如何用 Maven 和 Spring Cloud Alibaba 实现多环境部署。

3.2.1　初始化

本实例用 Spring Cloud Alibaba 作为基础框架，并添加了 Dubbo 和 RESTful API 接口。项目主要包含如下模块。

（1）user-server：包含一些业务核心代码的实现，比如 Dubbo 接口。

（2）user-client：包含 Dubbo 的 SDK、一些 Dubbo 的出参和入参，以及其他一些常量模块。

（3）user-distribution：包含项目工程的打包规则。

3.2.2　多环境配置

本实例的多环境配置主要包含两个业务产品：productA 和 productB。每个产品对应 3 个环境：development（开发）、test（测试）和 prod（生产）。项目的多环境代码结构如图 3-10 所示。

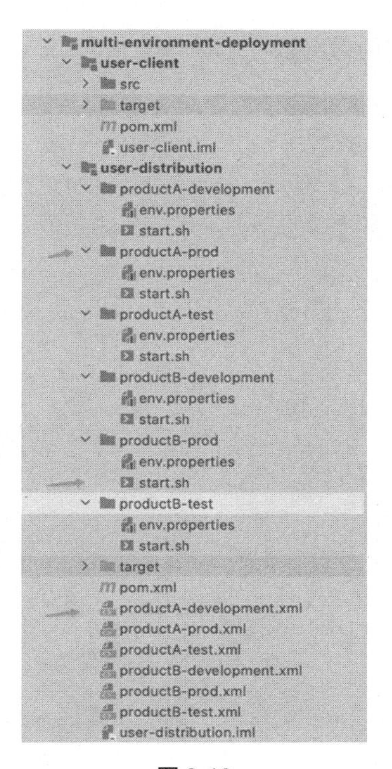

图 3-10

比如，productA-development 代表 productA 产品的 development（开发环境）；start.sh 为项目的启动脚本，运维人员可以直接运行它来启动应用；env.properties 为 productA 产品对应服务的个性化属性配置，具体代码如下：

```
#productA-development 应用的个性化参数
JAVA_HOME="/data/java/jdk1.8.0_161"
# JVM 参数配置在程序发生 OOM 时“杀死”进程，并通过保活机制重启程序
JAVA_PARAMS="$JAVA_PARAMS -XX:OnOutOfMemoryError='kill -9 %p'"
# 监控的配置
JAVA_PARAMS="$JAVA_PARAMS -DSW_AGENT_NAME=user-server"
JAVA_PARAMS="$JAVA_PARAMS
-javaagent:/data/skywalking/skywalking-agent.jar"
```

在以上代码中，在 env.properties 中配置了 JDK 路径、JVM 相关参数，以及 Skywalking 监控的配置信息。

3.2.3 构建

为了达到多环境部署的效果，须按照不同的环境标识去构建不同的 Jar 包。下面介绍如何构建项目。

（1）在 pom.xml 中添加 profile 标签，用于添加产品对应的环境配置。

productA 的 development 环境的具体代码如下：

```
<profile>
    //对应 MVN 的打包命令（-P）的入参
    <id>productA-development</id>
<properties>
    //对应产品及环境的文件夹名称
        <environment>productA-development</environment>
    </properties>
    <activation>
        <activeByDefault>true</activeByDefault>
    </activation>
<dependencies>
    //添加打包过程中所依赖的 Jar 包
        <dependency>
            <artifactId>user-server</artifactId>
            <groupId>com.alibaba.youxia</groupId>
            <version>1.0.0.release</version>
        </dependency>
        <dependency>
            <artifactId>user-client</artifactId>
            <groupId>com.alibaba.youxia</groupId>
```

```
            <version>1.0.0.release</version>
        </dependency>
    </dependencies>

    <build>
        <plugins>
            <plugin>
                <groupId>org.apache.maven.plugins</groupId>
                <artifactId>maven-assembly-plugin</artifactId>
                <version>2.4</version>
                <configuration>
                    //定义打包的名称

<finalName>user-distribution-${project.version}</finalName>
                    <appendAssemblyId>false</appendAssemblyId>
                    <descriptors>
                        //对应产品及环境的规则的 XML 文件
                        <descriptor>productA-development.xml</descriptor>
                    </descriptors>
                </configuration>
                <executions>
                    <execution>
                        <phase>package</phase>
                        <goals>
                            <goal>single</goal>
                        </goals>
                    </execution>
                </executions>
            </plugin>
        </plugins>
    </build>
</profile>
```

（2）配置对应产品及环境的打包规则。productA 的 development 环境对应的 XML 文件具体代码如下：

```
<assembly>
    //产品及环境名称
    <id>productA-development</id>
<includeBaseDirectory>true</includeBaseDirectory>
    //输出压缩包格式
    <formats>
        <format>dir</format>
        <format>tar.gz</format>
        <format>zip</format>
    </formats>
```

```
    <fileSets>
        <fileSet>
            <directory>productA-development</directory>
            <outputDirectory>bin</outputDirectory>
            <fileMode>0755</fileMode>
        </fileSet>

        <!-- 指定输出“target/classes”中的配置文件到 config 目录中 -->
        <fileSet>
            <directory>../user-server/target/classes</directory>
            //配置文件格式
            <includes>
                <include>*.yml</include>
                <include>*.properties</include>
                <include>*.xml</include>
            </includes>
            <outputDirectory>config</outputDirectory>
            <fileMode>0755</fileMode>
        </fileSet>

        <fileSet>
            <directory>${build.outputDirectory}/logs</directory>
            <outputDirectory>logs</outputDirectory>
            <fileMode>0644</fileMode>
        </fileSet>

        <!-- 将第三方依赖打包到 Lib 目录中 -->
        <fileSet>
            <directory>../user-client/target</directory>
            <includes>
                <include>*.jar</include>
            </includes>
            <outputDirectory>lib</outputDirectory>
            <fileMode>0755</fileMode>
        </fileSet>

        <fileSet>
            <directory>../user-server/target</directory>
            <includes>
                <include>*.jar</include>
            </includes>
            <outputDirectory>lib</outputDirectory>
            <fileMode>0755</fileMode>
        </fileSet>
```

```
            <!-- 将第三方依赖打包到 Lib 目录中 -->
            <fileSet>
                <directory>../user-server/target/starter-run/lib</directory>
                <includes>
                    <include>*.jar</include>
                </includes>
                <outputDirectory>lib</outputDirectory>
                <fileMode>0755</fileMode>
            </fileSet>
        </fileSets>
    </assembly>
```

3.2.4　效果演示

在搭建完多环境部署的项目后，需要通过 MVN 命令将其打包，生成对应环境的 Jar 包。

（1）通过命令“mvn clean package -PproductB-development”打包。

（2）在 user-distribution 项目的 target 目录下，可以找到对应的可以直接部署的 Jar 包：user-distribution-1.0.0.release.tar.gz 和 user-distribution-1.0.0.release.zip。

（3）检查对应的 env.properties 文件，可以看到它是 productB-development 的配置文件。具体效果如图 3-11 所示。

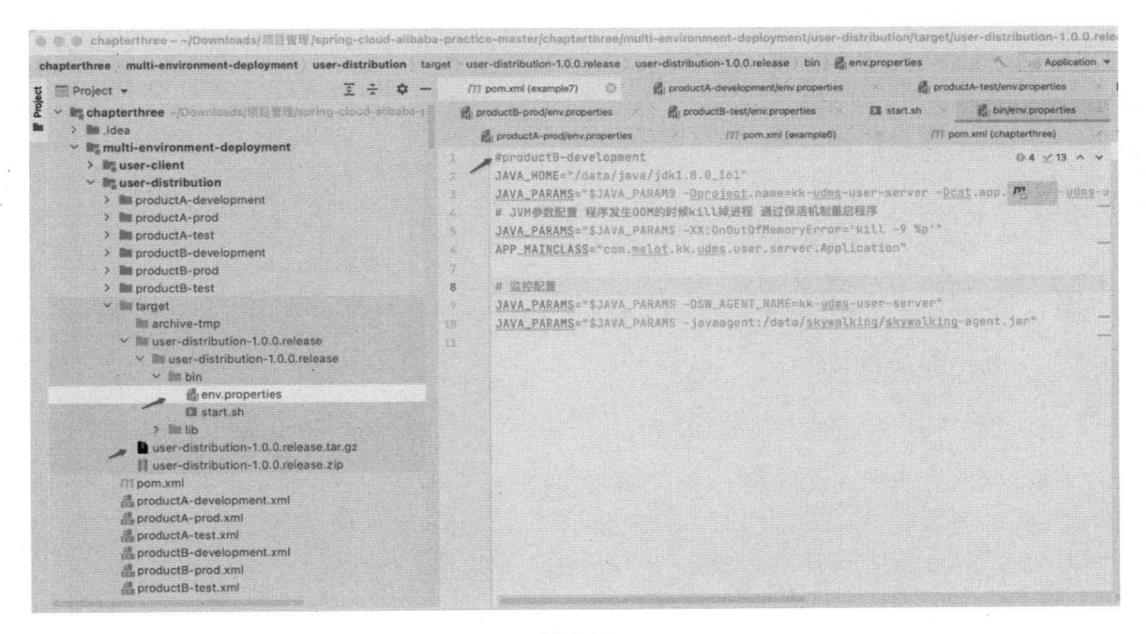

图 3-11

（4）在对应的 bin 目录下有对应产品的两个文件：env.properties 和 start.sh。如果在 env.properties 文件中存在属性值“#productB-development”，则说明已经将 productB-development 对应的应用参数加载到 Jar 文件指定的路径下了。

如果针对产品的不同环境的启动脚本也要做个性化，则可以通过定制化 start.sh 来实现，这里就不举例了。

3.3 【实例】用“MyBatis-Plus + Spring Cloud Alibaba”实现多租户架构

本实例的源码在本书配套资源的“/spring-cloud-alibaba-practice/chapterthree/multi-tenant-spring-cloud-alibaba/”目录下。

Spring Cloud Alibaba 是基础框架，开发者可以利用它和 MyBatis-Plus 来实现多租户架构。

3.3.1 多租户的概念

“多租户技术”又被称为“多重租赁技术”，是一种软件架构技术，即实现多用户（此处的多用户一般指企业用户）共用系统或程序组件，并确保各用户间数据的隔离性。简单来讲：在一台服务器上运行单个应用实例，它为多个租户提供服务。

从上面的定义中可以看出：多租户是一种架构，目的是让多个用户使用同一套程序，且保证用户间数据隔离。

3.3.2 多租户的原理

Mybatis-Plus 通过插件机制来实现多租户，其插件主体是 MybatisPlusInterceptor。

MybatisPlusInterceptor 是一个全局拦截器类，继承于 org.apache.ibatis.plugin.Interceptor，并重写了 intercept()方法。如果应用将 MybatisPlusInterceptor 托管到 IOC 容器中，则 Mybatis-Plus 插件机制会自动生效，并执行拦截处理。

Mybatis-Plus 单独封装了多租户拦截器 TenantLineInnerInterceptor。

在 MybatisPlusInterceptor 中，可以通过 addInnerInterceptor()方法添加拦截器。如果要在

应用中使用 Myatis 的多租户功能，则可以手动调用它——添加 TenantLineInnerInterceptor。

1. 多租户拦截器何时生效

MybatisPlusInterceptor 在 intercept()方法中遍历拦截器列表，并依次执行对应拦截器中的拦截处理方法。这样，被 MyBatis-Plus 封装的多租户拦截器就可以全局地执行拦截功能了。

Mybatis-Plus 非常灵活，可以通过注解配置白名单，有选择地对表及对应的 SQL 语句做多租户拦截处理。

2. 源码解读

MyBatis-Plus 通过 TenantLineInnerInterceptor 类实现租户拦截功能，部分源码如下：

```
    @Data
    @NoArgsConstructor
    @AllArgsConstructor
    @ToString(callSuper = true)
    @EqualsAndHashCode(callSuper = true)
    @SuppressWarnings({"rawtypes"})
    public class TenantLineInnerInterceptor extends JsqlParserSupport
implements InnerInterceptor {
     ...
     @Override
        protected void processInsert(Insert insert, int index, String sql, Object
obj) {
            if (tenantLineHandler.ignoreTable(insert.getTable().getName())) {
                //①如果配置了需要过滤掉表的 insert 操作，则直接返回
                return;
            }
            List<Column> columns = insert.getColumns();
            if (CollectionUtils.isEmpty(columns)) {
                //②不处理没有列名的 insert 操作
                return;
            }
            String tenantIdColumn = tenantLineHandler.getTenantIdColumn();
            if (columns.stream().map(Column::getColumnName).anyMatch(i ->
i.equals(tenantIdColumn))) {
                //③不处理非租户列的 insert 操作
                return;
            }
            columns.add(new Column(tenantLineHandler.getTenantIdColumn()));
            Select select = insert.getSelect();
            if (select != null) {
                this.processInsertSelect(select.getSelectBody());
            } else if (insert.getItemsList() != null) {
```

```
                // fixed github pull/295
                ItemsList itemsList = insert.getItemsList();
                if (itemsList instanceof MultiExpressionList) {
                    ((MultiExpressionList) itemsList).getExprList().forEach(el
-> el.getExpressions().add(tenantLineHandler.getTenantId()));
                } else {
                    ((ExpressionList)
itemsList).getExpressions().add(tenantLineHandler.getTenantId());
                }
            } else {
                throw ExceptionUtils.mpe("Failed to process multiple-table update,
please exclude the tableName or statementId");
            }
        }
        ...
    }
```

以上只是分析了 TenantLineInnerInterceptor 类中的 processInsert()方法的逻辑，还有很多其他的方法，这里就不分析了。如果读者感兴趣，可以查阅源码。

3.3.3 架构

多租户架构的技术难点是数据隔离。通常会采用物理隔离和逻辑隔离，前者的技术难度要远大于后者。

在多租户架构中，租户 ID 是一个很重要的概念。它可以用于单表数据的逻辑隔离，也可以用于分库分表之后的物理隔离。

1. 逻辑隔离的多租户架构

如果应用采用逻辑隔离来实现多租户，则对应的所有的租户会共享物理资源。物理资源包含服务资源和硬盘资源。

- 服务资源：租户要完成某一个业务功能所需要调用的应用服务。
- 硬盘资源：存储资源，比如数据库。

常规的逻辑隔离的多租户架构如图 3-12 所示。

业务流量通过负载均衡器（SLB），进入 API 网关。在 HTTP 请求头中设置租户 ID，租户 ID 会从网关层传递到微服务集群的微服务中。微服务在进行数据持久化之前，会重新解析 SQL 语句，并添加租户 ID，完成租户 ID 和对应数据的持久化。

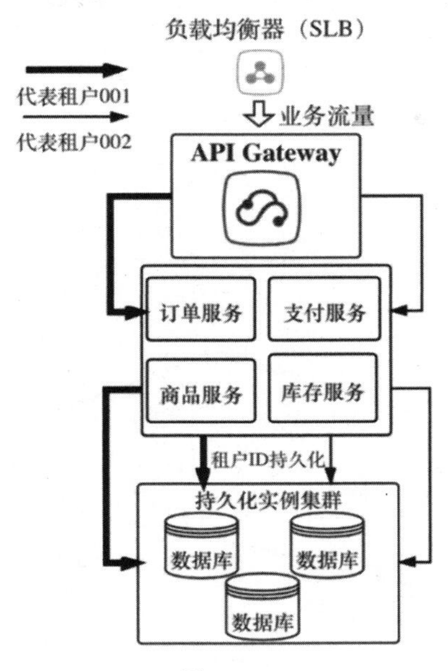

图 3-12

逻辑隔离主要的开发成本来源于：业务表都要增加额外的租户 ID 字段，以及在 ORM 层要做 SQL 拦截。比如，将租户 ID 为 001 和租户 ID 为 002 的请求都路由到同一个服务集群，再传递到数据库中。在图 3-12 中有 4 个微服务：订单服务、支付服务、商品服务和库存服务。需要在对应服务的订单表、支付流水表、库存表和商品表中添加租户 ID 字段，用于持久化租户 ID。如果服务查询带有租户 ID 字段的表，则需要在 SQL 语句中添加租户 ID 的查询条件。

2. 物理隔离的多租户架构

物理隔离通常是指，将不同租户产生的业务数据完全隔离。隔离的方式通常有：分表或者分库分表。分表相对比较简单，而分库分表则要考虑跨库查询和分布式事务的问题。

多租户架构通常都要考虑租户的 SLA（服务等级协议），即给不同的租户等级提供不同级别的 SLA。

常规的物理隔离的多租户架构如图 3-13 所示。流量通过负载均衡器 SLB 进入 API 网关，API 网关将租户 ID 和命名空间传递到微服务集群中。线上服务通常都是多实例部署。

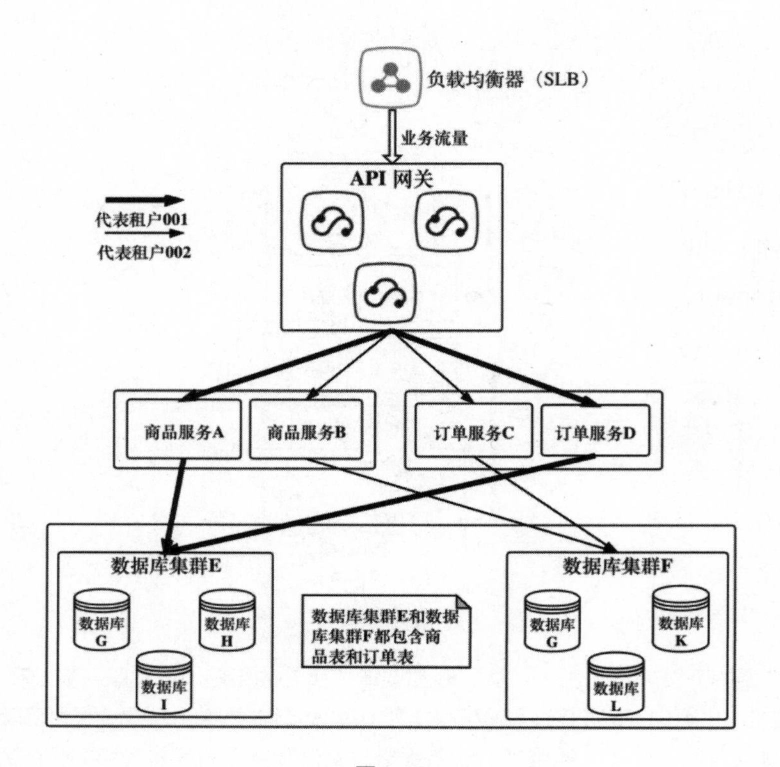

图 3-13

在订单服务和商品服务中都部署了两个实例，商品服务对应服务 A 和服务 B，订单服务对应服务 C 和服务 D。

- 通过租户 ID 可以做服务实例的物理隔离。比如：租户 ID 为 001 的数据库操作被路由到商品服务 A 和订单服务 D；租户 ID 为 002 的数据库操作被路由到商品服务 B 和订单服务 C。
- 在数据库持久层中，将商品表和订单表进行物理隔离。这样租户 ID 为 001 的数据库操作被请求路由到数据库集群 E，租户 ID 为 002 的数据库操作被请求路由到数据库集群 F。

3.3.4 搭建及效果演示

由于篇幅限制，这里只演示逻辑数据隔离的多租户。

1. 创建用于逻辑隔离的租户表

下面新建一张用户表并新增一个 tenant_id 字段，具体建表语句如下：

```
CREATE TABLE user
(
    id BIGINT(20) NOT NULL COMMENT '主键 ID',
```

```
    tenant_id BIGINT(20) NOT NULL COMMENT '租户 ID',
    name VARCHAR(30) NULL DEFAULT NULL COMMENT '姓名',
    PRIMARY KEY (id)
);
```

2. 搭建 Mybatis-Plus 租户环境

使用 mybatis-plus-boot-starter 快速搭建租户环境，具体可以参考本实例配套资源中的代码。

3. 执行带 tenant_id 的 SQL 语句

执行如下 SQL 语句，Mybatis-Plus 会在 SQL 语句中全局添加 tenant_id 字段：

```
SELECT u.id, u.name, a.name AS addr_name FROM user u LEFT JOIN user_addr a
ON a.user_id = u.id WHERE u.name LIKE concat(concat('%', ?), '%') AND u.tenant_id
= 001;
```

第 4 章 分布式服务治理——基于 Nacos

应用被水平拆分和垂直拆分后，在线上部署时，应用的实例数几何倍数增加，对开发人员和运维人员的技术挑战性非常高。

服务治理本质上是管理服务与服务之间的调用关系，并维持服务链路关系稳定的一系列架构方法论。

4.1 认识分布式服务治理

万丈高楼从地起。开发人员可以先认识分布式服务治理，再去熟悉分布式服务治理领域的核心技术。

4.1.1 什么是分布式服务治理

在分布式系统的构建中,一个好的服务治理平台可以大大降低协作开发的成本和版本的迭代效率。

服务治理主要包括如下几点。

- 服务注册与发现：在单体架构被拆分为微服务架构后，如果在微服务之间存在调用依赖，则进行服务治理需要得到目标服务的服务地址，即微服务治理的“服务发现”。要完成服务发现，则需要将服务信息存储在某个载体中。载体本身即微服务治理的“注册中心”，而存储到载体的动作即“服务注册”。

- 可观测性：在改为微服务架构之后，应用有了更多的部署载体，所以需要对众多服务间的调用关系、状态有清晰的掌控。可观测性包括调用拓扑关系、监控、日志、调用追踪等。
- 流量管理：由于微服务本身存在不同版本，所以在版本迭代过程中，需要对微服务间的调用进行控制，以完成微服务版本的平滑升级。在这个过程中，需要根据流量控制规则将流量路由到不同版本的服务，这也孵化出了灰度发布、蓝绿发布、A/B 测试等服务治理的细分主题。
- 安全性管理：不同微服务担当不同的业务角色，对于业务敏感的微服务，需要对其访问进行认证与鉴权（即安全问题）。
- 权限控制：在服务治理过程中，需要比较完善的权限控制体系，以控制服务之间的调用权限。

4.1.2　为什么需要分布式服务治理

在没有服务治理前，服务之间是通过直连的方式来相互访问的，如图 4-1 所示。比如，订单服务要访问支付服务的支付接口，所以订单服务需要在本地配置 IP 地址、端口号及支付接口 URL。

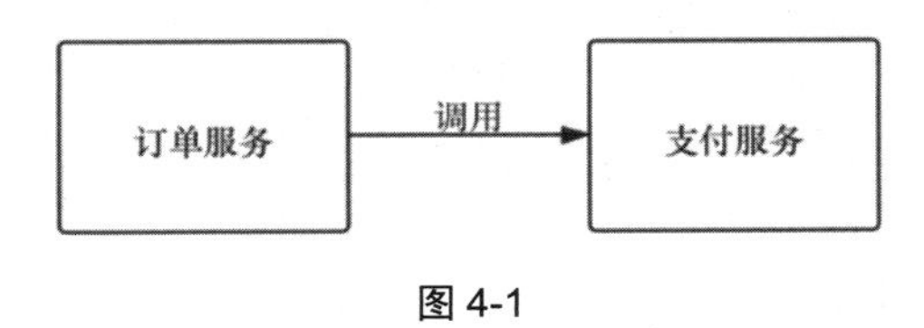

图 4-1

从图 4-1 可以看出，订单服务是直接调用支付服务。如果系统不复杂，则这样的调用是没什么问题的。但在采用微服务架构后，采用这种调用方式就会有问题了。

微服务架构中的服务非常多，并且都是采用集群部署的。每个服务都要维护一套服务之间的调用关系，非常烦琐。这时就需要一个中间代理来维护调用关系，注册中心应运而生。服务之间的调用如图 4-2 所示。

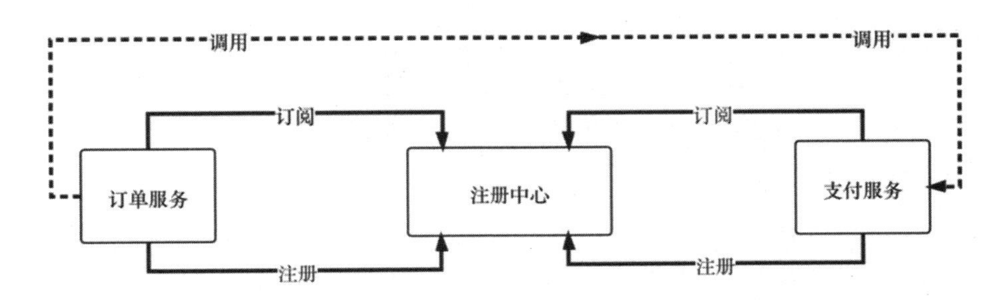

图 4-2

从图 4-2 可以看出，订单服务和支付服务都要注册到注册中心中，并且可以订阅注册中心中的所有的服务。订单服务通过订阅服务，可以间接地调用支付服务。

4.2 了解主流的注册中心

目前有很多注册中心的技术解决方案，Nacos、ZooKeeper、Sofa 等是主流的注册中心。

4.2.1 Nacos

Nacos 致力于发现、配置和管理微服务。它提供了一组简单易用的功能，帮助开发人员快速实现动态服务发现、服务配置、服务元数据及流量管理，从而更敏捷和更容易地构建、交付和管理微服务平台。

Nacos 是构建以“服务”为中心的现代应用架构（例如微服务范式、云原生范式）的基础设施。

Nacos 的主要特性以下。

1. 服务发现与服务健康检查

Nacos 支持基于 DNS 和 HTTP&API 的服务发现。服务提供者使用原生 SDK、OpenAPI 或一个独立的 Agent 注册 Service 后，服务消费者可以使用 DNS 或者 HTTP&API 查找和发现服务。

Nacos 提供了对服务的实时健康检查，阻止向不健康的主机或服务实例发送请求。Nacos 支持传输层和应用层的健康检查。

对于复杂的云环境和网络拓扑环境（如 VPC、边缘网络等）中的服务健康检查，Nacos 提供了“Agent 上报”和“服务器端主动检测”两种健康检查模式。

Nacos 还提供了统一的健康检查仪表盘，帮助用户根据健康状态来管理服务的可用性及流量。

2. 动态配置服务

动态配置服务可以让用户以中心化、外部化和动态化的方式，管理所有环境的应用配置和服务配置。动态配置消除了在配置变更时重新部署应用和服务的需要，让配置管理变得更加高效和敏捷。配置中心化管理让实现无状态服务变得更简单，让服务按需弹性扩展变得更容易。

3. 动态 DNS 服务

动态 DNS 服务支持权重路由，能更容易、更灵活地实现中间层负载均衡、路由策略、流量控制，以及数据中心内网的简单 DNS 解析。

利用动态 DNS 服务，能更容易地实现以 DNS 协议为基础的服务注册/订阅，以消除在客户端耦合到厂商私有服务时注册/订阅的可能风险。

4. 服务及其元数据管理

Nacos 能从微服务平台建设的视角，管理注册中心中的所有服务及元数据，包括管理服务的描述、生命周期、服务的静态依赖分析、服务的健康状态管理、服务的流量管理、路由及安全策略、服务的 SLA ，以及最重要的 metrics 统计数据。

4.2.2 ZooKeeper

ZooKeeper 从 Apache Hadoop 的子项目发展而来，于 2010 年 11 月正式成为 Apache 的顶级项目。ZooKeeper 为分布式应用提供了高效且可靠的分布式协调服务，以及统一命名服务、配置管理和分布式锁等分布式基础服务。

在解决分布式数据一致性方面，ZooKeeper 并没有直接采用 Paxos 算法，而是采用了一种被称为 ZAB 的一致性算法协议。

ZooKeeper 可以保证如下分布式特性。

- 顺序一致性：从同一个客户端发起的事务请求，将按照其发起顺序被严格地应用到 ZooKeeper 中。
- 原子性：所有事务请求的处理结果在整个集群中所有机器上的应用情况是一致的，即要么集群中的所有机器都成功应用了某一个事务，要么都没有应用，一定不会出现“集群中部分机器应用了该事务，而另外一部分没有应用”的情况，这是不符合数据一致性的。
- 单一视图：无论客户端连接的是哪个 ZooKeeper 服务器，其看到的服务器端数据模型都是一致的。
- 可靠性：一旦服务器成功应用了一个事务，并完成对客户端的响应，那么该事务所引起的服务器端状态变更会被一直保留下来，除非有另一个事务产生了状态变更。
- 实时性：通常人们看到实时性的第一反应是：一旦一个事务被成功应用，那么客户端能够立即从服务器端读取这个事务变更之后的最新的数据。注意，ZooKeeper 仅保证在一定的时间内，客户端能够从服务器端读取最新的数据。

利用 ZooKeeper 进行集群管理的框架有很多，但只有 Apache Dubbo （以下简称 Dubbo ）是基于 ZooKeeper 来实现分布式服务治理中的能力建设的，其他主流框架都只是用 ZooKeeper 来确保分布式环境下集群内数据的一致性的。

> 因为 ZooKeeper 能够确保 CAP 理论中的 C 和 P，即能够确保强一致性和分区容错性，所以，依赖 ZooKeeper 的主流框架几乎都是与数据处理相关的，对数据强一致性要求非常高。

4.2.3 Consul

Consul 是一种服务网格解决方案，提供服务发现和注册、分布式配置、健康检查、负载均衡等功能。将这些功能组合起来，则可以构建出服务治理中的服务网格控制平面。

Consul 集群的原理如图 4-3 所示。

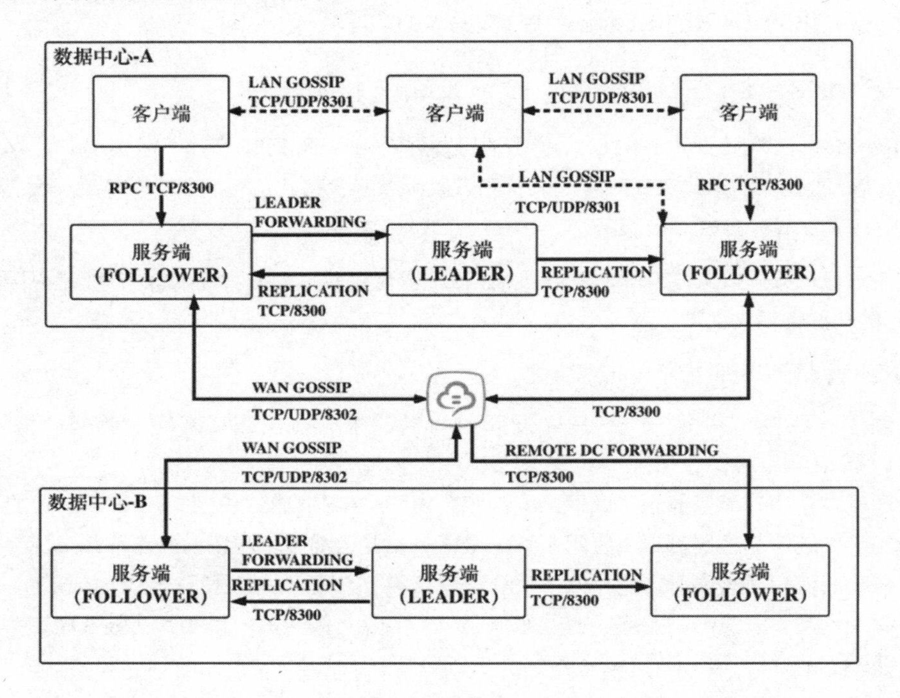

图 4-3

Consul 支持单个数据中心和多个数据中心。在图 4-3 中有两个数据中心，它们通过 Internet 互联。请注意，为了提高通信效率，只有 Server 节点才能加入跨数据中心的通信。

（1）Consul 集群选举。

- Consul 分为 Client 和 Server 两种节点（所有的节点都被称为 Agent）。Server 节点负责保存数据，Client 负责健康检查及转发数据请求到 Server。
- Server 节点分为 1 个 Leader 节点和多个 Follower。
- Leader 节点会将数据同步到 Follower 节点。推荐 Server 节点的数量是 3 个或者 5 个。
- 在 Leader 节点宕机时，会启动选举机制产生一个新的 Leader 节点。

（2）Consul 选举协议。

- 集群内的 Consul 节点通过 Gossip 协议维护成员关系。

- 单个数据中心的 Gossip 协议同时使用 TCP 和 UDP 进行通信，都使用 8301 端口。
- 跨数据中心的 Gossip 协议也同时使用 TCP 和 UDP 进行通信，都使用 8302 端口。
- 集群内数据的读写请求，既可以直接被发到 Server 节点，也可以通过 Client 节点使用 RPC 转发到 Server 节点，请求最终会到达 Leader 节点。

在允许数据轻微“陈旧”的情况下，“读”请求也可以在普通的 Server 节点完成，集群内数据的读写和复制都是通过 TCP 的 8300 端口完成的。

4.2.4 Sofa

Sofa 提供应用之间的点对点服务调用功能，具有高可伸缩、高容错的特性。

- 为保证高可用性，通常同一个应用或同一个服务提供方会被部署多份，以达到对等服务的目标。Sofa 是一个调度器，可以帮助服务消费方从这些对等的服务提供方中，合理地选择一个来执行相关的业务逻辑。
- 为保证应用的高容错性，需要服务消费方能够感知服务提供方的异常，并做出相应的处理，以减少应用出错后导致的“服务调用抖动”。在 Sofa 中，一切服务调用的容错机制均由软负载和配置中心控制。这样可以在应用系统无感知的情况下，帮助服务消费方正确选择“健康”的服务提供方，保障全站的稳定性。

Sofa 中的远程调用，是通过服务模型来定义服务调用双方的。服务分为服务消费方和服务提供方。对于 RPC 的调用端和被调用端，可以将它们理解为“调用客户端”和“调用服务器端”。对于 RPC 服务，服务提供方称之为“服务”（service），而服务消费方称之为“引用”（reference）。

4.2.5 Etcd

Etcd 是 CoreOS 团队于 2013 年 6 月发起的开源项目，其目标是构建一个高可用的分布式键值（Key-Value）数据库。在 Etcd 内部采用 Raft 协议实现一致性算法。Etcd 基于 Go 语言实现。

Etcd 主要分为以下四个部分。

- HTTP Server：用于处理用户发送的 API 请求，以及其他 Etcd 节点的同步与心跳信息请求。
- Store：用于处理 Etcd 支持的各类事务，包括数据索引、节点状态变更、监控与反馈、事件处理与执行等，是 Etcd 对用户提供的大多数 API 功能的具体实现。
- Raft：强一致性算法的具体实现，是 Etcd 的核心。
- WAL（Write Ahead Log，预写式日志）：Etcd 的数据存储方式。除在内存中存储所有数据的状态和节点的索引外，Etcd 还通过 WAL 进行持久化存储。在 WAL 中，所有数据在提交前都会事先记录日志。

- Snapshot：为了防止数据过多而进行的状态快照。
- Entry：存储的具体日志内容。

4.2.6 Eureka

Eureka 是基于 RESTful 的服务注册与发现的解决方案，最开始主要应用于 AWS 云上服务，以实现负载均衡和中间层服务器的故障转移。开源组织 Netflix 已经广泛应用 Eureka。

采用 Eureka 作为注册中心的服务治理框架有很多，包括 Spring Cloud 和 Dubbo。

Netflix 对 Eureka 1.x 还在维护，对 Eureka 2.0 已经停止维护。

4.2.7 对比 Nacos、ZooKeeper、Sofa、Consul、Etcd 和 Euraka

主流注册中心的对比见表 4-1。

表 4-1　对比主流注册中心

功能	Nacos	ZooKeeper	Consul	Sofa	Etcd	Eureka
开源时间	2018 年 6 月	2010 年 11 月	2014 年 4 月	2018 年 4 月	2013 年 6 月	2012 年 9 月
通信渠道	HTTP/gRPC	HTTP	HTTP/DNS	Netty	HTTP/gRPC	HTTP
编程语言	Java	Java	Go	Java	Go	Java
服务健康检查	TCP/HTTP/MySQL	长连接 Keepalive	提供详细的服务状态、内存，以及硬盘的健康检查	需要手动开启健康检查	连接心跳	需要手动开启健康检查
多数据中心	支持单机模式、集群模式，以及跨机房的多集群模式	无	通过 WAN 的 Gossip 协议支持	支持	无	无
Key-Value 存储	支持	支持	支持	支持	支持	无
数据一致性协议	Raft/Soft-Jraft	ZAB	Raft	Soft-Jraft	Raft	基于内存，只能保证最终一致性，不能保证强一致性
CAP	CP/AP	CP	AP	CP	CP	AP
监听器机制	支持	支持	支持	支持	支持	支持

续表

功能	Nacos	ZooKeeper	Consul	Sofa	Etcd	Eureka
自身监控	Metrics	需要使用第三方组件	Metrics	Metrics	Metrics	Metrics
安全及权限管理	ACL/HTTPS	ACL	ACL/HTTPS	支持	HTTPS	ACL
Spring Cloud Alibaba 支持	支持	支持	支持	支持	支持	只支持 Eureka 1.0

4.3　将应用接入 Nacos 注册中心

Nacos Naming 是 Nacos 注册中心的功能模块。开发者可以通过在 Nacos Client 中封装的 NacosNamingService 类将应用快速接入 Nacos 注册中心。

常规的接入方式有如下两种：

- 利用 Nacos 官方提供的“Nacos Client+Spring Boot”模式，这是一个针对 Spring Boot 应用提供的“开箱即用”解决方案。
- 利用 Spring Cloud Alibaba 提供的“Spring Cloud Alibaba Discovery”模式，这是针对 Spring Cloud Alibaba 应用提供的定制化解决方案。

4.3.1　【实例】用“Nacos Client + Spring Boot”接入

本实例的源码在本书配套资源的“/spring-cloud-alibaba-practice/chapterfour/use-nacos-spring-boot/”目录下。

本实例，先用 IDEA 创建一个 Spring Boot 项目，再在其中添加 Nacos Client 相关的依赖。

1. 添加 POM 依赖

```
<!—添加基于 Spring Boot 的 nacos-discovery-spring-boot-starter 的 Jar 包依赖-->
<dependency>
    <groupId>com.alibaba.boot</groupId>
    <artifactId>nacos-discovery-spring-boot-starter</artifactId>
    <version>0.2.7</version>
</dependency>
<!—添加基于 Spring Boot 的 dubbo-spring-boot-starter 的 Jar 包依赖-->
<dependency>
    <groupId>org.apache.dubbo</groupId>
    <artifactId>dubbo-spring-boot-starter</artifactId>
```

```
    <version>2.7.8</version>
</dependency>
```

2. 添加属性文件

在程序中添加配置文件 bootstrap.yaml，具体配置如下所示：

```
spring.application.name=use-nacos-spring-boot
server.port=7823
###添加 Dubbo 的配置信息
dubbo.application.name=use-nacos-spring-boot
dubbo.registry.address=nacos://127.0.0.1:8848
dubbo.protocol.name=dubbo
dubbo.protocol.port=20880
dubbo.application.parameters.namespace=c7ba173f-29e5-4c58-ae78-b102be11c4f9
dubbo.application.parameters.group=use-nacos-spring-boot
###添加 Nacos 的配置信息
nacos.discovery.server-addr=127.0.0.1:8848
nacos.discovery.namespace=c7ba173f-29e5-4c58-ae78-b102be11c4f9
nacos.discovery.register.group-name=use-nacos-spring-boot
nacos.discovery.auto-register=true
nacos.discovery.register.service-name=use-nacos-spring-boot
```

3. 添加 Dubbo 接口

在程序中定义一个 Dubbo 接口 PayService，具体代码如下所示：

```
@DubboService(group = "use-nacos-spring-boot",version = "1.0.0")
public class PayServiceImpl implements PayService {
    @Override
    public String pay() {
        return "payResult";
    }
}
```

4. 添加 RESTful API 接口

在程序中定义一个 RESTful API 接口，具体代码如下所示：

```
@RestController(value = "/pay")
public class PayController {
   @Resource
   private PayService payService;
   public String pay(){
      return payService.pay();
   }
}
```

5. 注册实例

启动服务后，在 Nacos 注册中心的控制台中可以看到服务“use-nacos-spring-boot”已经注册成功，如图 4-4 所示。

public | test | spring cloud alibaba practice | dev

Service List | spring cloud alibaba practice c7ba173f-29e5-4c58-ae78-b102be11c4f9

Service Name [Enter Service Name] Group Name [Enter Group Name] Hidden Empty Service: Search Create Service

Service Name	Group Name	Cluster Count	Instance Count	Healthy Instance Count	Trigger Protection Threshold	Operation
use-nacos-spring-boot	use-nacos-spring-boot	1	1	1	false	Details | Code Example | Delete
providers:com.alibaba.cloud.youxia.service.PayService:1.0.0:use-nacos-spring-boot	use-nacos-spring-boot	1	1	1	false	Details | Code Example | Delete

图 4-4

4.3.2 【实例】用 Spring Cloud Alibaba Discovery 接入

本实例的源码在本书配套资源的“/spring-cloud-alibaba-practice/chapterfour/use-spring-cloud-alibaba-discovery/”目录下。

本实例，先用 IDEA 创建一个 Spring Cloud Alibaba 项目，然后用 Spring Cloud Alibaba Discovery 将其接入 Nacos 注册中心。

1. 添加 POM 依赖

本程序中的 POM 依赖如下：

```
<!--添加 Spring Cloud Alibaba Nacos Discovery 的 Jar 包依赖-->
<dependency>
    <groupId>com.alibaba.cloud</groupId>
    <artifactId>spring-cloud-starter-alibaba-nacos-discovery</artifactId>
        <exclusions>
            <exclusion>
                <groupId>org.apache.httpcomponents</groupId>
                <artifactId>httpclient</artifactId>
            </exclusion>
        </exclusions>
</dependency>
<dependency>
    <groupId>com.alibaba.cloud</groupId>
```

```
    <artifactId>spring-cloud-starter-dubbo</artifactId>
    </dependency>
<dependency>
    <groupId>org.springframework.boot</groupId>
    <artifactId>spring-boot-starter-web</artifactId>
</dependency>
```

2. 添加属性配置

属性配置文件 bootstrap.yaml 中的配置信息如下：

```
dubbo:
  scan:
    base-packages: com.alibaba.cloud.youxia
  protocol:
    name: dubbo
    port: -1
spring:
  application:
    name: example11
  main:
    allow-bean-definition-overriding: true
  cloud:
    nacos:
      discovery:
        server-addr: 127.0.0.1:8848
        namespace: c7ba173f-29e5-4c58-ae78-b102be11c4f9
        group: use-spring-cloud-alibaba-discovery
server:
  port: 8089
```

3. 注册实例

应用在启动后会注册到 Nacos 注册中心中，Nacos 控制台中的效果如图 4-5 所示。

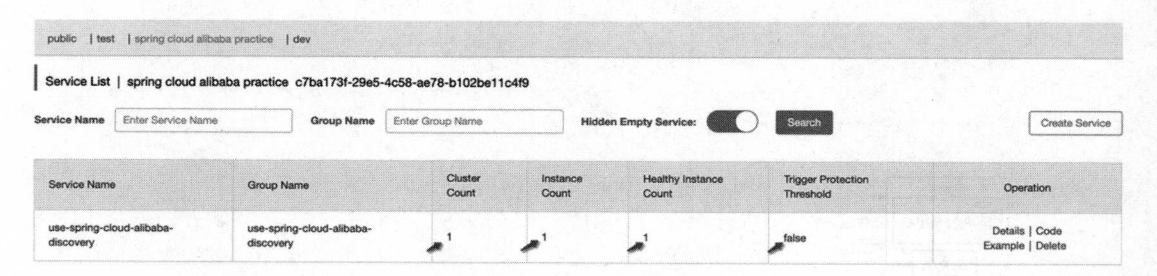

图 4-5

4.4 用“NacosNamingService 类 + @EnableDiscoveryClient”实现服务的注册/订阅

软件开发人员使用Spring Cloud Alibaba 开发Dubbo服务和Spring Cloud服务，其中Spring Cloud 服务依赖 Dubbo 服务，这样就可以在业务中用 Spring Cloud Alibaba 整合 Dubbo 服务和 Spring Cloud 服务。Spring Cloud Alibaba 支持如下两种通信协议。

- Dubbo 通信协议：用 Dubbo 作为最底层的 RPC 框架。
- Spring Cloud 通信协议：用 Spring Cloud 作为最底层的 RPC 框架。

Spring Cloud Alibaba 用“NacosNamingService 类 + @EnableDiscoveryClient”实现服务注册/订阅，具体包含如下两个模块。

- spring-cloud-starter-alibaba-nacos-discovery：封装了 Nacos Naming（注册中心），并通过 Nacos Client 调用与 Nacos 服务注册/订阅功能相关的 Open API。
- spring-cloud-starter-dubbo：封装了 Dubbo（RPC 框架）的部分功能。

下面我们来分析服务注册/订阅的核心原理。

4.4.1 服务注册的原理

在应用的服务接口注册到注册中心后，服务订阅者才能通过注册中心订阅该服务接口。

1. 服务注册的整体流程

Spring Cloud Alibaba 服务注册的整体流程如图 4-6 所示。

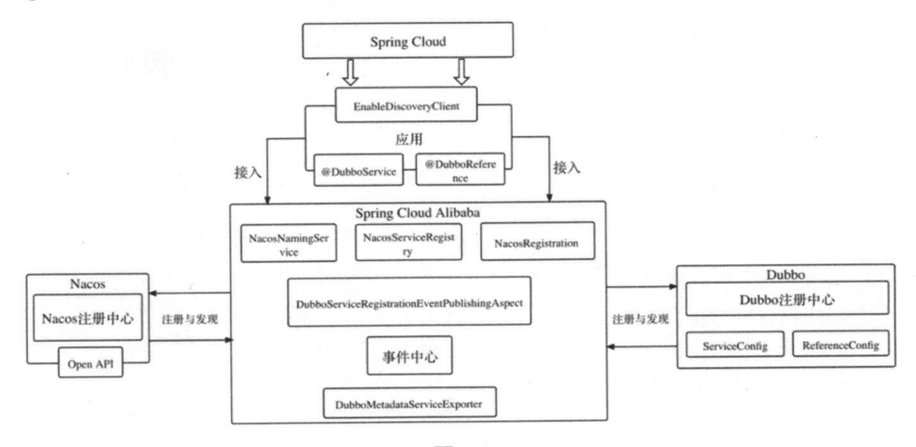

图 4-6

可以看出：

（1）Spring Cloud Alibaba 整合了 Spring Cloud 和 Dubbo 的服务注册功能，这样开发人员通过注解@EnableDiscoveryClient 开启 Spring Cloud 和 Dubbo 的服务注册。

（2）Spring Cloud Alibaba 完成 Nacos 注册中心的自动装配（装配 NacosServiceRegistry 类和 NacosRegistration 类）。

（3）Spring Cloud 触发自身的服务注册流程，调用 NacosServiceRegistry 类的 register() 方法。

（4）Spring Cloud Alibaba 通过前置切面拦截 NacosServiceRegistry 类的 register()方法，向事件中心发布一条注册预处理事件 ServiceInstancePreRegisteredEvent。

（5）事件处理器在监听到注册预处理事件后，执行导出当前应用服务元数据的操作。

（6）通过 Dubbo 的 API ServiceConfig 的 export()方法，可以导出所有被注解@DubboService 标记的服务接口对应的服务元数据，其实这也是服务提供者在应用端初始化的过程。熟悉 Dubbo 的开发者应该都知道，在服务初始化之后，会产生对应服务接口的全部元数据。

（7）应用在需要注册的服务接口上添加注解@DubboService，完成 Dubbo 接口的注册。

2. 通过注解@EnableDiscoveryClient 开启服务注册

注解@EnableDiscoveryClient 是开启 Spring Cloud Alibaba 服务注册/订阅的全局开关，是 Spring Cloud 的服务治理的基础注解。其代码如下所示：

```
@Target(ElementType.TYPE)
@Retention(RetentionPolicy.RUNTIME)
@Documented
@Inherited
@Import(EnableDiscoveryClientImportSelector.class)
public @interface EnableDiscoveryClient {
  boolean autoRegister() default true;
}
```

在上方代码中，通过注解@Import 自动注入 EnableDiscoveryClientImportSelector 类，属性 autoRegister 的默认值为 true。

EnableDiscoveryClientImportSelector 类的部分代码如下所示：

```
@Order(Ordered.LOWEST_PRECEDENCE - 100)
public class EnableDiscoveryClientImportSelector
        extends SpringFactoryImportSelector<EnableDiscoveryClient> {

    @Override
```

```
        public String[] selectImports(AnnotationMetadata metadata) {
            String[] imports = super.selectImports(metadata);
            AnnotationAttributes attributes = AnnotationAttributes.fromMap(
            metadata.getAnnotationAttributes(getAnnotationClass().getName(),
true));
            boolean autoRegister = attributes.getBoolean("autoRegister");

            if (autoRegister) {
                List<String> importsList = new
ArrayList<>(Arrays.asList(imports));
                    //如果 autoRegister 为 true，则加载
AutoServiceRegistrationConfiguration 类
                importsList.add(
        "org.springframework.cloud.client.serviceregistry.
AutoServiceRegistrationConfiguration");
                imports = importsList.toArray(new String[0]);
            }
            else {
                Environment env = getEnvironment();
                if (ConfigurableEnvironment.class.isInstance(env)) {
                    ConfigurableEnvironment configEnv = (ConfigurableEnvironment)
env;
                    LinkedHashMap<String, Object> map = new LinkedHashMap<>();

map.put("spring.cloud.service-registry.auto-registration.enabled", false);
                    MapPropertySource propertySource = new MapPropertySource(
                            "springCloudDiscoveryClient", map);
                    configEnv.getPropertySources().addLast(propertySource);
                }
            }
            return imports;
        }
        ...
    }
```

开发者可以通过如下方式来选择是否开启自动注册：

```
    //注解@EnableDiscoveryClient 默认开启自动注册。如果将 autoRegister 设置为 false，
则只订阅不注册服务
    @EnableDiscoveryClient(autoRegister = true)
    @SpringBootApplication
    public class DiscoveryApplication11 {
        public static void main(String[] args) {
            SpringApplication.run(DiscoveryApplication11.class, args);
        }
    }
```

3. 用 Spring Cloud Alibaba 自动装配 Nacos 注册中心

Spring Cloud Alibaba 依赖 Spring Cloud 来实现服务注册，于是 Spring Cloud Alibaba 定义了 NacosRegistration 类来实现 Spring Cloud 的 Registration 接口，也定义了 NacosServiceRegistry 类来实现 Spring Cloud 的 ServiceRegistry 和 ServiceInstance 接口。

NacosServiceRegistry 类注册逻辑的具体代码如下所示：

```
   public class NacosServiceRegistry implements ServiceRegistry<Registration>
{
       private final NacosDiscoveryProperties nacosDiscoveryProperties;
       ...
       @Autowired
       private NacosServiceManager nacosServiceManager;
       public NacosServiceRegistry(NacosDiscoveryProperties
nacosDiscoveryProperties) {
           this.nacosDiscoveryProperties = nacosDiscoveryProperties;
       }

       @Override
       public void register(Registration registration) {

           if (StringUtils.isEmpty(registration.getServiceId())) {
               log.warn("No service to register for nacos client……");
               return;
           }
           //①构造 NacosNamingService 类的实例对象
           NamingService namingService = namingService();
           //②从 NacosRegistration 中获取当前应用的唯一服务 ID
           String serviceId = registration.getServiceId();
           //③获取配置文件中的组名
           String group = nacosDiscoveryProperties.getGroup();
           //④将 NacosRegistration 类中的服务实例信息及元数据，转换为 Nacos 注册中心中
对应的实例信息及元数据
           Instance instance = getNacosInstanceFromRegistration(registration);
           try {
               //⑤调用 NacosNamingService 类的 registerInstance()方法完成服务注册
               namingService.registerInstance(serviceId, group, instance);
               log.info("nacos registry, {} {} {}:{} register finished", group,
serviceId,
                       instance.getIp(), instance.getPort());
           }
           catch (Exception e) {
               log.error("nacos registry, {} register failed……{},", serviceId,
                       registration.toString(), e);
```

```
                rethrowRuntimeException(e);
            }
        }
        ...
    }
```

NacosRegistration 类维护 Nacos 元数据具体逻辑的代码如下所示：

```
    public class NacosRegistration implements Registration, ServiceInstance {
        ...
        @PostConstruct
        public void init() {
            //①读取开发人员在应用的配置文件中配置的元数据
            Map<String, String> metadata =
nacosDiscoveryProperties.getMetadata();
            Environment env = context.getEnvironment();

            String endpointBasePath =
env.getProperty(MANAGEMENT_ENDPOINT_BASE_PATH);
            if (!StringUtils.isEmpty(endpointBasePath)) {
                metadata.put(MANAGEMENT_ENDPOINT_BASE_PATH, endpointBasePath);
            }
            Integer managementPort =
ManagementServerPortUtils.getPort(context);
            if (null != managementPort) {
                //②将端口号添加到 Nacos 的元数据中
                metadata.put(MANAGEMENT_PORT, managementPort.toString());
                //③将上下文添加到 Nacos 的元数据中
                String contextPath = env
                        .getProperty("management.server.servlet.context-path");
                String address = env.getProperty("management.server.address");
                //④将全局上下文路径添加到 Nacos 的元数据中
                if (!StringUtils.isEmpty(contextPath)) {
                    metadata.put(MANAGEMENT_CONTEXT_PATH, contextPath);
                }
                //⑤将服务地址添加到 Nacos 的元数据中
                if (!StringUtils.isEmpty(address)) {
                    metadata.put(MANAGEMENT_ADDRESS, address);
                }
            }
            //⑥将心跳周期添加到 Nacos 的元数据中
            if (null != nacosDiscoveryProperties.getHeartBeatInterval()) {
                metadata.put(PreservedMetadataKeys.HEART_BEAT_INTERVAL,

nacosDiscoveryProperties.getHeartBeatInterval().toString());
            }
```

```
            //⑦将心跳超时时间添加到 Nacos 的元数据中
            if (null != nacosDiscoveryProperties.getHeartBeatTimeout()) {
                metadata.put(PreservedMetadataKeys.HEART_BEAT_TIMEOUT,

nacosDiscoveryProperties.getHeartBeatTimeout().toString());
            }
            //⑧将删除 IP 地址的超时时间添加到 Nacos 的元数据中
            if (null != nacosDiscoveryProperties.getIpDeleteTimeout()) {
                metadata.put(PreservedMetadataKeys.IP_DELETE_TIMEOUT,

nacosDiscoveryProperties.getIpDeleteTimeout().toString());
            }
            customize(registrationCustomizers, this);
        }
        ...
    }
```

Nacos 注册中心的自动装配类 NacosServiceRegistryAutoConfiguration 的自动装配逻辑的代码如下所示：

```
    @Configuration(proxyBeanMethods = false)
    @EnableConfigurationProperties
    @ConditionalOnNacosDiscoveryEnabled
    @ConditionalOnProperty(value =
"spring.cloud.service-registry.auto-registration.enabled",
            matchIfMissing = true)
    @AutoConfigureAfter({ AutoServiceRegistrationConfiguration.class,
            AutoServiceRegistrationAutoConfiguration.class,
            NacosDiscoveryAutoConfiguration.class })
    public class NacosServiceRegistryAutoConfiguration {
        //①自动装配 Nacos 注册中心
        @Bean
        public NacosServiceRegistry nacosServiceRegistry(
                NacosDiscoveryProperties nacosDiscoveryProperties) {
            return new NacosServiceRegistry(nacosDiscoveryProperties);
        }
        //②自动装配 Nacos 的元数据
        @Bean
        @ConditionalOnBean(AutoServiceRegistrationProperties.class)
        public NacosRegistration nacosRegistration(
                ObjectProvider<List<NacosRegistrationCustomizer>>
registrationCustomizers,
                NacosDiscoveryProperties nacosDiscoveryProperties,
                ApplicationContext context) {
            return new
NacosRegistration(registrationCustomizers.getIfAvailable(),
```

```
                nacosDiscoveryProperties, context);
        }
        //③自动装配 NacosAutoServiceRegistration
        @Bean
        @ConditionalOnBean(AutoServiceRegistrationProperties.class)
        public NacosAutoServiceRegistration nacosAutoServiceRegistration(
              NacosServiceRegistry registry,
              AutoServiceRegistrationProperties
autoServiceRegistrationProperties,
              NacosRegistration registration) {
           return new NacosAutoServiceRegistration(registry,
                autoServiceRegistrationProperties, registration);
        }
    }
```

在自动装配完成后，在当前应用中就会存在 NacosServiceRegistry 类和 NacosRegistration 类的实例对象。

4. 通过 AOP 切面和事件中心拦截服务注册

Spring Cloud Alibaba 通过注解@EnableDiscoveryClient 在应用启动过程中触发 NacosServiceRegistry 类的 register()方法，执行服务注册逻辑；在执行服务注册之前，通过 AOP 切面拦截 NacosServiceRegistry 类的 register()方法。

（1）在 AOP 切面的 DubboServiceRegistrationEventPublishingAspect 类中，关于拦截 register()方法逻辑的具体代码如下所示：

```
    //①开启切面
    @Aspect
    public class DubboServiceRegistrationEventPublishingAspect
          implements ApplicationEventPublisherAware {
        //②定义切面，拦截 Spring Cloud 的 ServiceRegistry 接口的注册方法 register()，
最终拦截 NacosServiceRegistry 类的 register()方法
        public static final String REGISTER_POINTCUT_EXPRESSION = "execution(*
org.springframework.cloud.client.serviceregistry.ServiceRegistry.register(*))
&& target(registry) && args(registration)";
        //③Spring Framework 的事件中心对应的事件发布者
    private ApplicationEventPublisher applicationEventPublisher;
    //④在切面拦截之后，发布服务实例预注册事件
        @Before(value = REGISTER_POINTCUT_EXPRESSION, argNames = "registry,
registration")
    public void beforeRegister(ServiceRegistry registry, Registration
registration) {
        //⑤将 NacosRegistration 实例对象添加到事件中
          applicationEventPublisher.publishEvent(
```

```
                new ServiceInstancePreRegisteredEvent(registry,
registration));
        }
    }
```

（2）在 DubboServiceRegistrationAutoConfiguration 类中，监听事件中心的 ServiceInstancePreRegisteredEvent 事件的具体代码如下所示：

```
    @ConditionalOnProperty(value =
"spring.cloud.service-registry.auto-registration.enabled",
            matchIfMissing = true)
    @AutoConfigureAfter(name = { EUREKA_CLIENT_AUTO_CONFIGURATION_CLASS_NAME,
            CONSUL_AUTO_SERVICE_AUTO_CONFIGURATION_CLASS_NAME,
    "org.springframework.cloud.client.serviceregistry.AutoServiceRegistratio
nAutoConfiguration" },
            value = { DubboMetadataAutoConfiguration.class })
    public class DubboServiceRegistrationAutoConfiguration {
        ...
        //①监听 ServiceInstancePreRegisteredEvent 事件
        @EventListener(ServiceInstancePreRegisteredEvent.class)
        public void onServiceInstancePreRegistered
(ServiceInstancePreRegisteredEvent event) {
            //②从事件中获取全局的实例对象 NacosRegistration
            Registration registration = event.getSource();
            //③判断 DubboBootstrap 是否已经开启
            if (!DubboBootstrap.getInstance().isReady()
                    || !DubboBootstrap.getInstance().isStarted()) {
                ServiceRegistry<Registration> registry = event.getRegistry();
                synchronized (registry) {
                    registrations.putIfAbsent(registry, new HashSet<>());
                    registrations.get(registry).add(registration);
                }
            }
            else {
                //④将应用需要注册的服务接口的元数据添加到 Nacos 注册中心中
                attachDubboMetadataServiceMetadata(registration);
            }
        }
    }
```

5. 用 Spring Cloud Alibaba 导出被注解@DubboService 标注的服务接口的元数据

在事件中心监听到 ServiceInstancePreRegisteredEvent 事件后，Spring Cloud Alibaba 会添加新的元数据到 Nacos 注册中心中。

DubboServiceRegistrationAutoConfiguration 类的 attachDubboMetadataServiceMetadata()

方法的具体代码如下所示：

```
    private void attachDubboMetadataServiceMetadata(Registration registration) {
        if (registration == null) {
            return;
        }
    synchronized (registration) {
            //①从全局的 NacosRegistration 实例对象中获取存储元数据的对象
            Map<String, String> metadata = registration.getMetadata();
            //②添加新的元数据到 metadata 中
            attachDubboMetadataServiceMetadata(metadata);
        }
    }
    private void attachDubboMetadataServiceMetadata(Map<String, String>
metadata) {
        //③通过 DubboServiceMetadataRepository 类获取应用的最新元数据
        Map<String, String> serviceMetadata = dubboServiceMetadataRepository
                .getDubboMetadataServiceMetadata();
        if (!isEmpty(serviceMetadata)) {
            metadata.putAll(serviceMetadata);
        }
    }
```

DubboServiceMetadataRepository 类通过 getDubboMetadataServiceMetadata()方法获取元数据，具体代码如下所示：

```
    public Map<String, String> getDubboMetadataServiceMetadata() {
        //①用 DubboMetadataServiceExporter 类导出所有被注解@DubboService 标记的服务
接口对应的元数据
        List<URL> dubboMetadataServiceURLs =
dubboMetadataServiceExporter.export();
        //②元数据去重
        removeDubboMetadataServiceURLs(dubboMetadataServiceURLs);
    Map<String, String> metadata = newHashMap();
        //③重新复制一个新的元数据对象
        addDubboMetadataServiceURLsMetadata(metadata, dubboMetadataServiceURLs);
        //④添加 Dubbo 协议相关的元数据
    addDubboProtocolsPortMetadata(metadata);
        //⑤返回最新的元数据
        return Collections.unmodifiableMap(metadata);
    }
```

DubboMetadataServiceExporter 类的 export()方法用于导出最新的元数据，其具体代码如下所示：

```
    @Component
    public class DubboMetadataServiceExporter {
```

```
        private final Logger logger = LoggerFactory.getLogger(getClass());
        @Autowired
        private ApplicationConfig applicationConfig;
        @Autowired
        private ObjectProvider<DubboMetadataService> dubboMetadataService;
        @Autowired
        private Supplier<ProtocolConfig> protocolConfigSupplier;
          @Value("${spring.application.name:${dubbo.application.name:
application}}")
        private String currentApplicationName;
    private ServiceConfig<DubboMetadataService> serviceConfig;
    //①导出最新的元数据
    public List<URL> export() {
    //②如果 ServiceConfig 类已经被初始化，并且已经导出完毕，则直接返回
            if (serviceConfig == null || !serviceConfig.isExported()) {
                serviceConfig = new ServiceConfig<>();
                //③设置接口为 DubboMetadataService
                serviceConfig.setInterface(DubboMetadataService.class);
                //④设置元数据的版本号
                serviceConfig.setVersion(DubboMetadataService.VERSION);
                //⑤设置元数据组名为当前应用名称
                serviceConfig.setGroup(currentApplicationName);
                //⑥设置元数据对应的实例的可用性
                serviceConfig.setRef(dubboMetadataService.getIfAvailable());
                //⑦设置全局应用配置信息
                serviceConfig.setApplication(applicationConfig);
                //⑧设置协议信息
                serviceConfig.setProtocol(protocolConfigSupplier.get());
                //⑨导出元数据
                serviceConfig.export();
                if (logger.isInfoEnabled()) {
                    logger.info("The Dubbo service[{}] has been exported.",
                            serviceConfig.toString());
                }
            }
            return serviceConfig.getExportedUrls();
    }
    ...
    }
```

熟悉 Dubbo 的开发人员应该对 ServiceConfig 类不陌生，它是在 Dubbo 服务提供者初始化过程最核心的 API。Spring Cloud Alibaba 为了能够让开发者使用 Dubbo 的注解@DubboService 完成基于 Spring Cloud 的服务注册，执行了如下逻辑：

（1）用 ServiceConfig 类将需要注册的服务接口注册到 Dubbo 注册中心中。

（2）导出注册成功的接口元数据。

（3）将最新的接口元数据推送到 Nacos 元数据中心，这样应用就注册到 Nacos 注册中心中了。

关于 ServiceConfig 类的原理在这里就不再详细阐述，读者如果感兴趣请查询 Dubbo 相关源码。

6. 开发人员使用注解@DubboService 将服务接口注册到 Naccos 注册中心中

在经过 Spring Cloud Alibaba 的定制化封装后，开发人员可以使用注解@DubboService 快速将服务接口注册到 Nacos 注册中心中（开箱即用，非常方便），具体的代码样例如下所示：

```
//使用注解@DubboService 注册服务接口
@DubboService(group = "orderservice",version = "1.0.0")
public class OrderServiceProviderImpl implements OrderServiceProvider {
    @Override
    public String getOrderName() {
        return "orderName";
    }
}
```

综上所述，开发人员可以通过注解“@EnableDiscoveryClient + @DubboService”完成服务注册。

4.4.2　服务订阅的原理

在完成服务注册后，服务订阅者就可以从注册中心订阅服务，实现 RPC 调用了。

1. 应用初始化需要订阅的服务列表

应用可以通过 Spring Cloud Alibaba 的 DubboServiceMetadataRepository 类加载需要订阅的服务列表，如图 4-7 所示，

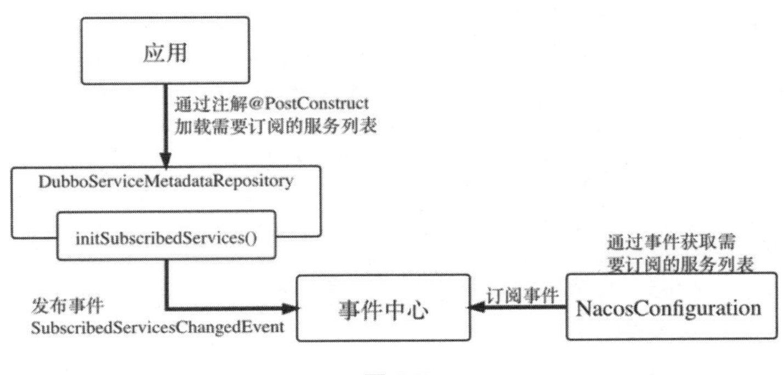

图 4-7

加载服务列表的具体逻辑如下：

（1）如果将 DubboCloudProperties.ALL_DUBBO_SERVICES 配置为“*”，则加载注册中心中的所有的服务。

（2）如果指定了需要订阅的服务（多个服务名之间用“,”分割），则按需加载服务。

具体代码逻辑可以参考 DubboServiceMetadataRepository 的源码。

2. 发布 SubscribedServicesChangedEvent 事件

SubscribedServicesChangedEvent 类是 Spring Cloud Alibaba 定义的订阅服务变更事件。在应用将需要订阅的服务列表中的服务初始化后，会将服务列表数据添加到事件中，并发布到事件中心中，具体代码如下所示：

```
    dispatchEvent(new SubscribedServicesChangedEvent(this,
oldSubscribedServices,
                newSubscribedServices));
    private void dispatchEvent(ApplicationEvent event) {
        applicationEventPublisher.publishEvent(event);
    }
```

3. 监听 SubscribedServicesChangedEvent 事件

在 DubboServiceDiscoveryAutoConfiguration 类中，Spring Cloud Alibaba 通过内部类 NacosConfiguration 来监听 SubscribedServicesChangedEvent 事件，具体代码如下所示：

```
    @EventListener(SubscribedServicesChangedEvent.class)
    public void
onSubscribedServicesChangedEvent(SubscribedServicesChangedEvent event)throws
Exception {
        //①遍历需要订阅的服务列表，添加监听器

event.getNewSubscribedServices().forEach(this::subscribeEventListener);
    }
    private void subscribeEventListener(String serviceName) {
        //②去重
        if (listeningServices.add(serviceName)) {
            try {
                String group = nacosDiscoveryProperties.getGroup();
                //③调用 NamingService 类的 subscribe()方法订阅对应的服务，并添加监听器
                namingService.subscribe(serviceName, group, event -> {
                    if (event instanceof NamingEvent) {
                        NamingEvent namingEvent = (NamingEvent) event;
                        //④将从 Nacos 注册中心返回的服务实例列表数据格式，转换为 Spring
Cloud 支持的服务列表数据格式
```

```
                    List<ServiceInstance> serviceInstances =
hostToServiceInstanceList(
                        namingEvent.getInstances(), serviceName);
                        //⑤向事件中心发布事件 ServiceInstancesChangedEvent
                        dispatchServiceInstancesChangedEvent(serviceName,
                                serviceInstances);
                    }
                });
            }catch (NacosException e) {
                ReflectionUtils.rethrowRuntimeException(e);
            }
        }
    }
```

事件发布的具体代码如下：

```
    private void dispatchServiceInstancesChangedEvent(String
serviceName,List<ServiceInstance> serviceInstances) {
        if (!hasText(serviceName) || Objects.equals(currentApplicationName,
serviceName)|| serviceInstances == null) {
            return;
        }

        //①将服务名称及对应的最新服务实例列表数据添加到事件对象中
        ServiceInstancesChangedEvent event = new
ServiceInstancesChangedEvent(serviceName,serviceInstances);
        if (logger.isInfoEnabled()) {
            logger.info("The event of the service instances[name : {} , size : {}]
change is about to be dispatched",
                    serviceName, serviceInstances.size());
    }
    //②向事件中心发布事件
        applicationEventPublisher.publishEvent(event);
    }
```

4. 监听 ServiceInstancesChangedEvent 事件

在 DubboCloudRegistry 类中，监听 ServiceInstancesChangedEvent 事件的具体代码如下所示：

```
    //①Spring Cloud Alibaba 重写了 Dubbo 注册中心的订阅方法 doSubscribe()
    @Override
    public final void doSubscribe(URL url, NotifyListener listener) {
        //②判断是否为 Dubbo 的后台管理协议
        if (isAdminURL(url)) {
            if (logger.isWarnEnabled()) {
```

```
            logger.warn("This feature about admin will be supported in the
future.");
          }
       //③判断是否为 DubboMetadataService 类
      }else if (isDubboMetadataServiceURL(url)) {
          subscribeDubboMetadataServiceURLs(url, listener);
    }else {
      //④标准的 Dubbo 服务订阅
          subscribeURLs(url, listener);
      }
    }
```

熟悉 Dubbo 框架的开发者应该知道 doSubscribe()方法。在应用通过注解@DubboReference 初始化服务消费者后，在 RPC 调用的过程中，应用会调用 doSubscribe()方法进行服务订阅。

下面来看看具体的订阅逻辑。subscribeURLs()方法的具体代码如下所示：

```
    private void subscribeURLs(URL url, NotifyListener listener) {
        //①同步订阅，并通知服务订阅者
    subscribeURLs(url, getServices(url), listener);
    //②注册 ServiceInstancesChangedEvent 事件的监听器
        registerServiceInstancesChangedListener(url,new
ApplicationListener<ServiceInstancesChangedEvent>() {
        private final URL url2subscribe = url;
            @Override
            @Order
            public void onApplicationEvent(ServiceInstancesChangedEvent event) {
                Set<String> serviceNames = getServices(url);
                String serviceName = event.getServiceName();
                if (serviceNames.contains(serviceName)) {
                   //③监听到服务实例变更事件，同步订阅，并通知服务订阅者
                    subscribeURLs(url, serviceNames, listener);
                }
            }
            @Override
            public String toString() {
                return "ServiceInstancesChangedEventListener:"+
url.getServiceKey();
            }
        });
    }
```

5. 通知服务订阅者最新的服务实例

Spring Cloud Alibaba 的 DubboCloudRegistry 类在监听到服务实例变更事件后，会用事件

通知机制通知服务订阅者，具体代码如下所示：

```
    private void subscribeURLs(URL url, Set<String> serviceNames,NotifyListener
listener) {
    List<URL> subscribedURLs = new LinkedList<>();
    //①获取对应服务名称的最新的元数据（Dubbo 注册中心）
        serviceNames.forEach(serviceName -> {
            subscribeURLs(url, subscribedURLs, serviceName,() ->
getServiceInstances(serviceName));
    });
    //②将最新的接口元数据通知到对应的服务订阅者
        notifyAllSubscribedURLs(url, subscribedURLs, listener);
    }
```

在以上代码中，关于获取对应服务名称的最新元数据的具体源码，读者可以查阅 DubboCloudRegistry 类。

为了保证服务实例信息的数据一致性，Spring Cloud Alibaba 并没有直接使用 ServiceInstancesChangedEvent 事件中的服务实例信息，而是通过 Nacos 的 NacosDiscoveryClient 类，重新从 Nacos 注册中心获取最新的服务实例信息，具体代码如下所示：

```
    private List<ServiceInstance> doGetServiceInstances(String serviceName) {
        List<ServiceInstance> serviceInstances = emptyList();
    try {
       //从 Nacos 配置中心获取最新的服务实例信息
            serviceInstances = discoveryClient.getInstances(serviceName);
        }catch (Exception e) {
            if (logger.isErrorEnabled()) {
                logger.error(e.getMessage(), e);
            }
        }
        return serviceInstances;
    }
```

通知服务订阅者的具体代码如下所示：

```
    private void notifyAllSubscribedURLs(URL url, List<URL> subscribedURLs,
                NotifyListener listener) {
        if (isEmpty(subscribedURLs)) {
            subscribedURLs.add(emptyURL(url));
        }
        if (logger.isDebugEnabled()) {
            logger.debug("The subscribed URL[{}] will notify all URLs : {}", url,
                    subscribedURLs);
```

```
    }
    //将最新的订阅的元数据推送给服务订阅者
        listener.notify(subscribedURLs);
    }
```

Spring Cloud Alibaba 利用 Dubbo 的事件监听机制，将 Dubbo 注册中心最新的元数据推送给服务订阅者。

6. Dubbo 的注解@DubboReference

开发人员使用注解@DubboReference 可以实时地订阅最新的服务信息，完成 RPC 通信。

具体代码样例如下：

```
@Component("orderServiceConsumer")
public class OrderServiceConsumer {
    @DubboReference(group = "orderservice",version = "1.0.0")
    private OrderServiceProvider orderServiceProvider;
    public String getOrderName() {
        return orderServiceProvider.getOrderName();
    }
}
```

4.4.3 【实例】通过服务幂等性设计验证服务的注册/订阅

本实例的源码在本书配套资源的"/spring-cloud-alibaba-practice/chapterfour/idempotent-design-spring-cloud-alibaba/"目录下。

所谓幂等性设计，即一次请求和多次请求某一个资源会得到同样的结果。用数学的语言来表达就是：$f(x) = f(f(x))$。比如，求绝对值的函数，abs(x) = abs(abs(x))。

为什么开发人员需要幂等性设计？在开发人员将系统解耦隔离后，服务间的调用可能会有 3 个状态：成功（Success）、失败（Failed）、超时（Timeout）。

"成功"和"失败"都是有明确原因的状态。但"超时"是没有原因的状态，可能是在网络传输过程中出现了丢包，也可能是请求没有到达应用服务，还有可能是"请求到达了，但没有正常返回结果"等。于是，调用方完全不知道下游系统是否存在网络问题、是否收到了请求、收到的请求是处理成功或者失败，还是在响应时遇到了网络的问题。

服务幂等性设计涉及的主要项目模块见表 4-2。

表 4-2　服务幂等性设计涉及的主要项目模块

模块名称	模块功能描述
distributed-uuid-server	分布式发号器服务，在分布式环境中，为服务产生跨实例的全局唯一性 ID
idempotent-design-user-api	公共 Dubbo 接口，对消费者暴露 SDK
idempotent-design-user-client	服务消费者
idempotent-design-user-server	服务提供者

1. 幂等性设计的思路

本实例模拟电商系统中商品库存被扣除的业务场景——如果用户下单成功，则扣除对应商品的库存。

本实例通过“UUID + Redis”判重来实现商品扣库存的幂等性设计，底层基础框架采用的是 Spring Cloud Alibaba，具体思路如下。

（1）在 idempotent-design-user-client 模块中封装 RESTful AP 作为消费者，来调用库存扣减服务 idempotent-design-user-server。

（2）开启消费者的 Dubbo 超时重试机制，并设置消费超时时间。

（3）通过动态开关控制是否开启幂等设计。

- 如果开启了幂等性设计，则消费者会通过分布式发号器服务生产一个全局唯一的 ID，并传递给服务提供者。服务提供者解析出全局唯一的 ID，并缓存这个 ID，且设置缓存的过期时间，比如 3s。这样，在 3s 内即使消费者一直重复请求，也不会影响数据的一致性，这样就能确保接口的幂等性。
- 如果没有开启幂等性设计，则消费者直接调用扣库存服务。如果出现服务调用超时，则 Dubbo 会自动重试，或者消费者主动重试，从而导致服务多次扣减库存。这样就会出现错误的业务执行过程：业务只发起了一次扣库存的请求，但是服务提供者实际上执行了多次扣库存的数据库变更。

2. 具体代码实现

（1）分布式发号器服务 idempotent-design-user-server。

分布式全局 ID 采用雪花算法，具体实现如下：

```
public class SnowFlake {
    //①起始的时间戳
    private final static long START_STMP = 1480166465631L;
    //②每一部分占用的位数
    private final static long SEQUENCE_BIT = 12;
    private final static long MACHINE_BIT = 5;
    private final static long DATACENTER_BIT = 5;
```

```
        //③每一部分的最大值
        private final static long MAX_DATACENTER_NUM = -1L ^ (-1L <<
DATACENTER_BIT);
        private final static long MAX_MACHINE_NUM = -1L ^ (-1L << MACHINE_BIT);
        private final static long MAX_SEQUENCE = -1L ^ (-1L << SEQUENCE_BIT);
        //④每一部分向左的位移
        private final static long MACHINE_LEFT = SEQUENCE_BIT;
        private final static long DATACENTER_LEFT = SEQUENCE_BIT + MACHINE_BIT;
        private final static long TIMESTMP_LEFT = DATACENTER_LEFT +
DATACENTER_BIT;
        private long datacenterId;
        private long machineId;
        private long sequence = 0L;
        private long lastStmp = -1L;
        public SnowFlake(long datacenterId, long machineId) {
            if (datacenterId > MAX_DATACENTER_NUM || datacenterId < 0) {
                throw new IllegalArgumentException("datacenterId can't be greater
than MAX_DATACENTER_NUM or less than 0");
            }
            if (machineId > MAX_MACHINE_NUM || machineId < 0) {
                throw new IllegalArgumentException("machineId can't be greater
than MAX_MACHINE_NUM or less than 0");
            }
            this.datacenterId = datacenterId;
            this.machineId = machineId;
        }
        //⑤产生下一个 ID
        public synchronized long nextId() {
            long currStmp = getNewstmp();
            if (currStmp < lastStmp) {
                throw new RuntimeException("Clock moved backwards.  Refusing to
generate id");
            }

            if (currStmp == lastStmp) {
                //⑥如果是相同毫秒，则序列号自动增加
                sequence = (sequence + 1) & MAX_SEQUENCE;
                //⑦相同毫秒的序列数已经达到最大
                if (sequence == 0L) {
                    currStmp = getNextMill();
                }
            } else {
                //⑧如果是不同毫秒，则将序列号设置为 0
                sequence = 0L;
            }
```

```
        lastStmp = currStmp;

        return (currStmp - START_STMP) << TIMESTMP_LEFT
            | datacenterId << DATACENTER_LEFT
            | machineId << MACHINE_LEFT
            | sequence;
    }

    private long getNextMill() {
        long mill = getNewstmp();
        while (mill <= lastStmp) {
            mill = getNewstmp();
        }
        return mill;
    }

    private long getNewstmp() {
        return System.currentTimeMillis();
    }
}
```

暴露分布式发号器服务的服务提供者的具体代码如下所示：

```
@DubboService(version = "1.0.0", group = "distributed-uuid-server")
public class DistributedServiceImpl implements DistributedService {
    @Autowired
    private SnowflakeConfig snowflakeConfig;
    Map <String,SnowFlake> snowFlakeHandlerMap = new ConcurrentHashMap <>();
    //①使用由分布式发号器调用方自定义的数据中心 ID 和机器 ID
@Override
public long nextId(final long datacenterId, final long machineId) {
        final long sdatacenterId = datacenterId;
        final long smachineId = machineId;
        final String handler = sdatacenterId + "_" + smachineId;
        SnowFlake snowFlake;
        //②从本地缓存中获取 SnowFlake 对象
        if (snowFlakeHandlerMap.containsKey(handler)) {
            snowFlake = snowFlakeHandlerMap.get(handler);
            return snowFlake.nextId();
        } else {
            //③如果没有命中缓存，则重新实例化一个 SnowFlake 对象，并将其存储在本地缓存中
            snowFlake = new SnowFlake(datacenterId, machineId);
            snowFlakeHandlerMap.putIfAbsent(handler, snowFlake);
            //④直接从缓存中获取 SnowFlake 对象
            snowFlake = snowFlakeHandlerMap.get(handler);
```

```
            return snowFlake.nextId();
        }
    }
    //⑤使用分布式发号器服务默认的数据中心 ID 和机器 ID
  @Override public long nextId() {
    //⑥从配置中心获取指定 IP 地址的数据中心 ID 和机器 ID，支持动态配置
        List <SnowflakeInfo> config = snowflakeConfig.getConfig();
        String localAddress = NetUtils.getLocalAddress();
        SnowflakeInfo snowflakeInfo = config.stream().filter(s
->Objects.equals(s.getIp(), localAddress)).findFirst().orElse(null);
        long dataCenterId =
Optional.ofNullable(snowflakeInfo).map(SnowflakeInfo: :getDataCenterId).orEl
se(0L);
        long machineId =
Optional.ofNullable(snowflakeInfo).map(SnowflakeInfo: :getMachineId).orElse(
0L);
        //⑥生成分布式 ID 并返回
        return nextId(dataCenterId, machineId);
    }
  }
```

（2）消费者服务 idempotent-design-user-client。

在服务 idempotent-design-user-client 中，通过 RESTful API 调用库存服务扣减商品的库存，具体代码如下：

```
  @RestController(value = "/good")
  public class GoodController {
    //①依赖服务提供者——商品服务，调用超时时间为 1s，超时之后重试次数为 4 次
    @DubboReference(version = "1.0.0", group =
"idempotent-design-user-server", retries = 4,timeout = 1)
    private GoodService goodService;
    //②依赖分布式发号器服务
    @DubboReference(version = "1.0.0", group = "distributed-uuid-server")
    private DistributedService distributedService;
    //③幂等性设计全局开关
    @Autowired
    private NacosConfig nacosConfig;

    @PostMapping(value = "/updataGoodNum")
    @ResponseBody
    public DefaultResult<GoodDTO> updateGoodNum(@RequestParam String goodId)
{
        long uuid = distributedService.nextId(7, 8);
        GoodServiceRequest<Example2ProductBo> request = new
GoodServiceRequest<Example2ProductBo>();
```

```
        Example2ProductBo example2ProductBo = new Example2ProductBo();
        example2ProductBo.setGoodId(Long.valueOf(goodId));
        request.setRequestData(example2ProductBo);
        //④如果开启了幂等性设计，则设置 UUID
        if (nacosConfig.isMideng()) {
            request.setUuid(uuid + "");
        }
        //⑤扣减库存
        return goodService.updateGoodNum(request);
    }
}
```

（3）服务提供者 idempotent-design-user-server。

在服务 idempotent-design-user-server 中，对比开启幂等性设计和关闭幂等性设计时接口处理过程的不同之处，具体代码如下：

```
@DubboService(version = "1.0.0",group = "idempotent-design-user-server")
public class GoodServiceImpl implements GoodService {

    @Autowired
private Example2ProductMapper example2ProductMapper;
    @Autowired
    private RedisTemplate redisTemplate;
    @Override
    public DefaultResult<GoodDTO> updateGoodNum(GoodServiceRequest
goodServiceRequest) {
        DefaultResult<GoodDTO> result=new DefaultResult<>();
        GoodDTO returnItem=new GoodDTO();
        Example2ProductBo
example2ProductBo=(Example2ProductBo)goodServiceRequest.getRequestData();
        //①如果开启了幂等性设计，则会利用 Redis 来校验 UUID 的唯一性；如果关闭了幂等性
设计，则过滤掉 UUID 的唯一性校验，直接执行库存扣减
        if(!StringUtils.isEmpty(goodServiceRequest.getUuid())){
            long uuid=Long.parseLong(goodServiceRequest.getUuid());
            if(null!=redisTemplate.opsForValue().get(uuid)){
                result.setData(new GoodDTO());
                result.setMessage("uuid:"+uuid+" 已经连续访问多次！");
                return result;
            }else{
                redisTemplate.opsForValue().set(uuid,true,2,
TimeUnit.SECONDS);
            }
        }
        try {
            //②设置执行延时时间为 2s
```

```
            Thread.sleep(2000);
        }catch (InterruptedException e){
            System.out.println(e.getMessage());
        }
        List<Example2ProductEntity> queryResult1=
example2ProductMapper.queryGoodInfoByGoodId(example2ProductBo);
        if(!CollectionUtils.isEmpty(queryResult1)){
            Example2ProductEntity item=queryResult1.get(0);
            System.out.println("开始扣减库存，扣除之前的商品库存为：
"+item.getNum()+" 商品 ID 为："+item.getGoodId());
        }
        example2ProductMapper.updateGoodNum(example2ProductBo);
        List<Example2ProductEntity> queryResult2=
example2ProductMapper.queryGoodInfoByGoodId(example2ProductBo);
        if(!CollectionUtils.isEmpty(queryResult2)){
            Example2ProductEntity item=queryResult2.get(0);
            System.out.println("开始扣减库存，扣除之后的商品库存为：
"+item.getNum()+" 商品 ID 为："+item.getGoodId());
        }
        returnItem.setGoodId(queryResult2.get(0).getGoodId());
        returnItem.setNum(queryResult2.get(0).getNum());
        result.setData(returnItem);
        result.setCode("200");
        result.setMessage("库存扣减成功！！！！！");
        return result;
    }
}
```

3. 场景验证

要验证幂等性设计，则需要准备相关的数据，并对比开启幂等性设计和关闭幂等性设计对数据的影响。

（1）测试数据准备。目前商品库存为 100，向商品表 example2_product 中插入一个测试商品，商品 ID 为 7878，具体代码如下：

```
insert into example2_product value (3477374334,7878,100,'苹果笔记本电脑');
commit ;
```

（2）按顺序启动服务 distributed-uuid-server、idempotent-design-user-server 和 idempotent-design-user-client，幂等服务的注册列表如图 4-8 所示，3 个服务完成注册。

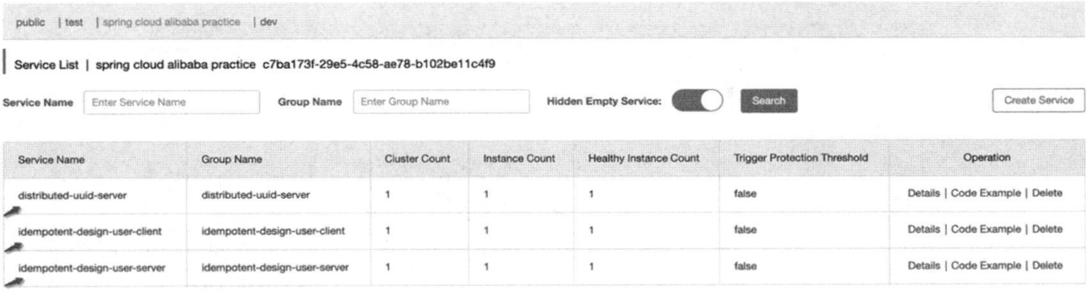

public | test | spring cloud alibaba practice | dev

Service List | spring cloud alibaba practice c7ba173f-29e5-4c58-ae78-b102be11c4f9

Service Name: Enter Service Name　Group Name: Enter Group Name　Hidden Empty Service:　Search　Create Service

Service Name	Group Name	Cluster Count	Instance Count	Healthy Instance Count	Trigger Protection Threshold	Operation
distributed-uuid-server	distributed-uuid-server	1	1	1	false	Details | Code Example | Delete
idempotent-design-user-client	idempotent-design-user-client	1	1	1	false	Details | Code Example | Delete
idempotent-design-user-server	idempotent-design-user-server	1	1	1	false	Details | Code Example | Delete

图 4-8

（3）验证没有开启幂等性设计的库存扣减的场景。

在扣减库存过程中，需要增加当前请求的耗时：用线程机制让当前请求对应的线程休眠 2s，并模拟在请求调用过程中的网络超时故障。

在增加耗时后，idempotent-design-user-client 会报超时异常，具体日志如下：

```
    org.apache.dubbo.remoting.TimeoutException: Waiting server-side response
timeout by scan timer. start time: 2021-03-12 23:20:50.604, end time: 2021-03-12
23:20:50.634, client elapsed: 0 ms, server elapsed: 30 ms, timeout: 1 ms, request:
Request [id=5, version=2.0.2, twoway=true, event=false, broken=false,
data=null], channel: /192.168.0.123:55884 -> /192.168.0.123:26756
```

通过 SQL 语句查询对应商品的库存值可以发现，库存值已经由 100 变为 95（如图 4-9 所示），这说明服务器端实际扣了 5 次。如果是正常扣减库存，则应该只扣减一次，库存值应该为 99。

Output　spring_cloud_alibaba...tice.example2_product

	id	good_id	num	good_name
1	3467374334	6778	95	手机
2	3477374334	7878	95	苹果笔记本电脑

图 4-9

（4）验证开启幂等性设计的库存扣减的场景。

还原测试商品的库存值为 100，并通过 Swagger 模拟用户调用。如图 4-10 所示，返回幂等性设计已经生效的提示信息。

通过 SQL 语句查询对应商品，发现库存值为 99，如图 4-11 所示。

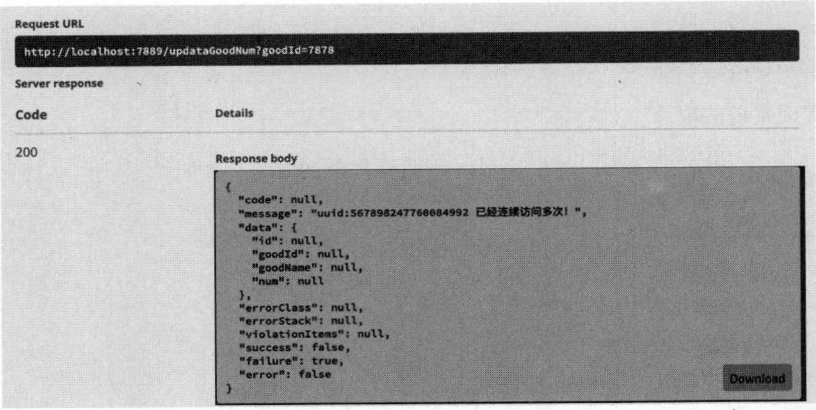

Request URL

`http://localhost:7889/updataGoodNum?goodId=7878`

Server response

Code　　Details

200

Response body


```
{
  "code": null,
  "message": "uuid:567898247760084992 已经连续访问多次！",
  "data": {
    "id": null,
    "goodId": null,
    "goodName": null,
    "num": null
  },
  "errorClass": null,
  "errorStack": null,
  "violationItems": null,
  "success": false,
  "failure": true,
  "error": false
}
```


Download

图 4-10

Output　spring_cloud_alibaba...tice.example2_product

2 rows　Tx: Auto　DDL　DML　Comma-...d (CSV)

	id	good_id	num	good_name
1	id: bigint 34	6778	95	手机
2	3477374334	7878	99	苹果笔记本电脑

图 4-11

idempotent-design-user-client 在调用 idempotent-design-user-server 时，配置了超时重试次数为 4，具体代码如下：

```
    @DubboReference(version = "1.0.0", group = "idempotent-design-user-server",
retries = 4,timeout = 1)
    private GoodService goodService;
```

如果超时，则理论上会执行 5 次库存扣减。在幂等性设计生效后，有效地拦截了 4 次库存扣减，保证了数据的一致性。

4.5 用“Ribbon + Nacos Client”实现服务发现的负载均衡

Spring Cloud Alibaba 可以用 Ribbon 和 Nacos Client 实现服务发现的负载均衡。开发人员只需要依赖它的 Jar 包，即可实现“开箱即用”。

4.5.1 为什么需要负载均衡

通过服务发现，可以发现一个能够提供服务的实例信息列表。但应用需要先采用策略从这个包含一组 IP 地址的服务列表中选出一个 IP 地址，然后对其发起 RPC 请求。负载均衡就是用来解决这个问题的，通常包括客户端负载均衡和服务器端负载均衡。

负载均衡器（Load Balancer，LB）的作用是：将用户的请求按照负载均衡算法分配到多个服务上，从而实现系统的高可用。

4.5.2 【实例】用“Ribbon + Nacos Client”实现负载均衡

本实例的源码在本书配套资源的“/spring-cloud-alibaba-practice/chapterfour/ribbon-discovery-spring-cloud-alibaba-provider/”和“/spring-cloud-alibaba-practice/chapterfour/ribbon-discovery-spring-cloud-alibaba-consumer/”目录下。

Ribbon 是一个负载均衡器。Ribbon 是基于客户端负载均衡算法的，所以需要开发人员主动开启负载均衡。

Spring Cloud Alibaba 采用 Ribbon 来实现负载均衡，从而提升服务订阅者调用的高可用性。

1. 添加负载均衡算法

在 Spring Cloud Alibaba 中，通过 NacosRule 类来定义负载均衡算法，具体代码如下所示：

```
public class NacosRule extends AbstractLoadBalancerRule {
    ...
    @Autowired
    private NacosDiscoveryProperties nacosDiscoveryProperties;

    @Autowired
    private NacosServiceManager nacosServiceManager;
    //重写 Spring Cloud 的抽象负载均衡规则类 AbstractLoadBalancerRule 的 choose()
方法
    @Override
    public Server choose(Object key) {
        try {
            //①获取集群名称
            String clusterName =
this.nacosDiscoveryProperties.getClusterName();
            //②获取组名称
            String group = this.nacosDiscoveryProperties.getGroup();
            DynamicServerListLoadBalancer loadBalancer =
(DynamicServerListLoadBalancer) getLoadBalancer();
            //③获取服务名称
            String name = loadBalancer.getName();
            //④获取 Nacos 的名字服务对象 NamingService
            NamingService namingService = nacosServiceManager
```

```
                    .getNamingService(nacosDiscoveryProperties.getNacosPrope
rties());
            //⑤从 Nacos 注册中心获取健康的服务实例列表
           List<Instance> instances = namingService.selectInstances(name,
group, true);
            if (CollectionUtils.isEmpty(instances)) {
                LOGGER.warn("no instance in service {}", name);
                return null;
            }
            //⑥剔除不属于本集群的服务实例，不允许跨集群负载均衡
            List<Instance> instancesToChoose = instances;
            if (StringUtils.isNotBlank(clusterName)) {
                List<Instance> sameClusterInstances = instances.stream()
                        .filter(instance -> Objects.equals(clusterName,
                                instance.getClusterName()))
                        .collect(Collectors.toList());
                if (!CollectionUtils.isEmpty(sameClusterInstances)) {
                    instancesToChoose = sameClusterInstances;
                }
                else {
                    LOGGER.warn(
                            "A cross-cluster call occurs, name = {}, clusterName
= {}, instance = {}",
                            name, clusterName, instances);
                }
            }
            //⑦按照服务实例的权重值，从服务列表中选出提供服务的实例
            Instance instance =
ExtendBalancer.getHostByRandomWeight2(instancesToChoose);
            //⑧返回选中的实例
            return new NacosServer(instance);
        }
        catch (Exception e) {
            LOGGER.warn("NacosRule error", e);
            return null;
        }
    }
    @Override
    public void initWithNiwsConfig(IClientConfig iClientConfig) {
    }
}
```

Spring Cloud Alibaba 在服务发现的过程中，会根据服务实例的权重选出合适的服务实例。详细源码可以查阅 ExtendBalancer 类的 getHostByRandomWeight2()方法。

2. 在应用中初始化 NacosRule

通过注解@Configuration 和@Bean 加载 NacosRule 类，具体代码如下所示：

```
//①初始化 NacosRule 实例
@Configuration
public class NacosRibbonRuleConfig {

    @Bean
    public NacosRule nacosRule(){
        return new NacosRule();
    }
}
//②初始化 RestTemplate 实例
@LoadBalanced
@Bean
public RestTemplate restTemplate() {
    return new RestTemplate();
}
//③配置全局负载均衡
@Configuration
@RibbonClient(name="ribbon-discovery-spring-cloud-alibaba-provider",conf
iguration =NacosRibbonRuleConfig.class)
public class NacosGlobalClientConfig {
}
```

3. 定义消费者和服务提供者

服务提供者的具体代码如下所示：

```
@RestController
@RequestMapping(value = "/provider")
public class ProviderController {

    @Autowired
private RibbonDiscoveryService ribbonDiscoveryService;

    @GetMapping(value = "/getRibbonConfig")
    @ResponseBody
public String testRestRibbon(){
    //①返回当前实例的节点信息、IP 地址和端口号
        return ribbonDiscoveryService.getRibbonConfig();
    }
}

@DubboService(version = "1.0.0",group = "ribbon-provider")
```

```
public class RibbonDiscoveryServiceImpl implements RibbonDiscoveryService
{

    @Autowired
    private Environment environment;

    @Override
public String getRibbonConfig() {
    //②获取当前应用的端口号
        String port = environment.getProperty("local.server.port");
        String ip = "";
        try {
            //③获取当前应用所在机器的 IP 地址
            ip = InetAddress.getLocalHost().toString();
        } catch (Exception e) {
            System.out.println(e.getMessage());
        }
        //④组装 IP 地址和端口号
        String result = "负载均衡成功！" + "当前机器节点 IP 地址为：" + ip + ":" +
port;
        return result;
    }
}
```

本实例用一个 RESTful API 接口作为服务消费者，调用具备负载均衡功能的服务提供者的接口，具体代码如下所示：

```
@RestController
@RequestMapping(value = "/ribbon")
public class NacosRibbonController {
    @Autowired
    private RestTemplate restTemplate;
    @GetMapping(value = "/test")
public String doRibbon(){
    //通过带有负载功能的 RestTemplate 访问服务提供者的服务实例
        return
restTemplate.getForObject("http://ribbon-discovery-spring-cloud-alibaba-prov
ider/provider/getRibbonConfig",String.class);
    }
}
```

4. 通过 Nacos 的控制台配置服务提供者对应实例的权重

服务提供者总共有 3 个实例，分别是 192.168.0.123:8079、192.168.0.123.8069、192.168.0.123:8099，如图 4-12 所示。

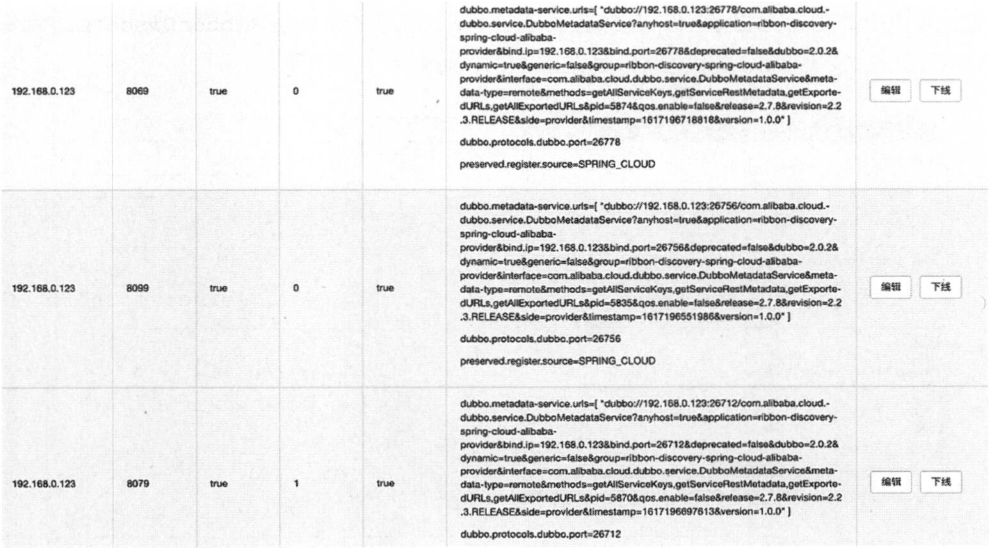

192.168.0.123	8069	true	0	true	dubbo.metadata-service.urls=["dubbo://192.168.0.123:26778/com.alibaba.cloud.-dubbo.service.DubboMetadataService?anyhost=true&application=ribbon-discovery-spring-cloud-alibaba-provider&bind.ip=192.168.0.123&bind.port=26778&deprecated=false&dubbo=2.0.2&dynamic=true&generic=false&group=ribbon-discovery-spring-cloud-alibaba-provider&interface=com.alibaba.cloud.dubbo.service.DubboMetadataService&meta-data-type=remote&methods=getAllServiceKeys,getServiceRestMetadata,getExporte-dURLs,getAllExportedURLs&pid=5874&qos.enable=false&release=2.7.8&revision=2.2.3.RELEASE&side=provider×tamp=1617196718818&version=1.0.0"] dubbo.protocols.dubbo.port=26778 preserved.register.source=SPRING_CLOUD	编辑 下线
192.168.0.123	8099	true	0	true	dubbo.metadata-service.urls=["dubbo://192.168.0.123:26756/com.alibaba.cloud.-dubbo.service.DubboMetadataService?anyhost=true&application=ribbon-discovery-spring-cloud-alibaba-provider&bind.ip=192.168.0.123&bind.port=26756&deprecated=false&dubbo=2.0.2&dynamic=true&generic=false&group=ribbon-discovery-spring-cloud-alibaba-provider&interface=com.alibaba.cloud.dubbo.service.DubboMetadataService&meta-data-type=remote&methods=getAllServiceKeys,getServiceRestMetadata,getExporte-dURLs,getAllExportedURLs&pid=5835&qos.enable=false&release=2.7.8&revision=2.2.3.RELEASE&side=provider×tamp=1617196551986&version=1.0.0"] dubbo.protocols.dubbo.port=26756 preserved.register.source=SPRING_CLOUD	编辑 下线
192.168.0.123	8079	true	1	true	dubbo.metadata-service.urls=["dubbo://192.168.0.123:26712/com.alibaba.cloud.-dubbo.service.DubboMetadataService?anyhost=true&application=ribbon-discovery-spring-cloud-alibaba-provider&bind.ip=192.168.0.123&bind.port=26712&deprecated=false&dubbo=2.0.2&dynamic=true&generic=false&group=ribbon-discovery-spring-cloud-alibaba-provider&interface=com.alibaba.cloud.dubbo.service.DubboMetadataService&meta-data-type=remote&methods=getAllServiceKeys,getServiceRestMetadata,getExporte-dURLs,getAllExportedURLs&pid=5870&qos.enable=false&release=2.7.8&revision=2.2.3.RELEASE&side=provider×tamp=1617196697613&version=1.0.0"] dubbo.protocols.dubbo.port=26712	编辑 下线

图 4-12

3 个实例的初始化权重都为 1。当服务消费者访问服务提供者时，3 个实例的访问流量是均衡的，验证如下情况：

（1）将 192.168.0.123:8079 的权重设置为 0，按照负载均衡策略，这个实例应该没有流量。

通过 Swagger 进行验证（本实例已经集成了 Swagger，读者可以通过对应的地址进行访问），请求只能到达其他两个节点，只返回 192.168.0.123:8069 和 192.168.0.123:8099，如图 4-13 所示。

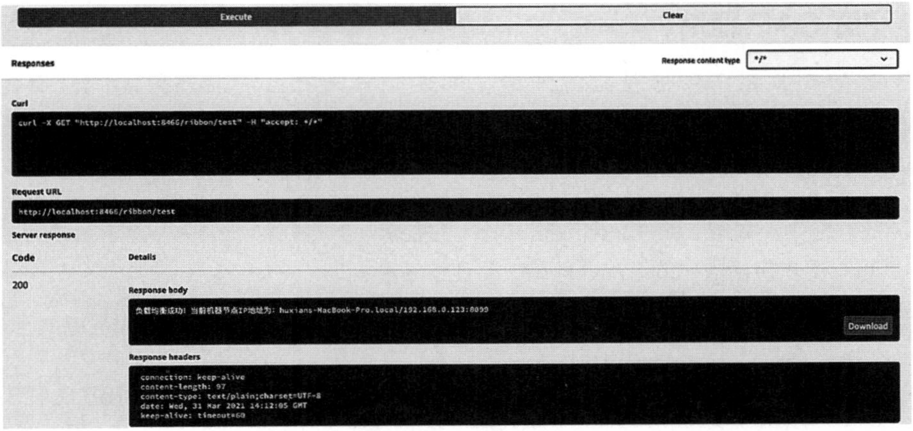

图 4-13

（2）将 192.168.0.123:8079 和 192.168.0.123:8099 的权重设置为 0，按照负载均衡策略，只有 192.168.0.123:8069 能被 Swagger 访问到，如图 4-14 所示。

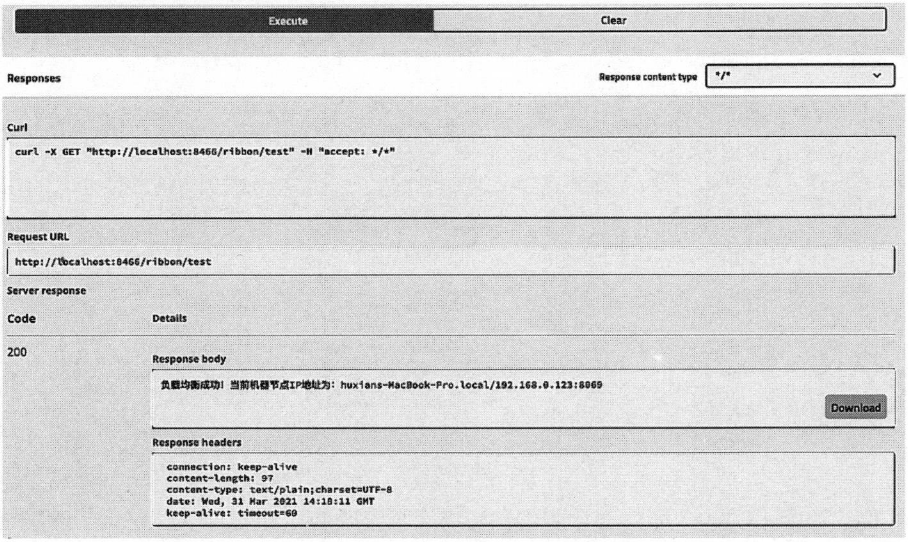

Execute　Clear

Responses　Response content type */*

Curl

```
curl -X GET "http://localhost:8466/ribbon/test" -H "accept: */*"
```

Request URL

```
http://localhost:8466/ribbon/test
```

Server response

Code　Details

200　Response body

```
负载均衡成功！当前机器节点IP地址为：huxians-MacBook-Pro.local/192.168.0.123:8069
```

Download

Response headers

```
connection: keep-alive
content-length: 97
content-type: text/plain;charset=UTF-8
date: Wed, 31 Mar 2021 14:18:11 GMT
keep-alive: timeout=60
```

图 4-14

4.6 用 CP 模式和 AP 模式来保持注册中心的数据一致性

实现数据一致性的技术方案有很多，Nacos 通过 Soft-Jraft 算法来实现注册中心数据的一致性。

4.6.1 了解 CAP 理论

CAP 是分布式系统中最基础的理论，即一个分布式系统最多只能同时满足一致性（Consistency）、可用性（Availability）和分区容错性（Partition tolerance）这三项中的两项。

1. 数据一致性

在 CAP 理论中，数据一致性是对数据“写”的要求。数据在收到“写”请求后才会发生变化。而数据的“写”请求包括数据的增加、删除、修改这三种。

在分布式环境中，对于数据的“写”请求，需要确保多个节点的数据（例如数据库存储的主从节点）保持一致。数据一致性分为以下两种。

- 强数据一致性：数据在各个节点之间随时保持完全一致，即数据的“写”操作要么成功，要么失败。

- 最终数据一致性：数据在各个节点之间在时间阈值内同步成功，保持数据的最终一致性。

2. 可用性

可用性在 CAP 理论里是对结果的要求。它要求系统内的节点无论是接收到“写”请求还是“读”请求，都能处理并给回响应结果。

可用性的两个前置条件：①返回结果必须在合理的时间内，这个“合理的时间”是根据业务来定的；②系统中能正常接收请求的所有节点都返回结果。

3. 分区容错性

分布式存储系统有很多的节点，这些节点都是通过网络进行通信的。而网络是不可靠的，如果节点间的通信出现了问题，则称当前分布式存储系统出现了分区。分区并不一定是由网络故障引起的，也可能是由机器故障引起的。

分区容错性是指：如果出现了分区问题，分布式存储系统还能继续运行；不能因为出现了分区问题，整个分布式节点都“罢工”了。

4.6.2　了解 Nacos 的 CP 模式和 AP 模式

Nacos 支持用 CP 模式和 AP 模式来确保其数据一致性。下面先介绍一下 CP 模式和 AP 模式。

1. AP 模式

在 AP 模式下，Nacos 采用 Nacos Server 之间互相的数据同步，来实现数据在集群之间的同步和复制。大致有以下几种情况。

AP 模式是基于 Nacos Server 的内存来存储数据的，所以很难达到强一致性。

（1）如果集群中有 5 个节点，由于硬件故障某个节点无法使用，那客户端可以通过负载均衡，将故障节点的服务注册请求重新路由到其他正常工作的 4 个节点中的一个，以保证 Nacos 集群的整体可用性。

（2）如果集群中有 5 个节点，有 3 个节点网络互通，其他 2 个节点网络出现了故障，形成了分区故障，则客户端被负载到 Nacos 集群中任意一个节点，可以正常完成服务注册/订阅，只是新注册的服务数据不能实时地同步到其他节点上。已经完成服务注册/订阅的应用，还可以正常通过自身的容错机制，保证在出现分区故障时 Nacos 集群的可用性。

（3）使用 HTTP 接口来同步变更服务信息，在排除网络故障之后，还是会有节点的资源达到阈值，并存在数据同步失败的风险。如果节点之间基于内存来维护服务信息，则会存在数据不一致的情况。可以通过重试机制来重试数据的同步，以实现一定阈值范围内的最终一致性。

2. CP 模式

在 CP 模式下，Nacos 采用 Raft 算法完成 Nacos Server 间数据的同步和复制操作。CP 模式主要通过 Raft 算法来保持数据的强一致性。大致有以下几种情况。

（1）如果集群中存在 5 个节点，在执行完成数据一致性算法之后，会产生一个 Leader 节点和 4 个 Follower 节点。如果 1 个 Follower 节点宕机，则 Leader 节点会把到该节点的请求路由到其他 3 个正常的节点上。如果 Leader 节点宕机，则集群会重新选举，产生新的 Leader 节点。在选举过程中，集群整体不可用。在选举过程中，如果发现当前集群中存活的节点数小于“*n*/2+1”（*n* 代表集群中总的节点数），则选举不能正常进行，集群整体不可用。

CP 模式是牺牲一定的可用性来确保数据的强一致性。

（2）集群中 5 个节点，出现了分区故障。1 个分区包含 1 个 Leader 节点和 2 个 Follower 节点，另外 1 个分区只包含 2 个 Follower 节点。只要 Leader 节点不宕机，那它还是可以正常地处理请求。如果 Leader 节点所在的分区的节点总数小于“*n*/2+1”，则影响集群的重新选举（这种概率非常小）。总体上来看，还是可以确保分布式系统的容忍性的。

（3）在 CP 模式下，节点间的数据同步只能是从 Leader 节点到 Followder 节点。并且，数据同步采用持久化方式复制日志文件。在同步过程中，如果有 1 个节点最终同步失败，则数据变更整体失败，并回滚。

总体上来看，采用基于 Raft 算法后的 CP 模型能够确保数据的强一致性。

4.6.3 了解 Raft 与 Soft-Jraft

1. 了解 Raft

Raft 是一种“共识”算法。“共识”是指，保证所有的参与者都有相同的认知。简单来说就是——在分布式环境下如何解决多个服务节点间的数据一致性。

“共识”包含“服务器端之间”和“客户端和服务器端之间”两方面。

- 在服务器端之间：所有服务器端要达成“共识”。比如 Nacos 集群，服务器端是 3 个节点——A、B、C 节点组成集群。服务提供者和消费者调用注册接口，完成服务元数据的注册。但是，服务提供者和服务消费者会依赖负载均衡将服务元数据同步到集群中的某一个节点。如果要达成服务器之间的共识，则其他两个节点也要同步，这样才能确保数据一致性。
- 在客户端与服务器端之间：客户端调用服务器端，并处理服务器端的响应。需要确保客户端能够正确识别响应结果，要么成功，要么失败，或者出现能够识别的异常。这样，在客户端与服务器端之间，相关联的数据就能保证一致性。

Raft 算法和 Paxos 算法都是“共识”算法。Raft 算法的主要优势如下：

- Paxos 算法偏理重于论；Raft 算法偏重于实践，Raft 算法比较容易实现。
- 采用 Raft 算法理论，开源社区已经落地了很多产品：Etcd、RocketMQ 及 Sofa-JRaft。

对于分布式算法的入门者来说，Raft 算法比 Paxos 算法更容易学习。

2. 了解 Soft-Jraft 中间件

Sofa-JRaft 中间件目前最新的版本是 1.3.5，社区活跃度还行，并且项目的 Issues 数量非常少，项目代码质量非常高，稳定性也非常高。

Sofa-JRaft 中间件具有以下优点。

- 工业级：它是一款稳定性非常高的中间件，已经达到企业级的标准。
- 高性能：其底层采用 Netty 完成 RPC。
- 高并发和低延迟：它能通过多个 Raft 组来应对高并发和低延迟的大流量业务场景。
- 开箱即用：它提供了很多 Demo 实例，通过这些实例，开发人员可以快速上手。

4.6.4　Nacos 注册中心 AP 模式的数据一致性原理

前面已经分析过 Spring Cloud Alibaba 的服务注册过程，下面从 NacosNamingService 的 registerInstance()方法开始分析 Nacos 注解中心 AP 模式的数据一致性原理。

1. 通过随机算法选出 Nacos 节点，发送服务注册请求

服务注册是通过 NamingProxy 类的 registerService()方法来完成 HTTP 请求的，具体代码如下所示：

```
    public void registerService(String serviceName, String groupName, Instance
instance) throws NacosException {
```

```
        NAMING_LOGGER.info("[REGISTER-SERVICE] {} registering service {} with
instance: {}", namespaceId, serviceName,
                instance);
    final Map<String, String> params = new HashMap<String, String>(16);
    //①组装服务注册的 HTTP 参数对象
        params.put(CommonParams.NAMESPACE_ID, namespaceId);
        params.put(CommonParams.SERVICE_NAME, serviceName);
        params.put(CommonParams.GROUP_NAME, groupName);
        params.put(CommonParams.CLUSTER_NAME, instance.getClusterName());
        params.put("ip", instance.getIp());
        params.put("port", String.valueOf(instance.getPort()));
        params.put("weight", String.valueOf(instance.getWeight()));
        params.put("enable", String.valueOf(instance.isEnabled()));
        params.put("healthy", String.valueOf(instance.isHealthy()));
        params.put("ephemeral", String.valueOf(instance.isEphemeral()));
    params.put("metadata", JacksonUtils.toJson(instance.getMetadata()));
        //②调用 Nacos 集群
        reqApi(UtilAndComs.nacosUrlInstance, params, HttpMethod.POST);
    }

    //③NamingProxy 类的 reqApi()方法的部分代码
    if (servers != null && !servers.isEmpty()) {
        Random random = new Random(System.currentTimeMillis());
        int index = random.nextInt(servers.size());
        //④按照服务节点数产生随机数
        for (int i = 0; i < servers.size(); i++) {
            String server = servers.get(index);
            try {
                //⑤调用接口路径"/nacos/v1/ns/instance"发起 POST 请求，注册服务
                return callServer(api, params, body, server, method);
            } catch (NacosException e) {
                exception = e;
                if (NAMING_LOGGER.isDebugEnabled()) {
                    NAMING_LOGGER.debug("request {} failed.", server, e);
                }
            }
                index = (index + 1) % servers.size();
        }
    }
```

2. Nacos 集群节点处理服务注册请求

Nacos 通过 InstanceController 类的 register()方法处理服务注册请求，具体代码如下所示：

```
@CanDistro
@PostMapping
```

```
    @Secured(parser = NamingResourceParser.class, action = ActionTypes.WRITE)
    public String register(HttpServletRequest request) throws Exception {

        final String namespaceId = WebUtils.optional(request,
CommonParams.NAMESPACE_ID, Constants.DEFAULT_NAMESPACE_ID);
    final String serviceName = WebUtils.required(request,
CommonParams.SERVICE_NAME);
    //①校验服务名称
    NamingUtils.checkServiceNameFormat(serviceName);
    //②解析注册请求
    final Instance instance = parseInstance(request);
    //③调用 ServiceManager 类的 registerInstance()方法完成注册
        serviceManager.registerInstance(namespaceId, serviceName, instance);
        return "ok";
    }
```

3. 用一致性算法处理服务注册请求

Nacos 通过 ServiceManager 类的 addInstance()方法处理服务注册数据的一致性请求，具体代码如下所示：

```
    public void addInstance(String namespaceId, String serviceName, boolean
ephemeral, Instance...... ips)
    throws NacosException {
        String key = KeyBuilder.buildInstanceListKey(namespaceId, serviceName,
ephemeral);
        Service service = getService(namespaceId, serviceName);
        synchronized (service) {
            List<Instance> instanceList = addIpAddresses(service, ephemeral,
ips);
            Instances instances = new Instances();
            instances.setInstanceList(instanceList);
            //调用一致性算法实现类完成服务注册
            consistencyService.put(key, instances);
        }
    }
```

ConsistencyService 接口通过如下代码完成初始化：

```
    //依赖当前应用中名称为“consistencyDelegate”的 bean 实例
    @Resource(name = "consistencyDelegate")
    private ConsistencyService consistencyService;

    @DependsOn("ProtocolManager")
    @Service("consistencyDelegate")
    public class DelegateConsistencyServiceImpl implements ConsistencyService {
```

```
        private final PersistentConsistencyServiceDelegateImpl
persistentConsistencyService;
        private final EphemeralConsistencyService ephemeralConsistencyService;

        @Override
        public void put(String key, Record value) throws NacosException {
            mapConsistencyService(key).put(key, value);
        }

    private ConsistencyService mapConsistencyService(String key) {
        //如果是临时节点，则用 EphemeralConsistencyService 接口完成注册；如果是持久化节点，则用 PersistentConsistencyServiceDelegateImpl 类完成注册
            return KeyBuilder.matchEphemeralKey(key) ?
ephemeralConsistencyService : persistentConsistencyService;
        }
        ...
    }
```

4. 用 DistroConsistencyServiceImpl 类（AP 模式）注册临时节点

（1）通过 DistroConsistencyServiceImpl 类的 put()方法注册临时节点，具体代码如下所示：

```
    @DependsOn("ProtocolManager")
    @org.springframework.stereotype.Service("distroConsistencyService")
    public class DistroConsistencyServiceImpl implements
EphemeralConsistencyService, DistroDataProcessor {
        private final DistroProtocol distroProtocol;
        private final GlobalConfig globalConfig;
        @Override
    public void put(String key, Record value) throws NacosException {
        //①用应用的本地缓存存储服务注册信息，并添加异步/同步任务
            onPut(key, value);
            //②调用一致性协议同步服务注册信息
            distroProtocol.sync(new DistroKey(key,
KeyBuilder.INSTANCE_LIST_KEY_PREFIX), DataOperation.CHANGE,
                    globalConfig.getTaskDispatchPeriod() / 2);
        }
        ...
    }
```

（2）在当前节点的内存中存储服务注册信息，具体代码如下所示：

```
    public void onPut(String key, Record value) {
        if (KeyBuilder.matchEphemeralInstanceListKey(key)) {
            Datum<Instances> datum = new Datum<>();
            datum.value = (Instances) value;
            datum.key = key;
```

```
            datum.timestamp.incrementAndGet();
            //①用本地缓存对象 DataStore 类存储服务注册信息
            dataStore.put(key, datum);
        }

        if (!listeners.containsKey(key)) {
            return;
    }
    //②通过监听器将服务注册信息异步地通知给服务订阅者
        notifier.addTask(key, DataOperation.CHANGE);
    }
```

（3）通过 DistroProtocol 类的 sync()方法完成 Nacos 集群节点间的数据同步，具体代码如下所示：

```
    @Component
    public class DistroProtocol {
        private final DistroTaskEngineHolder distroTaskEngineHolder;
        public void sync(DistroKey distroKey, DataOperation action, long delay) {
            for (Member each : memberManager.allMembersWithoutSelf()) {
                //①构造包含 Nacos 集群节点 IP 地址的 DistroKey 对象
                DistroKey distroKeyWithTarget = new
DistroKey(distroKey.getResourceKey(), distroKey.getResourceType(),
                        each.getAddress());
                //③构造数据同步的延迟任务
                DistroDelayTask distroDelayTask = new
DistroDelayTask(distroKeyWithTarget, action, delay);
                //④调用延迟任务的执行引擎添加延迟任务
                distroTaskEngineHolder.getDelayTaskExecuteEngine()
    .addTask(distroKeyWithTarget, distroDelayTask);
                if (Loggers.DISTRO.isDebugEnabled()) {
                    Loggers.DISTRO.debug("[DISTRO-SCHEDULE] {} to {}", distroKey,
each.getAddress());
                }
            }
        }
        ...
    }
```

DistroDelayTaskExecuteEngine 类延迟处理数据同步任务的详细代码可以查看相关源码，这里就不详细分析了。

综上所述，如果应用是临时节点，则 Nacos 会采用基于内存的 AP 模型来保证数据的一致性。在用事件模型和延迟任务同步数据时，AP 模型不能确保各个节点之间的数据能够实时地同步，所以只能保证数据的最终一致性。

4.6.5 Nacos 注册中心 CP 模式的数据一致性原理

如果应用是持久化节点，则 Nacos 会采用 CP 模式来保证数据的一致性。

1. 用 PersistentConsistencyServiceDelegateImpl 类处理服务注册请求

Nacos 在 1.4.0 之后用 Soft-Jraft 算法取代了 Jraft 算法。

（1）通过 Nacos 节点的版本号来确定是否开启了 Soft-Jraft 算法。

在 ClusterVersionJudgement 类中验证 Nacos 节点的版本号，具体代码如下所示：

```
    protected void judge() {
        Collection<Member> members = memberManager.allMembers();
        final String oldVersion = "1.4.0";
    boolean allMemberIsNewVersion = true;
    //①遍历所有节点，验证节点 Nacos 的版本号
        for (Member member : members) {
            final String curV = (String)
member.getExtendVal(MemberMetaDataConstants.VERSION);
            if (StringUtils.isBlank(curV) ||
VersionUtils.compareVersion(oldVersion, curV) > 0) {
                allMemberIsNewVersion = false;
            }
        }
        if (allMemberIsNewVersion && !this.allMemberIsNewVersion) {
            this.allMemberIsNewVersion = true;
            Collections.sort(observers);
            //②如果全部节点的版本号都大于或等于 1.4.0，则通知观察者使用 Soft-Jraft 算法
            for (ConsumerWithPriority consumer : observers) {
                consumer.consumer.accept(true);
            }
            observers.clear();
        }
    }
```

（2）用 PersistentConsistencyServiceDelegateImpl 类观察版本号，并初始化一致性选举算法，具体代码如下所示：

```
    @Component("persistentConsistencyServiceDelegate")
    public class PersistentConsistencyServiceDelegateImpl implements
PersistentConsistencyService {
        private volatile boolean switchNewPersistentService = false;

    private void init() {
        //①如果 isAllNewVersion 为 true,则将 switchNewPersistentService 设置为 true
```

```
        this.versionJudgement.registerObserver(isAllNewVersion ->
switchNewPersistentService = isAllNewVersion, -1);
    }

    private PersistentConsistencyService switchOne() {
        //②如果 switchNewPersistentService 为 true，则采用 Soft-Jraft 算法，否则采用
Jraft 算法
        return switchNewPersistentService ? newPersistentConsistencyService :
oldPersistentConsistencyService;
    }
    ...
}
```

2. 用 Soft-Jraft 算法处理数据一致性请求

（1）通过 PersistentServiceProcessor 类的 put()方法来处理服务注册的数据一致性请求，具体代码如下所示：

```
@Service
public class PersistentServiceProcessor extends LogProcessor4CP implements
PersistentConsistencyService {
    @Override
public void put(String key, Record value) throws NacosException {
    //①构造数据一致性请求对象
        final BatchWriteRequest req = new BatchWriteRequest();
        Datum datum = Datum.createDatum(key, value);
        req.append(ByteUtils.toBytes(key), serializer.serialize(datum));
        //②将请求对象序列化为 Raft 日志
        final Log log =
Log.newBuilder().setData(ByteString.copyFrom(serializer.serialize(req)))
                .setGroup(Constants.NAMING_PERSISTENT_SERVICE_GROUP).setOpe
ration(Op.Write.desc).build();
        try {
            //③调用一致性协议的 CPProtocol 类提交 Raft 日志
            protocol.submit(log);
        } catch (Exception e) {
            throw new NacosException(ErrorCode.ProtoSubmitError.getCode(),
e.getMessage());
        }
    }
    ...
}
```

（2）Nacos 的 CP 模式数据一致性协议的 CPProtocol 类，通过 ProtocolManager 类的 getCpProtocol()方法来完成初始化，具体代码如下所示：

```
    @Component(value = "ProtocolManager")
    @DependsOn("serverMemberManager")
    public class ProtocolManager extends MemberChangeListener implements
ApplicationListener<ContextStartedEvent>, DisposableBean {
        public CPProtocol getCpProtocol() {
            synchronized (this) {
                if (!cpInit) {
                    initCPProtocol();
                    cpInit = true;
                }
            }
            return cpProtocol;
        }
         //从当前运行的应用程序中，获取已经初始化完成的CPProtocol接口的实现类的实例对象
         private void initCPProtocol() {
            ApplicationUtils.getBeanIfExist(CPProtocol.class, protocol -> {
                Class configType =
ClassUtils.resolveGenericType(protocol.getClass());
                Config config = (Config) ApplicationUtils.getBean(configType);
                injectMembers4CP(config);
                protocol.init((config));
                ProtocolManager.this.cpProtocol = protocol;
            });
        }
    }
```

CPProtocol 接口的实现类的实例对象，通过 ConsistencyConfiguration 类实现初始化，具体代码如下所示：

```
    @Configuration
    public class ConsistencyConfiguration {

        @Bean(value = "strongAgreementProtocol")
    public CPProtocol strongAgreementProtocol(ServerMemberManager
memberManager) throws Exception {
        //①初始化Soft-Jraft选举协议的JRaftProtocol对象
            final CPProtocol protocol = getProtocol(CPProtocol.class, () -> new
JRaftProtocol(memberManager));
            //②返回JRaftProtocol对象
            return protocol;
        }

        private <T> T getProtocol(Class<T> cls, Callable<T> builder) throws
Exception  {
            ServiceLoader<T> protocols = ServiceLoader.load(cls);
```

```
        // Select only the first implementation

        Iterator<T> iterator = protocols.iterator();
        if (iterator.hasNext()) {
            return iterator.next();
        } else {
            return builder.call();
        }
    }
}
```

3. 用 JRaftProtocol 类处理数据一致性存储

在 JRaftProtocol 类中，通过 submitAsync()方法异步地提交数据一致性存储，具体代码如下所示：

```
public class JRaftProtocol extends AbstractConsistencyProtocol<RaftConfig,
LogProcessor4CP>
        implements CPProtocol<RaftConfig, LogProcessor4CP> {
    @Override
public Response submit(Log data) throws Exception {
    //①异步提交请求
        CompletableFuture<Response> future = submitAsync(data);
        //②返回异步结果，并设置结果响应的超时时间为 10s
        return future.get(10_000L, TimeUnit.MILLISECONDS);
    }

    @Override
public CompletableFuture<Response> submitAsync(Log data) {
    //③调用 RaftServer 类的 commit()方法提交 Raft 日志
        return raftServer.commit(data.getGroup(), data, new
CompletableFuture<>());
    }
    ...
}
```

4. 用 RaftServer 类存储 Raft 日志

Nacos 用 JRaftServer 类来存储 Raft 日志，具体代码如下所示：

```
public class JRaftServer {
    public CompletableFuture<Response> commit(final String group, final
Message data,
            final CompletableFuture<Response> future) {
        LoggerUtils.printIfDebugEnabled(Loggers.RAFT, "data requested this
time : {}", data);
        //①寻址 Raft 选举组中的节点
```

```
            final RaftGroupTuple tuple = findTupleByGroup(group);
            if (tuple == null) {
                future.completeExceptionally(new IllegalArgumentException("No
corresponding Raft Group found : " + group));
                return future;
            }
            FailoverClosureImpl closure = new FailoverClosureImpl(future);
            final Node node = tuple.node;
            //②判断节点的角色
            if (node.isLeader()) {
                //③如果是 Leader 节点，则直接处理日志请求
                applyOperation(node, data, closure);
            } else {
                //④如果是 Follower 节点，则将日志请求转发到集群的 Leader 节点
                invokeToLeader(group, data, rpcRequestTimeoutMs, closure);
            }
            return future;
        }
        ...
    }
```

5.用 Soft-Jraft 的 NodeImpl 类的 apply()方法异步地复制日志

如果是 Follower 节点，则执行 invokeToLeader()方法路由请求到 Leader 节点，具体代码如下所示：

```
    private void invokeToLeader(final String group, final Message request, final
int timeoutMillis,
                FailoverClosure closure) {
        try {
            final Endpoint leaderIp = Optional.ofNullable(getLeader(group))
                    .orElseThrow(() -> new
NoLeaderException(group)).getEndpoint();
            //①通过 RPC 客户端的 CliClientServiceImpl 类向 Leader 节点发送同步数据的路由
请求
            cliClientService.getRpcClient().invokeAsync(leaderIp, request, new
InvokeCallback() {
            @Override
            public void complete(Object o, Throwable ex) {
                if (Objects.nonNull(ex)) {
                        closure.setThrowable(ex);
                        closure.run(new Status(RaftError.UNKNOWN,
ex.getMessage()));
                        return;
                }
                //②处理异步请求的响应结果
```

```
                closure.setResponse((Response) o);
                closure.run(Status.OK());
            }

            @Override
            public Executor executor() {
                    return RaftExecutor.getRaftCliServiceExecutor();
                }
            }, timeoutMillis);
        } catch (Exception e) {
            closure.setThrowable(e);
            closure.run(new Status(RaftError.UNKNOWN, e.toString()));
        }
    }
```

如果是 Leader 节点，则执行 NodeImpl 类的 apply()方法，并通过队列异步地处理 Raft 日志的复制请求，具体代码如下所示：

```
    public class NodeImpl implements Node, RaftServerService {
        @Override
        public void apply(final Task task) {
            ...
            //①构造 LogEntry 对象
            final LogEntry entry = new LogEntry();
            entry.setData(task.getData());
            int retryTimes = 0;
            try {
                //②构造 EventTranslator 事件对象
                final EventTranslator<LogEntryAndClosure> translator = (event,
sequence) -> {
                    event.reset();
                    event.done = task.getDone();
                    event.entry = entry;
                    event.expectedTerm = task.getExpectedTerm();
                };
                //③如果 retryTimes 小于或等于 MAX_APPLY_RETRY_TIMES，则开始执行
                while (true) {
                    //④向环形队列 RingBuffer 中添加事件
                    if (this.applyQueue.tryPublishEvent(translator)) {
                        //⑤如果添加成功，则直接跳出循环
                        break;
                    } else {
                        //⑥如果添加失败，则重试并累加重试次数
                        retryTimes++;
                        //⑦如果重试次数大于重试次数的阈值，则退出循环
                        if (retryTimes > MAX_APPLY_RETRY_TIMES) {
```

```
                    Utils.runClosureInThread(task.getDone(),
                        new Status(RaftError.EBUSY, "Node is busy, has too
many tasks."));
                    LOG.warn("Node {} applyQueue is overload.",
getNodeId());
                    this.metrics.recordTimes("apply-task-overload-times",
1);
                    return;
                }
                ThreadHelper.onSpinWait();
            }
        }

    } catch (final Exception e) {
    ...
    }
  }
}
```

6. **用 Soft-Jraft 的 LogManagerImpl 类的 appendEntries()方法持久化日志。**

为了保证日志持久化的高性能，Soft-Jraft 会先将日志存储在内存中，然后通过事件机制异步地通知任务队列持久化日志。

持久化日志的具体代码如下所示：

```
public class LogManagerImpl implements LogManager {
    @Override
    public void appendEntries(final List<LogEntry> entries, final
StableClosure done) {
        Requires.requireNonNull(done, "done");
        if (this.hasError) {
            entries.clear();
            Utils.runClosureInThread(done, new Status(RaftError.EIO,
"Corrupted LogStorage"));
            return;
        }
        boolean doUnlock = true;
        //①加锁，确保线程安全
        this.writeLock.lock();
        try {
            if (!entries.isEmpty() && !checkAndResolveConflict(entries, done)) {
                entries.clear();
                return;
            }
            for (int i = 0; i < entries.size(); i++) {
```

```
                final LogEntry entry = entries.get(i);
                if (this.raftOptions.isEnableLogEntryChecksum()) {
                    entry.setChecksum(entry.checksum());
                }
                if (entry.getType() == EntryType.ENTRY_TYPE_CONFIGURATION) {
                    Configuration oldConf = new Configuration();
                    if (entry.getOldPeers() != null) {
                        oldConf = new Configuration(entry.getOldPeers(),
entry.getOldLearners());
                    }
                    final ConfigurationEntry conf = new
ConfigurationEntry(entry.getId(),
                        new Configuration(entry.getPeers(),
entry.getLearners()), oldConf);
                    this.configManager.add(conf);
                }
            }
            if (!entries.isEmpty()) {
                done.setFirstLogIndex(entries.get(0).getId().getIndex());
                //②添加到内存中
                this.logsInMemory.addAll(entries);
            }
            done.setEntries(entries);

            int retryTimes = 0;
            final EventTranslator<StableClosureEvent> translator = (event,
sequence) -> {
                event.reset();
                event.type = EventType.OTHER;
                event.done = done;
            };
            //③如果重试次数小于或等于最大重试次数，则执行日志复制
            while (true) {
                //④发布磁盘异步持久化事件，并开启节点间的日志复制
                if (tryOfferEvent(done, translator)) {
                    break;
                } else {
                    retryTimes++;
                    if (retryTimes > APPEND_LOG_RETRY_TIMES) {
                        reportError(RaftError.EBUSY.getNumber(), "LogManager
is busy, disk queue overload.");
                        return;
                    }
                    ThreadHelper.onSpinWait();
                }
```

```
            }
            doUnlock = false;
            if (!wakeupAllWaiter(this.writeLock)) {
                notifyLastLogIndexListeners();
            }
        } finally {
            if (doUnlock) {
                //⑤释放锁
                this.writeLock.unlock();
            }
        }
    }
    ...
}
```

关于 Soft-Jraft 算法日志持久化的详细过程，这里就不再一一分析了，如果读者感兴趣可以查阅相关源码。

4.6.6 【实例】用持久化的服务实例来验证注册中心的数据一致性

本实例的源码在本书配套资源的“/spring-cloud-alibaba-practice/chapterfour/persistence-discovery-spring-cloud-alibaba/”目录下。

Nacos 通过 Soft-Jraft 算法来持久化注册中心的数据，具体操作如下。

1. 用“Spring Boot + Spring Cloud Alibaba”初始化项目

（1）添加 Spring Boot 和 Spring Cloud Alibaba 的相关 Jar 包。

（2）添加配置文件 bootstrap.yaml，具体代码如下所示：

```
dubbo:
  scan:
    base-packages: com.alibaba.cloud.youxia
  protocol:
    name: dubbo
    port: 26734
  registry:
    address: nacos://127.0.0.1:8848
spring:
  application:
    name: persistence-discovery-spring-cloud-alibaba
  main:
    allow-bean-definition-overriding: true
```

```
    cloud:
      nacos:
        discovery:
          server-addr: 127.0.0.1:8848
          namespace: c7ba173f-29e5-4c58-ae78-b102be11c4f9
          group: persistence-discovery-spring-cloud-alibaba
          ##设置实例类型为持久化
          ephemeral: false
server:
  port: 7867
```

2. 运行程序，观察效果

运行程序后，可以在 Nacos 控制台上看到持久化的实例已经注册成功，如图 4-15 所示。

IP	端口	临时实例	权重	健康状态	元数据	操作
172.17.1.160	7867	false	1	true	dubbo.metadata-service.urls=["dubbo://172.17.1.160:26734/com.alibaba.cloud.dub- bo.service.DubboMetadataService?anyhost=true&application=persistence- discovery-spring-cloud- alibaba&bind.ip=172.17.1.160&bind.port=26734&deprecated=false&dubbo=2.0.2&dy namic=true&generic=false&group=persistence-discovery-spring-cloud-alibaba&inter- face=com.alibaba.cloud.dubbo.service.DubboMetadataService&metadata-type=re- mote&methods=getAllServiceKeys,getServiceRestMetadata,getExportedURLs,getAll- ExportedURLs&pid=7107&qos.enable=false&release=2.7.8&revision=2.2.3.RELEASE &side=provider×tamp=1617711549997&version=1.0.0"] dubbo.protocols.dubbo.port=26734 preserved.register.source=SPRING_CLOUD	编辑 下线

图 4-15

从图 4-14 中可以看出，临时实例字段被设置为 false，表示实例已被持久化到文件中了。

3. 关闭程序并更改 Dubbo 端口号，验证注册中心数据的一致性

（1）关闭程序并更改 Dubbo 的端口号，具体代码如下所示：

```
server:
  port: 7890
dubbo:
    port: 26767
```

（2）重新运行程序，在 Nacos 控制台中会出现两个实例，如图 4-16 所示。

已经下线的实例（172.17.1.160:26734）的健康状态为 false，新上线的实例（172.17.1.160:26767）的健康状态为 true。Nacos 集群会永久地保存被设置为持久化类型的实例的元数据。

在 Nacos 的安装目录的文件路径“/data/naming/data/{命名空间 ID}”下，可以看到持久化的实例元数据，如图 4-17 所示。

IP	端口	临时实例	权重	健康状态	元数据	操作
172.17.1.160	7867	false	1	false	dubbo.metadata-service.urls=["dubbo://172.17.1.160:26734/com.alibaba.cloud.dubbo.service.DubboMetadataService?anyhost=true&application=persistence-discovery-spring-cloud-alibaba&bind.ip=172.17.1.160&bind.port=26734&deprecated=false&dubbo=2.0.2&dynamic=true&generic=false&group=persistence-discovery-spring-cloud-alibaba&interface=com.alibaba.cloud.dubbo.service.DubboMetadataService&metadata-type=remote&methods=getAllServiceKeys,getServiceRestMetadata,getExportedURLs,getAllExportedURLs&pid=7107&qos.enable=false&release=2.7.8&revision=2.2.3.RELEASE&side=provider×tamp=1617711549997&version=1.0.0"] dubbo.protocols.dubbo.port=26734 preserved.register.source=SPRING_CLOUD	编辑 下线
172.17.1.160	7890	false	1	true	dubbo.metadata-service.urls=["dubbo://172.17.1.160:26767/com.alibaba.cloud.dubbo.service.DubboMetadataService?anyhost=true&application=persistence-discovery-spring-cloud-alibaba&bind.ip=172.17.1.160&bind.port=26767&deprecated=false&dubbo=2.0.2&dynamic=true&generic=false&group=persistence-discovery-spring-cloud-alibaba&interface=com.alibaba.cloud.dubbo.service.DubboMetadataService&metadata-type=remote&methods=getAllServiceKeys,getServiceRestMetadata,getExportedURLs,getAllExportedURLs&pid=7399&qos.enable=false&release=2.7.8&revision=2.2.3.RELEASE&side=provider×tamp=1617712449703&version=1.0.0"] dubbo.protocols.dubbo.port=26767 preserved.register.source=SPRING_CLOUD	编辑 下线

图 4-16

```
huxian@huxians-MacBook-Pro data % cd c7ba173f-29e5-4c58-ae78-b102be11c4f9
huxian@huxians-MacBook-Pro c7ba173f-29e5-4c58-ae78-b102be11c4f9 % ls
com.alibaba.nacos.naming.domains.meta.c7ba173f-29e5-4c58-ae78-b102be11c4f9##example10@@example10.datum
com.alibaba.nacos.naming.domains.meta.c7ba173f-29e5-4c58-ae78-b102be11c4f9##persistence-discovery-spring-cloud-alibaba
@@persistence-discovery-spring-cloud-alibaba.datum
com.alibaba.nacos.naming.iplist.c7ba173f-29e5-4c58-ae78-b102be11c4f9##example10@@example10.datum
com.alibaba.nacos.naming.iplist.c7ba173f-29e5-4c58-ae78-b102be11c4f9##persistence-discovery-spring-cloud-alibaba@@pers
istence-discovery-spring-cloud-alibaba.datum
```

图 4-17

4.7 用缓存和文件来存储 Nacos 的元数据

Nacos 的元数据是指 Nacos 服务注册与服务订阅的数据来源。Nacos 支持用缓存和文件来存储其元数据。

4.7.1 认识 Nacos 的元数据

Nacos 的元数据主要包含 Nacos 数据（如配置和服务）的描述信息，如服务版本、权重、容灾策略、负载均衡策略、鉴权配置、各种自定义的标签（label）。从作用范围来看，元数据分为服务元数据、集群元数据及实例元数据。

服务元数据、实例元数据及集群元数据之间的关系如图 4-18 所示。

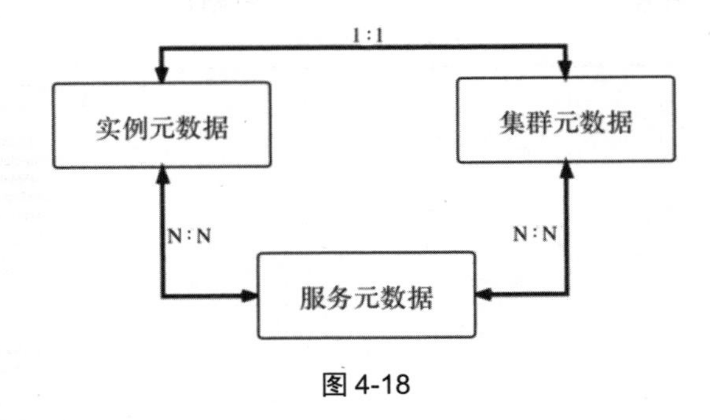

图 4-18

1. 服务元数据的字段

Nacos 通过 Service 类来定义服务元数据，字段的功能描述见表 4-3。

表 4-3　服务元数据字段及功能描述

元数据字段	字段功能描述
name	服务的名称
protectThreshold	保护阈值
fappName	服务对应的应用名称
groupName	服务对应的组名称
metadata	服务元数据
clusterMap	服务对应的集群列表
namespaceId	服务对应的命名空间 ID
enabled	服务的开关
resetWeight	重置服务权重的开关
owners	服务所有者
token	服务登录的 token

2. 实例元数据的字段

Nacos 通过 Instance 类来定义实例元数据，字段的功能描述见表 4-4。

表 4-4　实例元数据字段及功能描述

元数据字段	字段功能描述
instanceId	实例的 ID
ip	实例对应的 IP 地址
port	实例对应的端口号
weight	实例权重
healthy	实例健康状态
enabled	实例的开启开关

续表

元数据字段	字段功能描述
ephemeral	实例类型
clusterName	实例对应的集群名称
serviceName	实例对应的服务名称
metadata	实例元数据
tenant	实例对应的租户 ID
lastBeat	实例中最新的心跳时间

3. 集群元数据字段

Nacos 通过 Cluster 类来定义集群元数据，字段的功能描述见表 4-4。

表 4-4　集群元数据字段及功能描述

元数据字段	字段功能描述
serviceName	服务的名称
name	集群的名称
healthChecker	集群的健康检查
defaultPort	默认端口号
defaultCheckPort	默认检查端口号
metadata	集群的元数据
persistentInstances	持久化的实例列表
ephemeralInstances	临时的实例列表
service	集群对应的服务元数据

4.7.2　用缓存存储 Nacos 的元数据

用 Nacos 的控制台变更服务、实例及集群信息，需要存储元数据。如果实例为临时实例，则对应的元数据使用缓存来存储，具体代码如下所示：

```
@Component
public class DataStore {
    //①内存缓存对象
    private Map<String, Datum> dataMap = new ConcurrentHashMap<>(1024);
    //②添加 Nacos 的元数据
public void put(String key, Datum value) {
        dataMap.put(key, value);
    }
    //③删除 Nacos 的元数据
    public Datum remove(String key) {
        return dataMap.remove(key);
```

```
    }

    public Set<String> keys() {
        return dataMap.keySet();
    }
    //④获取 Nacos 的元数据
    public Datum get(String key) {
        return dataMap.get(key);
    }

    public boolean contains(String key) {
        return dataMap.containsKey(key);
    }
    //⑤批量获取 Nacos 的元数据
    public Map<String, Datum> batchGet(List<String> keys) {
        Map<String, Datum> map = new HashMap<>(128);
        for (String key : keys) {
            Datum datum = dataMap.get(key);
            if (datum == null) {
                continue;
            }
            map.put(key, datum);
        }
        return map;
    }

    public int getInstanceCount() {
        int count = 0;
        for (Map.Entry<String, Datum> entry : dataMap.entrySet()) {
            try {
                Datum instancesDatum = entry.getValue();
                if (instancesDatum.value instanceof Instances) {
                    count += ((Instances)
instancesDatum.value).getInstanceList().size();
                }
            } catch (Exception ignore) {
            }
        }
        return count;
    }

    public Map<String, Datum> getDataMap() {
        return dataMap;
    }
}
```

4.7.3 用文件存储 Nacos 的元数据

如果要持久化实例，则 Nacos 用文件存储其元数据，具体代码如下所示：

```
    @Component
    public class RaftStore implements Closeable {
        ...
        private final Properties meta = new Properties();
        private static final String META_FILE_NAME = DATA_BASE_DIR +
File.separator + "meta.properties";
        private static final String CACHE_DIR = DATA_BASE_DIR + File.separator
+ "data";
    public synchronized void write(final Datum datum) throws Exception {
        //①获取命名空间 ID
            String namespaceId = KeyBuilder.getNamespace(datum.key);
            //②生成缓存文件
            File cacheFile = new File(cacheFileName(namespaceId, datum.key));
            if (!cacheFile.exists() && !cacheFile.getParentFile().mkdirs()
&& !cacheFile.createNewFile()) {
                MetricsMonitor.getDiskException().increment();
                throw new IllegalStateException("can not make cache file: " +
cacheFile.getName());
            }
            FileChannel fc = null;
            ByteBuffer data;
            data = ByteBuffer.wrap(JacksonUtils.toJson(datum).
getBytes(StandardCharsets.UTF_8));
            try {
                fc = new FileOutputStream(cacheFile, false).getChannel();
                //③写元数据到缓存文件中
                fc.write(data, data.position());
                fc.force(true);
            } catch (Exception e) {
                MetricsMonitor.getDiskException().increment();
                throw e;
            } finally {
                if (fc != null) {
                    fc.close();
                }
            }
            if (StringUtils.isNoneBlank(namespaceId)) {
                if (datum.key.contains(Constants.DEFAULT_GROUP +
Constants.SERVICE_INFO_SPLITER)) {
                    String oldDatumKey = datum.key
```

```
                        .replace(Constants.DEFAULT_GROUP +
Constants.SERVICE_INFO_SPLITER, StringUtils.EMPTY);

                cacheFile = new File(cacheFileName(namespaceId,
oldDatumKey));
                if (cacheFile.exists() && !cacheFile.delete()) {
                    Loggers.RAFT.error("[RAFT-DELETE] failed to delete old
format datum: {}, value: {}", datum.key,
                            datum.value);
                    throw new IllegalStateException("failed to delete old
format datum: " + datum.key);
                }
            }
        }
    }
}
```

4.7.4　【实例】用 Spring Cloud Alibaba 整合 Nacos 和 Dubbo 的元数据

本实例的源码在本书配套资源的“/spring-cloud-alibaba-practice/chapterfour/integration-dubbo-alibaba-metadata/”目录下。

本实例用 Spring Cloud ALibaba 整合 Nacos 的元数据和 Dubbo 的元数据。

1. 属性配置

应用的属性配置如下（主要是增加了 Dubbo 的原生配置）：

```
dubbo:
  scan:
    base-packages: com.alibaba.cloud.youxia
  protocol:
    name: dubbo
    port: 28787
  application:
    name: integration-dubbo-alibaba-metadata
  registry:
    address: nacos://192.168.1.79:8848,nacos://192.168.1.79:8847,nacos://
192.168.1.79:8846
      parameters:
        namespace: c7ba173f-29e5-4c58-ae78-b102be11c4f9
        group: integration-dubbo-alibaba-metadata
spring:
  application:
```

```
    name: integration-dubbo-alibaba-metadata
  main:
    allow-bean-definition-overriding: true
  cloud:
    nacos:
      discovery:
        server-addr: 192.168.1.79:8848,192.168.1.79:8847,192.168.1.79:8846
        namespace: c7ba173f-29e5-4c58-ae78-b102be11c4f9
        group: integration-dubbo-alibaba-metadata
      config:
        server-addr: 192.168.1.79:8848,192.168.1.79:8847,192.168.1.79:8846
        namespace: c7ba173f-29e5-4c58-ae78-b102be11c4f9
        group: integration-dubbo-alibaba-metadata
```

2. 效果展示

（1）整合了 Dubbo 元数据中心后的 Nacos 服务列表如图 4-19 所示。

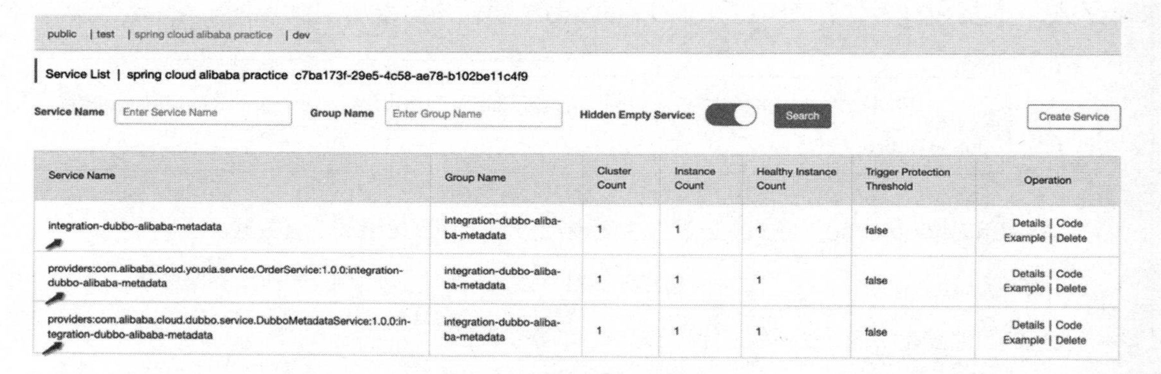

图 4-19

在本实例中暴露了一个 Dubbo 服务接口 com.alibaba.cloud.youxia.service.OrderService。启动应用后，订单接口会向注册中心注册 3 条元数据。如图 4-19 所示，在服务列表中已经注册成功 3 条元数据。

integration-dubbo-alibaba-metadata 元数据的样例如图 4-20 所示。

| Cluster: DEFAULT

Edit Cluster

IP	Port	Ephemeral	Weight	Healthy	Metadata	Operation
172.17.2.1	8080	false	1	true	dubbo.metadata-service.urls=["dubbo://172.17.2.1:28789/com.alibaba.-cloud.dubbo.service.DubboMetadataService?anyhost=true&application=integration-dubbo-alibaba-metadata&bind.ip=172.17.2.1&bind.port=28789&deprecated=false&dubbo=2.0.2&dynamic=true&generic=false&group=integration-dubbo-alibaba-metadata&interface=com.alibaba.cloud.dubbo.service.DubboMetadataService&metadata-type=remote&methods=getAllServiceKeys,getServiceRestMetadata,getExportedURLs,getAllExportedURLs&pid=8792&qos.enable=false&release=2.7.8&revision=2.2.3.RELEASE&side=provider×tamp=1629284563121&version=1.0.0"] dubbo.protocols.dubbo.port=28789 preserved.register.source=SPRING_CLOUD	Edit Offline

图 4-20

“providers:com.alibaba.cloud.dubbo.service.DubboMetadataService:1.0.0:integration-dubbo-alibaba-metadata”元数据的样例如图 4-21 所示。

| Cluster: DEFAULT

Edit Cluster

IP	Port	Ephemeral	Weight	Healthy	Metadata	Operation
172.17.2.1	28789	true	1	true	side=provider methods=getAllServiceKeys,getServiceRestMetadata,getExportedURLs,getAllExportedURLs release=2.7.8 deprecated=false dubbo=2.0.2 pid=8792 interface=com.alibaba.cloud.dubbo.service.DubboMetadataService version=1.0.0 generic=false revision=2.2.3.RELEASE path=com.alibaba.cloud.dubbo.service.DubboMetadataService protocol=dubbo metadata-type=remote application=integration-dubbo-alibaba-metadata dynamic=true category=providers anyhost=true group=integration-dubbo-alibaba-metadata timestamp=1629284563121	Edit Offline

图 4-21

“providers:com.alibaba.cloud.youxia.service.OrderService:1.0.0:integration-dubbo-alibaba-metadata”元数据的样例如图 4-22 所示。

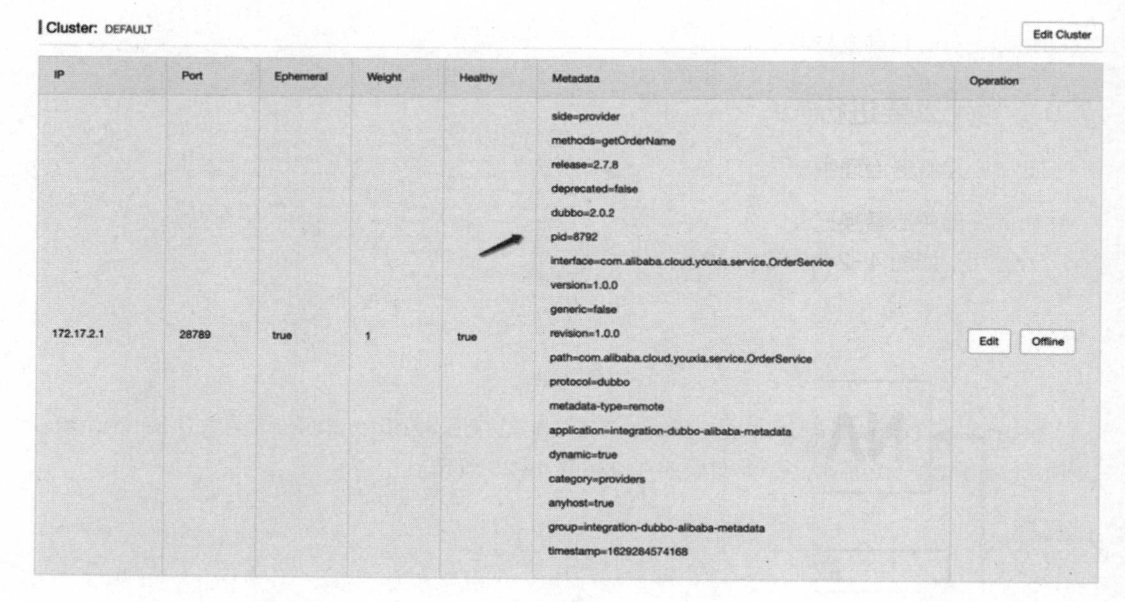

图 4-22

（2）整合了 Dubbo 元数据中心后的 Nacos 配置中心列表如图 4-23 所示。

图 4-23

4.8 用 Nacos Sync 来实现应用服务的数据迁移

Nacos Sync 是 Nacos 提供给开发者的一个应用服务数据迁移框架，开发者可以使用它完成从 Nacos 到指定注册中心的应用服务数据迁移。

Nacos Sync 目前支持 Nacos、ZooKeeper、Eureka 注册中心等。

4.8.1　为什么要进行应用服务的数据迁移

注册中心是服务治理中非常关键的功能，它维护服务提供者和服务订阅者间的订阅关系。

用新的注册中心替换已经在线上运行的注册中心，需要将订阅关系迁移到新的注册中心中，以确保服务可用，如图 4-24 所示。

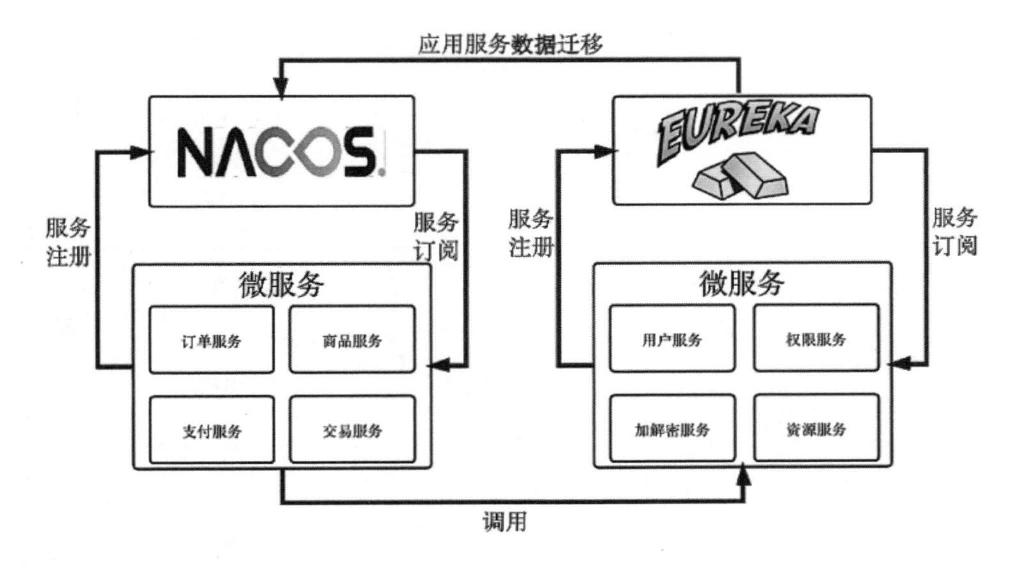

图 4-24

（1）如果 Eureka 注册中心已经上线，且大量的业务在使用 Eureka 作为注册中心，则业务中的大部分微服务应用都强依赖 Eureka 注册中心，所以在应用服务数据迁移的过程中，Eureka 注册中心服务不能下线。

（2）开发人员完成技术升级，并上线 Nacos 注册中心。订单服务、商品服务、支付服务和交易服务用 Nacos 作为注册中心，但它们需要调用 Eureka 注册中心中的用户服务、权限服务、加解密服务和资源服务。

（3）为了满足接入 Nacos 注册中心的服务（比如订单服务）也能够访问原有 Eureka 注册中心中的服务（用户服务），开发人员需要将 Eureka 注册中心中的服务迁移到 Nacos 注册中心中，以确保新的订单服务能够订阅用户服务，且用户服务不需要做技术升级/改造。通常，为了确保服务订阅的效率，在技术升级全量完成前，应用可以同时订阅 Nacos 和 Eureka 注册中心的服务。

4.8.2 如何完成应用服务的数据迁移

Nacos Sync 的架构如图 4-25 所示。开发人员可以在控制台中新增数据同步任务，来进行系统配置和集群配置。

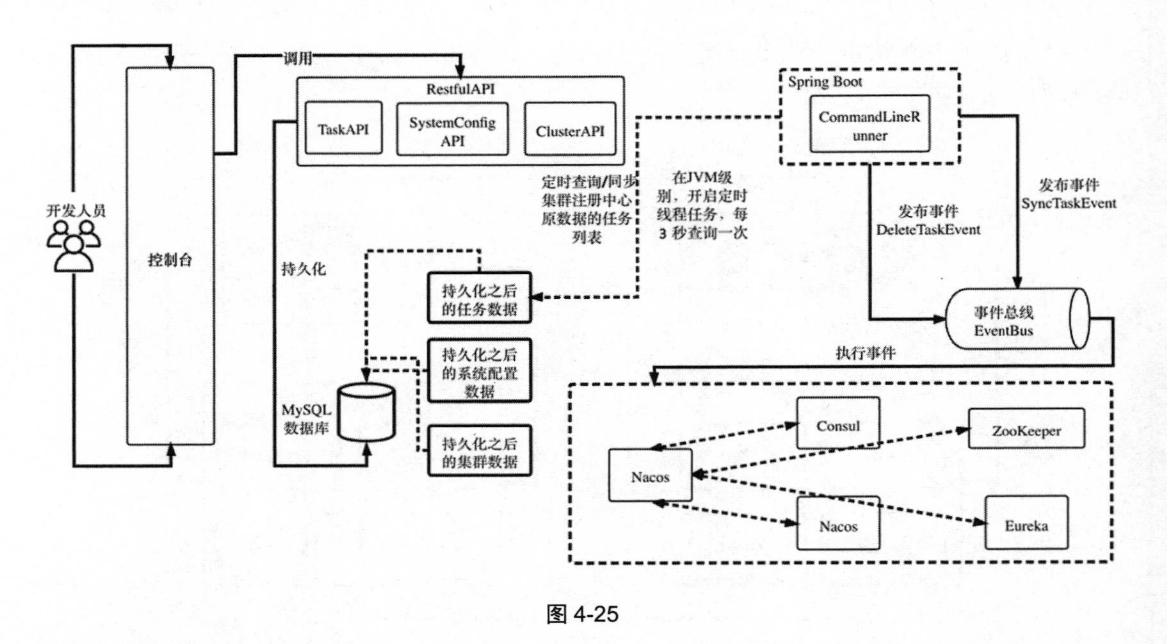

图 4-25

步骤如下：

（1）将数据持久化到 MySQL 数据库表、Cluster、system_config 及 Task 中。

（2）Nacos-Sync 会在 Spring Boot 的 CommandLineRunner 启动时，开启一个定时线程任务，每隔 3 秒轮询数据库，获取暂时没有执行的事件任务。

（3）事件类型主要分为 SyncTaskEvent 和 DeleteTaskEvent，前者是注册中心元数据同步事件，后者是对应的删除事件。SyncTaskEvent 会开启服务注册信息的同步，DeleteTaskEvent 会执行服务注册信息的删除。

（4）通过事件总线 EventBus 通知事件订阅者 EventListener，并执行 listenerSyncTaskEvent()和 listenerDeleteTaskEvent()方法，以及执行服务注册信息的双向同步逻辑。

4.8.3 【实例】将 Eureka 注册中心中的应用服务数据迁移到 Nacos 注册中心中

本实例的源码在本书配套资源的“/spring-cloud-alibaba-practice/chapterfour/nacos-sync/”目录下。

如果项目已经在使用 Spring Cloud 的 Eureka 作为线上应用服务的注册中心，线下又在将应用服务数据整体迁移到Naco 注册中心，则可以将“没有升级完成的老的应用服务数据”用Nacos Sync 迁移到 Nacos 注册中心中。

1. 搭建 Eureka 注册中心

本实例通过项目 eureka-server 来完成 Eureka 注册中心的搭建。

（1）添加 Jar 包依赖，具体代码如下所示：

```
<dependency>
    <groupId>org.springframework.cloud</groupId>
    <artifactId>spring-cloud-starter-netflix-eureka-server</artifactId>
</dependency>
<dependency>
     <groupId>org.springframework.boot</groupId>
     <artifactId>spring-boot-starter-web</artifactId>
     <version>2.0.5.RELEASE</version>
</dependency>
```

（2）添加配置文件 bootstrap.yml，具体代码如下所示：

```
server:
  port: 8761
eureka:
  instance:
    hostname: 127.0.0.1
  client:
    register-with-eureka: false
    fetch-registry: false
    service-url:
      defaultZone: http://${eureka.instance.hostname}:${server.port}/
eureka/
spring:
  application:
    name: eurka-server
```

（3）通过启动类 EurekaServerApplicatipn 运行程序，具体代码如下所示：

```
SpringBootApplication
@EnableEurekaServer
public class EurekaServerApplicatipn {
    public static void main(String[] args) {
        SpringApplication.run(EurekaServerApplicatipn.class, args);
    }
}
```

（4）输入 URL 127.0.0.1:8761，具体效果如图 4-26 所示。

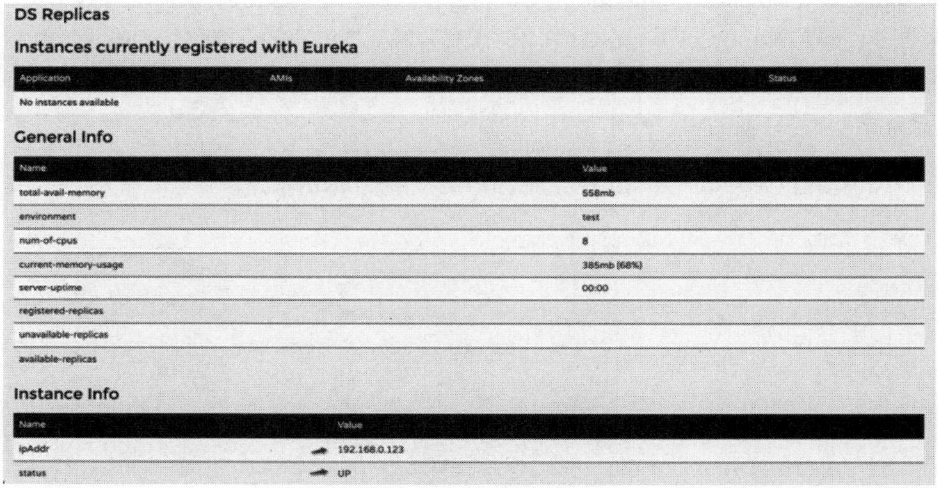

DS Replicas

Instances currently registered with Eureka

Application	AMIs	Availability Zones	Status
No instances available			

General Info

Name	Value
total-avail-memory	558mb
environment	test
num-of-cpus	8
current-memory-usage	385mb (68%)
server-uptime	00:00
registered-replicas	
unavailable-replicas	
available-replicas	

Instance Info

Name	Value
ipAddr	192.168.0.123
status	UP

图 4-26

2. 创建一个服务提供者，并将其注册到 Eureka 注册中心中

下面通过项目 eureka-provider 来实现一个服务提供者。

（1）添加 Jar 包依赖，具体代码如下所示。

```
<dependency>
    <groupId>org.springframework.cloud</groupId>
    <artifactId>spring-cloud-starter-netflix-eureka-client</artifactId>
</dependency>
<dependency>
    <groupId>org.springframework.boot</groupId>
    <artifactId>spring-boot-starter-web</artifactId>
</dependency>
```

（2）添加配置文件 bootstrap.yml，具体代码如下所示。

```
server:
  port: 8768
```

```
eureka:
  client:
    service-url:
      defaultZone: http://127.0.0.1:8761/eureka/
spring:
  application:
    name: eurka-provider
```

（3）定义一个 RESTful API 接口调用服务提供者，具体代码如下所示。

```
@RestController
@RequestMapping(value = "/data")
public class EurekaProviderController {
    @GetMapping(value = "/getdata")
    @ResponseBody
    public String getData(){
        System.out.println("通过 Eureka 注册中心暴露的 RESTful API 服务！");
        return "调用接口成功！";
    }
}
```

（4）通过启动类 EurekaProviderApplicatipn 运行项目，具体代码如下所示：

```
@SpringBootApplication
@EnableEurekaClient
public class EurekaProviderApplicatipn {
    public static void main(String[] args) {
        SpringApplication.run(EurekaProviderApplicatipn.class, args);
    }
}
```

（5）输入 URL 127.0.0.1:8761，具体效果如图 4-27 所示。可以看到，eurka-provider 已经被注册到 Eureka 注册中心中了。

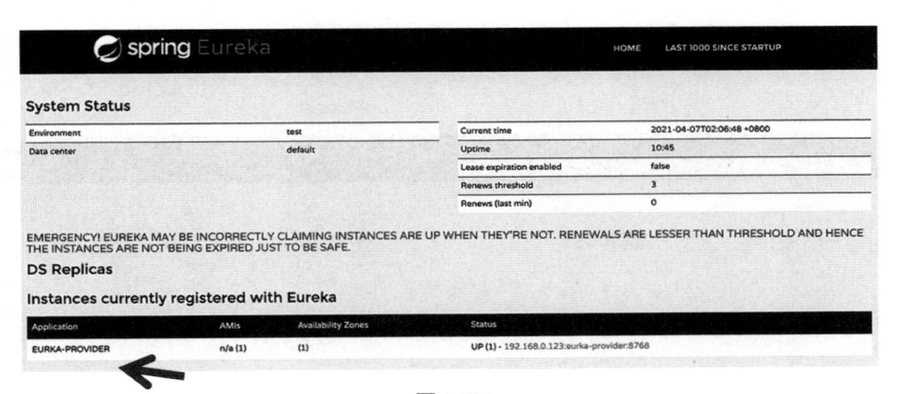

图 4-27

3. 搭建 Nacos Sync 环境

（1）从 Nacos Sync 的官网下载安装包，本程序使用的是 nacos-sync-0.4.4.tar.gz。

（2）修改安装包的配置文件 application.properties，具体代码如下所示。

```
server.port=8083
server.servlet.context-path=/
spring.jpa.properties.hibernate.dialect=org.hibernate.dialect.MySQL5Dial
ect
spring.jpa.hibernate.ddl-auto=update
spring.jpa.properties.hibernate.show_sql=false
##添加 MySQL 数据源
spring.datasource.url=jdbc:mysql://127.0.0.1:3306/nacos_sync?characterEn
coding=utf8
spring.datasource.username=root
spring.datasource.password=123456huxian
management.endpoints.web.exposure.include=*
management.endpoint.health.show-details=always
```

（3）切换到 bin 目录下并执行命令“sh startup.sh start”。

（4）通过“http://127.0.0.1:8083/#/serviceSync”访问 Nacos Sync 的控制台。

（5）在控制台中，进行集群配置和服务同步的配置。集群配置如图 4-28 所示；服务同步配置如图 4-29 所示。

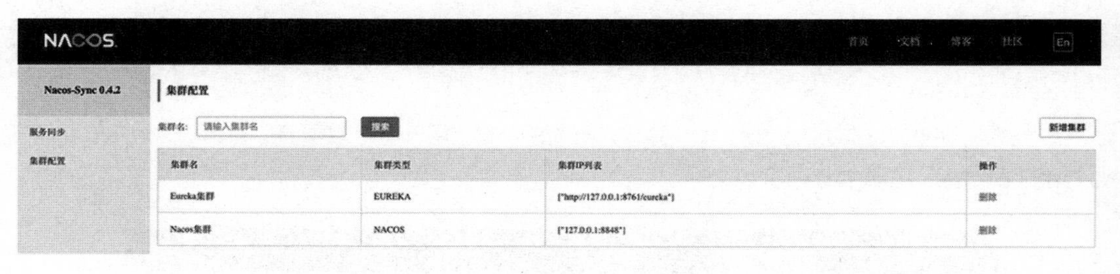

图 4-28

NACOS. 首页 文档 博客 社区 En

Nacos-Sync 0.4.2

服务同步

集群配置

同步服务

服务名: 请输入服务名 搜索 新增同步

服务名	分组	源集群	目标集群	操作
EURKA-PROVIDER	DEFAULT_GROUP	Eureka集群	Nacos集群	暂停
EURKA-PROVIDER	DEFAULT_GROUP	Eureka集群	Nacos集群	暂停

图 4-29

4. 观察 Nacos 注册中心是否同步成功

通过“127.0.0.1:8848/nacos”访问 Nacos 的控制台可以看出，服务提供者已经从 Eureka 注册中心中迁移到 Nacos 注册中心中了，如图 4-30 所示。

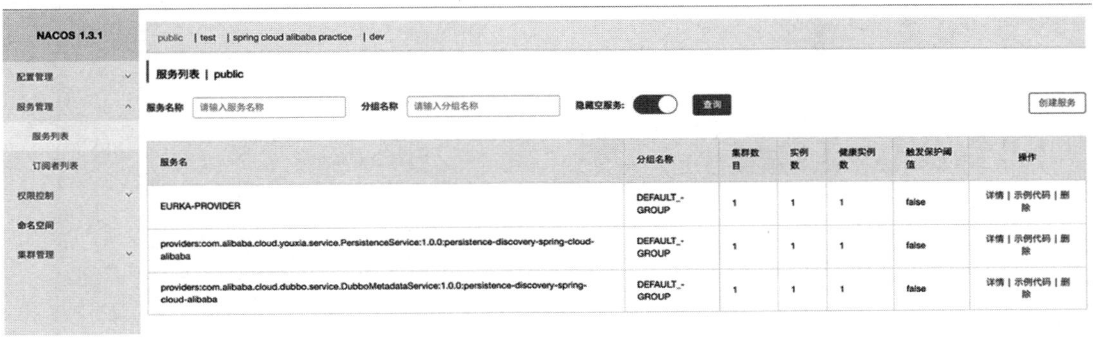

图 4-30

5. 创建一个服务订阅者，从 Nacos 注册中心订阅迁移过来的服务

下面通过项目 nacos-consumer 来实现服务订阅者。

（1）基于 Spring Cloud Alibaba 和 Spring Boot 创建项目。

（2）用 RestTemplate 类订阅服务提供者，具体代码如下所示。

```
@RestController
@RequestMapping(value = "nacosconsumer")
public class NacosConsumerController {
    @Autowired
    private RestTemplate restTemplate;

    @GetMapping(value = "/test")
public String doRibbon(){
    //通过 Nacos 注册中心订阅服务提供者 EURKA-PROVIDER
        return
restTemplate.getForObject("http://EURKA-PROVIDER/data/getdata",String.class)
;
    }
}
```

运行程序，并通过“127.0.0.1:7889/nacosconsumer/test”调用 nacos-consumer，发现可以正常订阅 eureka-provider。

第 5 章 分布式配置管理——基于 Nacos

分布式配置管理是分布式微服务架构中重要的基础功能。本章将结合 Nacos 来分析分布式配置管理的核心技术，以及关键原理。

5.1 认识分布式配置管理

分布式配置管理的本质：将分散的配置信息，通过公共组件集中起来管理。

5.1.1 什么是分布式配置管理

Nacos 控制台向开发者提供了管理配置信息的功能，可以将配置管理功能与应用的运行环境解耦。

如图 5-1 所示，有应用 A 和应用 B，它们分别读取本地的配置信息 A 和配置信息 B。如果将两个应用接入配置中心，则应用 A 和应用 B 就要通过远程调用从配置中心读取配置信息 A 和配置信息 B。

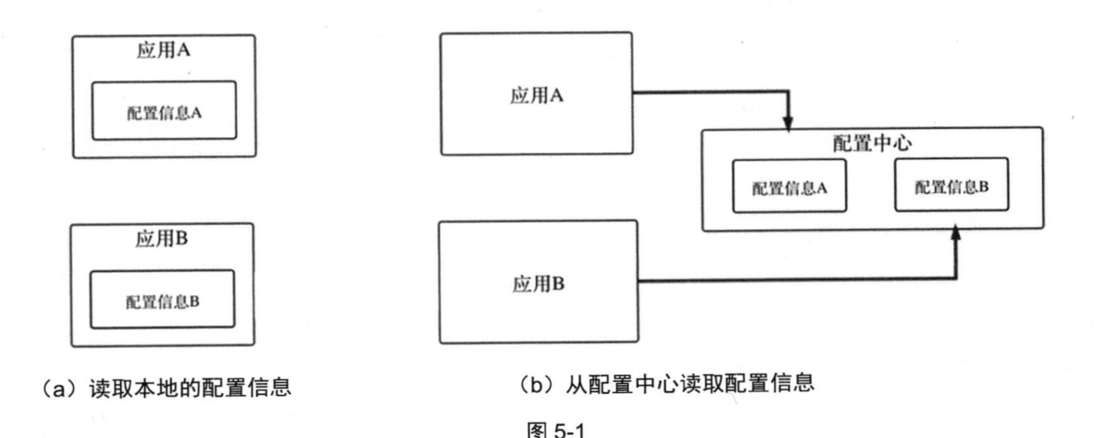

（a）读取本地的配置信息　　（b）从配置中心读取配置信息

图 5-1

应用和配置中心间的数据同步通常有如下 3 种模式。

- Pull 模式：让应用开启长轮询，即定时地从配置中心拉取最新的配置信息，并更新到应用的内存中。
- Push 模式：在配置中心配置数据变更之后，主动推送配置数据到指定的应用，应用更新到本地内存中。
- 混合模式：应用和配置中心通过“事件机制 + 监听器”模式保持长连接。如果应用监听的配置信息发生了变化，则配置中心发布对应类型的事件。应用只有在监听到该事件之后，才会处理对应的配置信息，并更新到应用本地。

通常将变更不是很频繁的应用初始化配置信息（比如数据库连接属性）放到配置中心中。应用在启动时，从配置中心加载属性，完成应用基础环境的初始化。

应用接入配置中心，会增加一次网络开销。配置中心采用“本地缓存 + 事件监听”模式，可以减少应用和配置中心间的网络开销。

配置中心通常会提供配置变更、配置推送、历史版本管理、灰度发布等功能。通过这些功能，可以降低分布式系统中管理配置信息的成本，降低因错误的配置信息变更带来的服务可用性下降和发生故障的风险。

5.1.2　为什么需要分布式配置管理

作为一名开发人员，对配置文件应该都不陌生。在 Spring Boot 项目中，默认提供了一个 application.properties 或者 application.yaml 文件。

可以把一些全局性的配置或者需要动态维护的配置（比如数据库连接、功能开关、降级、服务器地址等）写入该文件。

为了解决不同环境下服务连接配置信息的差异性，Spring Boot 还提供了基于 spring.profiles.active={profile}的机制来实现不同的环境切换。

单体架构向微服务架构演进已经是软件开发的必经之路。在服务被拆分之后，各个应用需要单独地维护本地配置文件，成本非常高。主要缺点如下。

- 不支持配置文件的动态更新：在实际的业务开发过程中，需要动态地更新配置文件，比如切换业务功能开关、变更图片服务器地址、变更数据库连接信息等。在传统配置模式下，需要修改本地配置文件并重新打包，然后重启应用并发布，这样才能保证配置文件能够生效。但这样会导致该服务在重启阶段不可用，从而影响服务整体的可用率。
- 不支持配置集中式管理：在微服务架构中，为了保证某些核心服务的高性能会部署几百个节点。如果在每个节点上都维护一个本地配置文件，则不管是对运维人员或者开发人员而言，成本都是巨大的。
- 不支持配置内容的安全性和权限校验：如果将配置文件随着源代码统一提交到代码库中，则配置文件内容会变得不可控。如果黑客主动攻击服务器并获取了源代码，即可直接拿到所有的配置信息。
- 不支持多环境部署：如果通过底层框架来维护不同环境的信息，则成本也是非常高的。

分布式配置管理就是弥补上述不足的方法。简单来说，其做法是：把各个服务的某些配置信息统一交给第三方中间件来维护；分布式配置管理上的数据变更，需要实时地推送到相对应的应用服务节点上。

5.2 了解主流的配置中心

主流的配置中心有以下四种，其中，Nacos 是 Alibaba 开源的分布式配置中心。本书将以 Nacos 来进行详细讲解。

5.2.1 Nacos

Nacos 的分布式配置非常简单和高效，并且部署简单，可以和注册中心共享一个集群。

1. Nacos 配置管理的维度

Nacos 支持以下维度的配置管理。

- 应用：这个比较好理解，每个配置文件中的配置信息都归属于某一个应用。
- Group（组）：如果服务配置了该项，则配置信息通过 Group（组）来隔离，该组中的所有应用都可以读取该配置信息，非本组的其他应用不能读取该配置信息。
- Namespace（命令空间）：如果服务配置了该项，则配置信息通过 Namespace（命令空间）来隔离，该命名空间中的所有应用都可以读取该配置信息，非本命令空间的其他应用不能读取该配置信息。

2. Nacos 配置管理的功能

Nacos 配置管理主要包括以下功能。

- 配置管理列表：开发人员可以在 Nacos 控制台上，实时地看到所有接入 Nacos 的应用的配置信息。
- 新建、编辑和删除配置文件：开发人员可以在 Nacos 控制台上操作指定的配置文件。
- 在不同命名空间之间复制指定的配置文件：如果开发环境和测试环境共用一个 Nacos 集群，则开发人员可以使用配置文件的复制功能，在不同命名空间之间复制指定的配置文件，完成配置文件的跨环境的数据迁移。
- 批量导出和导入配置文件：如果需要将测试环境的配置信息复制到预发布环境，但是测试环境和预发布环境不在一个集群，则开发人员可以使用批量导出和导入功能，完成配置文件的跨集群的数据迁移。
- 灰度环境的配置信息发布：开发人员可以将新增的配置信息发布到指定的客户端（用 IP 地址来区分），其余客户端配置保持不变。这样就可以用部分客户端验证新添加的配置信息对应的新的功能，以保证配置信息的平稳发布。灰度配置是生产环境中一个比较重要的功能，它能够确保生产环境运行的稳定。
- 支持多种配置文件格式：目前支持 TEXT、JSON、XML、YAML、HTML 和 Properties 格式。开发人员可以灵活地使用不同的文件格式将应用接入 Nacos 配置中心。

3. Nacos 配置管理支持 Open API

开发人员可以使用 Nacos 开放的 Open API 跨平台地管理配置文件。如果要将 Go 语言或者 Node.js 的应用接入 Nacos 配置中心，则可以通过 Open API 完成对应用配置文件的管理。

4. Nacos 配置管理的优势

Nacos 配置管理的主要优势如下。

（1）接入轻量级：开发者可以“开箱即用”地使用 Nacos 提供的 Starter 组件，几乎零技术成本地将应用接入 Nacos 配置中心。

（2）线上维护成本低：Nacos 本身就具备集群管理的能力，配置中心和注册中心可以共用一个集群，节省线上机器资源的成本。

（3）跨平台性：Nacos 通过 Open API 开放了它的配置管理功能，不同语言的客户端可以通过 Open API 接入 Nacos 配置中心。

（4）云原生：支持用 Docker 和 K8s 进行部署。

（5）开源社区活跃性高：目前 Nacos 开源社区的代码贡献者非常多，版本迭代速度非常快，并且 Nacos 的推广力度非常大。Nacos 线上和线下的技术分享活动非常多，能够解决开发者反馈的大部分技术问题。

Nacos 目前最新的版本 2.0.0，处于公测阶段。该版本最大的改动是：用 gRPC 框架替换了 HTTP 协议，提升了 Nacos 集群之间，以及集群与客户端之间的通信性能。

5.2.2 Spring Cloud Config

Spring Cloud Config 是 Spring Cloud 项目开源出来的分布式配置框架，最近一次版本更新的时间是 2021 年 1 月 9 日，最新的版本为 3.0.1。

Spring Cloud Config 就是我们通常意义上的配置中心——把应用原本放在本地文件中的配置信息抽取出来放在中心服务器中，从而能够提供统一的配置管理和发布。

1. 主要功能

Spring Cloud Config 提供以下 3 个维度的配置管理。

- 应用：这个比较好理解，每个配置文件中的配置信息都归属于某一个应用。
- 环境：每个配置信息都是归属于某一个环境，如开发环境、测试环境和生产环境等。
- 版本：这是一般配置中心所缺少的，即对同一份配置的不同版本进行管理，比如可以通过 Git 进行版本控制。Spring Cloud Config 提供了版本支持，即对于一个应用的不同部署实例，可以从服务器获取不同版本的配置。这对于一些特殊场景（如灰度发布、A/B 测试等）非常有用。

2. 主要优势

Spring Cloud Config 的主要优势如下。

- 基于应用、环境、版本这 3 个维度进行管理：支持 3 个维度的配置信息的版本管理。
- 配置信息可以持久化方式支持 Git：后端基于 Git 存储，一方面程序员非常熟悉，另一方面在部署上会非常简单。借助于 Git，能够非常友好地支持不同配置版本（Git 本身就是一个版本管理工具）。当然，Spring Cloud Config 还支持其他的存储模式，比如本地文件、SVN 等。

- 可以无缝地集成 Spring Cloud：因为 Spring Cloud 无缝地支持 Spring Framework 框架的 Environment 接口和 PropertySource 接口，所以它对于已有的 Spring Framwork 应用的迁移成本非常低，并且应用获取配置信息的方式完全一致。

5.2.3　Apollo

Apollo（简称阿波罗）的最新版本为 1.7.1，最近的更新时间是 2020 年 8 月 20 日。

Apollo 是携程框架部门研发的分布式配置中心框架，能够集中管理应用的不同环境、不同集群的配置信息。在配置信息被修改后，Apollo 能够将修改的内容实时地推送到应用端，并且具备规范的权限管理、数据推送流程治理等特性。

5.2.4　对比 Nacos、Spring Cloud Config、Apollo 和 Disconf

本书主要讲解 Nacos，它功能全面、强大，可以无缝地和 Spring Cloud Alibaba 体系结合，开箱即用。各个配置中心的对比见表 5-1。

表 5-1　对比主流的配置中心

功能	Nacos	Spring Cloud Config	Apollo
开源时间	2018 年 6 月	2014 年 9 月	2016 年 5 月
单机部署	Nacos 单节点	Config Server + Git（文件）+ Spring Cloud Bus	Apollo-quickstart + MySQL
配置实时推送	支持（HTTP 长轮询 1s 内）	支持（基于 Spring Cloud Bus）	支持(HTTP 长轮询 1s 内）
分布式部署	Nacos + MySQL	Config erver + Git + MQ + Spring Cloud Bus	Config + Admin + Portal + MySQL
版本管理	支持	支持	支持
配置回滚	支持	支持	支持
灰度发布	支持	支持	支持
权限管理	支持	支持	支持
集群	支持	支持	支持
多环境	支持	支持	支持
监听查询	支持	支持	支持
配置锁	不支持	支持	不支持
多语言	支持	支持	支持
配置格式校验	支持	不支持	支持

续表

功能	Nacos	Spring Cloud Config	Apollo
通信协议	HTTP 和 GRPC(Nacos 2.0 新特性)	HTTP 和 AMQP	HTTP
数据一致性	Raft 或者 Sofa-Jraft	Git 保证数据一致性	数据库模拟消息队列
配置界面	控制台	无	控制台
单点故障	支持 HA 部署，Nacos 自带高可用的集群管理功能	支持 HA 部署	支持 HA 部署
对 Spring Cloud Alibaba 的支持	深度支持	支持	支持
告警通知	通过 Open API 定制化，比如钉钉告警	不支持	不支持
配置限流和降级	支持	不支持	不支持

5.3 将应用接入 Nacos 配置中心

要将应用接入 Nacos 配置中心，就需要采用 Nacos 提供的接入方式。

5.3.1 接入方式

设计一个好配置中心，需要考虑接入方式的友好性。主流的配置中心通常包含以下接入方式。

- 原生的 SDK：提供统一的、稳定的、可维护的 Jar 包。
- Open API：提供符合 RESTful API 规范的 HTTP 接口。
- RPC：提供高性能的 RPC 接口，通过远程通信和事件机制来完成配置信息的交换。
- 第三方中间件：提供了更加稳定和高效的接入体验，并屏蔽了中间件底层的技术细节。

5.3.2 认识 Nacos 配置中心的配置信息模型

Nacos 配置中心的配置信息模型非常简单，具体见表 5-2。

表 5-2　Nacos Config 的配置信息模型

字段名称	语义描述
dataId	一条配置信息的唯一的名称，比如 order-server.yaml
groupId	配置信息所在组的 ID，比如 ORDER-SERVER
appName	配置信息对应的应用名称，通常和应用对应的名称保持一致，比如 example2-distributed-server
tenantId	租户的 ID，主要用于隔离对应租户的配置信息
content	配置内容，存储配置信息对应的数据
type	配置信息对应的文件格式，包括：TEXT、JSON、XML、YAML、HTML 和 Properties

续表

字段名称	语义描述
beatIp	发布灰度配置的灰度 IP 地址（客户端 IP 地址）
tagID	配置标签的 ID
namespace	Nacos 对应的命名空间

在 Nacos Config 中，tenantId 实际存储的是 namespace（命名空间）的值。

5.3.3 了解 NacosConfigService 类

NacosConfigService 类是 Nacos 配置中心的核心 API，它提供了表 5-3 中列出的方法。

表 5-3 NacosConfigService 类的方法

方法名称	方法功能描述
getConfig()	获取指定 dataId 和 group 的配置信息
publishConfig()	发布指定 dataId 和 group 的配置信息
removeConfig()	删除指定 dataId 和 group 的配置信息
getConfigAndSignListener()	获取指定 dataId 和 group 的配置信息，并且注册对应配置信息的监听器
addListener()	添加指定配置信息的监听器
removeListener()	删除指定配置信息的监听器

NacosConfigService 类的作用是：用开发人员在应用本地配置的 Nacos 集群的环境信息，初始化用于连接 Nacos 集群的 HTTP 代理。

在 Nacos 中，HTTP 代理的具体实现是 ServerHttpAgent 类。ServerHttpAgent 类只是一个代理，其代理的对象是 NacosRestTemplate 类。NacosRestTemplate 类是 Nacos 自研的 HTTP 框架，具体的原理会在 5.4 节介绍。

5.3.4 【实例】用 Nacos Client 接入应用

本实例的源码在本书配套资源的“/spring-cloud-alibaba-practice/chapterfive/use-nacos-client-nacos-config/”目录下。

Nacos Client 是 Nacos 暴露给开发者的原生 SDK。依赖 SDK，开发者可以用 NacosConfigService 类来调用 Nacos 配置中心。

1. 新建一个项目

（1）用 IDEA 快速创建一个 Spring Boot 项目，并添加启动项目的 Starter，具体代码如下所示：

```
    <dependency>
        <groupId>org.springframework.boot</groupId>
        <artifactId>spring-boot-actuator</artifactId>
    </dependency>
    <dependency>
        <groupId>org.springframework.boot</groupId>
        <artifactId>spring-boot-starter-web</artifactId>
    </dependency>
    <dependency>
        <groupId>org.springframework.boot</groupId>
        <artifactId>spring-boot-starter-logging</artifactId>
    </dependency>
```

（2）用注解@SpringBootApplication 快速启动项目：

```
    2021-03-21 08:42:10.876  INFO 59733 --- [           main]
o.s.b.w.embedded.tomcat.TomcatWebServer  : Tomcat started on port(s): 8080
(http) with context path ''
    2021-03-21 08:42:10.913  INFO 59733 --- [           main] c.a.cloud.youxia.
NacosClientApplication    : Started NacosClientApplication in 17.106 seconds
(JVM running for 23.597)
```

2. 引入 Nacos Client 依赖包

将 Nacos Client 引入项目中，代码如下：

```
    //根据 Nacos 集群对应的部署包的版本号，选择对应的 Nacos Client 的版本号
    <dependency>
        <groupId>com.alibaba.nacos</groupId>
        <artifactId>nacos-client</artifactId>
        <version>1.3.1</version>
    </dependency>
```

3. 用 NacosConfigService 从配置中心获取配置信息

下面使用注解@Configuration 和@Bean 初始化 NacosConfigService 对象，具体代码如下：

```
    @Configuration
    public class NacosConfig {
        @Bean
        public ConfigService configService() throws NacosException {
            final Properties properties = new Properties();
            //①设置 Nacos 节点对应的 IP 地址
           properties.setProperty(PropertyKeyConst.SERVER_ADDR,
"127.0.0.1:8848");
            //②设置命名空间
    properties.setProperty(PropertyKeyConst.NAMESPACE,"c7ba173f-29e5-4c58-ae
78-b102be11c4f9");
```

```
        //③如果开启了 Nacos 权限校验，则设置用户名
        properties.setProperty(PropertyKeyConst.USERNAME,"nacos");
        //④如果开启了 Nacos 权限校验，则设置用户密码
        properties.setProperty(PropertyKeyConst.PASSWORD,"nacos");
        //⑤设置获取配置信息的长轮训超时时间
    properties.setProperty(PropertyKeyConst.CONFIG_LONG_POLL_TIMEOUT,"3000");
        //⑥设置在获取配置信息失败后重试的次数
        properties.setProperty(PropertyKeyConst.CONFIG_RETRY_TIME,"5");
        //⑦设置是否开启客户端主动拉取最新的配置信息
     properties.setProperty(PropertyKeyConst.
ENABLE_REMOTE_SYNC_CONFIG,"true");
        //⑧设置最大重试次数
        properties.setProperty(PropertyKeyConst.MAX_RETRY,"5");
        //⑨构造一个 ConfigService 实例
      ConfigServiceconfigService=NacosFactory.
createConfigService(properties);
        return configService;
      }
    }
```

下面通过 NacosConfigManager 类，模拟业务调用方调用 NacosConfigService 类的 getConfig()方法获取配置信息：

```
    @Component
    public class NacosConfigManager {
       @Resource
       private ConfigService configService;
       @PostConstruct
       private void init() throws NacosException {
          String data=configService.getConfig
("use-nacos-client-nacos-config"," use-nacos-client-nacos-config",2000);
          System.out.print("从 Nacos 配置中心获取配置信息为: "+data);
       }
    }
```

在 Nacos 控制台中，添加 dataId 为“use-nacos-client-nacos-config”，group 为“use-nacos-client-nacos-config”的配置文件，如图 5-2 所示。

运行服务后可以看到，dataId 为“use-nacos-client-nacos-config”，group 为“use-nacos-client-nacos-config”的配置信息，可以将属性“spring.youxia.config.name”加载到当前应用中。

Edit Configuration

* Data ID: use-nacos-client-nacos-config

* Group: use-nacos-client-nacos-config

Advanced Options

Description:

Beta Publish: Not checked by default.

Format: TEXT JSON XML YAML HTML Properties

Configuration Content:

```
1 spring.application.name=use-nacos-client-nacos-config
2 spring.youxia.config.name=测试2
```

图 5-2

从 IDEA 控制台可以看到读取配置信息的日志如图 5-3 所示。

```
2021-08-18 20:42:42.342  INFO 49819 --- [           main] org.apache.catalina.core.StandardEngine  :
2021-08-18 20:42:42.458  INFO 49819 --- [           main] o.a.c.c.C.[Tomcat].[localhost].[/]       :
2021-08-18 20:42:42.459  INFO 49819 --- [           main] w.s.c.ServletWebServerApplicationContext :
通过Nacos Client 从Nacos配置中心获取配置信息为: spring.application.name=example13-nacosclient1234
spring.youxia.config.name=测试2
2021-08-18 20:42:43.633  INFO 49819 --- [           main] o.s.s.concurrent.ThreadPoolTaskExecutor  :
```

图 5-3

5.3.5 【实例】用 Open API 接入应用

本实例的源码在本书配套资源的“/spring-cloud-alibaba-practice/chapterfive/openapi-nacos-config/”目录下。

本实例使用 Nacos 的 Open API 将应用接入 Nacos 配置中心。

1. 新建一个项目

用 IDEA 新建一个项目，并添加相关的 Starter，具体代码如下：

```
<dependency>
   <groupId>org.springframework.boot</groupId>
   <artifactId>spring-boot-actuator</artifactId>
</dependency>
<dependency>
   <groupId>org.springframework.boot</groupId>
   <artifactId>spring-boot-starter-web</artifactId>
</dependency>
<dependency>
```

```
        <groupId>org.springframework.boot</groupId>
        <artifactId>spring-boot-starter-logging</artifactId>
    </dependency>
```

在上面的代码中，使用 Spring Framework 的 RestTemplate 类访问 Nacos Config 的 Open API。RestTemplate 是一个基础的 HTTP 客户端。

2. 初始化 RestTemplate

通过注解@Configuration 和@Bean 初始化 RestTemplate，具体代码如下：

```
    @Configuration
    public class OpenApiConfig {
        @Bean
        public RestTemplate restTemplate() {
            SimpleClientHttpRequestFactory requestFactory = new
SimpleClientHttpRequestFactory();
            //①设置连接超时时间
            requestFactory.setConnectTimeout(2000);
            //②设置读超时时间
            requestFactory.setReadTimeout(2000);
            RestTemplate restTemplate = new RestTemplate(requestFactory);
            return restTemplate;
        }
    }
```

3. 通过 RestTemplate 读取配置信息

通过注解@Component 和@PostConstruct，实现在应用启动的过程中调用 Open API 并读取配置信息，具体代码如下：

```
    @Component
    public class ReaderConfigManager {
        @Resource
        private RestTemplate restTemplate;
        //①用 RestTemplate 访问 Nacos Config 的 Open API
        @PostConstruct
        public void readConfig(){
            String url="http://127.0.0.1:8848/nacos/v1/cs/configs?dataId=
{dataId}&group={group}&tenant={tenant}";
            Map<String,String> params=new HashMap<String,String>(16);
            params.put("dataId", "openapi-nacos-config");
            params.put("group", "openapi-nacos-config");
            params.put("tenant","c7ba173f-29e5-4c58-ae78-b102be11c4f9");
            //②入参
            ResponseEntity<String>
responseEntity=restTemplate.getForEntity(url,String.class,params);
```

```
        String result=responseEntity.getBody();
        System.out.println("通过 Open API 从 Nacos 配置中心获取的配置信息为：
"+result);
    }
}
```

在 Nacos 控制台中，添加 dataId 为“openapi-nacos-config”，group 为“openapi-nacos-config”的配置文件，如图 5-4 所示。

Edit Configuration

* Data ID: openapi-nacos-config

* Group: openapi-nacos-config

Advanced Options

Description:

Beta Publish: Not checked by default.

Format: TEXT JSON XML YAML HTML Properties

Configuration Content:

```
1 spring.application.name=openapi-nacos-config
2 spring.youxia.config.name=测试2
```

图 5-4

运行服务后可以看到，dataId 为“openapi-nacos-config”，group 为“openapi-nacos-config”的配置信息，可以将属性“spring.youxia.config.name”加载到当前应用中。

从 IDEA 控制台，可以看到读取配置信息的日志如图 5-5 所示。

```
2021-08-19 03:48:11.457  INFO 38728 --- [           main] w.s.c.ServletWebServerApplicationContext :
通过Open API 从Nacos配置中心获取配置信息为: spring.application.name=openapi-nacos-config
spring.youxia.config.name=测试2
```

图 5-5

5.3.6 【实例】用 Spring Cloud Alibaba Config 接入应用

本实例的源码在本书配套资源的“/spring-cloud-alibaba-practice/chapterfive/spring-cloud-alibaba-config/”目录下。

Spring Cloud Alibaba 在 Nacos Client 的基础上，重新封装了一个“开箱即用”的基础组件 spring-cloud-starter-alibaba-nacos-config（简称 Spring Cloud Alibaba Config）。本实例使用 Spring Cloud Alibaba Config 接入应用。

1. 新建一个 Spring Cloud Alibaba 项目

用 IDEA 新建一个 Spring Boot 项目，并添加 Spring Cloud Alibaba Config 对应的 Jar 包依赖，具体代码如下：

```
<dependency>
    <groupId>com.alibaba.cloud</groupId>
    <artifactId>spring-cloud-starter-alibaba-nacos-config</artifactId>
</dependency>
```

2. 新增配置文件 bootstrap.yaml 和 application.properties

bootstrap.yaml 的具体代码如下所示：

```
spring:
  application:
    name: spring-cloud-alibaba-config
  cloud:
    nacos:
      config:
        namespace: c7ba173f-29e5-4c58-ae78-b102be11c4f9
        group: spring-cloud-alibaba-config
        password: nacos
        enable-remote-sync-config: true
        server-addr: 127.0.0.1:8848
        username: nacos
        name: spring-cloud-alibaba-config
        file-extension: properties
        extension-configs:
          - data-id: spring-cloud-alibaba-config-test.yaml
            group: spring-cloud-alibaba-config
            namespace: c7ba173f-29e5-4c58-ae78-b102be11c4f9
            refresh: true
```

application.properties 的具体代码如下所示：

```
spring.youxia.config.name=test2
spring.application.name=spring-cloud-alibaba-config
```

3. 解析配置信息

首先，在 application.properties 中新增属性键值对“spring.youxia.config.name=test2”；然后为了展示属性动态更新的效果，在配置中心中新增一个 Data ID 为 spring-cloud-alibaba-config.properties 的文件，配置内容如图 5-6 所示（其中配置了“spring.youxia.config.name=test20”）。

编辑配置

* Data ID: spring-cloud-alibaba-config.properties

* Group: spring-cloud-alibaba-config

更多高级选项

描述:

Beta发布: 默认不要勾选。

配置格式: TEXT JSON XML YAML HTML Properties

配置内容: 1 spring.youxia.config.name=test20

图 5-6

具体代码如下所示：

```
@Configuration
@RefreshScope
public class ConfigReader {
    @Autowired
    private NacosConfig nacosConfig;
    @PostConstruct
    public void init() {
        Executors.newCachedThreadPool().execute(new ConfigThread());
    }
    class ConfigThread implements Runnable {
        @Override
        public void run() {
            while (true) {
                //利用线程从 Nacos 配置中心读取配置信息
                System.out.println("使用 Spring Cloud Alibaba 接入 Nacos 配置中心，获取的配置信息 name 为：" + nacosConfig.getName());
                System.out.println("使用 Spring Cloud Alibaba 接入 Nacos 配置中心，获取的配置信息 value 为："+nacosConfig.getValue());
                try {
                    Thread.sleep(6000);
                } catch (InterruptedException e) {
                    System.out.println(e.getMessage());
                }
            }
        }
    }
}
```

启动程序，在控制台中会输出如下日志：

```
使用 Spring Cloud Alibaba 接入 Nacos 配置中心，获取的配置信息 name 为：test20
使用 Spring Cloud Alibaba 接入 Nacos 配置中心，获取的配置信息 value 为：7878
```

5.4　用 HTTP 协议和 gRPC 框架实现通信渠道

Nacos 在 2.0 版本及之后，底层通信渠道使用 gRPC 框架替换了原先的 HTTP 协议。

5.4.1　什么是 gRPC

HTTP 是基于 TCP/IP 协议的应用层协议。它经历过几个大的版本迭代：HTTP 1.0、HTTP 1.1、HTTP 2 及 HTTP 3。

开发人员使用 HTTP 协议，都会吐槽它的性能损耗。在应用层，业务接口之间的请求通常对性能的容忍度比较高，使用 HTTP 协议能够应对大部分的业务场景。但是，分布式配置管理是底层的基础服务，所以需要通信渠道具备高性能、高可用和高并发的质量属性。

gRPC 是一个高性能、开源和通用的 RPC 框架。它使用 Protocol Buffer（简称 Protobuf）作为序列化格式。

Protocol Buffer 是来自 Google 的序列化协议，比序列化协议 Jackson 更加轻便和高效。

另外，Protocol Buffer 基于 HTTP 2 标准设计，具有双向流、流量控制、头部压缩、单 TCP 连接上的多复用请求等特性。这些特性使得其在移动设备上表现更好、更省电和节省空间占用。

总体来说，gRPC 是一款高性能的 RPC 框架，支持长连接，可扩展性非常好。

5.4.2　“用 HTTP 实现 Nacos Config 通信渠道”的原理

在 Naco 配置中心和应用之间可以通过 HTTP 进行通信，那么 Nacos 是如何封装 HTTP 通信渠道的呢，下面逐步介绍。

1. 整体架构

用HTTP实现Nacos Config通信渠道的原理如图5-7所示，其中包含两个模块：Nacos Client 和 Open API。

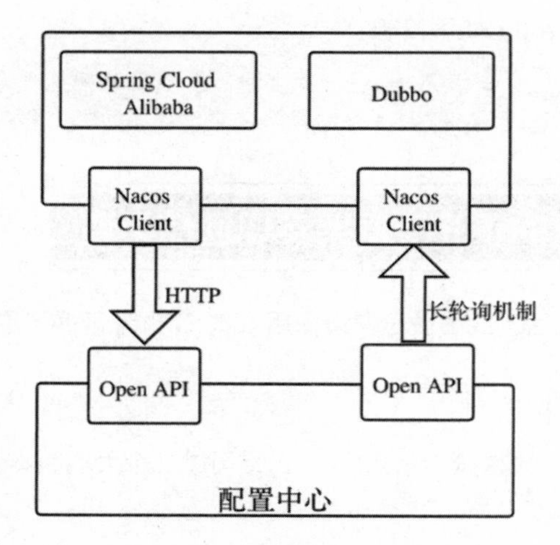

图 5-7

Spring Cloud Alibaba 和 Dubbo 框架可以通过 Nacos Client 调用配置中心的 Open API。Nacos Client 具备以下 3 个特性：

- Nacos Client 已经封装了 HTTP 客户端，将底层技术细节下沉在它的通信层。
- 配置中心将 Nacos Config 模块的 RESTful API（ConfigController 类、HistoryController 类等）以 Open API 的接入方式连接应用、Nacos Client 和配置中心。
- 通过长轮询机制，准实时地将配置中心变更后的数据，推送到使用 Nacos Client 的应用中，并更新到应用本地。Spring Cloud Alibaba 已经将这些技术细节封装在 Spring Cloud Alibaba Config 模块中，开发人员可以直接使用。

2. "用 HTTP 完成配置信息发布"的过程

下面介绍用 NacosConfigService 类中的 publishConfig()方法发布配置信息的过程。

（1）HTTP 客户端和 NacosConfigService 类的依赖关系。

NacosConfigService 类是 Nacos Client 中的核心 API，其提供的方法在 5.3.3 节已经介绍过。

在 Nacos Client 中，HTTP 客户端和 NacosConfigService 类的依赖关系如图 5-8 所示。

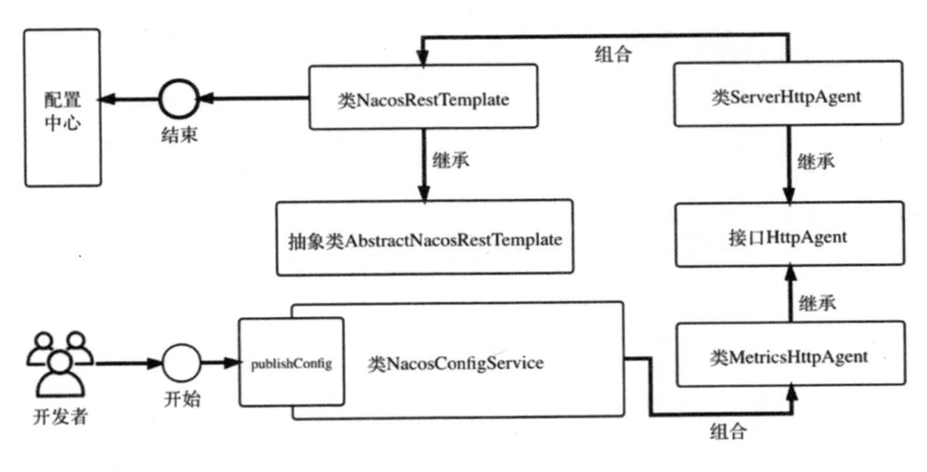

图 5-8

从图 5-8 中可以看到：

- 开发者通过 NacosConfigService 类的 publishConfig()方法发布配置信息。
- NacosConfigService 类的组合类 MetricsHttpAgent 是一个带有度量功能的 HTTP 客户端代理类（可以监控 HTTP 请求）。
- MetricsHttpAgent 类的代理 ServerHttpAgent 类是真正意义上的 HTTP 客户端。
- ServerHttpAgent 类的代理 NacosRestTemplate 类是 Nacos 自研的 HTTP Client 的实现。

（2）HTTP 客户端发布配置管理的过程。

HTTP 客户端发布配置管理的过程如图 5-9 所示。

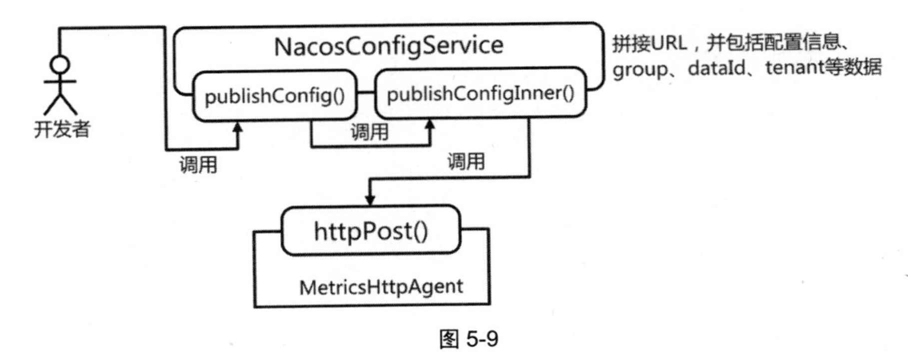

图 5-9

①在 NacosConfigService 类的 publishConfigInner()方法中拼接了 HTTP 请求的 URL，包括配置信息、group、dataId 和 tenant 等数据，具体代码如下所示：

```
    private boolean publishConfigInner(String tenant, String dataId, String
group, String tag, String appName,String betaIps, String content) throws
NacosException {
        group = null2defaultGroup(group);
    ParamUtils.checkParam(dataId, group, content);
    //①构造发布配置信息的 HTTP 请求参数
        ConfigRequest cr = new ConfigRequest();
        cr.setDataId(dataId);
        cr.setTenant(tenant);
        cr.setGroup(group);
        cr.setContent(content);
        configFilterChainManager.doFilter(cr, null);
        content = cr.getContent();
        String url = Constants.CONFIG_CONTROLLER_PATH;
        Map<String, String> params = new HashMap<String, String>(6);
        params.put("dataId", dataId);
        params.put("group", group);
        params.put("content", content);
        if (StringUtils.isNotEmpty(tenant)) {
            params.put("tenant", tenant);
        }
        if (StringUtils.isNotEmpty(appName)) {
            params.put("appName", appName);
        }
        if (StringUtils.isNotEmpty(tag)) {
            params.put("tag", tag);
        }
        Map<String, String> headers = new HashMap<String, String>(1);
        if (StringUtils.isNotEmpty(betaIps)) {
            headers.put("betaIps", betaIps);
        }
        HttpRestResult<String> result = null;
    try {
        //②调用 MetricsHttpAgent 的 httpPost()方法请求 Nacos Config 集群
            result = agent.httpPost(url, headers, params, encode, POST_TIMEOUT);
        } catch (Exception ex) {
            LOGGER.warn("[{}] [publish-single] exception, dataId={}, group={},
msg={}", agent.getName(), dataId, group,ex.toString());
            return false;
        }
        ...
    }
```

通过 HTTP 客户端发送“发布配置信息”的 HTTP 请求，调用 Nacos Config 的 Open API。

5.4.3 “用‘长轮询 + 注册监听器’机制将变更之后的配置信息同步到应用”的原理

Nacos 配置中心的配置信息会经常变更。在变更后，Nacos 会用“长轮询 + 监听器”机制保持应用和配置中心之间的数据一致。

下面介绍用“长轮询 + 注册监听器”机制将变更后的配置信息同步到应用的原理。

1. 开启配置信息的同步

Spring Cloud Alibaba Config 通过 NacosContextRefresher 类开启配置信息的同步，具体代码如下所示：

```
    @Configuration(proxyBeanMethods = false)
    @ConditionalOnProperty(name = "spring.cloud.nacos.config.enabled",
matchIfMissing = true)
    public class NacosConfigAutoConfiguration {
        //自动装配 NacosContextRefresher 对象
        @Bean
        public NacosContextRefresher nacosContextRefresher(NacosConfigManager
nacosConfigManager,NacosRefreshHistory nacosRefreshHistory) {
            return new NacosContextRefresher(nacosConfigManager,
nacosRefreshHistory);
        }
    }
```

2. 在应用启动后开启 Nacos Config 的注册监听器

NacosContextRefresher 类通过 onApplicationEvent()方法，将当前应用中所有的 dataID 注册到 Nacos Config 的注册监听器中，具体代码如下所示：

```
    public class NacosContextRefresher
            implements ApplicationListener<ApplicationReadyEvent>,
ApplicationContextAware {
        //①开启监听器的注册
        @Override
        public void onApplicationEvent(ApplicationReadyEvent event) {
            //②在应用中只注册一次
            if (this.ready.compareAndSet(false, true)) {
                //③注册 Nacos 监听器
                this.registerNacosListenersForApplications();
            }
        }
        //④循环遍历 NacosPropertySourceRepository 类中所有的 dataID，注册监听器
        private void registerNacosListenersForApplications() {
            //⑤如果开启了自动刷新，则添加监听器
```

```
        if (isRefreshEnabled()) {
            //⑥遍历所有的 dataID
            for (NacosPropertySource propertySource :
NacosPropertySourceRepository
                    .getAll()) {
                if (!propertySource.isRefreshable()) {
                    continue;
                }
                String dataId = propertySource.getDataId();
                //⑦注册 Nacos 监听器
                registerNacosListener(propertySource.getGroup(), dataId);
            }
        }
    }
    //⑧调用 Nacos Config 的 ConfigService 类的 addListener()方法注册监听器
    private void registerNacosListener(final String groupKey, final String
dataKey) {
        //⑨用组和 dataID 来组装本地监听器缓存的 key
        String key = NacosPropertySourceRepository.getMapKey(dataKey,
groupKey);
        //⑩构造 Nacos Config 的注册监听器 AbstractSharedListener 对象
        Listener listener = listenerMap.computeIfAbsent(key,
                lst -> new AbstractSharedListener() {
                    @Override
                    public void innerReceive(String dataId, String group,
                            String configInfo) {
                        refreshCountIncrement();
                        nacosRefreshHistory.addRefreshRecord(dataId, group,
configInfo);
                        applicationContext.publishEvent(
                                new RefreshEvent(this, null, "Refresh Nacos
config"));
                        if (log.isDebugEnabled()) {
                            log.debug(String.format(
                                    "Refresh Nacos config
group=%s,dataId=%s,configInfo=%s",
                                    group, dataId, configInfo));
                        }
                    }
                });
        try {
            configService.addListener(dataKey, groupKey, listener);
        }
        catch (NacosException e) {
            log.warn(String.format(
```

```
                    "register fail for nacos listener ,dataId=[%s],group=[%s]",
dataKey,
                    groupKey), e);
            }
        }
    }
```

3. 给 Nacos Config 开启定时任务，定时检查配置信息

Nacos Config 通过 ClientWorker 类开启定时任务，检查配置信息，具体代码如下所示：

```
    public class ClientWorker implements Closeable {

        public ClientWorker(final HttpAgent agent, final
ConfigFilterChainManager configFilterChainManager,
                final Properties properties) {
            //①开启定时器，执行周期为 1S
            this.executor.scheduleWithFixedDelay(new Runnable() {
                @Override
                public void run() {
                    try {
                        //②检查配置信息
                        checkConfigInfo();
                    } catch (Throwable e) {
                        LOGGER.error("[" + agent.getName() + "] [sub-check] rotate
check error", e);
                    }
                }
            }, 1L, 10L, TimeUnit.MILLISECONDS);
            ...
        }

    public void checkConfigInfo() {
        //③获取注册监听器总数
            int listenerSize = cacheMap.get().size();
            //④Nacos Config 默认最大的任务数为 3000,可以通过参数 PER_TASK_CONFIG_SIZE
配置
            int longingTaskCount = (int) Math.ceil(listenerSize /
ParamUtil.getPerTaskConfigSize());
            //⑤已经运行的线程任务数量没有超过阈值
            if (longingTaskCount > currentLongingTaskCount) {
                for (int i = (int) currentLongingTaskCount; i < longingTaskCount;
i++) {
                    //⑥开启线程，并执行轮询类
                    executorService.execute(new LongPollingRunnable(i));
                }
```

```
            currentLongingTaskCount = longingTaskCount;
        }
    }
}
```

4. **通过轮询实现配置信息变更之后的实时推送**

通过线程类 LongPollingRunnable 来轮询，将变更之后的信息实时地推送到客户端。

```
class LongPollingRunnable implements Runnable {
    private final int taskId;

    public LongPollingRunnable(int taskId) {
        this.taskId = taskId;
    }

    @Override
    public void run() {
        List<CacheData> cacheDatas = new ArrayList<CacheData>();
        List<String> inInitializingCacheList = new ArrayList<String>();
        try {
            //①遍历本地配置信息缓存 cacheMap
            for (CacheData cacheData : cacheMap.get().values()) {
                if (cacheData.getTaskId() == taskId) {
                    cacheDatas.add(cacheData);
                    try {
                        checkLocalConfig(cacheData);
                        if (cacheData.isUseLocalConfigInfo()) {
                            //②如果使用了本地缓存，则通知注册监听器，并发起远程回调
                            cacheData.checkListenerMd5();
                        }
                    } catch (Exception e) {
                        LOGGER.error("get local config info error", e);
                    }
                }
            }
            //③校验出变更的 dataID 列表
            List<String> changedGroupKeys = checkUpdateDataIds(cacheDatas,
inInitializingCacheList);
            if (!CollectionUtils.isEmpty(changedGroupKeys)) {
                LOGGER.info("get changedGroupKeys:" + changedGroupKeys);
            }
            //④更新本地配置信息的缓存 CacheData
            for (String groupKey : changedGroupKeys) {
                String[] key = GroupKey.parseKey(groupKey);
                String dataId = key[0];
```

```
                String group = key[1];
                String tenant = null;
                if (key.length == 3) {
                    tenant = key[2];
                }
                try {
                    //⑤从配置中心重新获取最新的配置信息
                    String[] ct = getServerConfig(dataId, group, tenant,
3000L);
                    CacheData cache =
cacheMap.get().get(GroupKey.getKeyTenant(dataId, group, tenant));
                    cache.setContent(ct[0]);
                    if (null != ct[1]) {
                        cache.setType(ct[1]);
                    }
                    LOGGER.info("[{}] [data-received] dataId={}, group={},
tenant={}, md5={}, content={}, type={}",
                                agent.getName(), dataId, group, tenant,
cache.getMd5(),
                                ContentUtils.truncateContent(ct[0]), ct[1]);
                } catch (NacosException ioe) {
                    String message = String
                                .format("[%s] [get-update] get changed config
exception. dataId=%s, group=%s, tenant=%s",
                                        agent.getName(), dataId, group, tenant);
                    LOGGER.error(message, ioe);
                }
            }
            //⑥遍历缓存，通知注册监听器
            for (CacheData cacheData : cacheDatas) {
                if (!cacheData.isInitializing() || inInitializingCacheList
                        .contains(GroupKey.getKeyTenant(cacheData.dataId,
cacheData.group, cacheData.tenant))) {
                    cacheData.checkListenerMd5();
                    cacheData.setInitializing(false);
                }
            }
            inInitializingCacheList.clear();
            executorService.execute(this);

        } catch (Throwable e) {
            LOGGER.error("longPolling error : ", e);
            executorService.schedule(this, taskPenaltyTime,
TimeUnit.MILLISECONDS);
        }
```

```
        }
    }
```

5. 轮询并回调客户端，同步变更了的配置信息

通过 CacheData 类的 checkListenerMd5()方法验证用 MD5 加密之后的配置信息，并对比加密前后。如果不一样，则回调客户端，同步变更的配置信息，具体代码如下所示：

```
     public class CacheData {
        ...
    void checkListenerMd5() {
        //①遍历注册监听器
            for (ManagerListenerWrap wrap : listeners) {
                //②对比缓存和 ManagerListenerWrap 类中同一个 dataID 通过密文加密前后的
字符串
                if (!md5.equals(wrap.lastCallMd5)) {
                    //③如果对比结果为不相同，则通知监听器
                    safeNotifyListener(dataId, group, content, type, md5, wrap);
                }
            }
        }
        private void safeNotifyListener(final String dataId, final String group,
final String content, final String type,
                final String md5, final ManagerListenerWrap listenerWrap) {
            final Listener listener = listenerWrap.listener;

            Runnable job = new Runnable() {
                @Override
                public void run() {
                    ClassLoader myClassLoader =
Thread.currentThread().getContextClassLoader();
                    ClassLoader appClassLoader =
listener.getClass().getClassLoader();
                    try {
                        if (listener instanceof AbstractSharedListener) {
                            AbstractSharedListener adapter =
(AbstractSharedListener) listener;
                            adapter.fillContext(dataId, group);
                            LOGGER.info("[{}] [notify-context] dataId={}, group={},
md5={}", name, dataId, group, md5);
                        }
                        //在执行回调之前，先将线程 classloader 设置为具体 webapp 的
classloader，以免在回调方法中调用 SPI 接口时出现异常或错用（多应用部署才会有该问题）
                        Thread.currentThread().setContextClassLoader
(appClassLoader);
                        //④构造最新的配置信息对象
```

```
                    ConfigResponse cr = new ConfigResponse();
                    cr.setDataId(dataId);
                    cr.setGroup(group);
                    cr.setContent(content);
                    //⑤过滤配置信息
                    configFilterChainManager.doFilter(null, cr);
                    String contentTmp = cr.getContent();
                    //⑥调用注册监听器，通知客户端接收最新的配置信息
                    listener.receiveConfigInfo(contentTmp);
                    if (listener instanceof AbstractConfigChangeListener) {
                        Map data = ConfigChangeHandler.getInstance()
                                .parseChangeData(listenerWrap.lastContent,
content, type);
                        ConfigChangeEvent event = new ConfigChangeEvent(data);
                        ((AbstractConfigChangeListener)
listener).receiveConfigChange(event);
                        listenerWrap.lastContent = content;
                    }

                    listenerWrap.lastCallMd5 = md5;
                    LOGGER.info("[{}] [notify-ok] dataId={}, group={}, md5={},
listener={} ", name, dataId, group, md5,
                            listener);
                } catch (NacosException ex) {
                    LOGGER.error("[{}] [notify-error] dataId={}, group={},
md5={}, listener={} errCode={} errMsg={}",
                            name, dataId, group, md5, listener, ex.getErrCode(),
ex.getErrMsg());
                } catch (Throwable t) {
                    LOGGER.error("[{}] [notify-error] dataId={}, group={},
md5={}, listener={} tx={}", name, dataId,
                            group, md5, listener, t.getCause());
                } finally {
                    Thread.currentThread().setContextClassLoader
(myClassLoader);
                }
            }
        };

        final long startNotify = System.currentTimeMillis();
        try {
            if (null != listener.getExecutor()) {
                listener.getExecutor().execute(job);
            } else {
                job.run();
```

```
            }
        } catch (Throwable t) {
            LOGGER.error("[{}] [notify-error] dataId={}, group={}, md5={},
listener={} throwable={}", name, dataId,
                    group, md5, listener, t.getCause());
        }
        final long finishNotify = System.currentTimeMillis();
        LOGGER.info("[{}] [notify-listener] time cost={}ms in ClientWorker,
dataId={}, group={}, md5={}, listener={} ",
                name, (finishNotify - startNotify), dataId, group, md5,
listener);
    }
}
```

5.4.4 "用 gRPC 框架实现客户端与 Nacos Config Server 之间通信渠道"的原理

在分析"用 gRPC 框架实现客户端与 Nacos Config Server 之间通信渠道"的原理前，可以先了解一下 Nacos Config 的整体架构。

1. 整体架构

Nacos 从版本 2.0 开始，支持 gRPC 通信。gRPC 通信按照功能总共分为 4 层，如图 5-10 所示。

- 接入层：主要包括需要接入 Nacos Config 的中间件，比如 Spring Cloud Alibaba 或者 Dubbo，可以用 Nacos Client 将它们接入。
- 通信层：主要指 RPC 框架 gRPC 和 Rsocket。
- 连接层：用来处理不同客户端及不同类型的 RPC 请求，以标准化输入和输出。
- 功能层：通过连接层访问功能层（配置中心），并返回指定类型请求的配置数据。

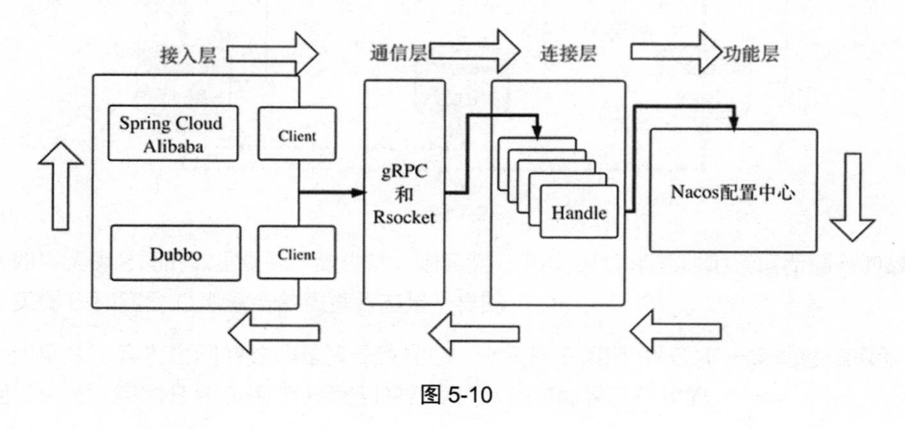

图 5-10

2. “用 gRPC 完成配置信息管理”的过程

在 Nacos 2.0 及之后的版本中，用 gRPC 作为客户端与 Nacos Config Server 之间通信渠道来完成配置信息管理（具体是用 NacosConfigService 类的 publishConfigCas()方法发布配置信息）。gRPC 客户端与 NacosConfigService 类的依赖关系如图 5-11 所示。

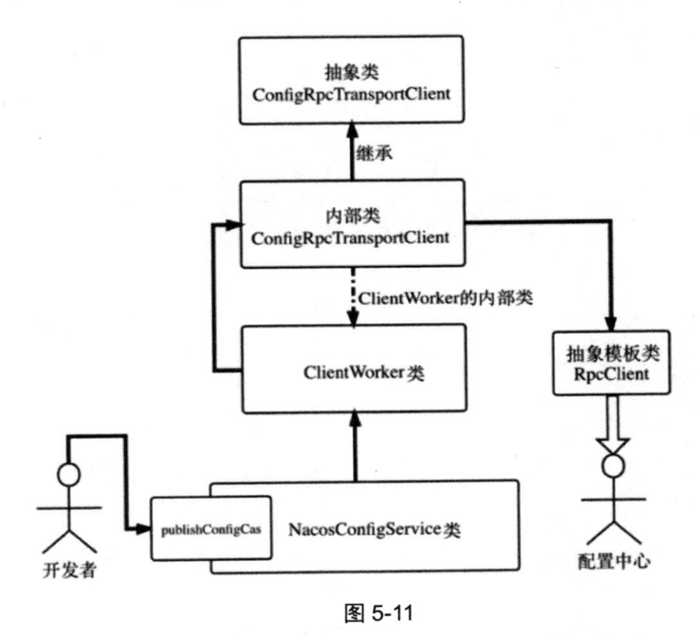

图 5-11

（1）RpcClient 是 gRPC 客户端的一个抽象模板类。开发人员先使用 NacosConfigService 类发布配置信息，然后用抽象模板类 RpcClient 与配置中心进行通信，完成配置信息的发布。

（2）ClientWorker 类是 Nacos 封装的长轮询功能类，NacosConfigService 类调用 ClientWorker 类的内部类 ConfigRpcTransportClient。

（3）用抽象模板类 RpcClient 的通信渠道连接开发者和配置中心。

ClientWorker 类的 publishConfig()方法的具体代码如下所示：

```
    @Override
    public boolean publishConfig(String dataId, String group, String tenant,
String appName, String tag,String betaIps, String content, String casMd5) throws
NacosException {
    try {
        //①构造 RPC 请求的对象（包含需要发布的配置信息 content）
            ConfigPublishRequest request = new ConfigPublishRequest(dataId,
group, tenant, content);
            request.setCasMd5(casMd5);
```

```
            request.putAdditionalParam("tag", tag);
            request.putAdditionalParam("appName", appName);
            request.putAdditionalParam("betaIps", betaIps);
            //②调用 gRPC 客户端发布配置信息
            ConfigPublishResponse response = (ConfigPublishResponse)
requestProxy(getOneRunningClient(), request);
            if (!response.isSuccess()) {
                LOGGER.warn("[{}] [publish-single] fail, dataId={}, group={},
tenant={}, code={}, msg={}",this.getName(), dataId, group, tenant,
response.getErrorCode(), response.getMessage());
            }
            //③返回结果
            return response.isSuccess();
        } catch (Exception e) {
            LOGGER.warn("[{}] [publish-single] error, dataId={}, group={},
tenant={}, code={}, msg={}",this.getName(), dataId, group, tenant, "unkonw",
e.getMessage());
            return false;
    }
    }
```

Nacos 是如何初始化 gRPC 客户端的呢？可以看一下 ClientWorker 类的 ensureRpcClient() 方法，具体代码如下所示：

```
    private synchronized RpcClient ensureRpcClient(String taskId) throws
NacosException {
        Map<String, String> labels = getLabels();
        Map<String, String> newLabels = new HashMap<String, String>(labels);
    newLabels.put("taskId", taskId);
    //①用 RpcClientFactory 类创建 RpcClient 客户端
        RpcClient rpcClient = RpcClientFactory.createClient("config-" + taskId
+ "-" + uuid, getConnectionType(), newLabels);
    if (rpcClient.isWaitInitiated()) {
        //②初始化处理器
            initRpcClientHandler(rpcClient);
            //③绑定租户
            rpcClient.setTenant(getTenant());
            rpcClient.clientAbilities(initAbilities());
            //④开启 RPC 客户端
            rpcClient.start();
        }
        return rpcClient;
    }
```

5.4.5 【实例】用“采用 gRPC 通信渠道的 Nacos Config”实现配置数据的动态更新

本实例的源码在本书配套资源的“/spring-cloud-alibaba-practice/chapterfive/nacos2.0-grpc-spring-cloud-alibaba-config/”目录下。

本实例需要连接 Nacos 2.0 配置中心，所以先搭建 Naocs 2.0 的环境。

1. 搭建 Nacos2.0 的环境

（1）从 Nacos 官网上下载 Nacos 2.0 的部署包“nacos-server-2.0.0.tar.gz”，解压缩文件。

（2）修改 conf/application.properties 文件，添加如下配置：

```
//配置 MySQL 数据源
db.num=1
db.url.0=jdbc:mysql://127.0.0.1:3306/nacos_config?characterEncoding=utf8
&connectTimeout=1000&socketTimeout=3000&autoReconnect=true&useUnicode=true&u
seSSL=false&serverTimezone=UTC
db.user=root
db.password=root
```

（3）启动脚本“bin/startup.sh”。

（4）创建命名空间“spring cloud alibaba practice”。

（5）搭建完成后 Nacos 2.0 的“单机部署+MySQL 数据源”环境如图 5-12 所示。

图 5-12

2. 创建 Spring Cloud Alibaba 项目，实现配置数据的动态更新

（1）添加 Spring Cloud Alibaba 的相关 Jar 包依赖，需要排除低版本的 Nacos-Client，并引

入 Nacos 2.0 对应的 Jar 包，具体的 POM 依赖可以参考本书配套资源中的代码。

（2）添加配置文件 bootstrap.yaml，具体的配置信息可以参考本书配套资源中的代码。

如果将应用采用 Spring Cloud Alibaba 作为基础框架去接入 Nacos 2.0，则开发者不需要对应用的配置文件做任何修改即可连接 Nacos 配置中心。

（3）启动应用。

如果看到如下日志，则说明 gRPC 通信渠道已经生效。

```
    2021-03-26 09:25:25.783  INFO 77061 --- [           main]
c.a.n.client.config.impl.ClientWorker    : [config_rpc_client] [subscribe]
nacos2.0-grpc-spring-cloud-alibaba-config-test.yaml+nacos2.0-grpc-spring-clo
ud-alibaba-config+c7ba173f-29e5-4c58-ae78-b102be11c4f9
    2021-03-26 09:25:25.784  INFO 77061 --- [           main]
c.a.nacos.client.config.impl.CacheData   : [config_rpc_client] [add-listener]
ok, tenant=c7ba173f-29e5-4c58-ae78-b102be11c4f9,
dataId=nacos2.0-grpc-spring-cloud-alibaba-config-test.yaml,
group=nacos2.0-grpc-spring-cloud-alibaba-config, cnt=1
```

对应的 Nacos 控制台效果如图 5-13 所示。

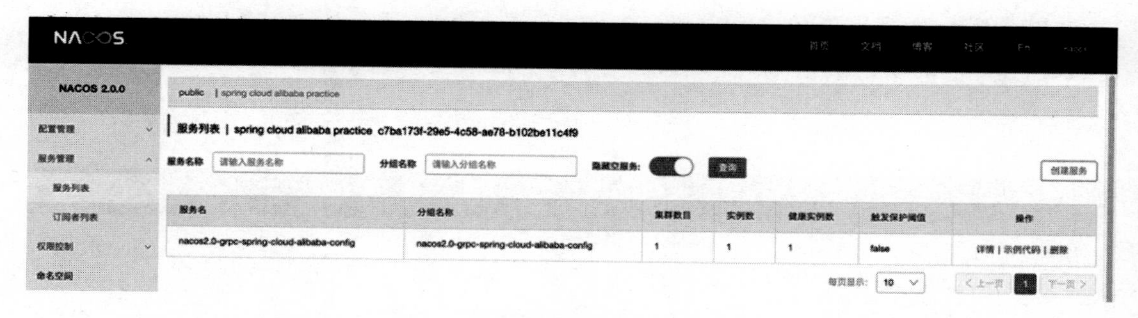

图 5-13

5.5 用“Sofa-Jraft + Apache Derby”保证配置中心的数据一致性

Nacos 在 1.4.0 版本及之后的版本中，用 Sofa-Jraft 框架替换了自研的 Raft 框架。Nacos Config 目前只支持通过“Soft-Jraft + Apache Derby”来保证数据一致性。其中，Apache Derby 是 Java 语言的轻量级数据库，支持嵌入式和集群模式部署。

Sofa-JRaft 是一个开源的 Raft 框架，目前最新的版本是 1.3.5，最近一次版本更新是 2020 年 11 月 26 日，社区活跃度还行，项目代码质量非常高，稳定性也非常高。

5.5.1　Nacos 配置中心的数据一致性原理

Nacos 配置中心的数据一致性原理如图 5-14 所示。

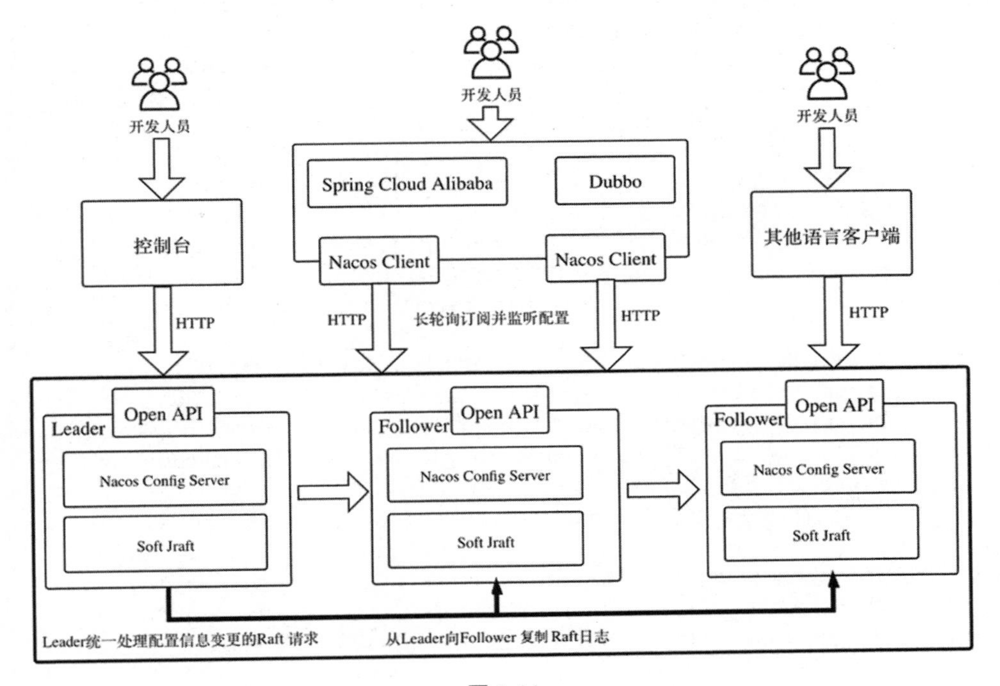

图 5-14

开发人员可以通过控制台、Nacos Client 或者其他语言的客户端，调用配置中心的 Open API 进行配置信息管理。具体流程如下：

（1）开发人员通过控制台新增配置信息，并给 Nacos Config Server 集群发送一个配置信息变更的请求。

（2）配置信息变更请求被同步到 Nacos Config Server 集群。

（3）如果当前节点是 Leader，则它会持久化 Raft 日志，并复制 Raft 日志到 Follower 节点。

（4）如果当前节点是 Followder 节点，则它会将请求转发到 Leader 节点，在 Leader 节点持久化 Raft 日志，并复制 Raft 日志到 Follower 节点中。

下面具体分析配置信息发布的过程。

1. Open API 处理配置信息发布的请求

通过 ConfigController 类的 publishConfig()方法处理请求，部分核心代码如下：

```
    if (StringUtils.isBlank(betaIps)) {
    if (StringUtils.isBlank(tag)) {
        //①如果标签为空，则处理 Raft 请求
            persistService.insertOrUpdate(srcIp, srcUser, configInfo, time,
configAdvanceInfo, true);
            ConfigChangePublisher
                .notifyConfigChange(new ConfigDataChangeEvent(false, dataId,
group, tenant, time.getTime()));
    } else {
        //②如果标签不为空，则处理 Raft 请求
            persistService.insertOrUpdateTag(configInfo, tag, srcIp, srcUser,
time, true);
            ConfigChangePublisher.notifyConfigChange(
                new ConfigDataChangeEvent(false, dataId, group, tenant, tag,
time.getTime()));
            }
        } else {
            //③如果是 beata 发布，则处理 Raft 请求
            persistService.insertOrUpdateBeta(configInfo, betaIps, srcIp,
srcUser, time, true);
                ConfigChangePublisher
                    .notifyConfigChange(new ConfigDataChangeEvent(true, dataId,
group, tenant, time.getTime()));
        }
    ConfigTraceService
        .logPersistenceEvent(dataId, group, tenant, requestIpApp,
time.getTime(),
InetUtils.getSelfIP(),ConfigTraceService.PERSISTENCE_EVENT_PUB, content);
```

2. 初始化 PersistService 类，并处理配置信息发布的请求

前面已经提到，只有 Nacos Config 开启了 Apache Derby 存储模式，才会采用 Sofa-Jraft 框架来保持数据一致性。

PersistService 类的初始化过程如下：

（1）在 Nacos 集群的配置文件 application.properties 中添加如下配置。

```
embeddedStorage=true
spring.datasource.platform=""
```

（2）通过条件注解类 ConditionOnEmbeddedStorage 读取变量 embeddedStorage 的值：

```
public class ConditionOnEmbeddedStorage implements Condition {
    @Override
public boolean matches(ConditionContext context, AnnotatedTypeMetadata
metadata) {
    //读取系统环境中变量 embeddedStorage 的值
        return PropertyUtil.isEmbeddedStorage();
    }
}
```

（3）通过条件变量控制 PersistService 类的初始化：

```
@Conditional(value = ConditionOnEmbeddedStorage.class)
@Component
public class EmbeddedStoragePersistServiceImpl implements PersistService {
    ...
}
```

如果条件注解的值为 true，则会在应用启动时实例化 PersistService 类的实现类 EmbeddedStoragePersistServiceImpl（主要用于访问 Apache Derby 数据库）。

3. 执行配置信息的发布

用 EmbeddedStoragePersistServiceImpl 类的 addConfigInfo()方法执行配置信息的发布，具体代码如下所示：

```
private void addConfigInfo(final String srcIp, final String srcUser, final
ConfigInfo configInfo,
    final Timestamp time, final Map<String, Object> configAdvanceInfo, final
boolean notify,BiConsumer<Boolean, Throwable> consumer) {
    try {
        final String tenantTmp =StringUtils.isBlank(configInfo.getTenant()) ?
StringUtils.EMPTY : configInfo.getTenant();
        configInfo.setTenant(tenantTmp);
        long configId = idGeneratorManager.nextId(RESOURCE_CONFIG_INFO_ID);
        long hisId = idGeneratorManager.nextId(RESOURCE_CONFIG_HISTORY_ID);
        //①将需要发布的配置信息插入表 config_info
        addConfigInfoAtomic(configId, srcIp, srcUser, configInfo, time,
configAdvanceInfo);
        String configTags = configAdvanceInfo == null ? null : (String)
configAdvanceInfo.get("config_tags");
        //②将需要发布的配置信息插入表 config_tags_relation
        addConfigTagsRelation(configId, configTags, configInfo.getDataId(),
configInfo.getGroup(),configInfo.getTenant());
```

```
            //③将需要发布的配置信息插入表 his_config_info
            insertConfigHistoryAtomic(hisId, configInfo, srcIp, srcUser, time,
"I");
            //④将需要发布的配置信息存储到当前线程的上下文中
            EmbeddedStorageContextUtils.onModifyConfigInfo(configInfo, srcIp,
time);
            //⑤处理 Raft 请求，并复制 Raft 日志
            databaseOperate.blockUpdate(consumer);
    } finally {
        //⑥清除当前线程上下文中所有的配置信息，以确保线程安全
            EmbeddedStorageContextUtils.cleanAllContext();
        }
    }
```

4. 处理 Raft 请求，并复制 Raft 日志

DatabaseOperate 类用于封装对 Apache Derby 数据库的操作。

Nacos Config 支持单机模式部署和集群模式部署，但 Sofa-Jraft 只有在集群模式才会生效，因为：

- Apache Derby 数据库采用嵌入式模式部署，支持和 Nacos Config 同时启动，并且只能当前 Nacos Config 节点访问 Apache Derby。
- 在单机模式部署时，Nacos Config 只有一个节点，不存在跨节点同步配置信息到其他节点的业务场景。
- 在集群模式部署时，Nacos Config 有多个节点。Nacos Config 通过 Soft-Jraft 确保节点中 Apache Derby 数据库中的配置信息的数据一致性。

处理 Raft 请求并复制 Raft 日志的具体过程如下：

（1）通过条件注解控制 Sofa-Jraft 的开启。

- 注解@ConditionDistributedEmbedStorage 的具体代码如下所示：

```
    public class ConditionDistributedEmbedStorage implements Condition {
        @Override
    public boolean matches(ConditionContext context, AnnotatedTypeMetadata
metadata) {
    //如果是采用 Apache Derby 存储，且 Nacos 是集群模式部署，则返回 true
    return PropertyUtil.isEmbeddedStorage()
&& !ApplicationUtils.getStandaloneMode();
        }
    }
```

- 注解@ConditionStandaloneEmbedStorag 的具体代码如下所示：

```
    public class ConditionStandaloneEmbedStorage implements Condition {
        @Override
    public boolean matches(ConditionContext context, AnnotatedTypeMetadata
metadata) {
        //如果是采用 Apache Derby 存储，且 Nacos 是单机模式部署，则返回 true
            return PropertyUtil.isEmbeddedStorage() &&
ApplicationUtils.getStandaloneMode();
        }
    }
```

（2）初始化 DatabaseOperate 类，并执行日志复制的请求。

初始化 DatabaseOperate 类的具体代码如下所示：

```
    //条件注解@ConditionDistributedEmbedStorage
    @Conditional(ConditionDistributedEmbedStorage.class)
    @Component
    @SuppressWarnings({"unchecked"})
    public class DistributedDatabaseOperateImpl extends LogProcessor4CP
implements BaseDatabaseOperate {
        ...
    }
    //条件注解@ConditionStandaloneEmbedStorage
    @Conditional(ConditionStandaloneEmbedStorage.class)
    @Component
    public class StandaloneDatabaseOperateImpl implements BaseDatabaseOperate
{
        ...
    }
```

代码中的两个条件注解，只能有一个生效，这样就能控制 DatabaseOperate 类的初始化。

（3）日志复制。

通过 DistributedDatabaseOperateImpl 类的 update()方法，开启集群节点之间的日志复制：

```
    @Override
    public Boolean update(List<ModifyRequest> sqlContext, BiConsumer<Boolean,
Throwable> consumer) {
        try {
            LoggerUtils.printIfDebugEnabled(LogUtil.DEFAULT_LOG,
"modifyRequests info : {}", sqlContext);
            final String key =System.currentTimeMillis() + "-" + group() + "-" +
memberManager.getSelf().getAddress() + "-"+
MD5Utils.md5Hex(sqlContext.toString(), Constants.ENCODE);
            //①从当前线程的上下文中获取需要发布的配置信息，并将其转换为 Raft 的 Log 日志文件
            Log log = Log.newBuilder().setGroup(group()).setKey(key)
```

```
                    .setData(ByteString.copyFrom(serializer.serialize(sqlCon
text)))
                    .putAllExtendInfo(EmbeddedStorageContextUtils.getCurrent
ExtendInfo())
                    .setType(sqlContext.getClass().getCanonicalName()).build
();
        if (Objects.isNull(consumer)) {
            //②调用 JRaftProtocol 类的 submit()方法同步提交 Log 日志
            Response response = this.protocol.submit(log);
            if (response.getSuccess()) {
                return true;
            }
            LogUtil.DEFAULT_LOG.error("execute sql modify operation failed :
{}", response.getErrMsg());
            return false;
        } else {
            //③调用 JRaftProtocol 类的 submitAsync()方法异步提交 Log 日志

this.protocol.submitAsync(log).whenComplete((BiConsumer<Response, Throwable>)
(response, ex) -> {
                String errMsg = Objects.isNull(ex) ? response.getErrMsg() :
ExceptionUtil.getCause(ex).getMessage();
                consumer.accept(response.getSuccess(),
                    StringUtils.isBlank(errMsg) ? null : new
NJdbcException(errMsg));
                });
        }
        return true;
    } catch (TimeoutException e) {
        LogUtil.FATAL_LOG.error("An timeout exception occurred during the
update operation");
        throw new NacosRuntimeException(NacosException.SERVER_ERROR,
e.toString());
    } catch (Throwable e) {
        LogUtil.FATAL_LOG.error("An exception occurred during the update
operation : {}", e);
        throw new NacosRuntimeException(NacosException.SERVER_ERROR,
e.toString());
    }
}
```

JRaftProtocol 类是 Nacos 的 Raft 算法的实现（基于 Soft-Jraft）。如果读者感兴趣，可以查阅相关源码。

5.5.2 【实例】用“切换所连接的 Nacos 节点”验证数据一致性

本实例的源码在本书配套资源的“/spring-cloud-alibaba-practice/chapterfive/soft-jraft-apache-derby-config/”目录下。

Nacos Config 采用集群部署，并开启了基于 Soft-Jraft 的数据一致性。为了验证数据一致性，下面分别通过控制台访问 Leader 节点和 Follower 节点，并变更配置信息。

1. 搭建“Soft-Jraft + Apache Derby”的集群环境

前面已经提到过，只有在集群模式下 Soft-Jraft 才会生效。

（1）下载安装包 nacos-server-1.4.0.zip，解压缩。

（2）配置集群信息。文件路径“conf/cluster.conf”，具体配置如下：

```
192.168.0.123:8846
192.168.0.123:8847
192.168.0.123:8849
```

（3）去掉 MySQL 数据源的配置信息

（4）启动脚本“bin/startup.sh”，启动命令如下：

```
sh startup.sh -p embedded
```

（5）用“192.168.0.123:8846/nacos/”访问 Nacos 集群，效果如图 5-15 所示。

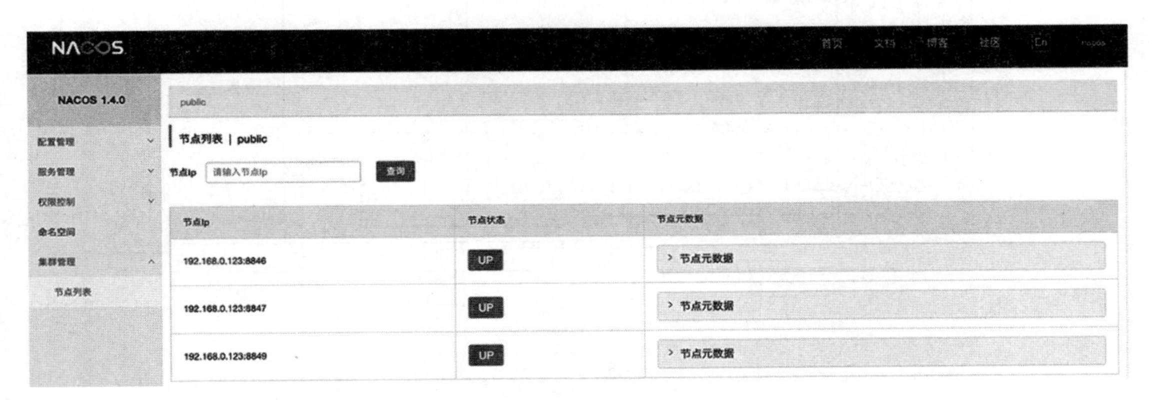

图 5-15

2. 搭建项目环境

用“Spring Cloud Alibaba Config + Spring Boot”快速构建一个项目。

3. 模拟场景一

下面通过控制台访问集群的 Leader 节点，添加配置信息并发布，然后更改配置信息；将应用通过 Spring Cloud Alibaba Config 接入配置中心的 Follower 节点，用线程实时监控配置信息的变化。如果能够实时获取最新的配置信息，则说明能够满足数据一致性。Leader 节点会通过 Soft-Jraft 算法复制日志文件到 Followder 节点。

（1）通过 Leader 节点访问控制台，创建命名空间 spring cloud Alibaba practice，并创建配置文件 soft-jraft-apache-derby-config-test.yaml：

```
spring:
   youxia:
      config:
         name: test50
```

（2）添加应用对应的配置文件：

```
spring:
  application:
    name: soft-jraft-apache-derby-config
  cloud:
    nacos:
      config:
        namespace: c7ba173f-29e5-4c58-ae78-b102be11c4f9
        group: soft-jraft-apache-derby-config
        password: nacos
        enable-remote-sync-config: true
        ###Follower 节点
        server-addr: 192.168.0.123:8847
        username: nacos
        name: soft-jraft-apache-derby-config
        file-extension: properties
        extension-configs:
          - data-id: soft-jraft-apache-derby-config-test.yaml
            group: soft-jraft-apache-derby-config
            namespace: c7ba173f-29e5-4c58-ae78-b102be11c4f9
            refresh: true
      discovery:
        namespace: c7ba173f-29e5-4c58-ae78-b102be11c4f9
        group: soft-jraft-apache-derby-config
        ###Follower 节点
        server-addr: 192.168.0.123:8847
```

（3）运行程序，在控制台中会输出如下日志：

```
使用 Spring Cloud Alibaba 将应用接入 Nacos 配置中心，获取配置信息 name 为：test50
```

将应用接入 Follower 节点，并通过 Nacos Client 获取配置信息，此时 Leader 节点上的数据“spring.youxia.config.name=test50”已经同步到 Follower 节点中了。

（4）通过 Leader 节点访问控制台，变更“spring.youxia.config.name”为“test51”，观察程序对应的控制台，输出如下日志：

```
使用 Spring Cloud Alibaba 将应用接入 Nacos 配置中心，获取配置信息 name 为：test51
```

通过 Leader 节点变更的配置信息，会实时地通过 Soft-Jraft 算法同步到 Follower 节点中。连接 Follower 节点的应用，能够实时获取最新的配置信息。

4. 模拟场景二

下面通过 Follower 节点访问控制台，并更改配置信息，监听 Leader 节点和另外一个 Follower 节点的配置信息。

（1）通过 Follower 节点访问控制台，变更“spring.youxia.config.name”为“test52”。

（2）将应用接入 Leader 节点，观察应用对应控制台的日志，如下：

```
使用 Spring Cloud Alibaba 接入 Nacos 配置中心，获取配置信息 name 为：test52
```

（3）将应用接入 Follower 节点，观察应用对应控制台的日志，如下：

```
使用 Spring Cloud Alibaba 接入 Nacos 配置中心，获取配置信息 name 为：test52
```

通过 Follower 节点访问控制台，在变更数据之后，接入的 Leader 节点和另外一个 Follower 节点都能实时访问最新的数据，3 个节点之间的数据保持一致。

5.6　用数据库持久化配置中心的数据

Nacos 配置中心存储的数据，是应用服务在启动过程中需要依赖的关键性数据。所以，需要用持久化技术保证配置中心的数据不会丢失。

5.6.1　为什么需要持久化

在配置中心中，关键的技术挑战之一就是做好配置信息的持久化。

常规的持久化有以下两种方式。

- 内存存储：采用内存存储能够确保数据存储的高性能，但是维护集群节点之间数据一致性的成本非常高。
- 数据库存储：在节点中不维护配置信息的数据状态，每个节点可以执行写操作。配置信息的数据状态的维护交给数据库，能够降低技术复杂度，但是不能确保高性能。

Nacos 1.4.0 版本及之后版本支持 MySQL 和“Soft-Jraft + Apache Derby”两种持久化方式。

5.6.2 持久化的基础配置

Nacos 配置中心持久化的基础配置见表 5-4。

表 5-4 Nacos 配置中心持久化的基础配置

配置项名称	功能描述
spring.datasource.platform	配置支持的数据库类型，默认为 MySQL
db.num	配置数据源
db.url	配置数据源对应连接池的 URL
db.user	配置数据源对应连接池的用户名
db.password	配置数据源对应连接池的密码
db.maxPoolSize	配置数据源对应连接池的最大线程数
db.minIdle	配置数据源对应连接池的最小空闲连接数

5.6.3 持久化的原理

Nacos 配置信息的持久化主要分为两种情况：①基于内嵌存储数据源，②基于外置数据源。

1. 内嵌存储数据源

Nacos 用 Apache Derby 框架来实现内嵌存储数据源，具体实现的核心类是 LocalDataSourceServiceImpl，通过它可以完成数据源的初始化。

下面来看看是如何初始化内嵌数据源的。

（1）通过注解@PostConstruct，将 LocalDataSourceServiceImpl 类和 Spring Framework 的 IOC 容器绑定。在 Spring Framework 的 IOC 容器初始化完成之后，开始执行数据源的初始化。其中，注解@PostConstruct 的语义是在 IOC 容器加载 Servlet 时运行被注解@PostConstruct 标注的方法，只会初始化一次。目前 Nacos 基于 Servlet 的服务器来启动整个服务。

（2）用系统属性参数 isUseExternalDB，来判断是否开启了内嵌存储数据源。如果开启了，则开始执行初始化。

（3）拼接 JDBC URL。其中，参数 derbyBaseDir 的值为“/data/derby-data”，参数 nacosHome 的值为“Nacos 集群部署的安装目录”。

（4）先使用（3）中拼接的 JDBC URL 初始化数据库连接池 HikariDataSource（包括设置 DriverClassName、JdbcUrl、Username、Password、IdleTimeout、MaximumPoolSize 和 ConnectionTimeout 等数据源常规的参数）；然后初始化 DataSourceTransactionManager 事

务管理器，并绑定到数据库连接池；接着初始化 JdbcTemplate 类和 TransactionTemplate 类，前者是 Spring JDBC 对数据库操作封装的一个模板类，后者是 Spring Framework 封装的事务模板类。

（5）加载数据表结构，完成 SQL 语句的 DDL 操作。主要是加载“META-INF/schema.sql”文件，并通过当前数据库连接池将文件中的 DDL 语句持久化到 Apache Derby 中。

2. 外置数据源

Nacos 外置数据源默认使用 MySQL，底层 ORM 框架采用 Spring JDBC，外置数据源的核心类是 ExternalDataSourceServiceImpl。

如果初始化外置数据源，则首先执行 DynamicDataSource 类的 getDataSource()方法，然后调用 ExternalDataSourceServiceImpl 类的 init()方法。具体过程如下：

（1）初始化 JdbcTemplate 类、DataSourceTransactionManager 类和 TransactionTemplate 类，其中，事务查询超时时间为 5s。

（2）如果当前服务使用外部数据源（通过 PropertyUtil.isUseExternalDB()方法来判断），则执行 reload()方法。

（3）开启定时器，执行任务 SelectMasterTask 和 CheckDbHealthTask。在这个过程中会间隔 10 s 去选择 Master 数据源，并进行数据源的健康检查。

3. 动态数据源

Nacos Config 动态数据源的核心类是 DynamicDataSource，它采用单例模式。

内嵌存储数据源和外置数据源都依赖 DynamicDataSource 类生成数据源，具体代码如下所示：

```
public synchronized DataSourceService getDataSource() {
    try {
        if (PropertyUtil.isEmbeddedStorage()) {
            if (localDataSourceService == null) {
                localDataSourceService = new LocalDataSourceServiceImpl();
                localDataSourceService.init();
            }
            return localDataSourceService;
            } else {
                if (basicDataSourceService == null) {
                    basicDataSourceService = new
ExternalDataSourceServiceImpl();
                    basicDataSourceService.init();
                }
                return basicDataSourceService;
```

```
            }
        } catch (Exception e) {
        throw new RuntimeException(e);
    }
}
```

4. 多数据源

只有外置数据源，才会存在多数据源的业务场景。

Nacos Config 用 Spring JDBC 作为其底层的 ORM 框架，并没有重复造轮子。开发人员可以在代码中配置 3 个数据源，具体代码如下所示：

```
//数据源总数为 3
db.num=3
//第 1 个数据源
db.url.0=jdbc:mysql://127.0.0.1:3306/nacos0?characterEncoding=utf8&connectTimeout=1000&socketTimeout=3000&autoReconnect=true&useUnicode=true&useSSL=false&serverTimezone=UTC
db.user.0=nacos
db.password.0=nacos
//第 2 个数据源
db.url.1=jdbc:mysql://127.0.0.1:3308/nacos1?characterEncoding=utf8&connectTimeout=1000&socketTimeout=3000&autoReconnect=true&useUnicode=true&useSSL=false&serverTimezone=UTC
db.user.1=nacos
db.password.1=nacos
//第 3 个数据源
db.url.2=jdbc:mysql://127.0.0.1:3307/nacos2?characterEncoding=utf8&connectTimeout=1000&socketTimeout=3000&autoReconnect=true&useUnicode=true&useSSL=false&serverTimezone=UTC
db.user.2=nacos
db.password.2=nacos
```

5. 数据表结构

Nacos 配置中心中与配置相关的数据表见表 5-5。

表 5-5　Nacos 配置中心中与配置相关的数据表

表　名	表功能描述
config_info	存储 nacos-config 对应 data_id 的配置信息，唯一性约束 'uk_configinfo_datagrouptenant' ('data_id','group_id','tenant_id')，配置信息主键 Key、服务组 ID 及租户 ID，存储应用直接使用的环境数据
config_info_beta	其表结构和 config_info 表一样，但是它存储的是测试环境的分布式配置数据，用于应用的测试版本发布

续表

表　　名	表功能描述
config_info_tag	数据表结构和 config_info 一样,但是它主要用于存储打上固定标签后的配置数据，tag 标签靠控制台入参
config_tags_relation	带有标签的配置信息
group_capacity	对应“组”的容量信息
his_config_info	历史版本的配置信息
tenant_capacity	租户容量信息
tenant_info	租户信息

5.6.4 【实例】用“配置信息的灰度发布”验证持久化

本实例的源码在本书配套资源的“/spring-cloud-alibaba-practice/chapterfive/beata-spring-cloud-alibaba-config/”目录下。

Nacos Config 支持配置信息的灰度发布，可以将配置信息发布到指定的客户端（不同的 IP 地址）。

1. 用“Spring Cloud Config + Spring Boot”快速创建项目

新增配置文件 bootstrap.yaml，具体代码如下所示：

```
spring:
  application:
    name: beata-spring-cloud-alibaba-config
  cloud:
    nacos:
      config:
        namespace: c7ba173f-29e5-4c58-ae78-b102be11c4f9
        group: beata-spring-cloud-alibaba-config
        password: nacos
        enable-remote-sync-config: true
        server-addr: 127.0.0.1:8848
        username: nacos
        name: beata-spring-cloud-alibaba-config
        file-extension: properties
        extension-configs:
          - data-id: beata-spring-cloud-alibaba-config-test.yaml
            group: beata-spring-cloud-alibaba-config
            namespace: c7ba173f-29e5-4c58-ae78-b102be11c4f9
            refresh: true
```

2. 在控制台中新增配置文件 beata-spring-cloud-alibaba-config-test.yaml

配置文件的具体内容如下所示：

```
spring:
  youxia:
    config:
      name: test56
```

3. 在灰度发布之前观察应用的运行日志

运行日志，效果如图 5-16 所示。

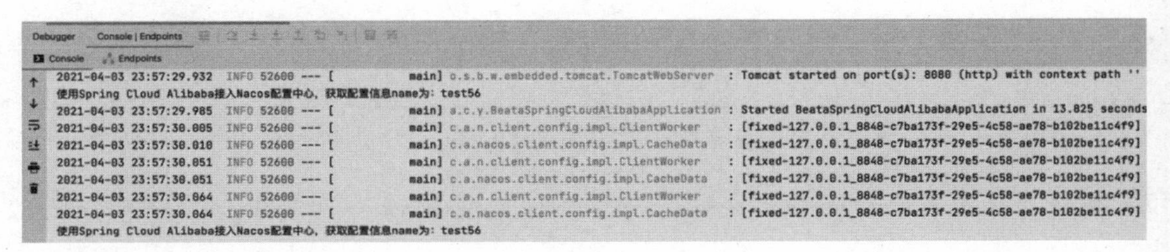

图 5-16

可以看出，使用 Spring Cloud Alibaba 将应用接入 Nacos 配置中心，能够获取“spring.youxia.config.name”属性的值为 test56。

4. 进行灰度发布

（1）通过控制台编辑 beata-spring-cloud-alibaba-config-test.yaml 文件进行灰度发布，如图 5-17 所示。

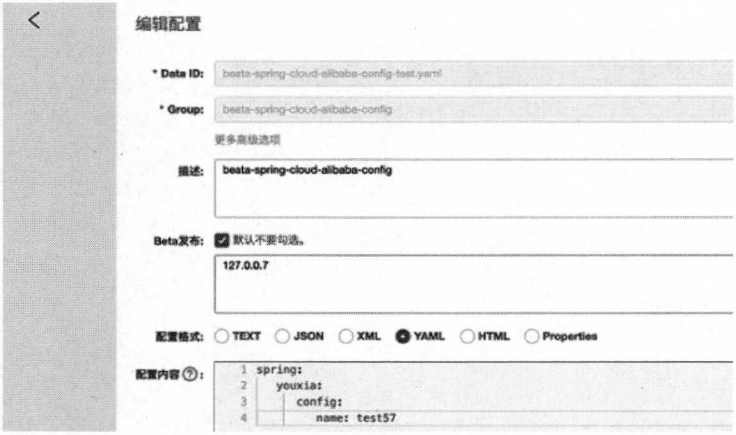

编辑配置

* Data ID: beata-spring-cloud-alibaba-config-test.yaml

* Group: beata-spring-cloud-alibaba-config

更多高级选项

描述: beata-spring-cloud-alibaba-config

Beta发布: 默认不要勾选。

127.0.0.7

配置格式: TEXT JSON XML YAML HTML Properties

配置内容:

```
1 spring:
2   youxia:
3     config:
4       name: test57
```

图 5-17

（2）更改“spring.youxia.config.name”为“test57”，并将其灰度发布给 IP 地址为 127.0.0.7 的客户端。当前应用对应的 IP 地址为 127.0.0.1，如果灰度发布生效，则应用监听到的值不会变。

（3）单击“灰度发布”按钮后，通过如下 SQL 语句查询对应的灰度表 config_info_beta。

```
    select * from nacos_config.config_info_beta t where
t.data_id="beata-spring-cloud-alibaba-config-test.yaml";
```

（4）查询结果如图 5-18 所示。可以看到，在灰度表中新增了一条记录。

	id	data_id	group_id	app_name	content	beta_ips
1	3	beata-spring-cloud-alibaba-config-test.yaml	beata-spring-cloud-alibaba-config		spring:↵ youxia:↵ …	127.0.0.7

图 5-18

5. 在灰度发布之后，观察运行日志。

使用 Spring Cloud Alibaba 将应用接入 Nacos 配置中心，获取的配置信息 name 为 test56。关闭灰度发布之后，重新发布。如图 5-19 所示，应用能够实时感知配置信息的变更，获取最新的配置数据。

```
2021-03-26 17:29:21.036  INFO 79686 --- [78-b102be11c4f9] c.a.n.client.config.impl.ClientWorker    : [fixed-127.0.0.1_8848-c7ba173f-29e5-4c58-ae78-b102be11c4f9] [p
2021-03-26 17:29:21.043  INFO 79686 --- [78-b102be11c4f9] c.a.n.client.config.impl.ClientWorker    : get changedGroupKeys:[beata-spring-cloud-alibaba-config-test.y
2021-03-26 17:29:21.091  INFO 79686 --- [78-b102be11c4f9] c.a.n.client.config.impl.ClientWorker    : [fixed-127.0.0.1_8848-c7ba173f-29e5-4c58-ae78-b102be11c4f9] [d
  youxia:
    config:
      name: test57, type=
2021-03-26 17:29:21.092  INFO 79686 --- [78-b102be11c4f9] c.a.nacos.client.config.impl.CacheData   : [fixed-127.0.0.1_8848-c7ba173f-29e5-4c58-ae78-b102be11c4f9] [n
2021-03-26 17:29:21.434  WARN 79686 --- [78-b102be11c4f9] c.a.c.n.c.NacosPropertySourceBuilder     : Ignore the empty nacos configuration and get it based on dataI
2021-03-26 17:29:21.443  WARN 79686 --- [78-b102be11c4f9] c.a.c.n.c.NacosPropertySourceBuilder     : Ignore the empty nacos configuration and get it based on dataI
2021-03-26 17:29:21.444  INFO 79686 --- [78-b102be11c4f9] b.c.PropertySourceBootstrapConfiguration : Located property source: [BootstrapPropertySource {name='boots
2021-03-26 17:29:21.451  INFO 79686 --- [78-b102be11c4f9] o.s.boot.SpringApplication               : No active profile set, falling back to default profiles: defau
2021-03-26 17:29:21.479  INFO 79686 --- [78-b102be11c4f9] o.s.boot.SpringApplication               : Started application in 0.375 seconds (JVM running for 1146.923
2021-03-26 17:29:21.513  INFO 79686 --- [78-b102be11c4f9] o.s.c.e.event.RefreshEventListener       : Refresh keys changed: [spring.youxia.config.name]
2021-03-26 17:29:21.513  INFO 79686 --- [78-b102be11c4f9] c.a.nacos.client.config.impl.CacheData   : [fixed-127.0.0.1_8848-c7ba173f-29e5-4c58-ae78-b102be11c4f9] [n
2021-03-26 17:29:21.513  INFO 79686 --- [78-b102be11c4f9] c.a.nacos.client.config.impl.CacheData   : [fixed-127.0.0.1_8848-c7ba173f-29e5-4c58-ae78-b102be11c4f9] [n
使用Spring Cloud Alibaba接入Nacos配置中心，获取配置信息name为：test57
使用Spring Cloud Alibaba接入Nacos配置中心，获取配置信息name为：test57
```

图 5-19

6. 查询数据库中已经持久化的灰度发布，以及正式发布的配置信息

通过如下 SQL 语句查询，发现在正式的配置信息表 config_info 中多了 1 条记录。

```
    select * from nacos_config.config_info t where
t.data_id="beata-spring-cloud-alibaba-config-test.yaml";
```

config_info 表的查询记录如图 5-20 所示。

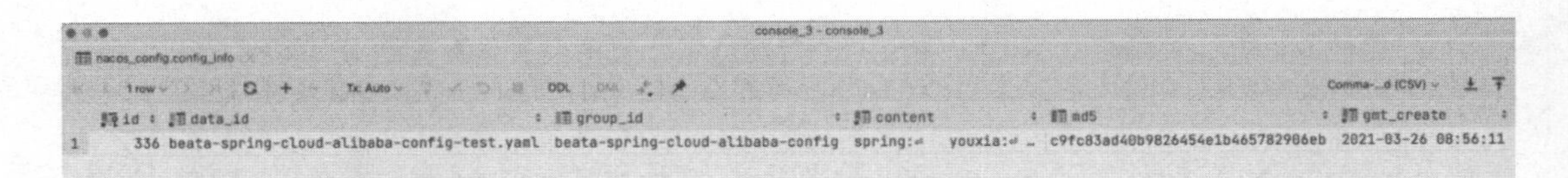

	id	data_id	group_id	content	md5	gmt_create
1	336	beata-spring-cloud-alibaba-config-test.yaml	beata-spring-cloud-alibaba-config	spring:↵ youxia:↵ …	c9fc83ad40b9826454e1b465782906eb	2021-03-26 08:56:11

图 5-20

5.7 用“Spring Cloud Alibaba Config + Nacos Config”实现配置管理（公共配置、应用配置和扩展配置）

在项目开发中，开发人员经常要利用配置信息优先级，来定义公共配置、应用配置和扩展配置。

Spring Cloud Alibaba 封装了 Nacos Config，可以实现配置信息按照优先级进行加载。

5.7.1 “按照优先级加载属性”的原理

Spring Cloud Alibaba Config 底层依赖 Spring Cloud，并通过 Nacos Config 的 Nacos Client 调用配置中心实现按照优先级加载属性。

1. 用 Spring Cloud 的 PropertySourceBootstrapConfiguration 类触发属性的加载

通过 PropertySourceBootstrapConfiguration 类的 initialize()方法，加载应用启动过程中的环境属性：

```
   @Configuration(proxyBeanMethods = false)
   @EnableConfigurationProperties(PropertySourceBootstrapProperties.class)
   public class PropertySourceBootstrapConfiguration implements
           ApplicationContextInitializer<ConfigurableApplicationContext>,
Ordered {

       @Override
       public void initialize(ConfigurableApplicationContext
applicationContext) {
           List<PropertySource<?>> composite = new ArrayList<>();
            //①排序属性资源加载器
           AnnotationAwareOrderComparator.sort(this.propertySourceLocators);
           boolean empty = true;
            //②获取配置属性环境对象
           ConfigurableEnvironment environment =
applicationContext.getEnvironment();
            //③循环遍历属性资源加载器列表
           for (PropertySourceLocator locator : this.propertySourceLocators) {
```

```
        //④通过属性资源加载器，加载在应用中自定义的属性（本地属性文件的属性和配置中心的属性）
        Collection<PropertySource<?>> source =
locator.locateCollection(environment);
            if (source == null || source.size() == 0) {
                continue;
            }
            List<PropertySource<?>> sourceList = new ArrayList<>();
            for (PropertySource<?> p : source) {
                if (p instanceof EnumerablePropertySource) {
                    EnumerablePropertySource<?> enumerable =
(EnumerablePropertySource<?>) p;
                    sourceList.add(new
BootstrapPropertySource<>(enumerable));
                }
                else {
                    sourceList.add(new SimpleBootstrapPropertySource(p));
                }
            }
            logger.info("Located property source: " + sourceList);
            composite.addAll(sourceList);
            empty = false;
        }
        if (!empty) {
            MutablePropertySources propertySources =
environment.getPropertySources();
            String logConfig =
environment.resolvePlaceholders("${logging.config:}");
            LogFile logFile = LogFile.get(environment);
            for (PropertySource<?> p : environment.getPropertySources()) {
                if (p.getName().startsWith(BOOTSTRAP_PROPERTY_SOURCE_NAME)) {
                    propertySources.remove(p.getName());
                }
            }
             //⑤将从应用中加载过来的属性，加载到配置属性环境对象中，属性加载生效
            insertPropertySources(propertySources, composite);
            reinitializeLoggingSystem(environment, logConfig, logFile);
            setLogLevels(applicationContext, environment);
            handleIncludedProfiles(environment);
        }
    }
    ...
}
```

2. 用属性加载器加载配置中心的属性信息

属性加载器 PropertySourceLocator 是 Spring Cloud 的一个通用接口。

Spring Cloud Alibaba 用 NacosPropertySourceLocator 类实现了 PropertySourceLocator 接口，具体代码如下所示：

```
@Order(0)
public class NacosPropertySourceLocator implements PropertySourceLocator {
    ...
    Override
    public PropertySource<?> locate(Environment env) {
        nacosConfigProperties.setEnvironment(env);
         //①获取 Nacos Config 的 ConfigService 对象
        ConfigService configService =
nacosConfigManager.getConfigService();
        if (null == configService) {
            log.warn("no instance of config service found, can't load config
from nacos");
            return null;
        }
        long timeout = nacosConfigProperties.getTimeout();
         //②初始化属性加载对象 NacosPropertySourceBuilder
        nacosPropertySourceBuilder = new
NacosPropertySourceBuilder(configService,
                timeout);
        String name = nacosConfigProperties.getName();
        String dataIdPrefix = nacosConfigProperties.getPrefix();
        if (StringUtils.isEmpty(dataIdPrefix)) {
            dataIdPrefix = name;
        }
        if (StringUtils.isEmpty(dataIdPrefix)) {
            dataIdPrefix = env.getProperty("spring.application.name");
        }
        CompositePropertySource composite = new CompositePropertySource(
                NACOS_PROPERTY_SOURCE_NAME);
          //③加载公共配置信息
        loadSharedConfiguration(composite);
          //④加载扩展配置信息
        loadExtConfiguration(composite);
          //⑤加载应用配置信息
        loadApplicationConfiguration(composite, dataIdPrefix,
nacosConfigProperties, env);
        return composite;
    }
}
```

加载配置信息的 3 个步骤如下：

（1）加载公共配置信息。

```
//加载公共配置信息
private void loadSharedConfiguration(
        CompositePropertySource compositePropertySource) {
    List<NacosConfigProperties.Config> sharedConfigs =
nacosConfigProperties
            .getSharedConfigs();
    if (!CollectionUtils.isEmpty(sharedConfigs)) {
        checkConfiguration(sharedConfigs, "shared-configs");
        loadNacosConfiguration(compositePropertySource, sharedConfigs);
    }
}
```

（2）加载扩展配置信息。

```
//加载扩展配置信息
private void loadExtConfiguration(CompositePropertySource
compositePropertySource) {
    List<NacosConfigProperties.Config> extConfigs = nacosConfigProperties
            .getExtensionConfigs();
    if (!CollectionUtils.isEmpty(extConfigs)) {
        checkConfiguration(extConfigs, "extension-configs");
        loadNacosConfiguration(compositePropertySource, extConfigs);
    }
}
```

（3）加载应用配置信息。

```
//加载应用配置信息
private void loadApplicationConfiguration(CompositePropertySource
compositePropertySource, String dataIdPrefix,
        NacosConfigProperties properties, Environment environment) {
    String fileExtension = properties.getFileExtension();
    String nacosGroup = properties.getGroup();
    loadNacosDataIfPresent(compositePropertySource, dataIdPrefix,
nacosGroup,fileExtension, true);
    loadNacosDataIfPresent(compositePropertySource,dataIdPrefix + DOT +
fileExtension, nacosGroup, fileExtension, true);
    for (String profile : environment.getActiveProfiles()) {
        String dataId = dataIdPrefix + SEP1 + profile + DOT + fileExtension;
        loadNacosDataIfPresent(compositePropertySource, dataId, nacosGroup,
                fileExtension, true);
    }
}
```

综上所述，扩展配置可以覆盖公共配置，应用配置可以覆盖扩展配置。

5.7.2 【实例】验证公共配置、应用配置和扩展配置的优先级顺序

本实例的源码在本书配套资源的“/spring-cloud-alibaba-practice/chapterfive/configuring-priorities-load-nacos/”目录下。

在项目开发过程中，开发人员会经常碰到一些数据库、缓存中间件等公共配置信息。同一命名空间中的所有应用，都可以访问这些配置信息。

扩展配置是指：在原有的应用配置的基础之上，增加一些定制化的扩展配置，解耦应用配置。

将 Spring Cloud Alibaba Config 接入 Nacos Config 后，即可使用 Nacos Config 的配置优先级的特性。优先级顺序为：

公共配置 ＜ 扩展配置 ＜ 应用配置

1. 新增配置文件

配置文件 bootstrap.yaml 的内容如下：

```
dubbo:
  scan:
    base-packages: com.alibaba.cloud.youxia
  protocol:
    name: dubbo
    port: -1
spring:
  application:
    //应用名称
    name: configuring-priorities-load-nacos
  main:
    allow-bean-definition-overriding: true
  cloud:
    nacos:
      discovery:
        //Nacos 注册中心集群的 IP 地址
        server-addr: 10.0.23.123:8848,10.0.23.123:8847,10.0.23.123:8846
        //Nacos 注册中心的命名空间
        namespace: c7ba173f-29e5-4c58-ae78-b102be11c4f9
        //nacos group
        group: configuring-priorities-load-nacos
      config:
        //Nacos 配置中心集群的 IP 地址
```

```
server-addr: 10.0.23.123:8848,10.0.23.123:8847,10.0.23.123:8846
//Nacos 配置中心的 Group
group: configuring-priorities-load-nacos
//Nacos 配置中心的命名空间
namespace: c7ba173f-29e5-4c58-ae78-b102be11c4f9
//扩展配置
extension-configs:
  - data-id: extension.yaml
    group: configuring-priorities-load-nacos
    refresh: true
//共享配置
sharedConfigs:
  - data-id: share.yaml
    group: configuring-priorities-load-nacos
    refresh: true
//应用级配置
name: configuring-priorities-load-nacos
file-extension: yaml
```

2. 启动应用，观察配置的优先级加载的顺序

启动程序，启动之后的 Nacos 配置中心如图 5-21 所示。

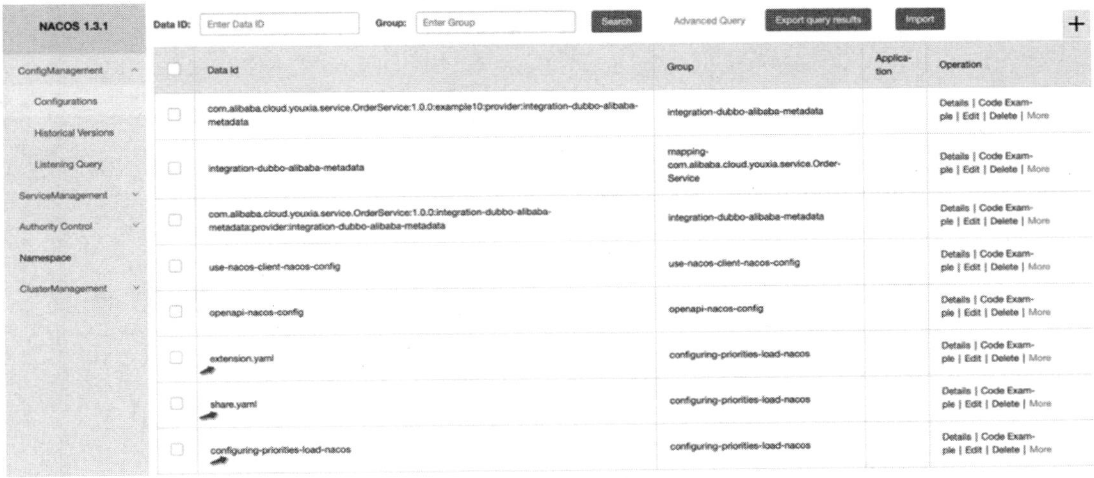

Data Id	Group	Applica-tion	Operation
com.alibaba.cloud.youxia.service.OrderService:1.0.0:example10:provider:integration-dubbo-alibaba-metadata	integration-dubbo-alibaba-metadata		Details | Code Example | Edit | Delete | More
integration-dubbo-alibaba-metadata	mapping-com.alibaba.cloud.youxia.service.OrderService		Details | Code Example | Edit | Delete | More
com.alibaba.cloud.youxia.service.OrderService:1.0.0:integration-dubbo-alibaba-metadata:provider:integration-dubbo-alibaba-metadata	integration-dubbo-alibaba-metadata		Details | Code Example | Edit | Delete | More
use-nacos-client-nacos-config	use-nacos-client-nacos-config		Details | Code Example | Edit | Delete | More
openapi-nacos-config	openapi-nacos-config		Details | Code Example | Edit | Delete | More
extension.yaml	configuring-priorities-load-nacos		Details | Code Example | Edit | Delete | More
share.yaml	configuring-priorities-load-nacos		Details | Code Example | Edit | Delete | More
configuring-priorities-load-nacos	configuring-priorities-load-nacos		Details | Code Example | Edit | Delete | More

图 5-21

Nacos 配置中心有 3 个配置文件：extension.yaml、share.yaml 和 example12，配置内容如下。

- extension.yaml：

```
spring:
  name: extensiontest 动态
```

- share.yaml：

```
spring:
   name: sharetest
```

- configuring-priorities-load-nacos：

```
spring:
  name: applicationname
```

输出日志如图 5-22 所示：打印出当前 Spring Name 为 applicationname，加载优先级最高的 example12 中的 spring.name=applicationname。

```
当前Spring Name是: applicationname
当前Spring Name是: applicationname
当前Spring Name是: applicationname
当前Spring Name是: applicationname
2021-02-03 00:44:19.329  WARN 3068 --- [ask-Scheduler-1] a.c.d.m.r.DubboServiceMetadataRepository : Current application will subscribe all services(size:1) in
当前Spring Name是: applicationname
当前Spring Name是: applicationname
当前Spring Name是: applicationname
当前Spring Name是: applicationname
当前Spring Name是: applicationname
当前Spring Name是: applicationname
2021-02-03 00:44:49.386  WARN 3068 --- [ask-Scheduler-1] a.c.d.m.r.DubboServiceMetadataRepository : Current application will subscribe all services(size:1) in
当前Spring Name是: applicationname
```

图 5-22

Nacos 注册中心的效果如图 5-23 所示。

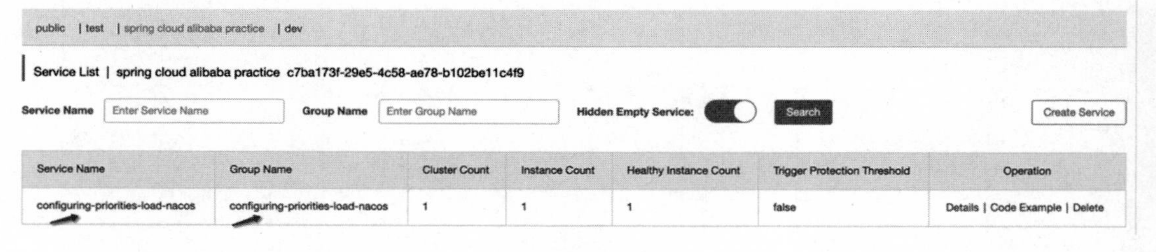

图 5-23

第 6 章 分布式流量防护——基于 Sentinel

流量防护是确保分布式系统高可用的关键手段之一。对应用做好流量防护（尤其是在复杂的调用链路关系中），能够减少很多线上故障。

Sentinel 是主流的分布式系统的高可用流量防护组件之一。本章将介绍如何利用 Sentinel 对应用进行高可用的流量防护。

6.1 认识分布式流量防护

在分布式系统中，服务之间的相互调用会生成分布式流量。如何通过组件进行流量防护，并有效控制流量，是分布式系统的技术挑战之一。

6.1.1 什么是分布式流量防护

在分布式系统中，一次 RPC 请求过程如图 6-1 所示。下面结合 RPC 请求，来分析流量防护组件在分布式系统中的角色：

- 请求被流量防护组件处理后，被路由到交易服务（见图 6-1 中的左侧）。
- 请求被交易服务处理完后，调用订单服务。同理，请求也会先被流量防护组件处理，再被路由到订单服务。
- 请求被订单服务处理完后，调用商品服务。同理，请求也会先被流量防护组件处理，再被商品服务处理。

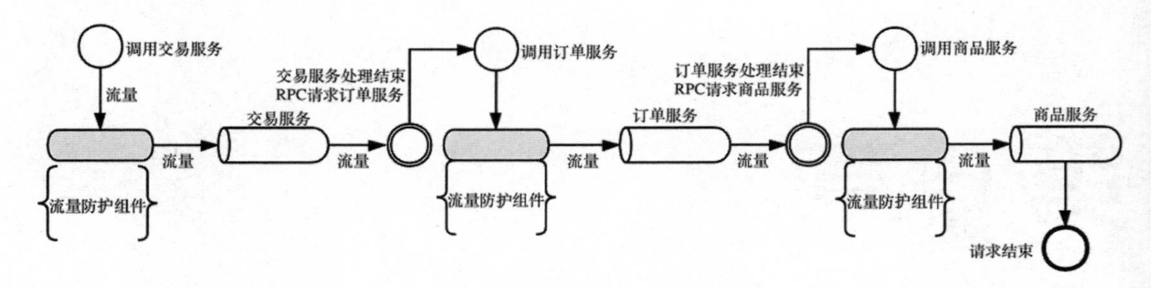

图 6-1

从分层架构的角度，可以将流量防护分为：前置式流量防护和嵌入式流量防护。

1. 前置式流量防护

如图 6-2 所示，在分层架构中，前置流量防护组件处于应用和负载均衡组件之前。可以在前置流量防护组件中添加流量防护的路由策略，使得流量还没有被路由到应用就先被前置流量防护组件有效地防护起来。常见的前置流量防护组件有：Spring Cloud Gateway、Zuul 等路由网关。

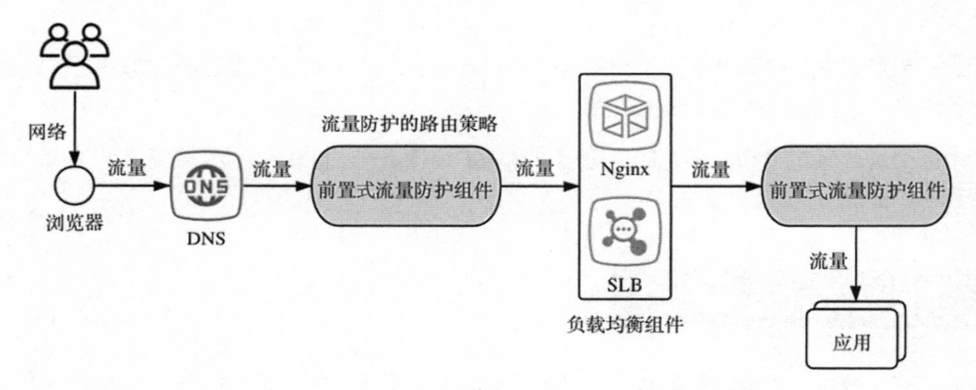

图 6-2

有些开发人员会将业务逻辑（比如安全校验等）放在负载均衡组件中。为了起到有效的流量防护作用，可以在前置式流量防护组件之前增加 DNS 的流量防护，真正做到前置式流量防护。

前置式流量防护具备如下优势：

- 对应用是零侵入的。
- 流量防护策略比较简单，不会给业务开发人员带来更多的技术成本。
- 开发人员能够非常方便地开启和关闭流量防护，而应用是无感知的。

2. 嵌入式流量防护

如图 6-3 所示，在分层架构中，嵌入式流量防护组件被内嵌到应用 A 和应用 B 中。应用需要依赖流量防护组件完成流量防护。

嵌入式流量防护组件通常都会和应用强耦合，但可以采用动态的全局开关做到流量防护功能的自动上下线。

嵌入式流量防护具备如下优势：

- 应用可以动态地调整流量防护策略。
- 流量防护的覆盖范围更广，可以从应用中的实例到应用中的任意代码块。
- 流量防护更加有效。

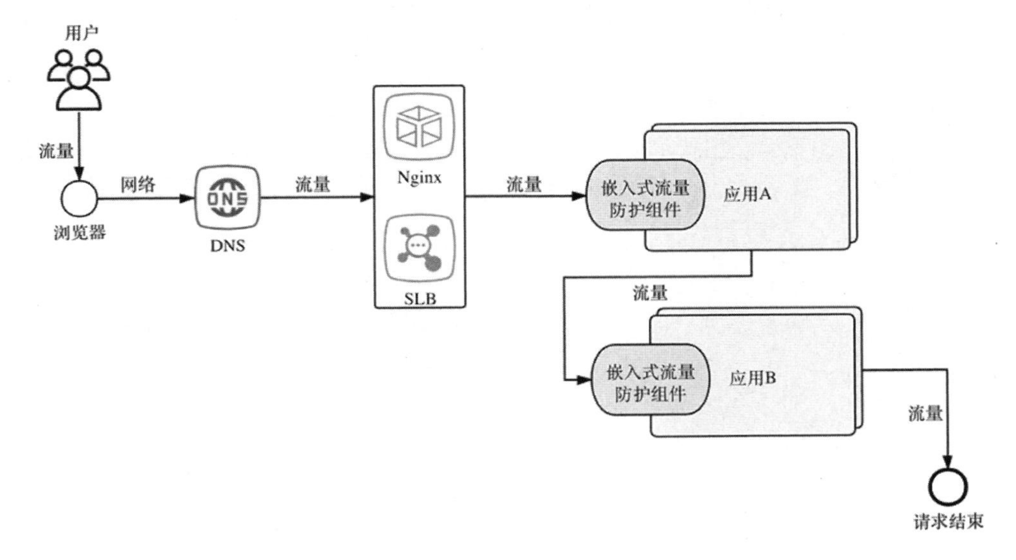

图 6-3

从图 6-3 中可以看出，在正常的流量请求到达应用 A 和应用 B 之前，嵌入式流量防护组件不会生效。

6.1.2　为什么需要分布式流量防护

在分布式系统中，服务被部署在不同的服务进程中，服务之间通过 RPC 通信完成服务请求。一次核心的业务请求通常需要调用几十个服务，从而产生上百次 RPC 请求。熟悉分布式系统的开发人员都知道，“网络抖动”会直接影响 RPC 请求的稳定性。

在分布式系统中，数据从服务 A 发送到服务 B 需要一段延迟时间，延迟时间由传输时间和处理时间组成。

通常“网络抖动”是指，数据从服务 A 传输到服务 B 的最大延迟时间和最小延迟时间之间的时间差。比如，最大延迟时间为 20ms，最小延迟时间为 5ms，那么“网络抖动”就是 15ms。“网络抖动”是衡量网络是否稳定的一个标志，“网络抖动”越大，网络越不稳定。

图 6-4 描述了电商领域中的交易服务、商品服务、订单服务、支付服务、库存服务、供应链服务、进销存货服务、仓库服务和物流服务之间的调用链路关系。

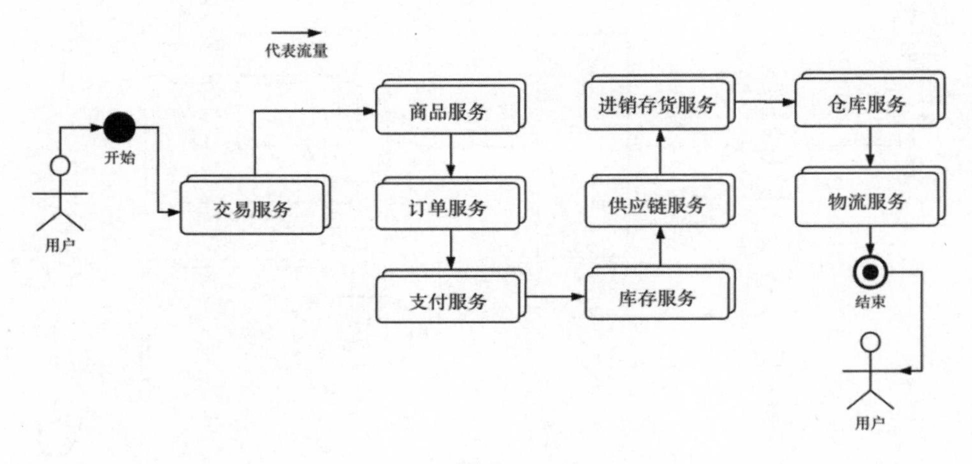

图 6-4

服务之间的调用链路关系介绍如下：

（1）用户发起订单交易请求，交易请求会先后调用交易服务、商品服务、订单服务等 9 个服务。

（2）在处理交易请求的过程中，会出现调用链路长、跨服务调用次数多的现象，从而导致 RPC 请求的次数增多，增加了服务调用不可用的可能性。

（3）如果某一个服务不规范，则需要将服务调用的网络超时时间设置得很长，比如 30s。如果出现超时异常，则这个服务在 30s 内只能处理这一个请求，严重地降低了服务的 QPS 和 TPS。

（4）如果在某一次的电商促销活动中（比如“双十一”），上百万个用户同时发起交易请求，形成了请求流量的“洪峰”，这时的请求流量是非“双十一”时间段的几十倍，此时支付服务集群中的某一个实例在处理大流量的支付请求的过程中出现了服务不可用的故障，若出现故障的支付服务实例没有接入流量防控组件，则支付服务集群中的其他稳定的实例也会受到影响，从而导致支付服务集群出现“整体不可用”。这样整个交易服务就会出现“雪崩”现象。

6.2 认识 Sentinel

在开源社区中主流的流量防护组件有 Sentinel、Nginx、Guava、Redis、Hystrix 和 Resilience4j。其中，Sentinel 是阿里巴巴开源的流量防护组件。本章重点介绍它。

Sentinel 是一款面向分布式服务架构的轻量级流量防控组件，主要以流量为切入点，从流量控制、熔断降级、系统自适应保护等多个维度，来帮助用户保障服务的稳定性。

Sentinel 的核心功能：根据资源配置流量防护规则，为资源执行相应的流量控制、降级和系统自适应保护等规则。

在 Sentinel 中，资源的定义和规则的配置是完全分开的。开发人员可以先通过 Sentinel API 将对应的业务代码定义为 Sentinel 能够识别的资源，然后将资源与流量防护规则绑定，从而达到流量防护的目的。

1. Sentinel 的发展历史

Sentinel 的发展历史如下：

- 2012 年，Sentinel 诞生，主要功能为入口流量控制。
- 2013—2017 年，Sentinel 在阿里巴巴内部迅速发展，成为基础技术模块，覆盖了所有的核心场景。
- 2018 年，Sentinel 开源，并持续演进。
- 2019 年，Sentinel 朝着多语言扩展的方向不断探索，推出 C++ 原生版本，同时针对 Service Mesh 场景也推出了 Envoy 集群流量控制支持，以解决 Service Mesh 架构下多语言限流的问题。
- 2020 年，推出 Sentinel Go 版本，继续朝着云原生方向演进。
- 2021 年，发布 1.8.1 版本。

2. Sentinel 的基本概念

Sentinel 的基本概念如下。

- 资源：它可以是 Java 应用中的任何内容，例如，由应用提供的服务，或由应用调用的其他应用提供的服务，甚至可以是一段代码。在接下来的文档中都会用“资源”来描述代码块。

只要是通过 Sentinel API 定义的代码，就是资源，能够被 Sentinel 保护起来。在大部分情况下，可以使用方法签名、URL，甚至服务名称，作为资源名来标识资源。

- 规则：根据资源的实时状态设定的规则，包括流量控制规则、熔断降级规则，以及系统保护规则。所有规则都可以被动态、实时地调整。

3. Sentinel 的优势

Sentinel 的优势如下。

- 轻量级：核心库无多余依赖，性能损耗小。
- 方便接入，开源生态广泛：Sentinel 为 Dubbo、Spring Cloud、Web Servlet、gRPC 等常用框架提供了适配模块，只需要引入相应依赖并进行简单配置，即可快速接入它们。Sentinel 还提供了低侵入性的注解资源定义方式，以方便开发人员灵活地将程序中的类或者方法接入 Sentinel 控制台。
- 丰富的流量控制场景：Sentinel 覆盖阿里巴巴多年“双十一”高流量的核心场景，流量控制的维度包括流量控制指标、流量控制效果（塑形）、调用关系、热点、集群等。针对系统维度，Sentinel 也提供了自适应保护机制。
- 易用的控制台：提供了实时监控、机器发现、规则管理等能力。
- 完善的扩展性设计：提供了多样化的 SPI 接口，方便用户根据需求给 Sentinel 添加自定义的逻辑。

6.3 将应用接入 Sentinel

要将应用接入 Sentinel，则需要先搭建 Sentinel 控制台。本节会使用两种方式将应用接入 Sentinel 中。

6.3.1 搭建 Sentinel 控制台

搭建 Sentinel 控制台的步骤如下：

（1）从官网下载 Sentinel 的安装包 sentinel-dashboard-1.8.0.jar。

（2）使用如下命令启动控制台。

```
    Java -Dserver.port=8080 -Dcsp.sentinel.dashboard.server=localhost:8080
-Dproject.name=sentinel-dashboard -jar sentinel-dashboard.jar
```

启动 Sentinel 控制台，需要 JDK 版本为 1.8 及以上。

其中，“-Dserver.port=8080”用于自定 Sentinel 控制台的端口号为 8080。

（3）在 Sentinel 控制台启动后，用“http://127.0.0.1:8080/#/login”访问控制台，输入默认的用户名和密码（sentinel/sentinel），之后在控制台中可以看到 sentinel-dashboard 监控的指标，如图 6-5 所示。提示：Sentinel 默认会开启 sentinel-dashboard。

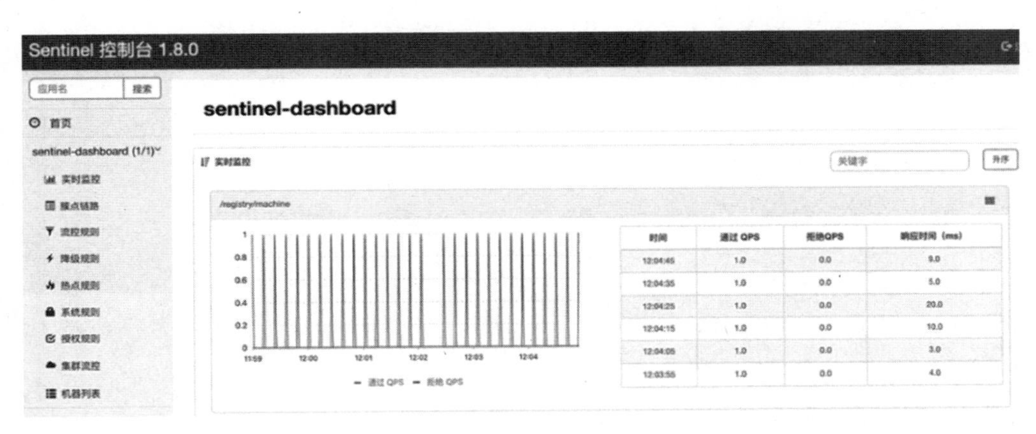

图 6-5

6.3.2 【实例】用 Sentinel Core 手动地将应用接入 Sentinel

本实例的源码在本书配套资源的“chaptersix/ sentinel-core-access”目录下。

在 Sentinel 控制台启动后，开发人员可以用 Sentinel Core 手动地将应用接入 Sentinel。

1. 创建一个 Spring Boot 项目，并引入 Sentinel 相关的 Jar 包

开发人员可以引入 Core 模块来定义流量控制规则，引入 Transport 模块来与 Sentinel 控制台进行通信。模块对应的 Jar 包依赖如下所示：

```
<dependency>
    <groupId>com.alibaba.csp</groupId>
    <artifactId>sentinel-core</artifactId>
    <version>1.8.0</version>
```

```
</dependency>
<dependency>
    <groupId>com.alibaba.csp</groupId>
    <artifactId>sentinel-transport-simple-http</artifactId>
    <version>1.8.0</version>
</dependency>
```

2. 添加配置文件 application.properties

在 application.properties 文件中，添加如下配置信息。

```
//应用的端口号
server.port=20000
//Sentinel 控制台的 IP 地址
csp.sentinel.dashboard.server=127.0.0.1:8080
//应用在 Sentinel 中的项目名称
project.name=sentinel-core-access
```

3. 配置应用服务的启动类

在启动类中，手动初始化 Sentinel 和应用之间的通信渠道，具体代码如下：

```
@SpringBootApplication
public class SentinelCoreAccessApplication {
public static void main(String[] args) {
    //手动初始化通信渠道，触发心跳机制，将应用接入 Sentinel 控制台
        InitExecutor.doInit();
        SpringApplication.run(SentinelCoreAccessApplication.class, args);
    }
}
```

4. 运行启动类，将应用接入 Sentinel

模拟流量去调用接口的具体效果如图 6-6 所示。

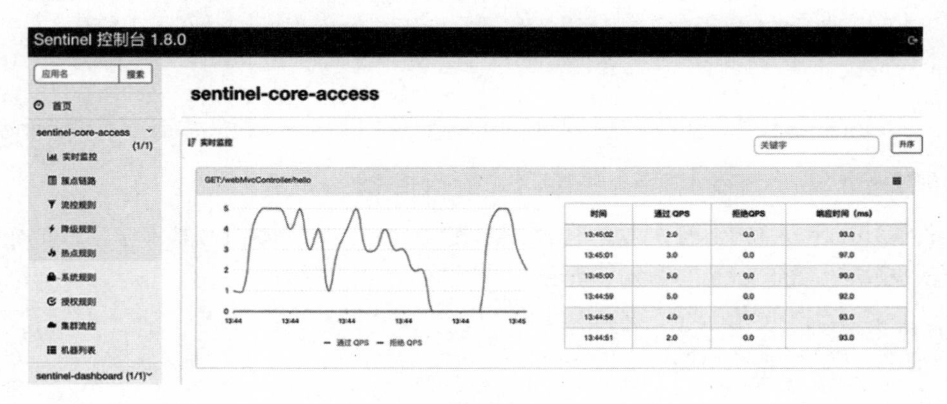

图 6-6

6.3.3 【实例】用 Spring Cloud Alibaba Sentinel 将应用接入 Sentinel

本实例的源码在本书配套资源的“chaptersix/ sentinel-spring-cloud-alibaba”目录下。

下面用 Spring Cloud Alibaba Sentinel 将应用接入 Sentinel。

1. 用 IDEA 快速搭建一个 Spring Boot 项目，并添加 Jar 包依赖

先用 IDEA 快速搭建一个 Spring Boot 项目，然后在程序的 pom.xml 文件中添加 Sentinel 相关的依赖关系，具体代码如下：

```
<dependency>
  <groupId>com.alibaba.cloud</groupId>
  <artifactId>spring-cloud-starter-alibaba-sentinel</artifactId>
</dependency>
```

2. 添加配置文件

在配置文件 application.properties 中，添加如下配置信息：

```
spring.application.name=sentinel-spring-cloud-alibaba-restfulapi
server.port=18083
management.endpoints.web.exposure.include=*
management.endpoint.health.show-details=always
management.health.diskspace.enabled=false
spring.cloud.sentinel.transport.dashboard=localhost:8080
spring.cloud.sentinel.eager=true
spring.cloud.sentinel.web-context-unify=true
```

3. 启动应用

在应用启动后，就可以将其接入 Sentinel，如图 6-7 所示。

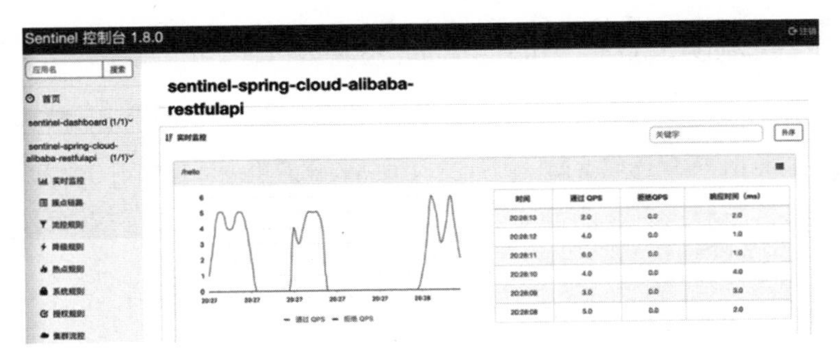

图 6-7

用 Spring Cloud Alibaba Sentinel 接入 Sentinel，需要在应用启动之前添加以下配置信息：

-Dcsp.sentinel.dashboard.server=127.0.0.1:8080 -Dproject.name=sentinel-core-access

之后，应用会正常接入 Sentinel 控制台。

6.4 用 HTTP 或者 Netty 实现通信渠道

在应用和 Sentinel 控制台之间，可以通过 HTTP 或者 Netty 进行通信。

6.4.1 认识 NIO 框架 Netty

1. 什么是 NIO

NIO（Non-blocking I/O，在 Java 领域也称之为 New I/O），是一种同步非阻塞的 I/O 模型，也是 I/O 多路复用的基础。它已经被越来越多地应用到大型应用服务器中，成为解决高并发与大量连接、I/O 处理问题的有效方式。

2. 什么是 Netty

Netty 被用来快速开发高性能、高可靠性的网络服务器和客户端程序。即 Netty 是一个基于 NIO 的客户和服务器端编程框架。

使用 Netty，用户可以快速和简单地开发出一个网络应用，例如实现某种协议的客户和服务器端应用。Netty 简化了网络应用的开发过程。

“快速”和“简单”并不意味着会让应用产生维护性或性能上的问题。Netty 借鉴了多种协议（包括 FTP、SMTP、HTTP，以及各种二进制和文本协议）的优点，并经过相当精心的设计，最终找到了一种方式：在保证高效开发的同时，还能保证应用的性能、稳定性和伸缩性。

6.4.2 用 SPI 机制实现插件化通信渠道的原理

Sentinel 支持用 Netty 和 HTTP 实现通信渠道。开发人员在生产环境中只能二选一。

- 如果使用 Netty 实现通信渠道，则开发人员需要在应用中添加如下 Jar 包依赖。

```
<dependency>
    <groupId>com.alibaba.csp</groupId>
    <artifactId>sentinel-transport-netty-http</artifactId>
    <version>1.8.0</version>
</dependency>
```

- 如果使用 HTTP 实现通信渠道，则开发人员需要在应用中添加如下 Jar 包依赖，并去掉 Netty 通信渠道的 Jar 包依赖。

```
<dependency>
    <groupId>com.alibaba.csp</groupId>
    <artifactId>sentinel-transport-simple-http</artifactId>
    <version>1.8.0</version>
</dependency>
```

Sentinel 是如何做到通过变更 Jar 包依赖零成本地切换应用的通信渠道的呢？基于 SPI 的插件化机制。

1. 用 Env 类开启 Sentinel 的 SPI 插件化机制的初始化

用 Env 类开启 Sentinel 的 SPI 插件化机制的初始化的具体代码如下：

```
public class Env {
    //①定义全局 Sph 对象
    public static final Sph sph = new CtSph();
static {
    //②用 InitExecutor 类的 doInit()方法开启初始化
        InitExecutor.doInit();
    }
}
```

2. 用 InitExecutor 类执行初始化

InitExecutor 类在 Java 的语法规范中是一个不可变的类。在应用启动的过程中，它只会初始化一次，以确保线程安全。具体代码如下：

```
//①用 Java 语法中的关键字 final 修饰 InitExecutor 类
public final class InitExecutor {
    //②定义初始化开关，默认为 false
    private static AtomicBoolean initialized = new AtomicBoolean(false);

public static void doInit() {
    //③用 Java 的 CAS 语法规范控制方法 doInit()只执行一次
        if (!initialized.compareAndSet(false, true)) {
            return;
        }
        try {
            //④用 JDK 自带的 SPI 加载 InitFunc 接口的所有实现类
            ServiceLoader<InitFunc> loader =
ServiceLoaderUtil.getServiceLoader(InitFunc.class);
            List<OrderWrapper> initList = new ArrayList<OrderWrapper>();
            //⑤设置接入 InitFunc 接口的所有实现类初始化的顺序
            for (InitFunc initFunc : loader) {
```

```
                RecordLog.info("[InitExecutor] Found init func: " +
initFunc.getClass().getCanonicalName());
                insertSorted(initList, initFunc);
            }
            for (OrderWrapper w : initList) {
                //⑥执行初始化
                w.func.init();
                RecordLog.info(String.format("[InitExecutor] Executing %s
with order %d",
                    w.func.getClass().getCanonicalName(), w.order));
            }
        } catch (Exception ex) {
            RecordLog.warn("[InitExecutor] WARN: Initialization failed", ex);
            ex.printStackTrace();
        } catch (Error error) {
            RecordLog.warn("[InitExecutor] ERROR: Initialization failed with
fatal error", error);
            error.printStackTrace();
        }
    }
}
```

3. 用通信渠道类 CommandCenterInitFunc 实现 InitFunc 接口

用通信渠道类 CommandCenterInitFunc 实现 InitFunc 接口的具体代码如下所示：:

```
//①设置初始化的顺序
@InitOrder(-1)
public class CommandCenterInitFunc implements InitFunc {
    @Override
public void init() throws Exception {
    //②加载通信渠道对应的通信事件中心接口的实现（Netty和HTTP）
        CommandCenter commandCenter =
CommandCenterProvider.getCommandCenter();
        if (commandCenter == null) {
            RecordLog.warn("[CommandCenterInitFunc] Cannot resolve
CommandCenter");
            return;
        }
        //③加载通信渠道对应的通信事件（Netty和HTTP）
        commandCenter.beforeStart();
       //④初始化通信渠道（Netty和HTTP）
        commandCenter.start();
        RecordLog.info("[CommandCenterInit] Starting command center: "
                + commandCenter.getClass().getCanonicalName());
    }
}
```

4. 用 HeartbeatSenderInitFunc 类实现 InitFunc 接口

在应用和 Sentinel 控制台之间，需要通过心跳机制保持连接。Sentinel 用 HeartbeatSenderInitFunc 类实现 InitFunc 接口，从而实现心跳机制的具体代码如下：

```
    //①设置初始化的顺序
    @InitOrder(-1)
    public class HeartbeatSenderInitFunc implements InitFunc {
        ...
        @Override
    public void init() {
        //②加载对应通信渠道的心跳发送的实现类（Netty 或者 HTTP）
            HeartbeatSender sender =
HeartbeatSenderProvider.getHeartbeatSender();
            if (sender == null) {
                RecordLog.warn("[HeartbeatSenderInitFunc] WARN: No
HeartbeatSender loaded");
                return;
            }
            //③初始化执行心跳任务的线程池
            initSchedulerIfNeeded();
            long interval = retrieveInterval(sender);
            //④设置发送心跳事件的周期
            setIntervalIfNotExists(interval);
            //⑤开启线程，按照固定周期执行心跳任务
            scheduleHeartbeatTask(sender, interval);
        }
    }
```

5. 用 CommandCenterProvider 类加载通信渠道对应的实现类

CommandCenterProvider 是一个公共的通信渠道加载类，具体代码如下：

```
    public final class CommandCenterProvider {
        private static CommandCenter commandCenter = null;
    static {
        //①通过静态方法触发通信事件中心类的初始化
            resolveInstance();
        }
        //②执行初始化，只执行一次
    private static void resolveInstance() {
        //③通过 SPI 加载 CommandCenter 接口的实现类（HTTP 或者 Netty）
            CommandCenter resolveCommandCenter =
SpiLoader.loadHighestPriorityInstance(CommandCenter.class);

            if (resolveCommandCenter == null) {
                //④如果在当前应用中没有引入 Netty 或者 HTTP 的 Jar 包，则 SPI 加载失败，输
```

```
出警告
            RecordLog.warn("[CommandCenterProvider] WARN: No existing
CommandCenter found");
        } else {
            //⑤实例全局通信事件中心类对应的对象
            commandCenter = resolveCommandCenter;
            RecordLog.info("[CommandCenterProvider] CommandCenter resolved:
" + resolveCommandCenter.getClass()
                .getCanonicalName());
        }
    }
    ...
    }
```

6. 用 HeartbeatSenderProvider 类，加载通信渠道心跳机制对应的实现类

HeartbeatSenderProvider 类是通信渠道心跳机制对应的实现类，是心跳机制的公共实现。具体代码如下：

```
    public final class HeartbeatSenderProvider {

    private static HeartbeatSender heartbeatSender = null;
    //①通过静态方法，触发心跳通信渠道心跳机制类的初始化
        static {
            resolveInstance();
        }
        //②执行初始化，只执行一次
    private static void resolveInstance() {
        //③通过 SPI 加载 HeartbeatSender 接口的实现类（HTTP 或者 Netty）
            HeartbeatSender resolved =
SpiLoader.loadHighestPriorityInstance(HeartbeatSender.class);
            if (resolved == null) {
                //④如果在当前应用中没有引入 Netty 或者 HTTP 的 Jar 包，则 SPI 加载失败，输
出警告
                RecordLog.warn("[HeartbeatSenderProvider] WARN: No existing
HeartbeatSender found");
            } else {
                //⑤实例全局心跳通信渠道心跳机制类对应的对象
                heartbeatSender = resolved;
                RecordLog.info("[HeartbeatSenderProvider] HeartbeatSender
activated: " + resolved.getClass()
                    .getCanonicalName());
            }
    }
    ...
    }
```

在通信渠道插件化后，开发人员就可以通过添加对应通信渠道的 Jar 包依赖，加载通信渠道插件，完成通信渠道的快速切换。

6.4.3 “用插件类 NettyHttpCommandCenter 实现通信渠道”的原理

在 Sentinel 中，主要用插件类 NettyHttpCommandCenter 来实现通信渠道（基于 Netty）。

1. 初始化 beforeStart()方法，注册通信事件

注册通信事件的具体代码如下：

```
@SpiOrder(SpiOrder.LOWEST_PRECEDENCE - 100)
public class NettyHttpCommandCenter implements CommandCenter {
   private final HttpServer server = new HttpServer();
   @Override
public void beforeStart() throws Exception {
   //①加载 Netty 通信事件的处理类
      Map<String, CommandHandler> handlers =
CommandHandlerProvider.getInstance().namedHandlers();
      //②向通信渠道中注册通信事件，以及对应的事件处理类
      server.registerCommands(handlers);
   }
   ...
}
```

2. 初始化 start()方法，开启 HTTP 通信渠道对应的连接

开启 HTTP 通信渠道对应的连接的具体代码如下：

```
@SpiOrder(SpiOrder.LOWEST_PRECEDENCE - 100)
public class NettyHttpCommandCenter implements CommandCenter {
   //①初始化 HTTP 通信渠道对应的实现类 HttpServer
private final HttpServer server = new HttpServer();
//②实例化一个线程池
   private final ExecutorService pool = Executors.newSingleThreadExecutor(
      new NamedThreadFactory("sentinel-netty-command-center-executor"));
   @Override
public void start() throws Exception {
   //③用线程池异步地开启 HTTP 通信渠道的连接
      pool.submit(new Runnable() {
         @Override
         public void run() {
            try {
               //④开启 HTTP 通信渠道的连接
               server.start();
            } catch (Exception ex) {
```

```
                    RecordLog.warn("[NettyHttpCommandCenter] Failed to start
Netty transport server", ex);
                    ex.printStackTrace();
                }
            }
        });
    }
    ...
}
```

3. 用 stop()方法关闭通信渠道

关闭通信渠道的具体代码如下：

```
@SpiOrder(SpiOrder.LOWEST_PRECEDENCE - 100)
public class NettyHttpCommandCenter implements CommandCenter {
private final HttpServer server = new HttpServer();
    private final ExecutorService pool = Executors.newSingleThreadExecutor(
        new NamedThreadFactory("sentinel-netty-command-center-executor"));
    @Override
public void stop() throws Exception {
    //①关闭通信渠道
        server.close();
        //②释放线程池资源
        pool.shutdownNow();
    }
    ...
}
```

4. 用 HttpServer 类实现 Netty 通信渠道

在 Sentinel 中，主要用 HttpServer 类封装 Netty 的通信 API，具体代码如下：

```
public final class HttpServer {
    private static final int DEFAULT_PORT = 8719;
    private Channel channel;
...
    //①开启 Netty 通信渠道
public void start() throws Exception {
    //②开启 Netty 的 boss 线程池
        EventLoopGroup bossGroup = new NioEventLoopGroup(1);
        //③开启 Netty 的工作线程池
        EventLoopGroup workerGroup = new NioEventLoopGroup();
        try {
            //④实例化 Netty 的通信服务启动类
            ServerBootstrap b = new ServerBootstrap();
            //⑤添加 boss 线程池和工作线程池到 Netty 通信服务启动类中，并添加 NIO 通道类
```

```
NioServerSocketChannel 和通信序列化类 HttpServerInitializer
            b.group(bossGroup, workerGroup)
                .channel(NioServerSocketChannel.class)
                .childHandler(new HttpServerInitializer());
            int port;
            try {
                if (StringUtil.isEmpty(TransportConfig.getPort())) {
                    CommandCenterLog.info("Port not configured, using default
port: " + DEFAULT_PORT);
                    port = DEFAULT_PORT;
                } else {
                    port = Integer.parseInt(TransportConfig.getPort());
                }
            } catch (Exception e) {
                throw new IllegalArgumentException("Illegal port: " +
TransportConfig.getPort());
            }
            int retryCount = 0;
            ChannelFuture channelFuture = null;
            while (true) {
                //⑥在通信渠道开启失败后，每重试 3 次就递增通信端口号
                int newPort = getNewPort(port, retryCount);
                try {
                    //⑦开启指定端口的 Netty 通信的连接
                    channelFuture = b.bind(newPort).sync();
                    TransportConfig.setRuntimePort(newPort);
                    CommandCenterLog.info("[NettyHttpCommandCenter] Begin
listening at port " + newPort);
                    break;
                } catch (Exception e) {
                    TimeUnit.MILLISECONDS.sleep(30);
                    RecordLog.warn("[HttpServer] Netty server bind error,
port={}, retry={}", newPort, retryCount);
                    retryCount ++;
                }
            }
            channel = channelFuture.channel();
            channel.closeFuture().sync();
        } finally {
            workerGroup.shutdownGracefully();
            bossGroup.shutdownGracefully();
        }
    }
}
```

6.4.4 “用 SimpleHttpCommandCenter 类实现通信渠道”的原理

在 Sentinel 中，主要用 SimpleHttpCommandCenter 类实现通信渠道（基于 HTTP）。

1. 初始化 beforeStart()方法，注册通信渠道事件

注册 HTTP 通信渠道事件的具体代码如下：

```
public class SimpleHttpCommandCenter implements CommandCenter {
    //①初始化通信渠道事件的缓存
    private static final Map<String, CommandHandler> handlerMap = new
ConcurrentHashMap<String, CommandHandler>();
    ...
    @Override
    @SuppressWarnings("rawtypes")
public void beforeStart() throws Exception {
    //②加载 HTTP 通信事件的处理类
        Map<String, CommandHandler> handlers =
CommandHandlerProvider.getInstance().namedHandlers();
        //③向 HTTP 通信渠道中注册通信事件
        registerCommands(handlers);
}
//④遍历通信渠道事件的缓存
    public static void registerCommands(Map<String, CommandHandler>
handlerMap) {
        if (handlerMap != null) {
            for (Entry<String, CommandHandler> e : handlerMap.entrySet()) {
                registerCommand(e.getKey(), e.getValue());
            }
        }
    }
    //⑤向缓存中添加通信渠道的事件
    @SuppressWarnings("rawtypes")
    public static void registerCommand(String commandName, CommandHandler
handler) {
        if (StringUtil.isEmpty(commandName)) {
            return;
        }
        if (handlerMap.containsKey(commandName)) {
            CommandCenterLog.warn("Register failed (duplicate command): " +
commandName);
            return;
        }
        handlerMap.put(commandName, handler);
    }
}
```

2. 初始化 start()方法，开启 HTTP 通信渠道对应的连接

开启 HTTP 通信渠道对应的连接的具体代码如下：

```
public class SimpleHttpCommandCenter implements CommandCenter {
    private static final int PORT_UNINITIALIZED = -1;
    private static final int DEFAULT_SERVER_SO_TIMEOUT = 3000;
    private static final int DEFAULT_PORT = 8719;
    @SuppressWarnings("PMD.ThreadPoolCreationRule")
    private ExecutorService executor = Executors.newSingleThreadExecutor(
        new NamedThreadFactory("sentinel-command-center-executor"));
    private ExecutorService bizExecutor;
    private ServerSocket socketReference;

    @Override
  public void start() throws Exception {
    //①根据 CPU 核心数算出最大线程数
        int nThreads = Runtime.getRuntime().availableProcessors();
        //②初始化通信事件处理的线程池
        this.bizExecutor = new ThreadPoolExecutor(nThreads, nThreads, 0L,
TimeUnit.MILLISECONDS,
        new ArrayBlockingQueue<Runnable>(10),
        new
NamedThreadFactory("sentinel-command-center-service-executor"),
        new RejectedExecutionHandler() {
                @Override
                public void rejectedExecution(Runnable r, ThreadPoolExecutor
executor) {
                    CommandCenterLog.info("EventTask rejected");
                    throw new RejectedExecutionException();
                }
            });
        //③初始化 HTTP 通信渠道的任务线程
        Runnable serverInitTask = new Runnable() {
            int port;

            {
                try {
                    port = Integer.parseInt(TransportConfig.getPort());
                } catch (Exception e) {
                    port = DEFAULT_PORT;
                }
            }

            @Override
            public void run() {
```

```
                boolean success = false;
                ServerSocket serverSocket =
getServerSocketFromBasePort(port);
                if (serverSocket != null) {
                    CommandCenterLog.info("[CommandCenter] Begin listening at
port " + serverSocket.getLocalPort());
                    socketReference = serverSocket;
                    //④用线程池 ExecutorService 异步地提交通信事件的处理线程
                    executor.submit(new ServerThread(serverSocket));
                    success = true;
                    port = serverSocket.getLocalPort();
                } else {
                    CommandCenterLog.info("[CommandCenter] chooses port fail,
http command center will not work");
                }

                if (!success) {
                    port = PORT_UNINITIALIZED;
                }

                TransportConfig.setRuntimePort(port);
                executor.shutdown();
            }

        };
        //⑤用线程异步地开启 HTTP 通信渠道的任务线程
        new Thread(serverInitTask).start();
    }
    ...
}
```

HTTP 通信渠道采用的是同步通信。Sentinel 通过线程异步处理请求，提高了通信渠道的吞吐量。通信事件处理线程的具体代码如下：

```
class ServerThread extends Thread {
    private ServerSocket serverSocket;
    ServerThread(ServerSocket s) {
        this.serverSocket = s;
        setName("sentinel-courier-server-accept-thread");
    }
    @Override
    public void run() {
        while (true) {
            Socket socket = null;
            try {
                //①初始化服务器端的套接字类 ServerSocket
```

```
                socket = this.serverSocket.accept();
                //②设置套接字通信的超时时间
                setSocketSoTimeout(socket);
                //③初始化套接字通信事件类 HttpEventTask
                HttpEventTask eventTask = new HttpEventTask(socket);
                //④用处理通信事件的线程池异步地执行线程
                bizExecutor.submit(eventTask);
            } catch (Exception e) {
                CommandCenterLog.info("Server error", e);
                if (socket != null) {
                    try {
                        socket.close();
                    } catch (Exception e1) {
                        CommandCenterLog.info("Error when closing an opened
socket",
                            e1);
                    }
                }
                try {
                    Thread.sleep(10);
                } catch (InterruptedException e1) {
                    break;
                }
            }
        }
    }
}
```

3. 用 stop()方法关闭通信渠道

关闭 HTTP 通信渠道的具体代码如下：

```
public class SimpleHttpCommandCenter implements CommandCenter {
    private static final Map<String, CommandHandler> handlerMap = new
ConcurrentHashMap<String, CommandHandler>();

    @SuppressWarnings("PMD.ThreadPoolCreationRule")
    private ExecutorService executor = Executors.newSingleThreadExecutor(
        new NamedThreadFactory("sentinel-command-center-executor"));
    private ExecutorService bizExecutor;
    private ServerSocket socketReference;
    public void stop() throws Exception {
        if (socketReference != null) {
            try {
                //①释放当前应用中套接字通信对象的线程池资源
                socketReference.close();
```

```
            } catch (IOException e) {
                CommandCenterLog.warn("Error when releasing the server socket",
e);
            }
        }
        //②释放通信事件线程池的资源
        bizExecutor.shutdownNow();
        //③释放线程池的资源
        executor.shutdownNow();
        TransportConfig.setRuntimePort(PORT_UNINITIALIZED);
        //④释放通信渠道事件的缓存资源
        handlerMap.clear();
    }
    ...
}
```

6.4.5 【实例】用 Netty 实现通信渠道，实现“从应用端到 Sentinel 控制台的流量控制规则推送”

本实例的源码在本书配套资源的“chaptersix/ netty-spring-cloud-alibaba-push-rule”目录下。

本实例用 Netty 实现通信渠道，将 Dubbo 接口对应的流量控制规则从应用端推送到 Sentinel 控制台。

1. 创建一个 Spring Cloud Alibaba 项目，并添加 Sentinel

创建一个 Spring Cloud Alibaba 项目，并添加 Sentinel，然后在 pom.xml 文件中添加相关的 Jar 包依赖，具体如下：

```
<dependency>
    <groupId>com.alibaba.cloud</groupId>
    <artifactId>spring-cloud-starter-alibaba-sentinel</artifactId>
</dependency>

<dependency>
    <groupId>com.alibaba.csp</groupId>
    <artifactId>sentinel-apache-dubbo-adapter</artifactId>
</dependency>

<dependency>
    <groupId>com.alibaba.csp</groupId>
    <artifactId>sentinel-transport-netty-http</artifactId>
```

```
    <version>1.8.0</version>
</dependency>
```

2. 在启动类中添加流量控制规则，并将其推送到 Sentinel 控制台

NettySpringCloudAlibabaPushApplication 是一个基于 Spring Boot 的启动类，具体代码如下：

```
@SpringBootApplication
@EnableDiscoveryClient
public class NettySpringCloudAlibabaPushApplication {

public static void main(String[] args) {
    //①用 Netty 实现通信渠道的初始化过程
        InitExecutor.doInit();
        //②开始构造流量控制规则，并推送
        initFlowRule(5,true);
        SpringApplicationBuilder providerBuilder = new
SpringApplicationBuilder();
        providerBuilder.web(WebApplicationType.NONE)
                .sources(NettySpringCloudAlibabaPushApplication.class).run(
args);
    }

    private static final String INTERFACE_RES_KEY =
RuleService.class.getName();
    private static final String RES_KEY = INTERFACE_RES_KEY + ":rulePush()";

    static void initFlowRule(int interfaceFlowLimit, boolean method) {
        //③定义对应 Dubbo 接口的流量控制规则
        FlowRule flowRule = new FlowRule(INTERFACE_RES_KEY)
                .setCount(interfaceFlowLimit)
                .setGrade(RuleConstant.FLOW_GRADE_QPS);
        List<FlowRule> list = new ArrayList<>();
        if (method) {
            FlowRule flowRule1 = new FlowRule(RES_KEY)
                    .setCount(5)
                    .setGrade(RuleConstant.FLOW_GRADE_QPS);
            list.add(flowRule1);
        }
        list.add(flowRule);
        //④调用 Netty 推送流量控制规则
        FlowRuleManager.loadRules(list);
    }
}
```

3. 启动程序并查看流量控制规则

在启动程序后，可以在 Sentinel 控制台中看到应用及对应的流量控制规则，如图 6-8 所示。

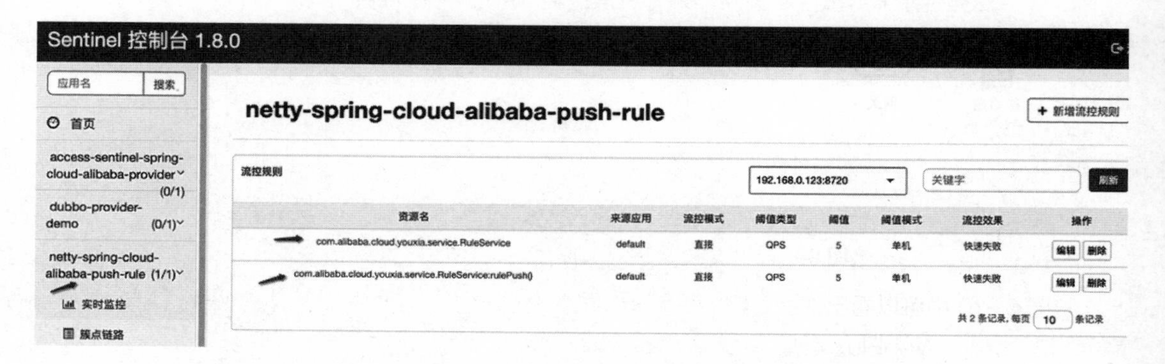

图 6-8

6.5 用过滤器和拦截器实现组件的适配

Sentinel 是嵌入式流量防护组件，它嵌入组件中才能实现流量防护。目前 Sentinel 能够适配主流的组件，包括 Spring Cloud Gateway、Dubbo、gRPC、Spring MVC 等。

Sentinel 主要通过定义对应组件的过滤器和拦截器，来实现对组件的适配。

6.5.1 什么是过滤器和拦截器

在 Java 领域，过滤器和拦截器是两种比较常用的技术。

1. 什么是过滤器

基于函数回调，过滤器可以对几乎所有的请求进行过滤。过滤器实例只能在容器初始化时被调用一次。使用过滤器的目的：做一些过滤操作，获取开发者想要获取的数据，以及处理一些过滤逻辑，如图 6-9 所示。

从图 6-9 中可以看出，过滤器通常用于处理在 Web 服务请求完成之后的 RPC 请求。Sentinel 通过过滤器可以适配 Dubbo、Sofa RPC、Spring Cloud Gateway 等。

2. 什么是拦截器

拦截器是基于特定请求进行拦截处理的。开发人员可以通过配置文件，配置需要拦截器进行拦截的请求路径和白名单路径，如图 6-10 所示。

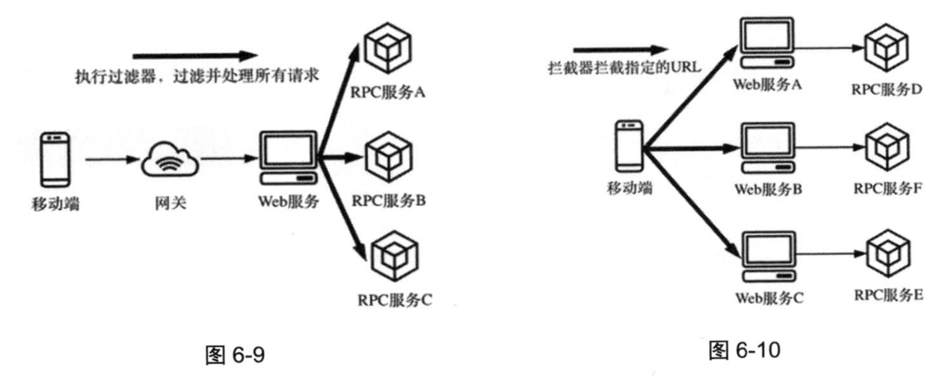

图 6-9　　　　　　　　图 6-10

从图 6-10 中可以看出，拦截器通常是用来拦截 URL 请求的，它通常和 Web 服务（比如 Spring MVC 和 Spring WebFlux 等组件）绑定在一起。

Sentinel 通过拦截器可以适配 Spring MVC、Spring WebFlux 等。

6.5.2 “Sentinel 通过过滤器适配 Dubbo”的原理

Sentinel 通过过滤器适配 Dubbo（同时支持 Apache 和 Alibaba 版本的 Dubbo）。本节只分析 Sentinel 适配 Apache 版本的 Dubbo 的原理。

> Dubbo 从 2.7.0 版本开始，将项目包的根路径从 com.alibaba.dubbo 变更为 org.apache。2019 年 5 月 21 日，Dubbo 框架正式成为 Apache 的顶级项目。

熟悉 Dubbo 框架的开发人员都知道，服务消费者和服务提供者是 Dubbo 中两个非常重要的角色。下面会从服务消费者和服务提供者的角度，去分析适配 Dubbo 的原理。

1. 用 SentinelDubboProviderFilter 类适配服务提供者

Sentinel 用 SentinelDubboProviderFilter 类实现 Dubbo 的 Filter 接口，具体代码如下：

```
@Activate(group = PROVIDER)
public class SentinelDubboProviderFilter extends BaseSentinelDubboFilter {
    ...
    @Override
public Result invoke(Invoker<?> invoker, Invocation invocation) throws
RpcException {
    //①从 Dubbo 的 RPC 请求对象 Invocation 中，解析出变量 dubboApplication 的值
        String origin =
DubboAdapterGlobalConfig.getOriginParser().parse(invoker, invocation);
        //②如果为空，则说明当前 RPC 请求为 Sentinel 适配之后的起点
```

```
        if (null == origin) {
            origin = "";
        }
        Entry interfaceEntry = null;
        Entry methodEntry = null;
        //③获取 Dubbo 服务提供者对应的资源名称的前缀
        String prefix =
DubboAdapterGlobalConfig.getDubboProviderResNamePrefixKey();
        //④用资源名称前缀，获取 Dubbo 接口名称
        String interfaceResourceName = getInterfaceName(invoker, prefix);
        //⑤用资源名称前缀，获取 Dubbo 接口中的方法名称
        String methodResourceName = getMethodName(invoker, invocation,
prefix);
        try {
            //⑥在当前线程上下文中，创建 Sentinel 入口流量的 Context 对象
            ContextUtil.enter(methodResourceName, origin);
            //⑦统计服务提供者的流量（流量类型为 EntryType.IN），统计的维度为接口，并
通过通信渠道将统计结果发送给 Sentinel 控制台进行持久化
            interfaceEntry = SphU.entry(interfaceResourceName,
ResourceTypeConstants.COMMON_RPC, EntryType.IN);
            //⑧统计服务提供者的流量（流量类型为 EntryType.IN），统计的维度为接口中的
方法，并通过通信渠道将统计结果发送给 Sentinel 控制台进行持久化
            methodEntry = SphU.entry(methodResourceName,
ResourceTypeConstants.COMMON_RPC, EntryType.IN,
                invocation.getArguments());
            //⑨执行 RPC 请求
            Result result = invoker.invoke(invocation);
            if (result.hasException()) {
                //⑩如果出现异常，则记录异常日志，并向控制台发送错误请求信息，方便
Sentinel 统计
                Tracer.traceEntry(result.getException(), interfaceEntry);
                Tracer.traceEntry(result.getException(), methodEntry);
            }
            return result;
        } catch (BlockException e) {
            return DubboAdapterGlobalConfig.getProviderFallback().handle
(invoker, invocation, e);
        } catch (RpcException e) {
            Tracer.traceEntry(e, interfaceEntry);
            Tracer.traceEntry(e, methodEntry);
            throw e;
        } finally {
            if (methodEntry != null) {
                methodEntry.exit(1, invocation.getArguments());
            }
```

```
            if (interfaceEntry != null) {
                interfaceEntry.exit();
            }
            ContextUtil.exit();
        }
    }
}
```

2. 用 SentinelDubboConsumerFilter 类适配服务消费者

Sentinel 用 DubboConsumerFilter 类实现了 Dubbo 的 Filter 接口，具体代码如下：

```
@Activate(group = CONSUMER)
public class SentinelDubboConsumerFilter extends BaseSentinelDubboFilter {
    ...
    @Override
public Result invoke(Invoker<?> invoker, Invocation invocation) throws
RpcException {
    //①解析 RPC 请求类型
        InvokeMode invokeMode = RpcUtils.getInvokeMode(invoker.getUrl(),
invocation);
        //②同步请求
        if (InvokeMode.SYNC == invokeMode) {
            return syncInvoke(invoker, invocation);
        } else {
            //③异步请求
            return asyncInvoke(invoker, invocation);
        }
    }
    private Result syncInvoke(Invoker<?> invoker, Invocation invocation) {
        Entry interfaceEntry = null;
        Entry methodEntry = null;
        String prefix =
DubboAdapterGlobalConfig.getDubboConsumerResNamePrefixKey();
        String interfaceResourceName = getInterfaceName(invoker, prefix);
        String methodResourceName = getMethodName(invoker, invocation,
prefix);
        try {
            //④统计服务消费者的流量（流量类型为 EntryType. OUT），统计的维度为接口，
并通过通信渠道将统计结果发送给 Sentinel 控制台进行持久化
            interfaceEntry = SphU.entry(interfaceResourceName,
ResourceTypeConstants.COMMON_RPC, EntryType.OUT);
            //⑤统计服务消费者的流量（流量类型为 EntryType.OUT），统计的维度为接口中的
方法，并通过通信渠道将统计结果发送给 Sentinel 控制台进行持久化
            methodEntry = SphU.entry(methodResourceName,
ResourceTypeConstants.COMMON_RPC, EntryType.OUT,
```

```
                invocation.getArguments());
            //⑥执行 RPC 请求
            Result result = invoker.invoke(invocation);
            if (result.hasException()) {
                Tracer.traceEntry(result.getException(), interfaceEntry);
                Tracer.traceEntry(result.getException(), methodEntry);
            }
            return result;
        } catch (BlockException e) {
            return DubboAdapterGlobalConfig.getConsumerFallback().
handle(invoker, invocation, e);
        } catch (RpcException e) {
            Tracer.traceEntry(e, interfaceEntry);
            Tracer.traceEntry(e, methodEntry);
            throw e;
        } finally {
            if (methodEntry != null) {
                methodEntry.exit(1, invocation.getArguments());
            }
            if (interfaceEntry != null) {
                interfaceEntry.exit();
            }
        }
    }

    private Result asyncInvoke(Invoker<?> invoker, Invocation invocation) {
        LinkedList<EntryHolder> queue = new LinkedList<>();
        String prefix =
DubboAdapterGlobalConfig.getDubboConsumerResNamePrefixKey();
        String interfaceResourceName = getInterfaceName(invoker, prefix);
        String methodResourceName = getMethodName(invoker, invocation,
prefix);
        try {
            //⑦统计服务消费者的流量（流量类型为 EntryType. OUT），统计的维度为接口，
将统计结果发送给缓存队列中
            queue.push(new EntryHolder(
                SphU.asyncEntry(interfaceResourceName,
ResourceTypeConstants.COMMON_RPC, EntryType.OUT), null));
            //⑧统计服务消费者的流量（流量类型为 EntryType. OUT），统计的维度为接口中
的方法，将统计结果发送给缓存队列中
            queue.push(new EntryHolder(
                SphU.asyncEntry(methodResourceName,
ResourceTypeConstants.COMMON_RPC,
                    EntryType.OUT, 1, invocation.getArguments()),
invocation.getArguments()));
```

```
            //⑨执行服务消费者的 RPC 调用
            Result result = invoker.invoke(invocation);
            result.whenCompleteWithContext((r, throwable) -> {
                Throwable error = throwable;
                if (error == null) {
                    error = Optional.ofNullable(r).map
(Result::getException).orElse(null);
                }
                while (!queue.isEmpty()) {
                    EntryHolder holder = queue.pop();
                    Tracer.traceEntry(error, holder.entry);
                    exitEntry(holder);
                }
            });
            //⑩返回 RPC 调用完成的结果
            return result;
        } catch (BlockException e) {
            while (!queue.isEmpty()) {
                exitEntry(queue.pop());
            }
            return DubboAdapterGlobalConfig.getConsumerFallback().handle
(invoker, invocation, e);
        }
    }
}
```

6.5.3 “Sentinel 通过拦截器适配 Spring MVC”的原理

RESTful 请求是服务之间常用的调用方式之一。Sentinel 通过拦截器可以适配 Spring MVC。但是，Sentinel 针对 Spring MVC 框架只定义了拦截器，如果开发者要开启拦截器，则需要在应用中手动添加拦截器并进行初始化。

1. 初始化拦截器的配置

在应用中定义 InterceptorConfig 类，具体代码如下：

```
@Configuration
public class InterceptorConfig implements WebMvcConfigurer {
    @Override
public void addInterceptors(InterceptorRegistry registry) {
    //①添加指定 URL 的 Spring MVC 拦截器
        addSpringMvcInterceptor(registry);
        //②添加全量 URL 的 Spring MVC 拦截器
        addSpringMvcTotalInterceptor(registry);
    }
    private void addSpringMvcInterceptor(InterceptorRegistry registry) {
```

```
            SentinelWebMvcConfig config = new SentinelWebMvcConfig();
            //③设置异常处理器
            config.setBlockExceptionHandler(new BlockExceptionHandler() {
                @Override
                public void handle(HttpServletRequest request,
HttpServletResponse response, BlockException e) throws Exception {
                    //④获取资源名称
                    String resourceName = e.getRule().getResource();
                    //⑤如果资源名称是"/hello"
                    if ("/hello".equals(resourceName)) {
                        //⑥则加载流量控制规则并触发流量防护
                        response.getWriter().write("/Blocked by sentinel");
                    } else {
                        throw e;
                    }
                }
            });
            config.setHttpMethodSpecify(false);
            config.setWebContextUnify(true);
            config.setOriginParser(new RequestOriginParser() {
                @Override
                public String parseOrigin(HttpServletRequest request) {
                    return request.getHeader("S-user");
                }
            });
            //⑦添加拦截器到 Spring MVC，拦截所有的 Web 请求
            registry.addInterceptor(new SentinelWebInterceptor(config)).
addPathPatterns("/**");
        }
        private void addSpringMvcTotalInterceptor(InterceptorRegistry registry)
{
            SentinelWebMvcTotalConfig config = new SentinelWebMvcTotalConfig();
config.setRequestAttributeName("my_sentinel_spring_mvc_total_entity_containe
r");
            config.setTotalResourceName("my_spring_mvc_total_url_request");
            //⑧添加全量 URL 拦截器到 Spring MVC，拦截所有的 Web 请求
            registry.addInterceptor(new SentinelWebTotalInterceptor(config)).
addPathPatterns("/**");
        }
    }
```

2. 用拦截器 SentinelWebInterceptor 类，进行可选 URL 的流量防护

SentinelWebInterceptor 类是 Sentinel 封装的拦截器，具体代码如下：

```
public class SentinelWebInterceptor extends AbstractSentinelInterceptor {
```

```
    private final SentinelWebMvcConfig config;
    public SentinelWebInterceptor() {
        this(new SentinelWebMvcConfig());
    }

    public SentinelWebInterceptor(SentinelWebMvcConfig config) {
        super(config);
        if (config == null) {
            this.config = new SentinelWebMvcConfig();
        } else {
            this.config = config;
        }
    }
    //①获取资源名称
    @Override
    protected String getResourceName(HttpServletRequest request) {
        //②从 HttpServletRequest 类中获取当前请求的资源对象
        Object resourceNameObject = request.getAttribute(HandlerMapping.
BEST_MATCHING_PATTERN_ATTRIBUTE);
        if (resourceNameObject == null || !(resourceNameObject instanceof
String)) {
            return null;
        }
        //③获取资源名称
        String resourceName = (String) resourceNameObject;
        UrlCleaner urlCleaner = config.getUrlCleaner();
        //④如果在应用中通过 UrlCleaner 自定义了统一资源名称的逻辑，则执行相关操作
        if (urlCleaner != null) {
            resourceName = urlCleaner.clean(resourceName);
        }
        if (StringUtil.isNotEmpty(resourceName) &&
config.isHttpMethodSpecify()) {
            resourceName = request.getMethod().toUpperCase() + ":" +
resourceName;
        }
        //⑤返回资源名称
        return resourceName;
    }
    ...
}
```

6.5.4 【实例】将 Spring Cloud Gateway 应用接入 Sentinel，管理流量控制规则

本实例的源码在本书配套资源的“chaptersix/ gateway-access-sentinel”目录下。

Sentinel 可以用过滤器适配基于 Spring Cloud Gateway 的应用。本实例使用 Sentinel 的适配器，将基于 Spring Cloud Gateway 的应用接入 Sentinel，并管理网关的流量控制规则。

1. 创建一个 Spring Cloud Alibaba 项目，并添加相关的 Jar 包

用 Spring Cloud Alibaba 初始化一个项目 gateway-access-sentinel，并添加 Sentinel 和 Spring Cloud Gateway 相关的 Jar 包，具体如下所示：

```
<dependency>
    <groupId>org.springframework.cloud</groupId>
    <artifactId>spring-cloud-starter-gateway</artifactId>
</dependency>
<dependency>
    <groupId>com.alibaba.cloud</groupId>
    <artifactId>spring-cloud-alibaba-sentinel-gateway</artifactId>
</dependency>
<dependency>
    <groupId>com.alibaba.cloud</groupId>
    <artifactId>spring-cloud-starter-alibaba-sentinel</artifactId>
</dependency>
<dependency>
    <groupId>com.alibaba.csp</groupId>
    <artifactId>sentinel-spring-cloud-gateway-adapter</artifactId>
    </dependency>
<dependency>
    <groupId>com.alibaba.csp</groupId>
    <artifactId>sentinel-api-gateway-adapter-common</artifactId>
</dependency>
```

2. 添加配置文件，读取本地的流量防护规则

部分配置内容如下，具体配置信息可以参考本书配套资源中的代码。

```
server:
  port: 18085
spring:
  application:
```

```
    sentinel:
      datasource.ds2.file:
        file: "classpath: gateway.json"
        ruleType: gw-flow
      datasource.ds1.file:
        file: "classpath: api.json"
        ruleType: gw-api-group
      transport:
        dashboard: localhost:8080
      filter:
        enabled: true
      scg.fallback:
          mode: response
          response-status: 444
          response-body: 1234
      scg:
        order: -100
...
```

在上面这段代码中，读取本地文件 api.json 和 gateway.json 中的规则，并将其同步到 Sentinel 控制台中。

3. 启动程序

程序在启动后将模拟用户调用服务接口，在控制台的实时监控模块中我们可以看到应用运行的监控指标，如图 6-11 所示。

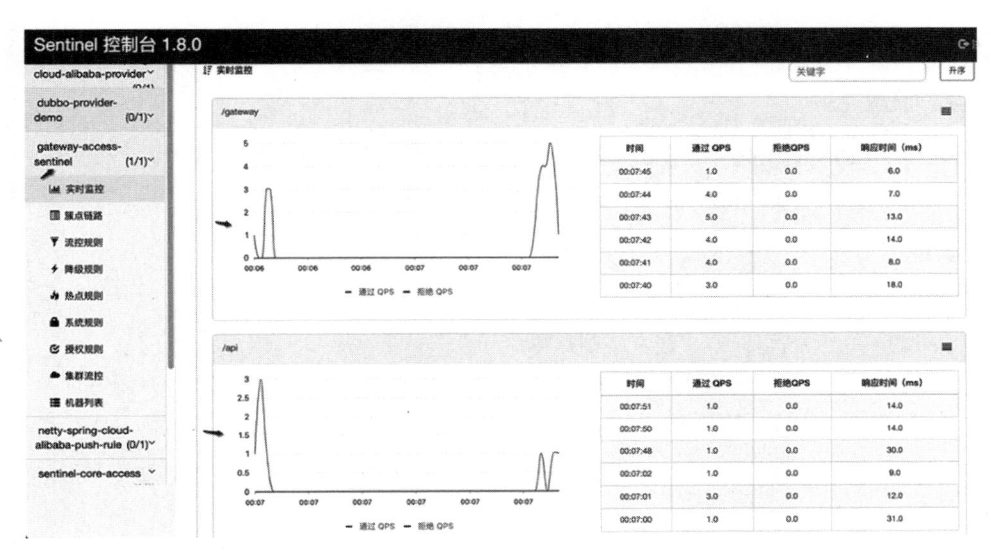

图 6-11

在“簇点链路”模块中能够看到从网关同步过来的簇点规则，如图 6-12 所示。

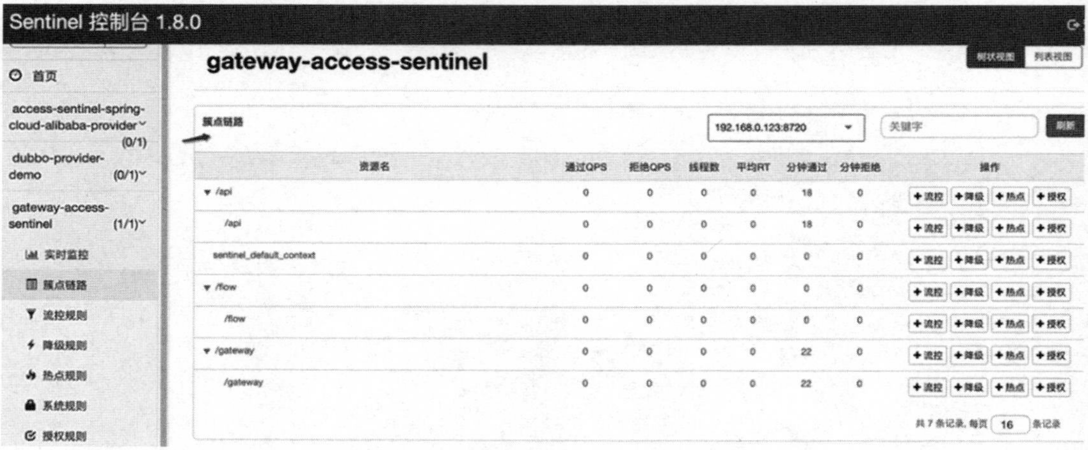

图 6-12

6.6 用“流量控制”实现流量防护

服务之间相互调用会产生流量。如果能够按照流量控制规则有效地控制流量，则可以实现流量防护。

6.6.1 什么是流量控制

1. 认识流量控制

流量控制（Flow Control）是指，根据流量、并发线程数、响应时间等指标，把随机到来的流量调整成合适的形状（即流量塑形），以避免应用被瞬时的流量高峰冲垮，从而保障应用的高可用性。

流量控制有以下几个角度。

- 运行指标：例如 QPS、并发线程数等。
- 资源的调用关系：例如资源的调用链路、资源和资源之间的关系及调用来源等。
- 控制效果：例如直接拒绝、Warm Up（预热）、排队等待等。

一条流量控制规则主要由以下几个元素组成，开发人员可以组合这些元素来实现不同的限流效果。

- 资源名：流量控制规则的作用对象。

- 阈值：流量控制阈值。
- 阈值类型：流量控制阈值类型（QPS 或并发线程数）。
- 来源应用：流量控制规则针对的调用来源，若为 default 则不区分调用来源。
- 流控模式：包括直接、关联及链路限流策略。
- 流控效果：包括快速失败、Warm Up 及排队等待模式。

2. 用 Sentinel 控制台新增流量控制规则

（1）在 Sentinel 控制台中单击“流控规则”菜单，输入资源名称为“/v1/flow/rules”，然后单击“新建”按钮，如图 6-13 所示。

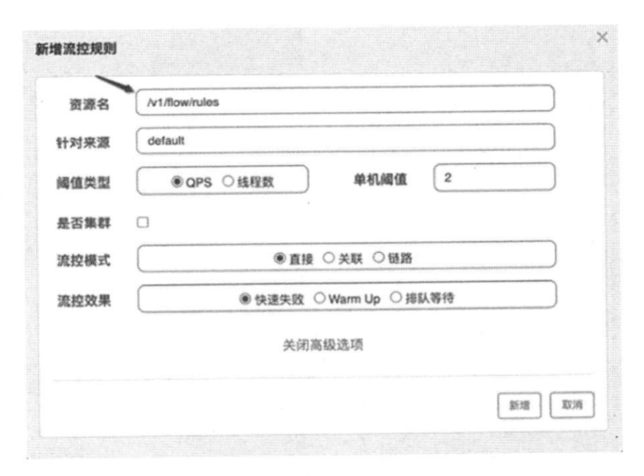

图 6-13

（2）新增具体的流量控制规则——将资源名称“/v1/flow/rules”的 QPS 的阈值设置为 4，如图 6-14 所示。如果 QPS 值超过 4，则资源会快速失败。

图 6-14

6.6.2 槽位（Slot）的动态加载机制

Sentinel 通过流量控制槽位类 FlowSlot 来校验流量控制规则。FlowSlot 类实现了 Sentinel 的接口 ProcessorSlot。

在 Sentinel 中，与流量防护相关的功能都是通过 Slot 的动态加载机制进行初始化的。Slot 的动态加载机制如图 6-15 所示。

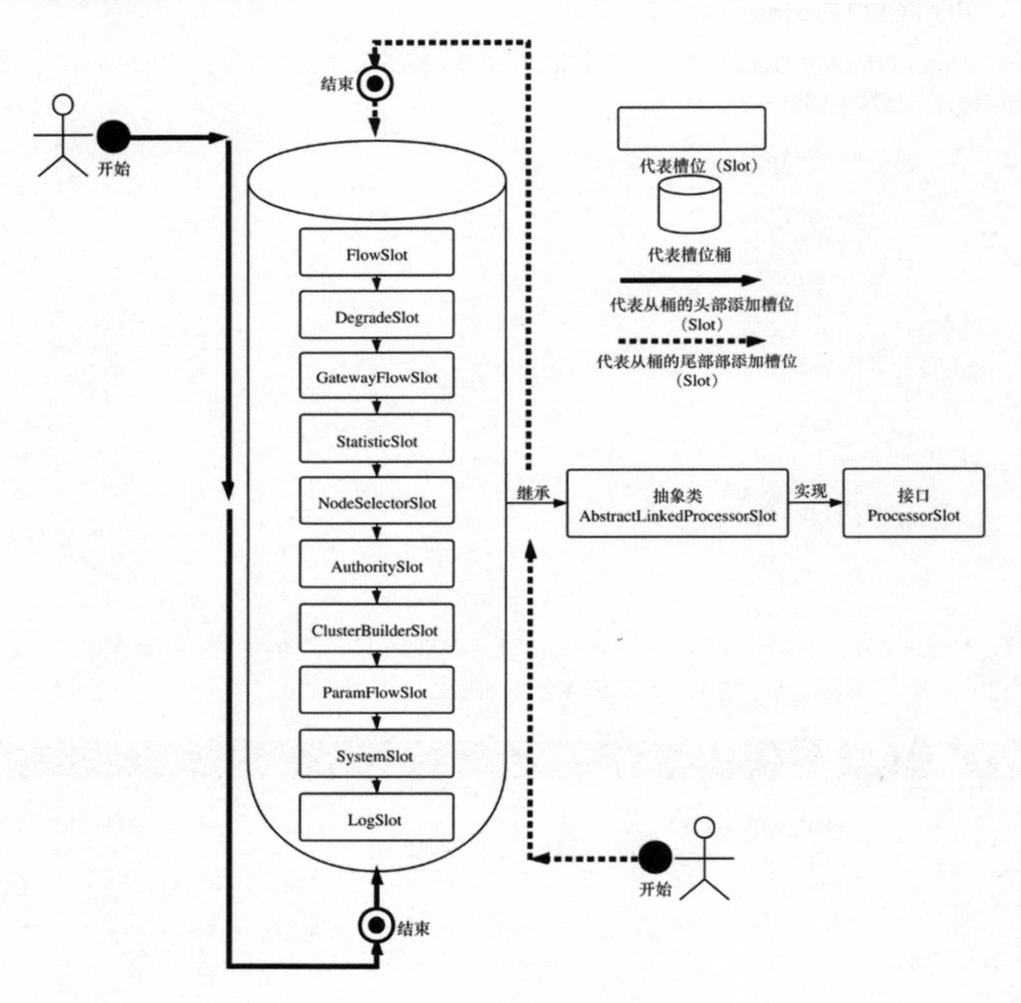

图 6-15

- Sentinel 定义了若干类型的槽位，比如 FlowSlot（流量控制规则槽位）、DegradeSlot（熔断降级槽位）、GatewayFlowSlot（网关流量控制槽位）、StatisticSlot（运行指标统计槽位）等。

- Sentinel 用槽位桶装载槽位。在槽位桶中，槽位会按照预先定义好的优先级顺序，完成入桶和出桶操作。
- ProcessorSlot 接口是所有槽位的父类。Sentinel 用 SPI 机制，将 ProcessorSlot 接口的实现类装载到槽位桶中。
- 应用在接入 Sentinel 后，会从槽位桶中按顺序取出槽位，执行对应槽位聚合的流量防护逻辑。

1. 用扩展接口 ProcessorSlot 实现 Slot 的动态加载

Sentinel 默认通过 DefaultSlotChainBuilder 类用 SPI 机制加载接口 ProcessorSlot，加载过程的具体代码如下所示：

```
public class DefaultSlotChainBuilder implements SlotChainBuilder {
    @Override
public ProcessorSlotChain build() {
    //①构造槽位链式处理对象 ProcessorSlotChain
        ProcessorSlotChain chain = new DefaultProcessorSlotChain();
        //②通过 SPI 机制加载扩展接口 ProcessorSlot 的所有实现类，返回排序后的列表（包含流量控制规则槽位类 FlowSlot），具体排序规则会在 6.6.3 节介绍
        List<ProcessorSlot> sortedSlotList =
SpiLoader.loadPrototypeInstanceListSorted(ProcessorSlot.class);
        //③循环遍历 ProcessorSlot 中的所有实现类的列表
        for (ProcessorSlot slot : sortedSlotList) {
            //④ProcessorSlot 的实现类需要继承抽象模板类，否则直接返回
            if (!(slot instanceof AbstractLinkedProcessorSlot)) {
                RecordLog.warn("The ProcessorSlot(" +
slot.getClass().getCanonicalName() + ") is not an instance of
AbstractLinkedProcessorSlot, can't be added into ProcessorSlotChain");
                continue;
            }
            //⑤将扩展接口 ProcessorSlot 的实现类添加到槽位的链式处理对象中
            chain.addLast((AbstractLinkedProcessorSlot<?>) slot);
        }
        return chain;
    }
}
```

2. 将槽位（Slot）添加到槽位桶中

Sentinel 用 DefaultProcessorSlotChain 类封装槽位桶，具体代码如下：

```
public class DefaultProcessorSlotChain extends ProcessorSlotChain {
    //①定义槽位桶的头部
AbstractLinkedProcessorSlot <?>first = new AbstractLinkedProcessorSlot <
Object > () {
```

```
    @Override public void entry(Context context, ResourceWrapper resourceWrapper,
Object t, int count, boolean prioritized, Object...args) throws Throwable {
                super.fireEntry(context, resourceWrapper, t, count, prioritized,
args);
            }

            @Override public void exit(Context context, ResourceWrapper
resourceWrapper, int count, Object...args) {
                super.fireExit(context, resourceWrapper, count, args);
            }
    };
    //②定义槽位桶的尾部
        AbstractLinkedProcessorSlot <?>end = first;
        //③向槽位桶的头部添加槽位
        @Override public void addFirst(AbstractLinkedProcessorSlot
<?>protocolProcessor) {
            protocolProcessor.setNext(first.getNext());
            first.setNext(protocolProcessor);
            if (end == first) {
                end = protocolProcessor;
            }
        }
        //④向槽位桶的尾部添加槽位
        @Override public void addLast(AbstractLinkedProcessorSlot
<?>protocolProcessor) {
            end.setNext(protocolProcessor);
            end = protocolProcessor;
        }
        //⑤向槽位链表中添加槽位
        @Override public void setNext(AbstractLinkedProcessorSlot < ?>next) {
            addLast(next);
        }
        //⑥获取槽位头部中的下一个槽位
        @Override public AbstractLinkedProcessorSlot <?>getNext() {
            return first.getNext();
        }
        //⑦进入槽位
        @Override public void entry(Context context, ResourceWrapper
resourceWrapper, Object t, int count, boolean prioritized, Object...args) throws
Throwable {
            first.transformEntry(context, resourceWrapper, t, count, prioritized,
args);
        }
        //⑧从槽位中出去
```

```
    @Override public void exit(Context context, ResourceWrapper
resourceWrapper, int count, Object...args) {
        first.exit(context, resourceWrapper, count, args);
    }
}
```

3. 实现槽位桶的链式存储

Sentinel 的槽位桶用链表来实现其链式存储，具体代码如下：

```
public abstract class AbstractLinkedProcessorSlot <T> implements
ProcessorSlot <T> {
    //①定义链表指针
    private AbstractLinkedProcessorSlot <?>next = null;

    @Override public void fireEntry(Context context, ResourceWrapper
resourceWrapper, Object obj, int count, boolean prioritized, Object...args)
throws Throwable {
        if (next != null) {
            next.transformEntry(context, resourceWrapper, obj, count,
prioritized, args);
        }
    }
    //②进入槽位
    @SuppressWarnings("unchecked") void transformEntry(Context context,
ResourceWrapper resourceWrapper, Object o, int count, boolean prioritized,
Object...args) throws Throwable {
        T t = (T) o;
        entry(context, resourceWrapper, t, count, prioritized, args);
    }
    //③从槽位中出去
    @Override public void fireExit(Context context, ResourceWrapper
resourceWrapper, int count, Object...args) {
        if (next != null) {
            next.exit(context, resourceWrapper, count, args);
        }
    }

    public AbstractLinkedProcessorSlot <?>getNext() {
        return next;
    }

    public void setNext(AbstractLinkedProcessorSlot <?>next) {
        this.next = next;
    }

}
```

通过 Slot 的动态加载，流量控制规则对应的槽位 FlowSlo 就被加载到槽位桶中了。

6.6.3 “加载应用运行的监控指标”的原理

对 Sentinel 实现流控，必须先获取应用运行的监控指标，然后加载指定资源的流量控制规则，用流量控制规则来校验运行的监控指标。如果达到阈值，则执行流量防控。

1. 排序槽位（Slot）

槽位（Slot）在槽位桶中是有顺序的。比如 FlowSlot 类的代码如下：

```
    //通过注解@SpiOrder 控制槽位的顺序
    @SpiOrder(-2000)
    public class FlowSlot extends AbstractLinkedProcessorSlot<DefaultNode> {
        ...
    }
```

Sentinel 在 SpiLoader 类中加载 SPI 扩展类时，先解析注解@SpiOrder 中的排序值，然后按照排序值进行排序。具体代码如下：

```
    //①SpiLoader 类的内部类
    private static class SpiOrderResolver {
        //②按照顺序插入
    private static <T> void insertSorted(List < SpiOrderWrapper <T>> list, T spi,
int order) {
        //③循环遍历链表
            int idx = 0;
            for (; idx < list.size(); idx++) {
                //④如果链表中槽位的排序值大于需要插入的槽位的排序值，则退出循环
                if (list.get(idx).getOrder() > order) {
                    break;
                }
            }
            //⑤将槽位直接插入下标为 idx+1 的位置
            list.add(idx, new SpiOrderWrapper <>(order, spi));
        }

    private static <T> int resolveOrder(T spi) {
        //⑥如果槽位没有使用注解@SpiOrder，则返回默认排序值
SpiOrder.LOWEST_PRECEDENCE
            if (!spi.getClass().isAnnotationPresent(SpiOrder.class)) {
                return SpiOrder.LOWEST_PRECEDENCE;
            } else {
                //⑦返回注解@SpiOrder 中的排序值
                return spi.getClass().getAnnotation(SpiOrder.class).value();
            }
```

```
        }
    }
    private static class SpiOrderWrapper <T> {
        private final int order;
        private final T spi;
        SpiOrderWrapper(int order, T spi) {
            this.order = order;
            this.spi = spi;
        }
        int getOrder() {
            return order;
        }
        T getSpi() {
            return spi;
        }
    }
```

SpiLoader 类中的 loadPrototypeInstanceListSorted()方法会调用排序类，具体代码如下：

```
    public static <T> List <T> loadPrototypeInstanceListSorted(Class <T> clazz) {
    try {
        //①加载指定接口的所有实现类
            ServiceLoader<T> serviceLoader =
ServiceLoaderUtil.getServiceLoader(clazz);
            List < SpiOrderWrapper <T>> orderWrappers = new ArrayList <>();
            for (T spi: serviceLoader) {
                int order = SpiOrderResolver.resolveOrder(spi);
                //②执行排序
                SpiOrderResolver.insertSorted(orderWrappers, spi, order);
                RecordLog.debug("[SpiLoader] Found {} SPI: {} with order {}",
clazz.getSimpleName(), spi.getClass().getCanonicalName(), order);
            }
            List <T> list = new ArrayList <>(orderWrappers.size());
            for (int i = 0; i < orderWrappers.size(); i++) {
                list.add(orderWrappers.get(i).spi);
            }
            return list;
        } catch(Throwable t) {
            RecordLog.error("[SpiLoader] ERROR: loadPrototypeInstanceListSorted
failed", t);
            t.printStackTrace();
            return new ArrayList <>();
        }
    }
```

2. 在 FlowSlot 槽位执行之前，获取应用运行的监控指标

Sentinel 用 NodeSelectorSlot、ClusterBuilderSlot 和 StatistcSlot 槽位来统计应用运行的实时监控指标信息，用注解@SpiOrder 来控制它们的顺序，具体代码如下：

```
    //排序值 NodeSelectorSlot<ClusterBuilderSlot<StatisticSlot<FlowSlot
    //槽位桶加载槽位的优先级从高到低：NodeSelectorSlot、ClusterBuilderSlot、
StatisticSlot 和 FlowSlot
    @SpiOrder(-10000)
    public class NodeSelectorSlot extends AbstractLinkedProcessorSlot<Object> {
        ...
    }
    @SpiOrder(-9000)
    public class ClusterBuilderSlot extends
AbstractLinkedProcessorSlot<DefaultNode> {
        ...
    }
    @SpiOrder(-7000)
    public class StatisticSlot extends AbstractLinkedProcessorSlot<DefaultNode> {
       ...
    }
    @SpiOrder(-2000)
    public class FlowSlot extends AbstractLinkedProcessorSlot<DefaultNode> {
       ...
    }
```

综上所述，在槽位桶加载流量控制规则对应的 FlowSlot 槽位时，应用运行的实时监控指标信息已被优先加载到内存中了。

6.6.4 “用 QPS/并发线程数实现流量控制”的原理

在流量控制规则类 FlowRule 中，用字段“grade”来定义阈值类型，具体代码如下：

```
    //阈值类型默认为 QPS，可以设置阈值类型为 FLOW_GRADE_THREAD（并发线程数）
    private int grade = RuleConstant.FLOW_GRADE_QPS;
```

（1）QPS 流量控制是指，当 QPS 值超过某个阈值时，采取措施进行流量控制。流量控制的手段包括下面 4 种，对应于 FlowRule 中的 controlBehavior 字段。

- 直接拒绝（RuleConstant.CONTROL_BEHAVIOR_DEFAULT）。这种方式是默认的流量控制方式。当 QPS 值超过规则的阈值后，新的请求会被立即拒绝，拒绝方式为抛出 FlowException。这种方式适用于对系统处理能力确切已知的情况下，比如通过压测确定了系统的准确水位。
- 冷启动（RuleConstant.CONTROL_BEHAVIOR_WARM_UP）。这种方式主要用于系统长期处于低水位的情况下，当流量突然增加时，直接把系统拉升到高水位。

- 匀速器（RuleConstant.CONTROL_BEHAVIOR_RATE_LIMITER）。这种方式严格控制请求通过的间隔时间，即让请求以均匀的速度通过，对应的是漏桶算法。这种方式主要用于处理间隔性突发流量，例如消息队列。想象一下这样的场景：系统在某一秒有大量的请求到来，而接下来的几秒处于空闲状态，开发人员希望系统能够在接下来的空闲期间逐渐处理这些请求，而不是在第 1 秒直接拒绝多余的请求。
- 匀速器和冷启动的混合模式（RuleConstant.CONTROL_BEHAVIOR_RATE_LIMITER）。

（2）并发线程数流量控制，可以保护业务线程数不被耗尽。例如，当应用所依赖的下游应用由于某种原因导致服务不稳定、响应延迟增加，对于调用者来说，这意味着吞吐量下降和更多的线程数占用（极端情况下甚至导致线程池耗尽）。

为应对高线程占用的情况，业内有使用隔离的方案，比如，通过不同业务逻辑使用不同线程池来隔离业务之间的资源争抢（线程池隔离），或者使用信号量来控制同时请求的个数（信号量隔离）。这些隔离方案虽然能够控制线程数量，但无法控制请求排队时间。当请求过多时，排队也是无益的，直接拒绝能够迅速降低系统压力。Sentinel 线程数流量控制不负责创建和管理线程池，而是简单统计当前请求上下文的线程个数，如果超出阈值，则新的请求会被立即拒绝。

1. 用 FlowSlot 类触发流量控制规则的校验

在 FlowSlot 类被初始化后，Sentinel 用它的 entry()方法校验流量控制规则，具体代码如下：

```
@SpiOrder(-2000)
public class FlowSlot extends AbstractLinkedProcessorSlot<DefaultNode> {
    //①初始化流量控制规则校验类 FlowRuleChecker
    private final FlowRuleChecker checker;
    public FlowSlot() {
        this(new FlowRuleChecker());
    }
    FlowSlot(FlowRuleChecker checker) {
        AssertUtil.notNull(checker, "flow checker should not be null");
        this.checker = checker;
    }
        @Override
        public void entry(Context context, ResourceWrapper resourceWrapper,
DefaultNode node, int count,
                    boolean prioritized, Object... args) throws Throwable {
            //②校验流量控制规则
            checkFlow(resourceWrapper, context, node, count, prioritized);
            fireEntry(context, resourceWrapper, node, count, prioritized, args);
    }
        void checkFlow(ResourceWrapper resource, Context context, DefaultNode
node, int count, boolean prioritized)
```

```
        throws BlockException {
        //③用流量控制规则校验类 FlowRuleChecker 校验流量控制规则
        checker.checkFlow(ruleProvider, resource, context, node, count,
prioritized);
    }
    private final Function<String, Collection<FlowRule>> ruleProvider = new
Function<String, Collection<FlowRule>>() {
        @Override
        public Collection<FlowRule> apply(String resource) {
            //④加载对应资源的流量控制规则
            Map<String, List<FlowRule>> flowRules =
FlowRuleManager.getFlowRuleMap();
            return flowRules.get(resource);
        }
    };
    ...
}
```

2. 用 FlowRuleChecker 类校验流量控制规则

FlowRuleChecker 类通过 checkFlow()方法处理校验请求，具体代码如下：

```
    public void checkFlow(Function < String, Collection < FlowRule >>
ruleProvider,  ResourceWrapper resource, Context context, DefaultNode node, int
count, boolean prioritized) throws BlockException {
        if (ruleProvider == null || resource == null) {
            return;
    }
    //①通过函数变量 ruleProvider 的回调方法 apply()获取指定资源的流量控制规则
        Collection < FlowRule > rules = ruleProvider.apply(resource.getName());
        if (rules != null) {
            for (FlowRule rule: rules) {
                //②遍历规则，并校验流量控制规则
                if (!canPassCheck(rule, context, node, count, prioritized)) {
                    throw new FlowException(rule.getLimitApp(), rule);
                }
            }
    }
    }
    public boolean canPassCheck(
    /*@NonNull*/
    FlowRule rule, Context context, DefaultNode node, int acquireCount, boolean
prioritized) {
        String limitApp = rule.getLimitApp();
        if (limitApp == null) {
            return true;
```

```
        }
    if (rule.isClusterMode()) {
        //③校验集群模式下的流量控制规则
            return passClusterCheck(rule, context, node, acquireCount, 
prioritized);
    }
    //④校验单机模式下的流量控制规则
        return passLocalCheck(rule, context, node, acquireCount, prioritized);
    }
```

3. 获取单机模式下需要流量防护的应用节点

获取单机模式下需要流量防护的应用节点的具体代码如下：

```
    private static boolean passLocalCheck(FlowRule rule, Context context, 
DefaultNode node, int acquireCount, boolean prioritized) {
        //①获取需要流量防护的应用节点
        Node selectedNode = selectNodeByRequesterAndStrategy(rule, context, 
node);
        if (selectedNode == null) {
            return true;
    }
    //②调用单机流量控制整形器，依据应用的运行指标校验流量控制规则
        return rule.getRater().canPass(selectedNode, acquireCount, 
prioritized);
    }

    static Node selectNodeByRequesterAndStrategy(
    FlowRule rule, Context context, DefaultNode node) {
    String limitApp = rule.getLimitApp();
    //③获取流量控制规则的流量控制模式
        int strategy = rule.getStrategy();
        String origin = context.getOrigin();
    if (limitApp.equals(origin) && filterOrigin(origin)) {
            if (strategy == RuleConstant.STRATEGY_DIRECT) {
                return context.getOriginNode();
            }
            return selectReferenceNode(rule, context, node);
        } else if (RuleConstant.LIMIT_APP_DEFAULT.equals(limitApp)) {
            if (strategy == RuleConstant.STRATEGY_DIRECT) {
                return node.getClusterNode();
            }
            return selectReferenceNode(rule, context, node);
        }else if(RuleConstant.LIMIT_APP_OTHER.equals(limitApp) && 
FlowRuleManager.isOtherOrigin(origin, rule.getResource())) {
            if (strategy == RuleConstant.STRATEGY_DIRECT) {
```

```
            return context.getOriginNode();
        }
        //④返回需要流量防护的应用节点
        return selectReferenceNode(rule, context, node);
    }
    return null;
}
```

4. 用流量整形器校验单机模式的流量控制规则

在单机模式下，Sentinel 定义了流量整形器的接口 TrafficShapingController。开发人员可以用 TrafficShapingController 接口来校验流量控制规则，并完成流量整形，具体代码如下：

```
public final class FlowRuleUtil {
private static TrafficShapingController generateRater(FlowRule rule) {
    //①基于 QPS 的流量控制规则，返回对应流量整形器
        if (rule.getGrade() == RuleConstant.FLOW_GRADE_QPS) {
            switch (rule.getControlBehavior()) {
                //②如果流量控制效果为冷启动，则返回整形器 WarmUpController
                case RuleConstant.CONTROL_BEHAVIOR_WARM_UP:
                    return new WarmUpController(rule.getCount(),
rule.getWarmUpPeriodSec(),
                        ColdFactorProperty.coldFactor);
                //③如果流量控制效果为匀速器，则返回整形器 RateLimiterController
                case RuleConstant.CONTROL_BEHAVIOR_RATE_LIMITER:
                    return new
RateLimiterController(rule.getMaxQueueingTimeMs(), rule.getCount());
                //④如果流量控制效果为匀速器和冷启动的混合模式，则返回整形器
WarmUpRateLimiterController
                case RuleConstant.CONTROL_BEHAVIOR_WARM_UP_RATE_LIMITER:
                    return new WarmUpRateLimiterController(rule.getCount(),
rule.getWarmUpPeriodSec(),
                        rule.getMaxQueueingTimeMs(),
ColdFactorProperty.coldFactor);
                //⑤如果流量控制效果为直接拒绝，则不返回流量整形器，直接失败
                case RuleConstant.CONTROL_BEHAVIOR_DEFAULT:
                default:
            }
        }
        //⑥基于并发线程数的流量控制规则，返回默认的流量整形器 DefaultController 类
        return new DefaultController(rule.getCount(), rule.getGrade());
    }
    ...
}
```

Sentinel 用不同的流量整形器来校验流量控制规则，具体原理可以参考 Sentinel 对应的流量整

形器，这里就不展开了。

5. 用 TokenService 类校验集群模式下的流量控制规则

用 TokenService 类校验集群模式下的流量控制规则的具体代码如下：

```
    private static boolean passClusterCheck(FlowRule rule, Context context,
DefaultNode node, int acquireCount, boolean prioritized) {
    try {
        //①获取集群模式下的 TokenService 类
            TokenService clusterService = pickClusterService();
            if (clusterService == null) {
                return fallbackToLocalOrPass(rule, context, node, acquireCount,
prioritized);
            }
            //②从“QPS/并发线程数”规则中获取规则 ID
            long flowId = rule.getClusterConfig().getFlowId();
            //③校验集群模式下的规则
            TokenResult result = clusterService.requestToken(flowId,
acquireCount, prioritized);
            return applyTokenResult(result, rule, context, node, acquireCount,
prioritized);
        } catch(Throwable ex) {
            RecordLog.warn("[FlowRuleChecker] Request cluster token unexpected
failed", ex);
        }
        return fallbackToLocalOrPass(rule, context, node, acquireCount,
prioritized);
    }

    private static boolean fallbackToLocalOrPass(FlowRule rule, Context context,
DefaultNode node, int acquireCount, boolean prioritized) {
        if (rule.getClusterConfig().isFallbackToLocalWhenFail()) {
            return passLocalCheck(rule, context, node, acquireCount,
prioritized);
        } else {
            return true;
        }
    }
    //④返回基于“QPS/并发线程数”的流量控制规则校验之后的结果
    private static boolean applyTokenResult(
    TokenResult result, FlowRule rule, Context context, DefaultNode node, int
acquireCount, boolean prioritized) {
        switch (result.getStatus()) {
        case TokenResultStatus.OK:
            return true;
```

```
        case TokenResultStatus.SHOULD_WAIT:
            try {
                Thread.sleep(result.getWaitInMs());
            } catch(InterruptedException e) {
                e.printStackTrace();
            }
            return true;
        case TokenResultStatus.NO_RULE_EXISTS:
        case TokenResultStatus.BAD_REQUEST:
        case TokenResultStatus.FAIL:
        case TokenResultStatus.TOO_MANY_REQUEST:
            return fallbackToLocalOrPass(rule, context, node, acquireCount,
prioritized);
        case TokenResultStatus.BLOCKED:
        default:
            return false;
        }
    }
```

6.6.5 “用调用关系实现流量控制”的原理

调用关系包括调用方和被调用方，方法又可能会调用其他方法，形成一个调用链路的层次关系。Sentinel 通过 NodeSelectorSlot 槽位类建立不同资源间的调用关系，并且通过 ClusterBuilderSlot 槽位类记录每个资源的实时统计信息。有了调用链路的统计信息，开发人员就可以衍生出多种流量控制手段。

1. 用 NodeSelectorSlot 槽位类关联服务的调用方与被调用方

Sentinel 用槽位类 NodeSelectorSlot 关联服务的调用方与被调用方，并形成拓扑调用关系的具体代码如下：

```
    //设置优先级
    @SpiOrder(-10000)
    public class NodeSelectorSlot extends AbstractLinkedProcessorSlot<Object> {
        private volatile Map<String, DefaultNode> map = new HashMap<String,
DefaultNode>(10);
        @Override
        public void entry(Context context, ResourceWrapper resourceWrapper,
Object obj, int count, boolean prioritized, Object... args)
            throws Throwable {
            DefaultNode node = map.get(context.getName());
            if (node == null) {
                synchronized (this) {
                    node = map.get(context.getName());
                    if (node == null) {
```

```
                    node = new DefaultNode(resourceWrapper, null);
                    HashMap<String, DefaultNode> cacheMap = new HashMap<String,
DefaultNode>(map.size());
                    cacheMap.putAll(map);
                    cacheMap.put(context.getName(), node);
                    map = cacheMap;
                    ((DefaultNode) context.getLastNode()).addChild(node);
                }
            }
        }
        context.setCurNode(node);
        fireEntry(context, resourceWrapper, node, count, prioritized, args);
    }
    ...
}
```

2. 用 ClusterBuilderSlot 槽位类实时统计链路信息

Sentinely 用槽位类 ClusterBuilderSlot 实时地统计链路信息的具体代码如下：

```
//设置优先级
@SpiOrder(-9000)
public class ClusterBuilderSlot extends
AbstractLinkedProcessorSlot<DefaultNode> {
    private static volatile Map<ResourceWrapper, ClusterNode> clusterNodeMap
= new HashMap<>();
    private static final Object lock = new Object();
    private volatile ClusterNode clusterNode = null;

    @Override
    public void entry(Context context, ResourceWrapper resourceWrapper,
DefaultNode node, int count,
                     boolean prioritized, Object... args)
        throws Throwable {
        if (clusterNode == null) {
            synchronized (lock) {
                if (clusterNode == null) {
                    clusterNode = new ClusterNode(resourceWrapper.getName(),
resourceWrapper.getResourceType());
                    HashMap<ResourceWrapper, ClusterNode> newMap = new
HashMap<>(Math.max(clusterNodeMap.size(), 16));
                    newMap.putAll(clusterNodeMap);
                    newMap.put(node.getId(), clusterNode);

                    clusterNodeMap = newMap;
                }
```

```
                }
            }
            node.setClusterNode(clusterNode);
            if (!"".equals(context.getOrigin())) {
                Node originNode =
node.getClusterNode().getOrCreateOriginNode(context.getOrigin());
                context.getCurEntry().setOriginNode(originNode);
            }

            fireEntry(context, resourceWrapper, node, count, prioritized, args);
        }
        ...
    }
```

槽位类 NodeSelectorSlot 的优先级要高于槽位类 ClusterBuilderSlot。先有调用链路关系，才会有基于链路的统计信息。

6.6.6 【实例】通过控制台实时地修改 QPS 验证组件的流量防控

本实例的源码在本书配套资源的“chaptersix/ verify-flow-rule-qps-thread”目录下。

本实例先在应用端配置 Dubbo 接口资源的流量控制规则，然后将流量控制规则同步到 Sentinel 控制台中。

1. 用 Spring Cloud Alibaba 创建项目 verify-flow-rule-qps-thread

创建项目 verify-flow-rule-qps-thread，并定义流量控制规则，具体代码如下：

```
@SpringBootApplication
@EnableDiscoveryClient
public class VerifyFlowRuleApplication {
    public static void main(String[] args) {
        InitExecutor.doInit();
        //①设置 QPS 的阈值为 5
        initFlowRule(5, true);
        SpringApplication.run(VerifyFlowRuleApplication.class, args);
    }
    private static final String INTERFACE_RES_KEY =
VerifyFlowService.class.getName();
    private static final String RES_KEY = INTERFACE_RES_KEY + ":verifyFlow()";
```

```
    static void initFlowRule(int interfaceFlowLimit, boolean method) {
        FlowRule flowRule = new FlowRule(INTERFACE_RES_KEY)
                .setCount(interfaceFlowLimit)
                .setGrade(RuleConstant.FLOW_GRADE_QPS);
        List<FlowRule> list = new ArrayList<>();
        if (method) {
            FlowRule flowRule1 = new FlowRule(RES_KEY)
                    .setCount(5)
                    .setGrade(RuleConstant.FLOW_GRADE_QPS);
            list.add(flowRule1);
        }
        list.add(flowRule);
        //②同步流量控制规则到 Sentinel 控制台中
        FlowRuleManager.loadRules(list);
    }
}
```

2. 启动程序

在程序运行后，在 Sentinel 控制台中可以看到应用已经接入，如图 6-16 所示。

图 6-16

基于 QPS 的流量控制规则是：QPS 的初始阈值为 5，流量控制模式为“直接失败”，流量控制效果为“快速失败”。

3. 模拟场景一：验证 QPS 的初始阈值 5

用命令“curl 127.0.0.1:7867/verify”模拟并发请求调用 Dubbo 接口资源 VerifyFlowService。

当并发请求超过阈值时，应用端会打印流量控制防护的错误日志，如图 6-17 所示。

```
Debug:  VerifyFlowRuleApplication
Debugger  Console | Endpoints
Console  Endpoints
2021-04-16 01:36:03.897 ERROR 51969 --- [nio-7867-exec-4] o.a.c.c.C.[.[.[/].[dispatcherServlet]    : Servlet.service() for servlet [dispatcherServlet] in context wi

java.lang.RuntimeException Create breakpoint : SentinelBlockException: FlowException
    at com.alibaba.csp.sentinel.slots.block.BlockException.block(BlockException.java:0) ~[sentinel-core-1.8.0.jar:1.8.0]

2021-04-16 01:36:04.263 ERROR 51969 --- [nio-7867-exec-7] o.a.c.c.C.[.[.[/].[dispatcherServlet]    : Servlet.service() for servlet [dispatcherServlet] in context wi

java.lang.RuntimeException Create breakpoint : SentinelBlockException: FlowException
    at com.alibaba.csp.sentinel.slots.block.BlockException.block(BlockException.java:0) ~[sentinel-core-1.8.0.jar:1.8.0]

2021-04-16 01:36:04.464 ERROR 51969 --- [nio-7867-exec-1] o.a.c.c.C.[.[.[/].[dispatcherServlet]    : Servlet.service() for servlet [dispatcherServlet] in context wi

java.lang.RuntimeException Create breakpoint : SentinelBlockException: FlowException
    at com.alibaba.csp.sentinel.slots.block.BlockException.block(BlockException.java:0) ~[sentinel-core-1.8.0.jar:1.8.0]
```

图 6-17

Sentinel 控制台会统计应用运行的监控指标，如图 6-18 所示。可以看到，超过阈值（QPS 的阈值为 5）的请求都被 Sentinel 直接拦截，并快速失败。

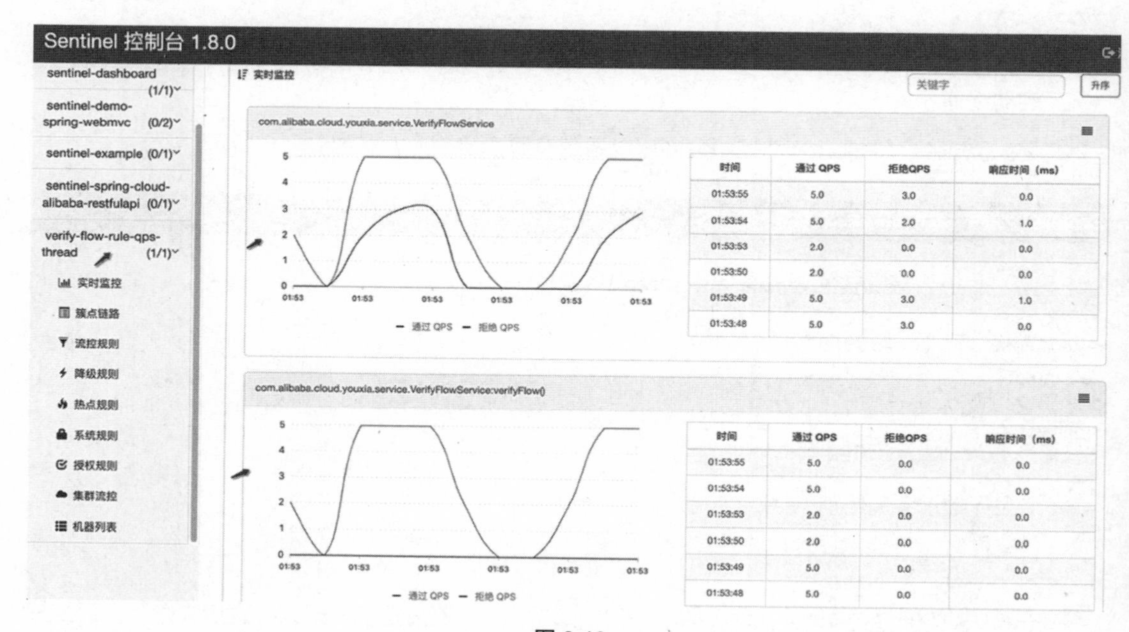

图 6-18

3. 模拟场景二：修改 QPS 的阈值为 10

在 Sentinel 控制台中，修改流量控制规则的 QPS 的阈值为 10，并验证流量控制规则，如图 6-19 所示。

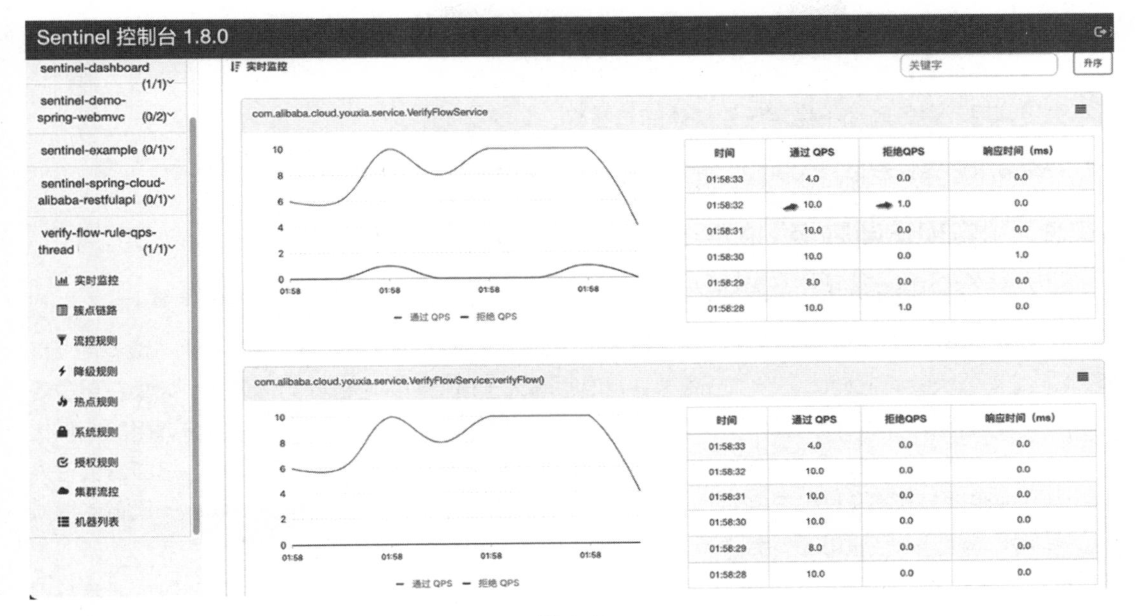

图 6-19

6.7　用“熔断降级”实现流量防护

除流量控制外，对调用链路中不稳定的资源进行熔断降级，也是保障系统高可用的重要措施之一。

6.7.1　什么是熔断降级

一个服务常常会调用别的模块——可能是另外的一个远程服务、数据库，或者第三方 API 等。例如，在支付时，可能需要远程调用银联提供的 API；在查询某个商品的价格时，可能需要进行数据库查询。然而，这个被依赖的服务的稳定性是不能保证的。

如果依赖的服务出现了不稳定的情况，则请求的响应时间会变长，那调用服务的方法的响应时间也会变长，线程会产生堆积，最终可能耗尽业务自身的线程池，服务本身也变得不可用。

现代微服务架构都是分布式的，由非常多的服务组成。不同服务之间相互调用，组成复杂的调用链路。

以上的问题在链路调用中会产生放大的效果：复杂链路上的某一环不稳定，可能会层层级联，最终导致整个链路都不可用。因此，开发者需要对不稳定的弱依赖服务调用进行熔断降级，暂时切断不稳定调用，避免局部不稳定导致整体的“雪崩”。

熔断降级作为保护自身的手段，通常在客户端（调用端）进行配置。

6.7.2 “实现熔断降级”的原理

在了解“实现熔断降级”的原理之前，可以先看一下 Sentinel 提供的几种熔断策略。

- 慢调用比例（SLOW_REQUEST_RATIO）：选择此项作为阈值，需要设置“慢调用 RT（即最大的响应时间）”，当请求的响应时间大于该值时，则统计为慢调用。如果在单位统计时长（statIntervalMs）内，请求数目大于设置的最小请求数目，并且慢调用的比例大于阈值，则在接下来的熔断时长内请求会自动被熔断。在超过熔断时长后，熔断器会进入探测恢复状态（HALF-OPEN 状态），即若接下来的一个请求响应时间小于设置的慢调用 RT，则结束熔断；若大于设置的慢调用 RT，则再次开始熔断。
- 异常比例（ERROR_RATIO）：如果在单位统计时长（statIntervalMs）内，请求数目大于设置的最小请求数目，并且异常的比例大于阈值，则在接下来的熔断时长内请求会自动被熔断。在超过熔断时长后，熔断器会进入探测恢复（HALF-OPEN）状态，即若接下来的一个请求成功完成（没有错误），则结束熔断，否则会再次开始熔断。异常比率的阈值范围是 [0.0,1.0]，代表 0 ～ 100%。
- 异常数（ERROR_COUNT）：如果在单位统计时长内的异常数目超过阈值，则自动进行熔断。在超过熔断时长后，熔断器会进入探测恢复（HALF-OPEN）状态，即若接下来的一个请求成功完成（没有错误），则结束熔断，否则会再次开始熔断。

1. 加载槽位桶中的熔断器槽位 DegradeSlot，并执行 entry()方法

DegradeSlot 类的 entry()方法的具体代码如下：

```
@SpiOrder(-1000)
public class DegradeSlot extends AbstractLinkedProcessorSlot<DefaultNode> {
    @Override
    public void entry(Context context, ResourceWrapper resourceWrapper, DefaultNode node, int count,
                      boolean prioritized, Object... args) throws Throwable {
        //①校验熔断器槽位对应的熔断降级规则
        performChecking(context, resourceWrapper);
        //②加载槽位桶中的其他槽位并执行
        fireEntry(context, resourceWrapper, node, count, prioritized, args);
    }
    void performChecking(Context context, ResourceWrapper r) throws BlockException {
```

```
        //③加载熔断降级规则
            List<CircuitBreaker> circuitBreakers =
DegradeRuleManager.getCircuitBreakers(r.getName());
            if (circuitBreakers == null || circuitBreakers.isEmpty()) {
                return;
            }
            //④遍历熔断降级规则，并校验规则
            for (CircuitBreaker cb : circuitBreakers) {
                if (!cb.tryPass(context)) {
                    throw new DegradeException(cb.getRule().getLimitApp(),
cb.getRule());
                }
            }
        }
        ...
    }
```

2. 用监听器动态更新熔断器的降级规则

用监听器通知应用加载降级规则，具体代码如下：

```
    public final class DegradeRuleManager {
        //①定义监听器
        private static final RulePropertyListener LISTENER = new
RulePropertyListener();
        private static SentinelProperty < List < DegradeRule >> currentProperty
= new DynamicSentinelProperty <>();
    static {
        //②添加监听器
            currentProperty.addListener(LISTENER);
        }
        //③加载新的熔断降级规则
        public static void loadRules(List <DegradeRule> rules) {
            try {
                //④遍历监听器列表，并触发监听器的通知机制
                currentProperty.updateValue(rules);
            } catch(Throwable e) {
                RecordLog.error("[DegradeRuleManager] Unexpected error when
loading degrade rules", e);
            }
    }

        private static class RulePropertyListener implements PropertyListener <
List <DegradeRule>> {
            private synchronized void reloadFrom(List <DegradeRule> list) {
                //⑤用新的降级规则构造容器降解器
```

```
            Map <String,List<CircuitBreaker>> cbs =
buildCircuitBreakers(list);
            Map <String,Set<DegradeRule>> rm = new HashMap <>(cbs.size());
            for (Map.Entry < String, List <CircuitBreaker>> e: cbs.entrySet()){
               assert e.getValue() != null && !e.getValue().isEmpty();
               Set <DegradeRule> rules = new HashSet <>(e.getValue().size());
               for (CircuitBreaker cb: e.getValue()) {
                  rules.add(cb.getRule());
               }
               rm.put(e.getKey(), rules);
            }
            DegradeRuleManager.circuitBreakers = cbs;
            DegradeRuleManager.ruleMap = rm;
         }
         //⑥触发监听器通知，应用更新本地缓存中的熔断降级规则
         @Override public void configUpdate(List <DegradeRule> conf) {
            reloadFrom(conf);
            RecordLog.info("[DegradeRuleManager] Degrade rules has been
updated to: " + ruleMap);
         }
   }
      ...
   }
```

3. 用熔断降级规则构造熔断降级器

Sentinel 用 DegradeRuleManager 类的 getExistingSameCbOrNew()方法构造熔断降级器，具体代码如下：

```
   private static CircuitBreaker getExistingSameCbOrNew(DegradeRule rule) {
      //①从缓存中获取对应规则的熔断降级器
      List <CircuitBreaker> cbs = getCircuitBreakers(rule.getResource());
      if (cbs == null || cbs.isEmpty()) {
         return newCircuitBreakerFrom(rule);
      }
      for (CircuitBreaker cb: cbs) {
         if (rule.equals(cb.getRule())) {
            return cb;
         }
      }
   }
   //②如果从缓存中获取不到，则构建新的熔断降级器
      return newCircuitBreakerFrom(rule);
   }
   private static CircuitBreaker newCircuitBreakerFrom(DegradeRule rule) {
      switch (rule.getGrade()) {
   case RuleConstant.DEGRADE_GRADE_RT:
```

```
        //③如果熔断降级规则的降级类型为 RuleConstant.DEGRADE_GRADE_RT（响应时间），则
返回熔断器 ResponseTimeCircuitBreaker
            return new ResponseTimeCircuitBreaker(rule);
            //④如果熔断降级规则的降级类型为（异常率和异常数），则返回熔断器
ExceptionCircuitBreaker
        case RuleConstant.DEGRADE_GRADE_EXCEPTION_RATIO:
        case RuleConstant.DEGRADE_GRADE_EXCEPTION_COUNT:
            return new ExceptionCircuitBreaker(rule);
        default:
            return null;
        }
    }
```

4. 遍历熔断器降级规则，校验规则

在 Sentinel 中，熔断降级器的状态有 3 种，用枚举类 State 实现，具体代码如下：

```
enum State {
    OPEN,
    HALF_OPEN,
    CLOSED
}
```

Sentinel 用熔断降级器的熔断状态判断是否降级，并将判定功能封装在抽象类 AbstractCircuitBreaker 中，具体代码如下：

```
public abstract class AbstractCircuitBreaker implements CircuitBreaker {
    protected final AtomicReference <State> currentState = new
AtomicReference <>(State.CLOSED);
    protected volatile long nextRetryTimestamp;
    //①判定熔断降级器的熔断状态
@Override
public boolean tryPass(Context context) {
    //②如果是关闭状态，则不触发熔断
        if (currentState.get() == State.CLOSED) {
            return true;
        }
        //③如果是开启状态，则触发熔断
        if (currentState.get() == State.OPEN) {
            //④如果开启了熔断，并且已经错过最近一次重试的时间窗口，则熔断降级器状态从
"开启"切换到"半开启"，不触发熔断
            return retryTimeoutArrived() && fromOpenToHalfOpen(context);
        }
        return false;
    }
    //⑤判断是否在最近一次重试的窗口期内，如果是，则返回 true，执行熔断
    protected boolean retryTimeoutArrived() {
```

```
            return TimeUtil.currentTimeMillis() >= nextRetryTimestamp;
        }
        //⑥熔断降级器的状态从“开启”切换到“半开启”
        protected boolean fromOpenToHalfOpen(Context context) {
            if (currentState.compareAndSet(State.OPEN, State.HALF_OPEN)) {
                notifyObservers(State.OPEN, State.HALF_OPEN, null);
                Entry entry = context.getCurEntry();
                entry.whenTerminate(new BiConsumer <Context, Entry> () {
                        @Override
                        public void accept(Context context, Entry entry) {
                        if (entry.getBlockError() != null) {
                            currentState.compareAndSet(State.HALF_OPEN,
State.OPEN);
                            notifyObservers(State.HALF_OPEN, State.OPEN, 1.0d);
                        }
                    }
                });
                return true;
            }
            return false;
        }
        //⑦通知熔断降级器的熔断状态观察者
        Private void notifyObservers(CircuitBreaker.State prevState,
CircuitBreaker.State newState, Double snapshotValue) {
            for(CircuitBreakerStateChangeObserver observer:
observerRegistry.getStateChangeObservers()) {
                observer.onStateChange(prevState, newState, rule,
snapshotValue);
            }
        }
        ...
    }
```

Sentinel 主要通过控制熔断降级器的熔断状态，来确定是否触发降级。

（1）如果降级类型为 RuleConstant.DEGRADE_GRADE_RT，则用 ResponseTimeCircuitBreaker 类校验规则，具体代码如下：

```
    public class ResponseTimeCircuitBreaker extends AbstractCircuitBreaker {
        private final long maxAllowedRt;
        private final double maxSlowRequestRatio;
        private final int minRequestAmount;
        //①定义基于环形数组的滑动窗口
        private final LeapArray <SlowRequestCounter> slidingCounter;

    @Override
```

```
    public void onRequestComplete(Context context) {
        //②从滑动窗口中获取慢请求统计类 SlowRequestCounter
            SlowRequestCounter counter =
slidingCounter.currentWindow().value();
            Entry entry = context.getCurEntry();
            if (entry == null) {
                return;
            }
            long completeTime = entry.getCompleteTimestamp();
            if (completeTime <= 0) {
                completeTime = TimeUtil.currentTimeMillis();
            }
            //③计算响应时间，如果响应时间大于规则中设置的最大响应时间
            long rt = completeTime - entry.getCreateTimestamp();
            if (rt > maxAllowedRt) {
                //④则在慢请求统计类中添加一个慢请求
                counter.slowCount.add(1);
            }
            //⑤慢请求统计类的总请求数加 1
            counter.totalCount.add(1);

            handleStateChangeWhenThresholdExceeded(rt);
        }
        private void handleStateChangeWhenThresholdExceeded(long rt) {
            if (currentState.get() == State.OPEN) {
                return;
            }
            if (currentState.get() == State.HALF_OPEN) {
                //⑥如果熔断降级器的状态为半开启，并且响应时间大于规则中设置的最大响应时间
                if (rt > maxAllowedRt) {
                    //⑦则熔断降级器的状态变更为“开启”
                    fromHalfOpenToOpen(1.0d);
                } else {
                    //⑧熔断降级器的状态变更为“关闭”
                    fromHalfOpenToClose();
                }
                return;
            }
            List <SlowRequestCounter> counters = slidingCounter.values();
            long slowCount = 0;
            long totalCount = 0;
            //⑨汇总慢请求和总的请求数
            for (SlowRequestCounter counter: counters) {
                slowCount += counter.slowCount.sum();
                totalCount += counter.totalCount.sum();
```

```
        }
        if (totalCount < minRequestAmount) {
            return;
        }
        //⑩计算慢请求的比例
        double currentRatio = slowCount * 1.0d / totalCount;
        if (currentRatio > maxSlowRequestRatio) {
            transformToOpen(currentRatio);
        }
}
...
}
```

（2）如果降级类型为 RuleConstant.DEGRADE_GRADE_EXCEPTION_RATIO 和 RuleConstant.DEGRADE_GRADE_EXCEPTION_COUNT，则用 ExceptionCircuitBreaker 类校验规则，具体代码如下：

```
public class ExceptionCircuitBreaker extends AbstractCircuitBreaker {
    ...
@Override
public void onRequestComplete(Context context) {
        ...
        Throwable error = entry.getError();
        SimpleErrorCounter counter = stat.currentWindow().value();
        if (error != null) {
            //①将错误统计器中的错误请求数加 1
            counter.getErrorCount().add(1);
        }
        //②增加总的请求数
        counter.getTotalCount().add(1);
        //③如果错误的次数达到预先设置的阈值，则变更资源的状态
        handleStateChangeWhenThresholdExceeded(error);
    }
    private void handleStateChangeWhenThresholdExceeded(Throwable error) {
        if (currentState.get() == State.OPEN) {
            return;
        }
        //④如果是半开状态，则直接返回
        if (currentState.get() == State.HALF_OPEN) {
            //⑤如果错误对象为空，则将“半开”状态更新为“关闭”状态
            if (error == null) {
                fromHalfOpenToClose();
            } else {
                //⑥如果错误对象不为空，则将“半开”状态更新为“开启”状态
                fromHalfOpenToOpen(1.0d);
```

```
            }
            return;
        }
        List <SimpleErrorCounter> counters = stat.values();
        long errCount = 0;
        long totalCount = 0;
        for (SimpleErrorCounter counter: counters) {
            //⑦统计总的错误请求次数和总的调用次数
            errCount += counter.errorCount.sum();
            totalCount += counter.totalCount.sum();
        }
        //⑧如果总的调用次数小于最小请求次数，则直接返回
        if (totalCount < minRequestAmount) {
            return;
        }
        double curCount = errCount;
        if (strategy == DEGRADE_GRADE_EXCEPTION_RATIO) {
            //⑨计算请求的错误率
            curCount = errCount * 1.0d / totalCount;
        }
        //⑩如果错误率大于预先设定的阈值，则将状态设置为“开启”
        if (curCount > threshold) {
            transformToOpen(curCount);
        }
}
...
}
```

6.7.3 【实例】用“模拟 Dubbo 服务故障”验证服务调用熔断降级的过程

本实例的源码在本书配套资源的“chaptersix/ verify-degrade-rule”目录下。

本实例从应用端同步降级规则到 Sentinel 控制台，模拟 Dubbo 服务故障，并触发熔断规则。

1. 用 Spring Cloud Alibaba 基础框架创建项目 verify-degrade-rule

用 IDEA 快速创建一个 Spring Cloud Alibaba 的基础项目，并初始化项目的基础环境。

在启动类中添加降级规则，具体代码如下：

```
@SpringBootApplication
@EnableDiscoveryClient
public class VerifyDegradeApplication {
```

```
    public static void main(String[] args) {
        InitExecutor.doInit();
        initDegradeRule(10,true);
        SpringApplication.run(VerifyDegradeApplication.class, args);
    }

    private static final String INTERFACE_RES_KEY =
DegradeService.class.getName();
    private static final String RES_KEY = INTERFACE_RES_KEY +
":verifyDegradeRule()";

  static void initDegradeRule(int interfaceDegradeLimit, boolean method) {
    //①设置 Dubbo 接口的熔断策略为“错误次数”
        DegradeRule degradeRule = new DegradeRule(INTERFACE_RES_KEY)
                .setGrade(RuleConstant.DEGRADE_GRADE_EXCEPTION_COUNT).setMi
nRequestAmount(interfaceDegradeLimit).setCount(8).setTimeWindow(2);
        List<DegradeRule> list = new ArrayList<>();
        if (method) {
            DegradeRule degradeRule1 = new DegradeRule(RES_KEY)
                    //②设置方法的熔断策略为“错误次数”
                    .setCount(8).setMinRequestAmount(interfaceDegradeLimit)
                    .setGrade(RuleConstant.DEGRADE_GRADE_EXCEPTION_COUNT).se
tTimeWindow(2);
            list.add(degradeRule1);
        }
        list.add(degradeRule);
        DegradeRuleManager.loadRules(list);
    }
  }
```

2. 准备熔断降级的环境

为了演示熔断降级的效果，本程序开启了 150 个线程来模拟并发请求，并通过随机数控制单次请求的出错率，具体代码如下：

```
  @DubboService(version = "1.0.0",group = "verify-degrade-rule")
  public class DegradeServiceImpl implements DegradeService{
    //①如果降级生效，则不会调用 verifyDegradeRule()方法
    @Override
    public String verifyDegradeRule() {
        int r= RandomUtils.nextInt(1,3);
        //②设置错误比率为 90%
        if(!("1".equals(r+""))) {
            //③随机抛出异常，让 Sentinel 统计错误的请求数
            throw new RuntimeException("错误");
        }
```

```
        for(int i=0;i<1000000000;i++){
        //④循环模拟 RT 延迟
        }
        return "verify degrade rule";
    }
}
```

在服务调用端开启 150 个线程，以模拟 150 个用户并发地调用接口，具体代码如下：

```
@RestController
public class DegradeController {
    @DubboReference(version = "1.0.0",group = "verify-degrade-rule")
    private DegradeService degradeService;
    @GetMapping(value = "/verify")
    @ResponseBody
public String verify(){
    //①开启 150 个线程
        Executors.newFixedThreadPool(50).execute(new ExecutorThread());
        Executors.newFixedThreadPool(50).execute(new ExecutorThread());
        Executors.newFixedThreadPool(50).execute(new ExecutorThread());
        return "success";
    }
    class ExecutorThread implements Runnable{
        @Override
        public void run() {
            while (true) {
                try {
                    //②调用接口
                    degradeService.verifyDegradeRule();
                }catch (Exception e){
                    System.out.println(e.getMessage());
                }
            }
        }
    }
}
```

通过如下命令开启服务调用：

```
curl 127.0.0.1:7867/verify
```

3. 启动程序，并验证熔断降级的效果

在程序运行后，在 Sentinel 控制台能看到应用已经接入，如图 6-20 所示。

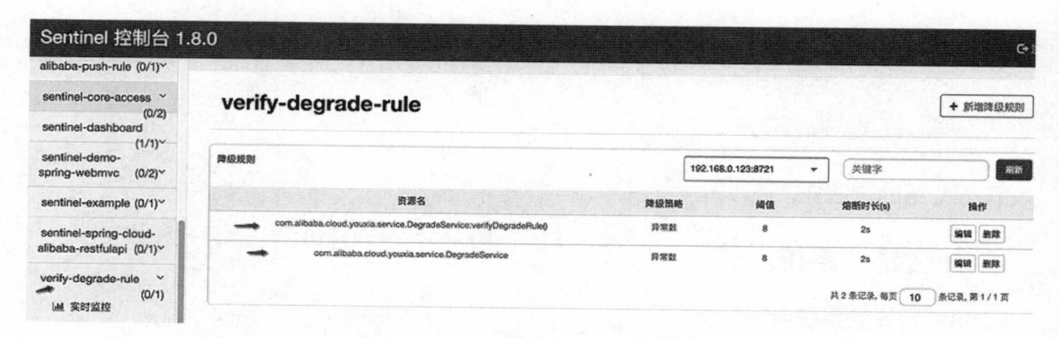

图 6-20

为了方便观察熔断降级的效果，这里调整熔断策略为“异常比例”，比例阈值为 0.4，最小请求次数为 3，熔断时长为 3，如图 6-21 所示。

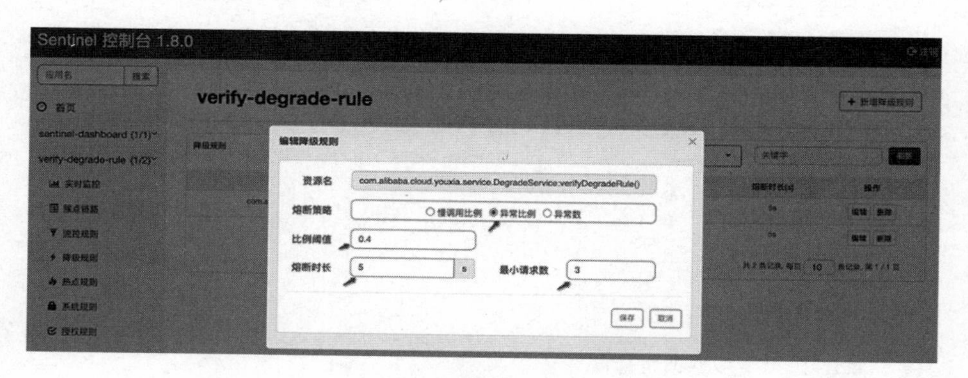

图 6-21

观察 Sentinel 控制台，接口会按照熔断降级规则进行降级处理（熔断窗口期为 5s），如图 6-22 所示。

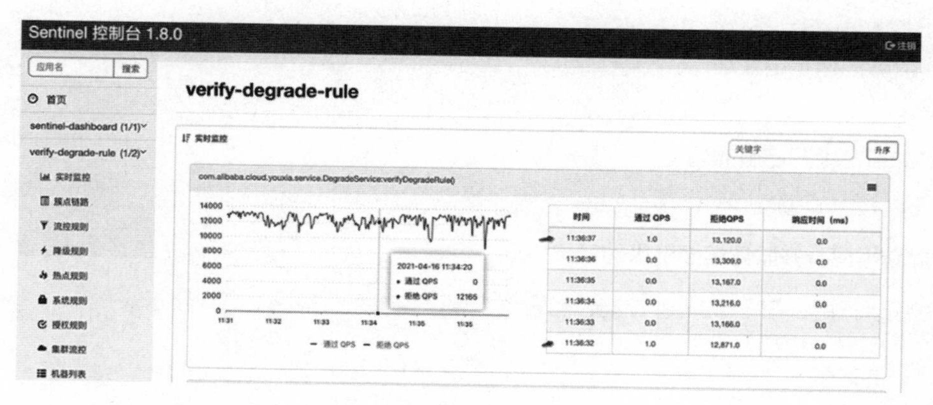

图 6-22

6.8 用“系统自适应保护”实现流量防护

除前面介绍的方法外，还可以用“系统自适应保护”实现流量防护。

6.8.1 什么是“系统自适应保护”

Sentinel 定义的系统自适应保护和传统的系统自适应保护是有区别的。

1. 传统的系统自适应保护

传统的系统自适应保护是根据硬件指标（即系统的负载）来做系统过载保护的：如果系统负载高于某个阈值，则禁止或者减少流量的进入；如果负载开始好转，则恢复流量的进入。

这个思路给应用系统带来了两个问题：

- 具有延迟性。从因果关系的角度来看，负载是一个“果”，如果应用根据负载的情况来调节流量的通过率，则在调节之后产生的流量整形效果始终是具有延迟性的（对通过率的任何调整，都会在过一段时间后才能看到效果）。如果当前系统的 CPU、内存、I/O 等资源非常丰富，并且应用服务此时一直处于“饥饿”状态，则传统的系统自适应保护不能实时地通过调整负载让应用充分利用当前系统的资源。这样，系统在运行期间产生的监控指标，在链路大屏上就会出现流量曲线的峰值和谷值波动很大的现象。
- 恢复慢。想象一下这样的一个场景：下游应用不可靠，导致应用 RT 很高，从而负载到了一个很高的点；过了一段时间后，下游应用恢复了，应用 RT 也相应减少。这时，其实应该大幅度增大流量的通过率；但由于这时负载仍然很高，所以通过率仍然不高。

2. Sentinel 的系统自适应保护

在 Sentinel 中，系统自适应保护策略是：用负载作为启动流量控制的值，且 Sentinel 允许通过的请求流量的值是由处理请求的能力（即请求的响应时间和当前系统正在处理的请求速率）来决定的。

Sentinel 根据“系统能够处理的请求”和“系统允许进来的请求”来做平衡，而不是根据一个间接的指标（系统负载）来做流量控制。

Sentinel 追求的目标是“在系统不被拖垮的情况下，提高系统的吞吐率”，而不是“负载一定要到低于某个阈值”。如果还是按照固有的思维——“超过特定的负载就禁止流量进入，系统负载恢复就放开流量”，则无论开发者怎么调节参数，调节比例，都是按照“果”来调节“因”，都无法取得良好的效果。

6.8.2 “系统自适应保护”的原理

Sentinel 在执行系统自适应保护时，会遵循如下系统规则。

- 负载：当系统负载超过阈值，且系统当前的并发线程数超过系统容量时，才会触发系统保护。系统容量由系统的“maxQps × minRt”计算得出。一般设定参考值是 “CPU 核数 × 2.5”。
- CPU 使用率：当系统的 CPU 使用率超过阈值时，触发系统保护（取值范围为 0.0 ～ 1.0）。
- RT：当单台机器上所有入口流量的平均 RT 达到阈值即触发系统保护，单位是 ms 。
- 线程数：当单台机器上所有入口流量的并发线程数达到阈值时，触发系统保护。
- 入口 QPS：当单台机器上所有入口流量的 QPS 达到阈值时，触发系统保护。

1. 用监听器动态地更新 SystemSlot 槽位中的系统规则

（1）调用 SystemRuleManager 类的 loadRules()方法加载新的系统规则，具体代码如下：

```
public final class SystemRuleManager {
//①定义系统规则监听器
private final static SystemPropertyListener listener = new
SystemPropertyListener();
//②定义 Sentinel 的动态属性加载类
    private static SentinelProperty <List<SystemRule>> currentProperty = new
DynamicSentinelProperty <List<SystemRule>> ();
static {
    //③添加监听器
        currentProperty.addListener(listener);
    }
public static void loadRules(List <SystemRule> rules) {
    //④用动态属性加载类 DynamicSentinelProperty 的 updateValue()方法更新系统规则
        currentProperty.updateValue(rules);
    }
    ...
}
```

（2）动态属性加载类 DynamicSentinelProperty 执行方法 updateValue()动态地更新系统规则，具体代码如下：

```
public class DynamicSentinelProperty<T> implements SentinelProperty<T> {
    //①定义监听器列表
    protected Set<PropertyListener<T>> listeners =
Collections.synchronizedSet(new HashSet<PropertyListener<T>>());
    private T value = null;
    //②添加监听器
    @Override
    public void addListener(PropertyListener<T> listener) {
```

```
            listeners.add(listener);
            listener.configLoad(value);
        }
        //③更新系统规则
        @Override
        public boolean updateValue(T newValue) {
            if (isEqual(value, newValue)) {
                return false;
            }
            RecordLog.info("[DynamicSentinelProperty] Config will be updated to:
" + newValue);

            value = newValue;
            //④遍历监听器列表，将最新的系统规则通过监听器通知到应用
            for (PropertyListener<T> listener : listeners) {
                listener.configUpdate(newValue);
            }
            return true;
        }
        ...
    }
```

（3）执行监听器类 SystemPropertyListener 的 configUpdate()方法，具体代码如下：

```
    static class SystemPropertyListener extends SimplePropertyListener <List<
SystemRule>> {

        @Override public synchronized void configUpdate(List <SystemRule> rules) {
            restoreSetting();
            if (rules != null && rules.size() >= 1) {
                //遍历系统规则列表，更新系统规则到应用中
                for (SystemRule rule: rules) {
                    loadSystemConf(rule);
                }
            } else {
                checkSystemStatus.set(false);
            }
    }
    ...
    }
```

2. 从槽位桶中取出 SystemSlot 槽位，执行系统自适应保护

Sentinel 通过槽位 SystemSlot 来实现系统自适应保护，具体代码如下：

```
    @SpiOrder(-5000) public class SystemSlot extends
AbstractLinkedProcessorSlot <DefaultNode> {
```

```
    @Override
    public void entry(Context context, ResourceWrapper resourceWrapper,
DefaultNode node, int count, boolean prioritized, Object...args) throws
Throwable {
        //①校验系统自适应保护的流量控制规则
            SystemRuleManager.checkSystem(resourceWrapper);
            //②执行槽位桶中其他的槽位
            fireEntry(context, resourceWrapper, node, count, prioritized, args);
    }
    ...
    }
```

3. 根据系统运行的指标，校验系统自适应保护的系统规则

Sentinel 通过 SystemRuleManager 类来校验系统自适应保护的流量控制规则，具体代码如下：

```
    public final class SystemRuleManager {
        public static void checkSystem(ResourceWrapper resourceWrapper) throws
BlockException {
            if (resourceWrapper == null) {
                return;
            }
            //①验证系统的状态
            if (!checkSystemStatus.get()) {
                return;
            }
            //②规则只对流量类型为 EntryType.IN 生效
            if (resourceWrapper.getEntryType()!= EntryType.IN) {
                return;
            }
            //③验证 QPS
            double currentQps = Constants.ENTRY_NODE == null ? 0.0 :
Constants.ENTRY_NODE.successQps();
            if (currentQps > qps) {
                throw new SystemBlockException(resourceWrapper.getName(),
"qps");
            }
            //④验证线程数
            int currentThread = Constants.ENTRY_NODE == null ? 0 :
Constants.ENTRY_NODE.curThreadNum();
            if (currentThread > maxThread) {
                throw new SystemBlockException(resourceWrapper.getName(),
"thread");
            }
            //⑤验证响应时间 rt
```

```
        double rt = Constants.ENTRY_NODE == null ? 0 :
Constants.ENTRY_NODE.avgRt();
        if (rt > maxRt) {
            throw new SystemBlockException(resourceWrapper.getName(), "rt");
        }
        //⑥验证系统负载和平均系统负载
        if (highestSystemLoadIsSet && getCurrentSystemAvgLoad() >
highestSystemLoad) {
            if (!checkBbr(currentThread)) {
                throw new SystemBlockException(resourceWrapper.getName(),
"load");
            }
        }
        //⑦验证系统 CPU 使用率
        if (highestCpuUsageIsSet && getCurrentCpuUsage() > highestCpuUsage) {
            throw new SystemBlockException(resourceWrapper.getName(),
"cpu");
        }
    }
    ...
    }
```

6.8.3 【实例】通过调整应用服务的入口流量和负载，验证系统自适应保护

本实例的源码在本书配套资源的“chaptersix/ verify-system-rule”目录下。

本实例将应用端的系统自适应保护规则同步到 Sentinel 的控制台中。

1. 用 Spring Cloud Alibaba 快速搭建应用项目 verify-system-rule

用 Spring Cloud Alibaba 快速搭建项目 verify-system-rule，然后在项目中添加 Sentinel 相关的 Jar 包依赖，具体如下：

```
<dependency>
    <groupId>com.alibaba.cloud</groupId>
    <artifactId>spring-cloud-starter-alibaba-sentinel</artifactId>
</dependency>
<dependency>
    <groupId>com.alibaba.csp</groupId>
    <artifactId>sentinel-apache-dubbo-adapter</artifactId>
</dependency>
<dependency>
```

```
    <groupId>com.alibaba.csp</groupId>
    <artifactId>sentinel-transport-netty-http</artifactId>
    <version>1.8.0</version>
</dependency>
```

2. 在启动类中添加系统规则

在启动类 VerifySystemRuleApplication 中添加系统自适应保护规则，并将其同步到 Sentinel 控制台中，具体代码如下：

```
@SpringBootApplication
@RefreshScope
@EnableDiscoveryClient
public class VerifySystemRuleApplication {
    public static void main(String[] args) {
        InitExecutor.doInit();
        initDegradeRule(10,true);
        SpringApplication.run(VerifySystemRuleApplication.class, args);
    }

    private static final String INTERFACE_RES_KEY =
SystemRuleService.class.getName();
    private static final String RES_KEY = INTERFACE_RES_KEY +
":verifySystemRule()";

    //初始化系统规则
    static void initDegradeRule(int interfaceDegradeLimit, boolean method) {
        SystemRule systemRule = new SystemRule();
        systemRule.setResource(INTERFACE_RES_KEY);
        systemRule.setAvgRt(2000);
        systemRule.setMaxThread(5000);
        systemRule.setQps(300000);
        systemRule.setHighestSystemLoad(1.5);
        systemRule.setHighestCpuUsage(0.8);
        List<SystemRule> list = new ArrayList<>();
        if (method) {
            SystemRule systemRule1 = new SystemRule();
            systemRule1.setResource(RES_KEY);
            systemRule1.setAvgRt(2000);
            systemRule1.setQps(300000);
            systemRule1.setHighestSystemLoad(1.5);
            systemRule1.setMaxThread(5000);
            systemRule1.setHighestCpuUsage(0.8);
            list.add(systemRule1);
        }
```

```
        list.add(systemRule);
        SystemRuleManager.loadRules(list);
    }
}
```

3. 实现动态变更接口流量和负载

通过线程和代码循环，分别模拟接口流量和负载。

模拟接口流量的具体代码如下：

```
@RestController
@RefreshScope
public class SystemRuleController {
    @Autowired
    private NacosConfig nacosConfig;
    @DubboReference(version = "1.0.0",group = "verify-system-rule")
    private SystemRuleService systemRuleService;
    @GetMapping(value = "/verify")
    @ResponseBody
    public String verifySystemRule(){
        Executors.newFixedThreadPool(50).execute(new Executor());
        Executors.newScheduledThreadPool(1).schedule(new Ext(),60*2,
TimeUnit.SECONDS);
        return "成功";
    }

    class Executor implements Runnable{
        @Override
        public void run() {
            while (true){
                systemRuleService.verifySystemRule();
            }
        }
    }
    class Ext implements Runnable{
        @Override
        public void run() {
            if(nacosConfig.isEnableaddthreadqps()){
                Executors.newFixedThreadPool(50).execute(new Executor());
            }
        }
    }
}
```

模拟负载的具体代码如下：

```
@DubboService(version = "1.0.0",group = "verify-system-rule")
public class SystemRuleServiceImpl implements SystemRuleService{
    @Autowired
    private NacosConfig nacosConfig;

    @Override
    public String verifySystemRule() {
        if(nacosConfig.isEnablecpuandrt()){
            for(int i=0;i<900000000;i++){}
            for(int i=0;i<900000000;i++){}
            for(int i=0;i<900000000;i++){}
            for(int i=0;i<900000000;i++){}
            for(int i=0;i<900000000;i++){}
            try {
                Thread.sleep(1000);
            }catch (InterruptedException e){

            }
        }
        return "验证系统自适应保护规则";
    }
}
```

4. 启动程序，验证动态流量和负载的效果

（1）启动程序，并通过如下命令（调用 HTTP 接口）模拟用户的并发请求。

```
curl 127.0.0.1:7867/verify
```

（2）系统自适应保护规则已经同步到 Sentinel 控制台中，如图 6-23 所示。

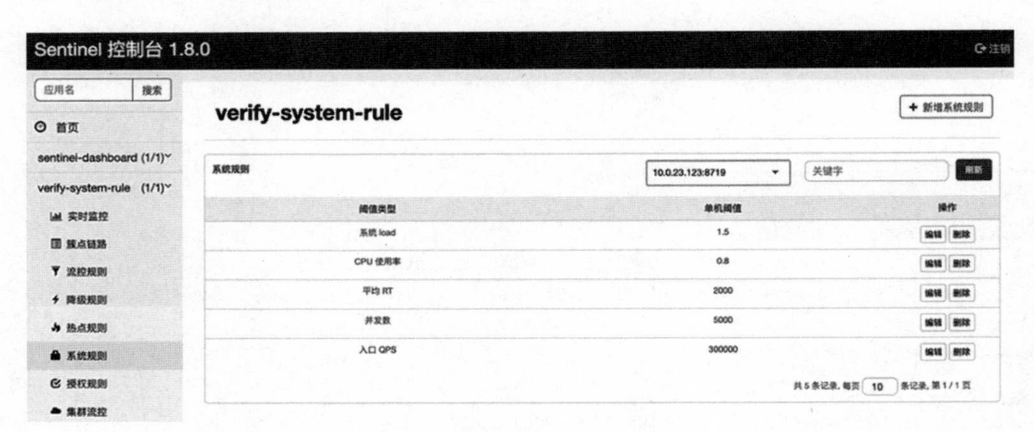

图 6-23

（3）开启负载，模拟调用延迟，系统自适应保护规则生效，如图 6-24 所示。

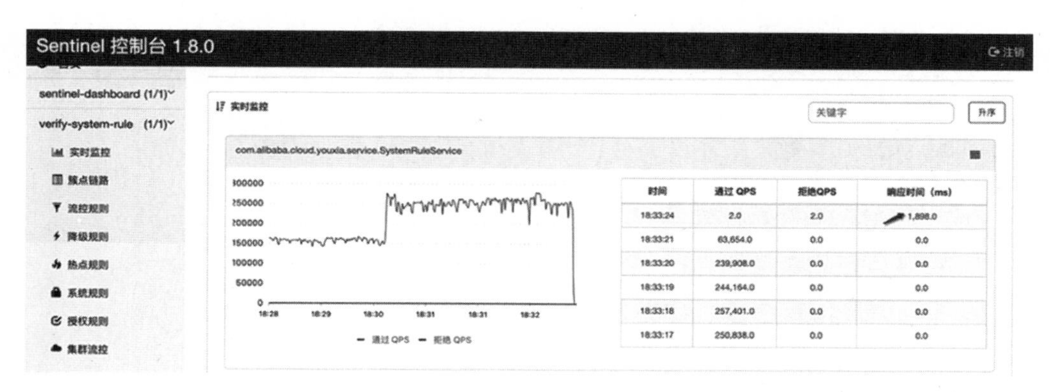

图 6-24

（4）开启流量模拟并发请求，当 QPS 超过阈值 260000 时，系统自适应保护规则生效，如图 6-25 所示。

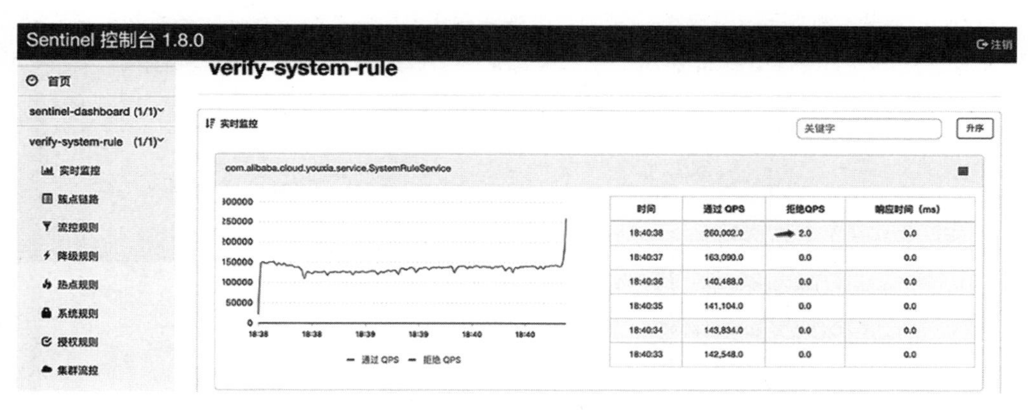

图 6-25

6.9　用 Nacos 实现规则的动态配置和持久化

Sentinel 可以用 Nacos、Apollo、ZooKeeper 等实现规则的动态配置和持久化。

6.9.1　为什么需要“规则的动态配置”

如果开发人员在 Sentinel 的控制台中操作资源对应的一条规则，则应用要能够实时地获取最新的规则。另外，开发人员在应用端操作资源对应的规则，则 Sentinel 也要能够实时地感知最新的规则，如图 6-26 所示。

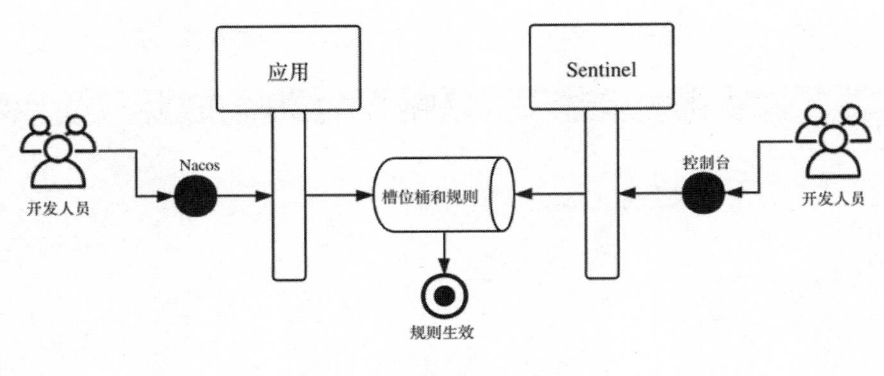

图 6-26

6.9.2 为什么需要“规则的持久化”

在图 6-27 中，应用和 Sentinel 的控制台会在内存中存储槽位桶和规则——如果应用和 Sentinel 控制台不重启，这是没有问题的。

Sentinel 需要将规则持久化到数据源中，以确保在其重启后能够重新获取最新的规则，并生成槽位桶，如图 6-27 所示。Sentinel 支持的数据源很多，包括 Nacos、Apollo 等。

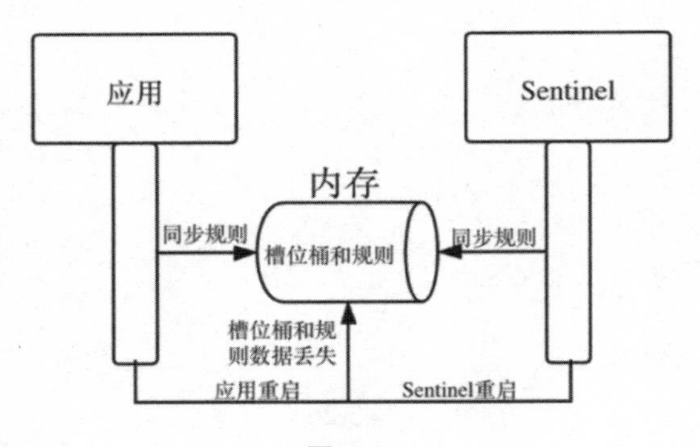

图 6-27

6.9.3 “规则的动态配置”的原理

在 6.4 节已经介绍过，Sentinel 和应用之间可以采用 HTTP 或者 Netty 实现通信渠道。下面以 Sentinel 流量控制规则的发布作为场景，分析“规则的动态配置”的原理。

1. 用控制台新增流量控制规则

Sentinel 用 FlowControllerV2 类来封装新增流量控制规则的 RESTful API，具体代码如下：

```
@RestController
@RequestMapping(value = "/v2/flow")
```

```
    public class FlowControllerV2 {
        private final Logger logger =
LoggerFactory.getLogger(FlowControllerV2.class);
        @Autowired
        private InMemoryRuleRepositoryAdapter<FlowRuleEntity> repository;
        @Autowired
        @Qualifier("flowRuleDefaultProvider")
        private DynamicRuleProvider<List<FlowRuleEntity>> ruleProvider;
        @Autowired
        @Qualifier("flowRuleDefaultPublisher")
        private DynamicRulePublisher<List<FlowRuleEntity>> rulePublisher;

        @PostMapping("/rule")
        @AuthAction(value = AuthService.PrivilegeType.WRITE_RULE)
        public Result<FlowRuleEntity> apiAddFlowRule(
            @RequestBody FlowRuleEntity entity) {
            Result<FlowRuleEntity> checkResult = checkEntityInternal(entity);
            if (checkResult != null) {
                return checkResult;
            }
            entity.setId(null);
            Date date = new Date();
            entity.setGmtCreate(date);
            entity.setGmtModified(date);
            entity.setLimitApp(entity.getLimitApp().trim());
            entity.setResource(entity.getResource().trim());
            try {
                //①将流量控制规则持久化到 Sentinel 控制台对应的内存中
                entity = repository.save(entity);
                //②将流量控制规则用通信渠道推送到应用端
                publishRules(entity.getApp());
            } catch (Throwable throwable) {
                logger.error("Failed to add flow rule", throwable);
                return Result.ofThrowable(-1, throwable);
            }
            return Result.ofSuccess(entity);
        }
    }
```

2. 用通信渠道推送流量控制规则

Sentinel 用 FlowRuleApiPublisher 类将资源对应的规则推送到应用端，具体代码如下：

```
    @Component("flowRuleDefaultPublisher")
    public class FlowRuleApiPublisher implements
DynamicRulePublisher<List<FlowRuleEntity>> {
```

```
    @Autowired
    private SentinelApiClient sentinelApiClient;
    @Autowired
    private AppManagement appManagement;

    @Override
    public void publish(String app, List<FlowRuleEntity> rules)
        throws Exception {
        if (StringUtil.isBlank(app)) {
            return;
        }
        if (rules == null) {
            return;
        }
        //①获取应用的机器实例列表
        Set<MachineInfo> set = appManagement.getDetailApp(app).getMachines();
        //②循环遍历机器实例列表
        for (MachineInfo machine : set) {
            if (!machine.isHealthy()) {
                continue;
            }
            //③用通信渠道推送流量控制规则
            sentinelApiClient.setFlowRuleOfMachine(app, machine.getIp(),
                machine.getPort(), rules);
        }
    }
}
```

调用 HTTP 接口“setRules”完成规则推送，具体代码如下：

```
@Component
public class SentinelApiClient {
    public boolean setFlowRuleOfMachine(String app, String ip, int port,
List<FlowRuleEntity> rules) {
        return setRules(app, ip, port, FLOW_RULE_TYPE, rules);
    }
    private boolean setRules(String app, String ip, int port, String type,
List<? extends RuleEntity> entities) {
        if (entities == null) {
            return true;
        }
        try {
            AssertUtil.notEmpty(app, "Bad app name");
            AssertUtil.notEmpty(ip, "Bad machine IP");
            AssertUtil.isTrue(port > 0, "Bad machine port");
            String data = JSON.toJSONString(
```

```
                    entities.stream().map(r ->
r.toRule()).collect(Collectors.toList()));
            Map<String, String> params = new HashMap<>(2);
            params.put("type", type);
            params.put("data", data);
            //用 CloseableHttpAsyncClient 来发起 HTTP 请求，并调用接口“flow”
            String result = executeCommand(app, ip, port, SET_RULES_PATH,
params, true).get();
            logger.info("setRules result: {}, type={}", result, type);
            return true;
        } catch (InterruptedException e) {
            logger.warn("setRules API failed: {}", type, e);
            return false;
        } catch (ExecutionException e) {
            logger.warn("setRules API failed: {}", type, e.getCause());
            return false;
        } catch (Exception e) {
            logger.error("setRules API failed, type={}", type, e);
            return false;
        }
    }
    ...
}
```

3. 应用端初始化 HTTP 接口，处理流量控制规则

应用端处理 Sentinel 请求的接口（包括流量控制规则接口“setRules”），主要是通过 CommandHandlerProvider 类来加载 CommandHandler 类，加载过程的具体代码如下：

```
public class CommandHandlerProvider implements Iterable<CommandHandler> {
    //①用 JDK 的 SPI 机制加载 CommandHandler 接口的实现类列表
    private final ServiceLoader<CommandHandler> serviceLoader =
ServiceLoaderUtil.getServiceLoader(
        CommandHandler.class);
    public Map<String, CommandHandler> namedHandlers() {
        Map<String, CommandHandler> map = new HashMap<String,
CommandHandler>();
        //②将 CommandHandler 接口的实现类列表转换成 HashMap
        for (CommandHandler handler : serviceLoader) {
            String name = parseCommandName(handler);
            if (!StringUtil.isEmpty(name)) {
                map.put(name, handler);
            }
        }
        return map;
}
```

```
    //③解析注解@CommandMapping 中的 name 字段的属性值，比如流量控制规则接口的名称
"setRules"
        private String parseCommandName(CommandHandler handler) {
            CommandMapping commandMapping =
handler.getClass().getAnnotation(CommandMapping.class);
            if (commandMapping != null) {
                return commandMapping.name();
            } else {
                return null;
            }
        }
        @Override
        public Iterator<CommandHandler> iterator() {
            return serviceLoader.iterator();
    }
        private static final CommandHandlerProvider INSTANCE = new
CommandHandlerProvider();
        public static CommandHandlerProvider getInstance() {
            return INSTANCE;
        }
    }
```

定义流量控制规则接口“setRules”，具体代码如下：

```
    @CommandMapping(name = "setRules", desc = "modify the rules, accept param:
type={ruleType}&data={ruleJson}")
    public class ModifyRulesCommandHandler implements CommandHandler<String> {
        private static final int FASTJSON_MINIMAL_VER = 0x01020C00;
        @Override
        public CommandResponse<String> handle(CommandRequest request) {
            if (VersionUtil.fromVersionString(JSON.VERSION) <
FASTJSON_MINIMAL_VER) {
                return CommandResponse.ofFailure(new RuntimeException("The
\"fastjson-" + JSON.VERSION
                        + "\" introduced in application is too old, you need
fastjson-1.2.12 at least."));
            }
            String type = request.getParam("type");
            String data = request.getParam("data");
            if (StringUtil.isNotEmpty(data)) {
                try {
                    data = URLDecoder.decode(data, "utf-8");
                } catch (Exception e) {
                    RecordLog.info("Decode rule data error", e);
                    return CommandResponse.ofFailure(e, "decode rule data error");
                }
```

```
        }
        RecordLog.info("Receiving rule change (type: {}): {}", type, data);
        String result = "success";
        //①如果是流量控制规则，则调用 FlowRuleManager 类的 loadRules()方法动态加载
规则到应用中
        if (FLOW_RULE_TYPE.equalsIgnoreCase(type)) {
            List<FlowRule> flowRules = JSONArray.parseArray(data,
FlowRule.class);
            FlowRuleManager.loadRules(flowRules);
            if (!writeToDataSource(getFlowDataSource(), flowRules)) {
                result = WRITE_DS_FAILURE_MSG;
            }
            return CommandResponse.ofSuccess(result);
        //②如果是鉴权规则，则调用 AuthorityRuleManager 类的 loadRules()方法动态加
载规则到应用中
        } else if (AUTHORITY_RULE_TYPE.equalsIgnoreCase(type)) {
            List<AuthorityRule> rules = JSONArray.parseArray(data,
AuthorityRule.class);
            AuthorityRuleManager.loadRules(rules);
            if (!writeToDataSource(getAuthorityDataSource(), rules)) {
                result = WRITE_DS_FAILURE_MSG;
            }
            return CommandResponse.ofSuccess(result);
        //③如果是降级规则，则调用 DegradeRuleManager 类的 loadRules()方法动态加载
规则到应用中
        } else if (DEGRADE_RULE_TYPE.equalsIgnoreCase(type)) {
            List<DegradeRule> rules = JSONArray.parseArray(data,
DegradeRule.class);
            DegradeRuleManager.loadRules(rules);
            if (!writeToDataSource(getDegradeDataSource(), rules)) {
                result = WRITE_DS_FAILURE_MSG;
            }
            return CommandResponse.ofSuccess(result);
        //④如果是系统规则，则调用 SystemRuleManager 类的 loadRules()方法动态加载规
则到应用中
        } else if (SYSTEM_RULE_TYPE.equalsIgnoreCase(type)) {
            List<SystemRule> rules = JSONArray.parseArray(data,
SystemRule.class);
            SystemRuleManager.loadRules(rules);
            if (!writeToDataSource(getSystemSource(), rules)) {
                result = WRITE_DS_FAILURE_MSG;
            }
            return CommandResponse.ofSuccess(result);
        }
```

```
            return CommandResponse.ofFailure(new
IllegalArgumentException("invalid type"));
        }
        ...
    }
```

4. 将初始化完成的 ModifyRulesCommandHandler 接口与通信渠道关联

通信渠道在初始化过程中会注册通信事件。一个 CommandHandler 接口的实现类处理一个类型的通信事件。

（1）用 HTTP 的通信事件中心的 SimpleHttpCommandCenter 类关联新增规则的 ModifyRulesCommandHandler 接口，具体代码如下：

```
    public class SimpleHttpCommandCenter implements CommandCenter {

        private static final Map<String, CommandHandler> handlerMap = new
ConcurrentHashMap<String, CommandHandler>();
        @Override
        @SuppressWarnings("rawtypes")
        public void beforeStart() throws Exception {
            //①加载所有的 CommandHandler 接口的实现类
            Map<String, CommandHandler> handlers =
CommandHandlerProvider.getInstance().namedHandlers();
            //②注册到本地缓存
            registerCommands(handlers);
        }
    }
```

（2）用异步事件任务 HttpEventTask 处理 HTTP 请求，具体代码如下：

```
    public class HttpEventTask implements Runnable {
        @Override
        public void run() {
        ...
        //①从请求中解析出事件名称，比如 setRules
        String commandName = HttpCommandUtils.getTarget(request);
        //②如果为空，则返回不合法
        if (StringUtil.isBlank(commandName)) {
            writeResponse(printWriter, StatusCode.BAD_REQUEST,
INVALID_COMMAND_MESSAGE);
            return;
        }
        //③从 SimpleHttpCommandCenter 的本地缓存中，获取对应事件名称的 CommandHandler
接口的实现类，比如 ModifyRulesCommandHandler
        CommandHandler<?> commandHandler =
SimpleHttpCommandCenter.getHandler(commandName);
```

```
        if (commandHandler != null) {
            //④处理 HTTP 请求，并返回结果
            CommandResponse<?> response = commandHandler.handle(request);
            handleResponse(response, printWriter);
        } else {
            writeResponse(printWriter, StatusCode.BAD_REQUEST, "Unknown command
`" + commandName + '`');
        }
        ...
        }
    }
```

用 Netty 实现通信渠道的关联过程和上面基本相同，感兴趣的读者可以查阅 NettyHttpCommandCenter 类的相关源码。

6.9.4 “规则的持久化”的原理

Sentinel 流量防护规则的持久化过程，主要是将规则从 Sentinel 的内存中同步到持久化数据源中。这样 Sentinel 在重启后能够从数据源中读取持久化后的规则。

1. 将规则持久化到 Nacos 配置中心

关于规则的持久化的原理，可以从 FlowRuleNacosPublisher 类开始分析，它是采用 Nacos 作为持久化数据源的起点，具体代码如下：

```
    @Component("flowRuleNacosPublisher")
    public class FlowRuleNacosPublisher implements
DynamicRulePublisher<java.util.List<FlowRuleEntity>> {

        @Autowired
        private ConfigService configService;
        @Autowired
        private Converter<java.util.List<FlowRuleEntity>, String> converter;
        @Override
        public void publish(String app, java.util.List<FlowRuleEntity> rules)
throws Exception {
            AssertUtil.notEmpty(app, "app name cannot be empty");
            if (rules == null) {
                return;
            }
            //①遍历流量控制规则，设置应用名称
            for (FlowRuleEntity rule : rules) {
                if (rule.getApp() == null) {
                    rule.setApp(app);
                }
```

```
            }
            //②将规则发布到 Nacos 配置中心
            configService.publishConfig(app +
NacosConfigUtil.FLOW_DATA_ID_POSTFIX,
                NacosConfigUtil.GROUP_ID, converter.convert(rules));
        }
    }
```

这样，Sentinel 维护的对应资源的规则就被持久化到 Nacos 配置中心中了。

2. Sentinel 在重启后，从 Nacos 配置中心中获取持久化后的规则

Sentinel 在重启后，会用 FlowRuleNacosProvider 类从 Nacos 配置中心获取最新的规则，具体代码如下：

```
    @Component("flowRuleNacosProvider")
    public class FlowRuleNacosProvider implements
DynamicRuleProvider<java.util.List<FlowRuleEntity>> {

        @Autowired
        private ConfigService configService;
        @Autowired
        private Converter<String, java.util.List<FlowRuleEntity>> converter;

        @Override
        public java.util.List<FlowRuleEntity> getRules(String appName) throws
Exception {
            //从配置中心获取规则
            String rules = configService.getConfig(appName +
NacosConfigUtil.FLOW_DATA_ID_POSTFIX,
                NacosConfigUtil.GROUP_ID, 3000);
            if (StringUtil.isEmpty(rules)) {
                return new ArrayList<>();
            }
            java.util.List<FlowRuleEntity> entityList =
converter.convert(rules);
            entityList.forEach(e -> e.setApp(appName));
            return entityList;
        }
    }
```

控制台通过 RESTful API 获取最新的规则，具体代码如下：

```
    @RestController
    @RequestMapping(value = "/v2/flow")
    public class FlowControllerV2 {
        @Autowired
```

```
        @Qualifier("flowRuleDefaultProvider")
        private DynamicRuleProvider<List<FlowRuleEntity>> ruleProvider;

        @GetMapping("/rules")
        @AuthAction(PrivilegeType.READ_RULE)
        public Result<List<FlowRuleEntity>> apiQueryMachineRules(@RequestParam
String app) {

            if (StringUtil.isEmpty(app)) {
                return Result.ofFail(-1, "app can't be null or empty");
            }
            try {
                //①使用 FlowRuleNacosProvider 类获取规则
                List<FlowRuleEntity> rules = ruleProvider.getRules(app);
                if (rules != null && !rules.isEmpty()) {
                    for (FlowRuleEntity entity : rules) {
                        entity.setApp(app);
                        if (entity.getClusterConfig() != null &&
entity.getClusterConfig().getFlowId() != null) {
                            entity.setId(entity.getClusterConfig().getFlowId());
                        }
                    }
                }
                //②将最新的规则更新到 Sentinel 的本地缓存中
                rules = repository.saveAll(rules);
                return Result.ofSuccess(rules);
            } catch (Throwable throwable) {
                logger.error("Error when querying flow rules", throwable);
                return Result.ofThrowable(-1, throwable);
            }
        }
        ...
    }
```

6.9.5 【实例】将 Dubbo 应用接入 Sentinel，实现规则的动态配置和持久化

本实例的源码在本书配套资源的“chaptersix/ verify-persistence-rule”目录下。

本实例将 Dubbo 应用接入 Sentinel，实现规则的动态配置和持久化，持久化的数据源为 Nacos 配置中心。Sentinel 默认是不支持规则的持久化的，需要手动改造其源码。

1. 改造 Sentinel Dashboard 源码，以支持 Nacos 数据源

（1）将能够发布和获取流量控制规则的 Nacos 相关类从 Dashboard 的单元测试包中，迁移到 Dashboard 的发布包中，如图 6-28 所示。

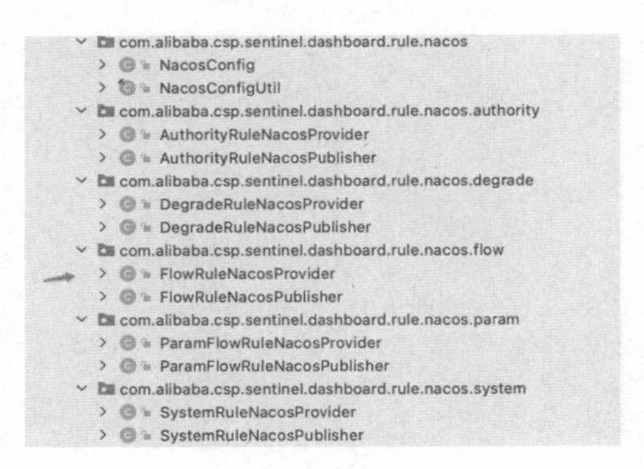

图 6-28

（2）用 Nacos 流量控制规则的发布器 FlowRuleNacosPublisher 类和获取器 FlowRuleNacosProvider 类，替换前台 API 中与通信渠道相关的类（比如 SentinelApiClient 类），具体代码如下：

```
@RestController
@RequestMapping(value = "/v1/flow")
public class FlowControllerV1 {
    @Autowired
    private InMemoryRuleRepositoryAdapter<FlowRuleEntity> repository;

    //①去掉通信渠道
    //@Autowired
    //private SentinelApiClient sentinelApiClient;
    //②添加 Nacos 的流量控制规则获取器
    @Autowired
    @Qualifier("flowRuleNacosProvider")
    private DynamicRuleProvider<List<FlowRuleEntity>> ruleProvider;

    //③添加 Nacos 的流量控制规则发布器
    @Autowired
    @Qualifier("flowRuleNacosPublisher")
    private DynamicRulePublisher<List<FlowRuleEntity>> rulePublisher;

    @PostMapping("/rule")
    @AuthAction(PrivilegeType.WRITE_RULE)
    public Result<FlowRuleEntity> apiAddFlowRule(
        @RequestBody
```

```
        FlowRuleEntity entity) {
            try {
                entity = repository.save(entity);
                //④修改新增流量控制规则的前台 API，通信渠道从 Netty 或者 HTTP 切换为 Nacos
                //publishRules(entity.getApp(), entity.getIp(),
entity.getPort()).get(5000, TimeUnit.MILLISECONDS);
                publishRules(entity.getApp());

                return Result.ofSuccess(entity);
            } catch (Throwable t) {
                Throwable e = (t instanceof ExecutionException) ? t.getCause() : t;
                logger.error("Failed to add new flow rule, app={}, ip={}",
                    entity.getApp(), entity.getIp(), e);

                return Result.ofFail(-1, e.getMessage());
            }
        }

        @DeleteMapping("/delete.json")
        @AuthAction(PrivilegeType.WRITE_RULE)
        public Result<Long> apiDeleteFlowRule(Long id) {
            try {
                //⑤修改删除流量控制规则的前台 API，通信渠道从 Netty 或者 HTTP 切换为 Nacos
                //publishRules(oldEntity.getApp(), oldEntity.getIp(),
oldEntity.getPort()).get(5000, TimeUnit.MILLISECONDS);
                publishRules(oldEntity.getApp());

                return Result.ofSuccess(id);
            } catch (Throwable t) {
                Throwable e = (t instanceof ExecutionException) ? t.getCause() : t;
                logger.error("Error when deleting flow rules, app={}, ip={}, id={}",
                    oldEntity.getApp(), oldEntity.getIp(), id, e);

                return Result.ofFail(-1, e.getMessage());
            }
        }

        @GetMapping("/rules")
        @AuthAction(PrivilegeType.READ_RULE)
        public Result<List<FlowRuleEntity>> apiQueryMachineRules(
            @RequestParam
        String app, @RequestParam
        String ip, @RequestParam
        Integer port) {
            try {
```

```
            //⑥修改获取流量控制规则的前台 API，通信渠道从 Netty 或者 HTTP 切换为 Nacos
            List<FlowRuleEntity> rules = ruleProvider.getRules(app);
            //List<FlowRuleEntity> rules = sentinelApiClient.
fetchFlowRuleOfMachine(app, ip, port);
            rules = repository.saveAll(rules);

            return Result.ofSuccess(rules);
        } catch (Throwable throwable) {
            logger.error("Error when querying flow rules", throwable);

            return Result.ofThrowable(-1, throwable);
        }
    }

    @PutMapping("/save.json")
    @AuthAction(PrivilegeType.WRITE_RULE)
    public Result<FlowRuleEntity> apiUpdateFlowRule(Long id, String app,
        String limitApp, String resource, Integer grade, Double count,
        Integer strategy, String refResource, Integer controlBehavior,
        Integer warmUpPeriodSec, Integer maxQueueingTimeMs) {
        try {
            entity = repository.save(entity);

            if (entity == null) {
                return Result.ofFail(-1, "save entity fail: null");
            }
            //⑦修改更新流量控制规则的前台 API，通信渠道从 Netty 或者 HTTP 切换为 Nacos
            //publishRules(entity.getApp(), entity.getIp(),
entity.getPort()).get(5000, TimeUnit.MILLISECONDS);
            publishRules(entity.getApp());

            return Result.ofSuccess(entity);
        } catch (Throwable t) {
            Throwable e = (t instanceof ExecutionException) ? t.getCause() : t;
            logger.error("Error when updating flow rules, app={}, ip={},
ruleId={}",
                entity.getApp(), entity.getIp(), id, e);

            return Result.ofFail(-1, e.getMessage());
        }
    }
    //⑧调用 Nacos 配置中心发布流量控制规则，并持久化
    private void publishRules( /*@NonNull*/
        String app) throws Exception {
        List<FlowRuleEntity> rules = repository.findAllByApp(app);
```

```
            rulePublisher.publish(app, rules);
        }
    }
```

在修改完流量控制规则的前台 API（FlowControllerV1）后，对应的前端的代码不需要调整即可生效。

前端对应的调用入口的具体代码如下（路径为 webapp/resources/app/scripts/directives/sidebar/sidebar.html）：

```
    <html>
     <head></head>
     <body>
      <li ui-sref-active="active" ng-if="!entry.isGateway"> <a
ui-sref="dashboard.flowV1({app: entry.app})"> <i class="glyphicon
glyphicon-filter"></i>  流量控制规则</a> </li>
     </body>
    </html>
    <html>
     <head></head>
     <body>
      .state('dashboard.flowV1', { templateUrl: 'app/views/flow_v1.html', url:
'/flow/:app', controller: 'FlowControllerV1', resolve: { loadMyFiles:
['$ocLazyLoad', function ($ocLazyLoad) { return $ocLazyLoad.load({ name:
'sentinelDashboardApp', files:
[ 'app/scripts/controllers/flow_v1.js', ] }); }] } }
     </body>
    </html>
```

（3）重新编译修改后的 Sentinel Dashboard，并部署。

2. 在控制台中创建 Dubbo 接口的流量控制规则，验证规则的持久化

创建基于流量控制的规则，如图 6-29 所示。重启 Sentinel 控制台后，规则可以从 Nacos 配置中心同步过来。

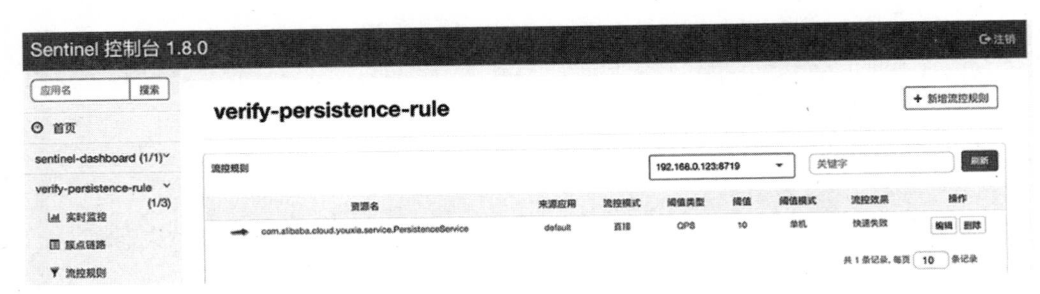

图 6-29

3. 模拟用户的并发请求，验证流量防护

在程序 verify-persistence-rule 中模拟用户并发请求，调用 Dubbo 接口 com.alibaba.cloud.youxia.service.PersistenceService。当 QPS 值超过阈值 10 时，流量防护生效，并快速失败，如图 6-30 所示。

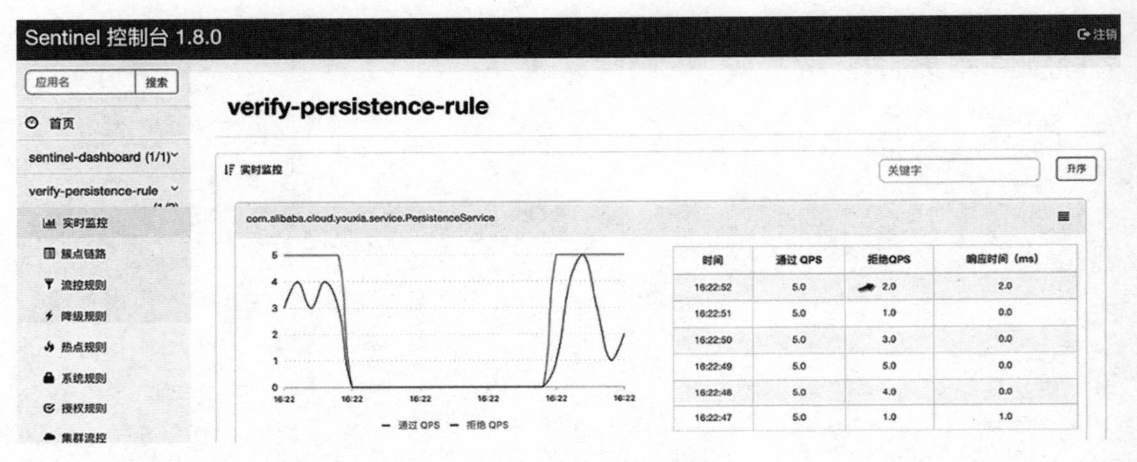

时间	通过 QPS	拒绝QPS	响应时间（ms）
16:22:52	5.0	2.0	2.0
16:22:51	5.0	1.0	0.0
16:22:50	5.0	3.0	0.0
16:22:49	5.0	5.0	0.0
16:22:48	5.0	4.0	0.0
16:22:47	5.0	1.0	1.0

图 6-30

中级篇

第 7 章 分布式事务处理——基于 Seata

随着业务的快速发展，以及业务复杂度越来越高，传统的单体应用逐渐暴露出一些问题——开发效率低、可维护性差、架构扩展性差、部署不灵活、健壮性差等。

微服务架构将单个服务拆分成一系列小服务，且这些小服务都拥有独立的进程，彼此独立，能很好地解决传统的单体应用的问题。但是服务被拆分后，需要开发人员来解决“确保在分布式环境下服务之间的数据一致性”这个技术难点。Seata 就是为了解决这个技术难点而生的。

7.1 认识分布式事务

7.1.1 什么是分布式事务

1. 事务的概念

事务是由多个计算任务构成的一个具有明确边界的工作集合。它包含以下两个目的：

- 为数据库操作提供从“失败态”恢复到“正常态”的方法，提供在数据库出现异常时确保数据一致性的方法。
- 当多个应用并发访问数据库时，可以隔离应用，确保应用能够安全地访问数据库。

事务具备如下 4 个特性。

- 原子性（Atomicity）：事务作为一个整体被执行，对数据库的操作要么都被执行，要么都不被执行。

- 一致性（Consistency）：事务应确保数据库的状态从一个“一致状态”转变为另一个“一致状态”。“一致状态”是指数据库中的数据满足完整性约束。
- 隔离性（Isolation）：在多个事务并发执行时，一个事务的执行不能影响其他事务的执行。
- 持久性（Durability）：一旦某个事务被提交，它对数据库的修改应该被永久保存在数据中。

2. 分布式事务的概念

分布式事务是指：事务的参与者、支持事务的服务器、资源服务器及事务管理器分别位于分布式系统的不同的实例上。

简单说，一个大的操作由不同的小操作组成，这些小的操作分布在不同的服务器上，且属于不同的应用。分布式事务需要保证这些小操作要么全部执行成功，要么全部执行失败。

从本质上来说，分布式事务就是为了确保不同应用间的数据一致性，包括 RPC 的数据一致性和数据存储的数据一致性。

如图 7-1 所示，将实例 A 中的本地事务拆分为 3 个部分：实例 B 中的事务 1、实例 C 中的事务 2 和实例 D 中的事务 3。事务之间的关系如下：

本地事务 = 事务 1 + 事务 2 + 事务 3

说明：事务 1、事务 2 和事务 3 利用 RPC 通信来传输事务数据并协调彼此的执行。

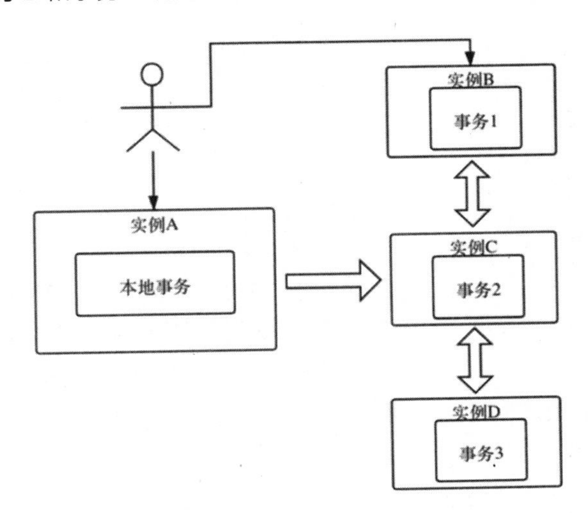

图 7-1

数据一致性可以分为如下 3 类。

- 强一致性：在数据更新成功后，任意时刻所有副本中的数据都是一致的。
- 弱一致性：在数据更新成功后，系统不承诺立即可以读到最新写入的数据，也不承诺具体多久后可以读取最新写入的数据。

在分布式环境中，弱一致性具有以下两个缺点。

- 如果更新数据时出现异常，则弱一致性不能确保数据最终能够被更新到数据副本中。
- 如果主副本更新成功并完成了数据的持久化，但是在主副本同步数据到从副本的过程中出现了数据同步异常，则弱一致性也不能确保主/从副本之间的数据一致性。

- 最终一致性：弱一致性的一种形式。在数据更新成功后，系统不承诺可以实时地返回最新写入的数据，但是可以保证一定会返回上一次更新操作之后的最新数据，只是会有一些时延。

如图 7-2 所示，实例 A、实例 B 及实例 C 同时对数据库进行读和写的操作，实例 A 执行 SQL 语句“insert into datasource0.t_order value (7878,1234,'测试订单');”，并插入一条数据，强一致性、弱一致性及最终一致性的表现如下：

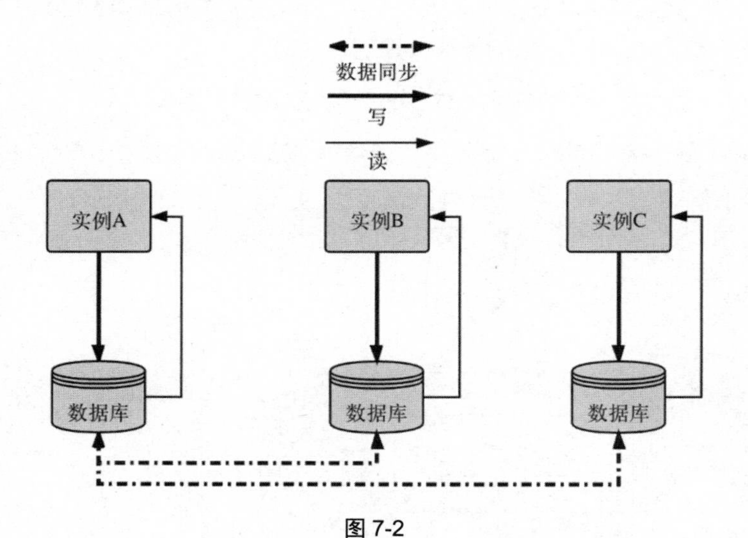

图 7-2

（1）如果实例之间的数据是强一致性的，则实例 B 和实例 C 能实时地读取新插入的数据，并且实例 A、实例 B 和实例 C 观察到的数据状态是一样的。

（2）如果 3 个实例之间的数据是弱一致性的，则实例 B 和实例 C 不一定能够读取新插入的数据，并且实例 A、实例 B 和实例 C 观察到的数据的状态可能是不一致的。

（3）如果 3 个实例之间的数据是最终一致性的，则实例 B 和实例 C 一定能够读取新插入的数据。在新插入的数据没有被同步到实例 B 和 C 之前，数据是不一致的；但是最终一致性会确保在一定时间之内，将新插入的数据从实例 A 对应的数据库同步到实例 B 和实例 C 对应的数据库。

7.1.2　为什么需要分布式事务

从系统整体的架构角度来看，分布式事务涉及的场景分为两类：

- 事务只涉及一个应用，但涉及多个数据存储。
- 事务涉及多个应用，且每个应用可能连接着一个或者多个数据存储。

图 7-3 所示的是电商领域最常见的两个服务：订单服务和商品服务。

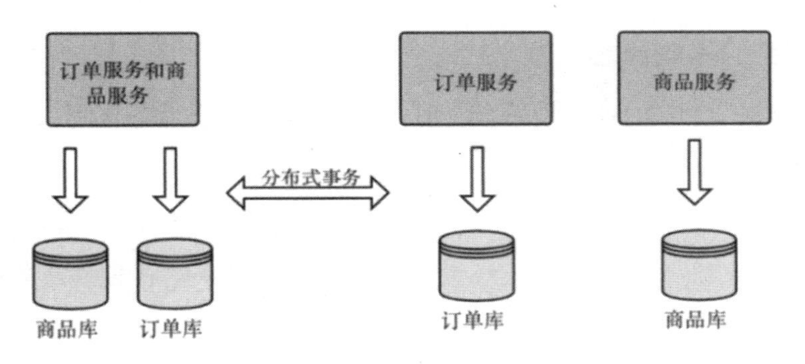

图 7-3

假设订单服务和商品服务没有被拆分，而是强耦合的，若将商品库和订单库进行水平拆分，则订单服务和商品服务将访问不同的实例数据库。如果在商品服务扣除商品库存成功后订单服务出现了故障，导致订单状态更新失败，若没有分布式事务，则商品库存和订单库存相关表数据就会不一致。假如，在订单服务出现故障时服务的流量非常大，比如 TPS 为 100 000，则会影响用户 100 000 次的正常购买商品。

服务之间的依赖关系越强，则故障造成的数据不一致性的影响越大。

如果将订单服务和商品服务进行水平拆分，则此时既要确保数据库层的数据一致性，还要确保 RPC 层的数据一致性。比如，在更新完商品库存后，通过 RPC 调用（比如 Dubbo）订单服务去更新订单状态，则存在以下两种情况：

- 如果在调用应用层的 RPC 接口时失败，则调用方可以通过重试或者补偿来确保业务功能的完整性。

- 如果在 RPC 接口更新订单库失败时已经成功调用 RPC 接口，但是数据库操作失败，则此时需要关联的 RPC 接口和数据库操作在一个全局事务里，并且具备分布式的特性。

由上可以看到，分布式事务可以解决 RPC 层和数据库层的分布式数据一致性问题。

7.2 认识 Seata

Seata 是一款开源的分布式事务解决方案，社区活跃度极高，它致力于在微服务架构中提供高性能和简单易用的分布式事务服务。Seata 为开发者提供了 AT、TCC、Saga 和 XA 共 4 种事务模式。

7.2.1 Seata 的基础概念

Seata 的基础概念主要包括：基础角色、事务分组、注册中心及配置中心。

1. 基础角色

Seata 的基础角色包括以下 3 个。

- TC（Transaction Coordinator）：事务协调者，它维护全局和分支事务的状态，驱动全局事务提交或回滚。
- TM（Transaction Manager）：事务管理器，它定义了全局事务的范围，主要包括开始全局事务、提交全局事务和回滚全局事务。
- RM（Resource Manager）：资源管理器，它管理分支事务处理的资源，与 TC 通信并注册分支事务和报告分支事务的状态，驱动分支事务的提交和回滚。

三者的关系如图 7-4 所示。

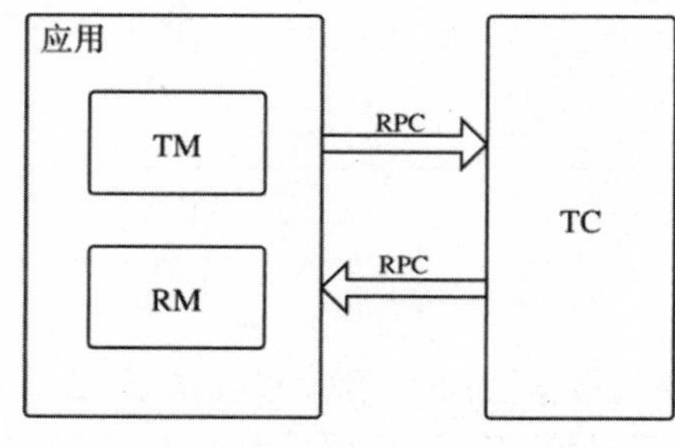

图 7-4

（1）TC 单独部署，TM 和 RM 与应用强绑定，共用一个虚拟机并一起部署。

（2）在 TM 和 RM 与 TC 之间用 RPC 完成资源数据的交换。

2. 事务分组

事务分组是 Seata 的资源逻辑，类似于服务实例。

3. 注册中心

Seata 可以将部署事务协调者的机器实例注册到注册中心，事务管理器和资源管理器可以从注册中心获取部署事务协调者的机器实例的集群列表。

Seata 支持的注册中心包括：Nacos、Sofa、Etcd、Eureka、Consul 等。

4. 配置中心

Seata 的服务器端（TC）和客户端（TM 和 RM）都支持分布式配置中心，开发人员可以将配置信息动态地存储在配置中心中。Seata 支持的配置中心包括：Nacos、Apollo、Etcd、ZooKeeper 等。

7.2.2 Seata 的事务模式

在 Seata 定义的全局事务基础框架中，不同角色的功能是不一样的。在全局事务基础框架中，Seata 处理分布式事务的整体流程如图 7-5 所示。

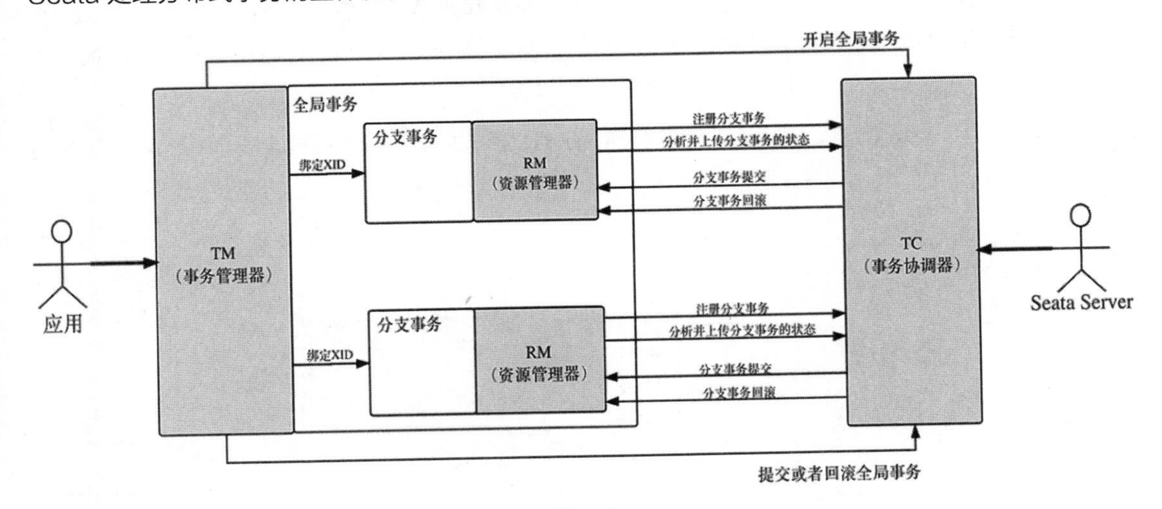

图 7-5

（1）TM 向 TC 发起请求，包括开启（Begin）、提交（Commit）和回滚（Rollback）全局事务。

（2）TM 把代表全局事务的 XID 绑定到分支事务。

（3）RM 向 TC 发起请求，并注册分事务，把分支事务关联到 XID 所代表的全局事务。

（4）RM 把分支事务的执行结果上报给 TC。

（5）TC 发送分支事务提交（Branch Commit）或分支事务回滚（Branch Rollback）命令给 RM。

在 Seata 中，全局事务的处理过程分为如下两个阶段。

（1）执行阶段：Seata 会执行与分支事务相关的逻辑，并保证执行结果是可以回滚和持久化的。

（2）完成阶段：应用会根据全局事务中所有的分支事务在执行阶段的结果，形成全局事务的决议，并通过 TM 向 TC 发起提交或者回滚全局事务的请求；之后，TC 向 RM 发起请求，提交或者回滚分支事务。

在 Seata 中，事务模式是指运行在 Seata 全局事务框架下的“分支事务的行为模式”（准确来讲应该是“分支事务模式”）。

不同的事务模式会使用不同的方式去完成全局事务的两个阶段，从而形成了 Seata 的 4 种事务模式：AT 模式、XA 模式、TCC 模式及 Saga 模式。

1. AT 模式

AT 模式是 Seata 主推的分布式事务解决方案，它最早来源于阿里中间件团队发布的 TXC 服务，“上云”后改名为 GTS。AT 模式屏蔽了底层 JDBC 数据层的细节，让应用能够无感知地使用分布式事务，自动代理应用数据源，并进行事务相关的操作。

图 7-6 是 Seata 的 AT 模式的架构示意图。在 AT 模式中，Seata 完成全局事务的两个阶段的过程如下。

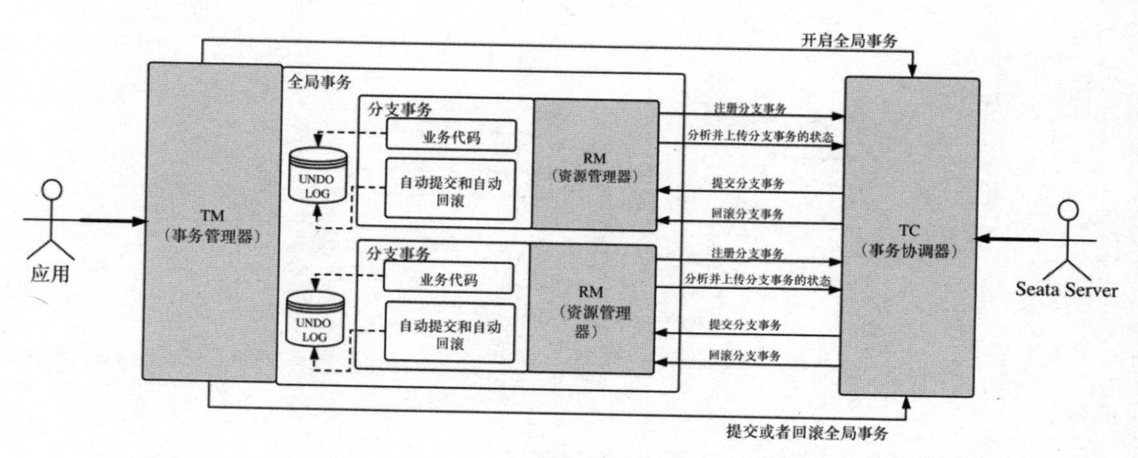

图 7-6

（1）第一阶段：应用只需要关注自己的“业务 SQL 代码”，Seata 将应用的“业务 SQL 代码的执行”作为第一阶段。Seata 框架会自动代理应用的数据源，并生成事务第二阶段的提交和回滚事务操作，记录在 UNDO LOG 日志表中。

（2）第二阶段：如果 TC 事务协调器通知分支事务处理成功，则 Seata 会提交分支事务（在 Seata 的 AT 模式中，提交分支事务就是“删除 UNDO LOG 日志表中的事务相关日志”）；如果 TC 事务协调器通知分支事务处理失败，则 Seata 会回滚分支事务（从 UNDO LOG 日志表中读取事物回滚的日志）。

2. XA 模式

XA 模式是 Seata 利用事务资源（数据库、消息服务等）来提供对 XA 协议的支持，以 XA 协议的机制来管理分支事务的一种事务模式。

图 7-7 是 Seata 的 XA 模式的架构示意图。在 XA 模式中，Seata 完成全局事务的两个阶段的过程如下：

（1）第一阶段，利用 RM 代理应用的数据源，并创建数据库的代理连接。通过开启 XA 模式事务拦截应用的 SQL 语句，执行 XA 模式预处理。为了防止应用宕机，造成数据丢失，第一阶段的 XA 模式操作会被 Seata 持久化。

（2）第二阶段，TC 通知 RM 执行提交或者回滚 XA 模式的分支事务。

（3）在 XA 模式中，应用在开启 XA 模式事务之后，会注册分支事务；在预处理 XA 模式事务之后，会分析并上传分支事务的状态。

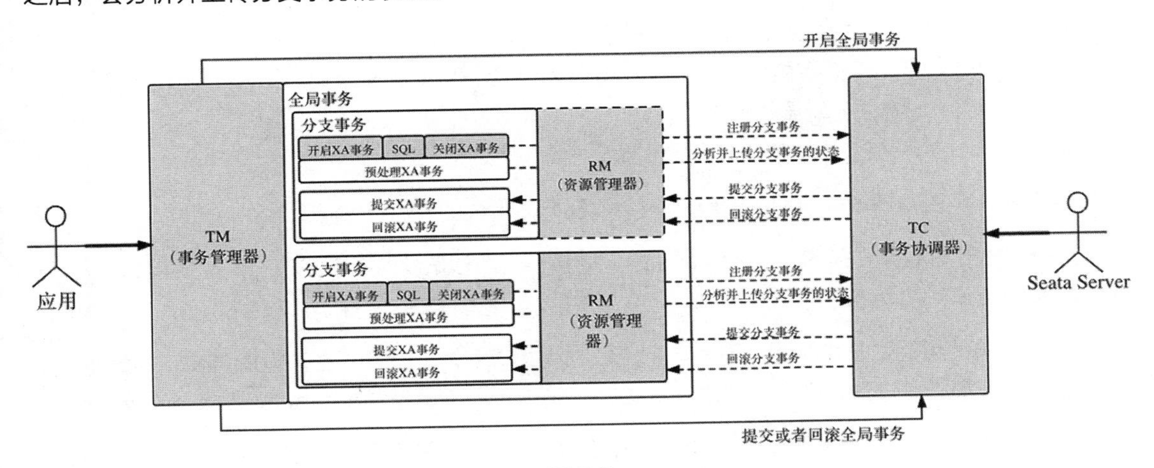

图 7-7

3. TCC 模式

2019 年 3 月，Seata 开源了 TCC 模式，它是一个分布式的全局事务，并且是一个两阶段的分布式事务。

在 TCC 模式中,全局事务是由若干分支事务组成的。分支事务要满足两阶段提交模型的要求(即需要每个分支事务都具备该模型），支持把自定义的分支事务纳入全局事务的管理中。

图 7-8 是 Seata 的 TCC 模式的架构示意图。

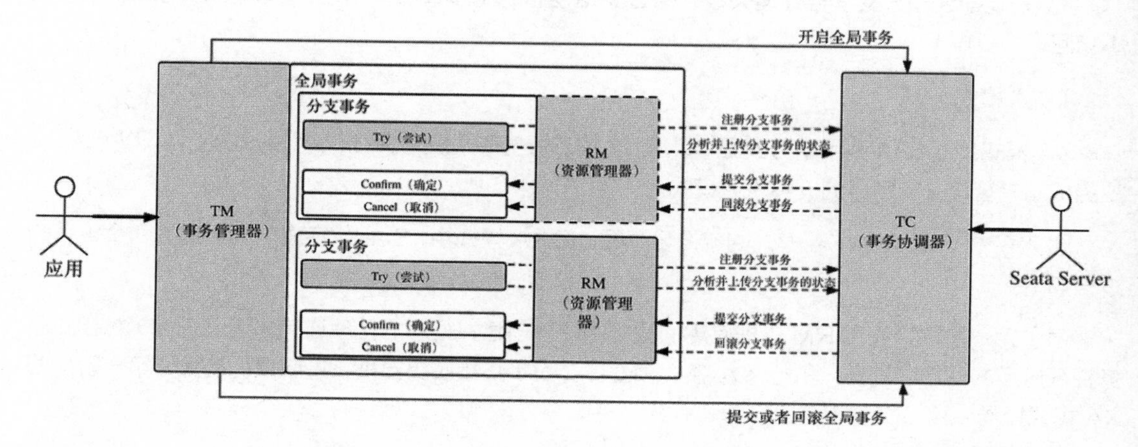

图 7-8

（1）在 TCC 模式中，用 Try()、Confirm()及 Cancel()这 3 个方法来实现事务的两阶段提交。3 个方法的描述如下。

- Try()：在应用中检测和预留系统资源。
- Confirm()：在应用中提交执行的业务操作。TCC 模式要求，如果 Try()方法执行成功，则 Confirm()方法一定要执行成功。
- Cancel()：在应用中释放预留的资源。

（2）RM 负责管理第一阶段的 Try 语句，以及第二阶段的 Confirm 和 Cancel 语句。

（3）TM 开启事务，在 RM 调用 TM 后，TM 执行一阶段的 Try()方法并注册分支事务，以及分析和上传分支事务的状态。

（4）如果调用链路已经完成，则 TM 向 TC 发起第二阶段的分支事务决议的请求；如果所有分支事务的决议都是通过的，则 TC 驱动 RM 去执行第二阶段的事务 Confirm 或者 Cancel 语句。

（5）应用需要在代码中用 Try()、Confirm()及 Cancel()方法实现分布式事务。即针对不同的代码，应用都需要重新实现分布式事务处理的业务逻辑。

4. Saga 模式

在 Saga 模式中，业务流程中的每个参与者都可以提交本地事务。如果某个参与者失败，则补偿本地事务提交成功的参与者，第一阶段正向服务和第二阶段补偿服务都由开发人员来实现。

Saga 的核心就是补偿：第一阶段就是服务的正常顺序调用（数据库事务正常提交），如果都执行成功，则第二阶段什么都不做；但如果在第一阶段中有执行发生异常，则在第二阶段中会依次调用其补偿服务（以保证整个交易的一致性）。

图 7-9 是 Saga 模式的架构示意图。

（1）在开启状态机后，会注册分支事务；在关闭状态机后，会分析并上传分支事务的状态。

（2）在 TC 通知 RM 去提交分支事务时，会转发状态机；在 TC 通知 RM 回滚分支事务时，会重新加载状态机。

（3）在运行状态机之后，将产生的数据存储在数据库中。

（4）在 Saga 模式中，主要用状态机来编排分布式事务中应用执行的顺序，以及事务执行的阶段。

（5）在 Saga 模式中，分布式事务通常是由事件驱动的，各个参与者之间是异步执行的。Saga 模式是一种长事务解决方案。

在 Saga 模式中，分布式事务具备如下优势：①在第一阶段提交本地数据库事务时，采用的是无锁设计，所以应用具备高性能；②参与者可以采用事务驱动来异步执行，所以应用具备高吞吐量；③补偿服务即正向服务的“反向”，所以应用非常容易实现，具备易用性。

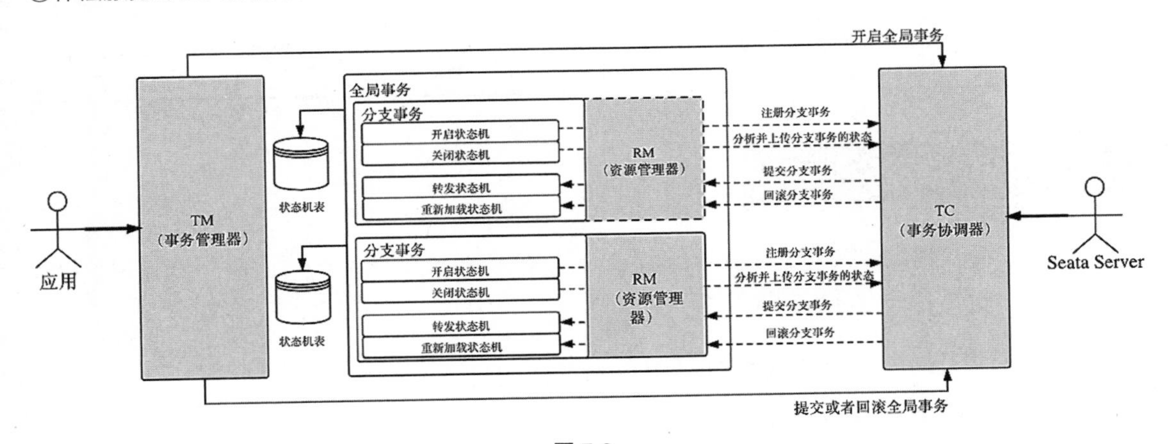

图 7-9

5. 对比 Seata 中的 AT、XA、TCC、Saga 四种事务模式

AT、XA、TCC、Saga 四种事务模式的对比见表 7-1。

表 7-1　AT、XA、TCC、Saga 四种事务模式的对比

对比项	AT 模式	XA 模式	TCC 模式	Saga 模式
数据源代理	支持代理应用的数据源	支持代理应用的数据源	不支持代理应用的数据源	不支持代理应用的数据源
零侵入	可以将 Seata 零侵入地接入应用	可以将 Seata 零侵入地接入应用	不支持	不支持
两阶段事务	是一个标准的两阶段事务	是一个标准的两阶段事务	是一个标准的两阶段事务	是一个标准的两阶段事务
状态机	不支持状态机	不支持状态机	不支持状态机	支持状态机
全局事务	支持全局事务	支持全局事务	支持全局事务	支持全局事务
数据库存储	需要数据库来存储	不支持数据库存储	不支持数据库存储	如果状态机是持久化的，则支持数据库存储
注册中心	支持用注册中心来管理 Seata Server 集群	支持用注册中心来管理 Seata Server 集群	支持用注册中心来管理 Seata Server 集群	支持用注册中心来管理 Seata Server 集群
配置中心	支持用配置中心来管理 Seata 的配置信息	支持用配置中心来管理 Seata 的配置信息	支持用配置中心来管理 Seata 的配置信息	支持用配置中心来管理 Seata 的配置信息
高性能的长事务	不支持高性能的长事务，性能非常差	不支持高性能的长事务，性能非常差	不支持高性能的长事务，性能非常差	支持高性能的长事务，性能非常高
支持本地 ACID 事务的关系型数据库	强支持	强支持	弱支持	弱支持
链路追踪	支持 Skywalking 的分布式链路追踪	支持 Skywalking 的分布式链路追踪	支持 Skywalking 的分布式链路追踪	支持 Skywalking 的分布式链路追踪

7.3　将应用接入 Seata

搭建高可用的服务器端环境（Seata Server），以及将应用接入 Seata，是开发人员使用 Seata 作为项目的分布式事务框架的前提条件。

7.3.1　搭建 Seata Server 的高可用环境

Seata 的高可用性依赖注册中心、配置中心和数据库来实现。

1. 下载 Seata 的安装包

从官网下载 Seata 的安装包 seata-server-1.4.1.zip，并解压缩。

2. 搭建注册中心和配置中心的环境

本书统一采用 Nacos 作为 Seata 的注册中心和配置中心。

将 Seata 接入 Nacos 的过程如下：

（1）搭建 Nacos 环境的具体细节可以参考本书的第 4 章和第 5 章。

（2）修改 Seata Server 的配置文件 registry.conf（文件路径为：/ seata1.4.1/config/）。在 registry.conf 中，注册中心和配置中心的配置信息如下。

```
//①注册中心的配置信息
registry {
  type = "nacos"
  loadBalance = "RandomLoadBalance"
  loadBalanceVirtualNodes = 10
  nacos {
    application = "seata-server"
    serverAddr = "127.0.0.1:8848"
    group = "SEATA_GROUP"
    namespace = "c7ba173f-29e5-4c58-ae78-b102be11c4f9"
    cluster = "default"
    username = "nacos"
    password = "nacos"
  }
  ...
}
//②配置中心的配置信息
config {
  type = "nacos"
  nacos {
    serverAddr = "127.0.0.1:8848"
    namespace = "c7ba173f-29e5-4c58-ae78-b102be11c4f9"
    group = "SEATA_GROUP"
    username = "nacos"
    password = "nacos"
  }
  ...
}
```

3. 搭建存储环境

Seata 支持文件、Redis 和关系型数据库（比如 MySQL）这 3 种全局锁的存储类型。这里采用 MySQL 来确保 Seata 存储的高可用性。

搭建过程如下：

（1）从 MySQL 官网下载安装包，并安装 MySQL 数据库。

（2）在 MySQL 数据库中创建全局 Session 表、分支 Session 表及全局锁表。对应的 SQL 语句如下：

```
//①创建全局 Session 表
CREATE TABLE IF NOT EXISTS `global_table`
(
    `xid`                       VARCHAR(128) NOT NULL,
    `transaction_id`            BIGINT,
    `status`                    TINYINT     NOT NULL,
    `application_id`            VARCHAR(32),
    `transaction_service_group` VARCHAR(32),
    `transaction_name`          VARCHAR(128),
    `timeout`                   INT,
    `begin_time`                BIGINT,
    `application_data`          VARCHAR(2000),
    `gmt_create`                DATETIME,
    `gmt_modified`              DATETIME,
    PRIMARY KEY (`xid`),
    KEY `idx_gmt_modified_status` (`gmt_modified`, `status`),
    KEY `idx_transaction_id` (`transaction_id`)
) ENGINE = InnoDB
  DEFAULT CHARSET = utf8;
//②创建分支 Session 表
CREATE TABLE IF NOT EXISTS `branch_table`
(
    `branch_id`         BIGINT       NOT NULL,
    `xid`               VARCHAR(128) NOT NULL,
    `transaction_id`    BIGINT,
    `resource_group_id` VARCHAR(32),
    `resource_id`       VARCHAR(256),
    `branch_type`       VARCHAR(8),
    `status`            TINYINT,
    `client_id`         VARCHAR(64),
    `application_data`  VARCHAR(2000),
    `gmt_create`        DATETIME(6),
    `gmt_modified`      DATETIME(6),
    PRIMARY KEY (`branch_id`),
    KEY `idx_xid` (`xid`)
) ENGINE = InnoDB
  DEFAULT CHARSET = utf8;
//③创建全局锁表
CREATE TABLE IF NOT EXISTS `lock_table`
(
```

```
    `row_key`       VARCHAR(128) NOT NULL,
    `xid`           VARCHAR(96),
    `transaction_id` BIGINT,
    `branch_id`     BIGINT      NOT NULL,
    `resource_id`   VARCHAR(256),
    `table_name`    VARCHAR(32),
    `pk`            VARCHAR(36),
    `gmt_create`    DATETIME,
    `gmt_modified`  DATETIME,
    PRIMARY KEY (`row_key`),
    KEY `idx_branch_id` (`branch_id`)
) ENGINE = InnoDB
  DEFAULT CHARSET = utf8;
```

（3）Seata 默认采用文件存储。在文件 file.conf 中配置 MySQL 存储的具体配置信息如下所示：

```
store {
  mode = "db"
  db {
    datasource = "druid"
    dbType = "mysql"
    driverClassName = "com.mysql.jdbc.Driver"
    url = "jdbc:mysql://127.0.0.1:3306/seata_server"
    user = "root"
    password = "123456huxian"
    minConn = 5
    maxConn = 100
    globalTable = "global_table"
    branchTable = "branch_table"
    lockTable = "lock_table"
    queryLimit = 100
    maxWait = 5000
  }
  ...
}
```

4. 将 Seata 的配置信息迁移到 Nacos 配置中心中

Seata 支持从配置中心读取配置信息。开发人员可以动态修改配置信息，配置信息会实时生效。

迁移过程如下：

（1）从 Seata 源码文件（文件路径：script/config-center/nacos/nacos-config.sh）中，将脚本文件 nacos-config.sh 复制到 Seata 安装目录的“/seata1.4.1/config /”中。

（2）从 Seata 源码文件（文件路径：script/config-center/config.txt）中，将文本文件 config.txt 复制到 Seata 安装目录的“/seata1.4.1/config /”中。

（3）切换到文件夹“/seata1.4.1/config”中执行如下命令，将文本文件 config.txt 中的内容推送的 Nacos 配置中心。

```
    sh nacos-config.sh -h localhost -p 8848 -g SEATA_GROUP -t
c7ba173f-29e5-4c58-ae78-b102be11c4f9 -u nacos -w nacos
```

（4）在 Nacos 的控制台中可以看到对应的配置信息，如图 7-10 所示。

图 7-10

5. 用命令启动 Seata Server

切换到 Seata 的安装目录文件“/seata1.4.1/config”，执行命令“sh seata-server.sh”即可启动 Seata Server。在启动成功之后，在 Nacos 控制台的服务列表中可以看到 Seata Server，如图 7-11 所示。

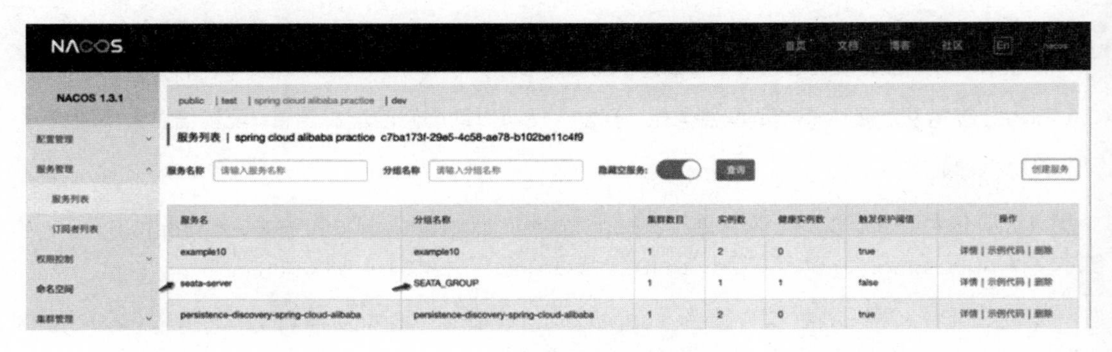

图 7-11

7.3.2 【实例】使用 seata-spring-boot-starter 将应用接入 Seata

本实例的源码在本书配套资源的"chapterseven/ use-spring-boot-access-seata"目录下。

本实例使用 seata-spring-boot-starter 将应用接入 Seata。

1. 创建一个 Spring Boot 项目，并添加 Seata 项目的 Jar 包依赖

用 IDEA 创建一个 Spring Boot 项目，然后添加 Seata 项目的 POM 文件的依赖，具体配置如下：

```
<!--添加 seata-spring-boot-starter 的版本-->
<dependency>
    <groupId>io.seata</groupId>
    <artifactId>seata-spring-boot-starter</artifactId>
    <exclusions>
        <exclusion>
           <groupId>org.springframework</groupId>
           <artifactId>spring-beans</artifactId>
        </exclusion>
    </exclusions>
    <version>1.4.1</version>
</dependency>
```

2. 添加程序启动的配置文件 bootstrap.yaml

在 bootstrap.yaml 中添加程序启动的配置信息，具体如下所示：

```
###①定义配置中心和注册中心的公共属性
nacos:
  server-addr: 127.0.0.1:8848
  namespace: c7ba173f-29e5-4c58-ae78-b102be11c4f9
  group: SEATA_GROUP
  seata:
    application: seata-server
    tx-service-group: use-spring-boot-access-seata-tx-group
spring:
  application:
    name: use-spring-boot-access-seata
  main:
allow-bean-definition-overriding: true
###②RM 对应的配置数据源
  datasource:
    url:
```

```
jdbc:mysql://127.0.0.1:3306/seata_server?characterEncoding=utf8&connectTimeo
ut=1000&socketTimeout=3000&autoReconnect=true&useUnicode=true&useSSL=false&s
erverTimezone=UTC
        username: root
        password: 123456huxian
    server:
      port: 7834
    ###③配置 Seata 的相关信息
    seata:
      registry:
        type: nacos
        nacos:
          group: ${nacos.group}
          application: ${nacos.seata.application}
          namespace: ${nacos.namespace}
          cluster: default
          server-addr: ${nacos.server-addr}
          username: "nacos"
          password: "nacos"
      tx-service-group: ${nacos.seata.tx-service-group}
      enabled: true
      enable-auto-data-source-proxy: true
      application-id: use-spring-boot-access-seata
      config:
        type: nacos
        nacos:
          group: ${nacos.group}
          server-addr: ${nacos.server-addr}
          namespace: ${nacos.namespace}
          username: "nacos"
          password: "nacos"
      data-source-proxy-mode: XA
```

3. 添加分布式事务注解，并开启分布式事务

开启分布式事务的代码如下所示：

```
    //通过注解@GlobalTransactional 启动分布式事务
    @GlobalTransactional(name = "spring-boot-access-pay-order", rollbackFor =
Exception.class, lockRetryInternal=5000
    ,lockRetryTimes=5,timeoutMills=30000,propagation=
Propagation.REQUIRES_NEW)
        public boolean placeOrder(Long userId, Long goodId, Long accountId,
Integer num) {
            Result orderResult = createOrder(userId, new BigDecimal(100),
goodId);
            WarehouseServiceBo warehouseServiceBo = new WarehouseServiceBo();
```

```
        warehouseServiceBo.setOrderId(orderResult.getId());
        warehouseServiceBo.setGoodId(goodId);
        warehouseServiceBo.setNum(num);
        WarehouseServiceDTO warehouseServiceDTO =
xaWarehouseService.notifyWarehouse(warehouseServiceBo);
        LogisticsServiceBo logisticsServiceBo = new LogisticsServiceBo();
        logisticsServiceBo.setOrderId(orderResult.getId());
        LogisticsServiceDTO logisticsServiceDTO =
xaLogisticsService.notifyLogistics(logisticsServiceBo);
        if (orderResult.isResult() && warehouseServiceDTO.getStatus() == 200
&& logisticsServiceDTO.getStatus() == 200) {
            boolean updataOrderStatusResult =
updataOrderStatus(orderResult.getId(), 1);
            if (updataOrderStatusResult) {
                return true;
            }
            return false;
        }
        return true;
    }
```

在上面的代码中，通过注解@GlobalTransactional 开启分布式事务，并设置事务回滚的异常类型为 Exception，超时之后的事务重试周期为 5s，超时重试次数为 5 次，事务执行的超时时间为 30s，全局事务隔离级别为 Propagation.REQUIRES_NEW（针对每一个新的连接都开启一个新的事务）。

4. 启动服务

在启动服务之后，可以在启动日志中看到 XA 模式的分布式事务已经初始化完成，如图 7-12 所示。

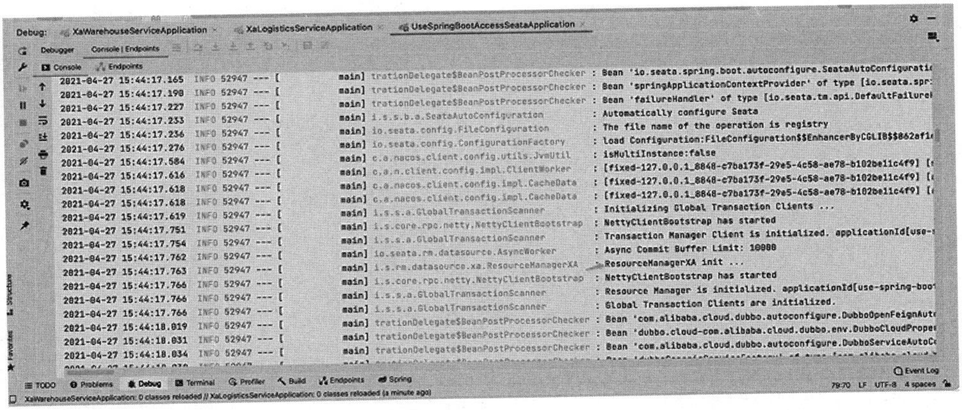

```
Debug: XaWarehouseServiceApplication  XaLogisticsServiceApplication  UseSpringBootAccessSeataApplication
Debugger  Console | Endpoints
Console  Endpoints
2021-04-27 15:44:17.165 INFO 52947 --- [ main] trationDelegate$BeanPostProcessorChecker : Bean 'io.seata.spring.boot.autoconfigure.SeataAutoConfigurati
2021-04-27 15:44:17.198 INFO 52947 --- [ main] trationDelegate$BeanPostProcessorChecker : Bean 'springApplicationContextProvider' of type [io.seata.spr
2021-04-27 15:44:17.227 INFO 52947 --- [ main] trationDelegate$BeanPostProcessorChecker : Bean 'failureHandler' of type [io.seata.tm.api.DefaultFailureH
2021-04-27 15:44:17.233 INFO 52947 --- [ main] i.s.s.b.a.SeataAutoConfiguration : Automatically configure Seata
2021-04-27 15:44:17.236 INFO 52947 --- [ main] io.seata.config.FileConfiguration : The file name of the operation is registry
2021-04-27 15:44:17.276 INFO 52947 --- [ main] io.seata.config.ConfigurationFactory : load Configuration:FileConfiguration$$EnhancerByCGLIB$$862af1
2021-04-27 15:44:17.584 INFO 52947 --- [ main] c.a.nacos.client.config.utils.JvmUtil : isMultiInstance:false
2021-04-27 15:44:17.616 INFO 52947 --- [ main] c.a.n.client.config.impl.ClientWorker : [fixed-127.0.0.1_8848-c7ba173f-29e5-4c58-ae78-b102be11c4f9] [
2021-04-27 15:44:17.618 INFO 52947 --- [ main] c.a.nacos.client.config.impl.CacheData : [fixed-127.0.0.1_8848-c7ba173f-29e5-4c58-ae78-b102be11c4f9] [
2021-04-27 15:44:17.618 INFO 52947 --- [ main] c.a.nacos.client.config.impl.CacheData : [fixed-127.0.0.1_8848-c7ba173f-29e5-4c58-ae78-b102be11c4f9] [
2021-04-27 15:44:17.619 INFO 52947 --- [ main] i.s.s.a.GlobalTransactionScanner : Initializing Global Transaction Clients ...
2021-04-27 15:44:17.751 INFO 52947 --- [ main] i.s.core.rpc.netty.NettyClientBootstrap : NettyClientBootstrap has started
2021-04-27 15:44:17.754 INFO 52947 --- [ main] i.s.s.a.GlobalTransactionScanner : Transaction Manager Client is initialized. applicationId[use-
2021-04-27 15:44:17.762 INFO 52947 --- [ main] io.seata.rm.datasource.AsyncWorker : Async Commit Buffer Limit: 10000
2021-04-27 15:44:17.763 INFO 52947 --- [ main] i.s.rm.datasource.xa.ResourceManagerXA : ResourceManagerXA init ...
2021-04-27 15:44:17.766 INFO 52947 --- [ main] i.s.core.rpc.netty.NettyClientBootstrap : NettyClientBootstrap has started
2021-04-27 15:44:17.766 INFO 52947 --- [ main] i.s.s.a.GlobalTransactionScanner : Resource Manager is initialized. applicationId[use-spring-boo
2021-04-27 15:44:17.766 INFO 52947 --- [ main] i.s.s.a.GlobalTransactionScanner : Global Transaction Clients are initialized.
2021-04-27 15:44:18.019 INFO 52947 --- [ main] trationDelegate$BeanPostProcessorChecker : Bean 'com.alibaba.cloud.dubbo.autoconfigure.DubboOpenFeignAut
2021-04-27 15:44:18.031 INFO 52947 --- [ main] trationDelegate$BeanPostProcessorChecker : Bean 'dubbo.cloud-com.alibaba.cloud.dubbo.env.DubboCloudPrope
2021-04-27 15:44:18.034 INFO 52947 --- [ main] trationDelegate$BeanPostProcessorChecker : Bean 'com.alibaba.cloud.dubbo.autoconfigure.DubboServiceAutoC
TODO  Problems  Debug  Terminal  Profiler  Build  Endpoints  Spring  Event Log
XaWarehouseServiceApplication: 0 classes reloaded // XaLogisticsServiceApplication: 0 classes reloaded (a minute ago)  79:70 LF UTF-8 4 spaces
```

图 7-12

在 Nacos 的控制台中，可以看到服务已经被注册到 Nacos 的注册中心中了，如图 7-13 所示。

NACOS. HOME DOCS BLOG COMMUNITY 中 nacos

NACOS 1.3.1

ConfigManagement
ServiceManagement
Service List
Subscribers
Authority Control
Namespace
ClusterManagement

public | test | spring cloud alibaba practice | dev

Service List | spring cloud alibaba practice c7ba173f-29e5-4c58-ae78-b102be11c4f9

Service Name Enter Service Name Group Name Enter Group Name Hidden Empty Service: Search Create Service

Service Name	Group Name	Cluster Count	Instance Count	Healthy Instance Count	Trigger Protection Threshold	Operation
xa-warehouse-service	SEATA_GROUP	1	1	1	false	Details | Code Example | Delete
example10	example10	1	2	0	true	Details | Code Example | Delete
seata-server	SEATA_GROUP	1	1	1	false	Details | Code Example | Delete
xa-logistics-service	SEATA_GROUP	1	1	1	false	Details | Code Example | Delete
use-spring-boot-access-seata	SEATA_GROUP	1	1	1	false	Details | Code Example | Delete
persistence-discovery-spring-cloud-alibaba	persistence-discovery-spring-cloud-alibaba	1	2	0	true	Details | Code Example | Delete

图 7-13

7.3.3 【实例】使用 Spring Cloud Alibaba 将应用接入 Seata

本实例的源码在本书配套资源的“chapterseven/ use-spring-cloud-alibaba-access-seata”目录下。

本实例使用 Spring Cloud Alibaba 作为基础框架将应用接入 Seata，具体过程如下。

1. 创建一个 Spring Cloud Alibaba 项目

使用 IDEA 创建一个 Spring Cloud Alibaba 项目。在项目创建完成后，添加 Seata 的相关 Jar 包依赖，具体如下所示：

```
  <dependency>
    <groupId>com.alibaba.cloud</groupId>
    <artifactId>spring-cloud-starter-alibaba-seata</artifactId>
        <exclusions>
            <exclusion>
                <groupId>io.seata</groupId>
                <artifactId>seata-spring-boot-starter</artifactId>
            </exclusion>
        </exclusions>
</dependency>
```

```
<dependency>
    <groupId>io.seata</groupId>
        <artifactId>seata-spring-boot-starter</artifactId>
        <exclusions>
            <exclusion>
                <groupId>org.springframework</groupId>
                <artifactId>spring-beans</artifactId>
            </exclusion>
        </exclusions>
    <version>1.4.1</version>
</dependency>
```

2. 启动服务

在启动服务之后，可以在启动日志中看到 XA 模式的分布式事务已经完成了初始化，如图 7-14 所示。

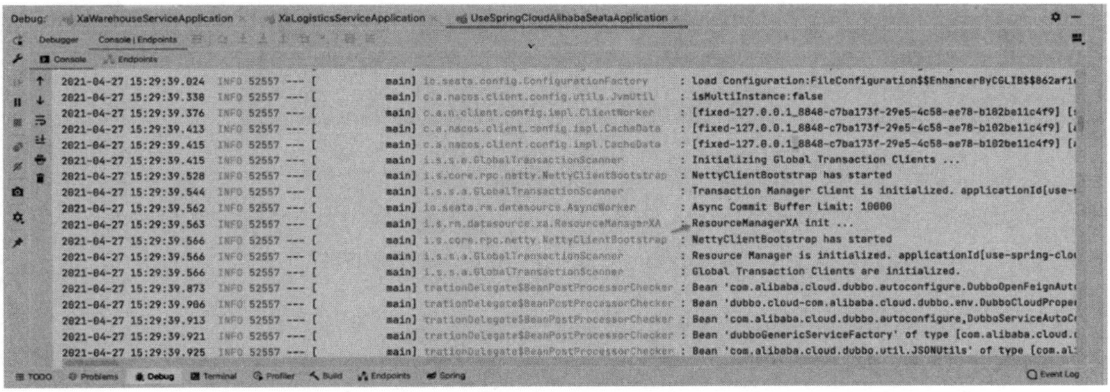

```
Debug: XaWarehouseServiceApplication  XaLogisticsServiceApplication  UseSpringCloudAlibabaSeataApplication
Debugger  Console | Endpoints
Console  Endpoints
2021-04-27 15:29:39.024 INFO 52557 --- [           main] io.seata.config.ConfigurationFactory     : load Configuration:FileConfiguration$$EnhancerByCGLIB$$862af1
2021-04-27 15:29:39.338 INFO 52557 --- [           main] c.a.nacos.client.config.utils.JvmUtil    : isMultiInstance:false
2021-04-27 15:29:39.376 INFO 52557 --- [           main] c.a.n.client.config.impl.ClientWorker    : [fixed-127.0.0.1_8848-c7ba173f-29e5-4c58-ae78-b102be11c4f9] [
2021-04-27 15:29:39.413 INFO 52557 --- [           main] c.a.nacos.client.config.impl.CacheData   : [fixed-127.0.0.1_8848-c7ba173f-29e5-4c58-ae78-b102be11c4f9] [
2021-04-27 15:29:39.415 INFO 52557 --- [           main] c.a.nacos.client.config.impl.CacheData   : [fixed-127.0.0.1_8848-c7ba173f-29e5-4c58-ae78-b102be11c4f9] [
2021-04-27 15:29:39.415 INFO 52557 --- [           main] i.s.s.a.GlobalTransactionScanner         : Initializing Global Transaction Clients ...
2021-04-27 15:29:39.528 INFO 52557 --- [           main] i.s.core.rpc.netty.NettyClientBootstrap  : NettyClientBootstrap has started
2021-04-27 15:29:39.544 INFO 52557 --- [           main] i.s.s.a.GlobalTransactionScanner         : Transaction Manager Client is initialized. applicationId[use-
2021-04-27 15:29:39.562 INFO 52557 --- [           main] io.seata.rm.datasource.AsyncWorker       : Async Commit Buffer Limit: 10000
2021-04-27 15:29:39.563 INFO 52557 --- [           main] i.s.rm.datasource.xa.ResourceManagerXA   : ResourceManagerXA init ...
2021-04-27 15:29:39.566 INFO 52557 --- [           main] i.s.core.rpc.netty.NettyClientBootstrap  : NettyClientBootstrap has started
2021-04-27 15:29:39.566 INFO 52557 --- [           main] i.s.s.a.GlobalTransactionScanner         : Resource Manager is initialized. applicationId[use-spring-clo
2021-04-27 15:29:39.566 INFO 52557 --- [           main] i.s.s.a.GlobalTransactionScanner         : Global Transaction Clients are initialized.
2021-04-27 15:29:39.873 INFO 52557 --- [           main] trationDelegate$BeanPostProcessorChecker : Bean 'com.alibaba.cloud.dubbo.autoconfigure.DubboOpenFeignAut
2021-04-27 15:29:39.906 INFO 52557 --- [           main] trationDelegate$BeanPostProcessorChecker : Bean 'dubbo.cloud-com.alibaba.cloud.dubbo.env.DubboCloudPrope
2021-04-27 15:29:39.913 INFO 52557 --- [           main] trationDelegate$BeanPostProcessorChecker : Bean 'com.alibaba.cloud.dubbo.autoconfigure.DubboServiceAutoC
2021-04-27 15:29:39.921 INFO 52557 --- [           main] trationDelegate$BeanPostProcessorChecker : Bean 'dubboGenericServiceFactory' of type [com.alibaba.cloud.
2021-04-27 15:29:39.925 INFO 52557 --- [           main] trationDelegate$BeanPostProcessorChecker : Bean 'com.alibaba.cloud.dubbo.util.JSONUtils' of type [com.al
TODO  Problems  Debug  Terminal  Profiler  Build  Endpoints  Spring                                    Event Log
```

图 7-14

在 Nacos 控制台中，可以看到服务已经成功地注册到 Nacos 的注册中心中了，如图 7-15 所示。

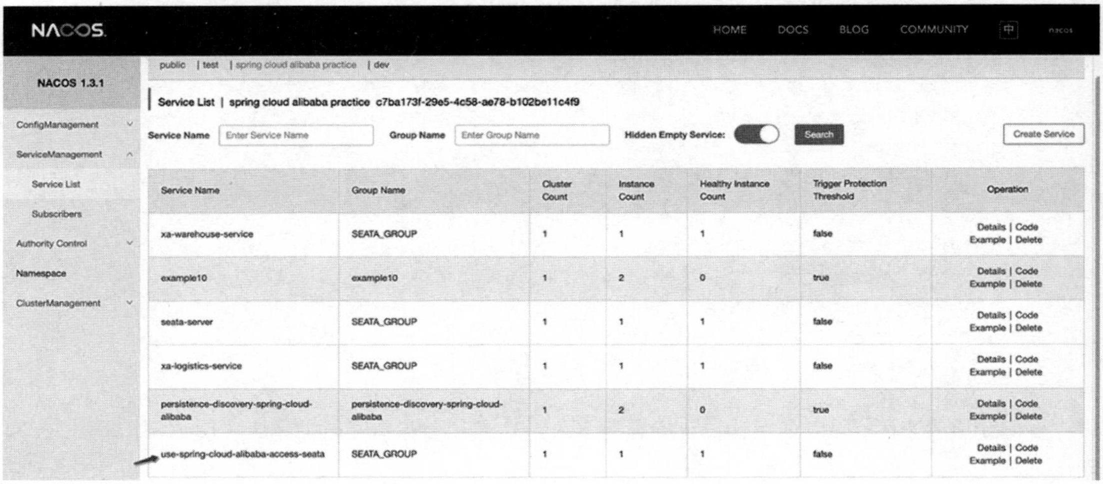

图 7-15

7.4 用 Netty 实现客户端与服务器端之间的通信渠道

在 Seata 中，TC 是服务器端，客户端（TM 和 RM）需要与它进行 RPC 通信，并完成全局事务和分支事务的数据传输，所以 Seata 采用 Netty 作为基础框架来实现客户端与服务器端之间的通信渠道。下面会从服务器端和客户端的角度分析 Seata 通信渠道的原理。

7.4.1 “用 Netty 实现通信渠道的服务器端”的原理

下面介绍在 Seata 中用 Netty 实现通信渠道的服务器端的原理。

1. 实例化 TC 的通信渠道类 NettyRemotingServer

在 Seata 中启动 Seata Server 的启动类 Server 时，会实例化 NettyRemotingServer 类，具体代码如下所示：

```
    public class Server {
        //①自定义工作线程池
        private static final ThreadPoolExecutor WORKING_THREADS = new
ThreadPoolExecutor(MIN_SERVER_POOL_SIZE,
            MAX_SERVER_POOL_SIZE, KEEP_ALIVE_TIME, TimeUnit.SECONDS,
            new LinkedBlockingQueue<>(MAX_TASK_QUEUE_SIZE),
            new NamedThreadFactory("ServerHandlerThread", MAX_SERVER_POOL_SIZE),
new ThreadPoolExecutor.CallerRunsPolicy());
        //②启动 Server 类
        public static void main(String[] args) throws IOException {
```

```
        //③实例化 NettyRemotingServer 类
        NettyRemotingServer nettyRemotingServer = new
NettyRemotingServer(WORKING_THREADS);
        //④实例化事务协调器，并组合实例化之后的 NettyRemotingServer 对象
        DefaultCoordinator coordinator = new
DefaultCoordinator(nettyRemotingServer);
        //⑤初始化事务协调器
        coordinator.init();
        nettyRemotingServer.setHandler(coordinator);
    }
    ...
}
```

2. 实例化服务器端的 NIO 通信类 NettyServerBootstrap

Seata 的 NettyRemotingServer 类在实例化的过程中，会实例化服务器端的 NIO 通信类 NettyServerBootstrap，具体代码如下所示：

```
public abstract class AbstractNettyRemotingServer extends
AbstractNettyRemoting
    implements RemotingServer {
    private final NettyServerBootstrap serverBootstrap;
    public AbstractNettyRemotingServer(ThreadPoolExecutor messageExecutor,
        NettyServerConfig nettyServerConfig) {
        super(messageExecutor);
        //①用 Netty 的配置文件 NettyServerConfig 实例化 NettyServerBootstrap 类
        serverBootstrap = new NettyServerBootstrap(nettyServerConfig);
        serverBootstrap.setChannelHandlers(new ServerHandler());
    }
}
public class NettyServerBootstrap implements RemotingBootstrap {
    private final ServerBootstrap serverBootstrap = new ServerBootstrap();
    private final EventLoopGroup eventLoopGroupWorker;
    private final EventLoopGroup eventLoopGroupBoss;
    private final NettyServerConfig nettyServerConfig;
    public NettyServerBootstrap(NettyServerConfig nettyServerConfig) {
        this.nettyServerConfig = nettyServerConfig;
        //②如果开启了 Netty 的 Epool 通信渠道，则使用 NIO 通信类
EpollServerSocketChannel
        if (NettyServerConfig.enableEpoll()) {
            //③初始化服务器端的 Boss 事件组 EpollEventLoopGroup
            this.eventLoopGroupBoss = new
EpollEventLoopGroup(nettyServerConfig.getBossThreadSize(),
                  new
NamedThreadFactory(nettyServerConfig.getBossThreadPrefix(),
                    nettyServerConfig.getBossThreadSize()));
```

```
            //④初始化服务器端的工作事件组 EpollEventLoopGroup
            this.eventLoopGroupWorker = new
EpollEventLoopGroup(nettyServerConfig.getServerWorkerThreads(),
                  new
NamedThreadFactory(nettyServerConfig.getWorkerThreadPrefix(),
                        nettyServerConfig.getServerWorkerThreads()));
        } else {
            //⑤初始化服务器端的 Boss 事件组 NioEventLoopGroup
            this.eventLoopGroupBoss = new
NioEventLoopGroup(nettyServerConfig.getBossThreadSize(),
                  new
NamedThreadFactory(nettyServerConfig.getBossThreadPrefix(),
                        nettyServerConfig.getBossThreadSize()));
            //⑥初始化服务器端的 Boss 事件组 NioEventLoopGroup
            this.eventLoopGroupWorker = new
NioEventLoopGroup(nettyServerConfig.getServerWorkerThreads(),
                  new
NamedThreadFactory(nettyServerConfig.getWorkerThreadPrefix(),
                        nettyServerConfig.getServerWorkerThreads()));
        }
    }
}
```

3. 注册 TC 支持的事件处理器

Seata 在初始化 NettyRemotingServer 类的过程中会注册消息事件处理器，具体代码如下所示：

```
public class NettyRemotingServer extends AbstractNettyRemotingServer {
    @Override
    public void init() {
        //①注册消息事件处理器
        registerProcessor();
        if (initialized.compareAndSet(false, true)) {
            super.init();
        }
    }
    private void registerProcessor() {
        //②注册请求消息处理器
        ServerOnRequestProcessor onRequestProcessor = new
ServerOnRequestProcessor(this,
                getHandler());
        super.registerProcessor(MessageType.TYPE_BRANCH_REGISTER,
            onRequestProcessor, messageExecutor);
        super.registerProcessor(MessageType.TYPE_BRANCH_STATUS_REPORT,
            onRequestProcessor, messageExecutor);
```

```
        super.registerProcessor(MessageType.TYPE_GLOBAL_BEGIN,
            onRequestProcessor, messageExecutor);
        super.registerProcessor(MessageType.TYPE_GLOBAL_COMMIT,
            onRequestProcessor, messageExecutor);
        super.registerProcessor(MessageType.TYPE_GLOBAL_LOCK_QUERY,
            onRequestProcessor, messageExecutor);
        super.registerProcessor(MessageType.TYPE_GLOBAL_REPORT,
            onRequestProcessor, messageExecutor);
        super.registerProcessor(MessageType.TYPE_GLOBAL_ROLLBACK,
            onRequestProcessor, messageExecutor);
        super.registerProcessor(MessageType.TYPE_GLOBAL_STATUS,
            onRequestProcessor, messageExecutor);
        super.registerProcessor(MessageType.TYPE_SEATA_MERGE,
            onRequestProcessor, messageExecutor);
        //③注册响应消息处理器
        ServerOnResponseProcessor onResponseProcessor = new
ServerOnResponseProcessor(getHandler(),
                getFutures());
        super.registerProcessor(MessageType.TYPE_BRANCH_COMMIT_RESULT,
            onResponseProcessor, messageExecutor);
        super.registerProcessor(MessageType.TYPE_BRANCH_ROLLBACK_RESULT,
            onResponseProcessor, messageExecutor);
        //④注册 RM 消息处理器
        RegRmProcessor regRmProcessor = new RegRmProcessor(this);
        super.registerProcessor(MessageType.TYPE_REG_RM, regRmProcessor,
            messageExecutor);
        //⑤注册 TM 消息处理器
        RegTmProcessor regTmProcessor = new RegTmProcessor(this);
        super.registerProcessor(MessageType.TYPE_REG_CLT, regTmProcessor,
null);
        //⑥注册心跳消息处理器
        ServerHeartbeatProcessor heartbeatMessageProcessor = new
ServerHeartbeatProcessor(this);
        super.registerProcessor(MessageType.TYPE_HEARTBEAT_MSG,
            heartbeatMessageProcessor, null);
    }
}
```

4. 开启 NIO 通信渠道

在实例化 NIO 通信类 NettyServerBootstrap 后，Seata 需要开通 NIO 通信渠道，以保证客户端能够建立与服务器端的 RPC 连接，具体代码如下所示：

```
public class NettyServerBootstrap implements RemotingBootstrap {
    @Override
public void start() {
```

```
        //①构造服务器端套接字对象ServerBootstrap，并设置TCP参数，比如开启长连接（对应参数ChannelOption.SO_KEEPALIVE）
        this.serverBootstrap.group(this.eventLoopGroupBoss,
            this.eventLoopGroupWorker)
                        .channel(NettyServerConfig.SERVER_CHANNEL_CLAZZ)
                        .option(ChannelOption.SO_BACKLOG,
            nettyServerConfig.getSoBackLogSize())
                        .option(ChannelOption.SO_REUSEADDR, true)
                        .childOption(ChannelOption.SO_KEEPALIVE, true)
                        .childOption(ChannelOption.TCP_NODELAY, true)
                        .childOption(ChannelOption.SO_SNDBUF,
            nettyServerConfig.getServerSocketSendBufSize())
                        .childOption(ChannelOption.SO_RCVBUF,
            nettyServerConfig.getServerSocketResvBufSize())
                        .childOption(ChannelOption.WRITE_BUFFER_WATER_MARK,
            new
WriteBufferWaterMark(nettyServerConfig.getWriteBufferLowWaterMark(),
                nettyServerConfig.getWriteBufferHighWaterMark()))
                        .localAddress(new
InetSocketAddress(listenPort)).childHandler(new
ChannelInitializer<SocketChannel>() {
                @Override
                public void initChannel(SocketChannel ch) {
                    ch.pipeline()
                      .addLast(new IdleStateHandler(
                          nettyServerConfig.getChannelMaxReadIdleSeconds(),
                          0, 0)).addLast(new ProtocolV1Decoder())
                      .addLast(new ProtocolV1Encoder());
                    if (channelHandlers != null) {
                        addChannelPipelineLast(ch, channelHandlers);
                    }
                }
            });
        try {
            //②绑定端口，并开启同步通信，等待客户端连接
            ChannelFuture future =
this.serverBootstrap.bind(listenPort).sync();
            LOGGER.info("Server started, listen port: {}", listenPort);
            //③将启动成功的服务器端注册到注册中心（比如Nacos）中
            RegistryFactory.getInstance()
                         .register(new
InetSocketAddress(XID.getIpAddress(),
                    XID.getPort()));
            initialized.set(true);
            future.channel().closeFuture().sync();
```

```
            } catch (Exception exx) {
                throw new RuntimeException(exx);
            }
        }
    }
```

7.4.2 “用 Netty 实现通信渠道的客户端”的原理

下面介绍在 Seata 中用 Netty 来实现通信渠道的客户端的原理。

1. 使用 Spring Boot 的自动配置模块加载 TM 和 RM 通信渠道

Seata 加载 TM 通信渠道的过程如下：

（1）给应用添加 seata-spring-boot-starter 的依赖，具体如下所示：

```
<dependency>
    <groupId>io.seata</groupId>
    <artifactId>seata-spring-boot-starter</artifactId>
</dependency>
```

（2）如果应用采用 Spring Boot 作为基础框架，则 Seata 的 SeataAutoConfiguration 类会依赖 Spring Boot 的自动配置，Seata 会自动初始化全局事务扫描类 GlobalTransactionScanner。

初始化全局事务扫描类的代码如下所示：

```
    @ComponentScan(basePackages =
"io.seata.spring.boot.autoconfigure.properties")
    @ConditionalOnProperty(prefix = SEATA_PREFIX, name = "enabled", havingValue
= "true", matchIfMissing = true)
    @Configuration
    @EnableConfigurationProperties({SeataProperties.class})
    public class SeataAutoConfiguration {
        @Bean
        @DependsOn({BEAN_NAME_SPRING_APPLICATION_CONTEXT_PROVIDER,
BEAN_NAME_FAILURE_HANDLER})
        @ConditionalOnMissingBean(GlobalTransactionScanner.class)
        public GlobalTransactionScanner
globalTransactionScanner(SeataProperties seataProperties, FailureHandler
failureHandler) {
            if (LOGGER.isInfoEnabled()) {
                LOGGER.info("Automatically configure Seata");
            }
            //自动装配 GlobalTransactionScanner 类（开启全局事务扫描）
            return new
GlobalTransactionScanner(seataProperties.getApplicationId(),
seataProperties.getTxServiceGroup(), failureHandler);
```

```
        }
    }
```

（3）Seata 用 Spring Framework 的 InitializingBean 类触发 TM 和 RM 通信渠道的初始化。

熟悉 Spring Framework 的开发人员应该都了解——业务功能可以通过继承 InitializingBean 类的 afterPropertiesSet()方法来完成初始化。

在 Seata 中，使用 InitializingBean 类的 afterPropertiesSet()方法触发 TM 和 RM 通信渠道初始化的代码如下所示：

```
    public class GlobalTransactionScanner extends AbstractAutoProxyCreator
        implements ConfigurationChangeListener, InitializingBean,
            ApplicationContextAware, DisposableBean {
        @Override
        public void afterPropertiesSet() {
    ConfigurationCache.addConfigListener(ConfigurationKeys.DISABLE_GLOBAL_TR
ANSACTION,(ConfigurationChangeListener) this);
            if (disableGlobalTransaction) {
                if (LOGGER.isInfoEnabled()) {
                    LOGGER.info("Global transaction is disabled.");
                }
                return;
            }
            //①开始初始化 TM 和 RM 之间的通信渠道，在当前 JVM 中只初始化一次
            if (initialized.compareAndSet(false, true)) {
                initClient();
            }
        }
        private void initClient() {
            if (LOGGER.isInfoEnabled()) {
                LOGGER.info("Initializing Global Transaction Clients ... ");
            }
            if (StringUtils.isNullOrEmpty(applicationId) ||
                    StringUtils.isNullOrEmpty(txServiceGroup)) {
                throw new IllegalArgumentException(String.format(
                        "applicationId: %s, txServiceGroup: %s", applicationId,
                        txServiceGroup));
            }
            //②调用 TMClient 的 init()方法初始化 TM 和 TC 之间的通信渠道
            TMClient.init(applicationId, txServiceGroup, accessKey, secretKey);
            if (LOGGER.isInfoEnabled()) {
                LOGGER.info("Transaction Manager Client is initialized.
applicationId[{}] txServiceGroup[{}]",
                    applicationId, txServiceGroup);
            }
```

```
        //③调用 RMClient 的 init()方法来初始化 RM 和 TC 之间的通信渠道
        RMClient.init(applicationId, txServiceGroup);

        if (LOGGER.isInfoEnabled()) {
            LOGGER.info("Resource Manager is initialized. applicationId[{}]
txServiceGroup[{}]",
                applicationId, txServiceGroup);
        }
        if (LOGGER.isInfoEnabled()) {
            LOGGER.info("Global Transaction Clients are initialized. ");
        }
        //④将用于关闭进程的回调方法注册到当前线程中
        registerSpringShutdownHook();
    }
}
```

2. 用 TMClient 类完成 TM 与 TC 之间的通信渠道的初始化

在 Seata 中，用 TMClient 类的 init()方法初始化通信渠道的代码如下所示：

```
public class TMClient {
    public static void init(String applicationId,
        String transactionServiceGroup, String accessKey, String secretKey) {
        //①实例化 TM 模块的 TmNettyRemotingClient 类
        TmNettyRemotingClient tmNettyRemotingClient =
TmNettyRemotingClient.getInstance(applicationId,
                transactionServiceGroup, accessKey, secretKey);
        //②初始化 TmNettyRemotingClient 类
        tmNettyRemotingClient.init();
    }
    ...
}
```

3. 用 RMClient 类完成 RM 通信渠道的初始化

用 RMClient 类的 init()方法初始化通信渠道的代码如下所示：

```
public class RMClient {
public static void init(String applicationId, String transactionServiceGroup) {
        //①实例化 RM 模块的 RmNettyRemotingClient 类
        RmNettyRemotingClient rmNettyRemotingClient =
RmNettyRemotingClient.getInstance(applicationId, transactionServiceGroup);
        //②设置资源管理器

rmNettyRemotingClient.setResourceManager(DefaultResourceManager.get());
rmNettyRemotingClient.setTransactionMessageHandler(DefaultRMHandler.get());
        //③初始化 RmNettyRemotingClient 类
```

```
            rmNettyRemotingClient.init();
        }
    }
```

4. 用抽象模板设计模式实现“基于 Netty 的通信渠道的客户端”

用抽象模板设计模式实现“基于 Netty 的通信渠道的客户端”，如图 7-16 所示。

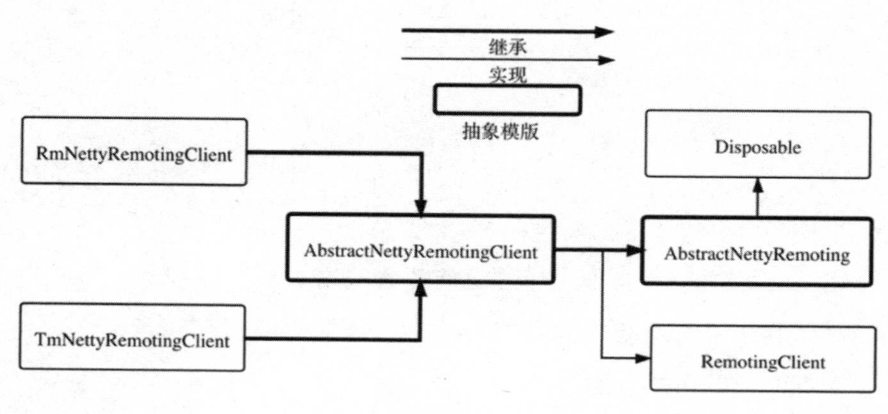

图 7-16

从图 7-16 中可以看出，主要包括如下几个部分。

- 两个抽象模板类：AbstractNettyRemotingClient 和 AbstractNettyRemoting。
- 两个个性化通信渠道类：TmNettyRemotingClient（TM）和 RmNettyRemotingClient（RM）。
- 一个公共的客户端通信渠道接口 RemotingClient。

在 Seata 中，建立客户端 TM 和 RM 之间的通信渠道的过程大致分为如下几个阶段：

（1）用抽象模板类 AbstractNettyRemotingClient 实例化客户端套接字类 NettyClientBootstrap，具体代码如下所示：

```
    public abstract class AbstractNettyRemotingClient extends
AbstractNettyRemoting
        implements RemotingClient {
        private final NettyClientBootstrap clientBootstrap;
        private NettyClientChannelManager clientChannelManager;
        public AbstractNettyRemotingClient(NettyClientConfig
nettyClientConfig,
            EventExecutorGroup eventExecutorGroup,
            ThreadPoolExecutor messageExecutor,
            NettyPoolKey.TransactionRole transactionRole) {
            //①初始化抽象模板类 AbstractNettyRemoting
```

```
        super(messageExecutor);
        this.transactionRole = transactionRole;
        //②初始化客户端套接字类 NettyClientBootstrap
        clientBootstrap = new NettyClientBootstrap(nettyClientConfig,
                eventExecutorGroup, transactionRole);
        clientBootstrap.setChannelHandlers(new ClientHandler());
        //③将通信渠道添加到 Netty 的客户端缓存中
        clientChannelManager = new NettyClientChannelManager(new
NettyPoolableFactory(
                    this, clientBootstrap), getPoolKeyFunction(),
                nettyClientConfig);
    }
}
```

（2）用 RmNettyRemotingClient 类的 init()方法开启 RM 客户端与 TC 之间的通信渠道，具体代码如下所示：

```
public final class RmNettyRemotingClient extends
AbstractNettyRemotingClient {
    @Override
public void init() {
    //①注册 RM 支持的事件处理器
        registerProcessor();
        if (initialized.compareAndSet(false, true)) {
            //②初始化模板类 AbstractNettyRemotingClient，并开启 Netty 通信渠道
            super.init();
            if ((resourceManager != null) &&
                    !resourceManager.getManagedResources().isEmpty() &&
                    StringUtils.isNotBlank(transactionServiceGroup)) {

getClientChannelManager().reconnect(transactionServiceGroup);
            }
        }
    }
    private void registerProcessor() {
        //③注册 RM 分支提交事件处理器
        RmBranchCommitProcessor rmBranchCommitProcessor = new
RmBranchCommitProcessor(getTransactionMessageHandler(),
                this);
        super.registerProcessor(MessageType.TYPE_BRANCH_COMMIT,
            rmBranchCommitProcessor, messageExecutor);
        //④注册 RM 分支回滚事件处理器
        RmBranchRollbackProcessor rmBranchRollbackProcessor = new
RmBranchRollbackProcessor(getTransactionMessageHandler(),
                this);
        super.registerProcessor(MessageType.TYPE_BRANCH_ROLLBACK,
```

```
            rmBranchRollbackProcessor, messageExecutor);
        //⑤注册 RM undolog 事件处理器
        RmUndoLogProcessor rmUndoLogProcessor = new
RmUndoLogProcessor(getTransactionMessageHandler());
        super.registerProcessor(MessageType.TYPE_RM_DELETE_UNDOLOG,
            rmUndoLogProcessor, messageExecutor);
        //⑥注册 TC 处理事件后的响应事件处理器
        ClientOnResponseProcessor onResponseProcessor = new
ClientOnResponseProcessor(mergeMsgMap,
                super.getFutures(), getTransactionMessageHandler());
        super.registerProcessor(MessageType.TYPE_SEATA_MERGE_RESULT,
            onResponseProcessor, null);
        super.registerProcessor(MessageType.TYPE_BRANCH_REGISTER_RESULT,
            onResponseProcessor, null);

super.registerProcessor(MessageType.TYPE_BRANCH_STATUS_REPORT_RESULT,
            onResponseProcessor, null);
        super.registerProcessor(MessageType.TYPE_GLOBAL_LOCK_QUERY_RESULT,
            onResponseProcessor, null);
        super.registerProcessor(MessageType.TYPE_REG_RM_RESULT,
            onResponseProcessor, null);
        //⑦注册心跳处理器
        ClientHeartbeatProcessor clientHeartbeatProcessor = new
ClientHeartbeatProcessor();
        super.registerProcessor(MessageType.TYPE_HEARTBEAT_MSG,
            clientHeartbeatProcessor, null);
    }
    ...
    }
```

（3）用 TmNettyRemotingClient 类的 init()方法开启 RM 客户端与 TC 之间的通信渠道，具体代码如下所示：

```
    public final class TmNettyRemotingClient extends
AbstractNettyRemotingClient {
        @Override
        public void init() {
            //①注册 TM 支持的事件处理器
            registerProcessor();
            if (initialized.compareAndSet(false, true)) {
                //②初始化模板类 AbstractNettyRemotingClient，并开启 Netty 通信渠道
                super.init();
            }
        }
        private void registerProcessor() {
            //③注册 TC 处理事件后的响应事件处理器
```

```
        ClientOnResponseProcessor onResponseProcessor = new
ClientOnResponseProcessor(mergeMsgMap,
                super.getFutures(), getTransactionMessageHandler());
        super.registerProcessor(MessageType.TYPE_SEATA_MERGE_RESULT,
            onResponseProcessor, null);
        super.registerProcessor(MessageType.TYPE_GLOBAL_BEGIN_RESULT,
            onResponseProcessor, null);
        super.registerProcessor(MessageType.TYPE_GLOBAL_COMMIT_RESULT,
            onResponseProcessor, null);
        super.registerProcessor(MessageType.TYPE_GLOBAL_REPORT_RESULT,
            onResponseProcessor, null);
        super.registerProcessor(MessageType.TYPE_GLOBAL_ROLLBACK_RESULT,
            onResponseProcessor, null);
        super.registerProcessor(MessageType.TYPE_GLOBAL_STATUS_RESULT,
            onResponseProcessor, null);
        super.registerProcessor(MessageType.TYPE_REG_CLT_RESULT,
            onResponseProcessor, null);
        //④注册心跳消息事件处理器
        ClientHeartbeatProcessor clientHeartbeatProcessor = new
ClientHeartbeatProcessor();
        super.registerProcessor(MessageType.TYPE_HEARTBEAT_MSG,
            clientHeartbeatProcessor, null);
    }
    ...
    }
```

7.5　用拦截器和过滤器适配主流的 RPC 框架

Seata 用拦截器和过滤器来适配主流的 RPC 框架（包括 Dubbo、gRPC、Motan 及 Soft RPC 等）。

7.5.1　“用过滤器适配 Dubbo”的原理

Seata 用 ApacheDubboTransactionPropagationFilter 类实现 Dubbo 的 Filter 接口。在 ApacheDubboTransactionPropagationFilter 类的 invoke()方法中，封装了 Seata 的分布式事务规范，并将事务规范侵入应用的 Dubbo 接口，具体代码如下所示：

```
    @Activate(group = {DubboConstants.PROVIDER, DubboConstants.CONSUMER}, order
= 100)
    public class ApacheDubboTransactionPropagationFilter implements Filter {

        private static final Logger LOGGER =
LoggerFactory.getLogger(ApacheDubboTransactionPropagationFilter.class);
```

```
    @Override
    public Result invoke(Invoker<?> invoker, Invocation invocation) throws
RpcException {
        //①获取分布式事务上下文中唯一的 XID
        String xid = RootContext.getXID();
        //②获取分布式事务上下文中的分支事务类型
        BranchType branchType = RootContext.getBranchType();
        //③获取当前 RPC 请求中的分布式事务 XID
        String rpcXid = getRpcXid();
        //④获取当前 RPC 请求中的支事务类型
        String rpcBranchType =
RpcContext.getContext().getAttachment(RootContext.KEY_BRANCH_TYPE);
        if (LOGGER.isDebugEnabled()) {
            LOGGER.debug("xid in RootContext[{}] xid in RpcContext[{}]", xid,
rpcXid);
        }
        boolean bind = false;
        //⑤如果分布式事务上下文存在 XID
        if (xid != null) {
            //⑥则将 XID 和分支事务类型设置到当前的 RPC 请求中
            RpcContext.getContext().setAttachment(RootContext.KEY_XID,
xid);

RpcContext.getContext().setAttachment(RootContext.KEY_BRANCH_TYPE,
branchType.name());
        } else {
            //⑦如果在当前 RPC 请求中存在 rpcXid
            if (rpcXid != null) {
                //⑧则将 rpcXid 绑定到分布式事务上下文中
                RootContext.bind(rpcXid);
                if (StringUtils.equals(BranchType.TCC.name(), rpcBranchType)) {
                    //⑨如果是 TCC 分布式事务类型，则设置分布式事务上下文的事务类型为
BranchType.TCC
                    RootContext.bindBranchType(BranchType.TCC);
                }
                bind = true;
                if (LOGGER.isDebugEnabled()) {
                    LOGGER.debug("bind xid [{}] branchType [{}] to RootContext",
rpcXid, rpcBranchType);
                }
            }
        }
        try {
            //⑩执行实际的 Dubbo RPC 请求
```

```
                return invoker.invoke(invocation);
            } finally {
                if (bind) {
                    BranchType previousBranchType = RootContext.getBranchType();
                    String unbindXid = RootContext.unbind();
                    if (BranchType.TCC == previousBranchType) {
                        RootContext.unbindBranchType();
                    }
                    if (LOGGER.isDebugEnabled()) {
                        LOGGER.debug("unbind xid [{}] branchType [{}] from
RootContext", unbindXid, previousBranchType);
                    }
                    if (!rpcXid.equalsIgnoreCase(unbindXid)) {
                        LOGGER.warn("xid in change during RPC from {} to
{},branchType from {} to {}", rpcXid, unbindXid,
                                rpcBranchType != null ? rpcBranchType : "AT",
previousBranchType);
                        if (unbindXid != null) {
                            RootContext.bind(unbindXid);
                            LOGGER.warn("bind xid [{}] back to RootContext",
unbindXid);
                            if (BranchType.TCC == previousBranchType) {
                                RootContext.bindBranchType(BranchType.TCC);
                                LOGGER.warn("bind branchType [{}] back to
RootContext", previousBranchType);
                            }
                        }
                    }
                }
            }
        }
        private String getRpcXid() {
            String rpcXid =
RpcContext.getContext().getAttachment(RootContext.KEY_XID);
            if (rpcXid == null) {
                rpcXid =
RpcContext.getContext().getAttachment(RootContext.KEY_XID.toLowerCase());
            }
            return rpcXid;
        }
    }
```

7.5.2 "用拦截器适配 gRPC"的原理

在 gRPC 中，开发人员可以通过拦截器拦截客户端和服务器端的请求。Seata 使用服务器端的拦截器类 ServerInterceptor 和客户端的拦截器类 ClientInterceptor 拦截请求，将 Seata 的分布式事务规范添加到 gRPC 请求中，具体代码如下所示：

```
    public class ServerTransactionInterceptor implements ServerInterceptor {
        @Override
        public <ReqT, RespT> ServerCall.Listener<ReqT> interceptCall(
            ServerCall<ReqT, RespT> serverCall,
            Metadata metadata,
            ServerCallHandler<ReqT, RespT> serverCallHandler) {
            //①从元数据中获取全局分布式事务的 XID
            String xid = getRpcXid(metadata);
            //②将从元数据中获取的全局分布式事务的 XID 设置到 gRPC 的当前请求中
            return new ServerListenerProxy<>(xid,
serverCallHandler.startCall(serverCall, metadata));
        }
    private String getRpcXid(Metadata metadata) {
            String rpcXid = metadata.get(GrpcHeaderKey.HEADER_KEY);
            if (rpcXid == null) {
                rpcXid = metadata.get(GrpcHeaderKey.HEADER_KEY_LOWERCASE);
            }
            return rpcXid;
        }
    }
    public class ClientTransactionInterceptor implements ClientInterceptor {
        @Override
        public <ReqT, RespT> ClientCall<ReqT, RespT> interceptCall(
            MethodDescriptor<ReqT, RespT> method, CallOptions callOptions,
            Channel next) {
            //③从分布式事务上下文中获取分布式事务的 XID
            String xid = RootContext.getXID();
            return new ForwardingClientCall.SimpleForwardingClientCall<ReqT,
RespT>(next.newCall(
                    method, callOptions)) {
                    @Override
                    public void start(Listener<RespT> responseListener,
                        Metadata headers) {
                        //④将分布式事务上下文对象中的分布式事务的 XID 设置到请求头中
                        if (xid != null) {
                            headers.put(GrpcHeaderKey.HEADER_KEY, xid);
                        }
                        super.start(new
ForwardingClientCallListener.SimpleForwardingClientCallListener<RespT>(
                                responseListener) {
                                @Override
                                public void onHeaders(Metadata headers) {
```

```
                        super.onHeaders(headers);
                    }
                }, headers);
            }
        };
    }
}
```

7.6　用 AT 模式实现分布式事务

Seata 的 AT 模式是 Seata 官方推荐的分布式事务模式，它有什么价值呢?

从技术原理的角度看，AT 模式平衡了非功能性的质量属性——高可用、高性能和易用性：

- 易用性，也被叫作低成本。AT 模式在需要实现两阶段协议的编程模型的前提下，能够做到轻量级并且零侵入应用。
- 高性能。AT 模式采用异步非阻塞的通信渠道，合理地利用了应用的系统资源，能保证应用的高吞吐量。
- 高可用。AT 模式自带熔断降级和线程隔离功能。在极端异常的情况下，可以比较友好地屏蔽异常，自动降级和隔离分布式事务。这样可以确保应用能够处理业务接口请求，但是会牺牲一部分数据一致性。

从功能的角度看，AT 模式并不是完美的，也存在缺陷：

- 在 AT 模式中，Seata 强依赖 SDK。如果能够将 SDK 下沉到代理层（Agent 或者 Sidecar 模式），则应用只需要简单的配置就能够使用 AT 模式。
- AT 模式的工作机制是基于本地事务的，所以应用依赖的分布式数据库必须具备事务能力。

7.6.1　"用数据源代理实现 AT 模式的零侵入应用"的原理

在 AT 模式中，Seata 通过代理应用的数据源，侵入应用的 SQL 语句的执行过程。在侵入过程中，应用无须更改任何业务逻辑的代码即可实现对应用的零侵入。

在 AT 模式中，Seata 代理应用的数据源的原理如下，如图 7-17 所示。

（1）Seata 利用 Spring Framework 的 AOP 框架，在应用启动的过程中动态地代理应用中的 Bean 对象（代理 JDK 中的数据源对象 DataSource），具体代码如下所示。Seata 不会代理应用中的所有对象，这样会使得应用接口的性能损耗比较大。

```
@Override
protected boolean shouldSkip(Class<?> beanClass, String beanName) {
```

```
    //数据源代理的白名单条件：① Bean 对象不是 DataSource 类，② Bean 对象是 SeataProxy
类，③ Bean 对象能够匹配在配置文件中配置的白名单类
    //如果满足①②③中的一个白名单条件，则取消 Bean 对象的代理
        return !DataSource.class.isAssignableFrom(beanClass) ||
            SeataProxy.class.isAssignableFrom(beanClass) ||
            excludes.contains(beanClass.getName());
    }
```

（2）应用调用 DataSource 类的 getConnection()方法，并传递对应的参数列表的值。

（3）方法拦截器 MethodInterceptor 拦截 getConnection()方法，并从当前 IOC 容器中获取 DataSource 类的代理类 DataSourceProxy。

（4）如果存在代理类，则执行代理类 DataSourceProxy 的 getConnection()方法并传递参数值。

（5）如果不存在代理类，则执行 DataSource 类的 getConnection()方法并传递参数值。

（6）Seata 在应用启动时，会自动注入 DataSource 类的代理类 DataSourceProxy。

（7）Seata 不仅会用 DataSourceProxy 类代理 DataSource 类，还会用 ConnectionProxy 类代理 java.sql.Connection 类，用 PreparedStatementProxy 类代理 java.sql.PreparedStatement 类。

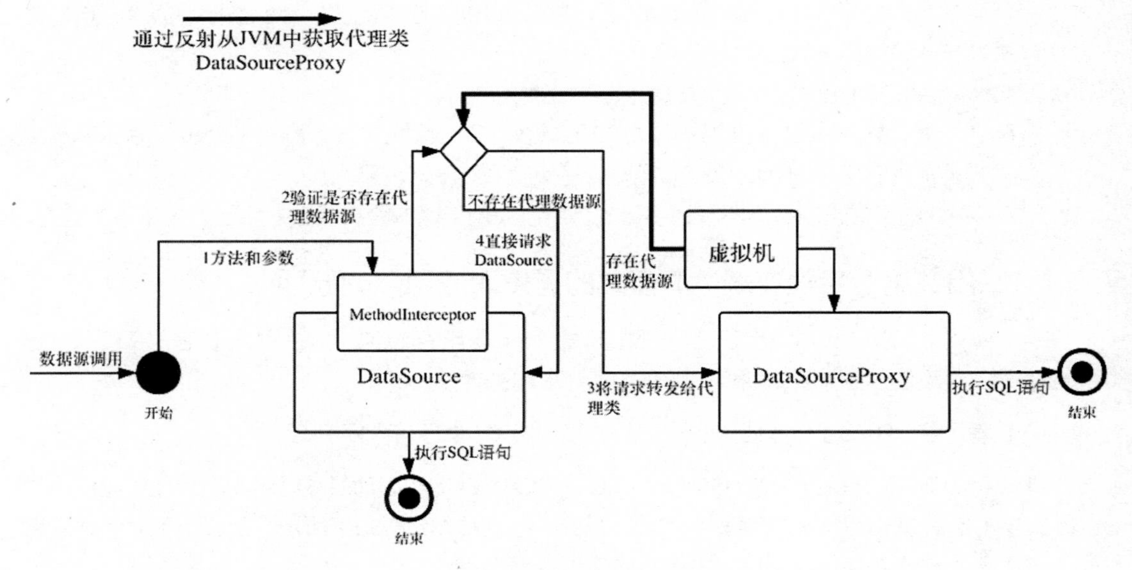

图 7-17

1. 自动注入 AT 模式的数据源代理 DataSourceProxy

在 Seata 中，可以用 SeataAutoDataSourceProxyAdvice 类来自动注入代理类

DataSourceProxy，具体代码如下所示：

```
public class SeataAutoDataSourceProxyAdvice implements MethodInterceptor,
IntroductionInfo {
//①数据源代理的分支事务类型
private final BranchType dataSourceProxyMode;
//②数据源代理
    private final Class<?extends SeataDataSourceProxy>
dataSourceProxyClazz;
public SeataAutoDataSourceProxyAdvice(String dataSourceProxyMode) {
    //③如果是 AT 模式
        if (BranchType.AT.name().equalsIgnoreCase(dataSourceProxyMode)) {
            //④则设置分支事务类型为 AT
            this.dataSourceProxyMode = BranchType.AT;
            //⑤设置数据源代理为 DataSourceProxy
            this.dataSourceProxyClazz = DataSourceProxy.class;
        } else if
(BranchType.XA.name().equalsIgnoreCase(dataSourceProxyMode))   {
            this.dataSourceProxyMode = BranchType.XA;
            this.dataSourceProxyClazz = DataSourceProxyXA.class;
        } else {
            throw new IllegalArgumentException("Unknown dataSourceProxyMode:
"   +dataSourceProxyMode);
        }
        //⑥设置分布式事务上下文中的分支事务类型
        RootContext.setDefaultBranchType(this.dataSourceProxyMode);
    }
}
```

2. 用方法拦截器拦截 Seata 的方法请求

在 Seata 中，可以用 SeataAutoDataSourceProxyAdvice 类实现方法拦截器 MethodInterceptor 的 invoke()方法，具体代码如下所示：

```
@Override
public Object invoke(MethodInvocation invocation) throws Throwable {
    //①如果在分布式事务上下文中已经开启了全局锁，且当前应用的事务类型和分布式事务上下
文中的事务类型不一样，则过滤拦截请求
    if (!RootContext.requireGlobalLock() && (dataSourceProxyMode !=
RootContext.getBranchType())) {
        return invocation.proceed();
}
//②获取拦截器的方法对象 Method
Method method = invocation.getMethod();
//③获取拦截方法对应的方法参数列表
Object[] args = invocation.getArguments();
```

```
    //④用代理类的接口定义、方法名及方法参数类型，从 IOC 容器中获取数据源代理类的实现对象
    Method m = BeanUtils.findDeclaredMethod(dataSourceProxyClazz,
method.getName(), method.getParameterTypes());
    //⑤如果在 JVM 中存在数据源代理类的实现对象
    if (m != null) {
        //⑥则用 DataSourceProxyHolder 类实例化数据源代理对象
            SeataDataSourceProxy dataSourceProxy =
DataSourceProxyHolder.get().putDataSource((DataSource) invocation.getThis(),
dataSourceProxyMode);
            //⑦使用数据源代理类的执行方法
            return m.invoke(dataSourceProxy, args);
    } else {
        //⑧如果在 JVM 中不存在数据源代理类的实现对象，则过滤拦截请求
            return invocation.proceed();
        }
    }
```

3. 调用 DataSourceProxyHolder 类实例化数据源代理对象

DataSourceProxyHolder 类是数据源代理的一个缓存容器。应用可以通过调用 DataSourceProxyHolder 类来实例化数据源代理对象，具体代码如下所示：

```
    public class DataSourceProxyHolder {
    private static final int MAP_INITIAL_CAPACITY = 8;
    //①数据源代理缓存
        private ConcurrentHashMap <DataSource,SeataDataSourceProxy>
dataSourceProxyMap;
    private DataSourceProxyHolder() {
        //②初始化数据源代理缓存，初始容量为 8
            dataSourceProxyMap = new ConcurrentHashMap < >(MAP_INITIAL_CAPACITY);
        }
        public SeataDataSourceProxy putDataSource(DataSource dataSource,
BranchType dataSourceProxyMode) {
            //③需要代理的源数据源
            DataSource originalDataSource;
            //④如果数据源已经是 SeataDataSourceProxy
            if (dataSource instanceof SeataDataSourceProxy) {
                SeataDataSourceProxy dataSourceProxy = (SeataDataSourceProxy)
dataSource;
                //⑤且数据源代理的分支事务类型和应用属性文件中配置的分支事务类型一致
                if (dataSourceProxyMode == dataSourceProxy.getBranchType()) {
                    //⑥则返回数据源代理
                    return (SeataDataSourceProxy) dataSource;
                }
                //⑦设置源数据源为数据源代理中的目标数据源
                originalDataSource = dataSourceProxy.getTargetDataSource();
```

```
            } else {
                //⑧设置源数据源为当前方法需要设置的数据源
                originalDataSource = dataSource;
            }
            //⑨如果分支事务类型为 AT，则在数据源缓存对象 DataSourceProxy 中，设置实例化
后的数据源代理 DataSourceProxy 和源数据源的映射关系
            return CollectionUtils.computeIfAbsent(this.dataSourceProxyMap,
originalDataSource, BranchType.XA == dataSourceProxyMode ?
DataSourceProxyXA: :new: DataSourceProxy: :new);
        }
        ...
    }
```

4. 用 ConnectionProxy 类代理数据库的连接

在 Seata 中，可以用 ConnectionProxy 类来代理数据库的连接，具体代码如下所示：

```
    public class DataSourceProxy extends AbstractDataSourceProxy implements
Resource {
        public DataSourceProxy(DataSource targetDataSource, String
resourceGroupId) {
            if (targetDataSource instanceof SeataDataSourceProxy) {
                LOGGER.info("Unwrap the target data source, because the type is:
{}", targetDataSource.getClass().getName());
                targetDataSource = ((SeataDataSourceProxy)
targetDataSource).getTargetDataSource();
            }
            this.targetDataSource = targetDataSource;
            //①初始化目标数据源
            init(targetDataSource, resourceGroupId);
        }
        private void init(DataSource dataSource, String resourceGroupId) {
            this.resourceGroupId = resourceGroupId;
            try (Connection connection = dataSource.getConnection()) {
                jdbcUrl = connection.getMetaData().getURL();
                dbType = JdbcUtils.getDbType(jdbcUrl);
                if (JdbcConstants.ORACLE.equals(dbType)) {
                    userName = connection.getMetaData().getUserName();
                }
            } catch(SQLException e) {
                throw new IllegalStateException("can not init dataSource", e);
            }
            //②将数据源代理注册到资源管理器（DefaultResourceManager 类）中
            DefaultResourceManager.get().registerResource(this);
            //③如果开启了表元数据的校验
            if (ENABLE_TABLE_META_CHECKER_ENABLE) {
```

```
                //④则用定时线程池定时地校验对应连接的元数据
                tableMetaExcutor.scheduleAtFixedRate(()->{
                    try (Connection connection = dataSource.getConnection()) {
    TableMetaCacheFactory.getTableMetaCache(DataSourceProxy.this.getDbType()
).refresh(connection, DataSourceProxy.this.getResourceId());
                    } catch(Exception ignore){}
                },
                0, TABLE_META_CHECKER_INTERVAL, TimeUnit.MILLISECONDS);
            }
            //⑤在分布式事务上下文中设置分支事务类型
            RootContext.setDefaultBranchType(this.getBranchType());
    }
    @Override
        public ConnectionProxy getConnection() throws SQLException {
            Connection targetConnection = targetDataSource.getConnection();
            //⑥用 ConnectionProxy 类代理数据库连接类 Connection
            return new ConnectionProxy(this, targetConnection);
        }
        @Override
        public ConnectionProxy getConnection(String username, String password)
throws SQLException {
            Connection targetConnection =
targetDataSource.getConnection(username, password);
            return new ConnectionProxy(this, targetConnection);
        }
    ...
    }
```

7.6.2 “用全局锁实现 AT 模式第二阶段的写隔离”的原理

在 Seata 的 AT 模式中，在第一阶段本地事务提交前需要确保已经拿到全局锁：

（1）如果拿不到全局锁，则不能提交本地事务；

（2）如果在重试次数的阈值范围内拿到全局锁，则提交本地事务，否则回滚本地事务并释放本地锁。

1. 认识全局锁

全局锁由“表名”和“操作记录的主键”两部分组成。在 Seata AT 模式中，使用服务器端（TC）存储全局锁。

Seata 支持文件、Redis 和关系型数据库这 3 种全局锁的存储类型。加载全局锁的过程如下：

- 在配置中心或者配置文件（包括 file.conf 和 file.yaml）中设置全局锁的存储类型，如下所示。

```
store.mode=db
```

- Seata 通过 LockerManagerFactory 类来加载指定类型的全局锁管理器，具体代码如下所示。

```
public class LockerManagerFactory {
    //①依据“store.mode”的值（存储类型），用 SPI 机制来加载指定存储类型的全局锁管理器
    private static final LockManager LOCK_MANAGER =
EnhancedServiceLoader.load(LockManager.class,
ConfigurationFactory.getInstance().getConfig(ConfigurationKeys.STORE_MODE));
    //②返回初始化完成的全局锁管理器
        public static LockManager getLockManager() {
            return LOCK_MANAGER;
        }
    }
```

- 在 Seata 中，用注解@LoadLevel 来完成全局锁管理器类 AbstractLockManager 的实现类的动态加载，具体代码如下所示：

```
    //①用注解@LoadLevel 设置全局锁的存储类型为关系型数据库
    @LoadLevel(name = "db")
    public class DataBaseLockManager extends AbstractLockManager implements
Initialize {
        ...
    }
    //②用注解@LoadLevel 设置全局锁的存储类型为 Redis
    @LoadLevel(name = "redis")
    public class RedisLockManager extends AbstractLockManager implements
Initialize {
        ...
    }
    //③用注解@LoadLevel 设置全局锁的存储类型为 File
    @LoadLevel(name = "file")
    public class FileLockManager extends AbstractLockManager {
        ...
    }
```

2. 注册全局锁

在 Seata 中，客户端在进行第一阶段的本地事务提交前，会先向服务器端注册分支事务，此时会将要修改行的表名和主键信息封装成全局锁，并发送到服务器端中进行保存；服务器端在保存全局锁时，如果发现已有其他的全局事务锁定了这些主键，则会抛出全局锁冲突异常，客户端循环等待并重试。

Seata 主要是通过注册、查询和释放等操作来操控全局锁的。

注册全局锁的过程如下。

（1）开发人员通过配置中心动态地配置全局锁对应的表。如果不配置，则默认为表 lock_table，如下所示：

```
store.db.lockTable=lock_table
```

（2）Seata 注册分支事务，开启分布式事务的全局 Session 并注册全局锁。AbstractCore 类中的代码如下所示：

```
 public abstract class AbstractCore implements Core {
     ...
     @Override
   public Long branchRegister(BranchType branchType, String resourceId, String
clientId, String xid, String applicationData, String lockKeys) throws
TransactionException {
         //①调用对应事务类型的全局锁注册，比如 AT 模式
            branchSessionLock(globalSession, branchSession);
            ...
         });
     }
     ...
   }
   public class ATCore extends AbstractCore {
     ...
     @Override
   protected void branchSessionLock(GlobalSession globalSession, BranchSession
branchSession) throws TransactionException {
     //②如果是 AT 模式，则执行 BranchSession 类的 lock()方法来注册全局锁
         if (!branchSession.lock()) {
            throw new BranchTransactionException(LockKeyConflict, String
                  .format("Global lock acquire failed xid = %s branchId = %s",
globalSession.getXid(),
                        branchSession.getBranchId()));
         }
     }
     ...
   }
```

（3）在 Seata 中，用 BranchSession 类的 lock()方法调用全局锁管理器，并利用全局锁管理器进行全局锁的注册，具体代码如下所示：

```
   public class BranchSession implements Lockable, Comparable<BranchSession>,
SessionStorable {
    @Override
   public boolean lock() throws TransactionException {
```

```
        //①目前 Seata 只在 AT 模式下支持全局锁
           if (this.getBranchType().equals(BranchType.AT)) {
              //②调用全局锁管理器 DataBaseLockManager 的 acquireLock()方法来注册全局锁
              return LockerManagerFactory.getLockManager().acquireLock(this);
           }
           return true;
        }
        ...
    }
```

（4）在 Seata 中，用全局锁管理器 DataBaseLockManager 的 acquireLock()方法调用 DataBaseLocker 类的 acquireLock()方法来操控数据库并注册全局锁，具体代码如下所示：

```
    @Override public boolean acquireLock(BranchSession branchSession) throws
TransactionException {
        if (branchSession == null) {
           throw new IllegalArgumentException("branchSession can't be null for
memory/file locker.");
        }
        String lockKey = branchSession.getLockKey();
        if (StringUtils.isNullOrEmpty(lockKey)) {
           return true;
        }
        List <RowLock> locks = collectRowLocks(branchSession);
        if (CollectionUtils.isEmpty(locks)) {
           return true;
    }
    //获取子类 DataBaseLockManager 中初始化的数据库锁的实现类 DataBaseLocker，并执行
acquireLock()方法
        return getLocker(branchSession).acquireLock(locks);
    }
```

（5）在 Seata 中，用 LockStoreDataBaseDAO 类的 acquireLock()方法来实现全局锁的注册（数据库存储类型），具体代码如下所示：

```
    @Override
    public boolean acquireLock(List <LockDO> lockDOs) {
        Connection conn = null;
        PreparedStatement ps = null;
        ResultSet rs = null;
        Set <String> dbExistedRowKeys = new HashSet < >();
        boolean originalAutoCommit = true;
        if (lockDOs.size() > 1) {
           lockDOs =
lockDOs.stream().filter(LambdaUtils.distinctByKey(LockDO: :getRowKey)).colle
ct(Collectors.toList());
```

```
        }
        try {
            //①获取数据库连接
            conn = lockStoreDataSource.getConnection();
            if (originalAutoCommit = conn.getAutoCommit()) {
                conn.setAutoCommit(false);
            }
            StringJoiner sj = new StringJoiner(",");
            for (int i = 0; i < lockDOs.size(); i++) {
                sj.add("?");
            }
            boolean canLock = true;
            //②拼接验证全局锁的 SQL 语句
            String checkLockSQL =
LockStoreSqlFactory.getLogStoreSql(dbType).getCheckLockableSql(lockTable,
sj.toString());
            ps = conn.prepareStatement(checkLockSQL);
            //③将需要锁住的数据拼接到全局锁的 SQL 语句中
            for (int i = 0; i < lockDOs.size(); i++) {
                ps.setString(i + 1, lockDOs.get(i).getRowKey());
            }
            rs = ps.executeQuery();
            String currentXID = lockDOs.get(0).getXid();
            //④遍历返回的数据，如果已经存在对应的行主键的锁，则注册锁失败
            while (rs.next()) {
                String dbXID =
rs.getString(ServerTableColumnsName.LOCK_TABLE_XID);
                if (!StringUtils.equals(dbXID, currentXID)) {
                    if (LOGGER.isInfoEnabled()) {
                        String dbPk =
rs.getString(ServerTableColumnsName.LOCK_TABLE_PK);
                        String dbTableName =
rs.getString(ServerTableColumnsName.LOCK_TABLE_TABLE_NAME);
                        Long dbBranchId =
rs.getLong(ServerTableColumnsName.LOCK_TABLE_BRANCH_ID);
                        LOGGER.info("Global lock on [{}:{}] is holding by xid {}
branchId {}", dbTableName, dbPk, dbXID, dbBranchId);
                    }
                    canLock &= false;
                    break;
                }
dbExistedRowKeys.add(rs.getString(ServerTableColumnsName.LOCK_TABLE_ROW_KEY)
);
            }
            //⑤如果不能注册全局锁，则回滚并直接返回注册失败
```

```
        if (!canLock) {
            conn.rollback();
            return false;
        }
        List <LockDO> unrepeatedLockDOs = null;
        if (CollectionUtils.isNotEmpty(dbExistedRowKeys)) {
            unrepeatedLockDOs = lockDOs.stream().filter(lockDO
->!dbExistedRowKeys.contains(lockDO.getRowKey())).collect(Collectors.toList(
));
        } else {
            unrepeatedLockDOs = lockDOs;
        }
        //⑥如果没有注册的行的主键为空，则说明该行的主键已经全部注册，直接返回注册成功
        if (CollectionUtils.isEmpty(unrepeatedLockDOs)) {
            conn.rollback();
            return true;
        }
        if (unrepeatedLockDOs.size() == 1) {
            LockDO lockDO = unrepeatedLockDOs.get(0);
            if (!doAcquireLock(conn, lockDO)) {
                if (LOGGER.isInfoEnabled()) {
                    LOGGER.info("Global lock acquire failed, xid {} branchId {}
pk {}", lockDO.getXid(), lockDO.getBranchId(), lockDO.getPk());
                }
                conn.rollback();
                return false;
            }
        } else {
            //⑦用递归算法执行全局锁的注册
            if (!doAcquireLocks(conn, unrepeatedLockDOs)) {
                if (LOGGER.isInfoEnabled()) {
                    LOGGER.info("Global lock batch acquire failed, xid {}
branchId {} pks {}", unrepeatedLockDOs.get(0).getXid(),
unrepeatedLockDOs.get(0).getBranchId(),
unrepeatedLockDOs.stream().map(lockDO
- >lockDO.getPk()).collect(Collectors.toList()));
                }
                conn.rollback();
                return false;
            }
        }
        conn.commit();
        return true;
    } catch(SQLException e) {
        throw new StoreException(e);
```

```
        } finally {
            IOUtil.close(rs, ps);
            if (conn != null) {
                try {
                    if (originalAutoCommit) {
                        conn.setAutoCommit(true);
                    }
                    conn.close();
                } catch(SQLException e) {}
            }
        }
    }
```

3. 校验全局锁

在 Seata 的 AT 模式中，被注解@GlobalLock 修饰的方法虽然不在某个全局事务下，但是在提交事务前也会进行全局锁查询。在执行当前事务之前，如果发现全局锁正在被其他全局事务持有，则当前事务会循环等待。

（1）在注解@GlobalLock 中，关于全局锁超时重试的周期和重试次数的代码如下所示：

```
@Retention(RetentionPolicy.RUNTIME)
@Target({ElementType.METHOD,ElementType.TYPE})
@Inherited
public @interface GlobalLock {
    //①设置超时重试的周期，比如 3s
    int lockRetryInternal() default 0;
    //②设置超时重试的次数，比如 3 次
    int lockRetryTimes() default -1;
}
```

（2）如果业务中的方法被注解@GlobalLock 修饰，则 Seata 会用拦截器类 GlobalTransactionalInterceptor 拦截方法并执行全局锁校验，具体代码如下所示：

```
public class GlobalTransactionalInterceptor
    implements ConfigurationChangeListener, MethodInterceptor {
    @Override
    public Object invoke(final MethodInvocation methodInvocation)
        throws Throwable {
        Class<?> targetClass = (methodInvocation.getThis() != null)
            ? AopUtils.getTargetClass(methodInvocation.getThis()) : null;
        Method specificMethod =
ClassUtils.getMostSpecificMethod(methodInvocation.getMethod(),
                targetClass);
        if ((specificMethod != null) &&
                !specificMethod.getDeclaringClass().equals(Object.class)) {
```

```
            final Method method =
BridgeMethodResolver.findBridgedMethod(specificMethod);
            final GlobalTransactional globalTransactionalAnnotation =
getAnnotation(method,targetClass, GlobalTransactional.class);
            //①如果方法被注解@GlobalLock（全局锁）修饰，则用getAnnotation()方法可以从方法的元数据中获取全局锁对象GlobalLock
            final GlobalLock globalLockAnnotation = getAnnotation(method,
                  targetClass, GlobalLock.class);
            //②disable代表是否开启全局事务（默认disable=false，表示开启全局事务）。如果开启了降级处理（degradeCheck=true），则对比降级的总次数和配置中心配置的降级次数的阈值次数
            boolean localDisable = disable ||
               (degradeCheck && (degradeNum >= degradeCheckAllowTimes));
            if (!localDisable) {
               if (globalTransactionalAnnotation != null) {
                  //如果globalTransactionalAnnotation不为空，则调用
handleGlobalTransaction()方法处理全局事务
                  return handleGlobalTransaction(methodInvocation,
                     globalTransactionalAnnotation);
               } else if (globalLockAnnotation != null) {
                  //如果globalLockAnnotation不为空，则调用handleGlobalLock()
方法处理全局锁
                  return handleGlobalLock(methodInvocation,
                     globalLockAnnotation);
               }
            }
         }
         //④返回实际的方法请求后的结果响应
         return methodInvocation.proceed();
      }
      Object handleGlobalLock(final MethodInvocation methodInvocation,
         final GlobalLock globalLockAnno) throws Throwable {
         //⑤用全局锁模板类GlobalLockTemplate执行全局锁的拦截请求处理
         return globalLockTemplate.execute(new GlobalLockExecutor() {
               //⑥如果全局锁检验通过，则返回实际的方法请求后的结果响应
               @Override
               public Object execute() throws Throwable {
                  return methodInvocation.proceed();
               }
               @Override
               public GlobalLockConfig getGlobalLockConfig() {
                  GlobalLockConfig config = new GlobalLockConfig();

config.setLockRetryInternal(globalLockAnno.lockRetryInternal());
```

```
config.setLockRetryTimes(globalLockAnno.lockRetryTimes());
                        return config;
                    }
                });
            }
        }
```

如果业务方法没有被@GlobalLock 修饰，但是被全局事务注解@GlobalTransactional 修饰，则也会开启全局锁的校验。

（1）在 Seata 中，可以在 AT 模式的代理连接类 ConnectionProxy 中的 commit()方法和 setAutoCommit()方法中开启全局锁的校验，具体代码如下所示：

```
    public class ConnectionProxy extends AbstractConnectionProxy {
        //①处理 SQL 语句的提交
        @Override
        public void commit() throws SQLException {
            try {
                LOCK_RETRY_POLICY.execute(() -> {
                    //②在 SQL 语句正常提交的过程中校验全局锁和全局事务
                    doCommit();
                    return null;
                });
            } catch (SQLException e) {
                if (targetConnection != null && !getAutoCommit()) {
                    rollback();
                }
                throw e;
            } catch (Exception e) {
                throw new SQLException(e);
            }
    }
    //③处理 SQL 语句的自动提交
        @Override
    public void setAutoCommit(boolean autoCommit) throws SQLException {
        //④如果开启了全局事务、全局锁，以及 SQL 的自动提交
            if ((context.inGlobalTransaction() || context.isGlobalLockRequire())
&& autoCommit && !getAutoCommit()) {
                //⑤则在 SQL 语句自动提交的过程中校验全局锁和全局事务
                doCommit();
            }
            targetConnection.setAutoCommit(autoCommit);
        }
    private void doCommit() throws SQLException {
        //⑥如果开启了全局事务，则执行全局事务提交请求
```

```
        if (context.inGlobalTransaction()) {
            processGlobalTransactionCommit();
        //⑦如果开启了全局锁，则执行全局锁提交的请求
        } else if (context.isGlobalLockRequire()) {
            processLocalCommitWithGlobalLocks();
        } else {
            //⑧如果都没开启，则直接提交 SQL 请求
            targetConnection.commit();
        }
    }
    ...
}
```

（2）如果开启了全局锁校验，则 Seata 会在 SQL 语句提交之前执行 ConnectionProxy 类的 processLocalCommitWithGlobalLocks()方法校验全局锁，具体代码如下所示：

```
private void processLocalCommitWithGlobalLocks() throws SQLException {
    //①校验全局锁
    checkLock(context.buildLockKeys());
    try {
        targetConnection.commit();
    } catch (Throwable ex) {
        throw new SQLException(ex);
    }
    context.reset();
}
public void checkLock(String lockKeys) throws SQLException {
    //②如果全局锁对应的 key 值为空，则直接返回
    if (StringUtils.isBlank(lockKeys)) {
        return;
    }
try {
    //③用 AT 模式的资源管理类 DataSourceManager 的 lockQuery()方法校验全局锁
        boolean lockable =
DefaultResourceManager.get().lockQuery(BranchType.AT,
            getDataSourceProxy().getResourceId(), context.getXid(),
lockKeys);
        if (!lockable) {
            throw new LockConflictException();
        }
    } catch (TransactionException e) {
        recognizeLockKeyConflictException(e, lockKeys);
    }
}
```

（3）在 Seata 的 AT 模式下，可以用 DataSourceManager 类的 lockQuery()方法校验全局

锁，具体代码如下所示：

```
    @Override
    public boolean lockQuery(BranchType branchType, String resourceId, String
xid, String lockKeys) throws TransactionException {
    try {
        //①构造全局锁校验的查询请求，设置 XID、全局锁 Key 及资源 ID
            GlobalLockQueryRequest request = new GlobalLockQueryRequest();
            request.setXid(xid);
            request.setLockKey(lockKeys);
            request.setResourceId(resourceId);
            GlobalLockQueryResponse response = null;
            if (RootContext.inGlobalTransaction() ||
RootContext.requireGlobalLock()) {
                //②向服务器端发起 RPC 请求
                response = (GlobalLockQueryResponse)
RmNettyRemotingClient.getInstance().sendSyncRequest(request);
            } else {
                throw new RuntimeException("unknow situation!");
            }
            if (response.getResultCode() == ResultCode.Failed) {
                throw new
TransactionException(response.getTransactionExceptionCode(), "Response[" +
response.getMsg() + "]");
            }
            //③返回校验全局锁的结果
            return response.isLockable();
        } catch(TimeoutException toe) {
            throw new RmTransactionException(TransactionExceptionCode.IO, "RPC
Timeout", toe);
        } catch(RuntimeException rex) {
            throw new
RmTransactionException(TransactionExceptionCode.LockableCheckFailed,
"Runtime", rex);
        }
    }
```

（4）Seata 在服务器端（TC）用 AbstractTCInboundHandler 类的 handle()方法处理客户端的 GlobalLockQueryRequest 请求，具体代码如下所示：

```
    public abstract class AbstractTCInboundHandler extends
AbstractExceptionHandler implements TCInboundHandler {
        ...
        @Override
        public GlobalLockQueryResponse handle(GlobalLockQueryRequest request,
            final RpcContext rpcContext) {
```

```
        GlobalLockQueryResponse response = new GlobalLockQueryResponse();
        exceptionHandleTemplate(new
AbstractCallback<GlobalLockQueryRequest, GlobalLockQueryResponse>() {
                @Override
                public void execute(GlobalLockQueryRequest request,
                    GlobalLockQueryResponse response)
                    throws TransactionException {
                    try {
                        //校验全局锁
                        doLockCheck(request, response, rpcContext);
                    } catch (StoreException e) {
                        throw new
TransactionException(TransactionExceptionCode.FailedStore,
                            String.format(
                                "global lock query request failed. xid=%s,
msg=%s",
                                request.getXid(), e.getMessage()), e);
                    }
                }
            }, request, response);

        return response;
    }
}
```

在 Seata 中，在 AbstractTCInboundHandler 类的实现类 DefaultCoordinator 中，可以用 doLockCheck()方法校验全局锁，具体代码如下所示：

```
    @Override
    protected void doLockCheck(GlobalLockQueryRequest request,
GlobalLockQueryResponse response, RpcContext rpcContext)
        throws TransactionException {
        response.setLockable(core.lockQuery(request.getBranchType(),
request.getResourceId(), request.getXid(), request.getLockKey()));
    }
```

（5）在 AT 模式下，可以用 ATCore 类的 lockQuery()方法调用全局锁管理器校验全局锁，具体代码如下所示：

```
    @Override
    public boolean lockQuery(BranchType branchType, String resourceId, String
xid, String lockKeys)throws TransactionException {
        return lockManager.isLockable(xid, resourceId, lockKeys);
    }
```

在服务器端，可以用 LockStoreDataBaseDAO 类的 checkLockable()方法检验全局锁，具体代码如下所示：

```
    protected boolean checkLockable(Connection conn, List<LockDO> lockDOs) {
        PreparedStatement ps = null;
        ResultSet rs = null;
        try {
            StringJoiner sj = new StringJoiner(",");
            for (int i = 0; i < lockDOs.size(); i++) {
                sj.add("?");
            }
            String checkLockSQL = LockStoreSqlFactory.getLogStoreSql(dbType).
getCheckLockableSql(lockTable, sj.toString());
            ps = conn.prepareStatement(checkLockSQL);
            for (int i = 0; i < lockDOs.size(); i++) {
                ps.setString(i + 1, lockDOs.get(i).getRowKey());
            }
            rs = ps.executeQuery();
            while (rs.next()) {
                String xid = rs.getString("xid");
                if (!StringUtils.equals(xid, lockDOs.get(0).getXid())) {
                    return false;
                }
            }
            return true;
        } catch (SQLException e) {
            throw new DataAccessException(e);
        } finally {
            IOUtil.close(rs, ps);
        }
    }
```

4. 释放全局锁

由于第二阶段提交是异步进行的，在服务器端向客户端发送 branch commit 请求后，客户端仅将分支提交信息插入内存队列即返回，所以，服务器端只要判断这个流程没有异常就会释放全局锁。因此可以说，如果第一阶段成功，则在第二阶段一开始就会释放全局锁，不会锁定到第二阶段提交流程结束。

7.6.3 【实例】搭建 Seata 的 AT 模式的环境，并验证 AT 模式的分布式事务场景

本实例的源码在本书配套资源的“chapterseven/spring-cloud-alibaba-access-seata-at”目录下。

如果要用 AT 模式接入 Seata 并实现分布式事务，则需要搭建 AT 模式的环境（包括客户端和服务器端）。其中，服务器端环境可以参考 7.3.1 节。

1. 搭建客户端数据库环境

开发人员要使用 Seata 的 AT 模式，则需要在应用的数据库中新建表 undo_log，DDL 语句如下所示：

```
create table seata_server.undo_log
(
    branch_id bigint not null comment 'branch transaction id',
    xid varchar(100) not null comment 'global transaction id',
    context varchar(128) not null comment 'undo_log context,such as
serialization',
    rollback_info longblob not null comment 'rollback info',
    log_status int not null comment '0:normal status,1:defense status',
    log_created datetime(6) not null comment 'create datetime',
    log_modified datetime(6) not null comment 'modify datetime',
    constraint ux_undo_log
        unique (xid, branch_id)
)
comment 'AT transaction mode undo table' charset=utf8;
```

2. 利用购买商品的业务场景验证 AT 模式

购买商品的业务逻辑如下：创建订单→扣减库存→完成支付→更改订单状态。

本程序包含 3 个子项目：交易订单服务（at-trade-service）、交易账户服务（at-account-service）和商品库存服务（at-storage-service）。

3. 初始化交易账户服务

使用 Spring Cloud Alibaba 框架初始化项目，具体步骤如下。

（1）在 pom.xml 中添加 Seata 相关的 POM 依赖，部分配置信息如下。

```
<!--省略部分 POM 依赖-->
<dependency>
```

```
  <groupId>com.alibaba.cloud</groupId>
  <artifactId>spring-cloud-starter-alibaba-seata</artifactId>
  <exclusions>
    <exclusion>
      <groupId>io.seata</groupId>
      <artifactId>seata-spring-boot-starter</artifactId>
    </exclusion>
  </exclusions>
</dependency>
<dependency>
  <groupId>io.seata</groupId>
  <artifactId>seata-spring-boot-starter</artifactId>
  <exclusions>
    <exclusion>
      <groupId>org.springframework</groupId>
      <artifactId>spring-beans</artifactId>
    </exclusion>
  </exclusions>
</dependency>
```

（2）在 bootstrap.yaml 中添加项目的配置信息，具体如下所示：

```
###省略部分配置信息
nacos:
  server-addr: 127.0.0.1:8848
  namespace: c7ba173f-29e5-4c58-ae78-b102be11c4f9
  group: SEATA_GROUP
  seata:
    application: seata-server
    tx-service-group: at-account-service-tx-group
spring:
  application:
    name: at-account-service
###RM 对应的配置数据源
  datasource:
    url:
jdbc:mysql://127.0.0.1:3306/seata_server?characterEncoding=utf8&connectTimeo
ut=1000&socketTimeout=3000&autoReconnect=true&useUnicode=true&useSSL=false&s
erverTimezone=UTC
    username: root
    password: 123456huxian
server:
  port: 7890
##配置 Seata 的相关信息
seata:
  registry:
```

```
    type: nacos
    nacos:
      group: ${nacos.group}
      application: ${nacos.seata.application}
      namespace: ${nacos.namespace}
      cluster: default
      server-addr: ${nacos.server-addr}
      username: "nacos"
      password: "nacos"
  tx-service-group: ${nacos.seata.tx-service-group}
  enabled: true
  enable-auto-data-source-proxy: false
  application-id: at-account-service
  config:
    type: nacos
    nacos:
      group: ${nacos.group}
      server-addr: ${nacos.server-addr}
      namespace: ${nacos.namespace}
      username: "nacos"
      password: "nacos"
```

（3）创建账户表，具体的 DDL 语句如下：

```
create table seata_server.at_account
(
   id bigint not null
      primary key,
   account_name varchar(50) null,
   account_id bigint null,
   amount bigint null,
   is_deleted int not null,
   gmt_create timestamp not null,
   gmt_modified timestamp null,
   user_id bigint null
);
```

（4）初始化交易账户服务，并模拟账户扣款的业务场景。用来扣款的 Dubbo 接口的代码如下所示：

```
DubboService(version = "1.0.0",group = "SEATA_GROUP")
public class AtAccountServiceImpl implements AtAccountService {
    @Autowired
    private AccountManager accountManager;
    @Override
public boolean deductAccountBalance(Long userId, BigDecimal amount,Long
accountId) {
```

```
        //执行账户扣款
            return accountManager.deductAccountBalance(userId,amount,accountId);
        }
    }
```

4. 初始化商品库存服务

下面用 Spring Cloud Alibaba 框架初始化商品库存服务项目。

（1）在 pom.xml 中添加 Seata 相关的 Jar 包依赖（和交易账户服务一致）。

（2）在 bootstrap.yaml 中添加项目的配置信息，具体如下所示：

```
###省略部分配置信息
nacos:
  server-addr: 127.0.0.1:8848
  namespace: c7ba173f-29e5-4c58-ae78-b102be11c4f9
  group: SEATA_GROUP
  seata:
    application: seata-server
    tx-service-group: at-storage-service-tx-group
spring:
  application:
    name: at-storage-service
  ###RM 对应的配置数据源
  datasource:
    url: jdbc:mysql://127.0.0.1:3306/seata_server?characterEncoding=
utf8&connectTimeout=1000&socketTimeout=3000&autoReconnect=true&useUnicode=tr
ue&useSSL=false&serverTimezone=UTC
    username: root
    password: 123456huxian
    maximumPoolSize: 10
    minimumIdle: 2
    idleTimeout: 600000
    connectionTimeout: 30000
    maxLifetime: 1800000
server:
  port: 7823
###配置 Seata 的相关信息
seata:
  registry:
    type: nacos
    nacos:
      group: ${nacos.group}
      application: ${nacos.seata.application}
      namespace: ${nacos.namespace}
      cluster: default
```

```
            server-addr: ${nacos.server-addr}
            username: "nacos"
            password: "nacos"
        tx-service-group: ${nacos.seata.tx-service-group}
        enabled: true
        enable-auto-data-source-proxy: false
        application-id: at-storage-service
        config:
          type: nacos
          nacos:
            group: ${nacos.group}
            server-addr: ${nacos.server-addr}
            namespace: ${nacos.namespace}
            username: "nacos"
            password: "nacos"
```

（3）创建商品库存表，具体的 DDL 语句如下：

```
create table seata_server.at_storage
(
    id bigint not null
        primary key,
    num bigint not null,
    good_id bigint null,
    is_deleted int null,
    gmt_create timestamp null,
    gmt_modified timestamp null
);
```

（4）初始化商品库存服务，并模拟商品扣减库存的业务场景。扣减库存的 Dubbo 接口的代码如下所示：

```
@DubboService(version = "1.0.0",group = "SEATA_GROUP")
public class AtStorageServiceImpl implements AtStorageService {
    @Resource
    private StorageManager storageManager;
    @Override
public boolean deductInventory(Long goodId,Integer num) {
    //执行商品库存扣减
        return storageManager.deductInventory(goodId,num);
    }
}
```

5. 初始化交易订单服务

下面用 Spring Cloud Alibaba 框架初始化交易订单服务项目。

（1）在 pom.xml 中添加 Seata 相关的 Jar 包依赖（和交易账户服务一致）。

（2）在 bootstrap.yaml 中添加项目的配置信息，具体如下所示：

```
###①省略部分配置信息
nacos:
  server-addr: 127.0.0.1:8848
  namespace: c7ba173f-29e5-4c58-ae78-b102be11c4f9
  group: SEATA_GROUP
  seata:
    application: seata-server
    tx-service-group: at-trade-service-tx-group
spring:
  application:
    name: at-trade-service
  main:
    allow-bean-definition-overriding: true
###②RM 对应的配置数据源
  datasource:
    url: jdbc:mysql://127.0.0.1:3306/seata_server?characterEncoding=
utf8&connectTimeout=1000&socketTimeout=3000&autoReconnect=true&useUnicode=tr
ue&useSSL=false&serverTimezone=UTC
    username: root
    password: 123456huxian
    maximumPoolSize: 10
    minimumIdle: 2
    idleTimeout: 600000
    connectionTimeout: 30000
    maxLifetime: 1800000
server:
  port: 7865
###③配置 Seata 的相关信息
seata:
  registry:
    type: nacos
    nacos:
      group: ${nacos.group}
      application: ${nacos.seata.application}
      namespace: ${nacos.namespace}
      cluster: default
      server-addr: ${nacos.server-addr}
      username: "nacos"
      password: "nacos"
  tx-service-group: ${nacos.seata.tx-service-group}
  enabled: true
  enable-auto-data-source-proxy: false
```

```
  application-id: at-trade-service
  config:
    type: nacos
    nacos:
      group: ${nacos.group}
      server-addr: ${nacos.server-addr}
      namespace: ${nacos.namespace}
      username: "nacos"
      password: "nacos"
```

（3）创建交易订单表，具体的 DDL 语句如下：

```
create table seata_server.at_order
(
    id bigint not null
        primary key,
    order_name varchar(50) null,
    order_id bigint null,
    order_amount bigint null,
    is_deleted int not null,
    gmt_create timestamp null,
    gmt_modified timestamp null,
    good_id bigint null,
    user_id bigint not null,
    order_status int default 0 not null
);
```

（4）初始化交易订单服务，并模拟订单创建的业务场景。订单创建接口的代码如下所示：

```
@Service
public class OrderManager {
    @Resource
    private OrderMapper orderMapper;
    @DubboReference(version = "1.0.0", group = "SEATA_GROUP")
    private AtAccountService atAccountService;
    @DubboReference(version = "1.0.0", group = "SEATA_GROUP")
    private AtStorageService atStorageService;
    //①创建订单
    public Result createOrder(Long userId, BigDecimal amount, Long goodId) {
        OrderEntity orderEntity = new OrderEntity();
        orderEntity.setOrderId(RandomUtils.nextLong());
        orderEntity.setId(RandomUtils.nextLong());
        orderEntity.setOrderAmount(amount);
        orderEntity.setOrderName("测试订单" + RandomUtils.nextInt());
        orderEntity.setGoodId(goodId);
        orderEntity.setUserId(userId);
        orderEntity.setIsDeleted(0);
```

```
        orderEntity.setGmtCreate(new Date());
        orderEntity.setGmtModified(new Date());
        Long result = orderMapper.createOrder(orderEntity);
        Result createResult = new Result();
        createResult.setResult(true ? (result > 0) : false);
        createResult.setId(orderEntity.getId());
        return createResult;
    }
    //②更新订单状态
    public boolean updataOrderStatus(Long id, Integer orderStatus) {
        OrderEntity orderEntity = new OrderEntity();
        orderEntity.setId(id);
        orderEntity.setOrderStatus(orderStatus);
        Integer result = orderMapper.updataOrderStatus(orderEntity);
        return true ? (result > 0) : false;
    }
    //③开启 Seata 的全局事务
    @GlobalTransactional(name = "buy-good", rollbackFor = Exception.class)
    public boolean buy(Long userId, Long goodId, Long accountId, Integer num) {
        //④创建订单
        Result orderResult = createOrder(userId, new BigDecimal(100), goodId);
        //⑤调用账户服务扣减金额
        boolean accountResult = atAccountService.deductAccountBalance(userId,
                new BigDecimal(100), accountId);
        //⑥调用库存服务扣减对应商品库存
        boolean storageResult = atStorageService.deductInventory(goodId, num);
        if (orderResult.isResult() && accountResult && storageResult) {
            //⑦更新订单状态
            boolean updataOrderStatusResult = updataOrderStatus(orderResult.getId(),
                    1);
            if (updataOrderStatusResult) {
                return true;
            }
            return false;
        }
        return false;
    }
    class Result {
        private Long id;
        private boolean result;
        public Long getId() {
            return id;
        }
        public void setId(Long id) {
            this.id = id;
```

```
        }
        public boolean isResult() {
            return result;
        }
        public void setResult(boolean result) {
            this.result = result;
        }
    }
}
```

6. 启动程序

启动后的交易账户服务、商品库存服务和交易订单服务如图 7-18 所示。

图 7-18

7. 模拟分布式事务回滚的业务场景

模拟过程如下。

（1）初始化账户表（at_account）,设置账户余额为 999 999 000 元，如图 7-19 所示。

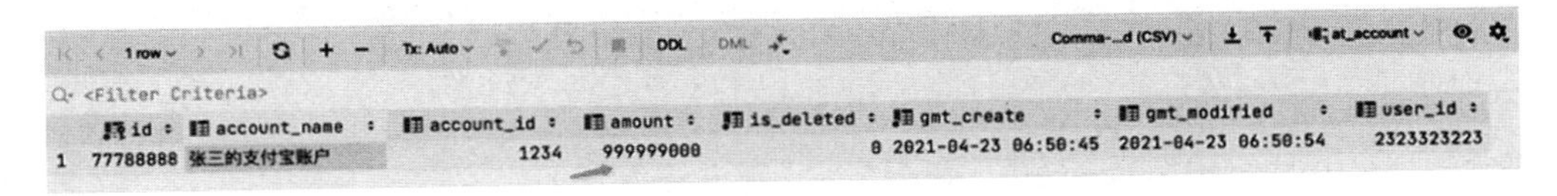

	id	account_name	account_id	amount	is_deleted	gmt_create	gmt_modified	user_id
1	77788888	张三的支付宝账户	1234	999999000	0	2021-04-23 06:50:45	2021-04-23 06:50:54	2323323223

图 7-19

（2）初始化商品库存表（at_storage）,设置商品的库存数量为 999 999，如图 7-20 所示。

	id	num	good_id	is_deleted	gmt_create	gmt_modified
1	665666	999999	34556656	0	2021-04-23 06:52:41	2021-04-23 06:52:43

图 7-20

（3）在购买商品过程中，服务之间调用的故障场景如图 7-21 所示。

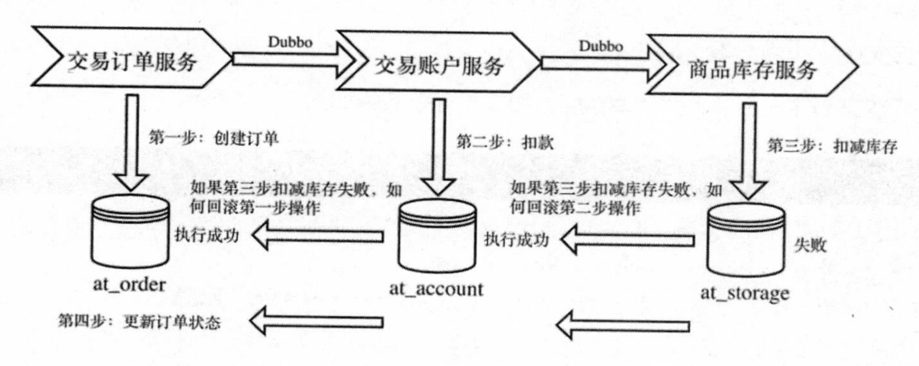

图 7-21

- 交易订单服务调用本地接口创建订单，通过 RPC（Dubbo）调用交易账户服务和商品库存服务，完成账户扣款和商品库存扣减。
- 在服务接口都返回“执行成功”后，更新订单状态为“交易成功”。
- 如果更新订单状态失败，则用户会看到购买失败的提示信息，但此时已经完成商品库存扣减和账户扣款，这样就形成了一个比较严重的服务之间接口状态不一致的故障。

（4）在购买商品的过程中，制造一个服务不可用的异常。

如果在交易订单服务完成账户扣款后关闭商品库存服务，则在交易服务中会出现回滚日志，如图 7-22 所示。

```
i.seata.tm.api.DefaultGlobalTransaction  : Begin new global transaction [192.168.0.123:8091:12913867805580
c.a.druid.pool.DruidAbstractDataSource   : discard long time none received connection. , jdbcUrl : jdbc:my
i.s.c.r.p.c.RmBranchRollbackProcessor    : rm handle branch rollback process:xid=192.168.0.123:8091:129138
io.seata.rm.AbstractRMHandler            : Branch Rollbacking: 192.168.0.123:8091:129138678055809024 12913
i.s.r.d.undo.AbstractUndoLogManager      : xid 192.168.0.123:8091:129138678055809024 branch 12913867839135
io.seata.rm.AbstractRMHandler            : Branch Rollbacked result: PhaseTwo_Rollbacked
```

图 7-22

在账户服务中也会出现回滚日志，如图 7-23 所示。

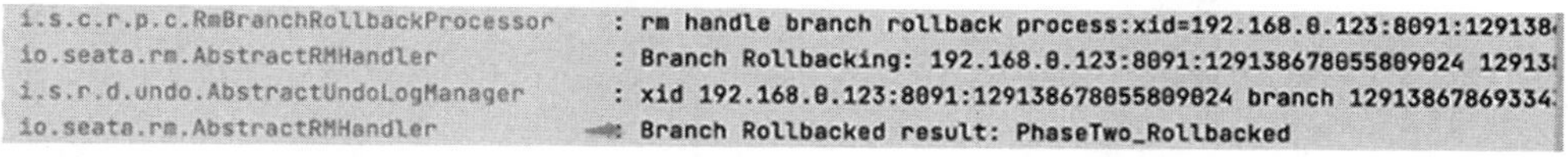

```
i.s.c.r.p.c.RmBranchRollbackProcessor    : rm handle branch rollback process:xid=192.168.0.123:8091:129138
io.seata.rm.AbstractRMHandler            : Branch Rollbacking: 192.168.0.123:8091:129138678055809024 12913
i.s.r.d.undo.AbstractUndoLogManager      : xid 192.168.0.123:8091:129138678055809024 branch 12913867869334
io.seata.rm.AbstractRMHandler            : Branch Rollbacked result: PhaseTwo_Rollbacked
```

图 7-23

综上所述，分布式事务回滚过程如下：

- 商品原始库存数量为 999999，账户原始余额为 999999000。
- 如果没有分布式事务，则会出现如下业务场景——商品库存扣减失败，但是账户扣款成功。在成功扣款 100 元后，最终余额为 999 998 900 元；在商品库存扣减失败后，最终商品库存还是 999 999。这时会提示用户交易失败，但是用户已经收到成功扣款 100 元的短信提示。

如果开启了 AT 模式的分布式事务，则会回滚账户表：用 undo_log 表中存储的回滚日志，将账户余额恢复到 999 999 000 元（用全局锁保证数据一致性）。undo_log 表的回滚日志如图 7-24 所示。

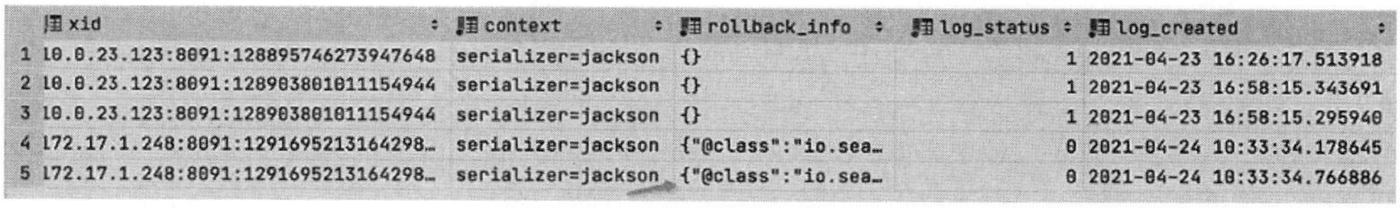

	xid	context	rollback_info	log_status	log_created
1	10.0.23.123:8091:128895746273947648	serializer=jackson	{}	1	2021-04-23 16:26:17.513918
2	10.0.23.123:8091:128903801011154944	serializer=jackson	{}	1	2021-04-23 16:58:15.343691
3	10.0.23.123:8091:128903801011154944	serializer=jackson	{}	1	2021-04-23 16:58:15.295940
4	172.17.1.248:8091:1291695213164298…	serializer=jackson	{"@class":"io.sea…	0	2021-04-24 10:33:34.178645
5	172.17.1.248:8091:1291695213164298…	serializer=jackson	{"@class":"io.sea…	0	2021-04-24 10:33:34.766886

图 7-24

7.7 用 TCC 模式实现分布式事务

在 Seata 中，如果应用需要使用 TCC 模式，则需要在类或者方法中添加 Seata 指定的注解（比如@LocalTCC 或者@TwoPhaseBusinessAction）。这样，Seata 就可以解析注解，并将类或者方法添加到 TCC 模式的分布式事务的作用域内。

7.7.1 用 GlobalTransactionScanner 类扫描客户端，开启 TCC 动态代理

在项目中，可以使用注解@LocalTCC 和@TwoPhaseBusinessAction 开启业务接口的 TCC 事务，这样 Seata 会用 GlobalTransactionScanner 类扫描这两个注解，具体代码如下所示：

```
    public class GlobalTransactionScanner extends AbstractAutoProxyCreator
        implements ConfigurationChangeListener, InitializingBean,
ApplicationContextAware, DisposableBean {
        private static final Set<String> PROXYED_SET = new HashSet<>();
        @Override
        protected Object wrapIfNecessary(Object bean, String beanName, Object
cacheKey) {
```

```
        try {
            synchronized (PROXYED_SET) {
                if (PROXYED_SET.contains(beanName)) {
                    return bean;
                }
                interceptor = null;
                //①验证是否是 TCC 代理的类或者方法
                if (TCCBeanParserUtils.isTccAutoProxy(bean, beanName,
applicationContext)) {
                    //②如果是，则初始化 TCC 模式的拦截器
                    interceptor = new
TccActionInterceptor(TCCBeanParserUtils.getRemotingDesc(beanName));
                    ConfigurationCache.addConfigListener
(ConfigurationKeys.DISABLE_GLOBAL_TRANSACTION,
                        (ConfigurationChangeListener)interceptor);
                } else {
                    ...
                }
                LOGGER.info("Bean[{}] with name [{}] would use interceptor
[{}]", bean.getClass().getName(), beanName,
interceptor.getClass().getName());
                if (!AopUtils.isAopProxy(bean)) {
                    bean = super.wrapIfNecessary(bean, beanName, cacheKey);
                } else {
                    AdvisedSupport advised =
SpringProxyUtils.getAdvisedSupport(bean);
                    Advisor[] advisor = buildAdvisors(beanName,
getAdvicesAndAdvisorsForBean(null, null, null));
                    for (Advisor avr : advisor) {
                        advised.addAdvisor(0, avr);
                    }
                }
                PROXYED_SET.add(beanName);
                return bean;
            }
        } catch (Exception exx) {
            throw new RuntimeException(exx);
        }
    }
}
```

在 Seata 中，用 TCCBeanParserUtils 类的 isTccProxyTargetBean()方法扫描注解@TwoPhaseBusinessAction，具体代码如下所示：

```
public static boolean isTccProxyTargetBean(RemotingDesc remotingDesc) {
    if (remotingDesc == null) {
```

```
            return false;
        }
        boolean isTccClazz = false;
        Class <?>tccInterfaceClazz = remotingDesc.getInterfaceClass();
        Method[] methods = tccInterfaceClazz.getMethods();
        TwoPhaseBusinessAction twoPhaseBusinessAction;
    for (Method method: methods) {
        //扫描应用代码中的方法是否被注解@TwoPhaseBusinessAction 修饰
            twoPhaseBusinessAction =
method.getAnnotation(TwoPhaseBusinessAction.class);
            if (twoPhaseBusinessAction != null) {
                isTccClazz = true;
                break;
            }
        }
        if (!isTccClazz) {
            return false;
        }
        short protocols = remotingDesc.getProtocol();
        if (Protocols.IN_JVM == protocols) {
            return true;
        }
        return remotingDesc.isReference();
    }
```

在 Seata 中，用 LocalTCCRemotingParser 类的 isReference()方法和 isService()方法扫描注解@ LocalTCC，具体代码如下所示：

```
    public class LocalTCCRemotingParser extends AbstractedRemotingParser {
        @Override
        public boolean isReference(Object bean, String beanName) {
            Class<?> classType = bean.getClass();
            Set<Class<?>> interfaceClasses =
ReflectionUtil.getInterfaces(classType);
            for (Class<?> interClass : interfaceClasses) {
                if (interClass.isAnnotationPresent(LocalTCC.class)) {
                    return true;
                }
            }
            return false;
        }
        @Override
        public boolean isService(Object bean, String beanName) {
            Class<?> classType = bean.getClass();
            Set<Class<?>> interfaceClasses = ReflectionUtil.getInterfaces(classType);
            for (Class<?> interClass : interfaceClasses) {
```

```
            if (interClass.isAnnotationPresent(LocalTCC.class)) {
                return true;
            }
        }
        return false;
    }
}
```

7.7.2 用拦截器 TccActionInterceptor 校验 TCC 事务

在 Saga 模式中，Seata 用 TccActionInterceptor 类实现了 Spring Framework 的方法拦截器 MethodInterceptor，具体代码如下所示：

```
public class TccActionInterceptor implements MethodInterceptor,
ConfigurationChangeListener {
    ...
    @Override
    public Object invoke(final MethodInvocation invocation) throws Throwable {
        //①如果分布式事务上下文中的全局事务开关为关闭状态，或者事务类型是 Saga，则过滤
拦截请求
        if (!RootContext.inGlobalTransaction() || disable ||
RootContext.inSagaBranch()) {
            return invocation.proceed();
        }
        Method method = getActionInterfaceMethod(invocation);
        TwoPhaseBusinessAction businessAction =
method.getAnnotation(TwoPhaseBusinessAction.class);
        if (businessAction != null) {
            String xid = RootContext.getXID();
            //②如果上一次请求的分支事务类型不是 TCC 事务，则绑定事务类型为 TCC
            BranchType previousBranchType = RootContext.getBranchType();
            if (BranchType.TCC != previousBranchType) {
                RootContext.bindBranchType(BranchType.TCC);
            }
            try {
                Object[] methodArgs = invocation.getArguments();
                //③调用 TCC 拦截器处理器校验 TCC 事务
                Map<String, Object> ret = actionInterceptorHandler.proceed(method,
    methodArgs, xid, businessAction,
                        invocation::proceed);
                return ret.get(Constants.TCC_METHOD_RESULT);
            }
            finally {
                if (BranchType.TCC != previousBranchType) {
                    RootContext.unbindBranchType();
```

```
                }
            }
        }
        return invocation.proceed();
    }
}
```

在 Saga 模式中，Seata 用 ActionInterceptorHandler 类的 proceed()方法校验 TCC 事务，具体代码如下所示：

```
public class ActionInterceptorHandler {
private static final Logger LOGGER =
LoggerFactory.getLogger(ActionInterceptorHandler.class);
//①处理 TCC 事务的切面
    public Map <String,Object> proceed(Method method, Object[] arguments,
String xid, TwoPhaseBusinessAction businessAction, Callback < Object >
targetCallback) throws Throwable {
        Map <String,Object> ret = new HashMap <>(4);
        String actionName = businessAction.name();
        //②构造一个业务执行器的上下文对象 BusinessActionContext
        BusinessActionContext actionContext = new BusinessActionContext();
        actionContext.setXid(xid);
        actionContext.setActionName(actionName);
        //③调用 doTccActionLogStore()方法来注册分支事务
        String branchId = doTccActionLogStore(method, arguments,
businessAction, actionContext);
        actionContext.setBranchId(branchId);
        Class <?>[] types = method.getParameterTypes();
        int argIndex = 0;
        for (Class <?>cls: types) {
            if (cls.getName().equals(BusinessActionContext.class.getName())) {
                arguments[argIndex] = actionContext;
                break;
            }
            argIndex++;
        }
        ret.put(Constants.TCC_METHOD_ARGUMENTS, arguments);
        ret.put(Constants.TCC_METHOD_RESULT, targetCallback.execute());
        return ret;
    }
    protected String doTccActionLogStore(Method method, Object[] arguments,
TwoPhaseBusinessAction businessAction, BusinessActionContext actionContext) {
        String actionName = actionContext.getActionName();
        //④获取分布式事务的 XID
        String xid = actionContext.getXid();
```

```
        Map<String,Object> context = fetchActionRequestContext(method,
arguments);
        context.put(Constants.ACTION_START_TIME,
System.currentTimeMillis());
        //⑤初始化业务上下文
        initBusinessContext(context, method, businessAction);
        //⑥初始化环境上下文
        initFrameworkContext(context);
        actionContext.setActionContext(context);
        Map <String,Object> applicationContext = new HashMap <>(4);
        applicationContext.put(Constants.TCC_ACTION_CONTEXT, context);
        String applicationContextStr =
JSON.toJSONString(applicationContext);
        try {
            //⑦调用 DefaultResourceManager 类的 branchRegister()方法来注册分支事务
            Long branchId =
DefaultResourceManager.get().branchRegister(BranchType.TCC, actionName, null,
xid, applicationContextStr, null);
            return String.valueOf(branchId);
        } catch(Throwable t) {
            String msg = String.format("TCC branch Register error, xid: %s",
xid);
            LOGGER.error(msg, t);
            throw new FrameworkException(t, msg);
        }
    }
}
```

7.7.3 【实例】搭建 Seata 的 TCC 模式的环境，并验证 TCC 模式的分布式事务场景

本实例的源码在本书配套资源的“chapterseven/spring-cloud-alibaba-access-seata-tcc”目录下。

下面使用 Spring Cloud Alibaba 框架来模拟用户下单购买商品的场景。

1. 初始化交易账户服务

下面用项目 tcc-account-service 初始化交易账户服务：

（1）用 Spring Cloud Alibaba 搭建基础环境（包括 Nacos、Seata 等），添加的 POM 依赖如下所示：

```xml
<dependency>
    <groupId>com.alibaba.cloud</groupId>
    <artifactId>spring-cloud-starter-alibaba-nacos-discovery</artifactId>
    <exclusions>
        <exclusion>
            <groupId>org.apache.httpcomponents</groupId>
            <artifactId>httpclient</artifactId>
        </exclusion>
    </exclusions>
</dependency>
<dependency>
    <groupId>com.alibaba.cloud</groupId>
    <artifactId>spring-cloud-starter-alibaba-nacos-config</artifactId>
</dependency>
<dependency>
    <groupId>com.alibaba.cloud</groupId>
    <artifactId>spring-cloud-starter-alibaba-seata</artifactId>
    <exclusions>
        <exclusion>
            <groupId>io.seata</groupId>
            <artifactId>seata-spring-boot-starter</artifactId>
        </exclusion>
    </exclusions>
</dependency>
<dependency>
    <groupId>io.seata</groupId>
    <artifactId>seata-spring-boot-starter</artifactId>
    <exclusions>
        <exclusion>
            <groupId>org.springframework</groupId>
            <artifactId>spring-beans</artifactId>
        </exclusion>
    </exclusions>
</dependency>
```

（2）添加配置文件 bootstrap.yaml，如下所示。

```yaml
###①省略部分配置信息
nacos:
  server-addr: 127.0.0.1:8848
  namespace: c7ba173f-29e5-4c58-ae78-b102be11c4f9
  group: SEATA_GROUP
  seata:
    application: seata-server
    tx-service-group: tcc-account-service-tx-group
spring:
```

```
    application:
      name: tcc-account-service
    main:
  allow-bean-definition-overriding: true
    ###②RM 对应的配置数据源
    datasource:
      url:
jdbc:mysql://127.0.0.1:3306/seata_server?characterEncoding=utf8&connectTimeo
ut=1000&socketTimeout=3000&autoReconnect=true&useUnicode=true&useSSL=false&s
erverTimezone=UTC
      username: root
      password: 123456huxian
  server:
    port: 7812
  ###③配置 Seata 的相关信息
  seata:
    registry:
      type: nacos
      nacos:
        group: ${nacos.group}
        application: ${nacos.seata.application}
        namespace: ${nacos.namespace}
        cluster: default
        server-addr: ${nacos.server-addr}
        username: "nacos"
        password: "nacos"
    tx-service-group: ${nacos.seata.tx-service-group}
    enabled: true
    enable-auto-data-source-proxy: false
    application-id: tcc-account-service
    config:
      type: nacos
      nacos:
        group: ${nacos.group}
        server-addr: ${nacos.server-addr}
        namespace: ${nacos.namespace}
        username: "nacos"
        password: "nacos"
```

（3）实现账户扣款的 TCC 功能，具体代码如下所示：

```
  @Service
  @Slf4j
  public class AccountTccActionImpl implements AccountTccAction {
      @Resource
  private AccountMapper accountMapper;
```

```
        @Transactional(rollbackFor = Exception.class)
        @Override
        public boolean prepareDecreaseMoney(BusinessActionContext
businessActionContext, Long userId, BigDecimal money) {
            log.info("减少账户金额，第一阶段锁定金额，userId=" + userId + ", money="
+ money);
            AccountDO account = accountMapper.findOneByUserId(userId);
            //①如果余额不足，则抛出“账户金额不足”异常
            if (account.getResidue().compareTo(money) < 0) {
                throw new RuntimeException("账户金额不足");
            }
            //②冻结账户
            accountMapper.updateFrozen(userId,
account.getResidue().subtract(money), account.getFrozen().add(money));
            ResultHolder.setResult(getClass(), businessActionContext.getXid(),
"p");
            return true;
        }
        @Transactional(rollbackFor = Exception.class)
        @Override
        public boolean commit(BusinessActionContext businessActionContext) {
            long userId =
Long.parseLong(businessActionContext.getActionContext("userId").toString());
            BigDecimal money = new
BigDecimal(businessActionContext.getActionContext("money").toString());
            log.info("减少账户金额，第二阶段，提交，userId=" + userId + ", money=" +
money);
            //③防止重复提交
            if (ResultHolder.getResult(getClass(),
businessActionContext.getXid()) == null) {
                return true;
            }
            accountMapper.updateFrozenToUsed(userId, money);
            //④删除标识
            ResultHolder.removeResult(getClass(),
businessActionContext.getXid());
            return true;
        }
        @Transactional(rollbackFor = Exception.class)
        @Override
        public boolean rollback(BusinessActionContext businessActionContext) {
            long userId =
Long.parseLong(businessActionContext.getActionContext("userId").toString());
            BigDecimal money = new
BigDecimal(businessActionContext.getActionContext("money").toString());
```

```
        if (ResultHolder.getResult(getClass(),
businessActionContext.getXid()) == null) {
            return true;
        }
        log.info("减少账户金额，第二阶段，回滚，userId=" + userId + ", money=" +
money);
        accountMapper.updateFrozenToResidue(userId, money);
        ResultHolder.removeResult(getClass(),
businessActionContext.getXid());
        return true;
    }
}
```

（4）暴露交易账户服务对应的 RPC 接口，具体代码如下所示：

```
@DubboService(version = "1.0.0",group = "SEATA_GROUP")
public class TccAccountServiceImpl implements TccAccountService {
    @Resource
    private AccountTccAction accountTccAction;
    @Override
    public boolean transformMoney(TccAccountServiceBo tccAccountServiceBo) {
        return false;
    }
    @Override
    public void decreaseMoney(Long userId, BigDecimal money) {
        accountTccAction.prepareDecreaseMoney(null, userId, money);
    }
}
```

2. 初始化商品库存服务

下面用项目 tcc-storage-service 初始化商品库存服务。

（1）用 Spring Cloud Alibaba 搭建基础环境，Jar 包的 POM 依赖与项目 tcc-account-service 一致。

（2）添加配置文件 bootstrap.yaml，如下所示。

```
###①省略部分配置信息
nacos:
  server-addr: 127.0.0.1:8848
  namespace: c7ba173f-29e5-4c58-ae78-b102be11c4f9
  group: SEATA_GROUP
  seata:
    application: seata-server
    tx-service-group: tcc-storage-service-tx-group
spring:
```

```
    application:
      name: tcc-storage-service
    main:
      allow-bean-definition-overriding: true
###②RM 对应的配置数据源
    datasource:
      url:
jdbc:mysql://127.0.0.1:3306/seata_server?characterEncoding=utf8&connectTimeo
ut=1000&socketTimeout=3000&autoReconnect=true&useUnicode=true&useSSL=false&s
erverTimezone=UTC
      username: root
      password: 123456huxian
server:
  port: 7848
###③配置 Seata 的相关信息
seata:
  registry:
    type: nacos
    nacos:
      group: ${nacos.group}
      application: ${nacos.seata.application}
      namespace: ${nacos.namespace}
      cluster: default
      server-addr: ${nacos.server-addr}
      username: "nacos"
      password: "nacos"
  tx-service-group: ${nacos.seata.tx-service-group}
  enabled: true
  enable-auto-data-source-proxy: false
  application-id: tcc-storage-service
  config:
    type: nacos
    nacos:
      group: ${nacos.group}
      server-addr: ${nacos.server-addr}
      namespace: ${nacos.namespace}
      username: "nacos"
      password: "nacos"
```

（3）实现商品扣减库存的 TCC 功能，具体代码如下所示：

```
@Slf4j
@Component
public class StorageTccActionImpl implements StorageTccAction {
    @Resource
private StorageMapper storageMapper;
```

```
    //①预扣减库存的数据库操作
        @Transactional(rollbackFor = Exception.class)
        @Override
        public boolean prepareDecreaseStorage(BusinessActionContext
businessActionContext, Long productId, Integer count) {
            log.info("减少商品库存，第一阶段，锁定减少的库存量，
productId="+productId+",  count="+count);
            StorageDO storage = storageMapper.findOneByProductId(productId);
            if (storage.getResidue()-count<0) {
                throw new RuntimeException("库存不足");
            }
            storageMapper.updateFrozen(productId, storage.getResidue()-count,
storage.getFrozen()+count);
            ResultHolder.setResult(getClass(), businessActionContext.getXid(),
"p");
            return true;
    }
    //②提交扣减库存的数据库操作
        @Transactional(rollbackFor = Exception.class)
        @Override
        public boolean commit(BusinessActionContext businessActionContext) {
            long productId =
Long.parseLong(businessActionContext.getActionContext("productId").toString(
));
            int count =
Integer.parseInt(businessActionContext.getActionContext("count").toString());
            log.info("减少商品库存,第二阶段提交,productId="+productId+", count="+count);
            if (ResultHolder.getResult(getClass(), businessActionContext.getXid())
== null) {
              return true;
            }
            storageMapper.updateFrozenToUsed(productId, count);
            ResultHolder.removeResult(getClass(),
businessActionContext.getXid());
            return true;
    }
    //③回滚扣减库存的数据库操作
        @Transactional(rollbackFor = Exception.class)
        @Override
        public boolean rollback(BusinessActionContext businessActionContext) {
            long productId = Long.parseLong
(businessActionContext.getActionContext("productId").toString());
            int count =
Integer.parseInt(businessActionContext.getActionContext("count").toString());
            log.info("减少商品库存,第二阶段,回滚,productId="+productId+", count="+count);
```

```
        if (ResultHolder.getResult(getClass(),
businessActionContext.getXid()) == null) {
            return true;
        }
        storageMapper.updateFrozenToResidue(productId, count);
        ResultHolder.removeResult(getClass(), businessActionContext.getXid());
        return true;
    }
}
```

（4）暴露 RPC 接口 TccStorageServiceImpl，具体代码如下所示：

```
@DubboService(version = "1.0.0",group = "SEATA_GROUP")
public class TccStorageServiceImpl implements TccStorageService {
    @Resource
    private StorageTccAction storageTccAction;
    @Override
    public void decreaseStorage(Long productId, Integer count) {
        storageTccAction.prepareDecreaseStorage(null, productId, count);
    }
}
```

3. 初始化交易订单服务

下面用项目 tcc-order-service 初始化交易订单服务。

（1）用 Spring Cloud Alibaba 搭建基础环境，Jar 包的 POM 依赖与项目 tcc-account-service 一致。

（2）添加配置文件 bootstrap.yaml，如下所示。

```
###①省略部分配置信息
nacos:
  server-addr: 127.0.0.1:8848
  namespace: c7ba173f-29e5-4c58-ae78-b102be11c4f9
  group: SEATA_GROUP
  seata:
    application: seata-server
    tx-service-group: tcc-order-service-tx-group
###②以上定义配置中心和注册中心的公共属性
spring:
  application:
    name: tcc-order-service
  main:
    allow-bean-definition-overriding: true
###③RM 对应的配置数据源
  datasource:
```

```
      url:
jdbc:mysql://127.0.0.1:3306/seata_server?characterEncoding=utf8&connectTimeo
ut=1000&socketTimeout=3000&autoReconnect=true&useUnicode=true&useSSL=false&s
erverTimezone=UTC
      username: root
      password: 123456huxian
  server:
    port: 7889
  ###④配置 Seata 的相关信息
  seata:
    registry:
      type: nacos
      nacos:
        group: ${nacos.group}
        application: ${nacos.seata.application}
        namespace: ${nacos.namespace}
        cluster: default
        server-addr: ${nacos.server-addr}
        username: "nacos"
        password: "nacos"
    tx-service-group: ${nacos.seata.tx-service-group}
    enabled: true
    enable-auto-data-source-proxy: false
    application-id: tcc-order-service
    config:
      type: nacos
      nacos:
        group: ${nacos.group}
        server-addr: ${nacos.server-addr}
        namespace: ${nacos.namespace}
        username: "nacos"
        password: "nacos"
```

（3）实现交易订单服务的 TCC 功能，具体代码如下所示：

```
  @Slf4j
  @Component
  public class OrderTccActionImpl implements OrderTccAction {
      @Resource
      private  OrderMapper orderMapper;
      //①预创建订单
      @Transactional(rollbackFor = Exception.class)
      @Override
      public boolean prepareCreateOrder(BusinessActionContext
businessActionContext, String orderNo,Long userId,
            Long productId,Integer amount,BigDecimal money) {
```

```
            orderMapper.save(new OrderDO(orderNo,userId, productId, amount,
money, 0));
            ResultHolder.setResult(OrderTccAction.class,
businessActionContext.getXid(), "p");
            return true;
        }
        //②提交事务，并执行 SQL 语句创建订单
        @Transactional(rollbackFor = Exception.class)
        @Override
        public boolean commit(BusinessActionContext businessActionContext) {
            String p = ResultHolder.getResult(OrderTccAction.class,
businessActionContext.getXid());
            if (p == null){
                return true;
            }
            String orderNo =
businessActionContext.getActionContext("orderNo").toString();
            orderMapper.updateStatusByOrderNo(orderNo, 1);
            ResultHolder.removeResult(OrderTccAction.class,
businessActionContext.getXid());
            return true;
    }
    //③如果出现异常，则回滚事务，删除已经入库的订单数据
        @Transactional(rollbackFor = Exception.class)
        @Override
        public boolean rollback(BusinessActionContext businessActionContext) {
            String p = ResultHolder.getResult(OrderTccAction.class,
businessActionContext.getXid());
            if (p == null){
                return true;
            }
            String orderNo =
businessActionContext.getActionContext("orderNo").toString();
            orderMapper.deleteByOrderNo(orderNo);
            ResultHolder.removeResult(OrderTccAction.class,
businessActionContext.getXid());
            return true;
        }

    }
```

（4）交易订单服务调用交易账户服务和商点库存服务，用注解@GlobalTransactional 开启 TCC 模式的分布式事务，具体代码如下所示：

```
    @Service
    public class OrderServiceImpl implements OrderService {
```

```
    //①添加 OrderTccAction 的实例对象
    @Resource
  private  OrderTccAction orderTccAction;
    //②添加 Dubbo 接口 TccAccountService 的实例对象
    @DubboReference(version = "1.0.0",group = "SEATA_GROUP")
  private TccAccountService tccAccountService;
    //③添加 Dubbo 接口 TccStorageService 的实例对象
    @DubboReference(version = "1.0.0",group = "SEATA_GROUP")
  private TccStorageService tccStorageService;
    //④使用注解@GlobalTransactional 开启全局事务
    @GlobalTransactional
    @Override
    public void createOrder(OrderDO orderDO) {
        String orderNo=this.generateOrderNo();
        orderTccAction.prepareCreateOrder(null,
                orderNo,
                orderDO.getUserId(),
                orderDO.getProductId(),
                orderDO.getAmount(),
                orderDO.getMoney());
     tccAccountService.decreaseMoney(orderDO.getUserId(),orderDO.getMoney());

tccStorageService.decreaseStorage(orderDO.getProductId(),orderDO.getAmount());
    }
    private String generateOrderNo(){
        return LocalDateTime.now()
                .format(
                        DateTimeFormatter.ofPattern("yyMMddHHmmssSSS")
                );
    }
  }
```

4. 启动程序，调用交易订单服务

分别启动交易账户服务、商品库存服务和交易订单服务，在启动成功后，3 个服务会被注册到 Nacos 注册中心中，如图 7-25 所示。

Service Name	Group Name	Cluster Count	Instance Count	Healthy Instance Count
tcc-account-service	SEATA_GROUP	1	1	1
tcc-storage-service	SEATA_GROUP	1	1	1
tcc-order-service	SEATA_GROUP	1	1	1

图 7-25

用 Swagger 控制台调用交易订单服务，并手动关闭商品库存服务。从运行日志中可以分析出 TCC 模式的分布式事务已经生效。从图 7-26 中可以看到，交易订单服务的分布式事务已经回滚成功；从图 7-27 中可以看到，交易账户服务的分布式事务已经回滚成功。

```
com.alibaba.nacos.client.naming             : [REGISTER-SERVICE] public registering service DEFAULT_GROUP@@consumers:c
o.a.d.r.t.netty4.NettyServerHandler         :  [DUBBO] The connection of /172.17.2.14:52900 -> /172.17.2.14:20882 is c
a.c.d.m.r.DubboServiceMetadataRepository    : Current application will subscribe all services(size:4) in registry, a 1
i.seata.tm.api.DefaultGlobalTransaction     : Begin new global transaction [172.17.2.14:8091:130413954703876096]
i.s.c.r.p.c.RmBranchRollbackProcessor       : rm handle branch rollback process:xid=172.17.2.14:8091:130413954703876096
io.seata.rm.AbstractRMHandler               : Branch Rollbacking: 172.17.2.14:8091:130413954703876096 130413954997477
io.seata.rm.AbstractResourceManager         : TCC resource rollback result : true, xid: 172.17.2.14:8091:13041395470387
io.seata.rm.AbstractRMHandler               : Branch Rollbacked result: PhaseTwo_Rollbacked
i.seata.tm.api.DefaultGlobalTransaction     : [172.17.2.14:8091:130413954703876096] rollback status: Rollbacked
ider available from registry 127.0.0.1:8848 for service SEATA_GROUP/com.alibaba.cloud.youxia.service.TccStorageServi
ctory.doList(RegistryDirectory.java:599)
ectory.list(AbstractDirectory.java:74)
erInvoker.list(AbstractClusterInvoker.java:292)
```

图 7-26

```
c.a.c.y.a.impl.AccountTccActionImpl         : 减少账户金额，第一阶段锁定金额，userId=12345678,  money=10
i.s.c.r.p.c.RmBranchRollbackProcessor       : rm handle branch rollback process:xid=172.17.2.14:8091:13
io.seata.rm.AbstractRMHandler               : Branch Rollbacking: 172.17.2.14:8091:130413954703876096 1
c.a.c.y.a.impl.AccountTccActionImpl         : 减少账户金额，第二阶段，回滚，userId=12345678,  money=10
io.seata.rm.AbstractResourceManager         : TCC resource rollback result : true, xid: 172.17.2.14:809
io.seata.rm.AbstractRMHandler               : Branch Rollbacked result: PhaseTwo_Rollbacked
a.c.d.m.r.DubboServiceMetadataRepository    : Current application will subscribe all services(size:4) i
o.a.d.r.t.netty4.NettyServerHandler         :  [DUBBO] The connection of /172.17.2.14:57870 -> /172.17.
com.alibaba.nacos.client.naming             : received push data: {"type":"dom","data":"{\"hosts\":[{\"
com.alibaba.nacos.client.naming             : new ips(1) service: SEATA_GROUP@@tcc-storage-service -> [
```

图 7-27

7.8 用 XA 模式实现分布式事务

从编程模型的角度来分析，Seata 的 XA 模式和 AT 模式的原理是完全一致的。开发人员要在项目中使用 XA 模式，则必须使用支持 XA 模式事务的数据库（比如 MySQL），并且要使用 JDBC 访问数据库。

XA 模式和 AT 模式一样，需要代理应用的数据源实现分布式事务的零侵入。

7.8.1　"用数据源代理实现 XA 模式的零侵入应用"的原理

在 XA 模式中，Seata 会先初始化数据源代理，然后注册这个数据源代理，并绑定一个通信渠道。

1. 初始化数据源代理

在 Seata 的 XA 模式中，初始化数据源代理的顺序和 AT 模式中的顺序是一致的。

（1）在 XA 模式中，Seata 用 DataSourceProxyXA 类来作为数据源代理，具体代码如下所示：

```
public class DataSourceProxyXA extends AbstractDataSourceProxyXA {
...
public DataSourceProxyXA(DataSource dataSource, String resourceGroupId) {
    //①如果 DataSource 匹配 SeataDataSourceProxy 对象
    if (dataSource instanceof SeataDataSourceProxy) {
        LOGGER.info("Unwrap the data source, because the type is: {}",
dataSource.getClass().getName());
        //②则将 dataSource 赋值为 SeataDataSourceProxy 对象中的目标数据源
        dataSource = ((SeataDataSourceProxy)
dataSource).getTargetDataSource();
    }
    this.dataSource = dataSource;
    //③设置数据源的事务类型为 BranchType.XA
    this.branchType = BranchType.XA;
    //④初始化数据源对象 dataSource
        JdbcUtils.initDataSourceResource(this, dataSource, resourceGroupId);
        RootContext.setDefaultBranchType(this.getBranchType());
        }
}
```

在 XA 模式中，Seata 的 DataSourceProxyXA 类是通过 SeataAutoDataSourceProxyAdvice 类来继承 MethodInterceptor 类并实现 invoke()方法，从而实现数据源的代理的。

（2）在 Seata 的 XA 模式中，Seata 用 JdbcUtils 类的 initDataSourceResource()方法初始化数据源代理，具体代码如下所示：

```
public final class JdbcUtils {
    public static void initDataSourceResource(BaseDataSourceResource
dataSourceResource, DataSource dataSource, String resourceGroupId) {
        dataSourceResource.setResourceGroupId(resourceGroupId);
        //①将 jdbcUrl（JDK 连接 URL）、driverClassName（驱动类名称）及 dbType（数
据库类型）设置到对象 dataSourceResource 中
        try (Connection connection = dataSource.getConnection()) {
            String jdbcUrl = connection.getMetaData().getURL();
            dataSourceResource.setResourceId(buildResourceId(jdbcUrl));
            String driverClassName =
com.alibaba.druid.util.JdbcUtils.getDriverClassName(jdbcUrl);
            dataSourceResource.setDriver(loadDriver(driverClassName));

dataSourceResource.setDbType(com.alibaba.druid.util.JdbcUtils.getDbType(jdbc
Url, driverClassName));
```

```
            } catch (SQLException e) {
                throw new IllegalStateException("can not init DataSourceResource
with " + dataSource, e);
            }
            //②调用 ResourceManagerXA 类的 registerResource()方法去注册数据源
            DefaultResourceManager.get().registerResource(dataSourceResource);
        }
        ...
    }
```

2. 注册数据源代理

在 XA 模式中，Seata 用 DefaultResourceManager 类的 registerResource()方法注册数据源代理，具体代码如下所示：

```
    public class DefaultResourceManager implements ResourceManager {
        ...
        @Override
    public void registerResource(Resource resource) {
        //①调用 ResourceManagerXA 类的 registerResource()方法注册数据源代理
            getResourceManager(resource.getBranchType()).
         registerResource(resource);
        }
    }

    //②其中，ResourceManagerXA 类继承抽象类
AbstractDataSourceCacheResourceManager
    public class ResourceManagerXA extends
AbstractDataSourceCacheResourceManager {
        ...
    }
    //③在抽象类 AbstractDataSourceCacheResourceManager 中，统一实现了注册数据源的
registerResource()方法
    public abstract class AbstractDataSourceCacheResourceManager extends
AbstractResourceManager implements Initialize {
        @Override
    public void registerResource(Resource resource) {
        //④将数据源设置到本地缓存中
            dataSourceCache.put(resource.getResourceId(), resource);
            //⑤调用父类 AbstractResourceManager 的 registerResource()方法注册数据源代理
            super.registerResource(resource);
        }
        ...
    }

    public abstract class AbstractResourceManager implements ResourceManager {
```

```
        @Override
    public void registerResource(Resource resource) {
    //⑥调用 RmNettyRemotingClient 类的 registerResource()方法绑定数据源代理和通信渠道
        RmNettyRemotingClient.getInstance().
    registerResource(resource.getResourceGroupId(),
resource.getResourceId());
        }
        ...
    }
```

3. 绑定数据源代理和通信渠道

在 XA 模式中，Seata 用 RmNettyRemotingClient 类的 registerResource()方法绑定数据源代理和通信渠道，具体代码如下所示：

```
    public final class RmNettyRemotingClient extends
AbstractNettyRemotingClient {
        public void registerResource(String resourceGroupId, String resourceId) {
            if (StringUtils.isBlank(transactionServiceGroup)) {
                return;
            }
            //①如果通信渠道列表为空，则直接返回
            if (getClientChannelManager().getChannels().isEmpty()) {
                getClientChannelManager().reconnect(transactionServiceGroup);
                return;
            }
            synchronized (getClientChannelManager().getChannels()) {
                //②遍历通信渠道列表，并发送注册消息
                for (Map.Entry<String, Channel> entry :
getClientChannelManager().getChannels().entrySet()) {
                    String serverAddress = entry.getKey();
                    Channel rmChannel = entry.getValue();
                    if (LOGGER.isInfoEnabled()) {
                        LOGGER.info("will register resourceId:{}", resourceId);
                    }
                    sendRegisterMessage(serverAddress, rmChannel, resourceId);
                }
            }
        }

        public void sendRegisterMessage(String serverAddress, Channel channel,
String resourceId) {
            //③构造注册消息的请求对象 RegisterRMRequest
            RegisterRMRequest message = new RegisterRMRequest(applicationId,
transactionServiceGroup);
            message.setResourceIds(resourceId);
```

```
            try {
                //④调用父类AbstractNettyRemotingClient的sendAsyncRequest()方法发
送异步消息的请求
                super.sendAsyncRequest(channel, message);
            } catch (FrameworkException e) {
                if (e.getErrcode() == FrameworkErrorCode.ChannelIsNotWritable &&
serverAddress != null) {
                    getClientChannelManager().releaseChannel(channel,
serverAddress);
                    if (LOGGER.isInfoEnabled()) {
                        LOGGER.info("remove not writable channel:{}", channel);
                    }
                } else {
                    LOGGER.error("register resource failed,
channel:{},resourceId:{}", channel, resourceId, e);
                }
            }
        }
        ...
    }
```

在 XA 模式中，Seata 用 AbstractNettyRemotingServer 类的 channelRead()方法读取通道中的消息，这里主要是指数据源和通信渠道绑定的消息对象 RegisterRMRequest，具体代码如下所示：

```
    public abstract class AbstractNettyRemotingServer extends
AbstractNettyRemoting implements RemotingServer {
        @Override
        public void channelRead(final ChannelHandlerContext ctx, Object msg)
throws Exception {
            if (!(msg instanceof RpcMessage)) {
                return;
            }
            //处理消息通道中的消息对象RegisterRMRequest
            processMessage(ctx, (RpcMessage) msg);
        }
        ...
    }
```

在 XA 模式中，Seata 用 RegRmProcessor 类的 process()方法处理消息对象 RegisterRMRequest，具体代码如下所示：

```
    public class RegRmProcessor implements RemotingProcessor {
        @Override
        public void process(ChannelHandlerContext ctx, RpcMessage rpcMessage)
throws Exception {
```

```
            onRegRmMessage(ctx, rpcMessage);
        }
        //①注册 RM 消息
        private void onRegRmMessage(ChannelHandlerContext ctx, RpcMessage
rpcMessage) {
            //②解析出消息对象 RegisterRMRequest
            RegisterRMRequest message = (RegisterRMRequest)
rpcMessage.getBody();
            String ipAndPort =
NetUtil.toStringAddress(ctx.channel().remoteAddress());
            boolean isSuccess = false;
            String errorInfo = StringUtils.EMPTY;
            try {
                if (null == checkAuthHandler ||
checkAuthHandler.regResourceManagerCheckAuth(message)) {
                    //③调用 ChannelManager 类的 registerRMChannel()方法注册 RM 通道
                    ChannelManager.registerRMChannel(message, ctx.channel());
                    Version.putChannelVersion(ctx.channel(),
message.getVersion());
                    isSuccess = true;
                    if (LOGGER.isDebugEnabled()) {
                        LOGGER.debug("checkAuth for
client:{},vgroup:{},applicationId:{} is OK", ipAndPort,
message.getTransactionServiceGroup(), message.getApplicationId());
                    }
                }
            } catch (Exception exx) {
                isSuccess = false;
                errorInfo = exx.getMessage();
                LOGGER.error("RM register fail, error message:{}", errorInfo);
            }
            RegisterRMResponse response = new RegisterRMResponse(isSuccess);
            if (StringUtils.isNotEmpty(errorInfo)) {
                response.setMsg(errorInfo);
            }
            //④异步地返回结果
            remotingServer.sendAsyncResponse(rpcMessage, ctx.channel(), response);
            if (LOGGER.isInfoEnabled()) {
                LOGGER.info("RM register success,message:{},channel:{},client
version:{}", message, ctx.channel(),
                    message.getVersion());
            }
        }
        ...
    }
```

（3）在 XA 模式中，Seata 用 ConnectionProxyXA 类代理数据库连接，具体代码如下所示：

```
public class ResourceManagerXA extends
AbstractDataSourceCacheResourceManager {
    //①在 XA 模式中提交分支事务
    @Override
  public BranchStatus branchCommit(BranchType branchType, String xid, long
branchId, String resourceId,String applicationData)
  throws TransactionException   {
      return finishBranch(true, branchType, xid, branchId, resourceId,
applicationData);
  }
  //②在 XA 模式中回滚分支事务
  @Override
  public BranchStatus branchRollback(BranchType branchType, String xid, long
branchId, String resourceId,
                              String applicationData) throws
TransactionException {
      return finishBranch(false, branchType, xid, branchId, resourceId,
applicationData);
  }
  //③关闭分支事务（提交或者回滚）
   private BranchStatus finishBranch(boolean committed, BranchType branchType,
String xid, long branchId, String resourceId,
                              String applicationData) throws
TransactionException {
        XAXid xaBranchXid = XAXidBuilder.build(xid, branchId);
        Resource resource = dataSourceCache.get(resourceId);
        if (resource instanceof AbstractDataSourceProxyXA) {
            //④获取代理数据库连接对象 ConnectionProxyXA
            try (ConnectionProxyXA connectionProxyXA =
((AbstractDataSourceProxyXA)resource).getConnectionForXAFinish(xaBranchXid)) {
                if (committed) {
                    //⑤用 ConnectionProxyXA 类的 xaCommit()方法来提交分支事务
                    connectionProxyXA.xaCommit(xid, branchId, applicationData);
                    LO BranchStatus.PhaseTwo_Rollbacked GGER.info(xaBranchXid
+ " was committed.");
                    //⑥返回分支事务的结果 BranchStatus.PhaseTwo_Committed
                    return BranchStatus.PhaseTwo_Committed;
                } else {
                    //⑦用类 ConnectionProxyXA 的 xaRollback()方法来回滚分支事务
                    connectionProxyXA.xaRollback(xid, branchId, applicationData);
                    LOGGER.info(xaBranchXid + " was rolled back.");
                    //⑧返回分支事务的结果 BranchStatus.PhaseTwo_Rollbacked
                    return BranchStatus.PhaseTwo_Rollbacked;
```

```
                    }
                } catch (XAException | SQLException sqle) {
                    if (sqle instanceof XAException) {
                        if (((XAException)sqle).errorCode == XAException.XAER_NOTA) {
                            if (committed) {
                                return BranchStatus.PhaseTwo_Committed;
                            } else {
                                return BranchStatus.PhaseTwo_Rollbacked;
                            }
                        }
                    }
                    if (committed) {
                        LOGGER.info(xaBranchXid + " commit failed since " +
sqle.getMessage(), sqle);
                        return BranchStatus.PhaseTwo_CommitFailed_Retryable;
                    } else {
                        LOGGER.info(xaBranchXid + " rollback failed since " +
sqle.getMessage(), sqle);
                        return BranchStatus.PhaseTwo_RollbackFailed_Retryable;
                    }
                }
            } else {
                LOGGER.error("Unknown Resource for XA resource " + resourceId +
" " + resource);
                if (committed) {
                    //⑨返回分支事务的结果 PhaseTwo_CommitFailed_Unretryable
                    return BranchStatus.PhaseTwo_CommitFailed_Unretryable;
                } else {
                    //⑩返回分支事务的结果 PhaseTwo_RollbackFailed_Unretryable
                    return BranchStatus.PhaseTwo_RollbackFailed_Unretryable;
                }
            }
        }
    }
```

7.8.2 “用 XACore 类处理 XA 模式的事务请求”的原理

在 XA 模式中，Seata 用 XACore 类继承抽象类 AbstractCore 来处理 XA 模式的事务请求，具体代码如下所示：

```
public class XACore extends AbstractCore {

    public XACore(RemotingServer remotingServer) {
        super(remotingServer);
    }
```

```
        @Override
        public BranchType getHandleBranchType() {
            return BranchType.XA;
        }
        @Override
    public void branchReport(BranchType branchType, String xid, long branchId,
BranchStatus status,String applicationData) throws TransactionException {
        //调用父类 AbstractCore 的 branchReport()方法去报告分支事务的状态
            super.branchReport(branchType, xid, branchId, status,
applicationData);
            if (BranchStatus.PhaseOne_Failed == status) {
            }
        }
    }
```

从XACore类的实现中可以看出，XA模式中的事务处理全部交给了抽象模板类AbstractCore，具体代码如下所示：

```
    public abstract class AbstractCore implements Core {
        //①注册分支事务
        @Override
        public Long branchRegister(BranchType branchType, String resourceId,
String clientId, String xid,String applicationData, String lockKeys) throws
TransactionException {
            GlobalSession globalSession = assertGlobalSessionNotNull(xid, false);
            return SessionHolder.lockAndExecute(globalSession, () -> {
                globalSessionStatusCheck(globalSession);
                        globalSession.addSessionLifecycleListener
(SessionHolder.getRootSessionManager());
                BranchSession branchSession =
SessionHelper.newBranchByGlobal(globalSession, branchType, resourceId,
                    applicationData, lockKeys, clientId);
                branchSessionLock(globalSession, branchSession);
                try {
                    globalSession.addBranch(branchSession);
                } catch (RuntimeException ex) {
                    branchSessionUnlock(branchSession);
                    throw new BranchTransactionException(
                        FailedToAddBranch,String.format("Failed to
                        store branch xid = %s branchId = %s",
                      globalSession.getXid(),branchSession.getBranchId()), ex);
                }
                if (LOGGER.isInfoEnabled()) {
                    LOGGER.info("Register branch successfully, xid = {}, branchId
= {}, resourceId = {} ,lockKeys = {}",
```

```
                    globalSession.getXid(), branchSession.getBranchId(),
resourceId, lockKeys);
            }
            return branchSession.getBranchId();
        });
    }
    //②分析并报告分支事务的状态
        @Override
        public void branchReport(BranchType branchType, String xid, long branchId,
BranchStatus status,
                              String applicationData) throws
TransactionException {
            GlobalSession globalSession = assertGlobalSessionNotNull(xid, true);
            BranchSession branchSession = globalSession.getBranch(branchId);
            if (branchSession == null) {
                throw new BranchTransactionException(BranchTransactionNotExist,
                    String.format("Could not found branch session xid = %s
branchId = %s", xid, branchId));
            }
            globalSession.addSessionLifecycleListener
(SessionHolder.getRootSessionManager());
            globalSession.changeBranchStatus(branchSession, status);

            if (LOGGER.isInfoEnabled()) {
                LOGGER.info("Report branch status successfully, xid = {}, branchId
= {}", globalSession.getXid(),branchSession.getBranchId());
            }
        }
        //③提交分支事务
        @Override
        public BranchStatus branchCommit(GlobalSession globalSession,
BranchSession branchSession) throws TransactionException {
            try {
                BranchCommitRequest request = new BranchCommitRequest();
                request.setXid(branchSession.getXid());
                request.setBranchId(branchSession.getBranchId());
                request.setResourceId(branchSession.getResourceId());

request.setApplicationData(branchSession.getApplicationData());
                request.setBranchType(branchSession.getBranchType());
                return branchCommitSend(request, globalSession, branchSession);
            } catch (IOException | TimeoutException e) {
                throw new
BranchTransactionException(FailedToSendBranchCommitRequest,String.format("Se
```

```
nd branch commit failed, xid = %s branchId = %s", branchSession.
getXid(),branchSession.getBranchId()), e);
            }
        }
        ...
    }
```

7.8.3 【实例】搭建 Seata 的 XA 模式的客户端运行环境，并验证 XA 模式的分布式事务回滚的效果

本实例的源码在本书配套资源的“chapterseven/ spring-cloud-alibaba-access-seata-xa”目录下。

本实例用电商供应链的业务场景来验证 XA 模式，包括 3 个项目：订单服务、仓库服务及物流服务。

1. 初始化仓库服务

在电商供应链中，仓库服务主要用于控制发货。其初始化过程如下。

（1）用 Spring Cloud Alibaba 快速创建项目，并添加 Seata 相关的 POM 依赖，具体配置如下所示：

```
<!—省略部分 POM 依赖->
<dependency>
    <groupId>com.alibaba.cloud</groupId>
    <artifactId>spring-cloud-starter-alibaba-seata</artifactId>
    <exclusions>
        <exclusion>
            <groupId>io.seata</groupId>
            <artifactId>seata-spring-boot-starter</artifactId>
        </exclusion>
    </exclusions>
</dependency>
<dependency>
    <groupId>io.seata</groupId>
    <artifactId>seata-spring-boot-starter</artifactId>
    <exclusions>
        <exclusion>
            <groupId>org.springframework</groupId>
            <artifactId>spring-beans</artifactId>
        </exclusion>
    </exclusions>
```

```
</dependency>
<dependency>
    <groupId>com.alibaba</groupId>
    <artifactId>druid-spring-boot-starter</artifactId>
    <version>1.2.5</version>
    <exclusions>
        <exclusion>
            <groupId>org.springframework.boot</groupId>
            <artifactId>spring-boot-autoconfigure</artifactId>
        </exclusion>
    </exclusions>
</dependency>
<dependency>
    <groupId>org.mybatis.spring.boot</groupId>
    <artifactId>mybatis-spring-boot-starter</artifactId>
    <version>2.1.4</version>
    <exclusions>
        <exclusion>
            <groupId>org.springframework.boot</groupId>
            <artifactId>spring-boot-autoconfigure</artifactId>
        </exclusion>
    </exclusions>
</dependency>
```

（2）创建仓库表，DDL 语句如下所示：

```
create table seata_server.at_warehouse
(
    id bigint not null,
    warehouse_name varchar(50) not null,
    status int default 0 not null,
    order_id bigint default 0 not null,
    good_id bigint default 0 not null,
    is_deleted int default 0 not null,
    gmt_create datetime null,
    gmt_modified datetime null,
    warehouse_id bigint null,
    num int default 0 not null,
    constraint at_warehouse_id_uindex
    unique (id)
);
alter table seata_server.at_warehouse
    add primary key (id);
```

（3）模拟仓库发货的过程，具体代码如下所示：

```
@DubboService(version = "1.0.0",group = "SEATA_GROUP")
public class XaWarehouseServiceImpl implements XaWarehouseService {
    @Resource
    private WarehouseManager warehouseManager;
    @Override
public WarehouseServiceDTO notifyWarehouse(WarehouseServiceBo
warehouseServiceBo) {
    //通知仓库去发货
        boolean result = warehouseManager.notify(warehouseServiceBo);
        WarehouseServiceDTO warehouseServiceDTO = new WarehouseServiceDTO();
        if (result) {
            warehouseServiceDTO.setStatus(200);
        } else {
            warehouseServiceDTO.setStatus(500);
        }
        return warehouseServiceDTO;
    }
}
```

2. 初始化物流服务

在电商供应链中，物流服务主要用于对接快递平台。其初始化过程如下。

（1）用 Spring Cloud Alibaba 快速创建项目，并添加 Seata 相关的 Jar 包依赖。

（2）创建物流表，DDL 语句如下所示：

```
create table seata_server.at_logistics
(
    id bigint not null,
    logistics_id bigint not null,
    logistics_name varchar(50) null,
    status int default 0 not null,
    order_id bigint not null,
    is_deleted int default 0 not null,
    gmt_create timestamp null,
    gmt_modified timestamp null,
    constraint at_logistics_id_uindex
    unique (id)
);

alter table seata_server.at_logistics
    add primary key (id);
```

（3）模拟物流信息，具体代码如下所示：

```
@DubboService(version = "1.0.0",group = "SEATA_GROUP")
public class XaLogisticsServiceImpl implements XaLogisticsService {
    @Resource
    private LogisticsManager logisticsManager;
    @Override
public LogisticsServiceDTO notifyLogistics(LogisticsServiceBo
logisticsServiceBo) {
    //通知物流服务更新物流信息
        Boolean result =
logisticsManager.notifyLogistics(logisticsServiceBo);
        LogisticsServiceDTO logisticsServiceDTO = new LogisticsServiceDTO();
        if (result) {
            logisticsServiceDTO.setStatus(200);
        } else {
            logisticsServiceDTO.setStatus(500);
        }
        return logisticsServiceDTO;
    }
}
```

3. 初始化订单服务

订单服务（xa-order-service）主要用于创建订单和修改订单状态。其初始化过程如下。

（1）用 Spring Cloud Alibaba 快速创建项目，并添加 Seata 相关的 Jar 包依赖。

（2）创建订单表，DDL 语句如下所示：

```
create table seata_server.at_order
(
    id bigint not null
        primary key,
    order_name varchar(50) null,
    order_id bigint null,
    order_amount bigint null,
    is_deleted int not null,
    gmt_create timestamp null,
    gmt_modified timestamp null,
    good_id bigint null,
    user_id bigint not null,
     order_status int default 0 not null
);
```

（3）模拟供应链的订单交易的业务场景，具体代码如下所示：

```
@Service
public class OrderManager {
    class Result{
        private Long id;
        private boolean result;
        public Long getId() {
            return id;
        }
        public void setId(Long id) {
            this.id = id;
        }
        public boolean isResult() {
            return result;
        }
        public void setResult(boolean result) {
            this.result = result;
        }
    }
    @Resource
    private OrderMapper orderMapper;
    @DubboReference(version = "1.0.0",group = "SEATA_GROUP")
    private XaWarehouseService xaWarehouseService;
    @DubboReference(version = "1.0.0",group = "SEATA_GROUP")
private XaLogisticsService xaLogisticsService;
//①创建测试订单
    public Result createOrder(Long userId, BigDecimal amount, Long goodId) {
        OrderEntity orderEntity = new OrderEntity();
        orderEntity.setOrderId(RandomUtils.nextLong());
        orderEntity.setId(RandomUtils.nextLong());
        orderEntity.setOrderAmount(amount);
        orderEntity.setOrderName("测试订单" + RandomUtils.nextInt());
        orderEntity.setGoodId(goodId);
        orderEntity.setUserId(userId);
        orderEntity.setIsDeleted(0);
        orderEntity.setGmtCreate(new Date());
        orderEntity.setGmtModified(new Date());
        Long result = orderMapper.createNewOrder(orderEntity);
        Result createResult = new Result();
        createResult.setResult(true ? result > 0 : false);
        createResult.setId(orderEntity.getId());
        return createResult;
}
//②更新订单的状态
    public boolean updataOrderStatus(Long id,Integer orderStatus) {
        OrderEntity orderEntity = new OrderEntity();
```

```
            orderEntity.setId(id);
            orderEntity.setOrderStatus(orderStatus);
            Integer result = orderMapper.updataOrderStatus(orderEntity);
            return true ? result > 0 : false;
    }
    //③开启全局事务，事务名称是“pay-order”，回滚的异常类型是 Exception
        @GlobalTransactional(name = "pay-order", rollbackFor = Exception.class)
        public boolean placeOrder(Long userId, Long goodId, Long accountId,
Integer num) {
            Result orderResult = createOrder(userId, new BigDecimal(100),
goodId);
            WarehouseServiceBo warehouseServiceBo = new WarehouseServiceBo();
            warehouseServiceBo.setOrderId(orderResult.getId());
            warehouseServiceBo.setGoodId(goodId);
            warehouseServiceBo.setNum(num);
            WarehouseServiceDTO warehouseServiceDTO =
xaWarehouseService.notifyWarehouse(warehouseServiceBo);
            LogisticsServiceBo logisticsServiceBo = new LogisticsServiceBo();
            logisticsServiceBo.setOrderId(orderResult.getId());
            LogisticsServiceDTO logisticsServiceDTO =
xaLogisticsService.notifyLogistics(logisticsServiceBo);
            if (orderResult.isResult() && warehouseServiceDTO.getStatus() == 200
&& logisticsServiceDTO.getStatus() == 200) {
                boolean updataOrderStatusResult =
updataOrderStatus(orderResult.getId(), 1);
                if (updataOrderStatusResult) {
                    return true;
                }
                return false;
            }
            return true;
        }
    }
```

4. 启动服务

启动仓库服务、物流服务、订单服务及 Seata Server。在服务启动完成后，可以发现这 4 个服务已经被注册到 Nacos 注册中心中了，如图 7-28 所示。

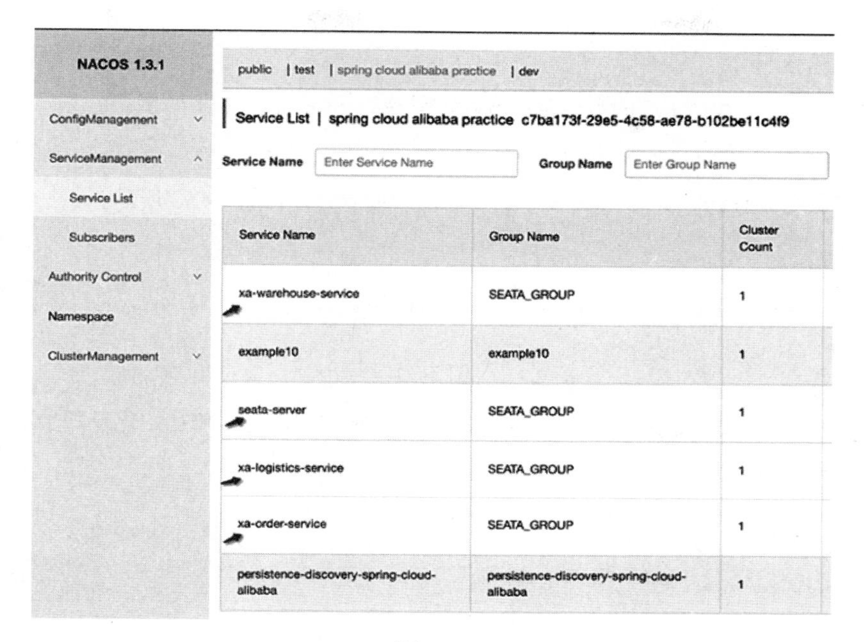

图 7-28

5. 用故障模拟分布式事务回滚的场景

如图 7-29 所示，订单服务调用仓库服务和物流服务，完成供应链的核心业务订单同步。

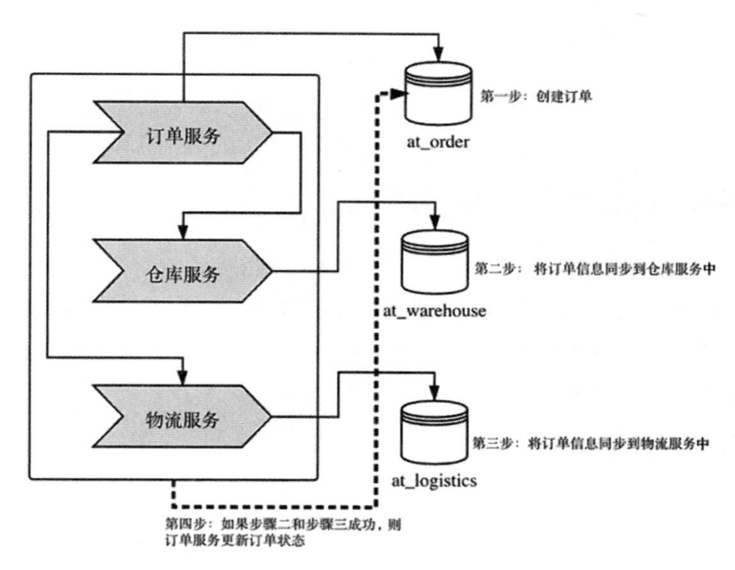

图 7-29

故障模拟过程如下：

（1）将订单表 at_order 中的数据全部删除，具体的 SQL 语句如下所示：

```
delete from at_order where 1=1;
```

执行完成后可以看到，at_order 表中的数据已经被清空，如图 7-30 所示。

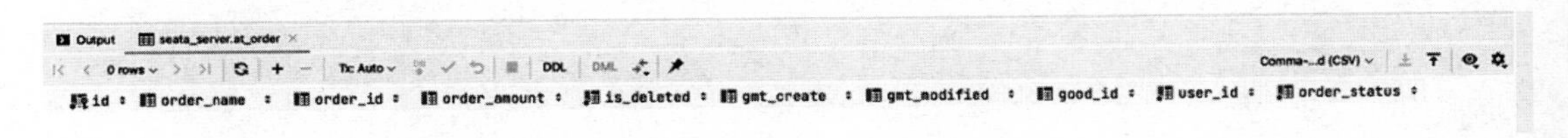

图 7-30

（2）本程序采用 Swagger 来产生请求的流量，并模拟供应链订单初始化及同步到仓库服务和物流服务的请求，如图 7-31 所示。

trade-controller Trade Controller
GET /order/placeOrder buy
Parameters
Cancel
Name
Description
accountId
integer($int64)
(query)
accountId
1234
goodId
integer($int64)
(query)
goodId
34556656
num
integer($int32)
(query)
num
1
userId
integer($int64)
(query)
userId
2323323223
Execute
Clear

图 7-31

（3）通过代码调试的断点功能和数据库连接的超时机制，制造出一个订单服务更新订单状态的技术故障。如果 XA 模式的分布式事务生效，则订单表（at_order）、物流表（at_logistics）和仓库表（at_warehouse）不会生成订单数据。

断点调试如图 7-32 所示。

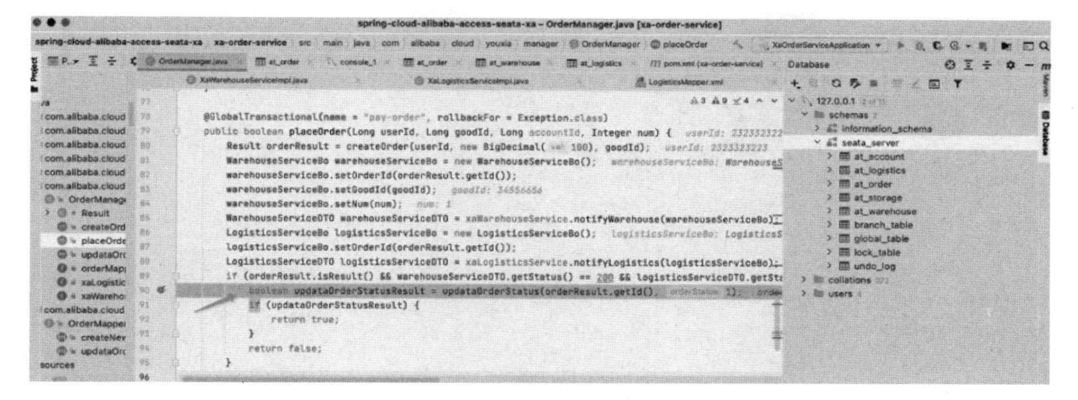

图 7-32

在订单服务中，如果出现数据库连接超时故障，则故障日志如图 7-33 所示。

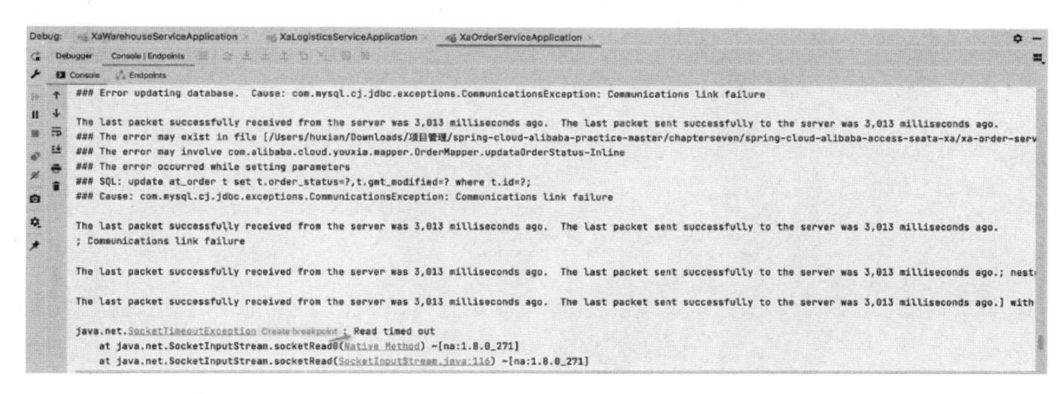

图 7-33

在订单服务中出现回滚日志，如图 7-34 所示。

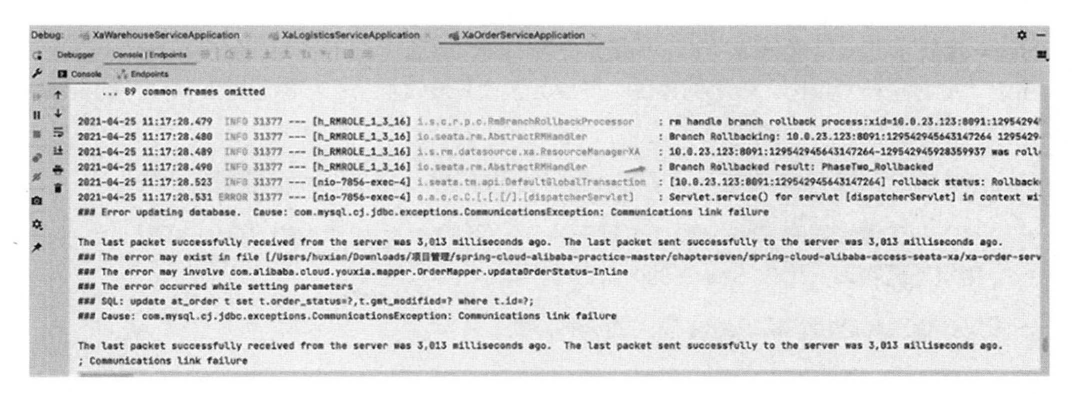

图 7-34

在仓库服务中出现回滚日志，如图 7-35 所示。

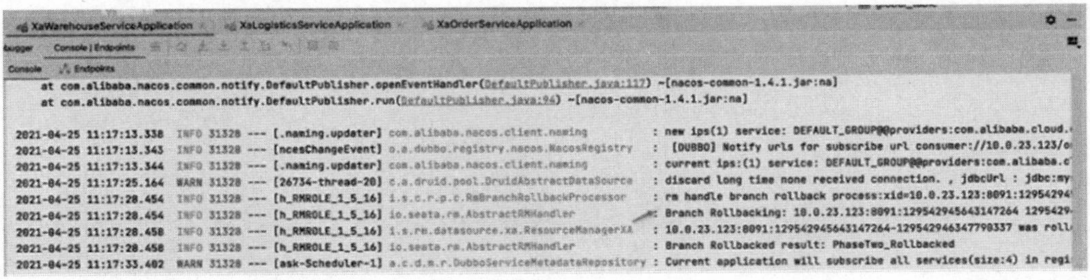

```
XaWarehouseServiceApplication    XaLogisticsServiceApplication    XaOrderServiceApplication
Debugger    Console | Endpoints
Console    Endpoints
    at com.alibaba.nacos.common.notify.DefaultPublisher.openEventHandler(DefaultPublisher.java:117) ~[nacos-common-1.4.1.jar:na]
    at com.alibaba.nacos.common.notify.DefaultPublisher.run(DefaultPublisher.java:94) ~[nacos-common-1.4.1.jar:na]

2021-04-25 11:17:13.338  INFO 31328 --- [.naming.updater] com.alibaba.nacos.client.naming          : new ips(1) service: DEFAULT_GROUP@@providers:com.alibaba.cloud.
2021-04-25 11:17:13.343  INFO 31328 --- [ncesChangeEvent] o.a.dubbo.registry.nacos.NacosRegistry   :  [DUBBO] Notify urls for subscribe url consumer://10.0.23.123/o
2021-04-25 11:17:13.344  INFO 31328 --- [.naming.updater] com.alibaba.nacos.client.naming          : current ips:(1) service: DEFAULT_GROUP@@providers:com.alibaba.c
2021-04-25 11:17:25.164  WARN 31328 --- [26734-thread-20] c.a.druid.pool.DruidAbstractDataSource   : discard long time none received connection. , jdbcUrl : jdbc:my
2021-04-25 11:17:28.454  INFO 31328 --- [h_RMROLE_1_5_16] i.s.c.r.p.c.RmBranchRollbackProcessor    : rm handle branch rollback process:xid=10.0.23.123:8091:12954294
2021-04-25 11:17:28.454  INFO 31328 --- [h_RMROLE_1_5_16] io.seata.rm.AbstractRMHandler            : Branch Rollbacking: 10.0.23.123:8091:129542945643147264 1295429
2021-04-25 11:17:28.458  INFO 31328 --- [h_RMROLE_1_5_16] i.s.rm.datasource.xa.ResourceManagerXA   : 10.0.23.123:8091:129542945643147264-129542946347798337 was roll
2021-04-25 11:17:28.458  INFO 31328 --- [h_RMROLE_1_5_16] io.seata.rm.AbstractRMHandler            : Branch Rollbacked result: PhaseTwo_Rollbacked
2021-04-25 11:17:33.402  WARN 31328 --- [ask-Scheduler-1] a.c.d.m.r.DubboServiceMetadataRepository : Current application will subscribe all services(size:4) in regi
```

图 7-35

在物流服务中出现回滚日志，如图 7-36 所示。

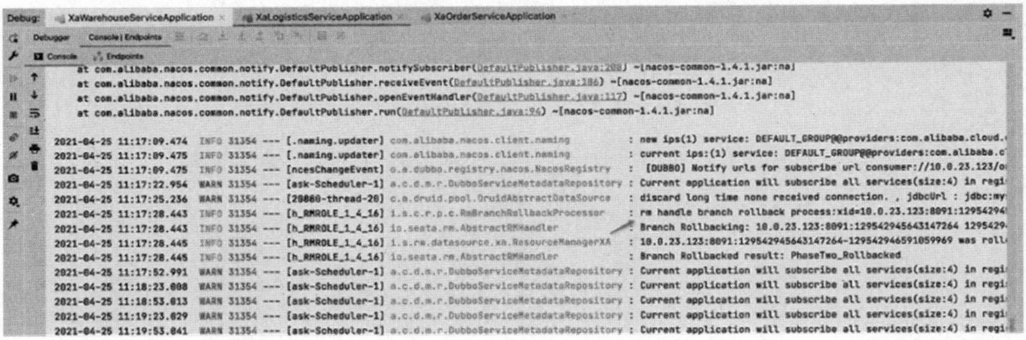

```
Debug:  XaWarehouseServiceApplication    XaLogisticsServiceApplication    XaOrderServiceApplication
Debugger    Console | Endpoints
Console    Endpoints
    at com.alibaba.nacos.common.notify.DefaultPublisher.notifySubscriber(DefaultPublisher.java:208) ~[nacos-common-1.4.1.jar:na]
    at com.alibaba.nacos.common.notify.DefaultPublisher.receiveEvent(DefaultPublisher.java:186) ~[nacos-common-1.4.1.jar:na]
    at com.alibaba.nacos.common.notify.DefaultPublisher.openEventHandler(DefaultPublisher.java:117) ~[nacos-common-1.4.1.jar:na]
    at com.alibaba.nacos.common.notify.DefaultPublisher.run(DefaultPublisher.java:94) ~[nacos-common-1.4.1.jar:na]

2021-04-25 11:17:09.474  INFO 31354 --- [.naming.updater] com.alibaba.nacos.client.naming          : new ips(1) service: DEFAULT_GROUP@@providers:com.alibaba.cloud.
2021-04-25 11:17:09.475  INFO 31354 --- [.naming.updater] com.alibaba.nacos.client.naming          : current ips:(1) service: DEFAULT_GROUP@@providers:com.alibaba.c
2021-04-25 11:17:09.475  INFO 31354 --- [ncesChangeEvent] o.a.dubbo.registry.nacos.NacosRegistry   :  [DUBBO] Notify urls for subscribe url consumer://10.0.23.123/o
2021-04-25 11:17:22.954  WARN 31354 --- [ask-Scheduler-1] a.c.d.m.r.DubboServiceMetadataRepository : Current application will subscribe all services(size:4) in regi
2021-04-25 11:17:25.236  WARN 31354 --- [20880-thread-20] c.a.druid.pool.DruidAbstractDataSource   : discard long time none received connection. , jdbcUrl : jdbc:my
2021-04-25 11:17:28.443  INFO 31354 --- [h_RMROLE_1_4_16] i.s.c.r.p.c.RmBranchRollbackProcessor    : rm handle branch rollback process:xid=10.0.23.123:8091:12954294
2021-04-25 11:17:28.443  INFO 31354 --- [h_RMROLE_1_4_16] io.seata.rm.AbstractRMHandler            : Branch Rollbacking: 10.0.23.123:8091:129542945643147264 1295429
2021-04-25 11:17:28.445  INFO 31354 --- [h_RMROLE_1_4_16] i.s.rm.datasource.xa.ResourceManagerXA   : 10.0.23.123:8091:129542945643147264-129542946591059969 was roll
2021-04-25 11:17:28.445  INFO 31354 --- [h_RMROLE_1_4_16] io.seata.rm.AbstractRMHandler            : Branch Rollbacked result: PhaseTwo_Rollbacked
2021-04-25 11:17:52.991  WARN 31354 --- [ask-Scheduler-1] a.c.d.m.r.DubboServiceMetadataRepository : Current application will subscribe all services(size:4) in regi
2021-04-25 11:18:23.008  WARN 31354 --- [ask-Scheduler-1] a.c.d.m.r.DubboServiceMetadataRepository : Current application will subscribe all services(size:4) in regi
2021-04-25 11:18:53.013  WARN 31354 --- [ask-Scheduler-1] a.c.d.m.r.DubboServiceMetadataRepository : Current application will subscribe all services(size:4) in regi
2021-04-25 11:19:23.029  WARN 31354 --- [ask-Scheduler-1] a.c.d.m.r.DubboServiceMetadataRepository : Current application will subscribe all services(size:4) in regi
2021-04-25 11:19:53.041  WARN 31354 --- [ask-Scheduler-1] a.c.d.m.r.DubboServiceMetadataRepository : Current application will subscribe all services(size:4) in regi
```

图 7-36

7.9 用 Saga 模式实现分布式事务

Saga 模式是 Seata 的长事务解决方案，使用场景：①业务流程长和业务流程复杂；②公司的遗留系统（或者是技术架构比较老的系统）在业务迭代开发过程中，不能通过业务重构的手段提供 TCC 模式要求的 3 个接口。

Saga 模式和 TCC 模式都是补偿事务。目前业界提供两种实现 Saga 模式的方式：

- 基于业务逻辑层的 Proxy 设计（基于 AOP 实现），比如华为的 ServiceComb。
- 用状态机实现，比如 Seata 的 Saga 模式。

下面只介绍用状态机实现 Saga 模式的这种方式。

7.9.1　“用状态机实现 Saga 模式”的原理

如果开发人员在指定的文件中定义了状态机，则 Seata 会去指定的文件目录进行解析，按照状态机规则驱动 Saga 分布式事务的执行。

（1）在项目中，开发人员可以用注解@Bean 初始化一个状态机对象 StateMachineEngine，具体代码如下：

```
    @Bean
    public StateMachineEngine process(DbStateMachineConfig
dbStateMachineConfig) {
    //①在当前 JVM 中初始化一个状态机
    ProcessCtrlStateMachineEngine processCtrlStateMachineEngine = new
ProcessCtrlStateMachineEngine();
        //②设置状态机的配置文件
        processCtrlStateMachineEngine.setStateMachineConfig
    (dbStateMachineConfig);
        return processCtrlStateMachineEngine;
    }
```

（2）将状态机和状态机引擎容器绑定，具体代码如下所示：

```
    @Bean
    public StateMachineEngineHolder
stateMachineEngineHolder(StateMachineEngine stateMachineEngine){
        StateMachineEngineHolder stateMachineEngineHolder=new
StateMachineEngineHolder();
    stateMachineEngineHolder.setStateMachineEngine(stateMachineEngine);
    //将绑定了状态机的状态机引擎设置到 IOC 容器中
        return stateMachineEngineHolder;
    }
```

在分支事务提交和回滚的过程中，用 RM 的实现类 SagaResourceManager 操作状态机，并控制业务接口的调用流程，从而保证分布式事务的一致性。

（3）在 Seata 中，用 StateMachineRepositoryImpl 类的 registryByResources()方法解析状态机规格文件，具体代码如下所示：

```
    public class StateMachineRepositoryImpl implements StateMachineRepository {
        ...
        Override
        public void registryByResources(Resource[] resources, String tenantId)
throws IOException {
            if (resources != null) {
                for (Resource resource : resources) {
                    String json = IOUtils.toString(resource.getInputStream(),
charset);
```

```
                StateMachine stateMachine =
StateMachineParserFactory.getStateMachineParser(jsonParserName).parse(json);
                if (stateMachine != null) {
                    stateMachine.setContent(json);
                    if (StringUtils.isBlank(stateMachine.getTenantId())) {
                        stateMachine.setTenantId(tenantId);
                    }
                    registryStateMachine(stateMachine);
                    if (LOGGER.isDebugEnabled()) {
                        LOGGER.debug("===== StateMachine Loaded: \n{}", json);
                    }
                }
            }
        }
    }
}
```

（4）在解析完状态机的规则后，将状态机的数据存储到数据库中，然后通过SpringBeanServiceInvoker类的invoke()方法处理业务接口的请求，具体代码如下所示：

```
public class SpringBeanServiceInvoker implements ServiceInvoker,
ApplicationContextAware {
    ...
    @Override
    public Object invoke(ServiceTaskState serviceTaskState, Object... input)
throws Throwable {
        ServiceTaskStateImpl state = (ServiceTaskStateImpl)
serviceTaskState;
        //①如果在状态机规格文件中配置了当前是异步请求，则用线程池异步执行
        if (state.isAsync()) {
            if (threadPoolExecutor == null) {
                if (LOGGER.isWarnEnabled()) {
                    LOGGER.warn(
                            "threadPoolExecutor is null, Service[{}.{}] cannot
execute asynchronously, executing "
                                    + "synchronously now. stateName: {}",
                            state.getServiceName(), state.getServiceMethod(),
state.getName());
                }
                return doInvoke(state, input);
            }

            if (LOGGER.isInfoEnabled()) {
                LOGGER.info("Submit Service[{}.{}] to asynchronously
executing. stateName: {}", state.getServiceName(),
                        state.getServiceMethod(), state.getName());
```

```
            }
            threadPoolExecutor.execute(new Runnable() {
                @Override
                public void run() {
                    try {
                        doInvoke(state, input);
                    } catch (Throwable e) {
                        LOGGER.error("Invoke Service[" + state.getServiceName()
+ "." + state.getServiceMethod() + "] failed.", e);
                    }
                }
            });
            return null;
        } else {
            //②同步执行请求
            return doInvoke(state, input);
        }
    }
    protected Object doInvoke(ServiceTaskStateImpl state, Object[] input)
throws Throwable {
        //③从当前的 IOC 容器中，获取状态机规格文件中配置的服务名称所对应的 Bean 的实现类
        Object bean = applicationContext.getBean(state.getServiceName());
        //④从状态机规格文件中，解析出执行的方法名称
        Method method = state.getMethod();
        if (method == null) {
            synchronized (state) {
                method = state.getMethod();
                if (method == null) {
                    method = findMethod(bean.getClass(),
state.getServiceMethod(), state.getParameterTypes());
                    if (method != null) {
                        state.setMethod(method);
                    }
                }
            }
        }
        if (method == null) {
            throw new EngineExecutionException(
                    "No such method[" + state.getServiceMethod() + "] on
BeanClass[" + bean.getClass() + "]",
                    FrameworkErrorCode.NoSuchMethod);

        }
        //⑤从状态机规格文件中，解析出方法的入参及参数类型
        Object[] args = new Object[method.getParameterCount()];
```

```
        try {
            Class[] paramTypes = method.getParameterTypes();
            if (input != null && input.length > 0) {
                int len = input.length < paramTypes.length ? input.length :
paramTypes.length;
                for (int i = 0; i < len; i++) {
                    args[i] = toJavaObject(input[i], paramTypes[i]);
                }
            }
        } catch (Exception e) {
            throw new EngineExecutionException(e,
                    "Input to java object error, Method[" +
state.getServiceMethod() + "] on BeanClass[" + bean.getClass()
                        + "]", FrameworkErrorCode.InvalidParameter);
        }
        if (!Modifier.isPublic(method.getModifiers())) {
            throw new EngineExecutionException("Method[" + method.getName()
+ "] must be public",
                    FrameworkErrorCode.MethodNotPublic);
        }
        Map<Retry, AtomicInteger> retryCountMap = new HashMap<>();
        //⑥用 retryCountMap 存储重试的次数总和。当出现异常时，如果重试的次数总和大于
或者等于“配置文件中配置的重试次数的阈值”，则直接抛出异常，否则继续重试
        while (true) {
            try {
                //⑦代理执行当前请求
                return invokeMethod(bean, method, args);
            } catch (Throwable e) {
                Retry matchedRetryConfig = matchRetryConfig(state.getRetry(), e);
                if (matchedRetryConfig == null) {
                    throw e;
                }
                AtomicInteger retryCount =
CollectionUtils.computeIfAbsent(retryCountMap, matchedRetryConfig,
                    key -> new AtomicInteger(0));
                if (retryCount.intValue() >=
matchedRetryConfig.getMaxAttempts()) {
                    throw e;
                }
                double intervalSeconds =
matchedRetryConfig.getIntervalSeconds();
                double backoffRate = matchedRetryConfig.getBackoffRate();
                //⑧计算当前请求的休眠时间
                long currentInterval = (long) (retryCount.intValue() > 0 ?
                        (intervalSeconds * backoffRate * retryCount.intValue()
```

```
* 1000) : (intervalSeconds * 1000));
                if (LOGGER.isWarnEnabled()) {
                    LOGGER.warn("Invoke Service[" + state.getServiceName() +
"." + state.getServiceMethod() + "] failed, will retry after "
                            + currentInterval + " millis, current retry count:
" + retryCount.intValue(), e);
                }
                //⑨用线程休眠机制让当前请求休眠
                try {
                    Thread.sleep(currentInterval);
                } catch (InterruptedException e1) {
                    LOGGER.warn("Retry interval sleep error", e1);
                }
                //⑩自增重试次数
                retryCount.incrementAndGet();
            }
        }
    }
}
```

7.9.2 【实例】搭建 Seata 的 Saga 模式的客户端运行环境，并验证 Saga 模式的分布式事务场景

本实例的源码在本书配套资源的“chapterseven/ spring-cloud-alibaba-access-seata-saga”目录下。

本实例用 Spring Cloud Alibaba 将应用接入 Saga。具体过程如下。

1. 搭建基础项目

在项目中添加 POM 文件的依赖，具体如下所示。

```
<!--省略项目中的部分 POM 依赖关系-->
<dependency>
    <groupId>com.alibaba.cloud</groupId>
    <artifactId>spring-cloud-starter-alibaba-seata</artifactId>
    <exclusions>
        <exclusion>
            <groupId>io.seata</groupId>
            <artifactId>seata-spring-boot-starter</artifactId>
        </exclusion>
    </exclusions>
</dependency>
```

2. 添加 Saga 模式的状态机

在 Saga 模式中，Seata 通过解析 JSON 文件中的状态机规格完成分布式事务的初始化。在项目的文件 saga-spring-cloud-alibaba.json 中，关于状态机规格定义的具体配置信息如下：

```
{
    //①状态机的名称
    "Name": "reduceInventoryAndBalance",
    "Comment": "reduce inventory then reduce balance in a transaction",
    //②状态机的初始状态
"StartState": "ReduceInventory",
    //③状态机的版本号
    "Version": "0.0.1",
"States": {
    //④配置状态机中扣减库存的正向服务接口 ReduceInventory
        "ReduceInventory": {
            "Type": "ServiceTask",
            "ServiceName": "inventoryAction",
            "ServiceMethod": "reduce",
            "CompensateState": "CompensateReduceInventory",
            "Next": "ChoiceState",
            "Input": [
                "$.[businessKey]",
                "$.[count]"
            ],
            "Output": {
                "reduceInventoryResult": "$.#root"
            },
            "Status": {
                "#root == true": "SU",
                "#root == false": "FA",
                "$Exception{java.lang.Throwable}": "UN"
            }
        },
        "ChoiceState":{
            "Type": "Choice",
            "Choices":[
                {
                    "Expression":"[reduceInventoryResult] == true",
                    "Next":"ReduceBalance"
                }
            ],
            "Default":"Fail"
        },
        //⑤配置状态机中扣减账户金额的正向服务接口 ReduceBalance
```

```
    "ReduceBalance": {
        "Type": "ServiceTask",
        "ServiceName": "balanceAction",
        "ServiceMethod": "reduce",
        "CompensateState": "CompensateReduceBalance",
        "Input": [
            "$.[businessKey]",
            "$.[amount]",
            {
                "throwException" : "$.[mockReduceBalanceFail]"
            }
        ],
        "Output": {
            "compensateReduceBalanceResult": "$.#root"
        },
        "Status": {
            "#root == true": "SU",
            "#root == false": "FA",
            "$Exception{java.lang.Throwable}": "UN"
        },
        "Catch": [
            {
                "Exceptions": [
                    "java.lang.Throwable"
                ],
                "Next": "CompensationTrigger"
            }
        ],
        "Next": "Succeed"
    },
    //⑥配置状态机中扣减库存的补偿服务接口
    "CompensateReduceInventory": {
        "Type": "ServiceTask",
        "ServiceName": "inventoryAction",
        "ServiceMethod": "compensateReduce",
        "Input": [
            "$.[businessKey]"
        ]
    },
    //⑦配置状态机中扣减账户余额的补偿服务接口
    "CompensateReduceBalance": {
        "Type": "ServiceTask",
        "ServiceName": "balanceAction",
        "ServiceMethod": "compensateReduce",
        "Input": [
```

```
                "$.[businessKey]"
            ]
        },
        //⑧配置状态机中执行事务补偿的触发条件为“Fail”
        "CompensationTrigger": {
            "Type": "CompensationTrigger",
            "Next": "Fail"
        },
        "Succeed": {
            "Type":"Succeed"
        },
        "Fail": {
            "Type":"Fail",
            "ErrorCode": "PURCHASE_FAILED",
            "Message": "purchase failed"
        }
    }
}
```

3. 模拟用户请求，调用状态机

在启动项目之后，模拟用户请求去调用状态机。模拟 Saga 分布式事务场景的代码如下所示：

```
@Configuration
@Order(Ordered.LOWEST_PRECEDENCE+1000000000)
public class SagaStarter {
    @Autowired
private ApplicationContext applicationContext;
    //①加载状态机对象 StateMachineEngine
    @Autowired
    private StateMachineEngine stateMachineEngine;
    @Resource
    private BalanceAction balanceAction;
    @Resource
    private InventoryAction inventoryAction;
    //②用注解@PostConstruct 开启 Saga 分布式事务
    @PostConstruct
    public void init(){
        run();
    }
public void run(){
        //③执行 Saga 事务的第一阶段
        transactionCommittedDemo(stateMachineEngine);
        //④执行 Saga 事务的第二阶段
        transactionCompensatedDemo(stateMachineEngine);
}
```

```
    //⑤在第一阶段提交本地事务
        private static void transactionCommittedDemo(StateMachineEngine
stateMachineEngine) {
            Map<String, Object> startParams = new HashMap<>(3);
            String businessKey = String.valueOf(System.currentTimeMillis());
            startParams.put("businessKey", businessKey);
            startParams.put("count", 10);
            startParams.put("amount", new BigDecimal("100"));
            StateMachineInstance inst =
stateMachineEngine.startWithBusinessKey("reduceInventoryAndBalance", null,
businessKey, startParams);
            Assert.isTrue(ExecutionStatus.SU.equals(inst.getStatus()), "saga
transaction execute failed. XID: " + inst.getId());
            System.out.println("saga transaction commit succeed. XID: " +
inst.getId());
            businessKey = String.valueOf(System.currentTimeMillis());
            inst =
stateMachineEngine.startWithBusinessKeyAsync("reduceInventoryAndBalance",
null, businessKey, startParams, CALL_BACK);
            waittingForFinish(inst);
            Assert.isTrue(ExecutionStatus.SU.equals(inst.getStatus()), "saga
transaction execute failed. XID: " + inst.getId());
            System.out.println("saga transaction commit succeed. XID: " +
inst.getId());
    }
    //⑥在第二阶段，如果出现异常，则执行事务的补偿
        private static void transactionCompensatedDemo(StateMachineEngine
stateMachineEngine) {
            Map<String, Object> startParams = new HashMap<>(4);
            String businessKey = String.valueOf(System.currentTimeMillis());
            startParams.put("businessKey", businessKey);
            startParams.put("count", 10);
            startParams.put("amount", new BigDecimal("100"));
            startParams.put("mockReduceBalanceFail", "true");
            StateMachineInstance inst =
stateMachineEngine.startWithBusinessKey("reduceInventoryAndBalance", null,
businessKey, startParams);
            businessKey = String.valueOf(System.currentTimeMillis());
            inst =
stateMachineEngine.startWithBusinessKeyAsync("reduceInventoryAndBalance",
null, businessKey, startParams, CALL_BACK);
            waittingForFinish(inst);
            Assert.isTrue(ExecutionStatus.SU.equals
(inst.getCompensationStatus()), "saga transaction compensate failed. XID: " +
inst.getId());
```

```
            System.out.println("saga transaction compensate succeed. XID: " +
inst.getId());
        }
        private static volatile Object lock = new Object();
        private static AsyncCallback CALL_BACK = new AsyncCallback() {
            @Override
            public void onFinished(ProcessContext context, StateMachineInstance
stateMachineInstance) {
                synchronized (lock){
                    lock.notifyAll();
                }
            }
            @Override
            public void onError(ProcessContext context, StateMachineInstance
stateMachineInstance, Exception exp) {
                synchronized (lock){
                    lock.notifyAll();
                }
            }
        };
        private static void waittingForFinish(StateMachineInstance inst){
            synchronized (lock){
                if(ExecutionStatus.RU.equals(inst.getStatus())){
                    try {
                        lock.wait();
                    } catch (InterruptedException e) {
                        e.printStackTrace();
                    }
                }
            }
        }
    }
```

4. 启动服务

启动服务 spring-cloud-alibaba-access-seata-saga 之后，在 Nacos 控制台中可以看到服务已经注册成功。

分析运行日志，可以看出 Saga 模式的分布式事务已经生效，如图 7-37 所示。

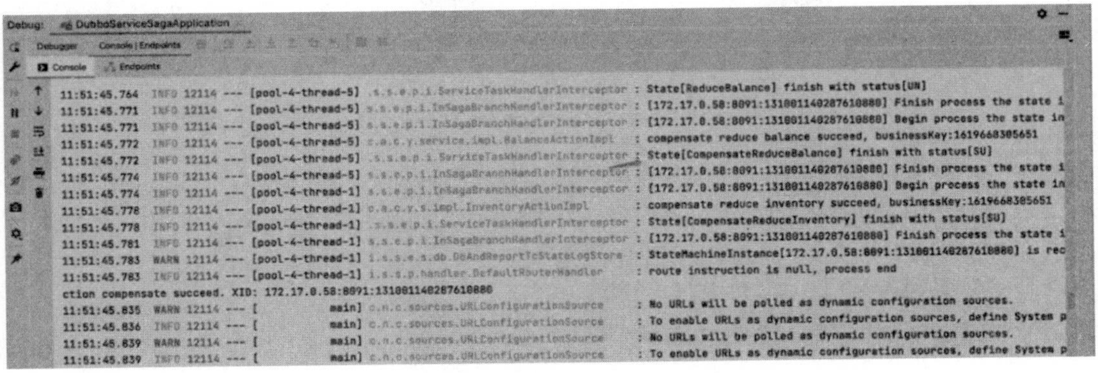

```
Debug:  DubboServiceSagaApplication
Debugger   Console | Endpoints
Console   Endpoints
11:51:45.764 INFO 12114 --- [pool-4-thread-5] .s.s.e.p.i.ServiceTaskHandlerInterceptor : State[ReduceBalance] finish with status[UN]
11:51:45.771 INFO 12114 --- [pool-4-thread-5] s.s.e.p.i.InSagaBranchHandlerInterceptor : [172.17.0.58:8091:131001140287610880] Finish process the state i
11:51:45.771 INFO 12114 --- [pool-4-thread-5] s.s.e.p.i.InSagaBranchHandlerInterceptor : [172.17.0.58:8091:131001140287610880] Begin process the state in
11:51:45.772 INFO 12114 --- [pool-4-thread-5] c.a.c.y.service.impl.BalanceActionImpl   : compensate reduce balance succeed, businessKey:1619668305651
11:51:45.772 INFO 12114 --- [pool-4-thread-5] .s.s.e.p.i.ServiceTaskHandlerInterceptor : State[CompensateReduceBalance] finish with status[SU]
11:51:45.774 INFO 12114 --- [pool-4-thread-5] s.s.e.p.i.InSagaBranchHandlerInterceptor : [172.17.0.58:8091:131001140287610880] Finish process the state i
11:51:45.774 INFO 12114 --- [pool-4-thread-1] s.s.e.p.i.InSagaBranchHandlerInterceptor : [172.17.0.58:8091:131001140287610880] Begin process the state in
11:51:45.778 INFO 12114 --- [pool-4-thread-1] c.a.c.y.s.impl.InventoryActionImpl       : compensate reduce inventory succeed, businessKey:1619668305651
11:51:45.778 INFO 12114 --- [pool-4-thread-1] .s.s.e.p.i.ServiceTaskHandlerInterceptor : State[CompensateReduceInventory] finish with status[SU]
11:51:45.781 INFO 12114 --- [pool-4-thread-1] s.s.e.p.i.InSagaBranchHandlerInterceptor : [172.17.0.58:8091:131001140287610880] Finish process the state i
11:51:45.783 WARN 12114 --- [pool-4-thread-1] i.s.s.e.s.db.DbAndReportTcStateLogStore  : StateMachineInstance[172.17.0.58:8091:131001140287610880] is rec
11:51:45.783 INFO 12114 --- [pool-4-thread-1] i.s.s.p.handler.DefaultRouterHandler     : route instruction is null, process end
ction compensate succeed. XID: 172.17.0.58:8091:131001140287610880
11:51:45.835 WARN 12114 --- [           main] c.n.c.sources.URLConfigurationSource     : No URLs will be polled as dynamic configuration sources.
11:51:45.836 INFO 12114 --- [           main] c.n.c.sources.URLConfigurationSource     : To enable URLs as dynamic configuration sources, define System p
11:51:45.839 WARN 12114 --- [           main] c.n.c.sources.URLConfigurationSource     : No URLs will be polled as dynamic configuration sources.
11:51:45.839 INFO 12114 --- [           main] c.n.c.sources.URLConfigurationSource     : To enable URLs as dynamic configuration sources, define System p
```

图 7-37

第 8 章 分布式消息处理——基于 RocketMQ

随着微服务架构的盛行，服务会被水平拆分和横向拆分，服务与服务之间的调用关系越来越复杂，服务治理的技术挑战性也越来越大。比如，将拆分之前和拆分之后的调用链路关系复杂度进行对比，后者通常都是前者的好几倍。架构师和技术专家面对复杂的业务场景带来的技术挑战，为了保证核心业务的高吞吐量，通常要考虑将核心业务的接口进行异步设计。

消息中间件作为异步设计的一种手段，不仅能够解耦服务之间的强依赖关系，还能够最大限度地提升服务的 QPS 处理能力。开源的消息中间件有很多， RocketMQ 是其中一款高可靠性的分布式消息中间件。

8.1 消息中间件概述

8.1.1 什么是消息中间件

1. 什么是中间件

如图 8-1 所示，系统 A 的服务（商品服务和交易服务）要访问系统 B 的服务（订单服务和库存服务），需要通过一种通信手段（比如 HTTP）完成数据的交换。但是，系统 A 中的服务有很多，系统 B 中的服务也有很多，并且类似系统 A 的系统又有很多（比如系统 C 也包含商品服务和交易服务），它们之间又有很多调用关系（比如 MQ 或者 RPC）。

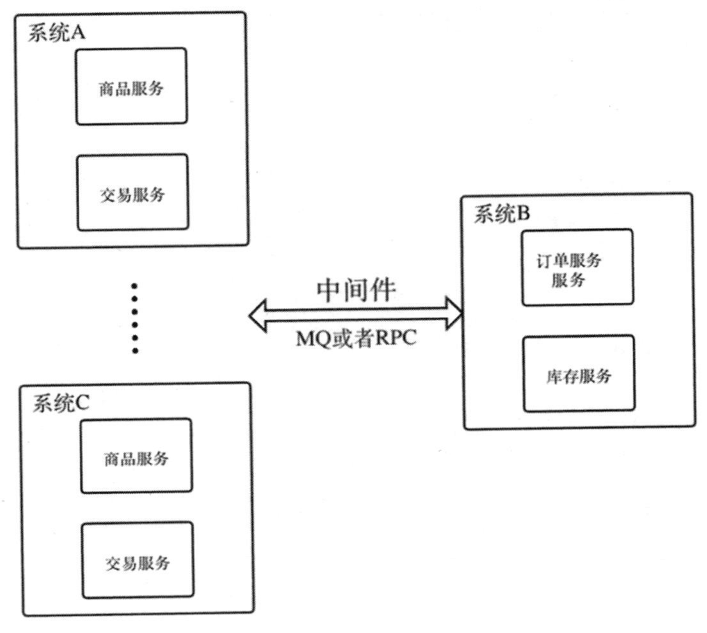

图 8-1

于是，开发人员将这些连接调用关系的代码沉淀下来，就形成了一套公用的代码库，并能够绑定不同系统的不同服务（甚至跨语言）。通常把这些能够解决跨语言并具备连接功能的解决方案叫作中间件。

2. 什么是消息中间件

消息中间件是基于队列与消息的网络传递技术，是在网络环境中为应用系统提供同步或异步、可靠的消息传输的支撑性软件系统。消息中间件利用高效、可靠的消息传递机制进行与平台无关的数据交流，并基于数据通信来进行分布式系统的集成。通过提供消息传递和消息排队模型，消息中间件可以在分布式环境下扩展进程间的通信。

开源社区中常用的消息中间件包括：RocketMQ、Kafka、Pulsar 等。

8.1.2　为什么需要消息中间件

图 8-2 所示为某直播平台主播的开播流程：

（1）主播访问直播应用，调用用户中心的用户服务登录应用。

（2）主播进入主播中心，调用主播服务加载与主播相关的资源。

（3）主播进入主播自己的房间，并调用房间中心加载房间相关的资源。

（4）主播上线之后，通知“粉丝”登录平台观看直播。

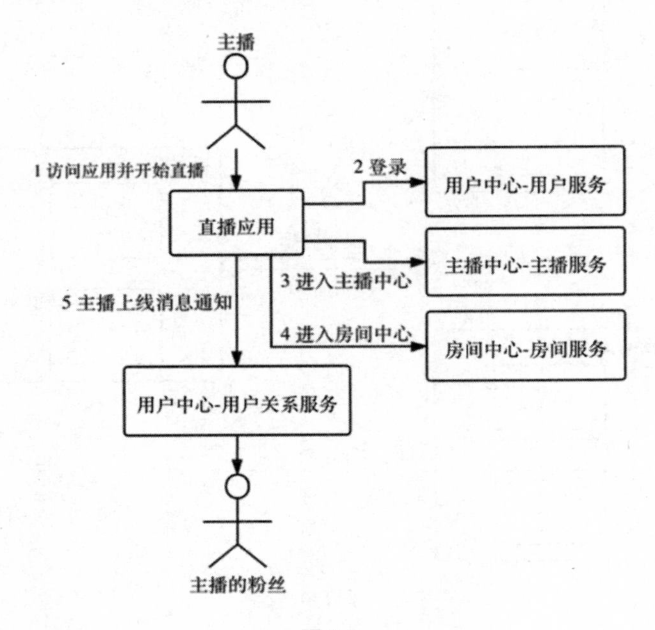

图 8-2

这里以直播平台基础功能“主播上线之后，通知‘粉丝’登录平台观看直播”作为业务场景来分析消息中间件的价值，主要是因为它具备接口异步设计的几个特征：

- 业务能够容忍一定程度的延迟；
- 业务瞬时流量洪峰大，正常时间段流量正常；
- 业务功能是可以降级的。

综上所述，如果主播是一个新人且“粉丝”数很少，则正常的同步消息通知是没有延迟问题的。如果主播是一个高流量的人气主播，“粉丝”数有几千万，则将消息同步地通知给“粉丝”，会产生很大的延迟。如果网络不稳定，则会出现“在主播上线了很久之后，“粉丝”才收到主播上线的消息”的情况。

消息中间件可以将消息通知从同步模式转为异步模式，这样可以提高消息通知接口的吞吐量，进行流量削峰，从而解决上述人气主播上线消息通知延迟的问题。

8.1.3 认识 RocketMQ

1. 什么是 RocketMQ

RocketMQ 主要由 Producer、Broker、Consumer 这 3 部分组成，其中，Producer 负责

生产消息，Consumer 负责消费消息，Broker 负责存储和转发消息。

Broker 在实际部署过程中对应一台服务器。每个 Broker 可以存储多个 Topic（主题）的消息，每个 Topic 的消息也可以分片存储于不同的 Broker 上。Message Queue 用于存储消息的物理地址，每个 Topic 中的消息地址存储于多个 Message Queue 中。ConsumerGroup 由多个 Consumer 实例构成。

2. RocketMQ 的基础术语

RocketMQ 主要包含。以下基础术语。

- 消息生产者（Producer）。

消息生产者负责生产消息，一般由业务系统负责生产消息。一个消息生产者会把业务应用系统产生的消息发送到 Broker 服务器。

RocketMQ 提供多种发送模式——同步、异步、最多发送一次。“同步”模式和“异步”模式均需要 Broker 服务器返回确认信息，“最多发送一次”模式则不需要。

- 消息消费者（Consumer）。

消息消费者负责消费消息，一般由后台系统负责异步消费。一个消息消费者会从 Broker 服务器拉取消息，并将其提供给应用程序。从用户应用的角度而言，RocketMQ 提供了两种消费形式：Pull 模式、Push 模式。

- 主题（Topic）。

主题表示一类消息的集合。每个主题包含若干条消息。每条消息只能属于一个主题，它是 RocketMQ 进行消息订阅的基本单位。

- 代理服务器（Broker Server）。

代理服务器负责存储和转发消息。在 RocketMQ 系统中，代理服务器负责接收从生产者发送来的消息并存储，同时为消费者的拉取请求做准备。代理服务器也存储消息相关的元数据，包括消费者组、消费进度偏移、主题和队列消息等。

- 名字服务器（Name Server）。

名字服务器充当路由消息的提供者。生产者或消费者能够通过名字服务器查找各主题相应的 Broker IP 地址列表。多个 Name Server 实例组成集群，但相互独立，没有信息交换。Name Server 是无状态的，其集群可以按照容量规划进行弹性扩容，对 Broker Server、Consumer 和 Producer 节点来讲是无感知的。

- 拉取式消费（Pull Consumer）。

拉取式消费是 Consumer 消费的一种类型。应用通常主动调用 Consumer 的拉消息方法从 Broker Server 拉取消息，拉取进度由应用来控制。一旦获取了批量消息，应用就会启动消费线程并消费消息。

- 推动式消费（Push Consumer）。

推模式消费是 Consumer 消费的一种类型。在该模式下，Broker Server 在收到数据后会主动推送给消费端。该消费模式一般实时性较高。

- 生产者组（Producer Group）。

生产者组是同一类 Producer 的集合，这类 Producer 发送同一类消息，且发送逻辑一致。如果发送的是事务消息，且原始生产者在发送之后崩溃，则 Broker Server 会联系同一生产者组的其他生产者实例以提交或回溯消费。

- 消费者组（Consumer Group）。

消费者组是同一类 Consumer 的集合，这类 Consumer 通常消费同一类消息，且消费逻辑一致。消费者组使得在消费消息方面实现负载均衡和容错变得非常容易。需要注意的是，消费者组的消费者实例必须订阅完全相同的 Topic。RocketMQ 支持两种消息模式：集群消费（Clustering）和广播消费（Broadcasting）。

- 集群消费（Clustering）。

在集群消费模式下，相同 Consumer Group 的每个 Consumer 实例平均地消费消息。

- 广播消费（Broadcasting）。

在广播消费模式下，相同 Consumer Group 的每个 Consumer 实例都能接收全量的消息。

- 普通顺序消息（Normal Ordered Message）。

在普通顺序消费模式下，消费者通过同一个消费队列收到的消息是有顺序的，不同消息队列收到的消息则可能是无顺序的。

- 严格顺序消息（Strictly Ordered Message）。

在严格顺序消息模式下，消费者收到的所有消息均是有顺序的。

- 消息（Message）。

消息是系统所传输信息的物理载体，也是生产和消费数据的最小单位。每条消息必须属于一个主题。RocketMQ 中的每个消息都拥有唯一的 Message ID，且可以携带具有业务标识的 Key。系统提供了通过 Message ID 和 Key 查询消息的功能。

- 标签（Tag）。

标签是 RocketMQ 为消息设置的标志，用于在同一个主题下区分不同类型的消息。来自同一个业务单元的消息，可以根据不同业务目的在同一个主题下设置不同标签。标签能够有效地保持代码的清晰度和连贯性，并优化 RocketMQ 提供的查询系统。消费者可以根据 Tag 对不同子主题的不同消费逻辑实现更好的扩展性。

8.2　搭建 RocketMQ 的运行环境

8.2.1　了解 RocketMQ 的安装包

RocketMQ 的安装包主要包括两部分：服务器端安装包和 UI 控制台安装包。

1. 服务器端安装包

从官网下载 RocketMQ 的服务器端安装包 rocketmq-all-4.8.0-source-release.zip，解压缩之后的文件目录结构如图 8-3 所示。

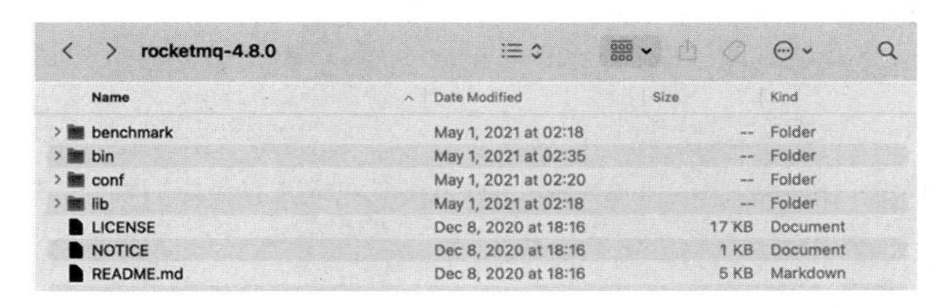

图 8-3

（1）bin 目录中存放的是服务器端的启动脚本，比如 Broker Server 的启动脚本 mqbroker.sh。

（2）conf 目录中存放的是服务器端的配置文件，比如 Broker Server 的配置文件 broker.conf。

（3）Lib 目录中存放的是服务器端的 Jar 包，比如 Broker Server 的 Jar 包 rocketmq-broker- 4.8.0.jar。

2. UI 控制台安装包

从官网下载 RocketMQ 的扩展安装包 rocketmq-externals-master，解压缩之后用 Maven 工具打包。在模块 rocketmq-console 的 target 目录下，会生成打包好的 UI 控制台安装包 rocketmq-console-ng-2.0.0.jar，如图 8-4 所示。

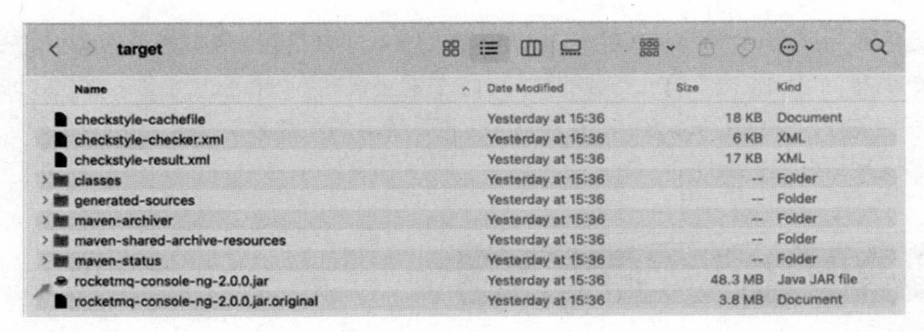

图 8-4

8.2.2 搭建单 Master 的单机环境

RocketMQ 支持单 Master 部署。单 Master 部署不需要修改配置信息，一切采用默认值即可。搭建过程如下。

（1）启动 Name Server。

在 bin 目录下，使用命令“nohup sh mqnamesrv &”启动 Name Server。

（2）启动 Broker Server。

在 bin 目录下，使用命令“nohup sh mqbroker −n localhost:9876 &”启动 Broker Server。

（3）使用命令“ps −ef| grep java”查看 RocketMQ 对应的 Java 进程。

（4）启动 UI 控制台。

在 rocketmq−console−ng−2.0.0.jar 文件对应的目录下，执行如下命令，启动 UI 控制台。

```
 java -jar rocketmq-console-ng-2.0.0.jar --server.port=12581
--rocketmq.config.namesrvAddr=127.0.0.1:9876
```

输入地址“127.0.0.1:12581”，可以看到一个 Master 节点（Broker Server）启动成功，并注册到 Name Server 中。

8.2.3 搭建多 Master 的集群环境

RocketMQ 支持多 Master 的集群部署，搭建过程如下：

（1）在 bin 目录下，使用命令“nohup sh mqnamesrv &”启动 Name Server。

（2）配置 Broker Server 节点 a 和 Broker Server 节点 b，具体配置如下：

```
###配置 Broker Server 节点 a
brokerClusterName=DefaultCluster
brokerName=broker-a
```

```
brokerId=0
deleteWhen=04
fileReservedTime=48
brokerRole=ASYNC_MASTER
flushDiskType=ASYNC_FLUSH
listenPort=10908
###配置 Broker Server 节点 b
brokerClusterName=DefaultCluster
brokerName=broker-b
brokerId=0
deleteWhen=04
fileReservedTime=48
brokerRole=ASYNC_MASTER
flushDiskType=ASYNC_FLUSH
listenPort=10911
```

（3）启动两个 Broker Server 节点：

```
###启动节点 a
nohup sh mqbroker -n 192.168.0.123:9876
-c ../conf/2m-noslave/broker-a.properties &
###启动节点 b
nohup sh mqbroker -n 192.168.0.123:9876
-c ../conf/2m-noslave/broker-b.properties &
```

（4）启动 UI 控制台。

在 rocketmq-console-ng-2.0.0.jar 文件对应的目录下，执行如下命令，启动 UI 控制台。

```
java -jar rocketmq-console-ng-2.0.0.jar --server.port=12581
--rocketmq.config.namesrvAddr=192.168.0.123:9876
```

输入地址“192.168.0.123:12581”，可以看到两个 Master 节点（Broker Server 节点）启动成功，并注册到 Name Server 中了，如图 8-6 所示。

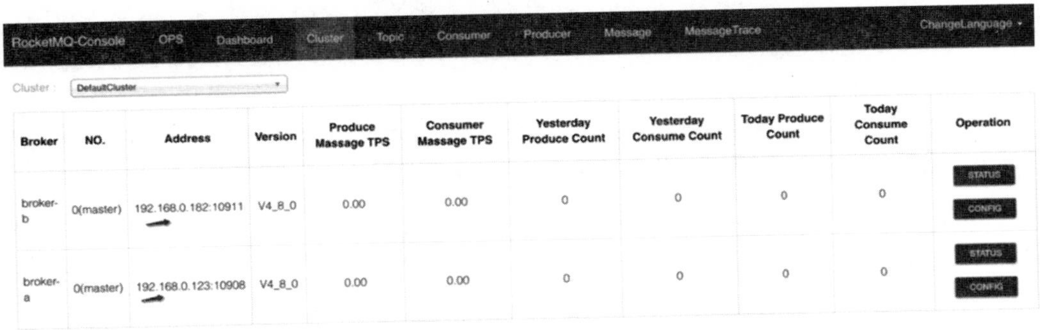

RocketMQ-Console　OPS　Dashboard　Cluster　Topic　Consumer　Producer　Message　MessageTrace　ChangeLanguage

Cluster : DefaultCluster

Broker	NO.	Address	Version	Produce Massage TPS	Consumer Massage TPS	Yesterday Produce Count	Yesterday Consume Count	Today Produce Count	Today Consume Count	Operation
broker-b	0(master)	192.168.0.182:10911	V4_8_0	0.00	0.00	0	0	0	0	STATUS CONFIG
broker-a	0(master)	192.168.0.123:10908	V4_8_0	0.00	0.00	0	0	0	0	STATUS CONFIG

图 8-6

8.2.4 搭建单 Master 和单 Slave 的集群环境

RocketMQ 支持多 Master 和多 Slave 的集群部署，比如一个 Master 节点可以对应多个 Slave 节点。

在 RocketMQ 中，Master 节点具备消息的“读”和“写”功能，Slave 节点只能具备消息的“读”功能。

如果开发人员的业务接口，对接口吞吐量要求非常高，对接口处理的时延性要求不是很高，则可以考虑使用“单 Master 对多 Slave 部署模式”：Master 节点负责写消息，Slave 节点负责读消息。这样可以做到消息的读写分离，Master 节点可以将资源全部给生产消息的应用客户端。

如果开发人员要提升消息的可用性，则可以考虑使用“多 Master 对多 Slave 部署模式”，如图 8-7 所示。

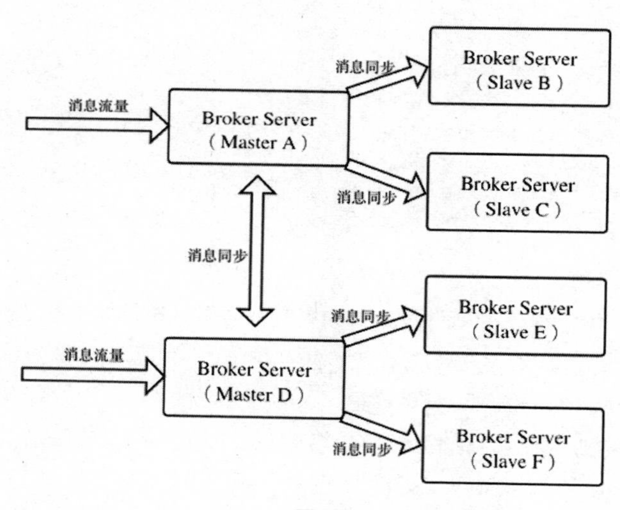

图 8-7

为了保持 RocketMQ 的高可用和吞吐量，线上通常都是部署 6 个节点。在消息流量被路由到 Master 节点 A 或者 Master 节点 D 后，会在 Master 节点 A 和 Master 节点 D 之间进行消息数据同步；Master 节点 A 会将消息数据同步到它对应的 Slave 节点 B 和 Slave 节点 C；Master 节点 D 会将消息数据同步到它对应的 Slave 节点 E 和 Slave 节点 F。

在 RocketMQ 中，Master 和 Slave 节点之间的消息数据主从同步（HA）支持如下两种模式。

- 异步复制。

如果消息数据主从同步采用异步复制模式，则主备之间有短暂的消息延迟（毫秒级）。采用异步复制模式，即使磁盘损坏，丢失的消息也非常少，且消息实时性不会受影响。另外，在 Master

节点宕机后，消费者仍然可以从 Slave 节点消费，而且此过程对应用是透明的。

- 同步双写。

HA 采用同步双写方式，那只有当主备都写成功后才会向应用返回成功。如果采用双写方式，在同步双写时，数据与服务都无单点故障；在 Master 宕机时，消息无延迟，服务可用性与数据可用性都非常高。

下面演示搭建单 Master 和单 Slave 的集群环境（异步复制）的过程。

（1）在 bin 目录下，使用命令“nohup sh mqnamesrv &”启动 Name Server。

（2）配置 Broker 节点。

配置 ASYNC_MASTER 节点的属性文件 broker-a.properties，具体配置如下：

```
brokerClusterName=DefaultCluster
brokerName=broker-a
brokerId=0
deleteWhen=04
fileReservedTime=48
brokerRole=ASYNC_MASTER
flushDiskType=ASYNC_FLUSH
```

配置 SLAVE 节点的属性文件 broker-a-s.properties，具体配置如下：

```
brokerClusterName=DefaultCluster
brokerName=broker-a
brokerId=1
deleteWhen=04
fileReservedTime=48
brokerRole=SLAVE
flushDiskType=ASYNC_FLUSH
```

（3）启动 ASYNC_MASTER 和 SLAVE 节点。

启动 ASYNC_MASTER 节点的命令如下所示：

```
nohup sh mqbroker -n 192.168.0.123:9876 -c ../conf
/2m-2s-async/broker-a.properties &
```

启动 SLAVE 节点的命令如下所示：

```
nohup sh mqbroker -n 192.168.0.123:9876 -c ../conf
/2m-2s-async/broker-a-s.properties &
```

启动之后，在 RocketmQ 控制台中可以看到两个 Broker 节点已经启动成功，如图 8-8 所示。

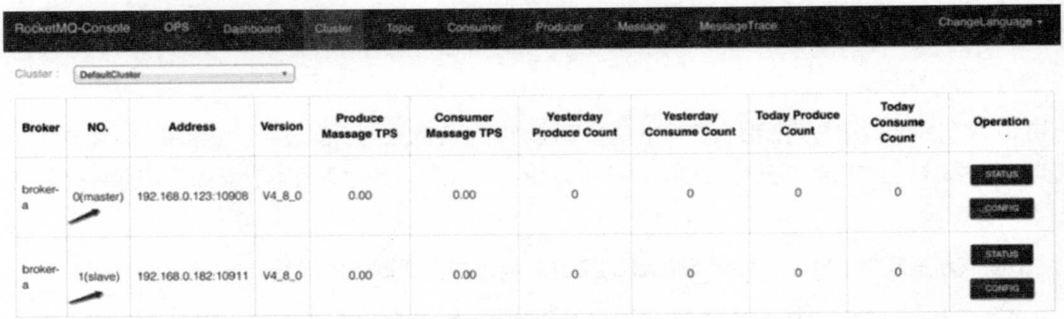

RocketMQ-Console OPS Dashboard Cluster Topic Consumer Producer Message MessageTrace ChangeLanguage

Cluster: DefaultCluster

Broker	NO.	Address	Version	Produce Massage TPS	Consumer Massage TPS	Yesterday Produce Count	Yesterday Consume Count	Today Produce Count	Today Consume Count	Operation
broker-a	0(master)	192.168.0.123:10908	V4_8_0	0.00	0.00	0	0	0	0	STATUS CONFIG
broker-a	1(slave)	192.168.0.182:10911	V4_8_0	0.00	0.00	0	0	0	0	STATUS CONFIG

图 8-8

8.2.5 搭建 Raft 集群环境

RocketMQ 从版本 4.5.0 开始，支持 Broker Server 的多副本机制。多副本机制主要是靠分布式一致性算法 Raft 来实现的。

下面演示搭建一组 Raft（3 个节点）集群环境的过程。

（1）在 bin 目录下，使用命令“nohup sh mqnamesrv &”启动 Name Server。

（2）配置 Raft 节点信息。

在文件 broker-n0.conf 中，配置 RaftNode00 节点，具体配置如下：

```
brokerClusterName = RaftCluster
brokerName=RaftNode00
listenPort=30911
namesrvAddr=192.168.0.123:9876
storePathRootDir=../../rmqstore/node00
storePathCommitLog=../../rmqstore/node00/commitlog
enableDLegerCommitLog=true
dLegerGroup=RaftNode00
dLegerPeers=n0-192.168.0.123:40911;n1-192.168.0.123:40912;n2-192.168.0.182:40913
dLegerSelfId=n0
sendMessageThreadPoolNums=16
```

在文件 broker-n1.conf 中，配置 RaftNode01 节点，具体配置如下：

```
brokerClusterName = RaftCluster
brokerName=RaftNode00
listenPort=30921
namesrvAddr=192.168.0.123:9876
storePathRootDir=../../rmqstore/node01
storePathCommitLog=../../rmqstore/node01/commitlog
```

```
enableDLegerCommitLog=true
dLegerGroup=RaftNode00
dLegerPeers=n0-192.168.0.123:40911;n1-192.168.0.123:40912;n2-192.168.0.182:40913
dLegerSelfId=n1
sendMessageThreadPoolNums=16
```

在文件 broker-n2.conf 中，配置 RaftNode02 节点，具体配置如下：

```
brokerClusterName = RaftCluster
brokerName=RaftNode00
listenPort=30931
namesrvAddr=192.168.0.123:9876
storePathRootDir=../../rmqstore/node01
storePathCommitLog=../../rmqstore/node01/commitlog
enableDLegerCommitLog=true
dLegerGroup=RaftNode00
dLegerPeers=n0-192.168.0.123:40911;n1-192.168.0.123:40912;n2-192.168.0.182:40913
dLegerSelfId=n2
sendMessageThreadPoolNums=16
```

（3）启动 Raft 节点。

使用如下命令分别启动 3 个节点。

```
nohup sh mqbroker -n 192.168.0.123:9876 -c ../conf/dledger/broker-n0.conf &
nohup sh mqbroker -n 192.168.0.123:9876 -c ../conf/dledger/broker-n1.conf &
nohup sh mqbroker -n 192.168.0.123:9876 -c ../conf/dledger/broker-n2.conf &
```

在 Raft 节点启动完成之后，在控制台中可以看到 3 个 Raft 节点已经启动，如图 8-9 所示。

RocketMQ-Console　OPS　Dashboard　Cluster　Topic　Consumer　Producer　Message　MessageTrace　ChangeLanguage

Cluster： RaftCluster

Broker	NO.	Address	Version	Produce Massage TPS	Consumer Massage TPS	Yesterday Produce Count	Yesterday Consume Count	Today Produce Count	Today Consume Count	Operation
RaftNode00	0(master)	192.168.0.123:30911	V4_8_0	0.00	0.00	0	0	0	0	STATUS CONFIG
RaftNode00	2(slave)	192.168.0.123:30921	V4_8_0	0.00	0.00	0	0	0	0	STATUS CONFIG
RaftNode00	3(slave)	192.168.0.182:30931	V4_8_0	0.00	0.00	0	0	0	0	STATUS CONFIG

图 8-9

8.2.6 【实例】用 RocketMQ Admin 控制台管控 RocketMQ

RocketMQ Broker 在启动的过程中，会启动 RocketMQ Admin 控制台。开发人员可以使用命令“sh mqadmin”查看控制台中的命令列表。

使用 RocketMQ 的控制台命令“updateTopic”，创建一个消息主题“useRocketMQCreateTopic”，具体命令如下：

```
sh mqadmin updateTopic -n 192.168.0.123:9876 -c RaftCluster -r 32 -t useRocketMQCreateTopic -w 32
```

在 RocketMQ 的控制台中会打印命令执行的日志，如图 8-10 所示。

```
huxian@huxians-MacBook-Pro bin % sh mqadmin updateTopic -n 192.168.0.123:9876 -c RaftCluster -t testTopic
RocketMQLog:WARN No appenders could be found for logger (io.netty.util.internal.PlatformDependent0).
RocketMQLog:WARN Please initialize the logger system properly.
create topic to 192.168.0.123:30911 success.
TopicConfig [topicName=testTopic, readQueueNums=8, writeQueueNums=8, perm=RW-, topicFilterType=SINGLE_TAG, topicSysFlag=0, order=false]
huxian@huxians-MacBook-Pro bin % sh mqadmin updateTopic -n 192.168.0.123:9876 -c RaftCluster -r 32 -t useRocketMQCreateTopic -w 32

RocketMQLog:WARN No appenders could be found for logger (io.netty.util.internal.PlatformDependent0).
RocketMQLog:WARN Please initialize the logger system properly.
create topic to 192.168.0.123:30911 success.
TopicConfig [topicName=useRocketMQCreateTopic, readQueueNums=32, writeQueueNums=32, perm=RW-, topicFilterType=SINGLE_TAG, topicSysFlag=0, order=false]
```

图 8-10

在 UI 控制台中可以看到，主题已经创建成功，如图 8-11 所示。

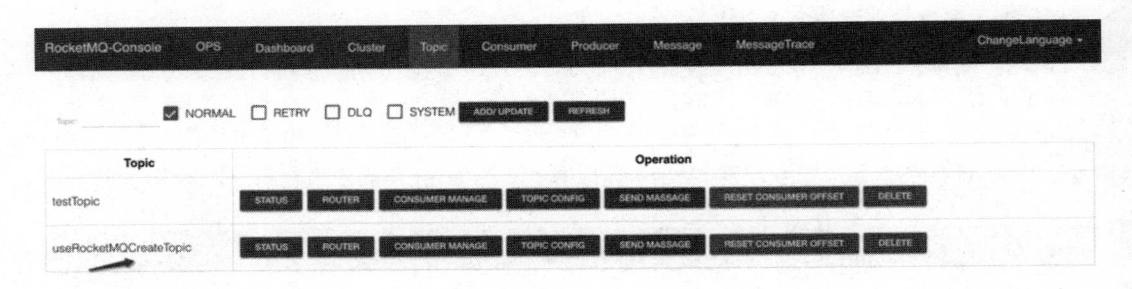

图 8-11

8.3 将应用接入 RocketMQ

RocketMQ 为开发人员提供了在客户端使用的 RocketMQ Client。开发人员可以直接依赖客户端，并利用生产者和消费者的 API 生产和消费消息。但这种方式对于开发人员来说成本非常高，需要开发人员自己维护客户端的版本，以及消费者和生产者的实例对象。

8.3.1 【实例】用 rocketmq-spring-boot-starter 框架将应用接入 RocketMQ

本实例的源码在本书配套资源的“chaptereight/ use-spring-boot-access-rocketmq”目录下。

本实例使用 RocketMQ 官方提供的 rocketmq-spring-boot-starter 框架将应用接入 RocketMQ。rocketmq-spring-boot-starter 是一个基于 Spring Boot 框架的“开箱即用”的 Starter 框架。

1. 初始化项目

本实例包含两个项目：use-spring-boot-produce（用于生产消息）和 use-spring-boot-consume（用于消费消息）。

（1）在两个项目中添加 rocketmq-spring-boot-starter 的依赖，具体配置如下：

```
<dependency>
    <groupId>org.apache.rocketmq</groupId>
    <artifactId>rocketmq-spring-boot-starter</artifactId>
    <version>2.2.0</version>
</dependency>
```

（2）在 use-spring-boot-produce 中配置连接 RocketMQ 的信息，具体配置如下：

```
rocketmq:
  name-server: 127.0.0.1:9876
  producer:
      group: use-spring-boot-access-rocketmq
      sendMessageTimeout: 300000
```

（3）在 use-spring-boot-consume 中配置连接 RocketMQ 的信息，具体配置如下：

```
rocketmq:
  name-server: 127.0.0.1:9876
  consumer:
    group: use-spring-boot-access-rocketmq
    topic: use-spring-boot-access-rocketmq
```

2. 生产和消费消息

在 use-spring-boot-produce 中启动一个线程来生产消息，具体代码如下所示：

```
@Configuration
public class UseSpringBootProduceConfig {
```

```
    @Resource
    private RocketMQTemplate rocketMQTemplate;
    @PostConstruct
  public void startProduce(){
    //①用线程触发消息的发送
        ExecutorService executorService= Executors.newCachedThreadPool();
        executorService.submit(new SendMessage());
    }
    class SendMessage implements Runnable{
        @Override
        public void run() {
            while (true) {
                //②用客户端 RocketMQTemplate 向 RocketMQ 的 Broker Server 发送消息
                SendResult sendResult =
rocketMQTemplate.syncSend("use-spring-boot-access-rocketmq", "hello word!");
                System.out.println("sendResult:" + sendResult.toString());
                try {
                    Thread.sleep(5000);
                } catch (InterruptedException e) {
                }
            }
        }
    }
  }
```

在 use-spring-boot-consume 中，启动一个线程消费消息，具体代码如下所示：

```
  @Configuration
  public class UseSpringBootConsumeConfig {
    @Resource
    private RocketMQTemplate rocketMQTemplate;

    @PostConstruct
  public void startConsume(){
    //①用线程触发消息的消费
        ExecutorService executorService=Executors.newCachedThreadPool();
        executorService.submit(new ConsumeMessage());
    }
    class ConsumeMessage implements Runnable {
        @Override
        public void run() {
            while (true) {
                //②用客户端 RocketMQTemplate 从 RocketMQ 的 Broker Server 消费消息，
默认为 Pull 消费模式
                List<String> result = rocketMQTemplate.receive(String.class);
                System.out.println("消费结果：" + result.toString());
```

```
                    try {
                        Thread.sleep(3000);
                    } catch (InterruptedException e) {
                    }
                }
            }
        }
    }
```

3. 启动项目

启动 use-spring-boot-produce 和 use-spring-boot-consume，如图 8-12 所示，生产者已经启动成功。

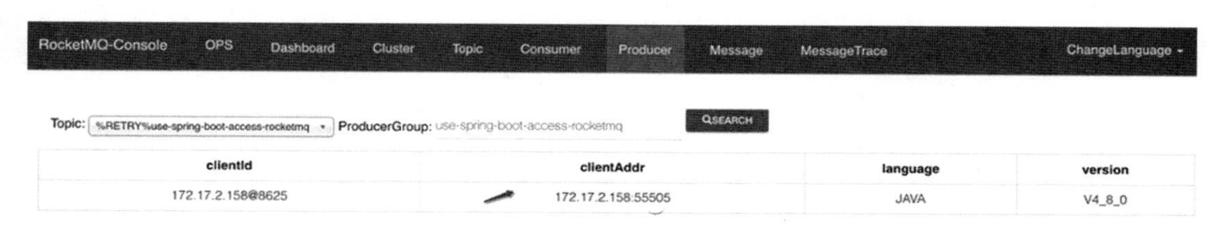

图 8-12

如图 8-13 所示，消费者已经启动成功，并开始消费消息。

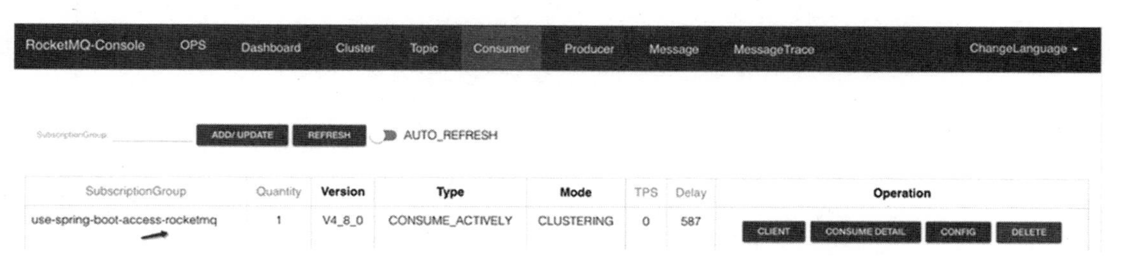

图 8-13

8.3.2 【实例】用 spring-cloud-starter-stream-rocketmq 框架将应用接入 RocketMQ

本实例的源码在本书配套资源的“chaptereight/use-spring-cloud-alibaba-access-rocketmq”目录下。

Spring Cloud Alibaba 官方提供了 spring-cloud-starter-stream-rocketmq 框架。本实例

使用 spring-cloud-starter-stream-rocketmq 框架将应用接入 RocketMQ。

1. 初始化项目

本实例主要包含两个项目：use-spring-cloud-alibaba-consume（消费者）和 use-spring-cloud-alibaba-produce（生产者）。

使用 Spring Cloud Alibaba 快速初始化两个项目，并添加 spring-cloud-starter-stream-rocketmq 相关的 Jar 包依赖，具体代码如下所示：

```
<dependency>
    <groupId>com.alibaba.cloud</groupId>
    <artifactId>spring-cloud-starter-stream-rocketmq</artifactId>
</dependency>
```

2. 生产和消费消息

在 use-spring-cloud-alibaba-produce 中定义一个线程生产消息，具体代码如下所示：

```
@Configuration
public class RocketmqConfig {
    @Autowired
    private SenderService senderService;

public void startRunSendMessage(){
    //①定义一个线程池
        ExecutorService executorService= Executors.newFixedThreadPool(1 );
        executorService.execute(new SendMessage());
    }
    class SendMessage implements Runnable{
        @Override
        public void run() {
            while (true) {
                try {
                    Thread.sleep(5000);
                    String msgContent = "msg-" + RandomUtils.nextLong();
                    //②生产消息
                    senderService.send(msgContent);
                }catch (InterruptedException e){
                    System.out.println(e.getMessage());
                }
            }
        }
    }
}
```

在 use-spring-cloud-alibaba-consume 中定义一个线程消费消息，具体代码如下所示：

```
@Configuration
public class RocketmqConsumeConfig {

    @Resource
    private MySink mySink;
    @PostConstruct
public void startConsume(){
    //①定义一个线程池
        ExecutorService executorService= Executors.newFixedThreadPool(20);
        executorService.execute(new ConsumeMessage());
    }
    class ConsumeMessage implements Runnable{
        @Override
        public void run() {
            while (true){
                //②消费消息
                mySink.input1();
            }
        }
    }
}
```

3. 启动项目

启动项目 use-spring-cloud-alibaba-produce 和 use-spring-cloud-alibaba-consume。如图 8-14 所示，生产者启动成功。

图 8-14

如图 8-15 所示消费者启动成功。

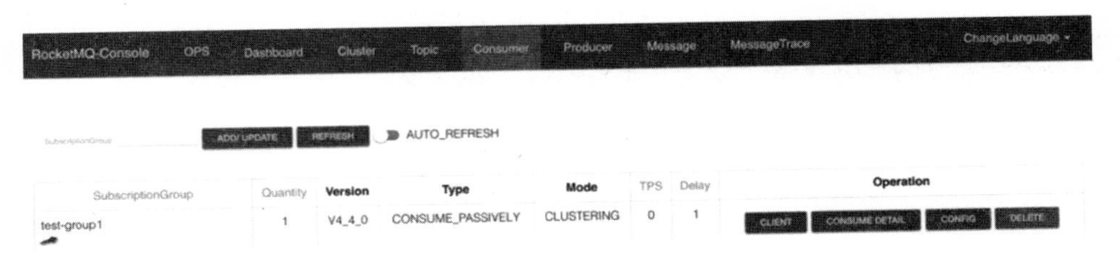

图 8-15

在项目 use-spring-cloud-alibaba-produce 中，用如下命令开启生产消息的线程：

```
curl 127.0.0.1:28081/produce/message
```

如图 8-16 所示，生产者成功地发送消息。

```
Debug: UseSpringCloudAlibabaProduceApplication  UseSpringCloudAlibabaConsumeApplication
Debugger  Console | Endpoints
Console  Endpoints
2021-05-12 07:18:53.437 DEBUG 12549 --- [pool-2-thread-1] c.a.c.s.b.r.i.RocketMQMessageHandler : com.alibaba.cloud.stream.binder.rocketmq.integration.RocketMQMe
2021-05-12 07:18:53.441 DEBUG 12549 --- [pool-2-thread-1] c.a.c.s.b.r.i.RocketMQMessageHandler : sync send to topic test-topic SendResult [sendStatus=SEND_OK, m
2021-05-12 07:18:58.444 DEBUG 12549 --- [pool-2-thread-1] c.a.c.s.b.r.i.RocketMQMessageHandler : com.alibaba.cloud.stream.binder.rocketmq.integration.RocketMQMe
2021-05-12 07:18:58.449 DEBUG 12549 --- [pool-2-thread-1] c.a.c.s.b.r.i.RocketMQMessageHandler : sync send to topic test-topic SendResult [sendStatus=SEND_OK, m
2021-05-12 07:19:03.455 DEBUG 12549 --- [pool-2-thread-1] c.a.c.s.b.r.i.RocketMQMessageHandler : com.alibaba.cloud.stream.binder.rocketmq.integration.RocketMQMe
2021-05-12 07:19:03.458 DEBUG 12549 --- [pool-2-thread-1] c.a.c.s.b.r.i.RocketMQMessageHandler : sync send to topic test-topic SendResult [sendStatus=SEND_OK, m
2021-05-12 07:19:08.464 DEBUG 12549 --- [pool-2-thread-1] c.a.c.s.b.r.i.RocketMQMessageHandler : com.alibaba.cloud.stream.binder.rocketmq.integration.RocketMQMe
2021-05-12 07:19:08.467 DEBUG 12549 --- [pool-2-thread-1] c.a.c.s.b.r.i.RocketMQMessageHandler : sync send to topic test-topic SendResult [sendStatus=SEND_OK, m
2021-05-12 07:19:13.472 DEBUG 12549 --- [pool-2-thread-1] c.a.c.s.b.r.i.RocketMQMessageHandler : com.alibaba.cloud.stream.binder.rocketmq.integration.RocketMQMe
2021-05-12 07:19:13.475 DEBUG 12549 --- [pool-2-thread-1] c.a.c.s.b.r.i.RocketMQMessageHandler : sync send to topic test-topic SendResult [sendStatus=SEND_OK, m
2021-05-12 07:19:18.480 DEBUG 12549 --- [pool-2-thread-1] c.a.c.s.b.r.i.RocketMQMessageHandler : com.alibaba.cloud.stream.binder.rocketmq.integration.RocketMQMe
2021-05-12 07:19:18.484 DEBUG 12549 --- [pool-2-thread-1] c.a.c.s.b.r.i.RocketMQMessageHandler : sync send to topic test-topic SendResult [sendStatus=SEND_OK, m
2021-05-12 07:19:23.488 DEBUG 12549 --- [pool-2-thread-1] c.a.c.s.b.r.i.RocketMQMessageHandler : com.alibaba.cloud.stream.binder.rocketmq.integration.RocketMQMe
2021-05-12 07:19:23.492 DEBUG 12549 --- [pool-2-thread-1] c.a.c.s.b.r.i.RocketMQMessageHandler : sync send to topic test-topic SendResult [sendStatus=SEND_OK, m
2021-05-12 07:19:28.496 DEBUG 12549 --- [pool-2-thread-1] c.a.c.s.b.r.i.RocketMQMessageHandler : com.alibaba.cloud.stream.binder.rocketmq.integration.RocketMQMe
2021-05-12 07:19:28.509 DEBUG 12549 --- [pool-2-thread-1] c.a.c.s.b.r.i.RocketMQMessageHandler : sync send to topic test-topic SendResult [sendStatus=SEND_OK, m
2021-05-12 07:19:33.514 DEBUG 12549 --- [pool-2-thread-1] c.a.c.s.b.r.i.RocketMQMessageHandler : com.alibaba.cloud.stream.binder.rocketmq.integration.RocketMQMe
2021-05-12 07:19:33.518 DEBUG 12549 --- [pool-2-thread-1] c.a.c.s.b.r.i.RocketMQMessageHandler : sync send to topic test-topic SendResult [sendStatus=SEND_OK, m
2021-05-12 07:19:38.521 DEBUG 12549 --- [pool-2-thread-1] c.a.c.s.b.r.i.RocketMQMessageHandler : com.alibaba.cloud.stream.binder.rocketmq.integration.RocketMQMe
2021-05-12 07:19:38.525 DEBUG 12549 --- [pool-2-thread-1] c.a.c.s.b.r.i.RocketMQMessageHandler : sync send to topic test-topic SendResult [sendStatus=SEND_OK, m
```

图 8-16

如图 8-17 所示，消费者成功地消费消息。

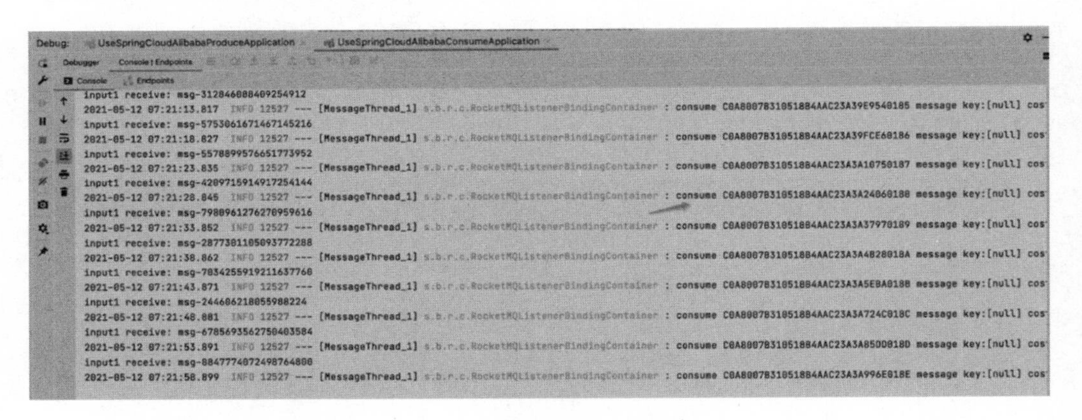

图 8-17

8.4 用 Netty 实现 RocketMQ 的通信渠道

在 RocketMQ 中，通信角色关系大致可以分为如下几种：

①消费者用通信渠道连接 Name Server 和 Broker Server；

②生产者用通信渠道连接 Name Server 和 Broker Server；

③Broker Server 连接 Name Server。

8.4.1　用 NettyRemotingClient 类实现客户端的通信渠道

1. 初始化 NettyRemotingClient 类

RocketMQ 在初始化客户端实例类 MQClientInstance 的过程中，初始化客户端通信渠道 NettyRemotingClient 类，具体代码如下所示：

```
public class MQClientManager {
    ...
    public MQClientInstance getOrCreateMQClientInstance(final ClientConfig
clientConfig, RPCHook rpcHook) {
        String clientId = clientConfig.buildMQClientId();
        MQClientInstance instance = this.factoryTable.get(clientId);
        if (null == instance) {
            //①初始化 MQClientInstance 类
            instance =
                new MQClientInstance(clientConfig.cloneClientConfig(),
                    this.factoryIndexGenerator.getAndIncrement(), clientId,
rpcHook);
            //②用本地缓存存储 MQClientInstance 类对应的实例对象
            MQClientInstance prev = this.factoryTable.putIfAbsent(clientId,
instance);
            if (prev != null) {
                instance = prev;
                log.warn("Returned Previous MQClientInstance for
clientId:[{}]", clientId);
            } else {
                log.info("Created new MQClientInstance for clientId:[{}]",
clientId);
            }
        }
        return instance;
    }
}
public class MQClientInstance {
    private final MQClientAPIImpl mQClientAPIImpl;
    ...
  public MQClientInstance(ClientConfig clientConfig, int instanceIndex,
String clientId, RPCHook rpcHook) {
     //③用构造函数初始化 MQClientAPIImpl 类
        this.mQClientAPIImpl = new MQClientAPIImpl(this.nettyClientConfig,
this.clientRemotingProcessor, rpcHook, clientConfig);
    }
```

```
    }
    public class MQClientAPIImpl {
        ...
        private final RemotingClient remotingClient;
        public MQClientAPIImpl(final NettyClientConfig nettyClientConfig,
            final ClientRemotingProcessor clientRemotingProcessor,
            RPCHook rpcHook, final ClientConfig clientConfig) {
            //④初始化客户端的通信渠道类 NettyRemotingClient
            this.remotingClient = new NettyRemotingClient(nettyClientConfig,
null);
        }
    }
```

2. 开启客户端通信渠道

RocketMQ 用 NettyRemotingClient 类的 start()方法开启客户端通信渠道，具体代码如下所示：

```
    @Override
    public void start() {
        //①开启事件执行器组
        this.defaultEventExecutorGroup = new DefaultEventExecutorGroup(
            nettyClientConfig.getClientWorkerThreads(),new ThreadFactory() {
                private AtomicInteger threadIndex = new AtomicInteger(0);
                @Override
                public Thread newThread(Runnable r) {
                    return new Thread(r, "NettyClientWorkerThread_" +
this.threadIndex.incrementAndGet());
                }
            });
        //②开启通信渠道启动类 Bootstrap
        Bootstrap handler =
this.bootstrap.group(this.eventLoopGroupWorker).channel(NioSocketChannel.cla
ss)
            .option(ChannelOption.TCP_NODELAY, true)
            .option(ChannelOption.SO_KEEPALIVE, false)
            .option(ChannelOption.CONNECT_TIMEOUT_MILLIS,
nettyClientConfig.getConnectTimeoutMillis())
            .option(ChannelOption.SO_SNDBUF,
nettyClientConfig.getClientSocketSndBufSize())
            .option(ChannelOption.SO_RCVBUF,
nettyClientConfig.getClientSocketRcvBufSize())
            .handler(new ChannelInitializer<SocketChannel>() {
            @Override
            public void initChannel(SocketChannel ch) throws Exception {
                ChannelPipeline pipeline = ch.pipeline();
```

```
            if (nettyClientConfig.isUseTLS()) {
                if (null != sslContext) {
                    pipeline.addFirst(defaultEventExecutorGroup,
"sslHandler", sslContext.newHandler(ch.alloc()));
                    log.info("Prepend SSL handler");
                } else {
                    log.warn("Connections are insecure as SSLContext is
null!");
                }
            }
            pipeline.addLast(
                defaultEventExecutorGroup,
                new NettyEncoder(),
                new NettyDecoder(),
                new IdleStateHandler(0, 0,
nettyClientConfig.getClientChannelMaxIdleTimeSeconds()),
                new NettyConnectManageHandler(),
                new NettyClientHandler());
            }
        });
        //③开启处理通信渠道结果的定时器
        this.timer.scheduleAtFixedRate(new TimerTask() {
            @Override
            public void run() {
                try {
                    NettyRemotingClient.this.scanResponseTable();
                } catch (Throwable e) {
                    log.error("scanResponseTable exception", e);
                }
            }
        }, 1000 * 3, 1000);
        if (this.channelEventListener != null) {
            this.nettyEventExecutor.start();
        }
    }
```

8.4.2　用 NettyRemotingServer 类实现服务器端的通信渠道

在 RocketMQ 中，Name Server 和 Broker Server 作为服务器端需要等待客户端连接，并完成 RPC 请求。在启动 Name Server 和 Broker Server 的过程中，会完成服务器端通信渠道的初始化，并开启对应的通信渠道。

1. 初始化 NettyRemotingServer 类

在 Broker Server 中的 BrokerController 类和 Name Server 中的 NamesrvController 类的

初始化过程中，会完成 NettyRemotingServer 类的初始化。其中，BrokerController 类中初始化的具体代码如下所示：

```
    public class BrokerController {
        ...
        private RemotingServer remotingServer;
        private RemotingServer fastRemotingServer;
        public boolean initialize() throws CloneNotSupportedException {
            boolean result = this.topicConfigManager.load();
            result = result && this.consumerOffsetManager.load();
            result = result && this.subscriptionGroupManager.load();
            result = result && this.consumerFilterManager.load();
            //如果主题配置信息、消费进度、订阅组及消费过滤器都加载完成，则初始化
NettyRemotingServer 类，否则不执行
            if (result) {
                this.remotingServer = new
NettyRemotingServer(this.nettyServerConfig, this.clientHousekeepingService);
                NettyServerConfig fastConfig = (NettyServerConfig)
this.nettyServerConfig.clone();
                fastConfig.setListenPort(nettyServerConfig.getListenPort() -
2);
                this.fastRemotingServer = new NettyRemotingServer(fastConfig,
this.clientHousekeepingService);
            }
        }
    }
```

NettyRemotingServer 类用构造函数初始化服务器端的通信渠道，具体代码如下所示：

```
    public NettyRemotingServer(final NettyServerConfig nettyServerConfig,
            final ChannelEventListener channelEventListener) {
        ...
    //①初始化 Netty 的 ServerBootstrap 类
        this.serverBootstrap = new ServerBootstrap();
    this.nettyServerConfig = nettyServerConfig;
    //②初始化通道事件的监听器
    this.channelEventListener = channelEventListener;
    //③从配置文件中获取公共线程数
        int publicThreadNums =
nettyServerConfig.getServerCallbackExecutorThreads();
        if (publicThreadNums <= 0) {
            publicThreadNums = 4;
    }
    //④定义一个线程池
        this.publicExecutor = Executors.newFixedThreadPool(publicThreadNums,
new ThreadFactory() {
```

```
            private AtomicInteger threadIndex = new AtomicInteger(0);
            @Override
            public Thread newThread(Runnable r) {
                return new Thread(r, "NettyServerPublicExecutor_" +
this.threadIndex.incrementAndGet());
            }
    });
    //⑤如果是 Epoll 类型的 NIO 通信，则初始化 EpollEventLoopGroup 类
        if (useEpoll()) {
            this.eventLoopGroupBoss = new EpollEventLoopGroup(1, new
ThreadFactory() {
                private AtomicInteger threadIndex = new AtomicInteger(0);
                @Override
                public Thread newThread(Runnable r) {
                    return new Thread(r, String.format("NettyEPOLLBoss_%d",
this.threadIndex.incrementAndGet()));
                }
            });
            //⑥初始化 EpollEventLoopGroup 类
            this.eventLoopGroupSelector = new EpollEventLoopGroup
(nettyServerConfig.getServerSelectorThreads(), new ThreadFactory() {
                private AtomicInteger threadIndex = new AtomicInteger(0);
                private int threadTotal =
nettyServerConfig.getServerSelectorThreads();
                @Override
                public Thread newThread(Runnable r) {
                    return new Thread(r,
String.format("NettyServerEPOLLSelector_%d_%d", threadTotal,
this.threadIndex.incrementAndGet()));
                }
            });
    } else {
        //⑦否则初始化 NioEventLoopGroup 类
            this.eventLoopGroupBoss = new NioEventLoopGroup(1, new
ThreadFactory() {
                private AtomicInteger threadIndex = new AtomicInteger(0);
                @Override
                public Thread newThread(Runnable r) {
                    return new Thread(r, String.format("NettyNIOBoss_%d",
this.threadIndex.incrementAndGet()));
                }
            });
            //⑧初始化 NioEventLoopGroup 类
```

```
            this.eventLoopGroupSelector = new
NioEventLoopGroup(nettyServerConfig.getServerSelectorThreads(), new
ThreadFactory() {
                private AtomicInteger threadIndex = new AtomicInteger(0);
                private int threadTotal =
nettyServerConfig.getServerSelectorThreads();
                @Override
                public Thread newThread(Runnable r) {
                    return new Thread(r,
String.format("NettyServerNIOSelector_%d_%d", threadTotal,
this.threadIndex.incrementAndGet()));
                }
            });
    }
    //⑨加载通信渠道的 SSL 上下文信息
        loadSslContext();
    }
```

2. 开启服务器端通信渠道

在初始化完 NettyRemotingServer 类之后，需要开启通信渠道以等待客户端连接。开启通信渠道的具体代码如下所示：

```
    ...
    @Override
    public void start() {
        //①初始化 Netty 的 DefaultEventExecutorGroup 类
        this.defaultEventExecutorGroup = new DefaultEventExecutorGroup(
        nettyServerConfig.getServerWorkerThreads(),
                new ThreadFactory() {
        private AtomicInteger threadIndex = new AtomicInteger(0);
        @Override
        public Thread newThread(Runnable r) {
            return new Thread(r, "NettyServerCodecThread_" +
this.threadIndex.incrementAndGet());
            }
    });
    //②预处理通信握手连接
    prepareSharableHandlers();
    //③初始化一个 Netty 的 ServerBootstrap 类
        ServerBootstrap childHandler =
            this.serverBootstrap.group(this.eventLoopGroupBoss,
this.eventLoopGroupSelector)
                .channel(useEpoll() ? EpollServerSocketChannel.class :
NioServerSocketChannel.class)
```

```
            //④添加底层的 TCP 连接的配置信息，比如参数 ChannelOption.SO_KEEPALIVE
（服务器端与客户端之间的探活机制，如果开启，则会验证通信渠道的有效性，否则不会）
            .option(ChannelOption.SO_BACKLOG, 1024)
            .option(ChannelOption.SO_REUSEADDR, true)
            .option(ChannelOption.SO_KEEPALIVE, false)
            .childOption(ChannelOption.TCP_NODELAY, true)
            .childOption(ChannelOption.SO_SNDBUF,
nettyServerConfig.getServerSocketSndBufSize())
            .childOption(ChannelOption.SO_RCVBUF,
nettyServerConfig.getServerSocketRcvBufSize())
            .localAddress(new
InetSocketAddress(this.nettyServerConfig.getListenPort()))
            //⑤向通信渠道中添加事件处理类
            .childHandler(new ChannelInitializer<SocketChannel>() {
                @Override
                public void initChannel(SocketChannel ch) throws Exception {
                    ch.pipeline()
                        .addLast(defaultEventExecutorGroup,
HANDSHAKE_HANDLER_NAME, handshakeHandler)
                        .addLast(defaultEventExecutorGroup,encoder,new
NettyDecoder(),
                        new IdleStateHandler(0, 0,
nettyServerConfig.getServerChannelMaxIdleTimeSeconds()),
                        connectionManageHandler,serverHandler
                        );
                }
            });

        if (nettyServerConfig.isServerPooledByteBufAllocatorEnable()) {
            childHandler.childOption(ChannelOption.ALLOCATOR,
PooledByteBufAllocator.DEFAULT);
        }
    try {
        //⑥绑定 IP 地址，等待客户端连接
            ChannelFuture sync = this.serverBootstrap.bind().sync();
            InetSocketAddress addr = (InetSocketAddress)
sync.channel().localAddress();
            this.port = addr.getPort();
        } catch (InterruptedException e1) {
            throw new RuntimeException("this.serverBootstrap.bind().sync()
InterruptedException", e1);
    }
    //⑦如果渠道事件监听器不为空，则开启监听器
        if (this.channelEventListener != null) {
            this.nettyEventExecutor.start();
```

```
    }
    //⑧使用定时器，定时地扫描通信渠道的响应结果
        this.timer.scheduleAtFixedRate(new TimerTask() {
            @Override
            public void run() {
            try {
                NettyRemotingServer.this.scanResponseTable();
            } catch (Throwable e) {
                    log.error("scanResponseTable exception", e);
                }
            }
        }, 1000 * 3, 1000);
    }
```

8.5 用“异步”“同步”和“最多发送一次”模式生产消息

RocketMQ 支持用“异步”“同步”和“最多发送一次”模式生产消息。

8.5.1 用“异步”模式生产消息的原理

在 RocketMQ 中，采用“异步”模式生产消息，可以提升消息发送的吞吐量。在了解“异步”模式生产消息的原理之前，要先认识“异步”模式。

1. 什么是“异步”模式

下面用电商支付的业务场景来描述“异步”模式，如图 8-18 所示。

（1）买家发起支付请求，金融支付网关只能确认是否收到了支付请求，并返回支付请求的结果。

（2）金融支付网关调用第三方支付公司的系统发起支付请求，并通过定时轮询拉取支付结果。

（3）第三方支付公司向买家开户行发起银行内部转账，买家开户行将支付结果推送给第三方支付公司。

（4）金融网关向买家异步推送支付结果，买家开户行将扣款短信发送到买家开户行对应的手机上。

通过上面分析电商支付的业务场景，可以将“异步”模式定义为：将“实时处理请求并响应处理结果”的方式，转换为“推拉结合的消息通知”的方式（调用方主动定时轮询结果，或者被调用方主动推送请求处理之后的结果）。

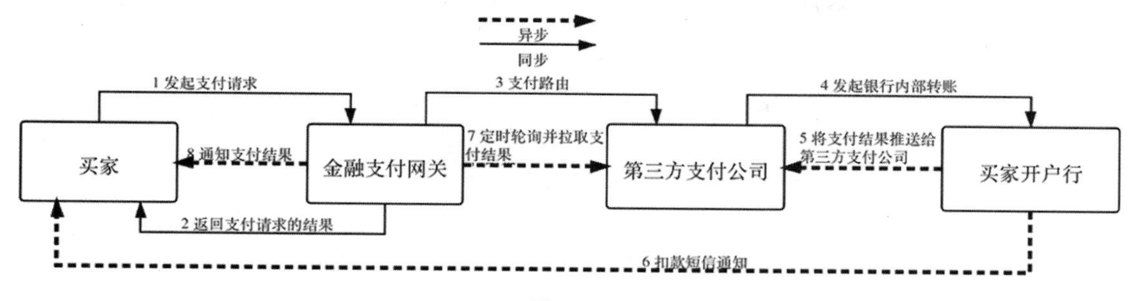

图 8-18

2. 异步生产消息

在 RocketMQ 中，开发人员可以比较灵活地使用不同的模式来生产消息，其中“异步”模式用枚举类型 CommunicationMode.ASYNC 来区分。

RocketMQ 在 DefaultMQProducerImpl 类的 sendKernelImpl()方法中处理异步消息，具体代码如下所示：

```
    private SendResult sendKernelImpl(final Message msg,
        final MessageQueue mq,
        final CommunicationMode communicationMode,
        final SendCallback sendCallback,
        final TopicPublishInfo topicPublishInfo,
        final long timeout) throws MQClientException, RemotingException,
MQBrokerException, InterruptedException {
        ...
        SendResult sendResult = null;
        switch (communicationMode) {
            //①如果是异步模式，则处理异步生产消息的逻辑
            case ASYNC:
                Message tmpMessage = msg;
                boolean messageCloned = false;
                if (msgBodyCompressed) {
                    tmpMessage = MessageAccessor.cloneMessage(msg);
                    messageCloned = true;
                    msg.setBody(prevBody);
                }
                if (topicWithNamespace) {
                    if (!messageCloned) {
                        tmpMessage = MessageAccessor.cloneMessage(msg);
                        messageCloned = true;
                    }
                    msg.setTopic(NamespaceUtil.withoutNamespace(msg.getTopic(),
this.defaultMQProducer.getNamespace()));
```

```
            }
            long costTimeAsync = System.currentTimeMillis() - beginStartTime;
            if (timeout < costTimeAsync) {
                throw new RemotingTooMuchRequestException("sendKernelImpl
call timeout");
            }
            //②调用 NettyRemotingClient 类的 invokeAsync()方法生产异步消息
            sendResult = this.mQClientFactory.getMQClientAPIImpl().
sendMessage(
                brokerAddr,
                mq.getBrokerName(),
                tmpMessage,
                requestHeader,
                timeout - costTimeAsync,
                communicationMode,
                sendCallback,
                topicPublishInfo,
                this.mQClientFactory,
                this.defaultMQProducer.getRetryTimesWhenSendAsyncFailed(),
                context,this);
            break;
        }
    }
```

3. 用回调机制实现用“异步”模式生产消息

（1）定义回调接口和消息发送结果接口的约束条件。

在 RocketMQ 中，用回调接口 SendCallback 来定义回调的约束条件，具体代码如下所示：

```
public interface SendCallback {
    //①异步发送成功之后的回调约束条件
void onSuccess(final SendResult sendResult);
    //②异步发送失败之后的回调约束条件
    void onException(final Throwable e);
}
```

在 RocketMQ 中，用枚举类 SendStatus 来定义生产消息的结果，具体代码如下所示：

```
public enum SendStatus {
//①消息发送成功
SEND_OK,
//②消息刷盘超时
FLUSH_DISK_TIMEOUT,
//③SLAVE 节点上的消息刷盘超时
FLUSH_SLAVE_TIMEOUT,
//④SLAVE 节点不可用
```

```
        SLAVE_NOT_AVAILABLE,
    }
```

（2）实现 onSuccess()和 onException()方法。

RocketMQ 为了方便开发人员异步地生产消息，已经在生产消息的客户端类 DefaultMQProducerImpl 的 request()方法中实现了异步消息回调功能，具体代码如下所示：

```
    public void request(Message msg, final RequestCallback requestCallback, long
timeout)
            throws RemotingException, InterruptedException, MQClientException,
MQBrokerException {
        //①记录开始生产消息的时间
    long beginTimestamp = System.currentTimeMillis();
        //②预处理需要生产的消息，并设置分布式唯一 ID
        prepareSendRequest(msg, timeout);
    final String correlationId =
msg.getProperty(MessageConst.PROPERTY_CORRELATION_ID);
        //③构造消息生产请求和响应的异步处理对象，用对应消息的分布式唯一 ID 来保证唯一性
        final RequestResponseFuture requestResponseFuture = new
RequestResponseFuture(correlationId, timeout, requestCallback);
        //④将异步处理对象设置到本地缓存中，方便定时任务异步地通知结果
        RequestFutureTable.getRequestFutureTable().put(correlationId,
requestResponseFuture);
    long cost = System.currentTimeMillis() - beginTimestamp;
        //⑤异步地生产消息
        this.sendDefaultImpl(msg, CommunicationMode.ASYNC, new SendCallback() {
            @Override
            public void onSuccess(SendResult sendResult) {
                //⑥如果生产消息成功，则将异步处理对象中的变量 sendRequestOk 设置为 true
                requestResponseFuture.setSendRequestOk(true);
            }
            @Override
            public void onException(Throwable e) {
                //⑦如果生产消息失败，则设置异常
                requestResponseFuture.setCause(e);
                requestFail(correlationId);
            }
        }, timeout - cost);
    }
```

8.5.2 用“同步”模式生产消息的原理

在 RocketMQ 中，“同步”模式用枚举类型 CommunicationMode.SYNC 来标识。具体代码如下所示：

"同步"模式和"异步"模式的主要区别是：前者没有采用回调机制，而是直接返回消息发送的结果。

```
    private SendResult sendKernelImpl(final Message msg,
            final MessageQueue mq,
            final CommunicationMode communicationMode,
            final SendCallback sendCallback,
            final TopicPublishInfo topicPublishInfo,
            final long timeout) throws MQClientException, RemotingException,
MQBrokerException, InterruptedException {
        ...
        SendResult sendResult = null;
        switch (communicationMode) {
            case SYNC:
                long costTimeSync = System.currentTimeMillis() - beginStartTime;
                if (timeout < costTimeSync) {
                    throw new RemotingTooMuchRequestException("sendKernelImpl
call timeout");
                }
                //用"同步"模式发送消息
                sendResult =
this.mQClientFactory.getMQClientAPIImpl().sendMessage(
                    brokerAddr,
                    mq.getBrokerName(),
                    msg,
                    requestHeader,
                    timeout - costTimeSync,
                    communicationMode,
                    context,
                this);
            break;
        }
    }
```

RocketMQ 用 MQClientAPIImpl 类的 sendMessageSync()方法处理同步消息，具体代码如下所示：

```
    private SendResult sendMessageSync(
            final String addr,
            final String brokerName,
            final Message msg,
            final long timeoutMillis,
            final RemotingCommand request
```

```
    ) throws RemotingException, MQBrokerException, InterruptedException {
        //①调用 NettyRemotingClient 的 invokeSync()方法发送同步消息
        RemotingCommand response = this.remotingClient.invokeSync(addr, request,
timeoutMillis);
    assert response != null;
        //②在返回发送消息之后，Broker Server 响应的结果
        return this.processSendResponse(brokerName, msg, response,addr);
    }
```

8.5.3 用“最多发送一次”模式生产消息的原理

在 RocketMQ 中，“同步”模式用枚举类型 CommunicationMode.ONEWAY 来标识。

所谓“最多发送一次”模式是指，RocketMQ 只能确保消息发送成功一次到 Broker Server，不能确保消息能够成功地被消费者消费。由于最多发送一次，并且不返回消息被消费的结果，所以该模式消息的吞吐量非常高，但是可靠性比“同步”模式和“异步”模式要低。

在 RocketMQ 中，用 NettyRemotingClient 类的 invokeOneway()方法来实现“最多发送一次”模式，具体代码如下所示：

```
    @Override
    public void invokeOneway(String addr, RemotingCommand request, long
timeoutMillis) throws InterruptedException,
    RemotingConnectException, RemotingTooMuchRequestException,
RemotingTimeoutException, RemotingSendRequestException {
    //①用 IP 地址获取对应的 Netty 通信通道
        final Channel channel = this.getAndCreateChannel(addr);
        if (channel != null && channel.isActive()) {
            try {
                doBeforeRpcHook①s(addr, request);
                //②发送消息
                this.invokeOnewayImpl(channel, request, timeoutMillis);
            } catch (RemotingSendRequestException e) {
                log.warn("invokeOneway: send request exception, so close the
channel[{}]", addr);
                this.closeChannel(addr, channel);
                throw e;
            }
        } else {
            this.closeChannel(addr, channel);
            throw new RemotingConnectException(addr);
        }
    }
    //③调用抽象类 NettyRemotingAbstract 的 invokeOnewayImpl()方法发送消息
```

```
    public void invokeOnewayImpl(final Channel channel, final RemotingCommand
request, final long timeoutMillis)
            throws InterruptedException, RemotingTooMuchRequestException,
RemotingTimeoutException, RemotingSendRequestException {
        request.markOnewayRPC();
        //④信号量 semaphoreOneway 默认的容量为 65535 个，可以通过参数
"com.rocketmq.remoting.clientOnewaySemaphoreValue"自定义容量大小
        boolean acquired = this.semaphoreOneway.tryAcquire(timeoutMillis,
TimeUnit.MILLISECONDS);
        //⑤从信号量池中获取信号量成功，acquired 为 true
        if (acquired) {
            //⑥用 SemaphoreReleaseOnlyOnce 类确保当前请求对应的信号量池只释放一次信号量
            final SemaphoreReleaseOnlyOnce once = new
SemaphoreReleaseOnlyOnce(this.semaphoreOneway);
            try {
                //⑦调用 Netty 通道，通知 Broker Server 处理消息发送的请求，并执行数据刷盘
                channel.writeAndFlush(request).addListener(new
ChannelFutureListener() {
                @Override
                public void operationComplete(ChannelFuture f) throws Exception {
                    //⑧直接释放信号量
                    once.release();
                    if (!f.isSuccess()) {
                        log.warn("send a request command to channel <" +
channel.remoteAddress() + "> failed.");
                    }
                }});
            } catch (Exception e) {
                once.release();
                log.warn("write send a request command to channel <" +
channel.remoteAddress() + "> failed.");
                throw new RemotingSendRequestException
(RemotingHelper.parseChannelRemoteAddr(channel), e);
            }
        } else {
            if (timeoutMillis <= 0) {
                //⑨如果远程请求的 TPS 已经超过 Broker 能够处理的阈值（Broker 处理请求的耗
时已经超过设定的阈值），则 Broker 自动降级，抛出异常
                throw new RemotingTooMuchRequestException("invokeOnewayImpl
invoke too fast");
            } else {
                String info = String.format(
                        "invokeOnewayImpl tryAcquire semaphore timeout, %dms,
waiting thread nums: %d semaphoreAsyncValue: %d",
                        timeoutMillis,
```

```
                    this.semaphoreOneway.getQueueLength(),
                    this.semaphoreOneway.availablePermits());
            log.warn(info);
            throw new RemotingTimeoutException(info);
        }
    }
}
```

8.5.4 【实例】在 Spring Cloud Alibaba 项目中生产同步消息和异步消息

本实例的源码在本书配套资源的"chaptereight/use-spring-cloud-alibaba-sync-async"目录下。

本实例使用 Spring Cloud Alibaba RocketMQ 来生产同步消息和异步消息，同步消息需要设置消息发送的超时时间，默认为 3s。

1. 初始化项目

本实例中集成 Spring Cloud Alibaba RocketMQ 的具体依赖如下：

```
<dependency>
    <groupId>com.alibaba.cloud</groupId>
    <artifactId>spring-cloud-starter-stream-rocketmq</artifactId>
</dependency>
```

在 application.properties 文件中，添加 RocketMQ 的配置信息，详细配置信息如下所示：

```
###配置异步通道
spring.cloud.stream.rocketmq.binder.name-server=127.0.0.1:9876
spring.cloud.stream.bindings.output1.destination=use-spring-cloud-alibab
a-async-topic
spring.cloud.stream.bindings.output1.content-type=text/plain
spring.cloud.stream.rocketmq.bindings.output1.producer.group=async-binde
r-group
###配置同步通道
spring.cloud.stream.bindings.output2.destination=use-spring-cloud-alibab
a-sync-topic
spring.cloud.stream.bindings.output2.content-type=text/plain
spring.cloud.stream.rocketmq.bindings.output2.producer.sync=true
spring.cloud.stream.rocketmq.bindings.output2.producer.group=sync-binder
-group
spring.cloud.stream.rocketmq.bindings.output2.producer.sendMessageTimeou
t=10000
```

2. 用同步通道和异步通道生产消息

本实例采用同步（output2）通道和异步通道（output1）分别生产同步消息和异步消息。如果要使用 Spring Cloud Alibaba 生产同步消息，则只需要添加配置信息“spring.cloud.stream.rocketmq.bindings.output2.producer.sync=true”即可，默认生产异步消息。

（1）定义 Spring Cloud Alibaba 的消息通道，具体代码如下所示：

```
public interface SyncAsyncMessageSource {
    //①异步消息通道
    @Output("output1")
MessageChannel output1();
    //②同步消息通道
    @Output("output2")
    MessageChannel output2();
}
```

（2）定义接口 SyncAsyncSendService，发送同步和异步消息，具体代码如下所示：

```
@Service
public class SyncAsyncSendService {
    @Resource
    private SyncAsyncMessageSource syncAsyncMessageSource;

    public void sendSyncMessage(String msg){
        System.out.println("同步发送消息: "+msg);
        //①用同步通道发送同步消息
        syncAsyncMessageSource.output2().send(MessageBuilder.
        withPayload(msg).build());
    }
    public void sendAsyncMessage(String msg){
        System.out.println("异步发送消息: "+msg);
        //②用异步通道发送异步消息
        syncAsyncMessageSource.output1().send(MessageBuilder.
        withPayload(msg).build());
    }
}
```

3. 运行项目，并用代码的 Debug 模式验证同步生产消息和异步生产消息

本实例采用开发人员最熟悉的代码的 Debug 模式来验证同步生产消息和异步生产消息。

（1）使用命令“curl 127.0.0.1:28082/syncAsyncSendMessage/syncSendMessage”开启同步地生产消息。

用 Debug 模式调试同步生产消息的过程如图 8-19 所示。用 Spring Cloud Alibaba RocketMQ 发送同步消息，最终调用 RocketMQTemplate 类的 syncSend()方法发送同步消息。

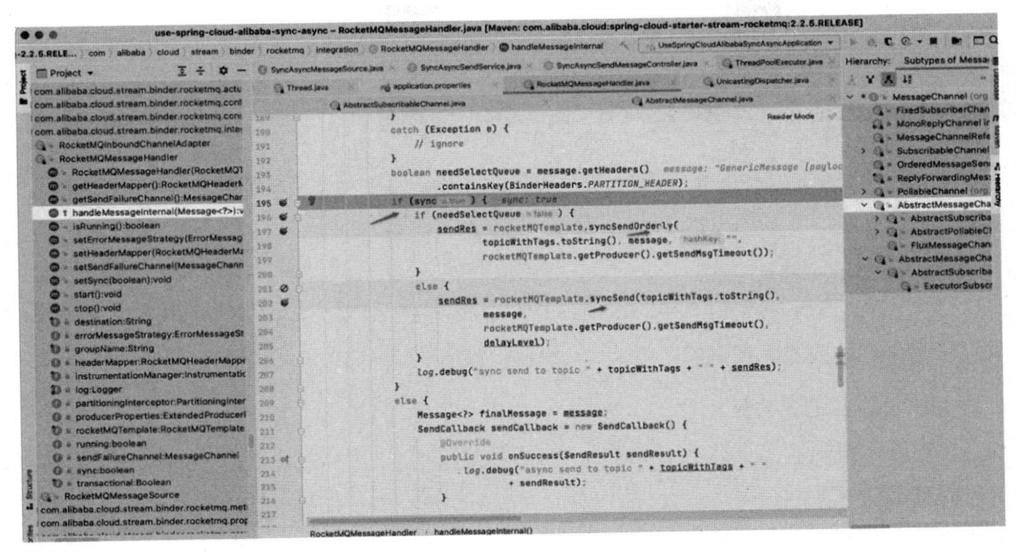

图 8-19

（2）使用命令“curl 127.0.0.1:28082/syncAsyncSendMessage/asyncSendMessage”开启异步的生产消息。

（3）用 Debug 模式调试同步生产消息的过程如图 8-20 所示。采用 Spring Cloud Alibaba RocketMQ 发送同步消息，然后调用 rocketMQTemplate 类的 asyncSend ()方法发送异步消息。

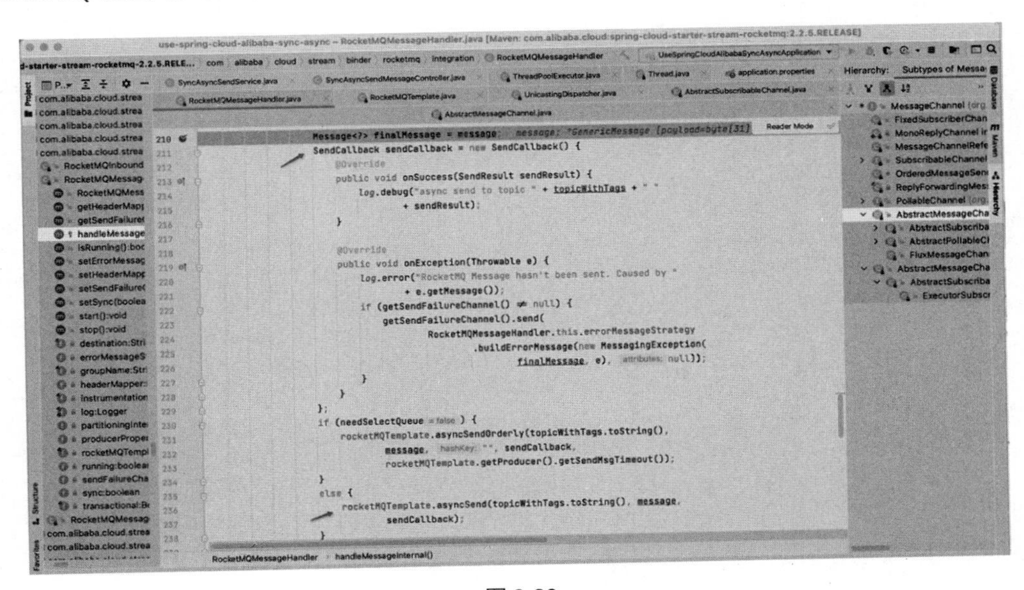

图 8-20

8.6 用 Push 模式和 Pull 模式消费消息

RocketMQ 支持用 Push 模式和 Pull 模式消费消息。

8.6.1 “用 Push 模式消费消息”的原理

在 RocketMQ 中，主要是用 DefaultMQPushConsumer 类实现 Push 模式的消息消费。

下面使用 Push 模式消费消息。

1. 开启 Push 模式并消费消息

开发人员可以在项目中添加 DefaultMQPushConsumer 类，并使用它来启动 Push 模式，具体示例代码实现如下：

```
    public class Consumer {
    public static void main(String[] args) throws InterruptedException,
MQClientException {
            //①消息主题
            String topic ="test-topic";
            //②消息标签
            String subscription ="test-subscription";
            //③消息组
            String group ="test-group";
            //④实例化 DefaultMQPushConsumer 对象
            DefaultMQPushConsumer consumer = new DefaultMQPushConsumer(group);
            //⑤添加消费者对应的 MQClientInstance 的实例名称
   consumer.setInstanceName(Long.toString(System.currentTimeMillis()));
            //⑥订阅消息
            consumer.subscribe(topic, subscription);
            //⑦注册消息监听器
            consumer.registerMessageListener(new MessageListenerConcurrently() {
            AtomicLong consumeTimes = new AtomicLong(0);
            @Override
            public ConsumeConcurrentlyStatus consumeMessage(List<MessageExt>
msgs,
                    ConsumeConcurrentlyContext context) {
                long currentTimes = this.consumeTimes.incrementAndGet();
                System.out.printf("%-8d %s%n", currentTimes, msgs);
                return ConsumeConcurrentlyStatus.CONSUME_SUCCESS;
            }});
            //⑧开启消费者线程，消费消息
            consumer.start();
```

```
        System.out.printf("Consumer Started.%n");
        }
    }
}
```

2. 用 PullMessageService 类实现 Push 模式的消息消费

RocketMQ 在底层用 PullMessageService 类实现 Push 模式的消息消费。

（1）启动 PullMessageService 类。

在 RocketMQ 中，每个 Consumer 在启动过程中都会初始化一个客户端实例类 MQClientInstance。

PullMessageService 类继承 ServiceThread 类，ServiceThread 类是 RocketMQ 统一封装的线程组件，所以，PullMessageService 类本质上是一个线程，它会在 MQClientInstance 类的初始化过程中启动线程并开启消息消费。RocketMQ 用“线程从 Broker Server 实时地拉取消息”来模拟 Push 模式消费消息（从 Broker Server 推送消息到 Consumer），具体代码如下所示：

```
public class MQClientInstance {
    ...
    public void start() throws MQClientException {
        synchronized (this) {
            switch (this.serviceState) {
                case CREATE_JUST:
                    //启动 PullMessageService 类
                    this.pullMessageService.start();
            }
        }
    }
}
```

（2）执行 PullMessageService 类中的 run()方法，开启消费线程

PullMessageService 类将消息消费的入口放在线程的 run()方法中，具体代码如下所示：

```
public class PullMessageService extends ServiceThread {
    ...
    @Override
    public void run() {
        log.info(this.getServiceName() + " service started");
        while (!this.isStopped()) {
            try {
                //①从 Pull 请求队列中获取拉取消息的请求
                PullRequest pullRequest = this.pullRequestQueue.take();
                //②执行拉取消息的请求
                this.pullMessage(pullRequest);
```

```
            } catch (InterruptedException ignored) {
            } catch (Exception e) {
                log.error("Pull Message Service Run Method exception", e);
            }
        }
        log.info(this.getServiceName() + " service end");
    }
    private void pullMessage(final PullRequest pullRequest) {
        //③从客户端实例的本地缓存中获取对应 Group 的消费者
        final MQConsumerInner consumer =
this.mQClientFactory.selectConsumer(pullRequest.getConsumerGroup());
        if (consumer != null) {
            //④如果消费者实例不为空，则将其强制转换为 DefaultMQPushConsumerImpl 对象
            DefaultMQPushConsumerImpl impl = (DefaultMQPushConsumerImpl)
consumer;
            //⑤执行 DefaultMQPushConsumerImpl 类的 pullMessage()方法拉取消息
            impl.pullMessage(pullRequest);
        } else {
            log.warn("No matched consumer for the PullRequest {}, drop it",
pullRequest);
        }
    }
}
```

3. 分析用 DefaultMQPushConsumerImpl 类的 pullMessage()方法拉取消息的过程

用 DefaultMQPushConsumerImpl 类的 pullMessage()方法拉取消息的原理比较复杂，这里将它拆分开，并逐一来分析。

（1）校验当前消费者状态。

RocketMQ 在执行拉取消息之前，需要校验当前消费者的状态，具体代码如下所示：

```
public void pullMessage(final PullRequest pullRequest) {
    //①从拉取消息的请求中获取处理请求的队列
final ProcessQueue processQueue = pullRequest.getProcessQueue();
    //②如果队列对应的 dropped 为 true，则直接返回
    if (processQueue.isDropped()) {
        log.info("the pull request[{}] is dropped.",
pullRequest.toString());
        return;
    }
    pullRequest.getProcessQueue().setLastPullTimestamp
(System.currentTimeMillis());
  try {
    //③验证当前消费者的状态，如果不是运行状态 ServiceState.RUNNING，则直接返回
```

```
            this.makeSureStateOK();
        } catch (MQClientException e) {
            log.warn("pullMessage exception, consumer state not ok", e);
            this.executePullRequestLater(pullRequest,
pullTimeDelayMillsWhenException);
            return;
    }
        //④如果是暂停消费者的消费进度（pause 被设置为 true），则直接返回
        if (this.isPause()) {
            log.warn("consumer was paused, execute pull request later.
instanceName={}, group={}", this.defaultMQPushConsumer.getInstanceName(),
this.defaultMQPushConsumer.getConsumerGroup());
            this.executePullRequestLater(pullRequest,
PULL_TIME_DELAY_MILLS_WHEN_SUSPEND);
            return;
    }
    //⑤获取处理队列中的消息总数
    long cachedMessageCount = processQueue.getMsgCount().get();
    //⑥获取请求队列中缓存消息的大小，单位为 MB
    long cachedMessageSizeInMiB = processQueue.getMsgSize().get() / (1024 *
1024);
    //⑦如果消息总数大于当前消费者设置的初始阈值，则延迟执行当前拉取请求，默认延迟 50ms
        if (cachedMessageCount >
this.defaultMQPushConsumer.getPullThresholdForQueue()) {
            this.executePullRequestLater(pullRequest,
PULL_TIME_DELAY_MILLS_WHEN_FLOW_CONTROL);
            if ((queueFlowControlTimes++ % 1000) == 0) {
                log.warn("the cached message count exceeds the threshold {}, so
do flow control, minOffset={}, maxOffset={}, count={}, size={} MiB,
pullRequest={},
flowControlTimes={}",this.defaultMQPushConsumer.getPullThresholdForQueue(),
processQueue.getMsgTreeMap().firstKey(),
processQueue.getMsgTreeMap().lastKey(), cachedMessageCount,
cachedMessageSizeInMiB, pullRequest, queueFlowControlTimes);
            }
            return;
    }
    //⑧如果“请求队列中缓存消息的大小”大于“当前消费者设置的初始阈值”，则延迟执行当前拉
取请求，默认延迟 50ms
        if (cachedMessageSizeInMiB >
this.defaultMQPushConsumer.getPullThresholdSizeForQueue()) {
            this.executePullRequestLater(pullRequest,
PULL_TIME_DELAY_MILLS_WHEN_FLOW_CONTROL);
            if ((queueFlowControlTimes++ % 1000) == 0) {
                log.warn(
```

```
                "the cached message size exceeds the threshold {} MiB, so do
flow control, minOffset={}, maxOffset={}, count={}, size={} MiB, pullRequest={},
flowControlTimes={}",
                this.defaultMQPushConsumer.getPullThresholdSizeForQueue(),
processQueue.getMsgTreeMap().firstKey(),
processQueue.getMsgTreeMap().lastKey(), cachedMessageCount,
cachedMessageSizeInMiB, pullRequest, queueFlowControlTimes);
            }
        return;
    }
    //⑨如果是非顺序的消费消息，若“请求队列中的消息树的最大深度”大于“当前消费者设置的初
始阈值”，则延迟执行当前拉取请求，默认延迟 50ms
      if (!this.consumeOrderly) {
          if (processQueue.getMaxSpan() >
this.defaultMQPushConsumer.getConsumeConcurrentlyMaxSpan()) {
              this.executePullRequestLater(pullRequest,
PULL_TIME_DELAY_MILLS_WHEN_FLOW_CONTROL);
              if ((queueMaxSpanFlowControlTimes++ % 1000) == 0) {
                  log.warn(
                      "the queue's messages, span too long, so do flow control,
minOffset={}, maxOffset={}, maxSpan={}, pullRequest={}, flowControlTimes={}",
                      processQueue.getMsgTreeMap().firstKey(),
processQueue.getMsgTreeMap().lastKey(), processQueue.getMaxSpan(),
                      pullRequest, queueMaxSpanFlowControlTimes);
              }
              return;
          }
    } else {
      //⑩如果请求队列上锁成功，则计算当前 pull 消费请求的起始位置 offset；如果上锁失败，
则延迟执行当前拉取请求，默认延迟时间为 50ms
          if (processQueue.isLocked()) {
              if (!pullRequest.isLockedFirst()) {
                  final long offset =
this.rebalanceImpl.computePullFromWhere(pullRequest.getMessageQueue());
                  boolean brokerBusy = offset < pullRequest.getNextOffset();
                  log.info("the first time to pull message, so fix offset from
broker. pullRequest: {} NewOffset: {} brokerBusy: {}",
                          pullRequest, offset, brokerBusy);
                  if (brokerBusy) {
                      log.info("[NOTIFYME]the first time to pull message, but pull
request offset larger than broker consume offset. pullRequest: {} NewOffset: {}",
                              pullRequest, offset);
                  }
                  pullRequest.setLockedFirst(true);
                  pullRequest.setNextOffset(offset);
```

```
            }
        } else {
            this.executePullRequestLater(pullRequest,
pullTimeDelayMillsWhenException);
            log.info("pull message later because not locked in broker, {}",
pullRequest);
            return;
        }
    }
    final SubscriptionData subscriptionData = this.rebalanceImpl.
getSubscriptionInner().get(pullRequest.getMessageQueue().getTopic());
    //从负载均衡器获取当前消息主题对应的订阅关系，如果为空，则延迟执行当前拉取请求，默认延
迟 50ms
        if (null == subscriptionData) {
            this.executePullRequestLater(pullRequest,
pullTimeDelayMillsWhenException);
            log.warn("find the consumer's subscription failed, {}", pullRequest);
            return;
        }
    }
```

（2）定义消息回调函数。

在 RocketMQ 中，用 PullCallback 类来定义消息处理后的回调函数，具体代码如下所示：

```
    final long beginTimestamp = System.currentTimeMillis();
    //①定义拉取消息的回调函数
    PullCallback pullCallback = new PullCallback() {
        @Override
        public void onSuccess(PullResult pullResult) {
            if (pullResult != null) {
                //②处理拉取消息的结果
                pullResult =
DefaultMQPushConsumerImpl.this.pullAPIWrapper.processPullResult(pullRequest.
getMessageQueue(), pullResult,subscriptionData);
                switch (pullResult.getPullStatus()) {
                    //③如果有新消息
                    case FOUND:
                        long prevRequestOffset = pullRequest.getNextOffset();
                        pullRequest.setNextOffset
(pullResult.getNextBeginOffset());
                        long pullRT = System.currentTimeMillis() - beginTimestamp;
                        //④则记录拉取消息的 RT 时间
            DefaultMQPushConsumerImpl.this.getConsumerStatsManager().
incPullRT(pullRequest.getConsumerGroup(),pullRequest.getMessageQueue().getTo
pic(), pullRT);
```

```
                    long firstMsgOffset = Long.MAX_VALUE;
                    if (pullResult.getMsgFoundList() == null ||
pullResult.getMsgFoundList().isEmpty()) {
                        //⑤如果拉取消息的列表为空，则重新执行拉取消息的请求
                        DefaultMQPushConsumerImpl.this.
executePullRequestImmediately(pullRequest);
                    } else {
                        firstMsgOffset = pullResult.getMsgFoundList().
get(0).getQueueOffset();
                        DefaultMQPushConsumerImpl.this.
getConsumerStatsManager().incPullTPS(pullRequest.getConsumerGroup(),pullRequ
est.getMessageQueue().getTopic(), pullResult.getMsgFoundList().size());
                        boolean dispatchToConsume =
processQueue.putMessage(pullResult.getMsgFoundList());
                        //⑥将消息列表推送给消费者，消费者提交消费消息的请求
                        DefaultMQPushConsumerImpl.this.
consumeMessageService.submitConsumeRequest(
                                pullResult.getMsgFoundList(),
                                processQueue,
                                pullRequest.getMessageQueue(),
                                dispatchToConsume);
                        //⑦如果当前消费者的消费策略为延迟消息，则执行延迟消费方法
executePullRequestLater()，延迟时间为消费者定义的 pullInterval
                        if
(DefaultMQPushConsumerImpl.this.defaultMQPushConsumer.getPullInterval() > 0)
{
                            DefaultMQPushConsumerImpl.this.
executePullRequestLater(pullRequest,DefaultMQPushConsumerImpl.this.defaultMQ
PushConsumer.getPullInterval());
                        } else {
                            //⑧如果消费者的消费策略为实时消费，则执行实时消费方法
executePullRequestImmediately()
                            DefaultMQPushConsumerImpl.this.
executePullRequestImmediately(pullRequest);
                        }
                    }
                    if (pullResult.getNextBeginOffset() < prevRequestOffset||
firstMsgOffset < prevRequestOffset) {
                        log.warn(
                            "[BUG] pull message result maybe data wrong,
nextBeginOffset: {} firstMsgOffset: {} prevRequestOffset: {}",
                            pullResult.getNextBeginOffset(),firstMsgOffset,
prevRequestOffset);
                    }
                    break;
```

```
                case NO_NEW_MSG:
                case NO_MATCHED_MSG:
                    //⑨如果没有匹配的消息，则拉取下一个 offset 的消息
                    pullRequest.setNextOffset(pullResult.
getNextBeginOffset());
                    DefaultMQPushConsumerImpl.this.
correctTagsOffset(pullRequest);
                    DefaultMQPushConsumerImpl.this.
executePullRequestImmediately(pullRequest);
                    break;
                case OFFSET_ILLEGAL:
                    log.warn("the pull request offset illegal, {} {}",
                    pullRequest.toString(), pullResult.toString());
                    pullRequest.setNextOffset(pullResult.
getNextBeginOffset());
                    pullRequest.getProcessQueue().setDropped(true);
                    //⑩如果拉取消息为非法请求
                    DefaultMQPushConsumerImpl.this.executeTaskLater(new
Runnable() {
                        @Override
                        public void run() {
                            try {
                                //则用定时器每隔 10s 更新消费进度 offset 并删除处理队列
                                DefaultMQPushConsumerImpl.this.offsetStore.
updateOffset(pullRequest.getMessageQueue(),pullRequest.getNextOffset(),
false);
                                DefaultMQPushConsumerImpl.this.offsetStore.
persist(pullRequest.getMessageQueue());
                                DefaultMQPushConsumerImpl.this.rebalanceImpl.
removeProcessQueue(pullRequest.getMessageQueue());
                                log.warn("fix the pull request offset, {}",
pullRequest);
                            } catch (Throwable e) {
                                log.error("executeTaskLater Exception", e);
                            }
                        }
                    }, 10000);
                    break;
                default:
                    break;
            }
        }
    }

    @Override
```

```
        public void onException(Throwable e) {
            if (!pullRequest.getMessageQueue().getTopic().startsWith
(MixAll.RETRY_GROUP_TOPIC_PREFIX)) {
                log.warn("execute the pull request exception", e);
            }
            //如果出现异常，则延迟一定的时间之后再执行当前拉取消息的请求
            DefaultMQPushConsumerImpl.this.executePullRequestLater(pullRequest,
pullTimeDelayMillsWhenException);
        }
    };
```

（3）开始消费消息。

RocketMQ 在校验完消费者状态，以及定义完成消费结果的回调函数之后，就开始消费消息，具体代码如下所示：

```
    boolean commitOffsetEnable = false;
    long commitOffsetValue = 0L;
    //①如果是集群消息模式，则需要从 Broker Server 读取全局的消费进度 offset
    if (MessageModel.CLUSTERING ==
this.defaultMQPushConsumer.getMessageModel()) {
        commitOffsetValue =
this.offsetStore.readOffset(pullRequest.getMessageQueue(),
ReadOffsetType.READ_FROM_MEMORY);
        if (commitOffsetValue > 0) {
            commitOffsetEnable = true;
        }
    }
    String subExpression = null;
    boolean classFilter = false;
    //②读取消费者的订阅信息
    SubscriptionData sd =
this.rebalanceImpl.getSubscriptionInner().get(pullRequest.getMessageQueue().
getTopic());
    if (sd != null) {
        if (this.defaultMQPushConsumer.isPostSubscriptionWhenPull()
&& !sd.isClassFilterMode()) {
            subExpression = sd.getSubString();
        }
        classFilter = sd.isClassFilterMode();
    }
    int sysFlag = PullSysFlag.buildSysFlag(commitOffsetEnable,
true,subExpression != null, classFilter);
    try {
        //③调用 PullAPIWrapper 类的 pullKernelImpl()方法开始消费消息
        this.pullAPIWrapper.pullKernelImpl(
```

```
            pullRequest.getMessageQueue(),
            subExpression,
            subscriptionData.getExpressionType(),
            subscriptionData.getSubVersion(),
            pullRequest.getNextOffset(),
            this.defaultMQPushConsumer.getPullBatchSize(),
            sysFlag,
            commitOffsetValue,
            BROKER_SUSPEND_MAX_TIME_MILLIS,
            CONSUMER_TIMEOUT_MILLIS_WHEN_SUSPEND,
            CommunicationMode.ASYNC,
            pullCallback);
    } catch (Exception e) {
        log.error("pullKernelImpl exception", e);
        this.executePullRequestLater(pullRequest,
pullTimeDelayMillsWhenException);
    }
```

（4）执行 PullAPIWrapper 类的 pullKernelImpl()方法消费消息。

RocketMQ 用 PullAPIWrapper 类的 pullKernelImpl()方法消费消息，具体代码如下所示：

```
    public PullResult pullKernelImpl(
        final MessageQueue mq,final String subExpression,final String
expressionType,
        final long subVersion,final long offset,final int maxNums,final int
sysFlag,
        final long commitOffset,final long brokerSuspendMaxTimeMillis,final long
timeoutMillis,
        final CommunicationMode communicationMode,final PullCallback
pullCallback
        ) throws MQClientException, RemotingException, MQBrokerException,
InterruptedException {
        //①判断 Broker 节点是否健康。在消息被消费的过程中，Broker 节点会存在不可用的风险
（因为时延性，消费者不能实时地感知 Broker 的健康状态）
        FindBrokerResult findBrokerResult
=this.mQClientFactory.findBrokerAddressInSubscribe(mq.getBrokerName(),
        this.recalculatePullFromWhichNode(mq), false);
        if (null == findBrokerResult) {

this.mQClientFactory.updateTopicRouteInfoFromNameServer(mq.getTopic());
            findBrokerResult
=this.mQClientFactory.findBrokerAddressInSubscribe(mq.getBrokerName(),
            this.recalculatePullFromWhichNode(mq), false);
        }
        if (findBrokerResult != null) {
```

```
        {
            if (!ExpressionType.isTagType(expressionType)
                    && findBrokerResult.getBrokerVersion() <
MQVersion.Version.V4_1_0_SNAPSHOT.ordinal()) {
                    throw new MQClientException("The broker[" +
mq.getBrokerName() + ", "
                        + findBrokerResult.getBrokerVersion() + "] does not
upgrade to support for filter message by " + expressionType, null);
                }
            }
            int sysFlagInner = sysFlag;
            if (findBrokerResult.isSlave()) {
                sysFlagInner = PullSysFlag.clearCommitOffsetFlag(sysFlagInner);
            }
            //②构造拉取消息的请求头 PullMessageRequestHeader
            PullMessageRequestHeader requestHeader = new
PullMessageRequestHeader();
            requestHeader.setConsumerGroup(this.consumerGroup);
            requestHeader.setTopic(mq.getTopic());
            requestHeader.setQueueId(mq.getQueueId());
            requestHeader.setQueueOffset(offset);
            requestHeader.setMaxMsgNums(maxNums);
            requestHeader.setSysFlag(sysFlagInner);
            requestHeader.setCommitOffset(commitOffset);
            requestHeader.setSuspendTimeoutMillis(brokerSuspendMaxTimeMillis);
            requestHeader.setSubscription(subExpression);
            requestHeader.setSubVersion(subVersion);
            requestHeader.setExpressionType(expressionType);
            String brokerAddr = findBrokerResult.getBrokerAddr();
            if (PullSysFlag.hasClassFilterFlag(sysFlagInner)) {
                brokerAddr = computePullFromWhichFilterServer(mq.getTopic(),
brokerAddr);
            }
            //③调用 MQClientAPIImpl 类的 pullMessage()方法从 Broker Server 拉取消息
            PullResult pullResult =
this.mQClientFactory.getMQClientAPIImpl().pullMessage(
                brokerAddr,requestHeader,timeoutMillis,communicationMode,
pullCallback);
            return pullResult;
        }
        throw new MQClientException("The broker[" + mq.getBrokerName() + "] not
exist", null);
    }
```

（5）调用 MQClientAPIImpl 类的 pullMessage()方法，向 Broker Server 发起拉取消息的

RPC 请求。

RocketMQ 会用 MQClientAPIImpl 类的 pullMessage()方法发起 RPC 请求，具体代码如下所示：

```
public PullResult pullMessage(
    final String addr,
    final PullMessageRequestHeader requestHeader,
    final long timeoutMillis,
    final CommunicationMode communicationMode,
    final PullCallback pullCallback)
    throws RemotingException, MQBrokerException, InterruptedException {
    //①构造消费消息的 RPC 请求
    RemotingCommand request =
RemotingCommand.createRequestCommand(RequestCode.PULL_MESSAGE,
requestHeader);
    switch (communicationMode) {
        case ONEWAY:
            assert false;
            return null;
        case ASYNC:
            //②如果是异步拉取消息，则执行 pullMessageAsync()方法
            this.pullMessageAsync(addr, request, timeoutMillis,
pullCallback);
            return null;
        case SYNC:
            //③如果是同步拉取消息，则执行 pullMessageSync()方法
            return this.pullMessageSync(addr, request, timeoutMillis);
        default:
            assert false;
            break;
    }
    return null;
}
```

8.6.2 “用 Pull 模式消费消息”的原理

在 RocketMQ 中，主要是用 DefaultLitePullConsumer 类实现 Pull 模式的消息消费。

1. 开启 Pull 模式并消费消息

在项目中添加 DefaultLitePullConsumer 类，并使用它来启动 Pull 模式并消费消息，具体示例代码实现如下：

```
//①初始化 DefaultLitePullConsumer 类
DefaultLitePullConsumer litePullConsumer = new
```

```
DefaultLitePullConsumer(consumerGroup + System.currentTimeMillis());
    //②设置 Name Server 的 IP 地址
    litePullConsumer.setNamesrvAddr("127.0.0.1:9876");
    //③设置消费模型
    litePullConsumer.setMessageModel(MessageModel.BROADCASTING);
    //④设置订阅消息的主题及 Tag 标签
    litePullConsumer.subscribe(topic, "*");
    //⑤开启 Pull 模式的消息消费
    litePullConsumer.start();
```

2. 设置订阅消息的主题、Tag 标签，以及消息选择器

在 RocketMQ 中，用 DefaultLitePullConsumer 类的 subscribe()方法来设置订阅消息的主题、Tag 标签，以及消息选择器，具体代码如下所示：

```
    //①设置订阅消息的主题及 Tag 标签
    public synchronized void subscribe(String topic, String subExpression)
throws MQClientException {
        try {
            if (topic == null || topic.equals("")) {
                throw new IllegalArgumentException("Topic can not be null or
empty.");
            }
            //②设置订阅类型为 SubscriptionType.SUBSCRIBE
            setSubscriptionType(SubscriptionType.SUBSCRIBE);
            //③构造订阅对象
            SubscriptionData subscriptionData =
FilterAPI.buildSubscriptionData(defaultLitePullConsumer.getConsumerGroup(),
                    topic, subExpression);
            //④向负载均衡器的本地缓存中设置订阅对象
            this.rebalanceImpl.getSubscriptionInner().put(topic,
subscriptionData);
            //⑤添加消息队列监听器
            this.defaultLitePullConsumer.setMessageQueueListener(new
MessageQueueListenerImpl());
            assignedMessageQueue.setRebalanceImpl(this.rebalanceImpl);
            if (serviceState == ServiceState.RUNNING) {
                this.mQClientFactory.sendHeartbeatToAllBrokerWithLock();
                updateTopicSubscribeInfoWhenSubscriptionChanged();
            }
        } catch (Exception e) {
            throw new MQClientException("subscribe exception", e);
        }
    }
    //⑥设置订阅消息的主题和消息选择器
```

```
    public synchronized void subscribe(String topic, MessageSelector
messageSelector) throws MQClientException {
        try {
            if (topic == null || topic.equals("")) {
                throw new IllegalArgumentException("Topic can not be null or
empty.");
            }
            setSubscriptionType(SubscriptionType.SUBSCRIBE);
            if (messageSelector == null) {
                subscribe(topic, SubscriptionData.SUB_ALL);
                return;
            }
            //⑦用消息选择器、消息主题构造订阅对象
            SubscriptionData subscriptionData = FilterAPI.build(topic,
                    messageSelector.getExpression(),
messageSelector.getExpressionType());
            this.rebalanceImpl.getSubscriptionInner().put(topic,
subscriptionData);
            //⑧设置消息队列监听器
            this.defaultLitePullConsumer.setMessageQueueListener(new
MessageQueueListenerImpl());
            assignedMessageQueue.setRebalanceImpl(this.rebalanceImpl);
            if (serviceState == ServiceState.RUNNING) {
                this.mQClientFactory.sendHeartbeatToAllBrokerWithLock();
                updateTopicSubscribeInfoWhenSubscriptionChanged();
            }
        } catch (Exception e) {
            throw new MQClientException("subscribe exception", e);
        }
    }
```

3. 加载用 Pull 模式消费消息所需要的资源

在 RocketMQ 中，用 DefaultLitePullConsumer 类的 start()方法来加载用 Pull 模式消费消息所需要的资源。

（1）执行 DefaultLitePullConsumer 类的 start()方法，具体代码如下所示：

```
    @Override
    public void start() throws MQClientException {
        setConsumerGroup(NamespaceUtil.wrapNamespace(this.getNamespace(),
this.consumerGroup));
        //调用 DefaultLitePullConsumer 的实现类 DefaultLitePullConsumerImpl 的
start()方法加载资源
        this.defaultLitePullConsumerImpl.start();
    }
```

（2）调用 DefaultLitePullConsumerImpl 类的 start()方法加载资源，具体代码如下所示：

```
    public synchronized void start() throws MQClientException {
        switch (this.serviceState) {
            case CREATE_JUST:
                this.serviceState = ServiceState.START_FAILED;
                this.checkConfig();
                if (this.defaultLitePullConsumer.getMessageModel() ==
MessageModel.CLUSTERING) {
                    this.defaultLitePullConsumer.changeInstanceNameToPID();
                }
                //①初始化 RocketMQ 客户端实例
                initMQClientFactory();
                //②加载负载均衡器
                initRebalanceImpl();
                //③加载 PullAPIWrapper 类
                initPullAPIWrapper();
                //④加载 LocalFileOffsetStore 存储类或者 RemoteBrokerOffsetStore 存储类
                initOffsetStore();
                //⑤开启 RocketMQ 客户端实例
                mQClientFactory.start();
                //⑥开启定时任务
                startScheduleTask();
                this.serviceState = ServiceState.RUNNING;
                log.info("the consumer [{}] start OK",
this.defaultLitePullConsumer.getConsumerGroup());
                operateAfterRunning();
                break;
            case RUNNING:
            case START_FAILED:
            case SHUTDOWN_ALREADY:
                throw new MQClientException("The PullConsumer service state not
OK, maybe started once, "
                        + this.serviceState
                        + FAQUrl.suggestTodo(FAQUrl.CLIENT_SERVICE_NOT_OK),
                        null);
            default:
            break;
        }
    }
```

4. 用监听器的 MessageQueueListenerImpl 类添加消息消费任务类 PullTaskImpl

在 RocketMQ 消息消费的 Pull 模式中，每个消费者都会注册一个监听器类 MessageQueueListenerImpl，用于监听 RocketMQ 的消息队列。如果消息队列有变动，则它会添加一个新的消息消费任务类 PullTaskImpl 到本地缓存，并发起消息消费的请求。

（1）用负载均衡器 RebalanceLitePullImpl 类和 RebalanceImpl 类触发监听器。

RocketMQ 用一个线程类 RebalanceService，每隔 20s（默认）进行消息消费的负载均衡处理，最终调用 RebalanceImpl 类的 rebalanceByTopic()方法触发监听器机制，具体代码如下所示：

```
    private void rebalanceByTopic(final String topic, final boolean isOrder) {
        switch (messageModel) {
            case BROADCASTING: {
                Set<MessageQueue> mqSet = this.topicSubscribeInfoTable.
get(topic);
                if (mqSet != null) {
                    boolean changed = this.updateProcessQueueTableInRebalance
(topic, mqSet, isOrder);
                    if (changed) {
                        //①在广播消息模式下，如果消息队列缓存有变更，则通知监听器
                        this.messageQueueChanged(topic, mqSet, mqSet);
                    }
                } else {
                        log.warn("doRebalance, {}, but the topic[{}] not exist.",
consumerGroup, topic);
                }
                break;
            }
            case CLUSTERING: {
                //②如果是集群模式，则从消息主题订阅关系的缓存中获取消息队列列表
                Set<MessageQueue> mqSet = this.topicSubscribeInfoTable.
get(topic);
                //③获取消费者 ID 列表
                List<String> cidAll = this.mQClientFactory.findConsumerIdList
(topic, consumerGroup);
                //④如果消息队列列表为空，则打印“消息主题不存在”的错误信息
                if (null == mqSet) {
                    if (!topic.startsWith(MixAll.RETRY_GROUP_TOPIC_PREFIX)) {
                            log.warn("doRebalance, {}, but the topic[{}] not exist.",
consumerGroup, topic);
                    }
                }
                //⑤如果消费者 ID 列表不为空，则打印“消费者不存在”的错误信息
                if (null == cidAll) {
                    log.warn("doRebalance, {} {}, get consumer id list failed",
consumerGroup, topic);
                }
                if (mqSet != null && cidAll != null) {
                    List<MessageQueue> mqAll = new ArrayList<MessageQueue>();
                    mqAll.addAll(mqSet);
                    Collections.sort(mqAll);
                    Collections.sort(cidAll);
```

```
                //⑥加载 RocketMQ 的负载均衡策略
                AllocateMessageQueueStrategy strategy =
this.allocateMessageQueueStrategy;
                List<MessageQueue> allocateResult = null;
                try {
                    //⑦执行负载均衡
                    allocateResult = strategy.allocate(
                            this.consumerGroup,
                            this.mQClientFactory.getClientId(),
                            mqAll,
                            cidAll);
                } catch (Throwable e) {
                    log.error("AllocateMessageQueueStrategy.allocate
Exception. allocateMessageQueueStrategyName={}", strategy.getName(),
                            e);
                    return;
                }
                Set<MessageQueue> allocateResultSet = new
HashSet<MessageQueue>();
                if (allocateResult != null) {
                    allocateResultSet.addAll(allocateResult);
                }
                //⑧将负载均衡后的结果更新到 RocketMQ 消息处理队列缓存中，这样
RocketMQ 就可以实时地使用最新的负载均衡的结果
                boolean changed =
this.updateProcessQueueTableInRebalance(topic, allocateResultSet, isOrder);
                //⑨如果负载均衡缓存有变更，则通知监听器
                if (changed) {
                    log.info(
                            "rebalanced result changed.
allocateMessageQueueStrategyName={}, group={}, topic={}, clientId={},
mqAllSize={}, cidAllSize={}, rebalanceResultSize={}, rebalanceResultSet={}",
                            strategy.getName(), consumerGroup, topic,
this.mQClientFactory.getClientId(), mqSet.size(), cidAll.size(),
                            allocateResultSet.size(), allocateResultSet);
                    //⑩在集群消息模式下，如果消息队列缓存有变更，则通知监听器
                    this.messageQueueChanged(topic, mqSet,
allocateResultSet);
                }
            }
            break;
        }
        default:
            break;
    }
}
```

（2）执行监听器。

在 RocketMQ 中，用 RebalanceLitePullImpl 类的 messageQueueChanged()方法执行监听器，具体代码如下所示：

```
    @Override
    public void messageQueueChanged(String topic, Set<MessageQueue> mqAll,
Set<MessageQueue> mqDivided) {
        MessageQueueListener messageQueueListener = this.litePullConsumerImpl.
getDefaultLitePullConsumer().getMessageQueueListener();
        if (messageQueueListener != null) {
            try {
                //如果消费者 DefaultLitePullConsumerImpl 存在监听器，则通知监听器消息队
列已经变更
                messageQueueListener.messageQueueChanged(topic, mqAll,
mqDivided);
            } catch (Throwable e) {
                log.error("messageQueueChanged exception", e);
            }
        }
    }
```

监听器在收到通知之后，执行 updatePullTask()方法更新拉取消息任务的本地缓存，并开启拉取消息任务，具体代码如下所示：

```
    //①监听器类 MessageQueueListenerImpl 是消费者 DefaultLitePullConsumerImpl 类的
内部类
    class MessageQueueListenerImpl implements MessageQueueListener {
        @Override
        public void messageQueueChanged(String topic, Set<MessageQueue> mqAll,
Set<MessageQueue> mqDivided) {
            MessageModel messageModel = defaultLitePullConsumer.
getMessageModel();
            switch (messageModel) {
                case BROADCASTING:
                    updateAssignedMessageQueue(topic, mqAll);
                    //②在广播消费模式下，更新拉取消息任务的本地缓存
                    updatePullTask(topic, mqAll);
                    break;
                case CLUSTERING:
                    updateAssignedMessageQueue(topic, mqDivided);
                    //③在集群消费模式下，更新拉取消息任务的本地缓存
                    updatePullTask(topic, mqDivided);
                    break;
                default:
                    break;
```

```
            }
        }
    }
```

（3）定时执行拉取消息的任务。

RocketMQ 执行 DefaultLitePullConsumerImpl 类的 startPullTask()方法以开始定时执行拉取消息的任务，具体代码如下所示：

```
    private void startPullTask(Collection<MessageQueue> mqSet) {
    for (MessageQueue messageQueue : mqSet) {
        //①遍历最新的消息队列列表，如果本地缓存中已经存在消息队列，则说明任务已经在执行中，
直接跳出当前循环，执行下一次循环
            if (!this.taskTable.containsKey(messageQueue)) {
                //②用消息队列构建拉取消息的任务
                PullTaskImpl pullTask = new PullTaskImpl(messageQueue);
                //③将任务设置到本地缓存中
                this.taskTable.put(messageQueue, pullTask);
                //④用定时器定时执行拉取消息的任务
                this.scheduledThreadPoolExecutor.schedule(pullTask, 0,
TimeUnit.MILLISECONDS);
            }
        }
    }
```

5. 从 Broker Server 拉取消息进行消费

PullTaskImpl 类是一个线程，它会通过 run()方法定时地从 Broker Server 拉取消息进行消费。

（1）校验当前消费者状态。

RocketMQ 在拉取消息之前，需要校验 DefaultLitePullConsumerImpl 类的状态，在 run()方法中具体代码如下所示：

```
    //①如果消费者被暂停，则延迟执行消费请求，延迟时间默认为 1s
    if (assignedMessageQueue.isPaused(messageQueue)) {
        scheduledThreadPoolExecutor.schedule(this,
PULL_TIME_DELAY_MILLS_WHEN_PAUSE, TimeUnit.MILLISECONDS);
        log.debug("Message Queue: {} has been paused!", messageQueue);
        return;
    }
    ProcessQueue processQueue =
assignedMessageQueue.getProcessQueue(messageQueue);
    //②如果消息处理队列为空或者队列被挂起，则直接返回
    if (null == processQueue || processQueue.isDropped()) {
```

```
        log.info("The message queue not be able to poll, because it's dropped.
group={}, messageQueue={}", defaultLitePullConsumer.getConsumerGroup(),
this.messageQueue);
        return;
    }
    //③如果“消费者请求缓存数×消息批量拉取数”大于“初始拉取阈值”，则延迟执行当前拉取消
息的请求
    if (consumeRequestCache.size() *
defaultLitePullConsumer.getPullBatchSize() >
defaultLitePullConsumer.getPullThresholdForAll()) {
        scheduledThreadPoolExecutor.schedule(this,
PULL_TIME_DELAY_MILLS_WHEN_FLOW_CONTROL, TimeUnit.MILLISECONDS);
        if ((consumeRequestFlowControlTimes++ % 1000) == 0)
            log.warn("The consume request count exceeds threshold {}, so do flow
control, consume request count={}, flowControlTimes={}",
consumeRequestCache.size(), consumeRequestFlowControlTimes);
            return;
    }
    long cachedMessageCount = processQueue.getMsgCount().get();
    long cachedMessageSizeInMiB = processQueue.getMsgSize().get() / (1024 *
1024);
    //④如果“消息缓存数”大于“队列阈值大小”，则延迟执行当前拉取消息的请求
    if (cachedMessageCount >
defaultLitePullConsumer.getPullThresholdForQueue()) {
        scheduledThreadPoolExecutor.schedule(this,
PULL_TIME_DELAY_MILLS_WHEN_FLOW_CONTROL, TimeUnit.MILLISECONDS);
        if ((queueFlowControlTimes++ % 1000) == 0) {
            log.warn(
            "The cached message count exceeds the threshold {}, so do flow control,
minOffset={}, maxOffset={}, count={}, size={} MiB, flowControlTimes={}",
            defaultLitePullConsumer.getPullThresholdForQueue(),
processQueue.getMsgTreeMap().firstKey(),
processQueue.getMsgTreeMap().lastKey(), cachedMessageCount,
cachedMessageSizeInMiB, queueFlowControlTimes);
        }
        return;
    }
    //⑤如果“消息大小”大于“队列中拉取消息的阈值”，则延迟执行当前拉取消息的请求
    if (cachedMessageSizeInMiB >
defaultLitePullConsumer.getPullThresholdSizeForQueue()) {
        scheduledThreadPoolExecutor.schedule(this,
PULL_TIME_DELAY_MILLS_WHEN_FLOW_CONTROL, TimeUnit.MILLISECONDS);
        if ((queueFlowControlTimes++ % 1000) == 0) {
            log.warn(
```

```
          "The cached message size exceeds the threshold {} MiB, so do flow control,
minOffset={}, maxOffset={}, count={}, size={} MiB, flowControlTimes={}",
          defaultLitePullConsumer.getPullThresholdSizeForQueue(),
processQueue.getMsgTreeMap().firstKey(),
processQueue.getMsgTreeMap().lastKey(), cachedMessageCount,
cachedMessageSizeInMiB, queueFlowControlTimes);
       }
       return;
   }
   //⑥如果“消息处理队列中的消息树的深度”大于“消费消息深度的阈值”，则延迟执行当前消费
消息的请求
   if (processQueue.getMaxSpan() >
defaultLitePullConsumer.getConsumeMaxSpan()) {
       scheduledThreadPoolExecutor.schedule(this,
PULL_TIME_DELAY_MILLS_WHEN_FLOW_CONTROL, TimeUnit.MILLISECONDS);
       if ((queueMaxSpanFlowControlTimes++ % 1000) == 0) {
          log.warn("The queue's messages, span too long, so do flow control,
minOffset={}, maxOffset={}, maxSpan={}, flowControlTimes={}",
          processQueue.getMsgTreeMap().firstKey(),
processQueue.getMsgTreeMap().lastKey(), processQueue.getMaxSpan(),
queueMaxSpanFlowControlTimes);
       }
       return;
   }
```

（2）开始消费消息。

在消费者状态校验通过之后，RocketMQ 就开始消费消息，具体代码如下所示：

```
   long offset = nextPullOffset(messageQueue);
   long pullDelayTimeMills = 0;
   try {
       //①构建订阅关系对象
       SubscriptionData subscriptionData;
       if (subscriptionType == SubscriptionType.SUBSCRIBE) {
          String topic = this.messageQueue.getTopic();
          subscriptionData = rebalanceImpl.getSubscriptionInner().get(topic);
       } else {
          String topic = this.messageQueue.getTopic();
          subscriptionData =
FilterAPI.buildSubscriptionData(defaultLitePullConsumer.getConsumerGroup(),
                          topic, SubscriptionData.SUB_ALL);
   }
   //②执行拉取消息的请求，并返回拉取结果
       PullResult pullResult = pull(messageQueue, subscriptionData, offset,
defaultLitePullConsumer.getPullBatchSize());
```

```
        switch (pullResult.getPullStatus()) {
            case FOUND:
                final Object objLock =
messageQueueLock.fetchLockObject(messageQueue);
                synchronized (objLock) {
                    if (pullResult.getMsgFoundList() != null
&& !pullResult.getMsgFoundList().isEmpty() &&
assignedMessageQueue.getSeekOffset(messageQueue) == -1) {
                        //③将拉取的消息存储在消息处理队列中
                        processQueue.putMessage(pullResult.getMsgFoundList());
                        //④提交消费消息的请求，异步地消费消息
                        submitConsumeRequest(new
ConsumeRequest(pullResult.getMsgFoundList(), messageQueue, processQueue));
                    }
                }
                break;
            case OFFSET_ILLEGAL:
                log.warn("The pull request offset illegal, {}",
pullResult.toString());
                break;
            default:
                break;
    }
    //⑤如果消费拉取是合法的，则更新 offset
        updatePullOffset(messageQueue, pullResult.getNextBeginOffset());
    } catch (Throwable e) {
        pullDelayTimeMills = pullTimeDelayMillsWhenException;
        log.error("An error occurred in pull message process.", e);
    }
    if (!this.isCancelled()) {
        //⑥如果取消消费消息的请求失败，则重新定时执行当前消费消息的请求
        scheduledThreadPoolExecutor.schedule(this, pullDelayTimeMills,
TimeUnit.MILLISECONDS);
    } else {
        log.warn("The Pull Task is cancelled after doPullTask, {}", messageQueue);
    }
```

8.6.3 【实例】生产者生产消息，消费者用 Pull 模式和 Push 模式消费消息

本实例的源码在本书配套资源的“chaptereight/ use-spring-cloud-alibaba-pull-consume”目录下。

Spring Cloud Alibaba RocketMQ 封装了 Spring Cloud Stream 组件。熟悉 Spring Cloud

Stream 的开发人员应该对 Spring Integration 项目非常熟悉，它是一种企业服务总线 ESB（Enterprise Service Bus）。在 Spring Integration 中，消费消息的通道被抽象成两种模式：PollableChannel 和 SubscribableChannel。

Spring Cloud Stream 为了提升开发者使用消息中间件的用户体验，将 Spring Integration 消费消息的 PollableChannel（轮询模式）和 SubscribableChannel（订阅模式）抽象为 Sink 模式，将消息生产抽象为 Source 模式。

1. 初始化项目

（1）用 Spring Cloud Alibaba 初始化一个项目，并添加 RocketMQ 相关的依赖。

```
<dependency>
    <groupId>com.alibaba.cloud</groupId>
    <artifactId>spring-cloud-starter-stream-rocketmq</artifactId>
</dependency>
```

（2）在工程 pull-message-produce 中，使用 Spring Cloud Alibaba RocketMQ 生产消息，具体代码如下所示：

```
//定义生产消息的 PullMessageSource 类，包含两个通道——output1 和 output2
public interface PullMessageSource {
    @Output("output1")
    MessageChannel output1();
    @Output("output2")
    MessageChannel output2();
}
@Service
public class PullMessageSendService {
    @Autowired
    private PullMessageSource pullMessageSource;
    public void sendMessage(String msg){
     //①使用通道 output1 生产消息
    pullMessageSource.output1().send(MessageBuilder
        .withPayload(msg).build());
    //②使用通道 output2 生产消息
    pullMessageSource.output2().send(MessageBuilder
        .withPayload(msg).build());
    }
}
@RestController
@RequestMapping(value = "pullmessage")
public class PullMessageController {
    @Autowired
    private PullMessageSendService pullMessageSendService;
```

```
    @GetMapping(value = "/sendMessage")
    public String sendMessage(){
        ExecutorService executorService= Executors.newFixedThreadPool(1);
        executorService.execute(new SendMessage());
        return "true";
    }
    class SendMessage implements Runnable{
        @Override
        public void run() {
            while (true){
                //③用线程触发 PullMessageSendService 类的 sendMessage()方法发送消息
                pullMessageSendService.sendMessage(RandomUtils.nextLong()+"
的消息");
                try{
                    Thread.sleep(3000);
                }catch (InterruptedException e){
                    System.out.println(e.getMessage());
                }
            }
        }
    }
}
```

（3）在工程 pull-message-consume 中，用 Spring Cloud Alibaba RocketMQ 的 Pull 和 Push 模式消费消息，具体代码如下所示：

```
public interface PullMessageConsumeSink {
    //①定义发布订阅模式（Push 模式）的消息通道
    @Input("input1")
SubscribableChannel input1();
//②定义轮询模式（Pull 模式）的消息通道
    @Input("input2")
    PollableMessageSource input2();
}
@Service
public class PushMessageReceiveService {
    //③用 Push 模式（发布订阅模式）消费消息，Broker Server 主动推送消息给消息消费者
    @StreamListener("input1")
    public void receiveInput1(String receiveMsg) {
        System.out.println("input1 receive: " + receiveMsg);
    }
}
//④定义 pull 模式的 PullLongPollMessageReceiveService 类去消费消息
@Service
public class PullLongPollMessageReceiveService {
    @Resource
```

```
    private PullMessageConsumeSink pullMessageConsumeSink;
  public void pollMessage(){
    //⑤定义线程池
      ExecutorService executorService= Executors.newFixedThreadPool(2);
      executorService.execute(new PollMessage());
    }
    class PollMessage implements Runnable{
      @Override
      public void run() {
        while (true){
          //⑥通过线程触发长轮询并消费消息，从消费者向Broker Server拉取消息
          pullMessageConsumeSink.input2().poll(m -> {
            String payload = (String) m.getPayload();
            System.out.println("pull msg: " + payload);
          }, new ParameterizedTypeReference<String>() {
          });
          try {
            Thread.sleep(5000);
          }catch (InterruptedException e){
            System.out.println(e.getMessage());
          }
        }
      }
    }
  }
  //⑦用RESTful API模拟长轮询的消费消息
  @RestController
  @RequestMapping(value = "/pullLongPollMessage")
  public class PullLongPollMessageController {

    @Resource
    private PullLongPollMessageReceiveService
pullLongPollMessageReceiveService;

    @GetMapping(value = "/poll")
  public String pollMessage(){
    //⑧调用PullLongPollMessageReceiveService类的pollMessage()方法长轮询消息
      pullLongPollMessageReceiveService.pollMessage();
      return "true";
    }
  }
```

2. 运行项目

分别运行项目 pull-message-produce 和 pull-message-consume。

（1）运行项目 pull-message-produce，在 UI 控制台中可以看到生产者信息。

调用 RESTful 接口并生产消息，具体命令如下：

```
curl 127.0.0.1:28082/pullmessage/sendMessage
```

如图 8-21 所示，pull-message-produce 已经在生产消息。

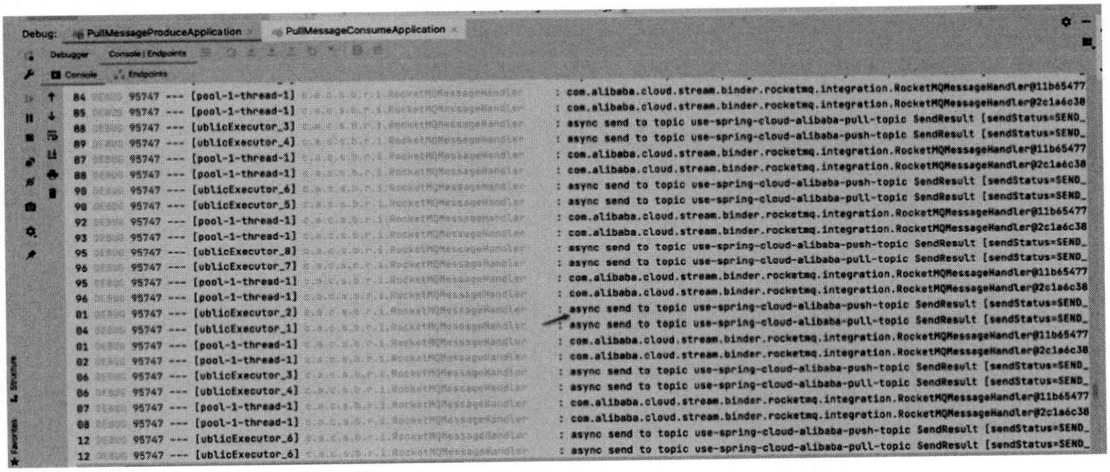

图 8-21

（2）运行项目 pull-message-consume，在 UI 控制台中可以看到消费者信息。

本实例默认开启 Push 模式去消费消息。当开发人员使用注解@StreamListener 监听 SubscribableChannel 模式的消息时，RocketMQ 会采用 Push 模式消费消息。

如图 8-22 所示，在实例运行的过程中，通过 Debug 调试，发现在 SubscribableChannel 模式下，RocketMQ 最终会调用 DefaultMQPushConsumerImpl 类来消费消息。

调用 RESTful 接口，开启 Pull 模式并消费消息，具体命令如下：

```
curl 127.0.0.1:28083/pullLongPollMessage/poll
```

如图 8-23 所示，RocketMQ 采用 Pull 模式消费消息。

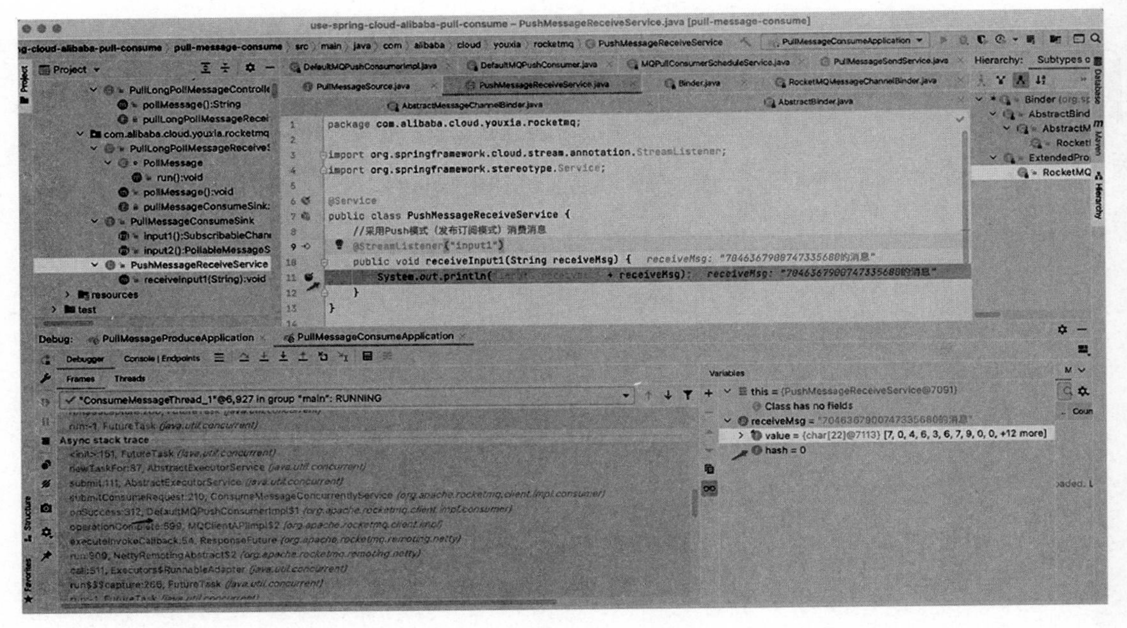
use-spring-cloud-alibaba-pull-consume – PushMessageReceiveService.java [pull-message-consume]
package com.alibaba.cloud.youxia.rocketmq;
import org.springframework.cloud.stream.annotation.StreamListener;
import org.springframework.stereotype.Service;
@Service
public class PushMessageReceiveService {
//采用Push模式（发布订阅模式）消费消息
@StreamListener("input1")
public void receiveInput1(String receiveMsg) {
System.out.println(
}
}
Debug: PullMessageProduceApplication PullMessageConsumeApplication
"ConsumeMessageThread_1"@6,927 in group "main": RUNNING
Async stack trace
Variables
this = {PushMessageReceiveService@7091}
Class has no fields
hash = 0

图 8-22

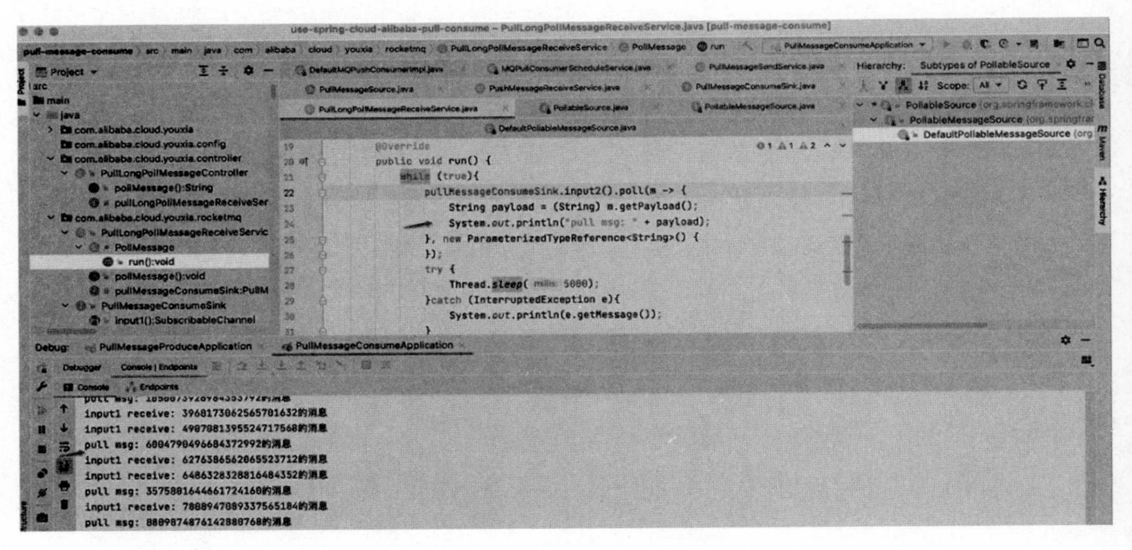
use-spring-cloud-alibaba-pull-consume – PullLongPollMessageReceiveService.java [pull-message-consume]
@Override
public void run() {
while (true){
pullMessageConsumeSink.input2().poll(m -> {
String payload = (String) m.getPayload();
System.out.println("pull msg: " + payload);
}, new ParameterizedTypeReference<String>() {
});
try {
Thread.sleep(millis: 5000);
}catch (InterruptedException e){
System.out.println(e.getMessage());
Debug: PullMessageProduceApplication PullMessageConsumeApplication
input1 receive: 3968173062565701632的消息
input1 receive: 4987081395524717568的消息
pull msg: 6004790496684372992的消息
input1 receive: 6276386562065523712的消息
input1 receive: 6486328328816484352的消息
pull msg: 3575801644661724160的消息
input1 receive: 7888947089337565184的消息
pull msg: 8809874876142880768的消息

图 8-23

8.7 用两阶段提交和定时回查事务状态实现事务消息

RocketMQ 用两阶段提交和定时回查事务状态实现了分布式事务消息，下面来分析其原理。

8.7.1 什么是事务消息

在搞清楚什么是事务消息之前，我们先来看看什么是消息。从广义上来看，消息就是数据，发送端发送消息，接收端解析消息并处理消息。消息可以分为很多种类，比如图片、文字、语音和视频等数据都是消息。

RocketMQ 在普通消息定义的基础上，对事务消息扩展了两个相关的概念：半消息（预处理消息）和消息状态回查。

1. Half（Prepare）Message：**半消息（预处理消息）**

半消息是一种特殊的消息类型，该类型的消息暂时不能被 Consumer 消费。当一条事务消息被成功投递到 Broker 上，但是 Broker 并没有接收到 Producer 发出的二次确认时，该事务消息就处于“暂时不可被消费”状态，该状态的事务消息被称为半消息。

2. Message Status Check：**消息状态回查**

网络抖动、Producer 重启等原因，可能会导致 Producer 向 Broker 发送的二次确认消息没有成功送达。如果 Broker 检测到某条事务消息长时间处于半消息状态，则会主动向 Producer 发起回查操作，查询该事务消息在 Producer 的事务状态（Commit 或 Rollback）。可以看出，Message Status Check 主要用来解决分布式事务中的超时问题。

8.7.2 两阶段提交的原理

在分析原理之前，可以先看看开发人员是如何发送一条事务消息的，具体代码实例如下：

```
    public static void main(String[] args) throws MQClientException,
UnsupportedEncodingException {
        //①构造一个事务监听器对象
        final TransactionListener transactionCheckListener = new
TransactionListenerImpl(statsBenchmark, config);
        //②构造事务消息的生产者
        final TransactionMQProducer producer =new
TransactionMQProducer("benchmark_transaction_producer", config.aclEnable ?
AclClient.getAclRPCHook():null);
        producer.setInstanceName(Long.toString(System.currentTimeMillis()));
        producer.setTransactionListener(transactionCheckListener);
```

```
        producer.setDefaultTopicQueueNums(1000);
        if (commandLine.hasOption('n')) {
            String ns = commandLine.getOptionValue('n');
            //③设置 Name Server 的 IP 地址
            producer.setNamesrvAddr(ns);
    }
    //④启动事务消息的生产者
        producer.start();
    try {
          //⑤发送事务消息
             SendResult sendResult =producer.sendMessageInTransaction
    (buildMessage(config), null);

success=sendResult!=null&&sendResult.getSendStatus()==SendStatus.SEND_OK;
        } catch (Throwable e) {
            success = false;
        }
    }
```

从事务消息实例中可以看出，RocketMQ 使用 TransactionMQProducer 类的 sendMessageInTransaction()方法来生产事务消息。RocketMQ 生产事务消息主要分为两阶段，下面来分析两阶段的原理。

1. 第一阶段

（1）TransactionMQProducer 类的 sendMessageInTransaction()方法是事务消息的起点，具体代码如下所示：

```
    public TransactionSendResult sendMessageInTransaction(final Message msg,
            final LocalTransactionExecuter localTransactionExecuter, final
Object arg)throws MQClientException {
        //①获取事务监听器
        TransactionListener transactionListener = getCheckListener();
        if (null == localTransactionExecuter && null == transactionListener) {
            throw new MQClientException("tranExecutor is null", null);
        }
        if (msg.getDelayTimeLevel() != 0) {

MessageAccessor.clearProperty(msg,MessageConst.PROPERTY_DELAY_TIME_LEVEL);
    }
    //②校验消息
        Validators.checkMessage(msg, this.defaultMQProducer);
    SendResult sendResult = null;
    //③设置事务消息属性值为 true
        MessageAccessor.putProperty(msg,
MessageConst.PROPERTY_TRANSACTION_PREPARED, "true");
```

```
        MessageAccessor.putProperty(msg, MessageConst.PROPERTY_PRODUCER_GROUP,
this.defaultMQProducer.getProducerGroup());
    try {
        //④第一阶段发送事务消息
            sendResult = this.send(msg);
        } catch (Exception e) {
            throw new MQClientException("send message Exception", e);
    }
    ...
    }
```

（2）用 SendMessageProcessor 类的 asyncSendMessage()方法异步处理事务消息，具体代码如下所示：

```
    private CompletableFuture<RemotingCommand>  asyncSendMessage(
        ChannelHandlerContext ctx, RemotingCommand request,
    SendMessageContext mqtraceContext,SendMessageRequestHeader requestHeader) {
    //①消息发送预处理，构造发送消息结果的响应头 SendMessageResponseHeader
        final RemotingCommand response = preSend(ctx, request, requestHeader);
        final SendMessageResponseHeader responseHeader =
(SendMessageResponseHeader)response.readCustomHeader();
        if (response.getCode() != -1) {
            return CompletableFuture.completedFuture(response);
        }
        final byte[] body = request.getBody();
        int queueIdInt = requestHeader.getQueueId();
        TopicConfig topicConfig =
this.brokerController.getTopicConfigManager().selectTopicConfig(requestHeade
r.getTopic());
        if (queueIdInt < 0) {
            queueIdInt = randomQueueId(topicConfig.getWriteQueueNums());
    }
    //②构造发送消息体 MessageExtBrokerInner
        MessageExtBrokerInner msgInner = new MessageExtBrokerInner();
        msgInner.setTopic(requestHeader.getTopic());
        msgInner.setQueueId(queueIdInt);
        if (!handleRetryAndDLQ(requestHeader, response, request, msgInner,
topicConfig)) {
            return CompletableFuture.completedFuture(response);
        }
        msgInner.setBody(body);
        msgInner.setFlag(requestHeader.getFlag());
        MessageAccessor.setProperties(msgInner,
MessageDecoder.string2messageProperties(requestHeader.getProperties()));
        msgInner.setPropertiesString(requestHeader.getProperties());
        msgInner.setBornTimestamp(requestHeader.getBornTimestamp());
```

```
        msgInner.setBornHost(ctx.channel().remoteAddress());
        msgInner.setStoreHost(this.getStoreHost());
        msgInner.setReconsumeTimes(requestHeader.getReconsumeTimes() == null ?
0 : requestHeader.getReconsumeTimes());
        String clusterName =
this.brokerController.getBrokerConfig().getBrokerClusterName();
        MessageAccessor.putProperty(msgInner, MessageConst.PROPERTY_CLUSTER,
clusterName);
msgInner.setPropertiesString(MessageDecoder.messageProperties2String(msgInne
r.getProperties()));
        CompletableFuture<PutMessageResult> putMessageResult = null;
        Map<String, String> origProps =
MessageDecoder.string2messageProperties(requestHeader.getProperties());
        String transFlag =
origProps.get(MessageConst.PROPERTY_TRANSACTION_PREPARED);
        if (transFlag != null && Boolean.parseBoolean(transFlag)) {
            if
(this.brokerController.getBrokerConfig().isRejectTransactionMessage()) {
                response.setCode(ResponseCode.NO_PERMISSION);
                response.setRemark(
                        "the broker[" + this.brokerController.
                                        getBrokerConfig().
                 getBrokerIP1()+ "] sending transaction message is forbidden");
                return CompletableFuture.completedFuture(response);
            }
            //③如果当前消息类型是事务消息，则执行 TransactionalMessageServiceImpl 类
的 asyncPrepareMessage()方法，发送 RocketMQ 事务消息的第一阶段的半消息（预处理消息）
            putMessageResult= this.brokerController.
                    getTransactionalMessageService().
                    asyncPrepareMessage(msgInner);
        } else {
            putMessageResult =
this.brokerController.getMessageStore().asyncPutMessage(msgInner);
    }
        //④在消息发送成功之后，将发送结果返给事务消息生产者
        return handlePutMessageResultFuture(putMessageResult, response, request,
msgInner, responseHeader, mqtraceContext, ctx, queueIdInt);
    }
```

（3）在 BrokerController 类中，RocketMQ 使用 SPI 机制加载事务消息服务 TransactionalMessageService 接口，如果开发人员没有自定义事务消息服务，则 RocketMQ 默认使用 TransactionalMessageServiceImpl 类，具体代码如下所示：

```
    public class BrokerController {
        private TransactionalMessageCheckService
```

```
transactionalMessageCheckService;
      private TransactionalMessageService transactionalMessageService;
      private AbstractTransactionalMessageCheckListener
transactionalMessageCheckListener;
   private void initialTransaction() {
      //①加载开发人员自定义的事务消息服务 TransactionalMessageService 接口的实现类
         this.transactionalMessageService = ServiceProvider.loadClass(
          ServiceProvider.TRANSACTION_SERVICE_ID,
TransactionalMessageService.class);
         if (null == this.transactionalMessageService) {
         //②如果没有自定义事务消息服务，则默认为 TransactionalMessageServiceImpl
            this.transactionalMessageService = new
TransactionalMessageServiceImpl(new TransactionalMessageBridge(this,
this.getMessageStore()));
            log.warn("Load default transaction message hook service: {}",
TransactionalMessageServiceImpl.class.getSimpleName());
         }
    //③加载开发人员自定义的事务消息
   校验监听器 AbstractTransactionalMessageCheckListener 的实现类
         this.transactionalMessageCheckListener = ServiceProvider.loadClass(
   ServiceProvider.TRANSACTION_LISTENER_ID,
AbstractTransactionalMessageCheckListener.class);
         if (null == this.transactionalMessageCheckListener) {
   //④如果开发人员没有自定义事务消息校验监听器，则默认是
DefaultTransactionalMessageCheckListener 类
            this.transactionalMessageCheckListener = new
DefaultTransactionalMessageCheckListener();
            log.warn("Load default discard message hook service: {}",
DefaultTransactionalMessageCheckListener.class.getSimpleName());
         }
         this.transactionalMessageCheckListener.setBrokerController(this);
         //⑤则初始化事务消息校验服务 TransactionalMessageCheckService
         this.transactionalMessageCheckService = new
TransactionalMessageCheckService(this);
      }
   }
```

（4）调用 TransactionalMessageBridge 类的 putHalfMessage()方法发送预处理消息，具体代码如下所示：

```
   public PutMessageResult putHalfMessage(MessageExtBrokerInner messageInner) {
      return store.putMessage(parseHalfMessageInner(messageInner));
   }
   private MessageExtBrokerInner parseHalfMessageInner(MessageExtBrokerInner
msgInner) {
   //①备份当前消息的主题
```

```
    MessageAccessor.putProperty(msgInner, MessageConst.PROPERTY_REAL_TOPIC,
msgInner.getTopic());
    //②备份当前消息的消息队列
        MessageAccessor.putProperty(msgInner,
MessageConst.PROPERTY_REAL_QUEUE_ID,
                String.valueOf(msgInner.getQueueId()));
        msgInner.setSysFlag(
                MessageSysFlag.resetTransactionValue(msgInner.getSysFlag(),
MessageSysFlag.TRANSACTION_NOT_TYPE));
        //③重新设置消息的主题为 RMQ_SYS_TRANS_HALF_TOPIC
        msgInner.setTopic(TransactionalMessageUtil.buildHalfTopic());
         //④重新设置消息的消息队列
         msgInner.setQueueId(0);
    msgInner.setPropertiesString(MessageDecoder.messageProperties2String(msg
Inner.getProperties()));
        return msgInner;
    }
```

2. 第二阶段

在第一阶段事务消息发送成功之后，RocketMQ 会在第二阶段处理本地事务，从而决定回滚或者提交事务消息。

（1）在 DefaultMQProducerImpl 类的 sendMessageInTransaction()方法中处理第二阶段事务，并从事务监听器中获取应用本地事务执行的状态，具体代码如下所示：

```
    LocalTransactionState localTransactionState =
LocalTransactionState.UNKNOW;
    Throwable localException = null;
    switch (sendResult.getSendStatus()) {
        case SEND_OK: {
            try {
                if (sendResult.getTransactionId() != null) {
                    msg.putUserProperty("__transactionId__",
sendResult.getTransactionId());
                }
                String transactionId =
msg.getProperty(MessageConst.PROPERTY_UNIQ_CLIENT_MESSAGE_ID_KEYIDX);
                if (null != transactionId && !"".equals(transactionId)) {
                    msg.setTransactionId(transactionId);
                }
                //①如果事务消息发送成功，则构建本地事务状态类 LocalTransactionState
                if (null != localTransactionExecuter) {
                    localTransactionState =
localTransactionExecuter.executeLocalTransactionBranch(msg, arg);
                } else if (transactionListener != null) {
```

```
                log.debug("Used new transaction API");
                //②从监听器中获取本地事务执行的状态
                localTransactionState =
transactionListener.executeLocalTransaction(msg, arg);
            }
            if (null == localTransactionState) {
                localTransactionState = LocalTransactionState.UNKNOW;
            }
            if (localTransactionState !=
LocalTransactionState.COMMIT_MESSAGE) {
                log.info("executeLocalTransactionBranch return {}",
localTransactionState);
                log.info(msg.toString());
            }
        } catch (Throwable e) {
            log.info("executeLocalTransactionBranch exception", e);
            log.info(msg.toString());
            localException = e;
        }
    }
        break;
    case FLUSH_DISK_TIMEOUT:
    case FLUSH_SLAVE_TIMEOUT:
  case SLAVE_NOT_AVAILABLE:
    //③如果发送消息的结果为 Broker Server 不可用，则将本地事务状态设置为回滚
        localTransactionState = LocalTransactionState.ROLLBACK_MESSAGE;
        break;
    default:
        break;
  }
```

（2）在 DefaultMQProducerImpl 类的 sendMessageInTransaction()方法中，结束事务或者回滚事务的具体代码如下所示：

```
  try {
      //①第二阶段的事务消息处理
      this.endTransaction(sendResult, localTransactionState,
localException);
  } catch (Exception e) {
      log.warn("local transaction execute " + localTransactionState + ", but
end broker transaction failed", e);
  }
  TransactionSendResult transactionSendResult = new TransactionSendResult();
  transactionSendResult.setSendStatus(sendResult.getSendStatus());
  transactionSendResult.setMessageQueue(sendResult.getMessageQueue());
  transactionSendResult.setMsgId(sendResult.getMsgId());
```

```
    transactionSendResult.setQueueOffset(sendResult.getQueueOffset());
    transactionSendResult.setTransactionId(sendResult.getTransactionId());
    transactionSendResult.setLocalTransactionState(localTransactionState);
    //②返回发送事务消息的结果
    return transactionSendResult;}
```

（3）在 DefaultMQProducerImpl 类的 endTransaction()方法中，通过第二阶段处理事务消息的具体代码如下所示：

```
    public void endTransaction(
        final SendResult sendResult,
        final LocalTransactionState localTransactionState,
        final Throwable localException) throws RemotingException,
MQBrokerException, InterruptedException, UnknownHostException {
        final MessageId id;
        if (sendResult.getOffsetMsgId() != null) {
            id = MessageDecoder.decodeMessageId(sendResult.getOffsetMsgId());
        } else {
            id = MessageDecoder.decodeMessageId(sendResult.getMsgId());
        }
        String transactionId = sendResult.getTransactionId();
    final String brokerAddr =
this.mQClientFactory.findBrokerAddressInPublish(sendResult.getMessageQueue()
.getBrokerName());
    //①构造第二阶段事务消息请求的请求头
    EndTransactionRequestHeader requestHeader = new
EndTransactionRequestHeader();
    //②设置事务 ID
    requestHeader.setTransactionId(transactionId);
    //③设置 CommitLog 的消费位置 offset
        requestHeader.setCommitLogOffset(id.getOffset());
        switch (localTransactionState) {
            case COMMIT_MESSAGE:
                //④如果本地事务为提交状态，则设置请求头中的 commitOrRollback 为提交状态
                requestHeader.setCommitOrRollback(MessageSysFlag.
    TRANSACTION_COMMIT_TYPE);
                break;
            case ROLLBACK_MESSAGE:
                 //⑤如果本地事务为回滚状态，则设置请求头中的 commitOrRollback 为回滚状态
                 requestHeader.setCommitOrRollback
                   (MessageSysFlag.TRANSACTION_ROLLBACK_TYPE);
                break;
            case UNKNOW:
                 //⑥如果本地事务为未知状态，则设置请求头中的 commitOrRollback 为未知状态
                 requestHeader.setCommitOrRollback(MessageSysFlag.
                  TRANSACTION_NOT_TYPE);
```

```
                break;
            default:
                break;
        }
requestHeader.setProducerGroup(this.defaultMQProducer.getProducerGroup());
        requestHeader.setTranStateTableOffset(sendResult.getQueueOffset());
        requestHeader.setMsgId(sendResult.getMsgId());
    String remark = localException != null ? ("executeLocalTransactionBranch
exception: " + localException.toString()) : null;
    //⑦调用 RocketMQ 的 MQClientAPIImpl 类的 endTransactionOneway()方法来结束事务
    this.mQClientFactory.getMQClientAPIImpl().endTransactionOneway(brokerAdd
r, requestHeader, remark,
                this.defaultMQProducer.getSendMsgTimeout());
    }
```

RocketMQ 用 MQClientAPIImpl 类的 endTransactionOneway()方法，向 Broker Server 发送 RPC 结束的事务的二阶段请求，请求类型为 RequestCode.END_TRANSACTION，具体代码如下所示：

```
    public void endTransactionOneway(
        final String addr,
        final EndTransactionRequestHeader requestHeader,
        final String remark,
        final long timeoutMillis
    ) throws RemotingException, MQBrokerException, InterruptedException {
        //①构造请求类型为 RequestCode.END_TRANSACTION 的 RPC 请求
        RemotingCommand request =
RemotingCommand.createRequestCommand(RequestCode.END_TRANSACTION,
requestHeader);
    request.setRemark(remark);
        //②调用 RemotingClient 类发起 RPC 请求
        this.remotingClient.invokeOneway(addr, request, timeoutMillis);
    }
```

（4）RocketMQ 在 Broker Server 中，用 EndTransactionProcessor 类的 processRequest() 方法处理二阶段事务消息的请求，具体代码如下所示：

```
    @Override
    public RemotingCommand processRequest(ChannelHandlerContext ctx,
RemotingCommand request) throws RemotingCommandException {...OperationResult
result = new OperationResult();
        //①如果结束事务的类型为提交事务
        if (MessageSysFlag.TRANSACTION_COMMIT_TYPE ==
requestHeader.getCommitOrRollback()) {
            //②则从 Broker Server 获取预处理的事务消息
```

```
            result =
this.brokerController.getTransactionalMessageService().commitMessage(request
Header);
            if (result.getResponseCode() == ResponseCode.SUCCESS) {
                //③校验预处理的事务消息
                RemotingCommand res =
checkPrepareMessage(result.getPrepareMessage(), requestHeader);
                if (res.getCode() == ResponseCode.SUCCESS) {
                //④在校验通过之后，构造 RocketMQ 的发送消息体 MessageExtBrokerInner，
用真实的消息主题 REAL_TOPIC 和消息队列 REAL_QID 替换临时的事务消息的消息主题和消息队列
                    MessageExtBrokerInner msgInner =
endMessageTransaction(result.getPrepareMessage());

msgInner.setSysFlag(MessageSysFlag.resetTransactionValue(msgInner.getSysFlag
(), requestHeader.getCommitOrRollback()));

msgInner.setQueueOffset(requestHeader.getTranStateTableOffset());

msgInner.setPreparedTransactionOffset(requestHeader.getCommitLogOffset());

msgInner.setStoreTimestamp(result.getPrepareMessage().getStoreTimestamp());
                    MessageAccessor.clearProperty(msgInner,
MessageConst.PROPERTY_TRANSACTION_PREPARED);
                    //⑤将消息体重新推送到 Broker Server 上，消费者可以直接消费
                    RemotingCommand sendResult = sendFinalMessage(msgInner);
                    if (sendResult.getCode() == ResponseCode.SUCCESS) {

this.brokerController.getTransactionalMessageService().deletePrepareMessage(
result.getPrepareMessage());
                    }
                    return sendResult;
                }
                return res;
            }
        }else if (MessageSysFlag.TRANSACTION_ROLLBACK_TYPE ==
requestHeader.getCommitOrRollback()) {
          //⑥如果结束事务的类型为回滚事务，则获取预处理的事务消息
            result =
this.brokerController.getTransactionalMessageService().rollbackMessage(reque
stHeader);
            if (result.getResponseCode() == ResponseCode.SUCCESS) {
                RemotingCommand res =
checkPrepareMessage(result.getPrepareMessage(), requestHeader);
                if (res.getCode() == ResponseCode.SUCCESS) {
              //⑦删除 Broker Server 上的预处理消息，这样消费者就消费不到这条消息
```

```
            this.brokerController.getTransactionalMessageService().
            deletePrepareMessage(result.getPrepareMessage());
                }
                return res;
            }
        }
        response.setCode(result.getResponseCode());
        response.setRemark(result.getResponseRemark());
        return response;
    }
```

8.7.3 定时回查事务状态的原理

RocketMQ 用线程的 TransactionalMessageCheckService 类定时地调用 TransactionalMessageServiceImpl 类的 check()方法回查事务状态，具体代码如下所示：

```
    public class TransactionalMessageCheckService extends ServiceThread {
        ...
        @Override
        public void run() {
            log.info("Start transaction check service thread!");
            long checkInterval =
brokerController.getBrokerConfig().getTransactionCheckInterval();
            while (!this.isStopped()) {
                //①设置定时执行的时间周期，默认为 60s
                this.waitForRunning(checkInterval);
            }
            log.info("End transaction check service thread!");
        }

        @Override
     protected void onWaitEnd() {
            //②获取事务消息的超时时间，默认为 6s
            long timeout =
brokerController.getBrokerConfig().getTransactionTimeOut();
            //③回查事务消息的次数，默认为 15 次
            int checkMax =
brokerController.getBrokerConfig().getTransactionCheckMax();
            long begin = System.currentTimeMillis();
            log.info("Begin to check prepare message, begin time:{}", begin);
            //④调用 TransactionalMessageServiceImpl 类的 check()方法回查事务消息
     this.brokerController.getTransactionalMessageService().check(timeout,
checkMax, this.brokerController.getTransactionalMessageCheckListener());
            log.info("End to check prepare message, consumed time:{}",
System.currentTimeMillis() - begin);
```

```
        }
    }
```

TransactionalMessageServiceImpl 类的 check()方法回查事务状态的逻辑非常复杂，下面将它拆分并逐一分析。

1. 获取当前事务消息主题所有的消息队列

RocketMQ 会从 Broker Server 获取当前事务消息主题的所有的消息队列，具体代码如下所示：

```
    @Override
    public void check(long transactionTimeout, int transactionCheckMax,
            AbstractTransactionalMessageCheckListener listener) {
        ...
    String topic = TopicValidator.RMQ_SYS_TRANS_HALF_TOPIC;
    //调用 TopicConfigManager 类，获取 Broker Server 本地缓存 topicConfigTable 中对应
事务消息的主题的消息队列
    Set<MessageQueue> msgQueues
                = transactionalMessageBridge.fetchMessageQueues(topic);
        if (msgQueues == null || msgQueues.size() == 0) {
            log.warn("The queue of topic is empty :" + topic);
            return;
        }
    }
```

2. 遍历消息队列，回查事务消息的事务状态

为了保证事务状态回查的可考性，RocketMQ 采用穷举法来保证功能的可靠性，具体代码如下所示：

```
    for (MessageQueue messageQueue : msgQueues) {
        long startTime = System.currentTimeMillis();
        //①获取 RMQ_SYS_TRANS_OP_HALF_TOPIC 主题的消息队列
        MessageQueue opQueue = getOpQueue(messageQueue);
        long halfOffset =
transactionalMessageBridge.fetchConsumeOffset(messageQueue);
        long opOffset = transactionalMessageBridge.fetchConsumeOffset(opQueue);
        log.info("Before check, the queue={} msgOffset={} opOffset={}",
messageQueue, halfOffset, opOffset);
        if (halfOffset < 0 || opOffset < 0) {
            log.error("MessageQueue: {} illegal offset read: {}, op offset:
{},skip this queue", messageQueue,
                        halfOffset, opOffset);
            continue;
        }
        List<Long> doneOpOffset = new ArrayList<>();
```

```
        HashMap<Long, Long> removeMap = new HashMap<>();
        PullResult pullResult = fillOpRemoveMap(removeMap, opQueue, opOffset,
halfOffset, doneOpOffset);
        if (null == pullResult) {
            log.error("The queue={} check msgOffset={} with opOffset={} failed,
pullResult is null",
                    messageQueue, halfOffset, opOffset);
            continue;
        }
        int getMessageNullCount = 1;
        long newOffset = halfOffset;
        long i = halfOffset;
        //②采用穷举算法
        while (true) {
            if (System.currentTimeMillis() - startTime > MAX_PROCESS_TIME_LIMIT) {
                log.info("Queue={} process time reach max={}", messageQueue,
MAX_PROCESS_TIME_LIMIT);
                break;
            }
            if (removeMap.containsKey(i)) {
                log.info("Half offset {} has been committed/rolled back", i);
                    Long removedOpOffset = removeMap.remove(i);
                    doneOpOffset.add(removedOpOffset);
            } else {
                //③获取预处理消息
                GetResult getResult = getHalfMsg(messageQueue, i);
                MessageExt msgExt = getResult.getMsg();
                if (msgExt == null) {
                    if (getMessageNullCount++ > MAX_RETRY_COUNT_WHEN_HALF_NULL) {
                        break;
                    }
                    if (getResult.getPullResult().getPullStatus() ==
PullStatus.NO_NEW_MSG) {
                        log.debug("No new msg, the miss offset={} in={}, continue
check={}, pull result={}", i,
                                messageQueue, getMessageNullCount,
getResult.getPullResult());
                        break;
                    } else {
                        log.info("Illegal offset, the miss offset={} in={},
continue check={}, pull result={}",
                                i, messageQueue, getMessageNullCount,
getResult.getPullResult());
                        i = getResult.getPullResult().getNextBeginOffset();
                        newOffset = i;
```

```
                    continue;
                }
            }
            //④删除预处理消息
            if (needDiscard(msgExt, transactionCheckMax) || needSkip(msgExt)) {
                listener.resolveDiscardMsg(msgExt);
                newOffset = i + 1;
                i++;
                continue;
            }
            if (msgExt.getStoreTimestamp() >= startTime) {
                log.debug("Fresh stored. the miss offset={}, check it later,
store={}", i,
                        new Date(msgExt.getStoreTimestamp()));
                break;
            }
            long valueOfCurrentMinusBorn = System.currentTimeMillis() -
msgExt.getBornTimestamp();
            long checkImmunityTime = transactionTimeout;
            String checkImmunityTimeStr = msgExt.getUserProperty
(MessageConst.PROPERTY_CHECK_IMMUNITY_TIME_IN_SECONDS);
            if (null != checkImmunityTimeStr) {
                checkImmunityTime = getImmunityTime(checkImmunityTimeStr,
transactionTimeout);
                if (valueOfCurrentMinusBorn < checkImmunityTime) {
                    if (checkPrepareQueueOffset(removeMap, doneOpOffset,
msgExt)) {
                        newOffset = i + 1;
                        i++;
                        continue;
                    }
                }
            } else {
                if ((0 <= valueOfCurrentMinusBorn) && (valueOfCurrentMinusBorn
< checkImmunityTime)) {
                    log.debug("New arrived, the miss offset={}, check it later
checkImmunity={}, born={}", i,
                            checkImmunityTime, new
Date(msgExt.getBornTimestamp()));
                    break;
                }
            }
            //⑤获取 RMQ_SYS_TRANS_OP_HALF_TOPIC 主题对应的事务消息
            List<MessageExt> opMsg = pullResult.getMsgFoundList();
```

```
            boolean isNeedCheck = (opMsg == null && valueOfCurrentMinusBorn >
checkImmunityTime)
                            || (opMsg != null && (opMsg.get(opMsg.size() -
1).getBornTimestamp() - startTime > transactionTimeout))
                            || (valueOfCurrentMinusBorn <= -1);
            //⑥如果需要校验消息，则将事务消息添加到预处理消息的列表中
            if (isNeedCheck) {
                if (!putBackHalfMsgQueue(msgExt, i)) {
                    continue;
                }
                listener.resolveHalfMsg(msgExt);
            } else {
                pullResult = fillOpRemoveMap(removeMap, opQueue,
pullResult.getNextBeginOffset(), halfOffset, doneOpOffset);
                log.debug("The miss offset:{} in messageQueue:{} need to get
more opMsg, result is:{}", i,
                        messageQueue, pullResult);
                continue;
            }
        }
        newOffset = i + 1;
        i++;
    }
    //⑦如果不是预处理消息，则更新 Broker Server 对应消息队列的消费进度
    if (newOffset != halfOffset) {
        transactionalMessageBridge.updateConsumeOffset(messageQueue,
newOffset);
    }
    long newOpOffset = calculateOpOffset(doneOpOffset, opOffset);
    if (newOpOffset != opOffset) {
        transactionalMessageBridge.updateConsumeOffset(opQueue,
newOpOffset);
    }
}
```

8.7.4 【实例】在 Spring Cloud Aliaba 项目中生产事务消息

本实例的源码在本书配套资源的“chaptereight/use-spring-cloud-alibaba-rocketmq-transaction-message”目录下。

本实例采用 Spring Cloud Alibaba，通过 Spring Cloud Alibaba RocketMQ 向 Broker Server 发送事务消息。

1. 初始化项目

使用 Spring Cloud Alibaba 初始化项目，并添加 RocketMQ 的依赖包，具体如下所示：

```
<dependency>
    <groupId>com.alibaba.cloud</groupId>
    <artifactId>spring-cloud-starter-stream-rocketmq</artifactId>
</dependency>
```

2. 生产事务消息

下面使用 RocketMQ 的 TransactionMQProducer 类生产事务消息：

```
@Configuration
public class ProducerConfig {
    //①初始化 TransactionMQProducer 类
    @Bean
    public TransactionMQProducer newTransactionMQProducer() throws
MQClientException {
        //②添加本地事务执行状态的监听器
    TransactionListener transactionListener =new TransactionListenerImpl();
        //③设置消息主题为 use-spring-cloud-alibaba-rocketmq-transaction-message
        TransactionMQProducer producer = new TransactionMQProducer
("use-spring-cloud-alibaba-rocketmq-transaction-message");
        ExecutorService executorService = new ThreadPoolExecutor(2, 5, 100,
TimeUnit.SECONDS, new ArrayBlockingQueue<Runnable>(2000), new ThreadFactory() {
            @Override
            public Thread newThread(Runnable r) {
                Thread thread = new Thread(r);
                thread.setName("检查客户端事务状态的线程");
                return thread;
            }
        });
        producer.setNamesrvAddr("127.0.0.1:9876");
        producer.setExecutorService(executorService);
        //④设置监听器
        producer.setTransactionListener(transactionListener);
        //⑤开启事务消息生产者
        producer.start();
        return producer;
    }
}

@Component
public class ProducerTransationService {
    //⑥加载 TransactionMQProducer 类
    @Resource
```

```
    private TransactionMQProducer transactionMQProducer;
    public String sendMessage() throws MQClientException,
InterruptedException{
        String[] tags = new String[]{"订单", "商品", "支付", "货品", "物流","
仓库"};
        SendResult sendResult=null;
        for (int i = 0; i < 200; i++) {
            try {
                Message msg =
                        new Message("use-spring-cloud-alibaba-
rocketmq-transaction-message", tags[i % tags.length], "KEY" + i,
                                ("事务测试消息" +
i).getBytes(RemotingHelper.DEFAULT_CHARSET));
                //⑦发送事务消息
                sendResult =
transactionMQProducer.sendMessageInTransaction(msg, null);
                System.out.printf("%s%n", sendResult);
                Thread.sleep(10);
            } catch (MQClientException | UnsupportedEncodingException e) {
                e.printStackTrace();
            }
        }
        for (int i = 0; i < 10; i++) {
            Thread.sleep(1000);
        }
        return sendResult.toString();
    }
}
```

本实例中的监听器的具体代码如下所示：

```
public class TransactionListenerImpl implements TransactionListener {
    //①原子更新
    private AtomicInteger transactionIndex = new AtomicInteger(0);
    //②缓存本地事务执行的结果
    private ConcurrentHashMap<String, Integer> localTransactionCache = new
ConcurrentHashMap<>();
    @Override
    public LocalTransactionState executeLocalTransaction(Message msg,
Object arg) {
        int value = transactionIndex.getAndIncrement();
        //③取模运算，模拟本地事务的 3 个状态
        int status = value % 3;
        localTransactionCache.put(msg.getTransactionId(), status);
        return LocalTransactionState.UNKNOW;
    }
```

```
    @Override
    public LocalTransactionState checkLocalTransaction(MessageExt msg) {
        //④从本地事务执行结果缓存中获取对应事务消息执行的结果
        Integer status = localTransactionCache.get(msg.getTransactionId());
        if (null != status) {
            switch (status) {
                case 0:
                    //⑤如果状态为 0，则返回 LocalTransactionState.UNKNOW
                    return LocalTransactionState.UNKNOW;
                case 1:
                    //⑥如果状态为 1，则返回
LocalTransactionState.COMMIT_MESSAGE，提交事务消息
                    return LocalTransactionState.COMMIT_MESSAGE;
                case 2:
                    //⑦如果状态为 2，则返回
LocalTransactionState.ROLLBACK_MESSAGE，回滚事务消息
                    return LocalTransactionState.ROLLBACK_MESSAGE;
                default:
                    //⑧默认返回 LocalTransactionState.COMMIT_MESSAGE
                    return LocalTransactionState.COMMIT_MESSAGE;
            }
        }
        return LocalTransactionState.COMMIT_MESSAGE;
    }
}
```

3. 运行实例，使用命令控制台验证事务消息

运行实例之后，在 RocketMQ 的 UI 控制台中可以看到消息主题为“RMQ_SYS_TRANS_HALF_TOPIC”的预处理消息，如图 8-24 所示。

通过 RocketMQ 的 Admin 控制台执行如下命令：

```
mqadmin queryMsgById -n 127.0.0.1:9876 -i C0A8010300002A9F000000000384C5F0
```

搜索事务消息，如图 8-25 所示，消息 ID 为“C0A8010300002A9F000000000384C5F0”的事务消息的主题为“RMQ_SYS_TRANS_HALF_TOPIC”，事务消息的 REAL_TOPIC 属性值为“use-spring-cloud-alibaba-rocketmq-transaction-message”。

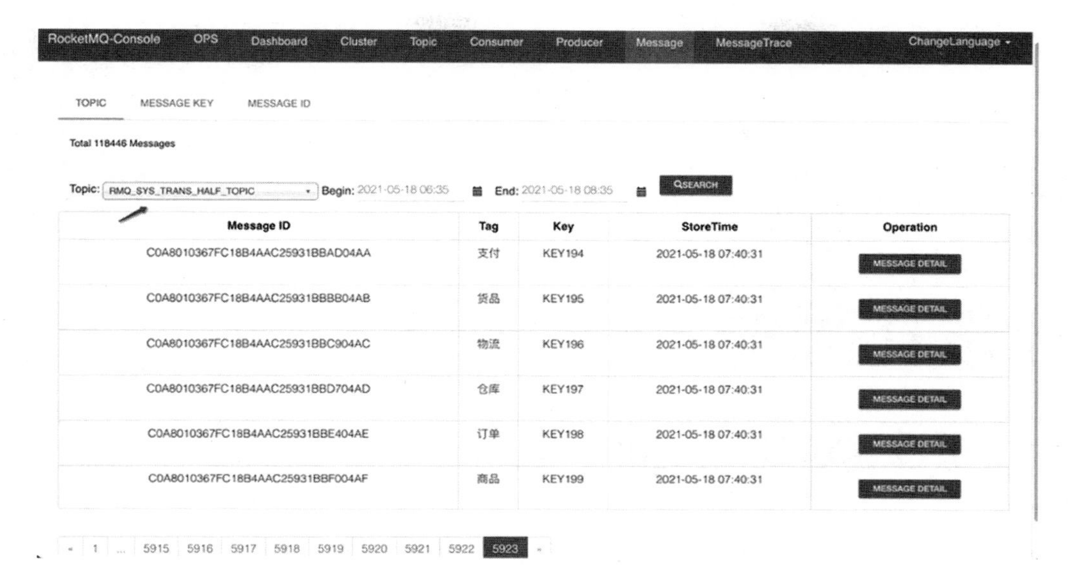

图 8-24

```
huxian@huxians-MacBook-Pro bin % sh mqadmin queryMsgById -n 127.0.0.1:9876 -i C0A8010300002A9F000000000384C5F0
RocketMQLog:WARN No appenders could be found for logger (io.netty.util.internal.PlatformDependent0).
RocketMQLog:WARN Please initialize the logger system properly.
OffsetID:              C0A8010300002A9F000000000384C5F0
Topic:                 RMQ_SYS_TRANS_HALF_TOPIC
Tags:                  [货品]
Keys:                  [KEY3]
Queue ID:              0
Queue Offset:          120659
CommitLog Offset:      59033072
Reconsume Times:       0
Born Timestamp:        2021-05-18 07:41:14,604
Store Timestamp:       2021-05-18 07:42:05,517
Born Host:             192.168.1.3:50072
Store Host:            192.168.1.3:10911
System Flag:           0
Properties:            {REAL_TOPIC=use-spring-cloud-alibaba-rocketmq-transaction-message, TRANSACTION_CHECK_TIMES=1, KEYS=KEY3, TRAN_MSG=true, UNIQ_KEY=C0A8010367FC18B4AAC25
93263EC057B, CLUSTER=DefaultCluster, WAIT=true, PGROUP=use-spring-cloud-alibaba-rocketmq-transaction-message, TAGS=货品, REAL_QID=1}
Message Body Path:   /tmp/rocketmq/msgbodys/C0A8010367FC18B4AAC2593263EC057B


MessageTrack [consumerGroup=CID_RMQ_SYS_TRANS, trackType=UNKNOWN, exceptionDesc=org.apache.rocketmq.client.exception.MQClientException: CODE: 17  DESC: No topic route info
 in name server for the topic: %RETRY%CID_RMQ_SYS_TRANS
See http://rocketmq.apache.org/docs/faq/ for further details., org.apache.rocketmq.client.impl.MQClientAPIImpl.getTopicRouteInfoFromNameServer(MQClientAPIImpl.java:1385)]
```

图 8-25

第 9 章

分布式网关

——基于 Spring Cloud Gateway

随着 Saas 付费软件的流行，企业在达到一定的规模后都会对外开放自身的 API，以实现技术变现，或增加企业在自身行业内的技术影响力。开放 API 需要有一个统一对接商户的业务网关，以处理不同商户的请求，并将请求路由到内部的微服务。

在企业实施业务中台战略后，公司的 App 产品会百花齐放。因为 App 产品需要统一调用后端的 API 服务，所以，开发人员在暴露后端的 API 给 App 时需要考虑很多非功能设计，其中，路由和流量控制就是其中比较重要的功能。

在引入业务网关后，可以将这些非功能设计从业务服务中统一抽取到网关，这样可以解耦后端的功能性和非功能性需求，让业务人员更加专注于理解自身的业务，以及按时地交付的业务功能。Spring Cloud Gateway 作为开源社区比较稳定和高性能的业务网关，已经被很多公司在业务的生产环境中使用。

9.1 认识网关

在分析业务网关 Spring Cloud Gateway 的原理前，先来认识一下网关。

9.1.1 什么是网关

开发人员可以从两个方面来理解网关。

1. 从字面含义的角度去理解

网关就是一个“关卡”，出入这个“关卡”的是外部的流量请求及其响应。如图 9-1 所示，网关作为“关卡”处理出入流量，并请求实际的资源，返回资源响应的结果。

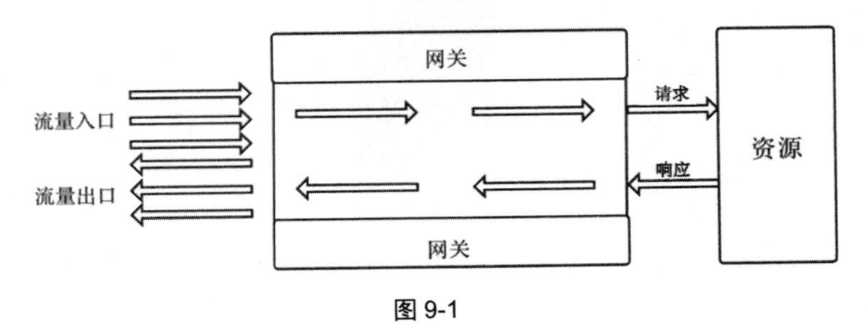

图 9-1

很多开发人员可能会认为网关和微服务是强相关的，其实网关并不是微服务独有的。

2. 从软件分层架构的角度去理解

网关是位于客户端与其依赖的服务之间的一个层，有时它也被称为“反向代理”。如图 9-2 所示，网关被作为从客户端到其服务的单一入口点。

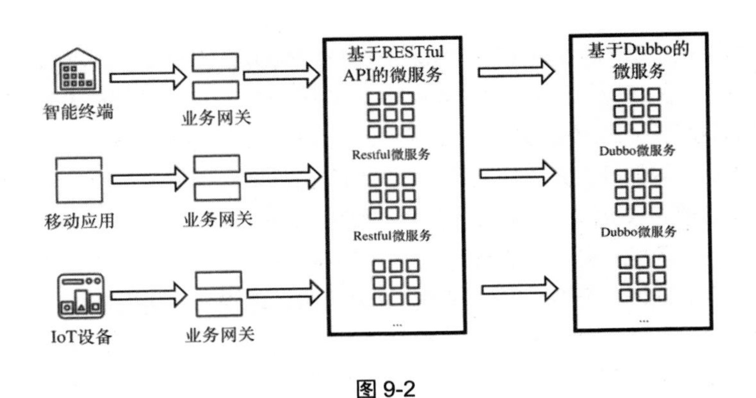

图 9-2

智能终端、移动应用和 IoT 设备连接业务网关，访问基于 RESTful API 微服务，再通过 RPC 框架访问基于 Dubbo 的微服务。

9.1.2　为什么需要网关

在微服务架构中，开发人员引入业务网关，主要是为了解决一些非功能性的问题，如图 9-3 所示。

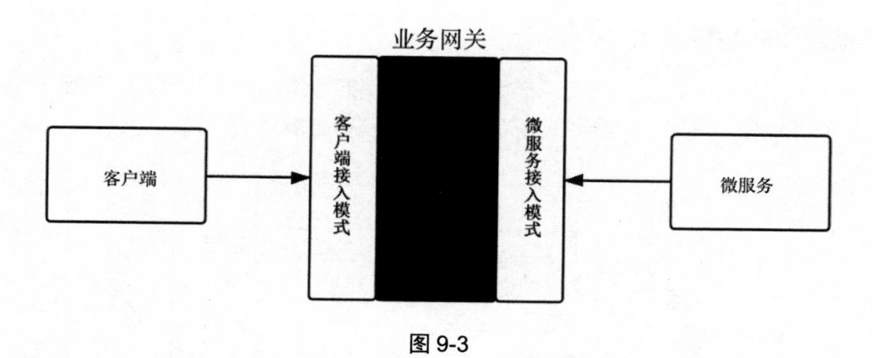

图 9-3

在图 9-3 中，开发人员可以将业务网关当作一个黑盒子：只需要按照约定的规则将应用接入业务网关，应用对应的 API 就可以暴露给第三方调用者。

1. **网关具体解决的哪些非功能性问题**

网关具体解决以下非功能性问题：

- 业务网关将外部公共 API 与内部微服务 API 进行隔离。内部微服务会持续迭代更新，但不会影响调用外部公共 API 的客户端。业务网关为微服务提供统一的入口点，向客户端屏蔽微服务的服务治理的和版本控制的技术细节。
- 业务网关能够统一地管理微服务的访问权限，解决微服务调用的安全性问题。
- 业务网关通常都支持多种通信协议，这样客户端可以使用外部 API（基于 HTTP 的 API 或者 RESTful API）完成和网关的 RPC 通信。在业务网关与内部微服务之间可以采用不同的通信协议，比如 ProtoBuf 或 AMQP，当然最常用的 RPC 协议是 RESTful 和 Dubbo 协议。业务网关可以跨这些不同的协议，提供一个外部的、统一的基于 RESTful 的 API，完成内部微服务 API 的暴露，并允许开发人员选择不同的内部通信协议来调用该 API。
- 业务网关能够降低开发人员实施和落地微服务的技术复杂度。在业务网关中，开发人员可以统一地治理业务 API，包括服务隔离、降级和统一认证等。
- 从业务中台的角度来分析，业务网关解决了不同产品线的业务功能，在线上强耦合部署的问题。在企业落地业务中台的过程中，必然会拆“烟囱”，将通用的业务功能下沉到业务中台，从而在上层会孵化出很多 API 产品及对应的 App 客户端，通过业务网关可以将不同 API 产品和对应的 App 客户端的流量完全隔离，但下层已经拆分的微服务是无感知的。

2. **一个自研业务网关实例**

图 9-4 是一个完整的自研的业务网关。

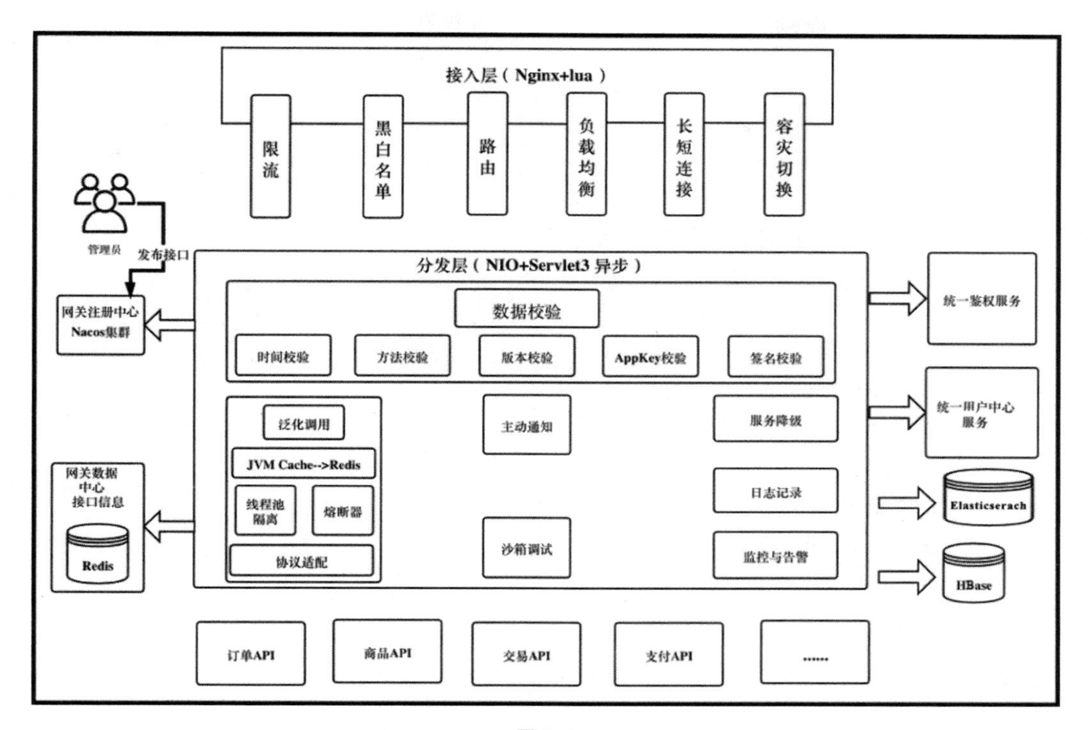

图 9-4

该网关需要具备的功能说明如下。

- 在接入层需要具备的功能主要包括：流量控制、黑/白名单、路由、负载均衡、长短连接及容灾切换等。
- 在分发层需要具备数据校验功能主要包括：时间校验、方法校验、版本校验、AppKey 校验、签名校验等。
- 业务网关需要具备的功能包括：泛化调用、线程池隔离、熔断器、通信协议适配、消息中心、服务降级和监控与告警等。其中，泛化调用的主要应用场景是业务网关对接 Dubbo API，业务网关需要使用 Dubbo 的泛化调用技术屏蔽掉对业务的 Dubbo API 的 Jar 包的强依赖（即在业务上线新功能后，需要调用方变更 Jar 包依赖关系才能生效）。业务网关需要具备在不重启服务的前提下，让外部客户端能够使用业务最新的业务功能。
- 如果是自研网关，则还需要统一的鉴权服务和统一的用户中心服务。
- 业务网关需要有自己独立的数据中心，用于存储在网关运行后需要持久化的一些非功能性数据。
- 业务网关要支持注册中心，以保证自身服务的高可用性和利用业务 API 的服务治理能力，增加自身服务路由的功能。
- 业务网关要支持配置中心，以确保网关配置信息的高可用，以及动态地更新指定的配置信息。

- 业务网关要具备“沙箱”环境，主要用于核心功能的功能测试。比如，要将支付接口开发给商户，商户按照接入文档，利用 SDK 将应用接入支付系统，但是，商户需要测试一下支付接口，确保能够跑通线上的支付业务。业务网关作为统一的入口，需要将“沙箱”环境当作一个稳定的产品来维护，以确保商户能够随时随地地调用，并保证服务接口的可用性。

9.1.3 认识 Spring Cloud Gateway

Spring Cloud Gateway 是 Spring Cloud 提供的高性能网关的技术解决方案，它是基于 Spring Framework 5.0 和 Spring Boot 2.0 构建的 API 网关，它的主要特性如下：

- JDK 要求最低版本为 1.8；
- 支持 Spring Framework 5.0；
- 支持动态路由；
- 用过滤器链来拦截业务网关的 API 流量请求；
- 支持配置中心和注册中心；
- 能够集成 Hystrix 断路器；
- 支持基于“Redis + Lua”的分布式流量控制。

Spring Cloud Gateway 主要提供路由、断言和过滤的功能。

1. 路由

路由是网关的基本组件，它由 ID、目标 URI、谓词集合和过滤器集合定义。如果聚合谓词为 true，则匹配路由。

图 9-5 描述了传统的 Nginx 请求路由的流程。当 URI 请求到达 Nginx 层后，Nginx 会触发路由算法（路由算法是在 Nginx 启动时配置的，不支持动态修改），将 URI 请求路由到指定的服务实例上。如果采用的是传统的路由模式，则开发人员不能动态地控制路由规则。

图 9-6 描述了基于网关路由的流程。当 URI 请求到达网关后，网关会根据开发人员预先设置的路由规则完成服务路由，将 URI 请求路由到指定的服务实例。

如果开发人员觉得路由规则不是很合理，可以动态地修改路由规则，在不重启网关的前提下让路由规则实时地生效。

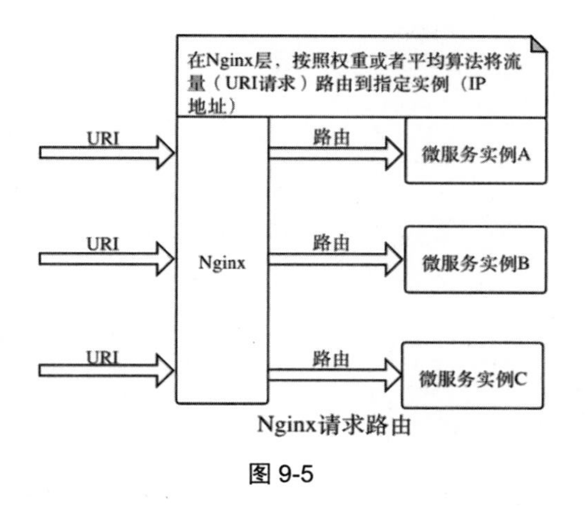

图 9-5

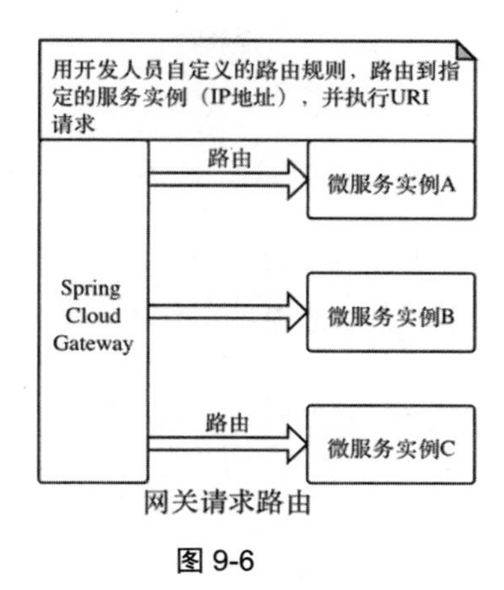

图 9-6

2. 断言

Java 8 引入了函数式编程，断言也是 Java 8 新增的语义。断言语义的核心是“输入一个条件，返回一个 Bolean 类型”，这一点非常适合作为网关路由规则的触发条件。

Spring Cloud Gateway 已经支持断言，在 Spring Cloud Gateway 中将断言作为路由转发的判断条件。目前 SpringCloud Gateway 支持多种断言方式，常见如基于 Path、Query、Method、Header 等。

在 Spring Cloud Gateway 中，路由是可以组合的。如图 9-7 所示，比如路由 A、路由 B 和路由 C，组合后的路由会按照优先级执行；但如果断言不成功，则 Spring Cloud Gateway 会不处理路由请求。

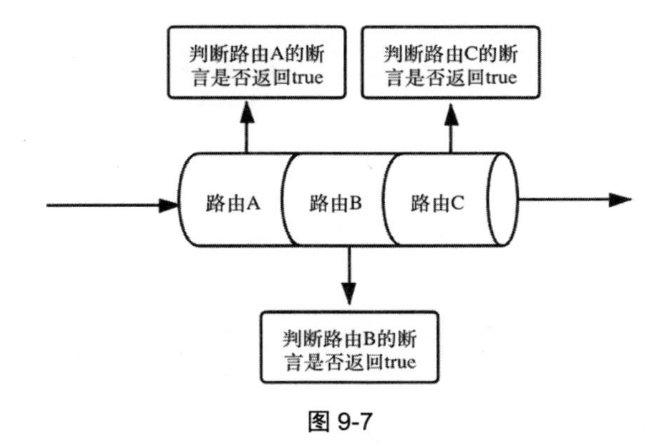

图 9-7

基于 Path 的断言如下所示，如果 Path 路径匹配不成功，则路由规则不生效。

```
routes:
  - id: user_api_route
```

```
        uri: http://127.0.0.1:28089
        predicates:
          - Path=/user/getUserInfo
```

3. 过滤器

过滤器是 Spring Cloud Gateway 中的核心功能。在 Spring Cloud Gateway 中，处理过滤器中 HTTP 请求的核心流程如图 9-8 所示。

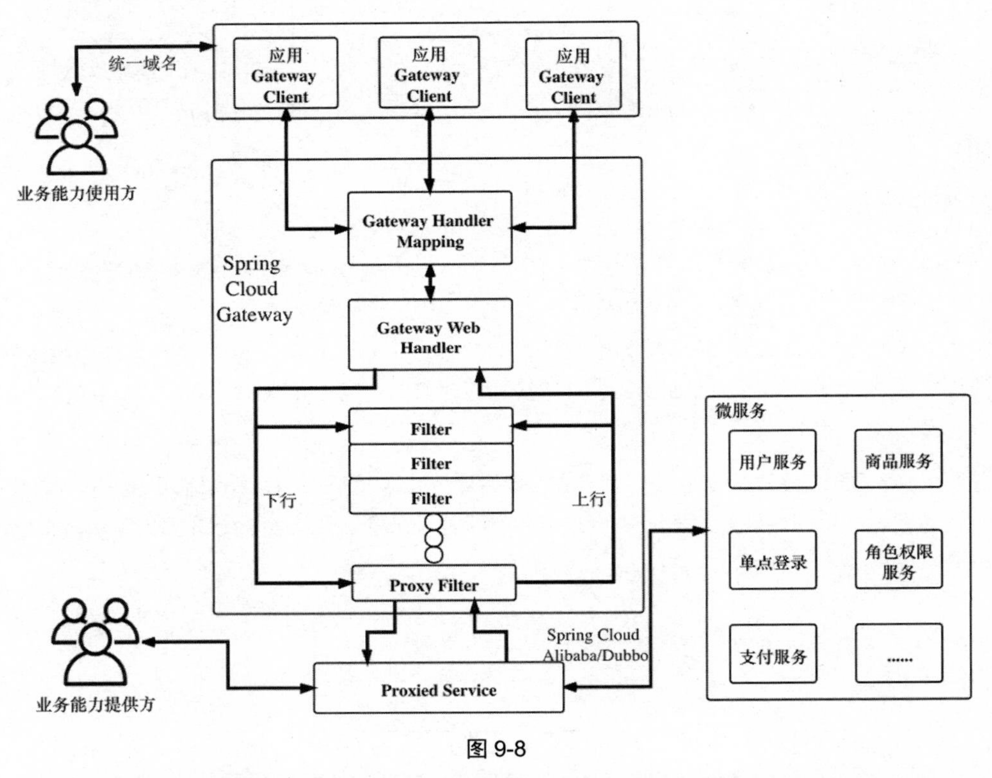

图 9-8

（1）“业务能力使用方”使用统一的域名调用应用的 Gateway Client。Gateway Client 通常都是集群部署的。

（2）Gateway Client 调用 Spring Cloud Gateway 的 Gateway Handler Mapping（网关映射处理类）。

（3）Gateway Handler Mapping 调用 Gateway Web Handler（网关的 Web 请求处理类）。

（4）Web 请求处理类会将请求转发到过滤器链。在 Spring Cloud Gateway 中，真正处理 URI 请求的是过滤器链中的过滤器。

（5）在按照预先设定的优先级执行完过滤器后，Spring Cloud Gateway 会将请求转发到代理过滤器（比如 NettyRoutingFilter）。通过代理过滤器执行 URI 的 RPC 请求，Spring Cloud Gateway 在代理请求执行的过程中，会变更原始请求的端口号。比如，业务人员统一访问访问 IP 地址“127.0.0.1:8080/user-api/getUserInfo”，经过代理过滤器转发后，业务 API 服务收到的请求会变为“127.0.0.1:26785/user-api/getUserInfo”。

（6）业务 API 服务在处理完成后，会通过过滤器将响应结果返回给业务 API 的调用者。

9.2　用 Reactor Netty 实现 Spring Cloud Gateway 的通信渠道

Spring Cloud Gateway 是一个高性能的业务网关。为什么说它是高性能的呢?因为其底层通信采用的是 Reactor Netty，通过它来集成高性能的 RPC 框架 Netty。

9.2.1　什么是 Reactor Netty

为了满足微服务架构中服务之间多场景的 RPC 调用，Reactor Netty 支持基于 HTTP（WebSocket）、TCP 或 UDP 的通信渠道，其底层基于 Netty 框架。

Spring Cloud Gateway 运行 Reactor Netty 的软件环境是：①JDK 版本为 1.8 及以上；②Reactive Streams API 的版本为 1.0 及以上；③Reactor Core 的版本为 3.x 及以上；④Netty 的版本为 4.x 及以上。

Reactor Netty 是高性能的 RPC 框架，主要包含如下几个核心的概念。

1. TCP Server

Reactor Netty 提供了一个易于使用和配置的 TCP Server。它隐藏了创建 TCP 服务器所需的大部分 Netty 功能，并添加了流式处理的功能。

TCP Server 的使用示例如下：

```
import reactor.netty.DisposableServer;
import reactor.netty.tcp.TcpServer;
public class Application {
  public static void main(String[] args) {
    //①用 TcpServer 类创建一个 DisposableServer 对象
    DisposableServer server =
        TcpServer.create()
          .bindNow();
    //②启动 TCP Server，等待客户端连接
```

```
    server.onDispose()
          .block();
  }
}
```

2. TCP Client

Reactor Netty 提供了易于使用和配置的 TCP Client。它隐藏了创建 TCP 客户端所需的大部分 Netty 功能，并添加了流式处理的功能。

TCP Client 的使用示例如下：

```
import reactor.netty.Connection;
import reactor.netty.tcp.TcpClient;
public class Application {
  public static void main(String[] args) {
    //①用 TcpClient 类创建一个 Connection 对象
    Connection connection =
        TcpClient.create()
                .connectNow();
    //②启动 TCP Client 去连接 TCP Server
    connection.onDispose()
            .block();
  }
}
```

3. HTTP Server

Reactor Netty 提供了易于使用和易于配置的 HTTP Server 类。它隐藏了创建 HTTP 服务器所需的大部分 Netty 功能，并添加了流式处理的功能。

HTTP Server 的使用示例如下：

```
import reactor.netty.DisposableServer;
import reactor.netty.http.server.HttpServer;
public class Application {
  public static void main(String[] args) {
    //①用 HttpServer 类创建一个 DisposableServer 对象
    DisposableServer server =
        HttpServer.create()
                .bindNow();
    //②启动 HTTP Server，等待客户端连接
    server.onDispose()
          .block();
  }
}
```

4. HTTP Client

Reactor Netty 提供了易于使用和易于配置的 HTTP Client。它隐藏了创建 HTTP 客户端所需的大部分 Netty 功能，并添加了流式处理的功能。

HTTP Client 的使用示例如下：

```
import reactor.netty.http.client.HttpClient;
public class Application {
  public static void main(String[] args) {
//①用 HttpClient 类创建一个 HttpClient 对象
HttpClient client = HttpClient.create();
//②用 get()方法调用 HTTP Server
    client.get()
          .uri("https://127.0.0.1:8080/")
          .response()
          .block();
  }
}
```

5. UDP Server

Reactor Netty 提供了易于使用和易于配置的 UDP Server，它包含 Netty 的大部分功能，并且屏蔽掉了底层的技术细节，这样开发人员可以利用 Reactor Netty 的 API 创建一个具备流式处理功能的 UDP Server。

UDP Server 的使用示例如下：

```
import reactor.netty.Connection;
import reactor.netty.udp.UdpServer;
import java.time.Duration;
public class Application {
  public static void main(String[] args) {
    //①用 UdpServer 类创建一个 Connection 对象
    Connection server =
        UdpServer.create()
                  .bindNow(Duration.ofSeconds(30));
    //②启动 UDP Server，等待客户端连接
    server.onDispose()
          .block();
  }
}
```

6. UDP Client

Reactor Netty 提供了易于使用和易于配置的 UDP Client，它包含 Netty 的大部分功能，并且屏蔽掉了底层的技术细节，这样开发人员可以利用 Reactor Netty 的 API 创建一个具备流式处理功

能的 UDP Client。

UDP Client 的使用示例如下：

```
import reactor.netty.Connection;
import reactor.netty.udp.UdpClient;
import java.time.Duration;
public class Application {
  public static void main(String[] args) {
    //①用 UdpClient 类创建一个 Connection 对象
    Connection connection =
        UdpClient.create()
                .connectNow(Duration.ofSeconds(30));
    //②用 onDispose()方法去连接 UDP Server
    connection.onDispose()
            .block();
  }
}
```

9.2.2 “用过滤器代理网关请求”的原理

在 Spring Cloud Gateway 中，通常都用过滤器来拦截所有业务的 API 请求，所以，Spring Cloud Gateway 通信渠道也是通过“定义一个过滤器（NettyRoutingFilter）并将其添加到过滤器链中”来实现的，具体原理如下。

1. 用 NettyConfiguration 类加载与 Netty 相关的资源

在自动配置类 GatewayAutoConfiguration 中，用 NettyConfiguration 类来加载 Netty 相关的资源。

（1）用 HttpClientProperties 类配置通信渠道的连接参数，具体代码如下所示：

```
@ConfigurationProperties("spring.cloud.gateway.httpclient")
@Validated
public class HttpClientProperties {
    //连接超时时间
    private Integer connectTimeout;
    //响应超时时间
    private Duration responseTimeout;
    //配置 HTTP 请求解码的最大报头大小
    private DataSize maxHeaderSize;
    //配置 HTTP 请求的初始行解码的最大长度
    private DataSize maxInitialLineLength;
    //配置 Netty HttpClient 的连接池信息
    private Pool pool = new Pool();
    //配置 Netty HttpClient 的代理连接信息
```

```
    private Proxy proxy = new Proxy();
    //配置Netty HttpClient的SSl信息
    private Ssl ssl = new Ssl();
    //配置Netty HttpClient的Websocket信息
    private Websocket websocket = new Websocket();
    //配置是否启用监听调试Netty HttpClient，true表示启用，false表示不启用
    private boolean wiretap;
    //配置Netty HttpClient是否启用压缩，true表示启动，false表示不启动
    private boolean compression;
    ...
  }
  public static class Pool {
    //HttpClient要使用的池类型，默认为ELASTIC（弹性）
    private PoolType type = PoolType.ELASTIC;
    //连接池的名称，默认为“proxy”
    private String name = "proxy";
    //连接池中连接数的最大值
    private Integer maxConnections =
ConnectionProvider.DEFAULT_POOL_MAX_CONNECTIONS;
    //仅对于FIXED类型，等待获取的最大时间（以ms为单位）
    private Long acquireTimeout =
ConnectionProvider.DEFAULT_POOL_ACQUIRE_TIMEOUT;
    //连接的最大空闲时间
    private Duration maxIdleTime = null;
    //连接的最大活跃时间
    private Duration maxLifeTime = null;
    //以指定的时间间隔在后台执行定期逐出检查
    private Duration evictionInterval = Duration.ZERO;
    public enum PoolType {
        //弹性
        ELASTIC,
        //固定
        FIXED,
        //禁用
        DISABLED
        }
    ...
  }
  public static class Proxy {
    //配置Netty HttpClient的代理类型，默认为HTTP
    private ProxyProvider.Proxy type = ProxyProvider.Proxy.HTTP;
    //配置Netty HttpClient的IP地址
    private String host;
    //配置Netty HttpClient的端口号
    private Integer port;
```

```
        //配置 Netty HttpClient 的用户名
        private String username;
        //配置 Netty HttpClient 的密码
        private String password;
        //配置 Netty HttpClient 不需要代理的主机匹配的正则表达式
        private String nonProxyHostsPattern;
        ...
    }
    public static class Ssl {
        //是否使用不安全的认证证书管理，默认为 false
        private boolean useInsecureTrustManager = false;
        //证书列表
        private List<String> trustedX509Certificates = new ArrayList<>();
        //TCP 握手超时时间，默认为 10s
        private Duration handshakeTimeout = Duration.ofMillis(10000);
        //关闭 SSL 连接后的通知刷新的超时时间，默认为 3s
        private Duration closeNotifyFlushTimeout = Duration.ofMillis(3000);
        //关闭 SSL 连接后的通知读的超时时间，默认为 0s
        private Duration closeNotifyReadTimeout = Duration.ZERO;
        //SSL 连接配置类型，默认为 TCP
        private SslProvider.DefaultConfigurationType defaultConfigurationType =
SslProvider.DefaultConfigurationType.TCP;
        //Netty HttpClient 的 keyStore 的存储路径
        private String keyStore;
        //Netty HttpClient 的 keyStore 的类型，默认为 JKS
        private String keyStoreType = "JKS";
        //Netty HttpClient 的 Keystore 的提供者
        private String keyStoreProvider;
        //KeyStore 的密码
        private String keyStorePassword;
        //Key 的密码，默认和 KeyStore 一样
        private String keyPassword;
        ...
    }
    //实例化 HttpClientProperties 对象
    @Bean
    public HttpClientProperties httpClientProperties() {
        return new HttpClientProperties();
    }
```

（2）初始化 Reactor Netty Http 的 HttpClient 类，用于和微服务完成 RPC 通信。在 NettyConfiguration 类中，用注解@Bean 和条件注解@ConditionalOnMissingBean 将初始化后 HttpClient 对象加载到 Spring Framework 的 IOC 容器，具体代码如下所示：

```
    @Bean
    @ConditionalOnMissingBean
```

```
    public HttpClient gatewayHttpClient(HttpClientProperties properties,
List<HttpClientCustomizer> customizers) {
    //①定义通信连接的生产者 ConnectionProvider
    ConnectionProvider connectionProvider =
buildConnectionProvider(properties);
    //②定义 HttpClient 对象
        HttpClient httpClient =
HttpClient.create(connectionProvider).httpResponseDecoder(spec -> {
            if (properties.getMaxHeaderSize() != null) {
                spec.maxHeaderSize((int)
properties.getMaxHeaderSize().toBytes());
            }
            if (properties.getMaxInitialLineLength() != null) {
                spec.maxInitialLineLength((int)
properties.getMaxInitialLineLength().toBytes());
            }
            return spec;
    //③配置 TcpClient
        }).tcpConfiguration(tcpClient -> {
            if (properties.getConnectTimeout() != null) {
                tcpClient =
tcpClient.option(ChannelOption.CONNECT_TIMEOUT_MILLIS,
                    properties.getConnectTimeout());
            }
            HttpClientProperties.Proxy proxy = properties.getProxy();
            if (StringUtils.hasText(proxy.getHost())) {
                tcpClient = tcpClient.proxy(proxySpec -> {
                    ProxyProvider.Builder builder =
proxySpec.type(proxy.getType()).host(proxy.getHost());
                    PropertyMapper map = PropertyMapper.get();
                    map.from(proxy::getPort).whenNonNull().to(builder::port);

map.from(proxy::getUsername).whenHasText().to(builder::username);
                    map.from(proxy::getPassword).whenHasText()
                        .to(password -> builder.password(s -> password));

map.from(proxy::getNonProxyHostsPattern).whenHasText().to(builder::nonProxyH
osts);
                });
            }
            return tcpClient;
    });
    //④配置 SSL
        HttpClientProperties.Ssl ssl = properties.getSsl();
          if ((ssl.getKeyStore() != null && ssl.getKeyStore().length() > 0)
```

```
                || ssl.getTrustedX509CertificatesForTrustManager().length > 0
|| ssl.isUseInsecureTrustManager()) {
            httpClient = httpClient.secure(sslContextSpec -> {
                SslContextBuilder sslContextBuilder =
SslContextBuilder.forClient();
                X509Certificate[] trustedX509Certificates =
ssl.getTrustedX509CertificatesForTrustManager();
                if (trustedX509Certificates.length > 0) {
                    sslContextBuilder =
sslContextBuilder.trustManager(trustedX509Certificates);
                }else if (ssl.isUseInsecureTrustManager()) {
                    sslContextBuilder =
sslContextBuilder.trustManager(InsecureTrustManagerFactory.INSTANCE);
                }
                try {
                    sslContextBuilder =
sslContextBuilder.keyManager(ssl.getKeyManagerFactory());
                }catch (Exception e) {
                    logger.error(e);
                }

sslContextSpec.sslContext(sslContextBuilder).defaultConfiguration(ssl.getDef
aultConfigurationType())
                    .handshakeTimeout(ssl.getHandshakeTimeout())
                    .closeNotifyFlushTimeout(ssl.getCloseNotifyFlushTimeout())
                    .closeNotifyReadTimeout(ssl.getCloseNotifyReadTimeout());
            });
        }
        if (properties.isWiretap()) {
            httpClient = httpClient.wiretap(true);
        }
        if (properties.isCompression()) {
            httpClient = httpClient.compress(true);
        }
        //⑤加载开发人员自定义的通信渠道
        if (!CollectionUtils.isEmpty(customizers)) {
            customizers.sort(AnnotationAwareOrderComparator.INSTANCE);
            for (HttpClientCustomizer customizer : customizers) {
                httpClient = customizer.customize(httpClient);
            }
        }
    //⑥返回 HttpClient 对象
    return httpClient;
}
```

2. 加载过滤器 NettyRoutingFilter

在 Spring Cloud Gateway 中，用过滤器 NettyRoutingFilter 来处理 API 请求，具体代码如下所示：

```
    @Bean
    @ConditionalOnEnabledGlobalFilter
    public NettyRoutingFilter routingFilter(HttpClient httpClient,
    ObjectProvider<List<HttpHeadersFilter>> headersFilters,
HttpClientProperties properties) {
    //定义过滤器 NettyRoutingFilter，并将其加载到 Spring Framework 的 IOC 容器中
        return new NettyRoutingFilter(httpClient, headersFilters, properties);
    }
```

3. 定义过滤器 NettyRoutingFilter 的优先级

在 Spring Cloud Gateway 中，不同的过滤器用优先级来确定加载顺序，NettyRoutingFilter 的优先级的具体实现如下：

```
    public class NettyRoutingFilter implements GlobalFilter, Ordered {
      public static final int ORDER = Ordered.LOWEST_PRECEDENCE;
      ...
    }
```

4. 用 filter()方法拦截 API 的 RPC 请求

在 NettyRoutingFilter 类中，用 filter()方法拦截 API 的 RPC 请求，具体代码如下所示：

```
    @Override
    @SuppressWarnings("Duplicates")
    public Mono<Void> filter(ServerWebExchange exchange,
          GatewayFilterChain chain) {
        //①获取请求路径
        URI requestUrl =
exchange.getRequiredAttribute(GATEWAY_REQUEST_URL_ATTR);
        //②解析出请求路径的 scheme
    String scheme = requestUrl.getScheme();
        if (isAlreadyRouted(exchange) || (!"http".equals(scheme)
&& !"https".equals(scheme))) {
            return chain.filter(exchange);
    }
    //③记录已经执行路由的 API 请求，防止重复路由
        setAlreadyRouted(exchange);
    ServerHttpRequest request = exchange.getRequest();
    //④解析出 API 请求的方法对象 HttpMethod
        final HttpMethod method = HttpMethod.valueOf(request.getMethodValue());
        final String url = requestUrl.toASCIIString();
```

```
        HttpHeaders filtered = filterRequest(getHeadersFilters(), exchange);
        final DefaultHttpHeaders httpHeaders = new DefaultHttpHeaders();
        filtered.forEach(httpHeaders::set);
    boolean preserveHost =
exchange.getAttributeOrDefault(PRESERVE_HOST_HEADER_ATTRIBUTE, false);
    //⑤从 Exchange 对象中读取路由规则
    Route route = exchange.getAttribute(GATEWAY_ROUTE_ATTR);
    //⑥获取提前初始化完成的 HttpClient 对象，如果连接超时，则重新初始化 HttpClient 对象
    Flux<HttpClientResponse> responseFlux = getHttpClient(route,
    //⑦构造请求头
    exchange).headers(headers -> {
            headers.add(httpHeaders);
        headers.remove(HttpHeaders.HOST);
            if (preserveHost) {
                String host = request.getHeaders().getFirst(HttpHeaders.HOST);
                headers.add(HttpHeaders.HOST, host);
            }
        //⑧执行 RPC 请求
        }).request(method).uri(url).send((req, nettyOutbound) -> {
            if (log.isTraceEnabled()) {
                nettyOutbound.withConnection(connection -> log.trace("outbound
route: "
                        + connection.channel().id().asShortText() + ", inbound:
" + exchange.getLogPrefix()));
            }
            return
nettyOutbound.send(request.getBody().map(this::getByteBuf));
        //⑨处理“请求-响应”的结果
        }).responseConnection((res, connection) -> {
            exchange.getAttributes().put(CLIENT_RESPONSE_ATTR, res);
            exchange.getAttributes().put(CLIENT_RESPONSE_CONN_ATTR,
connection);
            ServerHttpResponse response = exchange.getResponse();
            HttpHeaders headers = new HttpHeaders();
            res.responseHeaders().forEach(entry -> headers.add(entry.getKey(),
entry.getValue()));
            String contentTypeValue =
headers.getFirst(HttpHeaders.CONTENT_TYPE);
            if (StringUtils.hasLength(contentTypeValue)) {

exchange.getAttributes().put(ORIGINAL_RESPONSE_CONTENT_TYPE_ATTR,
contentTypeValue);
            }
            setResponseStatus(res, response);
```

```
        HttpHeaders filteredResponseHeaders =
HttpHeadersFilter.filter(getHeadersFilters(), headers, exchange,
                Type.RESPONSE);
        if
(!filteredResponseHeaders.containsKey(HttpHeaders.TRANSFER_ENCODING)
                &&
filteredResponseHeaders.containsKey(HttpHeaders.CONTENT_LENGTH)) {
            response.getHeaders().remove(HttpHeaders.TRANSFER_ENCODING);
        }
        exchange.getAttributes().put(CLIENT_RESPONSE_HEADER_NAMES,
filteredResponseHeaders.keySet());
        response.getHeaders().putAll(filteredResponseHeaders);
        return Mono.just(res);
      });
    Duration responseTimeout = getResponseTimeout(route);
    //⑩如果请求结果响应超时，则执行错误处理
      if (responseTimeout != null) {
        responseFlux = responseFlux
                .timeout(responseTimeout,
                        Mono.error(new TimeoutException("Response took
longer than timeout: " + responseTimeout)))
                .onErrorMap(TimeoutException.class,
                        th -> new ResponseStatusException
(HttpStatus.GATEWAY_TIMEOUT, th.getMessage(), th));
    }
      return responseFlux.then(chain.filter(exchange));
    }
```

9.3 用“路由规则定位器”（RouteDefinitionLocator）加载网关的路由规则

Spring Cloud Gateway 支持不同类型的路由规则定位器，主要包括注册中心、内存、Redis 及配置中心。

9.3.1 “基于注册中心的路由规则定位器”的原理

Spring Cloud Gateway 可以动态地从注册中心获取服务实例信息，目前主要支持 Nacos、Consul、Eureka 等注册中心。

1. 从注册中心获取服务路由规则的流程

Spring Cloud Gateway 支持从注册中心获取路由规则，并完成服务路由，如图 9-9 所示。

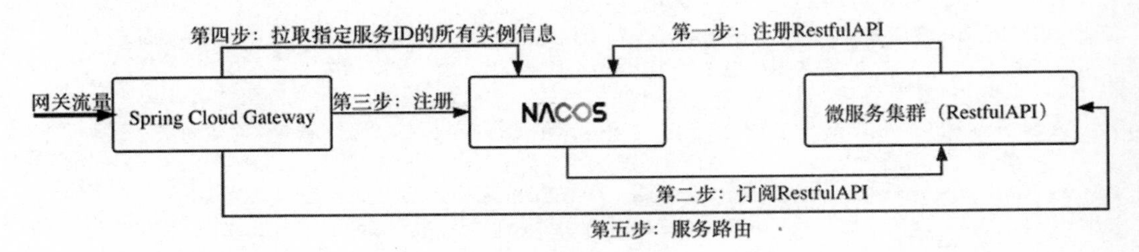

图 9-9

（1）基于 RESTful API 的微服务集群，将 API 注册到 Nacos 注册中心中。

（2）微服务可以订阅注册中心的 API，完成 RPC。

（3）Spring Cloud Gateway 在启动时注册到 Nacos 注册中心中。

（4） Spring Cloud Gateway 拉取指定服务 ID 的所有健康的实例信息，并设置到本地缓存中。

（5）Spring Cloud Gateway 解析实例信息并生成服务路由信息，进行服务路由。

2. 初始化注册中心定位器

Spring Cloud Gateway 用自动配置类 GatewayDiscoveryClientAutoConfiguration 来初始化注册中心定位器（DiscoveryClientRouteDefinitionLocator 类），具体代码如下所示：

```
@Configuration(proxyBeanMethods = false)
@ConditionalOnProperty(name = "spring.cloud.gateway.enabled",
matchIfMissing = true)
@AutoConfigureBefore(GatewayAutoConfiguration.class)
@AutoConfigureAfter(CompositeDiscoveryClientAutoConfiguration.class)
@ConditionalOnClass({ DispatcherHandler.class })
@EnableConfigurationProperties
public class GatewayDiscoveryClientAutoConfiguration {
    //①初始化断言信息
    public static List<PredicateDefinition> initPredicates() {
        ArrayList<PredicateDefinition> definitions = new ArrayList<>();
        PredicateDefinition predicate = new PredicateDefinition();
        predicate.setName(normalizeRoutePredicateName
        (PathRoutePredicateFactory.class));
        predicate.addArg(PATTERN_KEY, "'/'+serviceId+'/**'");
        definitions.add(predicate);
        return definitions;
    }
    //②初始化过滤器
    public static List<FilterDefinition> initFilters() {
        ArrayList<FilterDefinition> definitions = new ArrayList<>();
        FilterDefinition filter = new FilterDefinition();
```

```
        filter.setName(normalizeFilterFactoryName
        (RewritePathGatewayFilterFactory.class));
        String regex = "'/' + serviceId + '/(?<remaining>.*)'";
        String replacement = "'/${remaining}'";
        filter.addArg(REGEXP_KEY, regex);
        filter.addArg(REPLACEMENT_KEY, replacement);
        definitions.add(filter);
        return definitions;
  }
     @Bean
  public DiscoveryLocatorProperties discoveryLocatorProperties() {
        //③加载应用中关于注册中心路由的配置信息
        DiscoveryLocatorProperties properties = new
DiscoveryLocatorProperties();
        //④加载断言信息
        properties.setPredicates(initPredicates());
        //⑤加载过滤器
        properties.setFilters(initFilters());
        return properties;
     }
     @Configuration(proxyBeanMethods = false)
     @ConditionalOnProperty(value =
"spring.cloud.discovery.reactive.enabled",
           matchIfMissing = true)
  public static class ReactiveDiscoveryClientRoute
  DefinitionLocatorConfiguration {
        @Bean
        @ConditionalOnProperty(name =
"spring.cloud.gateway.discovery.locator.enabled")
        public DiscoveryClientRouteDefinitionLocator
discoveryClientRouteDefinitionLocator(
              ReactiveDiscoveryClient discoveryClient,
              DiscoveryLocatorProperties properties) {
         //⑥如果应用中配置“spring.cloud.gateway.discovery.locator.enabled”
为 true，则初始化注册中心定位器的 DiscoveryClientRouteDefinitionLocator 类
           return new
DiscoveryClientRouteDefinitionLocator(discoveryClient, properties);
        }
     }
     ...
  }
```

DiscoveryClientRouteDefinitionLocator 类在初始化的过程中，会从注册中心获取服务实例信息，具体代码如下所示：

```
    public class DiscoveryClientRouteDefinitionLocator implements
RouteDefinitionLocator {
        private final DiscoveryLocatorProperties properties;
        private Flux<List<ServiceInstance>> serviceInstances;

        public DiscoveryClientRouteDefinitionLocator(ReactiveDiscoveryClient
discoveryClient,
                DiscoveryLocatorProperties properties) {
            this(discoveryClient.getClass().getSimpleName(), properties);
            //从注册中心获取服务实例信息，并缓存在变量 serviceInstances 中
            serviceInstances = discoveryClient.getServices()
           .flatMap(service -> discoveryClient.getInstances(
           service).collectList());
        }
        ...
    }
```

3. 解析服务实例信息并生成服务路由规则

Spring Cloud Gateway 在获取服务实例信息后，需要将服务实例信息转换为服务路由规则，具体代码如下所示：

```
    public class DiscoveryClientRouteDefinitionLocator
                        implements RouteDefinitionLocator {
        @Override
        public Flux<RouteDefinition> getRouteDefinitions() {
            ...
            //①遍历服务实例信息列表
            return serviceInstances.filter(instances -> !instances.isEmpty())
                    .map(instances -> instances.get(0)).filter(includePredicate)
                    .map(instance -> {
                        RouteDefinition routeDefinition
                           =buildRouteDefinition(urlExpr,instance);
                        final ServiceInstance instanceForEval
                           = new DelegatingServiceInstance(instance, properties);
                        //②解析配置文件中的断言信息
                        for (PredicateDefinition original
                              : this.properties.getPredicates()) {
                            PredicateDefinition predicate = new
PredicateDefinition();
                            predicate.setName(original.getName());
                            for (Map.Entry<String, String> entry :
original.getArgs()
                                    .entrySet()) {
                                String value = getValueFromExpr(evalCtxt, parser,
                                        instanceForEval, entry);
```

```
                        predicate.addArg(entry.getKey(), value);
                    }
                    routeDefinition.getPredicates().add(predicate);
                }
                //③解析配置文件中的过滤器
                for (FilterDefinition original : this.properties
                     .getFilters()) {
                    FilterDefinition filter = new FilterDefinition();
                    filter.setName(original.getName());
                    for (Map.Entry<String, String> entry :
original.getArgs()
                            .entrySet()) {
                        String value = getValueFromExpr(evalCtxt, parser,
                                instanceForEval, entry);
                        filter.addArg(entry.getKey(), value);
                    }
                    routeDefinition.getFilters().add(filter);
                }
                //④返回路由规则
                return routeDefinition;
            });
    }
    ...
}
```

9.3.2 “基于内存的路由规则定位器”的原理

Spring Cloud Gateway 支持用内存来存储路由规则。开发人员可以使用 Open API 来维护内存中的路由规则，具体代码如下所示：

```
public class InMemoryRouteDefinitionRepository implements
RouteDefinitionRepository {
    //①基于内存的路由规则
    private final Map<String, RouteDefinition> routes = synchronizedMap(
            new LinkedHashMap<String, RouteDefinition>());
    //②新增路由规则
    @Override
    public Mono<Void> save(Mono<RouteDefinition> route) {
        return route.flatMap(r -> {
            if (StringUtils.isEmpty(r.getId())) {
                return Mono.error(new IllegalArgumentException("id may not be
empty"));
            }
            routes.put(r.getId(), r);
            return Mono.empty();
```

```
            });
        }
        //③删除路由规则
        @Override
        public Mono<Void> delete(Mono<String> routeId) {
            return routeId.flatMap(id -> {
                if (routes.containsKey(id)) {
                    routes.remove(id);
                    return Mono.empty();
                }
                return Mono.defer(() -> Mono.error(
                        new NotFoundException("RouteDefinition not found: " +
routeId)));
            });
        }
        //④返回内存中的路由规则
        @Override
        public Flux<RouteDefinition> getRouteDefinitions() {
            return Flux.fromIterable(routes.values());
        }
    }
```

Spring Cloud Gateway 用 AbstractGatewayControllerEndpoint 类暴露 Open API，具体代码如下所示：

```
    public class AbstractGatewayControllerEndpoint implements
ApplicationEventPublisherAware {
        //①新增路由规则
        @PostMapping("/routes/{id}")
        @SuppressWarnings("unchecked")
        public Mono<ResponseEntity<Object>> save(@PathVariable String id,
                @RequestBody RouteDefinition route) {
            return Mono.just(route).filter(this::validateRouteDefinition)
                    .flatMap(routeDefinition -> this.routeDefinitionWriter
                            .save(Mono.just(routeDefinition).map(r -> {
                                r.setId(id);
                                log.debug("Saving route: " + route);
                                return r;
                            }))
                            .then(Mono.defer(() -> Mono.just(ResponseEntity
                                    .created(URI.create("/routes/" +
id)).build()))))
                    .switchIfEmpty(
                            Mono.defer(() ->
Mono.just(ResponseEntity.badRequest().build())));
        }
```

```
        //②删除路由规则
        @DeleteMapping("/routes/{id}")
        public Mono<ResponseEntity<Object>> delete(@PathVariable String id) {
            return this.routeDefinitionWriter.delete(Mono.just(id))
                    .then(Mono.defer(() ->
Mono.just(ResponseEntity.ok().build())))
                    .onErrorResume(t -> t instanceof NotFoundException,
                            t -> Mono.just(ResponseEntity.notFound().build()));
        }
        ...
    }
```

9.3.3 “基于 Redis 缓存的路由规则定位器”的原理

Spring Cloud Gateway 默认不支持“基于 Redis 缓存来缓存路由规则”，需要开发人员改造规则定位器，具体代码如下所示：

```
    @Component
    public class RedisRouteDefinitionRepository
            implements RouteDefinitionRepository {
        public static final String GATEWAY_ROUTES_PREFIX = "GETEWAY_ROUTES";

        @Resource
        private StringRedisTemplate stringRedisTemplate;
        //①用 HashSet 来缓存路由规则
        private Set<RouteDefinition> routeDefinitions = new HashSet<>();

        @Override
        public Flux<RouteDefinition> getRouteDefinitions() {
            routeDefinitions.clear();
            BoundHashOperations<String, String, String> boundHashOperations =
stringRedisTemplate.boundHashOps(GATEWAY_ROUTES_PREFIX);
            //②从 Redis 中读取路由规则
            Map<String, String> map = boundHashOperations.entries();
            Iterator<Map.Entry<String, String>> it = map.entrySet().iterator();
            //③将路由规则更新到本地缓存中
            while (it.hasNext()) {
                Map.Entry<String, String> entry = it.next();
                routeDefinitions.add(JSON.parseObject(entry.getValue(),
RouteDefinition.class));
            }
            return Flux.fromIterable(routeDefinitions);
    }
        //④开发人员可以通过 save()方法新增路由规则
        @Override
```

```
    public Mono<Void> save(Mono<RouteDefinition> route) {
        return route.flatMap(routeDefinition -> {
            routeDefinitions.add(routeDefinition);
            return Mono.empty();
        });
    }
    //⑤开发人员可以通过 delete()方法删除路由规则
    @Override
    public Mono<Void> delete(Mono<String> routeId) {
        return routeId.flatMap(id -> {
            List<RouteDefinition> routeDefinitionList =
routeDefinitions.stream().filter(
                    routeDefinition ->
StringUtils.equals(routeDefinition.getId(), id)
            ).collect(Collectors.toList());
            routeDefinitions.removeAll(routeDefinitionList);
            return Mono.empty();
        });
    }
}
```

9.3.4 “基于属性文件的路由规则定位器”的原理

Spring Cloud Gateway 默认支持“基于属性文件的路由规则定位器”，具体实现方式是用 PropertiesRouteDefinitionLocator 类从属性文件中读取路由规则，具体代码如下所示：

```
public class PropertiesRouteDefinitionLocator
   implements RouteDefinitionLocator {
    private final GatewayProperties properties;

    public PropertiesRouteDefinitionLocator(
        GatewayProperties properties) {
        this.properties = properties;
    }

    @Override
    public Flux<RouteDefinition> getRouteDefinitions() {
        //从网关属性类 GatewayProperties 中读取所有的路由规则，不支持动态路由
        return Flux.fromIterable(this.properties.getRoutes());
    }
}
```

9.3.5 【实例】用“基于注册中心和配置中心的路由规则定位器”在网关统一暴露 API

本实例的源码在本书配套资源的“chapternine/user-api”“chapternine /order-api/”和“chapternine/use-spring-cloud-alibaba-nacos-config-gateway”目录下。

本实例用“基于注册中心和配置中心的路由规则定位器”在网关统一暴露 API。

1. 初始化项目

本实例主要包括 3 个项目：业务网关 use-spring-cloud-alibaba-nacos-config-gateway、user-api 服务和 order-api 服务。

（1）使用 Spring Cloud Alibaba 初始化业务网关，部分 POM 文件的依赖如下：

```
<!—添加 webflux 的依赖包-->
<dependency>
    <groupId>org.springframework.boot</groupId>
    <artifactId>spring-boot-starter-webflux</artifactId>
</dependency>
<!—添加负载均衡 loadbalancer 的依赖包-->
<dependency>
    <groupId>org.springframework.cloud</groupId>
    <artifactId>spring-cloud-starter-loadbalancer</artifactId>
</dependency>
...
```

（2）在本实例的 3 个项目中，添加项目启动的配置信息。其中，业务网关的配置信息如下所示，其他服务的配置信息可以参考本书配套资源中的代码。

```
spring:
  cloud:
    gateway:
      discovery:
        locator:
          lower-case-service-id: true
          enabled: false
    nacos:
      ###①连接 Nacos 配置中心的配置信息
      config:
        namespace: c7ba173f-29e5-4c58-ae78-b102be11c4f9
        group: gateway-dynamic-route-rule
        enable-remote-sync-config: true
        server-addr: 127.0.0.1:8848
```

```
        file-extension: yaml
        ###②从 Nacos 配置中心的文件 gateway-dynamic-route-rule.json 中读取路由规则
        prefix: gateway-dynamic-route-rule
  ###③Redis 的连接信息
  redis:
    host: localhost
    port: 6379
    database: 0
```

2. 在配置中心中添加业务网关的路由规则

在 Nacos 控制台中添加配置文件 gateway-dynamic-route-rule.json，并配置如下信息：

```
[{
   "id": "user-api-router",
   "order": 0,
   "predicates": [{
       "args": {
           "pattern": "/user/**"
       },
       "name": "Path"
   }],
   "uri": "lb://user-api"
}]
```

在 Nacos 配置中心中，服务 user-api 的路由规则已经生效，如图 9-10 所示。

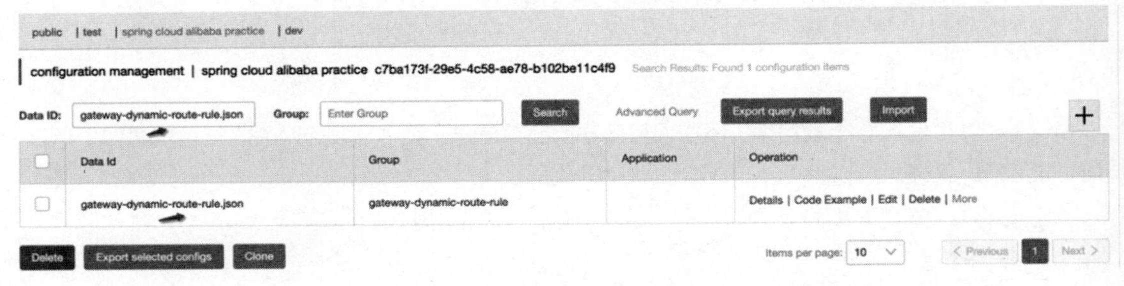

图 9-10

4. 动态读取路由规则，并将其加载到业务网关中

定义一个 DynamicRouteService 类并实现 ApplicationEventPublisherAware 接口，具体代码如下所示：

```
@Slf4j
@Service
public class DynamicRouteService implements ApplicationEventPublisherAware {
    //①引入路由规则写入器 RouteDefinitionWriter
    @Autowired
```

```
    private RouteDefinitionWriter routeDefinitionWriter;
    //②引入路由规则定位器
        @Autowired
        private RouteDefinitionLocator routeDefinitionLocator;
        @Autowired
        private ApplicationEventPublisher publisher;
        //③设置 Spring Framework 的事件发布器，用于在路由规则变更后通知应用
        @Override
        public void setApplicationEventPublisher(ApplicationEventPublisher
applicationEventPublisher) {
            this.publisher = applicationEventPublisher;
        }
        //④删除路由规则
        public String delete(String id) {
            try {
                log.info("gateway delete route id {}",id);
                this.routeDefinitionWriter.delete(Mono.just(id)).subscribe();
                this.publisher.publishEvent(new RefreshRoutesEvent(this));
                return "delete success";
            } catch (Exception e) {
                return "delete fail";
            }
        }
        //⑤更新路由规则
        public String updateList(List<RouteDefinition> definitions) {
            log.info("gateway update route {}",definitions);
            List<RouteDefinition> routeDefinitionsExits =
routeDefinitionLocator.getRouteDefinitions().buffer().blockFirst();
            if (!CollectionUtils.isEmpty(routeDefinitionsExits)) {
                routeDefinitionsExits.forEach(routeDefinition -> {
                    log.info("delete routeDefinition:{}", routeDefinition);
                    delete(routeDefinition.getId());
                });
            }
            definitions.forEach(definition -> {
                updateById(definition);
            });
            return "success";
        }

        public String updateById(RouteDefinition definition) {
            try {
                log.info("gateway update route {}",definition);

this.routeDefinitionWriter.delete(Mono.just(definition.getId()));
```

```
        } catch (Exception e) {
            return "update fail,not find route  routeId:
"+definition.getId();
        }
        try {
            routeDefinitionWriter.save(Mono.just(definition)).subscribe();
            this.publisher.publishEvent(new RefreshRoutesEvent(this));
            return "success";
        } catch (Exception e) {
            return "update route fail";
        }
    }
    //⑥添加路由规则
    public String add(RouteDefinition definition) {
        log.info("gateway add route {}",definition);
        routeDefinitionWriter.save(Mono.just(definition)).subscribe();
        this.publisher.publishEvent(new RefreshRoutesEvent(this));
        return "success";
    }
}
```

用 NacosRouteDynamicDataSource 类实现 ApplicationRunner 接口，从配置中心动态地读取路由规则，具体代码如下所示：

```
@Component
@RefreshScope
@Slf4j
public class NacosRouteDynamicDataSource implements ApplicationRunner {
    //①引入 Spring Cloud Alibaba 提供的配置中心管理器 NacosConfigManager
    @Autowired
    private NacosConfigManager nacosConfigManager;
    private ConfigService configService;
    @Override
    public void run(ApplicationArguments args) throws Exception {
        ExecutorService executorService= Executors.newFixedThreadPool(1);
        //②开启线程执行动态的读取路由规则
        executorService.execute(new RefreshRouteCache());
    }
    class RefreshRouteCache implements Runnable{
        @Override
        public void run() {
            while (true) {
                log.info("gateway route init...");
                try {
                    if (null == configService) {
```

```
                    configService =
nacosConfigManager.getConfigService();
                }
             //③使用 Nacos Config 的 ConfigService 接口读取
gateway-dynamic-route-rule.json 的路由规则
                String configInfo =
configService.getConfig("gateway-dynamic-route-rule.json",
"gateway-dynamic-route-rule"
                        , 4000);
                log.info("获取网关当前配置:\r\n{}", configInfo);
                List<RouteDefinition> definitionList =
JSON.parseArray(configInfo, RouteDefinition.class);
                //④调用 DynamicRouteService 类添加路由规则
                for (RouteDefinition definition : definitionList) {
                    log.info("update route : {}", definition.toString());
                    dynamicRouteService.add(definition);
                }
                Thread.sleep(20000);
             } catch (Exception e) {
                log.error("初始化网关路由时发生错误", e);
             }
             //⑤添加 gateway-dynamic-route-rule.json 的监听器
             dynamicRouteByNacosListener("gateway-dynamic-route-
rule.json", "gateway-dynamic-route-rule");
         }
      }
   }

   public void dynamicRouteByNacosListener (String dataId, String group){
      try {
         configService.addListener(dataId, group, new Listener()  {
            @Override
            public void receiveConfigInfo(String configInfo) {
               log.info("进行网关更新:\n\r{}",configInfo);
               //⑥如果配置中心中 gateway-dynamic-route-rule.json 文件的配置
信息发生变更，则主动地更新业务网关中的路由规则
               List<RouteDefinition> definitionList =
JSON.parseArray(configInfo, RouteDefinition.class);
               log.info("update route : {}",definitionList.toString());
               dynamicRouteService.updateList(definitionList);
            }
            @Override
            public Executor getExecutor() {
               log.info("getExecutor\n\r");
               return null;
```

```
                }
            });
        } catch (NacosException e) {
            log.error("从 nacos 接收动态路由配置出错!!!",e);
        }
    }
    @Resource
    private DynamicRouteService dynamicRouteService;
}
```

5. 验证业务网关到用户 API 的路由效果

执行命令“curl http://127.0.0.1:28082/user-api/user/getUserInfo”（如图 9-11 所示），这样之后对业务网关 IP 地址的访问就会直接被注册中心路由到用户 API。

```
huxian@huxians-MacBook-Pro bin % curl http://127.0.0.1:28082/user-api/user/getUserInfo
192.168.0.123:50713 已经通过网关路由到了用户服务%
huxian@huxians-MacBook-Pro bin %
```

图 9-11

6. 验证动态路由效果

在没有添加订单 API 的路由规则前，执行命令“curl http://127.0.0.1:28082/order-api/order/getOrderInfo”，订单 API 响应 404 错误，如图 9-12 所示。

```
huxian@huxians-MacBook-Pro bin % curl http://127.0.0.1:28082/order-api/order/getOrderInfo
{"timestamp":"2021-05-26T17:42:25.674+0000","path":"/order-api/order/getOrderInfo","status":404,"error":"Not Found","message":null
,"requestId":"a0303526-30"}%
huxian@huxians-MacBook-Pro bin %
```

图 9-12

在 Nacos 配置中心的配置文件 gateway-dynamic-route-rule.json 中，添加订单 API 的路由规则，如图 9-13 所示。

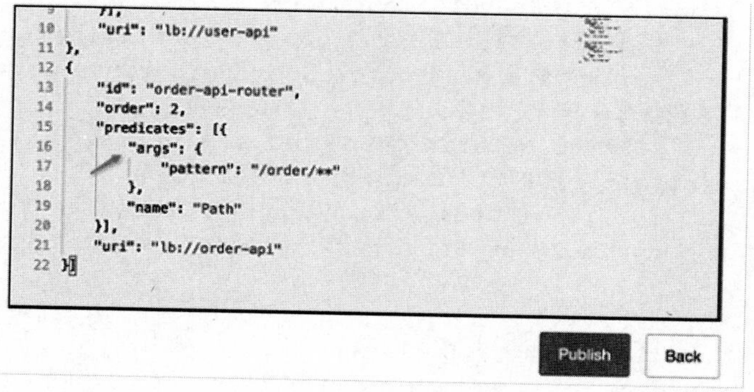

```
10      "uri": "lb://user-api"
11  },
12  {
13      "id": "order-api-router",
14      "order": 2,
15      "predicates": [{
16          "args": {
17              "pattern": "/order/**"
18          },
19          "name": "Path"
20      }],
21      "uri": "lb://order-api"
22  }]
```

Publish　Back

图 9-13

在订单 API 的路由规则的配置信息添加成功后，在不重启业务网关的前提下，执行命令“curl http://127.0.0.1:28082/order-api/order/getOrderInfo”，调用订单 API 成功，如图 9-14 所示。

```
huxian@huxians-MacBook-Pro bin % curl http://127.0.0.1:28082/order-api/order/getOrderInfo
192.168.0.123:54264 已经通过网关路由到了订单服务
huxian@huxians-MacBook-Pro bin %
```

图 9-14

9.4 用“Redis + Lua”进行网关 API 的限流

在微服务架构中，分布式限流是确保服务稳定性的一项关键性技术。Spring Cloud Gateway 集成 Spring Data Redis，支持开发人员采用“Redis + Lua”模式，并利用 Redis 的分布式限流算法，实现网关 API 的限流。

9.4.1 “网关用 Redis + Lua 实现分布式限流”的原理

下面分析“网关用 Redis + Lua 实现分布式限流”的原理。

1. 加载限流算法

Spring Cloud Gateway 默认实现了“Redis + Lua 模式”的限流算法——令牌桶算法。如图 9-15 所示，在 Spring Cloud Gateway Core 工程的 request_rate_limiter.lua 文件中，用 Lua 脚本语言定义了令牌桶算法。开发人员只需要按照 Spring Cloud Gateway 制定的配置语法，在配置信息中添加令牌桶算法对应的路由限流规则，即可使用 Spring Cloud Gateway 默认的限流算法。

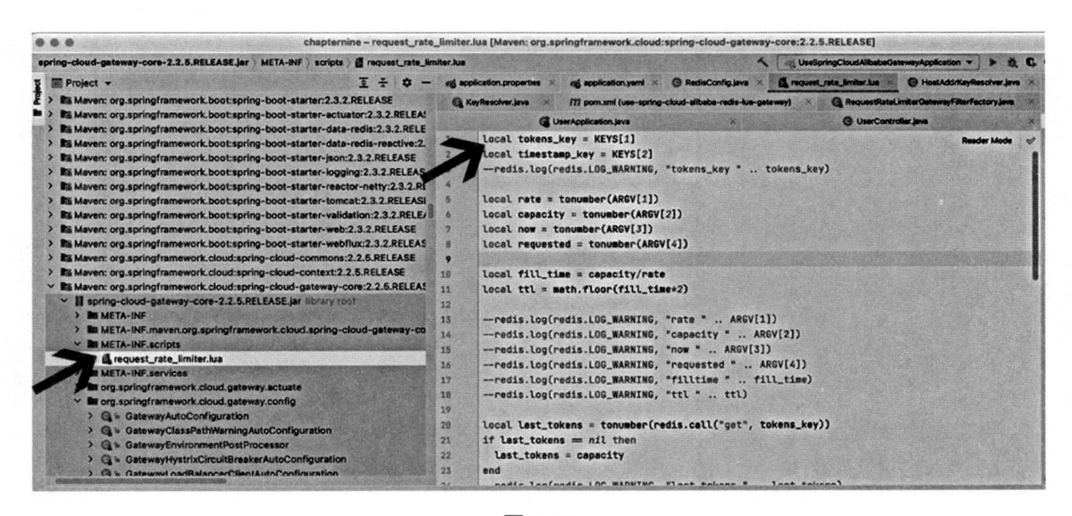

图 9-15

在 Spring Cloud Gateway 中，用自动配置类 GatewayRedisAutoConfiguration 的 redisRequestRateLimiterScript()方法来加载限流算法，具体代码如下所示：

```
@Configuration(proxyBeanMethods = false)
@AutoConfigureAfter(RedisReactiveAutoConfiguration.class)
@AutoConfigureBefore(GatewayAutoConfiguration.class)
@ConditionalOnBean(ReactiveRedisTemplate.class)
@ConditionalOnClass({ RedisTemplate.class, DispatcherHandler.class })
class GatewayRedisAutoConfiguration {
    @Bean
    @SuppressWarnings("unchecked")
    public RedisScript redisRequestRateLimiterScript() {
         //①定义 Redis 脚本解析器
        DefaultRedisScript redisScript = new DefaultRedisScript<>();
         //②设置 Lua 脚本的存储路径，用于读取限流算法
        redisScript.setScriptSource(
         new ResourceScriptSource(new ClassPathResource
("META-INF/scripts/request_rate_limiter.lua")));
Aa        redisScript.setResultType(List.class);
        return redisScript;
    }
    ...
}
```

2. 加载 Redis 限流器

用自动配置类 GatewayRedisAutoConfiguration 的 redisRateLimiter()方法来加载 Redis 限流器，具体代码如下所示：

```
    @Bean
    @ConditionalOnMissingBean
    public RedisRateLimiter redisRateLimiter(ReactiveStringRedisTemplate
redisTemplate,
        @Qualifier(RedisRateLimiter.REDIS_SCRIPT_NAME)
RedisScript<List<Long>> redisScript,
        ConfigurationService configurationService) {
         //加载 Redis 限流器 RedisRateLimiter，并设置脚本解析器的 DefaultRedisScript
对象和 Redis 客户端的 ReactiveStringRedisTemplate 对象
        return new RedisRateLimiter(redisTemplate, redisScript,
configurationService);
    }
```

3. 初始化请求限流过滤器工厂类 RequestRateLimiterGatewayFilterFactory

在网关启动的过程中，会执行自动配置类 GatewayAutoConfiguration，并初始化请求限流过滤器工厂类 RequestRateLimiterGatewayFilterFactory，具体代码如下所示：

```
@Configuration(proxyBeanMethods = false)
@ConditionalOnProperty(name = "spring.cloud.gateway.enabled",
matchIfMissing = true)
@EnableConfigurationProperties
@AutoConfigureBefore({ HttpHandlerAutoConfiguration.class,
WebFluxAutoConfiguration.class })
@AutoConfigureAfter({ GatewayReactiveLoadBalancerClientAutoConfiguration
.class,
        GatewayClassPathWarningAutoConfiguration.class })
@ConditionalOnClass(DispatcherHandler.class)
public class GatewayAutoConfiguration {
    //①如果开发人员没有自定义限流键的解析器（KeyResolver 接口的实现类），则网关默认
的是限流键解析器 PrincipalNameKeyResolver
    @Bean(name = PrincipalNameKeyResolver.BEAN_NAME)
    @ConditionalOnBean(RateLimiter.class)
    @ConditionalOnMissingBean(KeyResolver.class)

@ConditionalOnEnabledFilter(RequestRateLimiterGatewayFilterFactory.class)
    public PrincipalNameKeyResolver principalNameKeyResolver() {
        return new PrincipalNameKeyResolver();
    }
    //②如果在应用中存在 Redis 限流器和限流键的解析器，则初始化请求限流过滤器工厂类
RequestRateLimiterGatewayFilterFactory
    @Bean
    @ConditionalOnBean({ RateLimiter.class, KeyResolver.class })
    @ConditionalOnEnabledFilter
    public RequestRateLimiterGatewayFilterFactory
requestRateLimiterGatewayFilterFactory(RateLimiter rateLimiter,
            KeyResolver resolver) {
        return new RequestRateLimiterGatewayFilterFactory(rateLimiter,
resolver);
    }
    ...
}
```

4. 用 GatewayProperties 类读取开发人员配置的限流规则

读取开发人员配置的限流规则的 GatewayProperties 类的具体代码如下所示：

```
@ConfigurationProperties("spring.cloud.gateway")
@Validated
public class GatewayProperties {
    @NotNull
    @Valid
    private List<RouteDefinition> routes = new ArrayList<>();
    public List<RouteDefinition> getRoutes() {
```

```
        return routes;
    }
    public void setRoutes(List<RouteDefinition> routes) {
        this.routes = routes;
        if (routes != null && routes.size() > 0 && logger.isDebugEnabled()) {
            logger.debug("Routes supplied from Gateway Properties: " +
routes);
        }
    }
    ...
}
```

Spring Cloud Gateway 会读取以“spring.cloud.gateway.*”为前缀的路由配置信息。application.yaml 中的具体配置信息示例如下：

```
spring:
  cloud:
 gateway:
   ###对应 GatewayProperties 类的属性字段 routes。routes 是一个数组链表，用于存储路
由信息 RouteDefinition
      routes:
        - id: redis_limit_route
          uri: http://127.0.0.1:28067
          predicates:
            - Path=/**
          filters:
            - name: RequestRateLimiter
              args:
                key-resolver: '#{@hostAddrKeyResolver}'
                redis-rate-limiter.replenishRate: 5
                redis-rate-limiter.burstCapacity: 10
```

在 Spring Cloud Gateway 中，路由配置信息与 RouteDefinition 类中的属性字段是一一对应的，具体代码如下所示：

```
@Validated
public class RouteDefinition {
    //①对应 application.yaml 中的配置属性字段 id
    private String id;
    //②对应 application.yaml 中的配置属性字段 predicates
   @NotEmpty
   @Valid
   private List<PredicateDefinition> predicates = new ArrayList<>();
    //③对应 application.yaml 中的配置属性字段 filters
   @Valid
```

```
    private List<FilterDefinition> filters = new ArrayList<>();
    //④对应 application.yaml 中的配置属性字段 uri
    @NotNull
    private URI uri;
    //⑤对应 application.yaml 中的配置属性字段 metadata
    private Map<String, Object> metadata = new HashMap<>();
    private int order = 0;
    ...
}
```

通过 Debug 模式调试 Spring Cloud Gateway 读取限流规则的过程如图 9-16 所示。

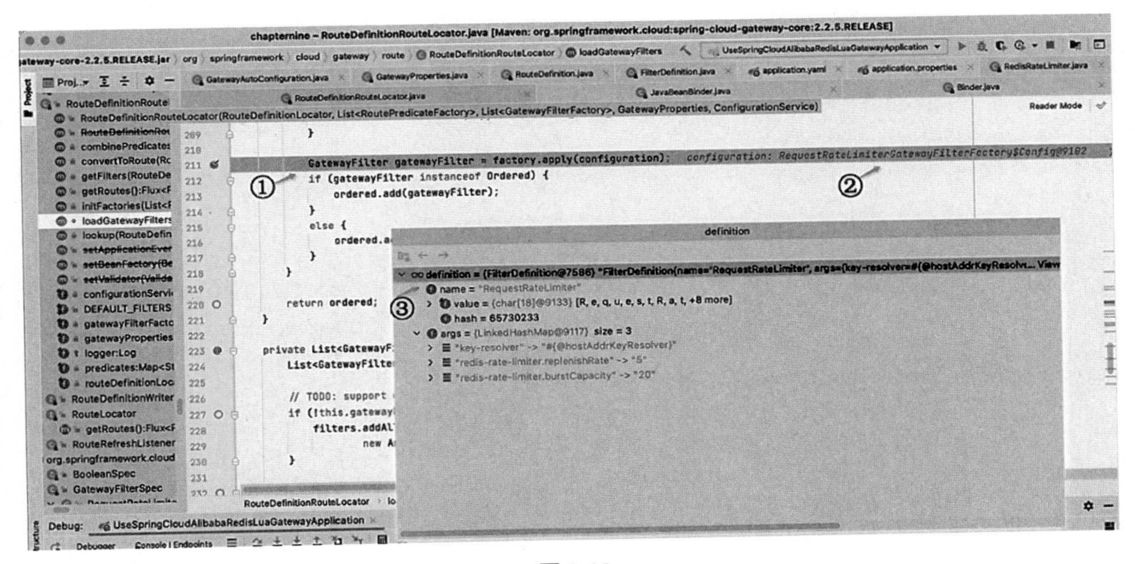

图 9-16

5. 用 RouteDefinitionRouteLocator 类加载请求路由限流过滤器

在 Spring Cloud Gateway 中，用 RouteDefinitionRouteLocator 类加载请求路由限流过滤器。

（1）用 getRoutes()方法获取网关的路由信息，具体代码如下所示：

```
public class RouteDefinitionRouteLocator implements RouteLocator,
        BeanFactoryAware, ApplicationEventPublisherAware {
    @Override
    public Flux<Route> getRoutes() {
        //①从 RouteDefinitionRouteLocator 类中获取路由定义，并将其转换为路由器
        Flux<Route> routes = this.routeDefinitionLocator.
         getRouteDefinitions().map(this::convertToRoute);
        if (!gatewayProperties.isFailOnRouteDefinitionError()) {
            routes = routes.onErrorContinue((error, obj) -> {
```

```
                if (logger.isWarnEnabled()) {
                    logger.warn("RouteDefinition id " + ((RouteDefinition)
obj).getId()
                            + " will be ignored. Definition has invalid configs, "
                            + error.getMessage());
                }
            });
        }
         //②返回路由规则
        return routes.map(route -> {
            if (logger.isDebugEnabled()) {
                logger.debug("RouteDefinition matched: " + route.getId());
            }
            return route;
        });
    }
    ...
}
```

（2）用 getFilters()方法获取路由过滤器，具体代码如下所示：

```
//①将路由规则转换为路由器
private Route convertToRoute(RouteDefinition routeDefinition) {
        AsyncPredicate<ServerWebExchange> predicate =
combinePredicates(routeDefinition);
         //②将路由规则转换为 GatewayFilter 类
        List<GatewayFilter> gatewayFilters = getFilters(routeDefinition);
    return Route.async(routeDefinition).asyncPredicate(predicate)
            .replaceFilters(gatewayFilters).build();
}
private List<GatewayFilter> getFilters(RouteDefinition routeDefinition) {
    List<GatewayFilter> filters = new ArrayList<>();
    if (!this.gatewayProperties.getDefaultFilters().isEmpty()) {
        filters.addAll(loadGatewayFilters(DEFAULT_FILTERS,
            new ArrayList<>(this.gatewayProperties.getDefaultFilters())));
    }
     //③加载指定路由 ID 的路由规则
    if (!routeDefinition.getFilters().isEmpty()) {
        filters.addAll(loadGatewayFilters(routeDefinition.getId(),
                new ArrayList<>(routeDefinition.getFilters())));
    }
    AnnotationAwareOrderComparator.sort(filters);
    return filters;
}
```

（3）用 loadGatewayFilters()方法将路由 ID 对应的路由规则转换为 GatewayFilter 类，具体

代码如下所示：

```
    List<GatewayFilter> loadGatewayFilters(String id,
            List<FilterDefinition> filterDefinitions) {
        ArrayList<GatewayFilter> ordered =
         new ArrayList<>(filterDefinitions.size());
         //①遍历配置文件中定义的路由规则中的过滤器
        for (int i = 0; i < filterDefinitions.size(); i++) {
            FilterDefinition definition = filterDefinitions.get(i);
             //②从本地的网关过滤器工厂缓存中，获取指定过滤器名称的网关过滤器工厂对象
RequestRateLimiterGatewayFilterFactory
            GatewayFilterFactory factory = this.gatewayFilterFactories
                    .get(definition.getName());
            //③如果当前应用中没有过滤器的工厂类 GatewayFilterFactory，则直接抛出异常
             if (factory == null) {
                throw new IllegalArgumentException(
                        "Unable to find GatewayFilterFactory with name "
                                + definition.getName());
            }
            if (logger.isDebugEnabled()) {
                logger.debug("RouteDefinition " + id + " applying filter "
                        + definition.getArgs() + " to " + definition.getName());
            }
            //④获取 Spring Cloud Gateway 的配置服务类 ConfigurationService
            Object configuration = this.configurationService.with(factory)
                .name(definition.getName())
                .properties(definition.getArgs())
                .eventFunction((bound, properties) -> new FilterArgsEvent(
                RouteDefinitionRouteLocator.this, id, (Map<String, Object>)
properties))
                .bind();
            //⑤如果 ConfigurationService 类的类型匹配 HasRouteId 类的类型，则将路由 ID
设置到 HasRouteId 类中
            if (configuration instanceof HasRouteId) {
                HasRouteId hasRouteId = (HasRouteId) configuration;
                hasRouteId.setRouteId(id);
            }
            //⑥用网关过滤器工厂类 GatewayFilterFactory 的方法 apply()从
ConfigurationService 类中的获取网关过滤器 GatewayFilter 类
            GatewayFilter gatewayFilter = factory.apply(configuration);
            //⑦如果 GatewayFilter 类是有优先级，则将 GatewayFilter 类添加到链表容器
ordered 中
            if (gatewayFilter instanceof Ordered) {
                ordered.add(gatewayFilter);
            }else {
```

```
            //⑧如果 GatewayFilter 类是没有优先级，则使用 OrderedGatewayFilter 类定
义一个默认的优先级
            ordered.add(new OrderedGatewayFilter(gatewayFilter, i + 1));
        }
    }
    //⑨返回包含网关过滤器的链表容器
    return ordered;
}
```

9.4.2 【实例】将 Spring Cloud Alibaba 应用接入网关，用“Redis +Lua”进行限流

本实例的源码在本书配套资源的“chapternine/ use-spring-cloud-alibaba-redis-lua-gateway”和“chapternine/user-api”目录下。

本实例包含两个项目：业务网关 use-spring-cloud-alibaba-redis-lua-gateway 和业务项目 user-api。

1. 初始化项目

下面使用 Spring Cloud Alibaba 初始化业务网关和业务项目。

（1）在业务网关中添加 Spring Cloud Gateway 相关依赖包，部分 POM 文件的依赖如下：

```
<!—Spring Cloud Gateway 的依赖包-->
<dependency>
    <groupId>org.springframework.cloud</groupId>
    <artifactId>spring-cloud-starter-gateway</artifactId>
</dependency>
<!—基于 reactive 的 Redis 的依赖包-->
<dependency>
    <groupId>org.springframework.boot</groupId>
    <artifactId>spring-boot-starter-data-redis-reactive</artifactId>
</dependency>
```

（2）在业务网关中添加路由和限流的配置信息，部分配置如下：

```
spring:
  cloud:
gateway:
  ###①路由规则
      routes:
        - id: limit_route
          uri: http://127.0.0.1:28089
```

```
          predicates:
            - Path=/**
          ###②限流过滤器
          filters:
            - name: RequestRateLimiter
              ###③限流规则
              args:
                key-resolver: '#{@hostAddrKeyResolver}'
                redis-rate-limiter.replenishRate: 1
                redis-rate-limiter.burstCapacity: 3
  ###④Redis 的连接信息
  redis:
    host: localhost
    port: 6379
    database: 0
```

2. 添加“基于 IP 地址限流键的解析器”

本实例定义了 HostAddrKeyResolver 类实现 Spring Cloud Gateway 的 KeyResolver 接口，具体代码如下所示：

```
public class HostAddrKeyResolver implements KeyResolver {
    @Override
public Mono<String> resolve(ServerWebExchange exchange) {
    //①返回网关请求中对应的 IP 地址
        return Mono.just(exchange.getRequest().getRemoteAddress().
        getAddress().getHostAddress());
    }
}
@Configuration
public class RedisConfig {
    //②实例化 HostAddrKeyResolver 对象
    @Bean
    public HostAddrKeyResolver hostAddrKeyResolver() {
        return new HostAddrKeyResolver();
    }
}
```

3. 启动项目

下面用多线程模拟业务网关流量去调用用户 API 服务，以验证“Redis + Lua”分布式限流的效果。

（1）用多线程模拟网关流量，具体代码如下所示：

```
@RestController
@RequestMapping("/gateway")
public class GatewayController {
```

```
        @Resource
        private RestTemplate restTemplate;
        @GetMapping(value = "/thread")
        public String executeThread(){
            ExecutorService executorService= Executors.newFixedThreadPool(20);
            //开启一个容量为 20 个的线程池，并使用 HTTP 客户端 RestTemplate 调用网关
            executorService.execute(new GatewayThread());
            return "成功";
        }
        class GatewayThread implements Runnable{
            @Override
            public void run() {
                while (true){
                    restTemplate.getForObject
                     ("http://127.0.0.1:28082/user/getUserInfo",String.class);
                    try{
                        Thread.sleep(2000);
                    }catch (InterruptedException e){
                        System.out.println(e.getMessage());
                    }
                }
            }
        }
    }
```

（2）启动业务网关和用户 API 项目，在 Nacos 控制台能够看到服务已经启动并注册成功，如图 9-17 所示。

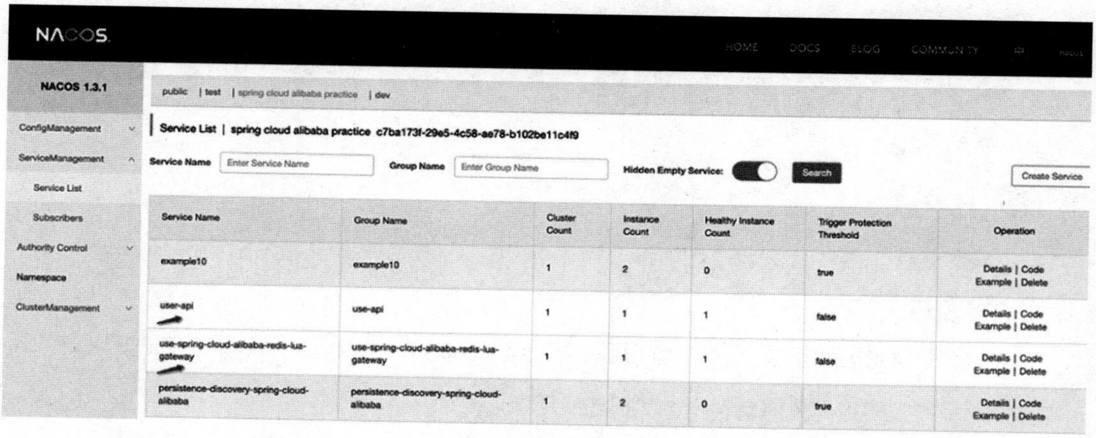

图 9-17

（3）执行命令“curl 127.0.0.1:28082/gateway/thread”开启多线程调用网关。在 Redis 后

台执行命令“./redis−cli −h 127.0.0.1 −p 6379”登录 Redis 的命令控制台。使用 Redis 客户端命令“keys *”查看当前 Redis 实例中的主键 Key 的列表。如图 9−18 所示，在 Redis 实例中已经生成了分布式限流 Key。

```
huxian@huxians-MacBook-Pro bin % ./redis-cli -h 127.0.0.1 -p 6379
127.0.0.1:6379> keys *
                    1) "request_rate_limiter.{127.0.0.1}.timestamp"
                                                          2) "7878"
                                                                 3) "request_rate_limiter.{127.0.0.1}.tokens"
                                                                                                   127.0.0.1:6379>
```

图 9-18

Spring Cloud Alibaba
微服务架构实战派

（下册）

胡弦◎著

電子工業出版社
Publishing House of Electronics Industry
北京•BEIJING

内 容 简 介

本书覆盖了微服务架构的主要技术点，包括分布式服务治理、分布式配置管理、分布式流量防护、分布式事务处理、分布式消息处理、分布式网关、分布式链路追踪、分布式Job、分库分表、读写分离、分布式缓存、服务注册/订阅路由、全链路蓝绿发布和灰度发布。这些技术点采用“是什么→怎么用→什么原理（源码解析）”的主线来讲解。

为了方便读者在企业中落地Spring Cloud Alibaba项目，本书还包括几个相对完整的项目实战：全链路日志平台、中台架构、数据迁移平台、业务链路告警平台。

本书的目标是：①让读者在动手中学习，而不是“看书时好像全明白了，一动手却发现什么都不会”；②读者可以掌握微服务全栈技术，而不仅仅是Spring Cloud Alibaba框架，对于相关的技术（Seata、RocketMQ），基本都是从零讲起，这样避免了读者为了学会微服务技术，得找Spring Cloud Alibaba的书、Seata的书、RocketMQ的书……本书是一站式解决方案。

本书适合对微服务架构感兴趣的开发人员。无论读者是否接触过微服务开发，只要具备一定的Java开发基础，都能通过本书的学习快速掌握微服务开发技能，快速搭建出可以在企业中应用的微服务架构。

图书在版编目（CIP）数据

Spring Cloud Alibaba 微服务架构实战派. 上下册 / 胡弦著. —北京：电子工业出版社，2022.1
ISBN 978-7-121-42313-0

Ⅰ. ①S… Ⅱ. ①胡… Ⅲ. ①互联网络－网络服务器 Ⅳ. ①TP368.5

中国版本图书馆 CIP 数据核字（2021）第 226217 号

责任编辑：吴宏伟
印　　刷：三河市良远印务有限公司
装　　订：三河市良远印务有限公司
出版发行：电子工业出版社
　　　　　北京市海淀区万寿路 173 信箱　　邮编：100036
开　　本：787×980　1/16　印张：60.25　字数：1446 千字
版　　次：2022 年 1 月第 1 版
印　　次：2024 年 5 月第 5 次印刷
定　　价：236.00 元（上下册）

凡所购买电子工业出版社图书有缺损问题，请向购买书店调换。若书店售缺，请与本社发行部联系，联系及邮购电话：（010）88254888，88258888。

质量投诉请发邮件至 zlts@phei.com.cn，盗版侵权举报请发邮件至 dbqq@phei.com.cn。

本书咨询联系方式：010-51260888-819，faq@phei.com.cn。

前言

在写这本书之前，我先后在两家杭州的“独角兽”公司担任技术负责人，并负责推进公司核心业务的“中台化”改造。在落地业务中台和技术中台的过程中，我督促并指导开发人员统一使用 Spring Cloud Alibaba 作为中台服务最底层的基础框架。为了快速推进业务服务 Spring Cloud Alibaba 化的进度，我冲在业务的第一线，收集和整理开发人员在使用 Spring Cloud Alibaba 过程中反馈的技术问题，并提供有效的技术解决方案，直至项目落地。

我每周都会做技术复盘，通过分析大量的问题总结出一个结论：开发人员反馈的问题大部分都是由于 Spring Cloud Alibaba 使用不合理所造成的。也就是说，很多开发人员并不了解 Spring Cloud Alibaba 的原理及如何落地实践。于是，我就产生了把我这几年落地 Spring Cloud Alibaba 的经验通过图书的方式输出的想法。

1. 本书特色

本书聚焦于 Spring Cloud Alibaba 微服务架构实战，全面分析了基于 Spring Cloud Alibaba 的微服务架构全栈技术原理。本书有如下特色：

（1）技术新。

Spring Cloud Alibaba 是一个将 Spring Cloud“阿里巴巴化”的微服务架构框架，它具备 Spring Cloud 所有的能力，并添加了 Nacos、Dubbo、RocketMQ 等 Spring Cloud 不具备的微服务架构能力。简单来说就是：搭建微服务架构，使用 Spring Cloud Alibaba 比使用 Spring Cloud 更高效，更简单，开发的技术成本更低。

本书中所有代码采用 Spring Cloud Alibaba 目前的最新版本（2.2. 5.RELEASE）来编写，与 Spring Cloud Alibaba 相关的微服务技术（Seata、RocketMQ 等）也采用的是目前最新的稳定版本。

（2）精心设计的主线：零基础入门，循序渐进，直至项目实战。

本书精心研究了程序类、架构类知识的认知规律，全书总共分为五个部分：入门篇、基础篇、中级篇、高级篇及项目实战篇，设计了一条相对科学的主线“它是什么→怎么搭建基础开发环境→怎么进行单项技术开发→怎么完成一个完整的项目”，让读者快速从“菜鸟”向微服务架构实战高手迈进。

（3）不只介绍 Spring Cloud Alibaba 框架本身，而是微服务架构全栈技术。

有的同类书只介绍 Spring Cloud Alibaba 框架本身，假定读者对支撑 Spring Cloud Alibaba 微服务架构相关技术（比如 Seata、Skywalking、Redis、RocketMQ 等）是了解的。这样就会存在一个问题——为了学会微服务技术，得找 Spring Cloud Alibaba 的书、Seata 的书、RocketMQ 的书……而这恰恰是难点所在——怎样将它们平滑地衔接起来学习。

本书是以“实现完整的 Spring Cloud Alibaba 微服务架构”为目标，为了这个目标，除介绍 Spring Cloud Alibaba 这个“主角”外，对于支撑 Spring Cloud Alibaba 微服务架构的技术（比如 Seata、Skywalking、Redis、RocketMQ 等）也基本都是从零讲起，保证读者能够平滑地学习。本书是“一站到底”的解决方案：读者只需从这里上车，中途无需转乘，读者需要什么，本书就提供什么，直达终点。

（4）绘制了大量的图，便于理解原理、架构、流程。

一图胜千文，书中在涉及原理、架构、流程的地方都尽量配有插图，以便读者有直观的理解。

（5）实战性强。

本书介绍了大量的实战案例，能让读者“动起来”，在实践中体会功能，而不只是一种概念上的理解。

在讲解每一个知识模块时，我们都在思考：在这个知识模块中，哪些是读者必须实现的“标准动作”（实例）；哪些“标准动作”是可以先完成的，以求读者能快速有一个感知；哪些“标准动作”是有一定难度，需要放到后面完成的。读者在跟随书中实例一个个实践之后，再去理解那些抽象的概念和原理就是水道渠成了。

本书的目标之一是，让读者在动手中学习，而不是“看书时好像全明白了，一动手却发现什都不会”。本书相信“知行合一”理念，不是“只知，而无行”，避免眼高手低。

（6）深入剖析原理。

本书以系统思维的方式，从业务功能视角剖析微服务架构中技术的底层原理，使读者具备快速阅读新框架源码的能力。读者只有具备了这种功能，才能举一反三，实现更复杂的功能，应对更复杂的应用场景。

（7）采用真实项目，实现“从树木到森林”的突破。

本书的“项目实战篇”是从架构、代码和业务的视角，在真实项目中验证“Spring Cloud Alibaba 微服务架构”的架构方法论及核心技术原理，让读者有身临生产级场景的感觉。

（8）衔接运维，一键部署。

本书中所有的技术框架都有详细的“搭建技术框架运维环境的步骤”，读者只需要按照本书的安装步骤，就可以快速搭建运维环境，从而在本地环境中快速运行本书的实例。

（9）干货丰富，知识的“巨无霸”。

本书共分为上、下两册，总计 18 章，近 1000 页，内容非常丰富，算得上是相关领域图书中的“巨无霸”。

2. 阅读本书，你能学到什么

- 掌握 Spring Cloud Alibaba 的核心原理及微服务架构项目实战经验；
- 掌握 Nacos 注册中心和配置中心的核心原理及微服务架构项目实战经验；
- 掌握 Sentinel 的核心原理及微服务架构项目实战经验；
- 掌握 Seata 的核心原理及微服务架构项目实战经验；
- 掌握 RocketMQ 的核心原理及微服务架构项目实战经验；
- 掌握 Skywalking 的核心原理及微服务架构项目实战经验；
- 掌握 Elastic Job 的核心原理及微服务架构项目实战经验；
- 掌握 ShardingSphere 的核心原理及微服务架构项目实战经验；
- 掌握 Spring Cloud Gateway 的核心原理及微服务架构项目实战经验；
- 掌握分布式缓存 Redis 的集群管理和分布式锁的原理及微服务架构项目实战经验；
- 掌握 Discovery 的核心原理及微服务架构项目实战经验；
- 掌握在业务中台和技术中台中落地“基于 Spring Cloud Alibaba 微服务架构”的项目实战经验；
- 掌握在微服务架构中“基于 DataX 的异构数据迁移”的项目实战经验；
- 掌握在微服务架构中“基于 Skywalking 的链路告警平台”的项目实战经验；
- 掌握在微服务架构中“基于 ELK 和 Sywalking 的全链路日志平台”的项目实战经验。

3. 读者对象

本书读者对象如下：

◎ 初学 Java 的自学者；
◎ 软件开发工程师；
◎ Java 语言中高级开发人员；
◎ 编程爱好者；
◎ 中间件爱好者；
◎ 技术总监；
◎ 培训机构的老师和学员；
◎ 高等院校计算机相关专业学生；
◎ Spring Cloud Alibaba 初学者；
◎ DevOps 运维人员；
◎ 技术经理；
◎ 其他对 Spring Cloud Alibaba 感兴趣的 IT 人员。

4. 致谢

Spring Cloud Alibaba 是我深度使用的微服务框架之一，也是我实现职业生涯飞跃的敲门砖。虽然我在微服务架构领域中有很多技术沉淀，但是作为一个技术人员，将自己懂的技术通过文字输出给读者，还是需要很强的技术布道能力及文字编排能力的。

感谢我的家人，特别是我太太陈益超和我三岁的儿子胡辰昱，在我写书期间对我的支持。同时也要感谢电子工业出版社的编辑吴宏伟老师，将我带进“通过文字进行技术知识输出”的大门。

胡弦

2021.09.17

读者服务

微信扫码回复：42313

- 获取本书配套代码
- 加入本书读者交流群，与作者互动
- 获取【百场业界大咖直播合集】（持续更新），仅需 1 元

目录

高级篇

项目实战篇

高级篇

第 10 章 分布式链路追踪——基于 Skywalking

在微服务架构中，开发人员利用各种中间件将不同的服务连接起来，比如用 Dubbo 完成服务之间的 RPC。随着引入的中间件越来越多，业务调用关系越来越复杂，开发人员就会意识到中间件本身也会影响业务的稳定性。

在服务运行过程中，除需要依赖基础框架（比如 Spring MVC、事务框架等）外，还需要依赖各类中间件（比如利用 RocketMQ 发送消息），所以，从技术或者业务的角度都需要分布式链路追踪技术。

10.1 认识分布式链路追踪

10.1.1 什么是分布式链路追踪

分布式链路追踪可以拆分为 3 个概念：分布式、链路和追踪。

1. 分布式

分布式链路追踪的前提是分布式。分布式链路追踪，主要用于解决不同进程之间的链路追踪问题。如果应用都强耦合在一个进程中，则不能发挥链路追踪的价值，并且成本会非常大。

什么是分布式？通俗来讲就是将一个进程拆分为多个进程，并各自独享系统资源，如图 10-1 所示。

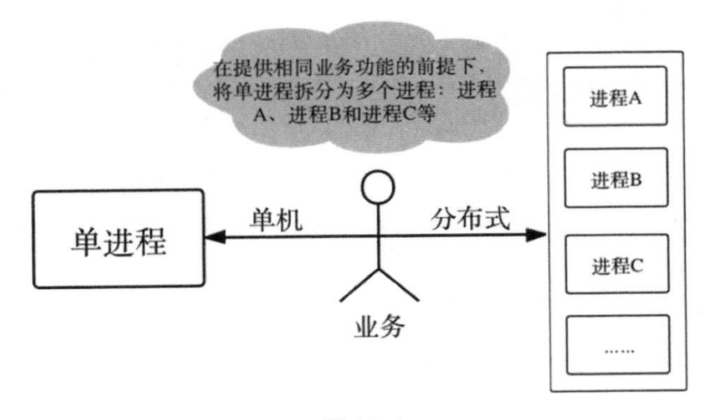

图 10-1

2. 链路

（1）对比理解。

可以将链路比作物理学中的电路图，具体如下：

- 链路和电路图都有确定的起点和终点。
- 链路和电路图都通过可视化的技术手段，来模拟真实组件的运行轨迹。
- 链路和电路图都非常复杂，并且随着系统复杂程度的增加，其可视化复杂度也会成几何倍数增加。

什么是链路？通俗地讲，它是一个网，利用该网可以将组件运行的轨迹可视化。

（2）举例。

如图 10-2 所示，在直播业务中，用户服务会调用会员服务、账户服务、主播服务和房间服务，并利用数据库、文件、缓存和消息中间件，形成一套完整的直播服务体系。

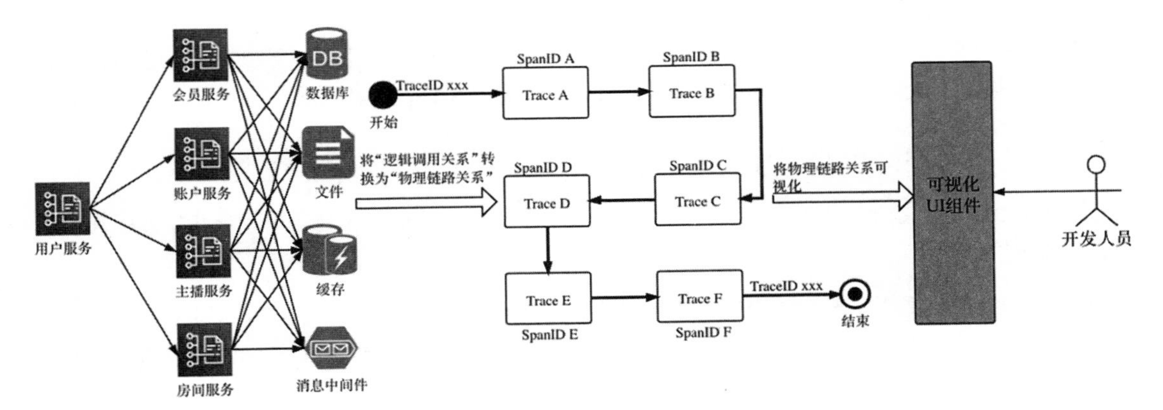

图 10-2

- 在直播服务体系中，开发人员可以梳理出服务的“逻辑调用关系”，但“逻辑调用关系”是静态的，并且没有和“服务之间调用产生的数据”衔接起来，所以并没有很大的实用价值。
- 在分布式链路追踪中，将“逻辑调用关系”和“服务之间调用产生的数据”衔接起来，形成一条网状链路。每一条链路都有一个唯一的 TraceID xxx，子链路都有唯一的 SpanID，并且 TraceID xxx 和 SpanID 通过 TraceID 串联起来。
- 在将“逻辑调用关系”转换为“链路”之后，分布式链路追踪将“链路”转换为“可视化的 UI 组件”，这样开发人员可以实时地观察链路。

3. 追踪

如果分布式是分布式链路追踪的前置条件，链路是分布式链路追踪的技术实现，那么追踪就是分布式链路追踪的目的，如图 10-3 所示。

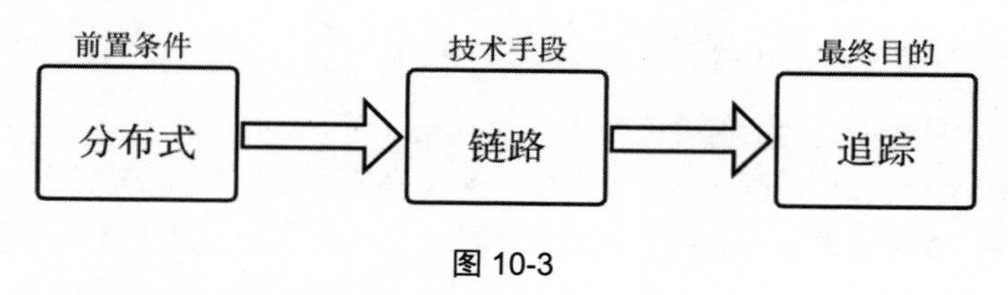

图 10-3

在分布式链路中，通过追踪可以实现以下目的。

- 业务完整性校验：比如，完成一次完整的业务请求需要 10 个子链路，但由于出现异常，其中的 1 个链路追踪失败，只追踪到 9 个子链路（追踪到的子链路能够持久化到存储设备中），那么业务请求就不完整的，这时可以添加业务不完整的预警。
- 快速地定位业务和技术问题：如果业务运行数据是可以追踪的，则可以通过可视化 UI 工具，为开发人员提供故障预判数据，以及实时的故障定位预警。
- 性能监控：将“业务运行数据”和“链路”转换为度量监控指标，开发人员可以利用监控指标实时地评估接口的性能，并评估服务的容量，实现性能和成本的平衡。

10.1.2 认识 Skywalking

SkyWalking 是一个开源的可观测平台，主要用于收集、分析、聚合和可视化来自服务和云原生基础设施的数据。

SkyWalking 提供了一种简单的方式来维护分布式系统的视图关系，它甚至能够支持跨云的服务。SkyWalking 是 APM（应用性能管理）领域的一种技术解决方案，是专门为基于云原生容器的分布式系统而设计的。

SkyWalking 提供了在不同场景下观察和监控分布式系统的解决方案。SkyWalking 为服务提供了自动探针代理，服务的语言包括 Java、C#、JavaScript、Go、PHP 等。在多语言、可持续部署的环境中，云原生基础设施变得越来越强大，但也越来越复杂。SkyWalking 的服务网格接收器允许 SkyWalking 从 Istio/Envoy 和 Linkerd 等服务网格框架中接收遥测数据，使用户能够鸟瞰完整的分布式系统环境。

SkyWalking 描述了服务、服务实例和端点的语义，具体如下。

- 服务：通常指接入 Skywalking 的应用名称。
- 服务实例：通常指应用部署机器的 IP 地址。
- 端点：通常指服务中用于传入请求的路径，如 URI 路径或“gRPC 服务类+方法签名”。

1. Skywalking 的核心概念

Skywalking 中的核心的概念如图 10-4 所示。

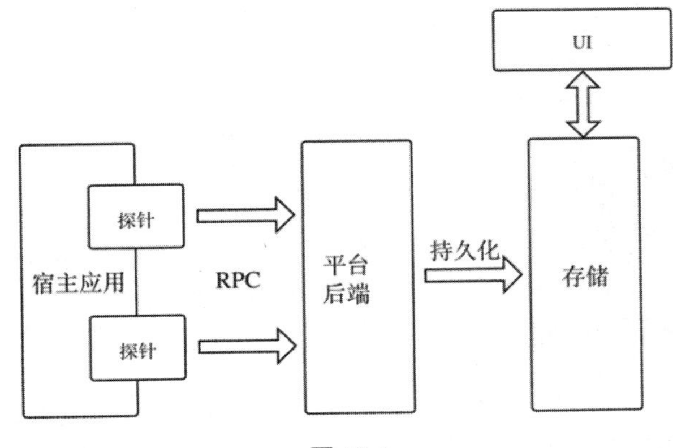

图 10-4

- 宿主应用：被探针通过字节码技术侵入的应用。从探针的视角来看，应用是探针寄生的宿主。
- 探针：收集从应用采集到的链路数据，并将其转换成 Skywalking 能够识别的数据格式。
- RPC：宿主应用和平台后端之间的通信渠道。
- 平台后端：支持数据聚合、分析和流处理，包括跟踪、指标和日志等。
- 存储：通过开放的、可插拔的接口来存储 SkyWalking 的链路数据。Skywalking 目前支持 ElasticSearch、H2、MySQL、TiDB、InfluxDB。
- UI：一个可定制的基于 Web 的界面，允许 SkyWalking 终端用户管理和可视化 SkyWalking 的链路数据。

2. Skywalking 能够解决什么问题

分析一项技术能够解决什么问题，需要从全局的视角去观察引入技术对应用的正向影响和反向的影响。如果正向的影响程度要远大于反向的影响程度，则说明引入的新技术是有利于应用稳定运行的。

（1）引入 Skywalking 的正向影响。

- Skywalking 提供了非常完善的分布式链路追踪功能。开发人员可以在其 UI 界面中，实时地查询指定服务、服务实例和端点的链路追踪信息。开发人员可以实时地观察到接口运行的轨迹和耗时。
- Skywalking 提供了非常完善的错误诊断功能。如果开发人员在接口中封装了系统异常和业务异常，则 Skywalking 可以捕获异常，将这些异常的节点标记为红色，并记录错误节点的完整异常信息。
- Skywalking 提供了非常完善的实时链路大屏功能。如果将应用接入 Skywalking，则在应用运行时产生的指标数据会被准实时地收集到 Skywalking 的后台。Skywalking 支持的指标类型非常多，比如 JVM、慢查询、P99、P95 等。
- Skywalking 提供了非常完善的服务拓扑关系分析功能，以及调用依赖关系功能。在微服务架构中，梳理服务之间的调用关系和解决强依赖关系，是开发人员经常要做的事情。Skywalking 能解决这个技术痛点。
- Skywalking 提供了非常完善的链路指标及错误告警功能。开发人员可以自定义告警规则，从而实时地收到告警，并及时地处理故障。
- Skywalking 提供了在线性能诊断功能。如果线上接口存在性能问题，但又不能在本地环境中模拟性能问题，则可以借助线上环境（真实流量和真实的数据）实时地排查性能问题。

（2）引入 Skywalking 的反向影响。

- Skywalking 需要借助 Java Agent（字节码技术）侵入使用的中间件。既然是侵入，肯定会增加一定的性能开销。那些高流量的核心接口对性能损耗的容忍度是非常低的，是不能接受这种性能开销的。
- Skywalking 是一个技术复杂度和业务复杂度都非常高的分布式链路追踪系统。另外，从架构的角度来看，它也是一个平台。引入 Skywalking 肯定会增加架构的复杂度，开发人员需要投入更多的精力去熟悉和维护 Skywalking。
- Skywalking 是通过侵入应用的中间件从而达到侵入应用的。引入 Skywalking，会增加中间件出现故障的可能性。

通过上面的分析可以发现：技术本身都是有风险的。开发人员能够做的是熟悉技术的原理和使用场景。只有这样才能提高技术的正向影响，降低反向影响。

10.2 搭建 Skywalking 环境

Skywalking 支持单机部署和集群部署。开发人员可以在开发测试环境中搭建单机环境，在生产环境中搭建集群环境。

10.2.1 搭建单机环境

从官方下载 Skywalking 的部署包 apache-skywalking-apm-es7-8.5.0.tar.gz，并解压缩部署包。

1. 修改配置信息

打开 Skywalking 的部署包的 config 目录下的文件 application.yml，配置单机环境信息。

关于环境的配置信息如下所示（其他配置信息可以默认）。

```
###①配置管理集群的注册中心，本书采用 Nacos
cluster:
  selector: ${SW_CLUSTER:nacos}
  standalone:
  nacos:
    serviceName: ${SW_SERVICE_NAME:"SkyWalking_OAP_Cluster"}
    hostPort: ${SW_CLUSTER_NACOS_HOST_PORT:127.0.0.1:8848}
    namespace:
${SW_CLUSTER_NACOS_NAMESPACE:"c7ba173f-29e5-4c58-ae78-b102be11c4f9"}
    username: ${SW_CLUSTER_NACOS_USERNAME:""}
    password: ${SW_CLUSTER_NACOS_PASSWORD:""}
    accessKey: ${SW_CLUSTER_NACOS_ACCESSKEY:""}
secretKey: ${SW_CLUSTER_NACOS_SECRETKEY:""}
###②配置通信渠道
core:
  selector: ${SW_CORE:default}
  default:
role: ${SW_CORE_ROLE:Mixed} # Mixed/Receiver/Aggregator
###③配置 HTTP 通信渠道
    restHost: ${SW_CORE_REST_HOST:127.0.0.1}
    restPort: ${SW_CORE_REST_PORT:12800}
    restContextPath: ${SW_CORE_REST_CONTEXT_PATH:/}
    restMinThreads: ${SW_CORE_REST_JETTY_MIN_THREADS:1}
    restMaxThreads: ${SW_CORE_REST_JETTY_MAX_THREADS:200}
    restIdleTimeOut: ${SW_CORE_REST_JETTY_IDLE_TIMEOUT:30000}
    restAcceptorPriorityDelta: ${SW_CORE_REST_JETTY_DELTA:0}
restAcceptQueueSize: ${SW_CORE_REST_JETTY_QUEUE_SIZE:0}
###④配置 gRPC 通信渠道
```

```
      gRPCHost: ${SW_CORE_GRPC_HOST:127.0.0.1}
      gRPCPort: ${SW_CORE_GRPC_PORT:11800}
      maxConcurrentCallsPerConnection:
${SW_CORE_GRPC_MAX_CONCURRENT_CALL:0}
      maxMessageSize: ${SW_CORE_GRPC_MAX_MESSAGE_SIZE:0}
      gRPCThreadPoolQueueSize: ${SW_CORE_GRPC_POOL_QUEUE_SIZE:-1}
      gRPCThreadPoolSize: ${SW_CORE_GRPC_THREAD_POOL_SIZE:-1}
      gRPCSslEnabled: ${SW_CORE_GRPC_SSL_ENABLED:false}
      gRPCSslKeyPath: ${SW_CORE_GRPC_SSL_KEY_PATH:""}
      gRPCSslCertChainPath: ${SW_CORE_GRPC_SSL_CERT_CHAIN_PATH:""}
   gRPCSslTrustedCAPath: ${SW_CORE_GRPC_SSL_TRUSTED_CA_PATH:""}
   ###⑤配置 Skywalking 在单机环境下可以直接使用内存数据库 H2
   torage:
     selector: ${SW_STORAGE:h2}
     h2:
       driver: ${SW_STORAGE_H2_DRIVER:org.h2.jdbcx.JdbcDataSource}
       url:
${SW_STORAGE_H2_URL:jdbc:h2:mem:skywalking-oap-db;DB_CLOSE_DELAY=-1}
       user: ${SW_STORAGE_H2_USER:sa}
       metadataQueryMaxSize: ${SW_STORAGE_H2_QUERY_MAX_SIZE:5000}
       maxSizeOfArrayColumn: ${SW_STORAGE_MAX_SIZE_OF_ARRAY_COLUMN:20}
       numOfSearchableValuesPerTag:
${SW_STORAGE_NUM_OF_SEARCHABLE_VALUES_PER_TAG:2}
   ###⑥配置管理集群的配置中心，本书采用 Nacos
   configuration:
     selector: ${SW_CONFIGURATION:nacos}
     none:
     nacos:
       serverAddr: ${SW_CONFIG_NACOS_SERVER_ADDR:127.0.0.1}
       port: ${SW_CONFIG_NACOS_SERVER_PORT:8848}
       group: ${SW_CONFIG_NACOS_SERVER_GROUP:skywalking}
       namespace:
${SW_CONFIG_NACOS_SERVER_NAMESPACE:c7ba173f-29e5-4c58-ae78-b102be11c4f9}
       period: ${SW_CONFIG_NACOS_PERIOD:60}
       username: ${SW_CONFIG_NACOS_USERNAME:"nacos"}
       password: ${SW_CONFIG_NACOS_PASSWORD:"nacos"}
       accessKey: ${SW_CONFIG_NACOS_ACCESSKEY:""}
       secretKey: ${SW_CONFIG_NACOS_SECRETKEY:""}
```

2. 执行脚本，启动单机环境

（1）在 Linux 或者 Mac 环境下，执行脚本“bin/startup.sh”（如果是 Windows 环境，则执行“bin/startup.bat”）。

（2）在 Nacos 配置中心中能够看到 Skywalking 的节点。

（3）访问“http://127.0.0.1:8080/”可以看到，Skywalking 的可视化 UI 控制台已经启动成功，如图 10-5 所示。

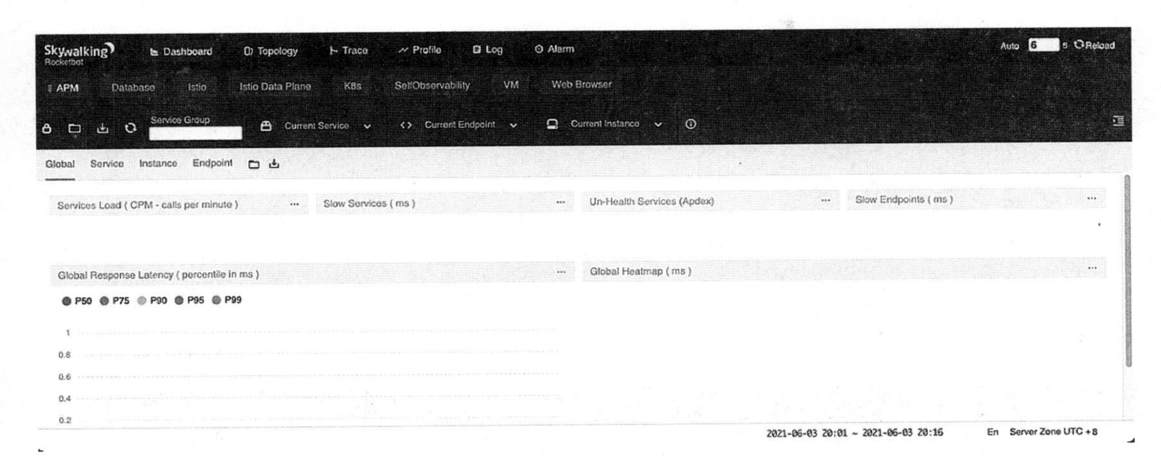

图 10-5

10.2.2　搭建集群环境

如果要确保线上 Skywalking 的高可用，则建议开发人员线上环境采用集群部署，底层存储采用 Elasticsearch。

通过更改 Skywalking 的部署包的 config 目录下的文件 application.yml 中的 gRPC 和 HTTP 通信渠道的端口号，可以在同一台机器上部署 Skywaking 后端的多个进程，用于模拟线上集群的部署。

1. 搭建 Elasticsearch 环境

从 Elasticsearch 的官网下载部署包 elasticsearch-7.13.1-darwin-x86_64.tar.gz，并解压缩部署包。

（1）修改解压缩后的 Elasticsearch 部署包中的配置文件“/config/elasticsearch.yml”，具体配置信息如下：

```
cluster.name: Skywaking-Elasticsearch
network.host: 127.0.0.1
http.port: 9200
```

（2）执行命令“nohup bash bin/elasticsearch &”启动 Elasticsearch。

（3）从 Kibana 的官网下载部署包 kibana-7.13.1-darwin-x86_64.tar.gz（Kibana 的部署

包要和 Elasticsearch 的版本一致）。

（4）修改解压缩后的 Kibana 部署包中的配置文件“/config/kibana.yml”，具体配置信息如下：

```
server.port: 5601
server.host: "127.0.0.1"
server.name: "skywalking-kibana"
elasticsearch.hosts: ["http://127.0.01:9200"]
ops.interval: 5000
i18n.locale: "zh-CN"
```

（5）执行命令“nohup bash bin/kibana &”启动 Kibana。

（6）用 IP 地址“127.0.0.1:5601”访问 Kibana 的 UI 控制台，可以发现 Kibana 已经启动成功，如图 10-6 所示。

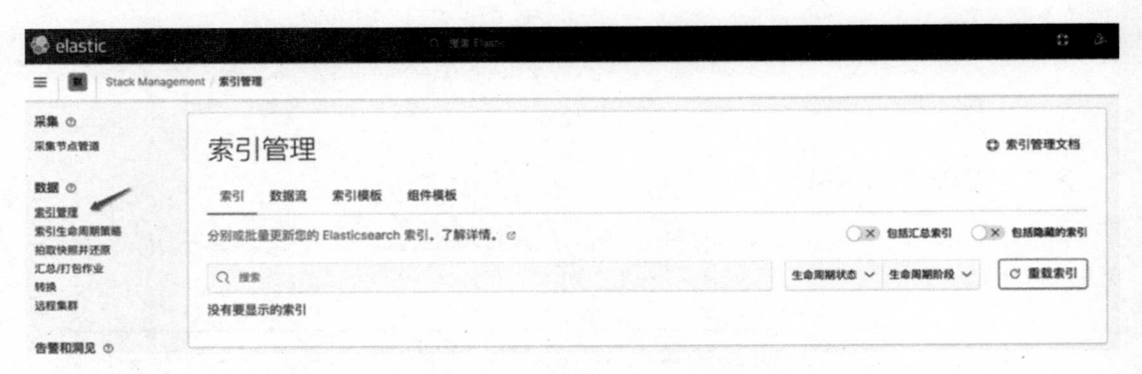

图 10-6

2. 在同一台机器上，新建 3 个部署目录

这里在同一台机器上新建了 3 个部署目录，用于存放 3 个 Skywalking 节点的部署包，分别是：apache-skywalking-apm-bin-es7-one 、 apache-skywalking-apm-bin-es7-two 和 apache-skywalking-apm-bin-es7-three，如图 10-7 所示。

skywalking-cluster

Name	Date Modified	Size	Kind
apache-skywalking-apm-bin-es7-one	Yesterday at 22:02	--	Folder
apache-skywalking-apm-bin-es7-three	Yesterday at 22:02	--	Folder
apache-skywalking-apm-bin-es7-two	Yesterday at 22:02	--	Folder

图 10-7

3. 在单机环境的基础之上，修改集群部署目录中相关的配置信息

打开 Skywalking 部署包的配置文件“config/application.yml”，修改配置信息。

（1）修改 apache-skywalking-apm-bin-es7-one 的配置信息，用 Elasticsearch 7 替换 H2，部分配置信息如下：

```
###用 elasticsearch7 替换 H2
storage:
  selector: ${SW_STORAGE:elasticsearch7}
  elasticsearch7:
    nameSpace: ${SW_NAMESPACE:""}
    clusterNodes: ${SW_STORAGE_ES_CLUSTER_NODES:127.0.1:9200}
    protocol: ${SW_STORAGE_ES_HTTP_PROTOCOL:"http"}
    trustStorePath: ${SW_STORAGE_ES_SSL_JKS_PATH:""}
    trustStorePass: ${SW_STORAGE_ES_SSL_JKS_PASS:""}
    dayStep: ${SW_STORAGE_DAY_STEP:1}
    indexShardsNumber: ${SW_STORAGE_ES_INDEX_SHARDS_NUMBER:1}
    indexReplicasNumber: ${SW_STORAGE_ES_INDEX_REPLICAS_NUMBER:1}
    ...
```

同理，将 apache-skywalking-apm-bin-es7-two 和 apache-skywalking-apm-bin-es7-three 中的 H2 替换为 Elasticsearch 7。

（2）为了模拟在单机器实例上搭建 Skywalking 集群，这里需要更改通信渠道默认的端口号，以避免端口冲突。

- apache-skywalking-apm-bin-es7-one 的配置文件“config/application.yml”中的 HTTP 和 gRPC 分别采用默认的端口 12800 和 11800；
- apache-skywalking-apm-bin-es7-two 中的 HTTP 和 gRPC 分别采用端口 13800 和 14800；
- apache-skywalking-apm-bin-es7-three 中的 HTTP 和 gRPC 分别采用端口 15800 和 16800。

（3）Skywalking 在启动服务平台的过程中，会一起启动 UI 控制台，但 UI 控制台默认的端口号为 8080。为了保证在单机器实例上能够启动集群，需要更改 UI 控制台的默认端口。

在部署包“apache-skywalking-apm-bin-es7-one”中，配置文件“webapp/webapp.yml”的配置信息如下：

```
server:
  port: 8080
collector:
  path: /graphql
  ribbon:
```

```
    ReadTimeout: 10000
    listOfServers: 127.0.0.1:12800,127.0.0.1:13800,127.0.0.1:15800
```

在部署包“apache-skywalking-apm-bin-es7-two”中，配置文件“webapp/webapp.yml”的配置信息如下：

```
server:
  port: 8081
collector:
  path: /graphql
  ribbon:
    ReadTimeout: 10000
    listOfServers: 127.0.0.1:12800,127.0.0.1:13800,127.0.0.1:15800
```

在部署包“apache-skywalking-apm-bin-es7-three”中，配置文件“webapp/webapp.yml”的配置信息如下：

```
server:
  port: 8082
collector:
  path: /graphql
  ribbon:
    ReadTimeout: 10000
    listOfServers: 127.0.0.1:12800,127.0.0.1:13800,127.0.0.1:15800
```

3. 分别执行 3 个部署包中的脚本，启动集群环境

在 3 个部署包中执行脚本“bin/startup.sh”，在执行完成后在 Nacos 注册中心中会注册 3 个新的 Skywalking 节点，分别是 127.0.0.1:11800、127.0.0.1:14800 和 127.0.0.1:16800，如图 10-8 所示。

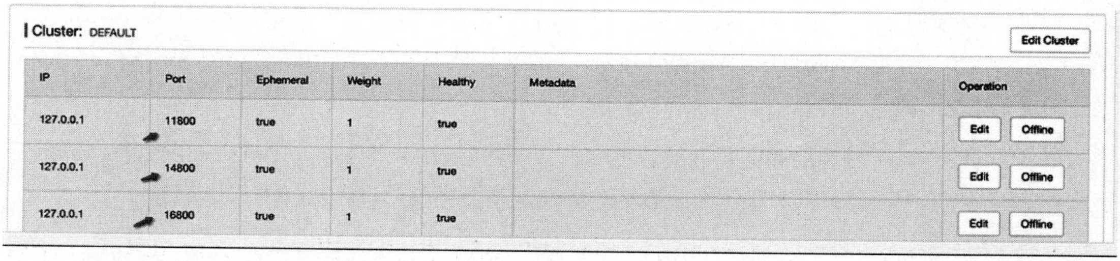

Cluster: DEFAULT　　Edit Cluster

IP	Port	Ephemeral	Weight	Healthy	Metadata	Operation
127.0.0.1	11800	true	1	true		Edit Offline
127.0.0.1	14800	true	1	true		Edit Offline
127.0.0.1	16800	true	1	true		Edit Offline

图 10-8

4. 访问 Kibana 控制台，验证存储设备 Elasticsearch 索引中的表结构已经完成初始化

访问 Kibana 控制台的 IP 地址“127.0.0.1:5601”，可以发现 Elasticsearch 索引中的表结构已经完成了初始化，如图 10-9 所示。

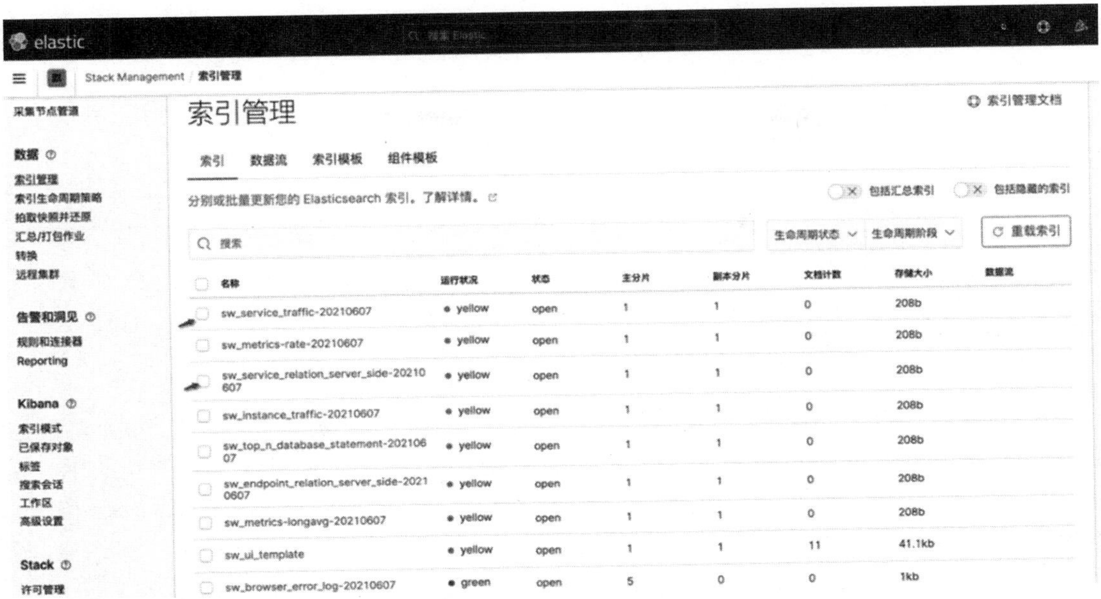

elastic
Stack Management / 索引管理
采集节点管理
数据
索引管理
索引生命周期策略
拍取快照并还原
汇总/打包作业
转换
远程集群
告警和洞见
规则和连接器
Reporting
Kibana
索引模式
已保存对象
标签
搜索会话
工作区
高级设置
Stack
许可管理
索引管理
索引管理文档
索引
数据流
索引模板
组件模板
分别或批量更新您的 Elasticsearch 索引。了解详情。
包括汇总索引
包括隐藏的索引
搜索
生命周期状态
生命周期阶段
重载索引
名称
运行状况
状态
主分片
副本分片
文档计数
存储大小
数据流
sw_service_traffic-20210607 yellow open 1 1 0 208b
sw_metrics-rate-20210607 yellow open 1 1 0 208b
sw_service_relation_server_side-20210607 yellow open 1 1 0 208b
sw_instance_traffic-20210607 yellow open 1 1 0 208b
sw_top_n_database_statement-20210607 yellow open 1 1 0 208b
sw_endpoint_relation_server_side-20210607 yellow open 1 1 0 208b
sw_metrics-longavg-20210607 yellow open 1 1 0 208b
sw_ui_template yellow open 1 1 11 41.1kb
sw_browser_error_log-20210607 green open 5 0 0 1kb

图 10-9

10.3　用 Java Agent 将 Spring Cloud Alibaba 应用接入 Skywalking

Skywalking 采用 Java Agent（字节码技术）将 Spring Cloud Alibaba 应用接入其后台中。在将 Spring Cloud Alibaba 应用接入 Skywaking 之前，需要先了解一下 Java Agent 的原理。

10.3.1　什么是 Java Agent

1. Java Agent 产生的背景

在开发的过程中，开发人员经常会使用 IDEA 的 Debug 功能（包含本地和远程 Debug）调试应用，并在 JVM 进程运行期间获取应用运行的 JVM 信息、变量信息等。如果开发人员需要应用线上运行期间实时地备份在 JVM 中运行的数据，也可以通过 JDK 自带的 Jmap、Jstack 等工具获取在 JVM 中运行的数据。

在 JVM 设计之初，JVM 就考虑了虚拟机的状态监控、debug、线程和内存分析等功能。在 JDK 1.5 之前，JVM 规范定义了 JVMPI（Java Virtual Machine Profiler Interface）语义，JVMPI 提供了一批 JVM 分析接口。

JVM 规范还定义了 JVMDI（Java Virtual Machine Debug Interface）语义，JVMDI 提供了一批 JVM 调试接口。JDK 1.5 及以后的版本将 JVMPI 和 JVMDI 合并，形成一套 JVM 语义 JVMTI，

包含 JVM 分析接口和 JVM 调试接口。

JVMTI 是一套 JVM 的接口规范，在不同的 JVM 中其实现方式可以不一样。JVMTI 提供的是 Native API 调用方式（即常用的 JNI 调用方式）。JVMTI 接口用 C/C++语言来暴露 Native API，并最终以动态链路库的形式被 JVM 加载并运行。

在 Java Agent 产生之前，开发人员只能通过调用 JVMTI 的 Native API 调用方式完成代码的动态侵入。这种调用方式非常不友好，并且技术门槛很高。

2. 在 JDK 1.5 之后，JVM 提供了 Java Agent

在 JDK 1.5 之后，JVM 开始提供了探针接口（Instrumentation 接口），这样开发人员可以使用探针接口编写 Java Agent。但是，探针接口底层依然依赖 JVMTI 语义的 Native API。在 JDK 1.6 及之后的版本中，JVM 提供了 Attach 接口，这样开发人员可以使用 Attach 接口来实现 Java Agent。Attach 接口底层也依赖 JVMTI 语义的 Native API。

Java Agent 提供了一种在加载字节码时对字节码进行修改的方式。它共有两种方式执行：

- 在应用运行之前，通过 premain()方法来实现“在应用启动时侵入并代理应用”（Instrumentation 接口）；
- 在应用运行之后，通过 Attach API 和 agentmain()方法来实现“在应用启动后的某一个运行阶段中侵入并代理应用”（Attach 接口）。

具体的使用方式如下：

（1）用 Instrumentation 接口和 premain()方法来实现字节码在“应用运行之前”加载。

开发人员通过 JVM 参数“-javaagent:**.jar”启动应用。应用在启动时，会优先加载 Java Agent，并执行 premain()方法，这时大部分的类都还没有被加载，此时可以实现对新加载的类进行字节码修改，但如果 premain()方法执行失败或抛出异常，则 JVM 会被中止，这是很致命的问题。

（2）用 Attach 接口和 agentmain()方法来实现字节码在“应用运行之后”加载。

开发人员在通过 JVM 参数“javaagent:**.jar”启动应用之后，可以通过 Java JVM 的 Attach 接口加载 Java Agent。Attach 接口其实是 JVM 进程之间的沟通桥梁，底层通过 Socket 进行通信，JVM A 可以发送一些指令给 JVM B，JVM B 在收到指令之后执行相应的逻辑。比如在命令行中经常使用的 Jstack、Jcmd、Jps 等，很多都是基于这种机制实现的。

10.3.2 “Skywalking 使用 Java Agent 零侵入应用”的原理

Skywalking 选择使用技术成本相对较小的 Instrumentation 接口和 premain()方法来实现字节码在应用运行前加载字节码，并实现对应用的零侵入。

1. 用 SkyWalkingAgent 类实现 Java Agent

Skywalking 用 SkyWalkingAgent 类定义 Java Agent 的 SkyWalkingAgent 类的逻辑比较复杂，下面将它拆分为几个部分来分析。

（1）实现 premain()方法，并调用 Instrumentation 接口。

```
public class SkyWalkingAgent {
    ...
    public static void premain(String agentArgs, Instrumentation
instrumentation) throws PluginException {
        final PluginFinder pluginFinder;
        try {
            //①读取应用中 Java Agent 配置文件 agent.config 中的配置信息
            SnifferConfigInitializer.initializeCoreConfig(agentArgs);
        } catch (Exception e) {
            LogManager.getLogger(SkyWalkingAgent.class)
                    .error(e, "SkyWalking agent initialized failure. Shutting
down.");
            return;
        } finally {
            LOGGER = LogManager.getLogger(SkyWalkingAgent.class);
        }

        try {
            //②加载 Skywalking 支持的所有中间件插件（中间件探针）
            pluginFinder = new PluginFinder(new
PluginBootstrap().loadPlugins());
        } catch (AgentPackageNotFoundException ape) {
            LOGGER.error(ape, "Locate agent.jar failure. Shutting down.");
            return;
        } catch (Exception e) {
            LOGGER.error(e, "SkyWalking agent initialized failure. Shutting
down.");
            return;
        }
        //③集成字节码框架 byte-buddy
        final ByteBuddy byteBuddy = new
ByteBuddy().with(TypeValidation.of(Config.Agent.IS_OPEN_DEBUGGING_CLASS));
        //④添加字节码侵入的白名单(比如以“javassist”开头的 Class 类)，则 Skywalking
就不会进行字节码侵入
```

```
        AgentBuilder agentBuilder = new
AgentBuilder.Default(byteBuddy).ignore(
                nameStartsWith("net.bytebuddy.")
                        .or(nameStartsWith("org.slf4j."))
                        .or(nameStartsWith("org.groovy."))
                        .or(nameContains("javassist"))
                        .or(nameContains(".asm."))
                        .or(nameContains(".reflectasm."))
                        .or(nameStartsWith("sun.reflect"))
                        .or(allSkyWalkingAgentExcludeToolkit())
                        .or(ElementMatchers.isSynthetic()));

        JDK9ModuleExporter.EdgeClasses edgeClasses = new
JDK9ModuleExporter.EdgeClasses();
        try {
            //⑤抽取 Skywalking 中优先级最高的公共的插件
            agentBuilder = BootstrapInstrumentBoost.inject(pluginFinder,
instrumentation, agentBuilder, edgeClasses);
        } catch (Exception e) {
            LOGGER.error(e, "SkyWalking agent inject bootstrap
instrumentation failure. Shutting down.");
            return;
        }
        try {
            agentBuilder = JDK9ModuleExporter.openReadEdge(instrumentation,
agentBuilder, edgeClasses);
        } catch (Exception e) {
            LOGGER.error(e, "SkyWalking agent open read edge in JDK 9+ failure.
Shutting down.");
            return;
        }
        //⑥如果在 agent.config 文件中将 IS_CACHE_ENHANCED_CLASS 设置为“true”,
则 Skywalking 会执行缓存字节码文件的操作，将增加之后的字节码文件缓存在本地
        if (Config.Agent.IS_CACHE_ENHANCED_CLASS) {
            try {
                agentBuilder = agentBuilder.with(new
CacheableTransformerDecorator(Config.Agent.CLASS_CACHE_MODE));
                LOGGER.info("SkyWalking agent class cache [{}] activated.",
Config.Agent.CLASS_CACHE_MODE);
            } catch (Exception e) {
                LOGGER.error(e, "SkyWalking agent can't active class cache.");
            }
        }
        //⑦用集成字节码框架 byte-buddy 的 AgentBuilder 类，将 Skywalking 支持的中
间件插件转换为字节码
```

```
        agentBuilder.type(pluginFinder.buildMatch())
                    .transform(new Transformer(pluginFinder))
                    .with(AgentBuilder.RedefinitionStrategy.RETRANSFORMATION)
                    .with(new RedefinitionListener())
                    .with(new Listener())
                    .installOn(instrumentation);
        try {
            //⑧启动 Skywalking 的 Java Agent，并侵入应用
            ServiceManager.INSTANCE.boot();
        } catch (Exception e) {
            LOGGER.error(e, "Skywalking agent boot failure.");
        }
        //⑨开启 Skywalking 的 Java Agent 的优雅关闭
        Runtime.getRuntime()
                .addShutdownHook(new Thread(ServiceManager.
INSTANCE::shutdown, "skywalking service shutdown thread"));
    }
}
```

（2）在 SkyWalkingAgent 类中定义一个静态内部类 Transformer，并实现 Byte Buddy（字节码框架）中的 AgentBuilder 类的内部接口 Transformer，具体代码如下所示：

```
private static class Transformer implements AgentBuilder.Transformer {
    private PluginFinder pluginFinder;
    Transformer(PluginFinder pluginFinder) {
        this.pluginFinder = pluginFinder;
    }
    @Override
    public DynamicType.Builder<?> transform(final DynamicType.Builder<?>
builder,
            final TypeDescription typeDescription,final ClassLoader
classLoader,
            final JavaModule module) {
        //①从插件缓存中获取指定的插件规格对象列表
        List<AbstractClassEnhancePluginDefine> pluginDefines =
pluginFinder.find(typeDescription);
        if (pluginDefines.size() > 0) {
            DynamicType.Builder<?> newBuilder = builder;
            EnhanceContext context = new EnhanceContext();
         //②遍历插件规格对象列表并解析插件规格对象，得到 Byte Buddy 能够识别的字节码
对象
            for (AbstractClassEnhancePluginDefine define : pluginDefines) {
                DynamicType.Builder<?> possibleNewBuilder = define.define(
                    typeDescription, newBuilder, classLoader, context);
                if (possibleNewBuilder != null) {
                    newBuilder = possibleNewBuilder;
```

```
                }
            }
            if (context.isEnhanced()) {
                LOGGER.debug("Finish the prepare stage for {}.",
typeDescription.getName());
            }
            return newBuilder;
        }
        LOGGER.debug("Matched class {}, but ignore by finding mechanism.",
typeDescription.getTypeName());
        //③返回字节码增强构造器对象
        return builder;
    }
}
```

2. 使用 PluginBootstrap 类加载中间件插件（中间件探针）

Skywalking 使用 PluginBootstrap 类加载中间件插件，具体代码如下所示：

```
public class PluginBootstrap {
    private static final ILog LOGGER =
LogManager.getLogger(PluginBootstrap.class);
    public List<AbstractClassEnhancePluginDefine> loadPlugins() throws
AgentPackageNotFoundException {
        AgentClassLoader.initDefaultLoader();
        PluginResourcesResolver resolver = new PluginResourcesResolver();
        //①获取存储插件的路径列表 URL，路径为“resources/skywalking-plugin.def”
        List<URL> resources = resolver.getResources();
        if (resources == null || resources.size() == 0) {
            LOGGER.info("no plugin files (skywalking-plugin.def) found,
continue to start application.");
            return new ArrayList<AbstractClassEnhancePluginDefine>();
        }
        //②遍历存储插件的路径 URL 列表，获取插件规格的元数据 PluginDefine 类，主要包
含插件名称和插件类的名称
        for (URL pluginUrl : resources) {
            try {
                PluginCfg.INSTANCE.load(pluginUrl.openStream());
            } catch (Throwable t) {
                LOGGER.error(t, "plugin file [{}] init failure.", pluginUrl);
            }
        }
        List<PluginDefine> pluginClassList =
PluginCfg.INSTANCE.getPluginClassList();
        List<AbstractClassEnhancePluginDefine> plugins = new
ArrayList<AbstractClassEnhancePluginDefine>();
```

```
        //③遍历存储插件元数据列表，将元数据转换为字节码框架能够识别的数据对象
TypeDescription，并将其存储在 plugins 中
        for (PluginDefine pluginDefine : pluginClassList) {
            try {
                LOGGER.debug("loading plugin class {}.",
pluginDefine.getDefineClass());
                AbstractClassEnhancePluginDefine plugin =
(AbstractClassEnhancePluginDefine)
Class.forName(pluginDefine.getDefineClass(), true, AgentClassLoader
                    .getDefault()).newInstance();
                plugins.add(plugin);
            } catch (Throwable t) {
                LOGGER.error(t, "load plugin [{}] failure.",
pluginDefine.getDefineClass());
            }
        }
        //④加载开发人员自定义的插件（InstrumentationLoader 接口的实现）
        plugins.addAll(DynamicPluginLoader.INSTANCE
         .load(AgentClassLoader.getDefault()));
        return plugins;
    }
}
```

10.3.3 【实例】将“基于 Spring Cloud Alibaba 的服务消费者和服务订阅者”接入 Skywalking

本实例的源码在本书配套资源的“access-skywalking-spring-cloud-alibaba-provider/”和“access-skywalking-spring-cloud-alibaba-consumerr/”目录下。

本实例包含服务提供者和服务消费者。

1. 初始化服务提供者 access-skywalking-spring-cloud-alibaba-provider

使用 Spring Cloud Alibaba 初始化一个项目，并添加一个 Dubbo 接口，具体代码如下所示：

```
@DubboService(version = "1.0.0",group
 = "access-skywalking-spring-cloud-alibaba-provider")
public class SkywalkingServiceImpl implements SkywalkingService {
    @Override
    public String skywalkingServiceProvider() {
        return "access skywalking!";
    }
}
```

2. 初始化服务消费者 access-skywalking-spring-cloud-alibaba-consumer

使用 Spring Cloud Alibaba 初始化一个项目，并定义一个服务订阅者接口，具体代码如下所示：

```
@Service
public class ConsumerService {
    @DubboReference(version = "1.0.0",group =
"access-skywalking-spring-cloud-alibaba-provider")
    private SkywalkingService skywalkingService;
    public String consumer(){
        return skywalkingService.skywalkingServiceProvider();
    }
}
@RequestMapping(value = "/skywalking")
public class SkywalkingController {
    @Resource
    private ConsumerService consumerService;
    @GetMapping(value = "")
    public String skywalking(){
        return consumerService.consumer();
    }
}
```

3. 启动服务提供者和服务消费者

在启动服务提供者和服务消费者时，在 IDEA 启动配置项的 JVM 配置中添加如下配置信息：

```
###①在 access-skywalking-spring-cloud-alibaba-consumer 中添加的配置信息
-javaagent:/Users/huxian/Downloads/skywalking-env/agent/skywalking-agent
.jar-Dskywalking.agent.service_name=access-skywalking-spring-cloud-alibaba-
consumer
###②在 access-skywalking-spring-cloud-alibaba-provider 中添加的配置信息
-javaagent:/Users/huxian/Downloads/skywalking-env/agent/skywalking-agent
.jar
-Dskywalking.agent.service_name=access-skywalking-spring-cloud-alibaba-provi
der
```

启动服务提供者和服务消费者，在 Nacos 配置中心中可以看到服务已经注册成功，如图 10-10 所示。

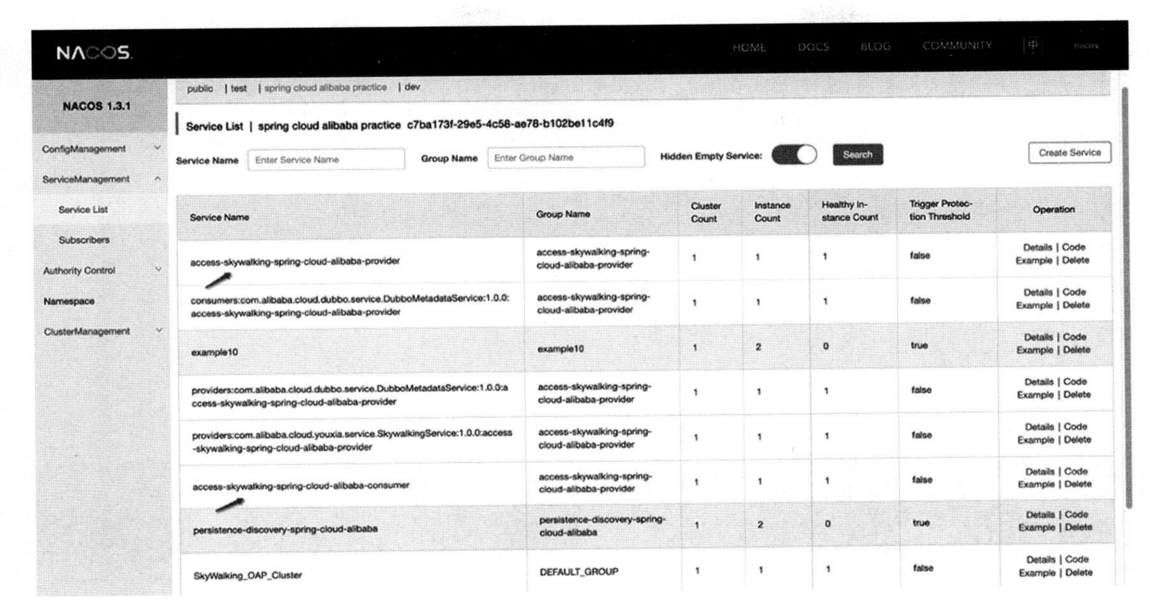

图 10-10

4. 模拟用户请求访问服务消费者，观察 Skywaking 上的链路关系

执行完命令“127.0.0.1:8078/topology”后，可以在 Skywalking 的 UI 控制台中看到服务消费者调用服务提供者的拓扑关系，如图 10–11 所示。

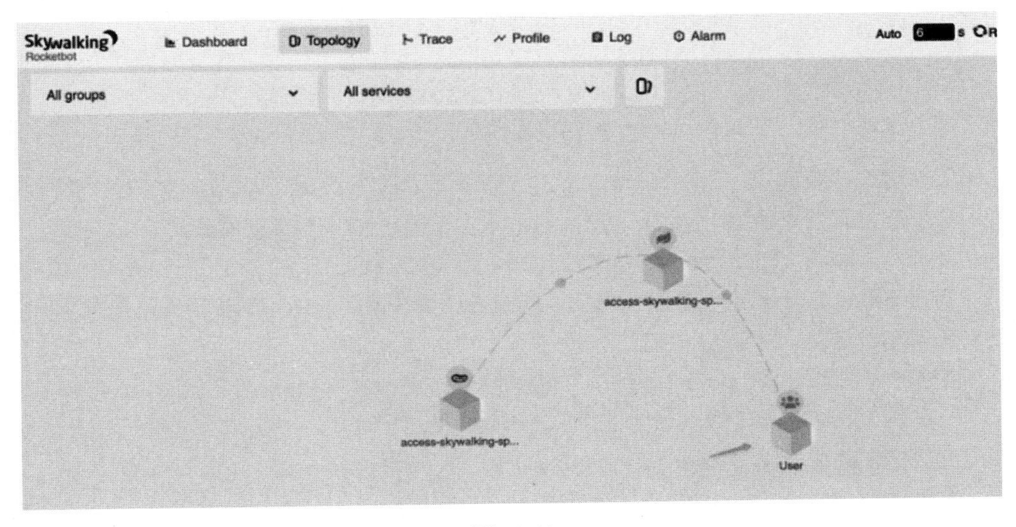

图 10-11

还可以在 Trace 页面中看到链路调用关系，如图 10–12 所示。

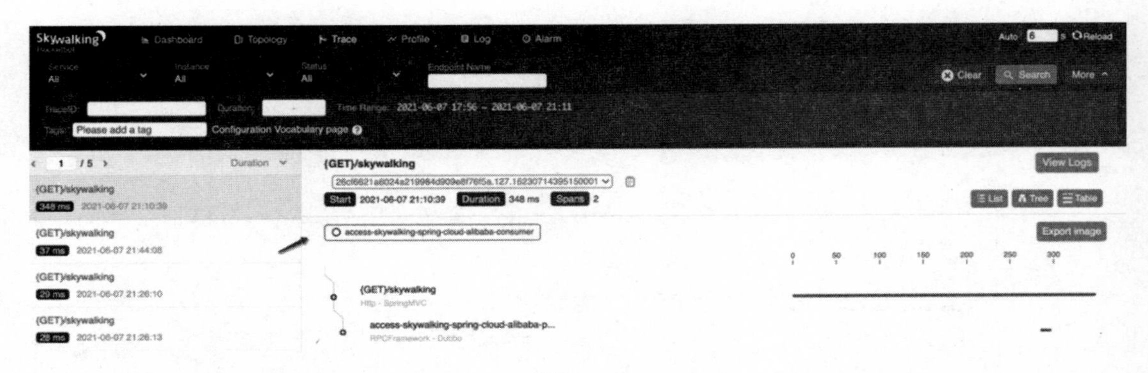

图 10-12

10.4 用 ModuleProvider 和 ModuleDefine 将 Skywalking 的功能进行模块化设计

在 Skywalking 中，所有与分布式链路相关的功能都是采用模块化来完成的。这样，模块内部是高内聚的，模块之间是低耦合的。

10.4.1 为什么需要模块化设计

Skywalking 是一个针对分布式系统的应用性能监控（Application Performance Monitor，APM）和可观察性分析平台（Observability Analysis Plaform）。

1. 包含的功能模块

为了能够实现平台化，Skywalking 提供了很多与分布式链路追踪相关的功能模块。

- 服务健康检查模块：Skywalking 从遥测模块收集指标数据，根据指标数据分析服务的健康状态，开发人员可以通过 Skywalking 暴露的 API 查询服务的健康状态。
- 注册中心模块：Skywalking 提供了注册中心（Nacos、Consul、Eureka 等）功能，支持将 Skywalking Server 的服务实例信息注册到注册中心中，应用可以从注册中心中获取服务实例信息，并完成 RPC。
- 配置中心模块：Skywalking 提供了配置中心（Nacos、Apollo、ZooKeeper 等）功能，支持将 Skywalking Server 的配置信息存储到配置中心中，应用可以动态地配置数据，开发人员不需要重启 Skywalking Server。
- 存储模块：Skywalking 提供了分布式链路数据存储（ElasticSearch、Influxdb、MySQL、Postgresql、Zipkin 等）功能，支持将从应用端采集的分布式链路数据持久化到不同类型的存储设备。

- GraphQL 查询模块：Skywalking 通过 GraphQL API 为 UI 层提供链路指标数据。
- JVM 收集模块：Skywalking 可以收集应用端运行的 JVM 指标数据。
- Kafka 收集模块：Skywalking 支持异步采集链路追踪数据（Kafka），在 Skywalking 的服务器端用 Kafka 消费主题消息。
- 告警模块：Skywalking 支持将链路告警规则（结合链路监控指标数据）转换为告警通知，并持久化到存储设备中。
- Agent 分析模块：Skywalking 支持对从应用端采集的数据进行分析，并持久化到存储设备中。
- Zipkin 收集模块：Skywalking 支持用 Zipkin 作为链路数据收集的数据源，将应用从 Zipkin 链路平台上的链路数据收集到 Skywalking 中，并持久化到存储设备中。
- Trace 收集模块：Skywalking 支持收集从应用端采集的 Trace 数据。
- Prometheus 收集模块：Skywalking 支持将链路监控指标数据同步到 Prometheus 系统中（Skywalking 暴露 HTTP 回调接口给 Prometheus）。
- 浏览器收集模块：Skywalking 能够监听浏览器，并收集浏览器的链路监控指标数据。
- Service Mesh 收集模块：在该模块中，SkyWalking 会利用从 Istio 中收集的监控指标来进行聚合分析。

2. 采用模块化设计的主要原因

Skywalking 采用模块化设计的主要原因如下：

- Skywalking 是 Apache 的顶级项目，在 Skywalking 项目上贡献代码的开发者非常多，但是开发人员之间的代码风格和质量存在很大的差异性。Skywalking 为了确保代码质量，要求将新功能模块化，做到功能的“高内聚，低耦合”。
- Skywalking 是一个平台，所以它的功能非常多。如果 Skywalking 要实现一个功能，这个功能可以用技术 A、技术 B 或者技术 C 中的一种来实现，那么 Skywalking 需要利用模块化设计来兼容这三种技术。这样，如果开发人员需要使用这个功能，则可以结合项目的实际情况来使用技术 A、技术 B 或者技术 C 中的一种。比如，Skywalking 提供了 Nacos、Eureka、Consul 等注册中心模块，开发人员可以根据自己项目注册中心的技术栈，合理地选择某个注册中心模块（本书统一使用 Nacos）。
- Skywalking 是可扩展的。在 Skywalking 中，不同的功能都需要采用插件的形式侵入业务。如果要插件化，则需要开发人员将功能模块化，将功能对应的配置信息模块化。
- Skywalking 是支持功能降级的。在 Skywalking 中，需要将核心功能和非核心功能隔离开。如果要隔离功能，则需要将功能核心与非核心功能模块化。

10.4.2 Skywalking 模块化设计的原理

1. Skywalking 的模块化设计

在 Skywalking 中，所有的链路功能都采用统一的模块化设计，如图 10-13 所示。

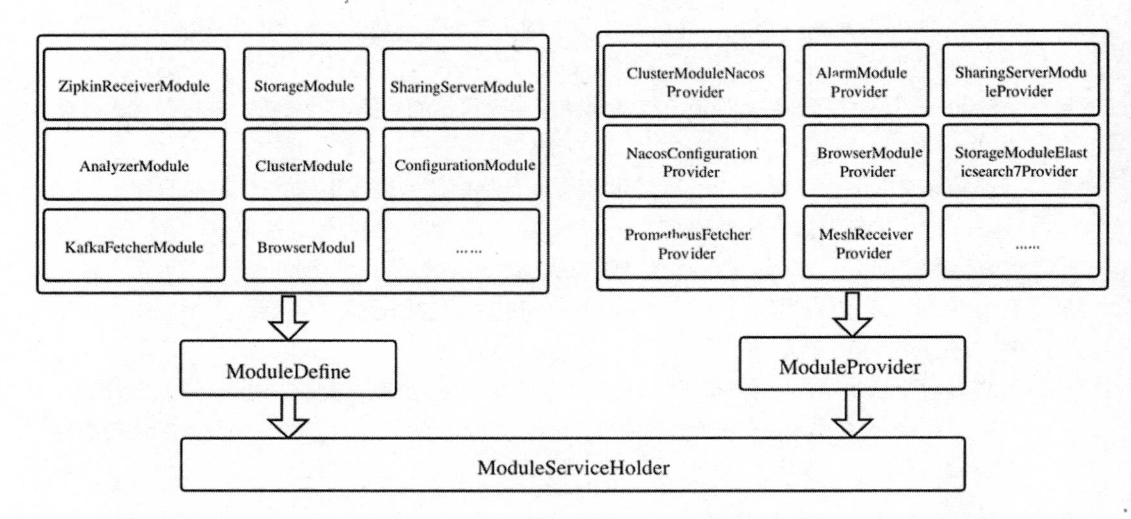

图 10-13

从图 10-13 可以看出，抽象模板 ModuleProvider 和 ModuleDefine 类是 Skywalking 模块化设计的核心类，它们的主要功能如下。

（1）抽象模板 ModuleDefine 类的主要功能包括：①用 services()方法返回模块支持的功能服务列表（主要是实现 Service 接口的功能服务）；②定义模块名称；③用 prepare()方法预加载模块。抽象模板类 ModuleDefine 的具体代码如下所示：

```
    public abstract class ModuleDefine implements ModuleProviderHolder {
    ...
    //①定义模块名称
        public final String name() {
            return name;
    }
    //②返回功能服务列表
    public abstract Class[] services();
    //③预加载模块
        void prepare(ModuleManager moduleManager,
ApplicationConfiguration.ModuleConfiguration configuration,
            ServiceLoader<ModuleProvider> moduleProviderLoader)
            throws ProviderNotFoundException, ServiceNotProvidedException,
            ModuleConfigException, ModuleStartException {
```

```
        //④遍历模块提供者列表，如果匹配当前需要加载模块类的类型，则执行预加载
        for (ModuleProvider provider : moduleProviderLoader) {
            if (!configuration.has(provider.name())) {
                continue;
            }
            if (provider.module().equals(getClass())) {
                if (loadedProvider == null) {
                    loadedProvider = provider;
                    loadedProvider.setManager(moduleManager);
                    loadedProvider.setModuleDefine(this);
                } else {
                    throw new DuplicateProviderException(this.name() + "
module has one " + loadedProvider.name() + "[" + loadedProvider
                        .getClass()
                        .getName() + "] provider already, " + provider.name()
+ "[" + provider.getClass()

      .getName() + "] is defined as 2nd provider.");
                }
            }
        }
        if (loadedProvider == null) {
            throw new ProviderNotFoundException(this.name() + " module no
provider found.");
        }
        LOGGER.info("Prepare the {} provider in {} module.",
loadedProvider.name(), this.name());
        try {
            copyProperties(loadedProvider.createConfigBeanIfAbsent(),
configuration.getProviderConfiguration(loadedProvider
                .name()), this.name(), loadedProvider.name());
        } catch (IllegalAccessException e) {
            throw new ModuleConfigException(this.name() + " module config
transport to config bean failure.", e);
        }
        //⑤执行 ModuleProvider 类的 prepare()方法完成指定模块的预加载
        loadedProvider.prepare();
    }
    //⑥返回已经加载完成的模块 loadedProvider
    @Override
    public final ModuleProvider provider() throws DuplicateProviderException,
ProviderNotFoundException {
        if (loadedProvider == null) {
            throw new ProviderNotFoundException("There is no module provider
in " + this.name() + " module!");
```

```
        }
        return loadedProvider;
    }
}
```

（2）抽象模板类 ModuleProvider 的主要功能包括：①返回模块管理器（ModuleManager）；②返回模块提供者的名称；③预加载模块提供者；④开启模块；⑤如果模块已经开启，则通知其他模块；⑥设置当前需要初始化的模块对其他模块的依赖关系；⑦注册模块需要支持的功能服务。抽象模板类 ModuleProvider 的具体代码如下所示：

```
public abstract class ModuleProvider implements ModuleServiceHolder {
...
//①设置管理模块的 ModuleManager 类
    @Setter
private ModuleManager manager;
//②设置定义模块的 ModuleDefine 类
    @Setter
    private ModuleDefine moduleDefine;
    private final Map<Class<? extends Service>, Service> services = new
HashMap<>();
    public ModuleProvider() {
    }
    protected final ModuleManager getManager() {
        return manager;
    }
    public abstract String name();
    public abstract Class<? extends ModuleDefine> module();
    //③定义预加载模块提供者的抽象方法 prepare()
    public abstract void prepare() throws ServiceNotProvidedException,
ModuleStartException;
    //④定义启动模块的抽象方法 start()
public abstract void start() throws ServiceNotProvidedException,
ModuleStartException;
//⑤定义模块加载完成后的通知方法 notifyAfterCompleted()
public abstract void notifyAfterCompleted() throws
ServiceNotProvidedException, ModuleStartException;
//⑥定义当前模块对其他模块的依赖关系
public abstract String[] requiredModules();
//⑦在当前需要初始化的模块中，注册它提供的服务功能列表，并将其缓存在变量 services 中
    @Override
    public final void registerServiceImplementation(Class<? extends Service>
serviceType,
        Service service) throws ServiceNotProvidedException {
        if (serviceType.isInstance(service)) {
            this.services.put(serviceType, service);
```

```
        } else {
            throw new ServiceNotProvidedException(serviceType + " is not
implemented by " + service);
        }
    }
    //⑧校验当前模块提供的服务功能列表中的服务的可用性，如果服务没有完成初始化则直接抛出异常
    void requiredCheck(Class<? extends Service>[] requiredServices) throws
ServiceNotProvidedException {
        if (requiredServices == null)
            return;
        for (Class<? extends Service> service : requiredServices) {
            if (!services.containsKey(service)) {
                throw new ServiceNotProvidedException("Service:" +
service.getName() + " not provided");
            }
        }
        if (requiredServices.length != services.size()) {
            throw new ServiceNotProvidedException("The " + this.name() + "
provider in " + moduleDefine.name() + " moduleDefine provide more service
implementations than ModuleDefine requirements.");
        }
    }
    //⑨返回当前模块能够提供的服务列表
    @Override
    public @SuppressWarnings("unchecked")
    <T extends Service> T getService(Class<T> serviceType) throws
ServiceNotProvidedException {
        Service serviceImpl = services.get(serviceType);
        if (serviceImpl != null) {
            return (T) serviceImpl;
        }
        throw new ServiceNotProvidedException("Service " +
serviceType.getName() + " should not be provided, based on moduleDefine
define.");
    }
    ModuleDefine getModule() {
        return moduleDefine;
    }
    //⑩获取在 ModuleDefine 类中定义的模块名称
    String getModuleName() {
        return moduleDefine.name();
    }
}
```

2. 全局扫描 ModuleProvider 类和 ModuleDefine 类

在 Skywalking 将功能模块化后，就会有很多原子模块。那么在启动的过程中，Skywalking 需要全局扫描应用中的 ModuleProvider 类和 ModuleDefine 类，并加载 ModuleProvider 类和 ModuleDefine 类对应的实现类，从而初始化原子模块，具体代码如下所示：

```
    public class ModuleManager implements ModuleDefineHolder {
        ...
    private boolean isInPrepareStage = true;
    //①缓存应用中所有的 ModuleDefine 类
        private final Map<String, ModuleDefine> loadedModules = new HashMap<>();
        public void init(
            ApplicationConfiguration applicationConfiguration) throws
ModuleNotFoundException,
            ProviderNotFoundException, ServiceNotProvidedException,
CycleDependencyException,
            ModuleConfigException, ModuleStartException {
            //②加载配置文件中的所有模块名称
            String[] moduleNames = applicationConfiguration.moduleList();
            //③用 JDK 自带的 SPI 加载应用中所有 ModuleDefine 类的实现类，并将其存储在变量
moduleServiceLoader 中
            ServiceLoader<ModuleDefine> moduleServiceLoader =
ServiceLoader.load(ModuleDefine.class);
           //④用 JDK 自带的 SPI 加载应用中所有的 ModuleProvider 类的实现类
            ServiceLoader<ModuleProvider> moduleProviderLoader =
ServiceLoader.load(ModuleProvider.class);
            HashSet<String> moduleSet = new
HashSet<>(Arrays.asList(moduleNames));
            //⑤遍历变量 moduleServiceLoader 中的 ModuleDefine 元素，并和配置文件模块名
称对比
            for (ModuleDefine module : moduleServiceLoader) {
                //⑥如果配置文件中的模块名称和遍历变量 moduleServiceLoader 中的
ModuleDefine 元素的模块名称一样，则加载这个模块，并执行模块的预加载操作
                if (moduleSet.contains(module.name())) {
                    module.prepare(this,
applicationConfiguration.getModuleConfiguration(module.name()),
moduleProviderLoader);
                    loadedModules.put(module.name(), module);
                    //⑦删除缓存中已经完成初始化的模块名称
                    moduleSet.remove(module.name());
                }
            }
            isInPrepareStage = false;
            if (moduleSet.size() > 0) {
```

```
            throw new ModuleNotFoundException(moduleSet.toString() + "
missing.");
        }
        BootstrapFlow bootstrapFlow = new BootstrapFlow(loadedModules);
        bootstrapFlow.start(this);
        bootstrapFlow.notifyAfterCompleted();
    }
}
```

3. 分析配置中心模块的加载过程

在 Skywalking 的配置文件 application.yml 中，配置了配置中心模块的配置信息，具体配置信息如下：

```
configuration:
  selector: ${SW_CONFIGURATION:nacos}
  none:
  ...
  nacos:
    serverAddr: ${SW_CONFIG_NACOS_SERVER_ADDR:127.0.0.1}
    port: ${SW_CONFIG_NACOS_SERVER_PORT:8848}
    group: ${SW_CONFIG_NACOS_SERVER_GROUP:skywalking}
    namespace: ${SW_CONFIG_NACOS_SERVER_NAMESPACE:}
    period: ${SW_CONFIG_NACOS_PERIOD:60}
    username: ${SW_CONFIG_NACOS_USERNAME:""}
    password: ${SW_CONFIG_NACOS_PASSWORD:""}
    accessKey: ${SW_CONFIG_NACOS_ACCESSKEY:""}
    secretKey: ${SW_CONFIG_NACOS_SECRETKEY:""}
```

Skwalking 解析配置文件 application.yml 的步骤如下。

（1）Skywalking 用 ApplicationConfigLoader 类来加载 application.yml 中的模块，具体代码如下所示：

```
public class ApplicationConfigLoader
public class ApplicationConfigLoader    implements
ConfigLoader<ApplicationConfiguration> {
    private static final String DISABLE_SELECTOR = "-";
    private static final String SELECTOR = "selector";
    private final Yaml yaml = new Yaml();
@Override
public ApplicationConfiguration load() throws
        ConfigFileNotFoundException {
        ApplicationConfiguration configuration
                = new ApplicationConfiguration();
        //①加载配置文件 application.yml 中的配置信息
        this.loadConfig(configuration);
```

```
        //②用在应用启动时配置的系统参数覆盖配置文件中的配置信息
        this.overrideConfigBySystemEnv(configuration);
        //③返回加载完成的配置信息对象 ApplicationConfiguration
        return configuration;
    }

    private void loadConfig(ApplicationConfiguration configuration) throws
ConfigFileNotFoundException {
        try {
        //④读取 application.yml 中的配置信息
        Reader applicationReader = ResourceUtils.read("application.yml");
        //⑤将配置信息转换为 Map 对象
        Map<String, Map<String, Object>> moduleConfig =
yaml.loadAs(applicationReader, Map.class);
        if (CollectionUtils.isNotEmpty(moduleConfig)) {
            //⑥在配置文件中用选择标签（selector）配置的子模块名称的配置信息（比如
nacos），如果不配置，则默认解析“default”的配置信息
            selectConfig(moduleConfig);
            moduleConfig.forEach((moduleName, providerConfig) -> {
                if (providerConfig.size() > 0) {
                    log.info("Get a module define from application.yml, module
                        name: {}", moduleName);
                    ApplicationConfiguration.ModuleConfiguration
                    moduleConfiguration =
                        configuration.addModule(moduleName);
                    providerConfig.forEach((providerName, config) -> {
                    ...
                    //⑦将需要初始化的模块及配置信息，添加到
                        ApplicationConfiguration 对象中
                   moduleConfiguration.addProviderConfiguration(
                      providerName, properties);
                    });
                } else {
                    ...
                }
            });
        }
    } catch (FileNotFoundException e) {
        throw new ConfigFileNotFoundException(e.getMessage(), e);
    }
    }
}
```

（2）Skywalking 通过 ApplicationConfigLoader 类的 selectConfig() 方法解析 application.yml 中的选择标签（selector），具体代码如下所示：

```
   private void selectConfig(final Map<String, Map<String, Object>>
moduleConfiguration) {
   final Set<String> modulesWithoutProvider = new HashSet<>();
   //①遍历模块信息
      for (final Map.Entry<String, Map<String, Object>> entry :
moduleConfiguration.entrySet()) {
         final String moduleName = entry.getKey();
         final Map<String, Object> providerConfig = entry.getValue();
         if (!providerConfig.containsKey(SELECTOR)) {
            continue;
         }
         //②获取在 selector 标签中配置的子模块名称，如果是配置中心模块，则子模块名称可
以配置为 Nacos、gRPC、Apollo、ZooKeeper 等
         final String selector = (String) providerConfig.get(SELECTOR);
         final String resolvedSelector =
PropertyPlaceholderHelper.INSTANCE.replacePlaceholders(
            selector, System.getProperties());
         //③如果在 selector 标签中配置了 Nacos,则删除已经加载到配置类 providerConfig
中的其他类型的配置中心（比如 Apollo、ZooKeeper 等）
         providerConfig.entrySet().removeIf(e
-> !resolvedSelector.equals(e.getKey()));
         if (!providerConfig.isEmpty()) {
            continue;
         }
         //④如果没有配置任何标签，则直接报“没有模块提供者”错误
         if (!DISABLE_SELECTOR.equals(resolvedSelector)) {
            throw new ProviderNotFoundException(
                  "no provider found for module " + moduleName + ", " +
                     "if you're sure it's not required module and want to
remove it, " +
                     "set the selector to -"
            );
         }
         modulesWithoutProvider.add(moduleName);
      }
   }
```

10.4.3 Skywalking 启动的原理

Skywalking 是一个轻量级的模块化和插件化的分布式链路追踪框架，它并没有依赖 Spring Framework，所以它的启动速度非常快。下面来分析它启动的原理。

1. 用脚本 oapService.sh 启动 Skywalking

Skywalking 用脚本文件 oapService.sh 来启动服务。如图 10-14 所示，在脚本文件

oapService.sh 中，Skywalking 会加载 org.apache.skywalking.oap.server.starter.OAPServerStartUp 类。

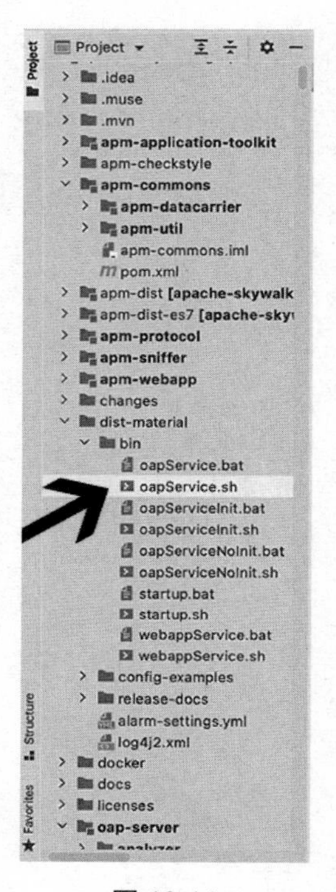

图 10-14

2. 加载 org.apache.skywalking.oap.server.starter.OAPServerStartUp 类

Skywalking 的存储支持 Elasticsearch，但是 Elasticsearch 的 6.x 和 7.x 版本差异非常大，所以，Skywalking 的存储表结构在 Elasticsearch 6.x 和 7.x 上也存在很大的差异。

在构建代码时，Skywalking 能够根据应用的 Elasticsearch 的版本构建出不同的部署包，所以，Skywalking 的启动工程有两个：server-starter 和 server-starter-es7。

开发人员可以结合项目中实际使用的 Elasticsearch 版本，灵活地选择不同的部署包进行线上部署。

如图 10-15 所示，两个工程都是先启动 OAPServerStartUp 类，然后启动 org.apache.skywalking.oap.server.starter.OAPServerBootstrap 类。在 OAPServerBootstrap 类中定义了一个静态方法 start()，OAPServerStartUp 类的 main()方法会调用静态方法 start()从而启动 Skywalking 服务。

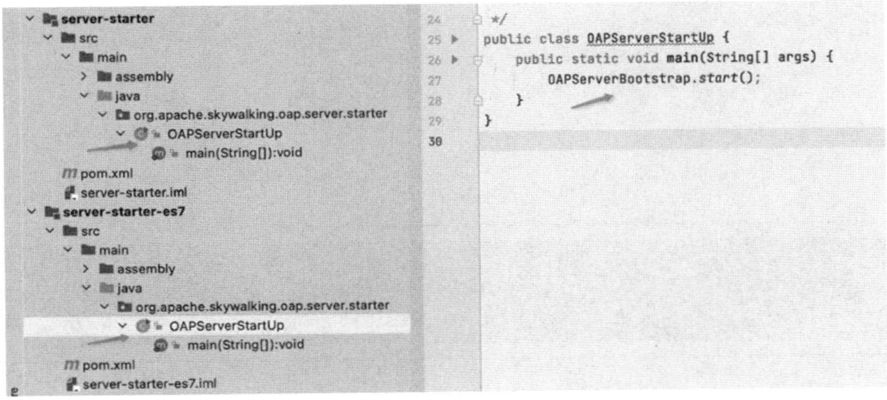

图 10-15

Skywalking 的启动类 OAPServerBootstrap 的具体代码如下所示：

```
@Slf4j
public class OAPServerBootstrap {
public static void start() {
    //①从系统参数中读取系统启动模式，比如 init 和 no-init
        String mode = System.getProperty("mode");
        RunningMode.setMode(mode);
        ApplicationConfigLoader configLoader = new
ApplicationConfigLoader();
        ModuleManager manager = new ModuleManager();
        try {
          //②加载模块的配置信息
            ApplicationConfiguration applicationConfiguration =
configLoader.load();
            //③初始化模块
            manager.init(applicationConfiguration);
            //④用 MetricsCreator 类收集 Skywalking 服务器端运行的度量指标信息，用于
健康检查
            manager.find(TelemetryModule.NAME)
                  .provider()
                  .getService(MetricsCreator.class)
                  .createGauge("uptime", "oap server start up time",
MetricsTag.EMPTY_KEY, MetricsTag.EMPTY_VALUE)
                  .setValue(System.currentTimeMillis() / 1000d);
            //⑤如果启动模式为 init，则正常关闭 Java 虚拟机并退出进程
            if (RunningMode.isInitMode()) {
                log.info("OAP starts up in init mode successfully, exit
now...");
                System.exit(0);
            }
```

```
        } catch (Throwable t) {
            log.error(t.getMessage(), t);
            System.exit(1);
        }
    }
}
```

10.5 用 HTTP、gRPC 和 Kafka 实现“应用与 Skywalking 之间的通信渠道”

Skywalking 用核心模块（core）来加载“应用与 Skywalking 之间的通信渠道”。

10.5.1 “基于 HTTP 实现通信渠道”的原理

Skywalking 用 Jetty Server 框架来封装 HTTP 服务器端。开发人员可以使用支持 HTTP 的客户端访问 Skywalking，从而完成 RPC。

如果开发人员直接使用 Skywalking 的 Java Agent 将应用接入（零侵入），则 Skywalking 默认会采用 gRPC 和 Kafka 作为通信渠道。HTTP 通信渠道主要服务于前端、第三方平台（Zipkin）及其他语言的客户端，能够使用 HTTP 通信渠道将链路数据从应用传输到 Skywalking。

1. 读取服务器端 HTTP 的配置信息

在 Skywalking 的配置文件 application.yml 中，可以设置 HTTP 服务器端的配置信息，具体配置信息如下：

```
core:
  selector: ${SW_CORE:default}
  default:
role: ${SW_CORE_ROLE:Mixed}
###①配置 HTTP 服务器端的 IP 地址
restHost: ${SW_CORE_REST_HOST:127.0.0.1}
###②配置 HTTP 服务器端的端口号
    restPort: ${SW_CORE_REST_PORT:12800}
    restContextPath: ${SW_CORE_REST_CONTEXT_PATH:/}
    restMinThreads: ${SW_CORE_REST_JETTY_MIN_THREADS:1}
    restMaxThreads: ${SW_CORE_REST_JETTY_MAX_THREADS:200}
    restIdleTimeOut: ${SW_CORE_REST_JETTY_IDLE_TIMEOUT:30000}
    restAcceptorPriorityDelta: ${SW_CORE_REST_JETTY_DELTA:0}
    restAcceptQueueSize: ${SW_CORE_REST_JETTY_QUEUE_SIZE:0}
```

2. 初始化 HTTP 服务器端

Skywalking 用核心模块类 CoreModuleProvider 的 prepare()方法来初始化 HTTP 服务器端，具体代码如下所示：

```
public class CoreModuleProvider extends ModuleProvider {
    ...
    private JettyServer jettyServer;
    @Override
    public void prepare() throws ServiceNotProvidedException,
ModuleStartException {
        //①读取配置文件中 HTTP 通信渠道的配置信息，并加载 JettyServerConfig 对象
        JettyServerConfig jettyServerConfig = JettyServerConfig.builder()
            //②配置 IP 地址
            .host(moduleConfig.getRestHost())
            //③配置端口号
            .port(moduleConfig.getRestPort())
            //④配置 Web 请求的上下文路径
            .contextPath(moduleConfig.getRestContextPath())
            //⑤配置通信连接的空闲超时时间
            .jettyIdleTimeOut(moduleConfig.getRestIdleTimeOut())
            .jettyAcceptorPriorityDelta(moduleConfig.getRestAcceptorPriori
tyDelta())
            //⑥配置通信连接的最小线程数
            .jettyMinThreads(moduleConfig.getRestMinThreads())
            //⑦配置通信连接的最大线程数
            .jettyMaxThreads(moduleConfig.getRestMaxThreads())
            //⑧配置通信连接的缓冲队列的容量
            .jettyAcceptQueueSize(moduleConfig.getRestAcceptQueueSize()).b
uild();
        //⑨加载服务器端的 JettyServer 对象
        jettyServer = new JettyServer(jettyServerConfig);
        //⑩调用 JettyServer 类的 initialize()方法初始化服务器端的通信渠道
        jettyServer.initialize();
        this.registerServiceImplementation(JettyHandlerRegister.class, new
JettyHandlerRegisterImpl(jettyServer));
    }
}
```

3. 用 JettyServer 类的 initialize()方法初始化 org.eclipse.jetty.server.Server 类

Skywalking 用 JettyServer 类的 initialize()方法初始化 org.eclipse.jetty.server.Server 类，具体代码如下所示：

```
public class JettyServer implements Server {
    ...
```

```
        private org.eclipse.jetty.server.Server server;
        private ServletContextHandler servletContextHandler;
        private JettyServerConfig jettyServerConfig;
        @Override
    public void initialize() {
        //①定义通信连接对应的队列线程池
            QueuedThreadPool threadPool = new QueuedThreadPool();
            threadPool.setMinThreads(jettyServerConfig.getJettyMinThreads());
            threadPool.setMaxThreads(jettyServerConfig.getJettyMaxThreads());
            //②初始化 org.eclipse.jetty.server.Server 类
            server = new org.eclipse.jetty.server.Server(threadPool);
            //③初始化 ServerConnector 类，并绑定 IP 地址
            ServerConnector connector = new ServerConnector(server);
            connector.setHost(jettyServerConfig.getHost());
            connector.setPort(jettyServerConfig.getPort());
            connector.setIdleTimeout(jettyServerConfig.getJettyIdleTimeOut());
    connector.setAcceptorPriorityDelta(jettyServerConfig.getJettyAcceptorPri
orityDelta());
     connector.setAcceptQueueSize(jettyServerConfig.
getJettyAcceptQueueSize());
            //④绑定 ServerConnector 类和 Server 类
            server.setConnectors(new Connector[] {connector});
            servletContextHandler = new
ServletContextHandler(ServletContextHandler.NO_SESSIONS);
      servletContextHandler.setContextPath(jettyServerConfig.
getContextPath());
            LOGGER.info("http server root context path: {}",
jettyServerConfig.getContextPath());
            //⑤绑定 Servlet 上下文处理类和 Server 类
            server.setHandler(servletContextHandler);
        }
    }
```

4. 注册 HTTP 服务器端的服务处理类

Skywalking 用 registerServiceImplementation()方法，向核心模块的 CoreModuleProvider 类中注册 Jetty 的服务处理类 JettyHandlerRegisterImpl，具体代码如下所示：

```
    @Override
    public final void registerServiceImplementation(Class<? extends Service>
serviceType,
    Service service) throws ServiceNotProvidedException {
    //①注册 Jetty 服务处理类，比如注册 serviceType 为 JettyHandlerRegister 类的服务处
理类 JettyHandlerRegisterImpl
        if (serviceType.isInstance(service)) {
            this.services.put(serviceType, service);
```

```
        } else {
            throw new ServiceNotProvidedException(serviceType + " is not
implemented by " + service);
        }
    }
    public class JettyHandlerRegisterImpl implements JettyHandlerRegister {
        private final JettyServer server;
        public JettyHandlerRegisterImpl(JettyServer server){
            this.server = server;
    }
    //②添加 JettyHandler 类
        @Override
        public void addHandler(JettyHandler serverHandler){
            server.addHandler(serverHandler);
        }
    }
```

10.5.2 “基于 gRPC 实现通信渠道”的原理

gRPC 是一个高性能的 RPC 框架，Skywalking 支持应用（Agent 客户端）采用 gRPC 与 Skywalking 平台进行通信，并完成 RPC。

1. 在应用的 Agent 客户端中完成 gRPC 客户端的初始化

在 Skywalking 中，gRPC 通信渠道会在探针插件侵入应用的过程中完成初始化，并确保链路追踪数据能够准实时地发送给 Skywalking。

（1）在应用启动过程中，Skywalking 通过字节码技术侵入应用，并加载 gRPC 渠道相关的配置信息。

在 Skywalking 中，gRPC 通信渠道的配置类 Config 的具体代码如下所示：

```
    public class Config {
        ...
        public static class Collector {
            //①gRPC 通道状态检查的周期
            public static long GRPC_CHANNEL_CHECK_INTERVAL = 30;
            //②Agent 向后端上报心跳的周期
            public static long HEARTBEAT_PERIOD = 30;
            //③客户端将实例属性发送到后端，周期为
HEARTBEAT_PERIOD*PROPERTIES_REPORT_ PERIOD_FACTOR,
            其中，PROPERTIES_REPORT_PERIOD_FACTOR 为实例属性报告周期的影响因子
            public static int PROPERTIES_REPORT_PERIOD_FACTOR = 10;
            //④Skywalking 的链路采集器的服务地址
            public static String BACKEND_SERVICE = "";
            //⑤gRPC 客户端的 UPSTREAM 超时时间
```

```
        public static int GRPC_UPSTREAM_TIMEOUT = 30;
        //⑥获取配置文件任务列表的周期
        public static int GET_PROFILE_TASK_INTERVAL = 20;
        //⑦客户端动态解析配置信息的周期
        public static int GET_AGENT_DYNAMIC_CONFIG_INTERVAL = 20;
        //⑧如果为 true，则 Skywalking Agent 将启用周期性解析 DNS 的功能，用于更新
Skywalking 的链路采集器的服务地址
        public static boolean IS_RESOLVE_DNS_PERIODICALLY = false;
    }
}
```

（2）Skywalking 用 SnifferConfigInitializer 类解析应用端的配置文件 agent.config。应用在启动之后，会主动加载配置文件 agent.config 中的配置信息。

SnifferConfigInitializer 类加载配置信息的核心代码如下：

```
public class SnifferConfigInitializer {
    private static ILog LOGGER =
LogManager.getLogger(SnifferConfigInitializer.class);
    private static final String SPECIFIED_CONFIG_PATH = "skywalking_config";
    //①默认的配置文件 agent.config 的存储路径
    private static final String DEFAULT_CONFIG_FILE_NAME =
"/config/agent.config";
    private static final String ENV_KEY_PREFIX = "skywalking.";
    private static Properties AGENT_SETTINGS;
    private static boolean IS_INIT_COMPLETED = false;
     //②加载文件 agent.config 中的配置信息
     public static void initializeCoreConfig(String agentOptions) {
        AGENT_SETTINGS = new Properties();
        try (final InputStreamReader configFileStream = loadConfig()) {
            AGENT_SETTINGS.load(configFileStream);
            for (String key : AGENT_SETTINGS.stringPropertyNames()) {
                String value = (String) AGENT_SETTINGS.get(key);
                AGENT_SETTINGS.put(key,
PropertyPlaceholderHelper.INSTANCE.replacePlaceholders(value,
AGENT_SETTINGS));
            }
        } catch (Exception e) {
            LOGGER.error(e, "Failed to read the config file, skywalking is
going to run in default config.");
        }
        try {
            //③用系统属性中的配置信息覆盖当前应用的配置信息
            overrideConfigBySystemProp();
        } catch (Exception e) {
            LOGGER.error(e, "Failed to read the system properties.");
        }
        agentOptions = StringUtil.trim(agentOptions, ',');
        if (!StringUtil.isEmpty(agentOptions)) {
            try {
```

```
                agentOptions = agentOptions.trim();
                LOGGER.info("Agent options is {}.", agentOptions);
                //④用 Agent 启动命令中的配置信息覆盖当前应用的配置信息
                overrideConfigByAgentOptions(agentOptions);
            } catch (Exception e) {
                LOGGER.error(e, "Failed to parse the agent options, val is {}.",
agentOptions);
            }
        }
        //⑤将配置信息转换为 Config 对象
        initializeConfig(Config.class);
        configureLogger();
        LOGGER = LogManager.getLogger(SnifferConfigInitializer.class);
        if (StringUtil.isEmpty(Config.Agent.SERVICE_NAME)) {
            throw new ExceptionInInitializerError("`agent.service_name` is
missing.");
        }
        if (StringUtil.isEmpty(Config.Collector.BACKEND_SERVICE)) {
            throw new
ExceptionInInitializerError("`collector.backend_service` is missing.");
        }
        if (Config.Plugin.PEER_MAX_LENGTH <= 3) {
            LOGGER.warn(
                "PEER_MAX_LENGTH configuration:{} error, the default value of
200 will be used.",
                Config.Plugin.PEER_MAX_LENGTH
            );
            Config.Plugin.PEER_MAX_LENGTH = 200;
        }
        IS_INIT_COMPLETED = true;
    }
    //⑥解析 Agent 启动命令中的配置信息
    private static void overrideConfigByAgentOptions(String agentOptions)
throws IllegalAccessException {
        for (List<String> terms : parseAgentOptions(agentOptions)) {
            if (terms.size() != 2) {
                throw new IllegalArgumentException("[" + terms + "] is not a
key-value pair.");
            }
            AGENT_SETTINGS.put(terms.get(0), terms.get(1));
        }
    }
    //⑦解析系统属性中的配置信息
    private static void overrideConfigBySystemProp() throws
IllegalAccessException {
        Properties systemProperties = System.getProperties();
        for (final Map.Entry<Object, Object> prop :
systemProperties.entrySet()) {
            String key = prop.getKey().toString();
            if (key.startsWith(ENV_KEY_PREFIX)) {
                String realKey = key.substring(ENV_KEY_PREFIX.length());
```

```
                AGENT_SETTINGS.put(realKey, prop.getValue());
            }
        }
    }
}
```

（3）在 Skywaking 中，主要是用 GRPCChannelManager 类初始化 gRPC 的客户端，具体代码如下所示：

```
@DefaultImplementor
public class GRPCChannelManager implements BootService, Runnable {
    private volatile GRPCChannel managedChannel = null;
    private volatile ScheduledFuture<?> connectCheckFuture;
    private volatile boolean reconnect = true;
    private final Random random = new Random();
    private final List<GRPCChannelListener> listeners =
Collections.synchronizedList(new LinkedList<>());
    private volatile List<String> grpcServers;
    private volatile int selectedIdx = -1;
    private volatile int reconnectCount = 0;
    @Override
    public void boot() {
        //①读取配置文件中的 BACKEND_SERVICE 属性值，比如 127.0.0.1:12800,
127.0.0.2: 12800, 127.0.0.3:12800
        if (Config.Collector.BACKEND_SERVICE.trim().length() == 0) {
            LOGGER.error("Collector server addresses are not set.");
            LOGGER.error("Agent will not uplink any data.");
            return;
        }
        grpcServers
        =   Arrays.asList(Config.Collector.BACKEND_SERVICE.split(","));
        connectCheckFuture = Executors.newSingleThreadScheduledExecutor(
            new DefaultNamedThreadFactory("GRPCChannelManager")
        ).scheduleAtFixedRate(
            new RunnableWithExceptionProtection(
                this,
                t -> LOGGER.error("unexpected exception.", t)
            ), 0, Config.Collector.GRPC_CHANNEL_CHECK_INTERVAL,
TimeUnit.SECONDS
        );
    }
    @Override
    public void run() {
        LOGGER.debug("Selected collector grpc service running,
reconnect:{}.", reconnect);
```

```
        //②如果 Skywalking Agent 启用了周期性解析 DNS，并启用了重连开关，则重新初始
化 gRPC 通道
        if (IS_RESOLVE_DNS_PERIODICALLY && reconnect) {
            String backendService =
Config.Collector.BACKEND_SERVICE.split(",")[0];
            try {
                String[] domainAndPort = backendService.split(":");

                List<String> newGrpcServers = Arrays
                        .stream(InetAddress.getAllByName(domainAndPort[0]))
                        .map(InetAddress::getHostAddress)
                        .map(ip -> String.format("%s:%s", ip,
domainAndPort[1]))
                        .collect(Collectors.toList());
                //③重新初始化最新的 Skywalking 集群
                grpcServers = newGrpcServers;
            } catch (Throwable t) {
                LOGGER.error(t, "Failed to resolve {} of backend service.",
backendService);
            }
        }
        if (reconnect) {
            if (grpcServers.size() > 0) {
                String server = "";
                try {
                    //④采用随机算法从 Skywalking 实例的集群列表中选出一个实例
                    int index = Math.abs(random.nextInt()) %
grpcServers.size();
                    if (index != selectedIdx) {
                        selectedIdx = index;
                        server = grpcServers.get(index);
                        String[] ipAndPort = server.split(":");
                        if (managedChannel != null) {
                            managedChannel.shutdownNow();
                        }
                        //⑤初始化客户端的 gRPC 通道 GRPCChannel
                        managedChannel = GRPCChannel.newBuilder(ipAndPort[0],
Integer.parseInt(ipAndPort[1]))
                    .addManagedChannelBuilder(new StandardChannelBuilder())
                    .addManagedChannelBuilder(new TLSChannelBuilder())
                    .addChannelDecorator(new AgentIDDecorator())
                    .addChannelDecorator(new AuthenticationDecorator())
                    .build();
                        //⑥通知监听器客户端的 gRPC 通道已经连接成功
                        notify(GRPCChannelStatus.CONNECTED);
```

```
                        reconnectCount = 0;
                        reconnect = false;
                    } else if (managedChannel.isConnected(++reconnectCount >
Config.Agent.FORCE_RECONNECTION_PERIOD)) {
                        reconnectCount = 0;
                        notify(GRPCChannelStatus.CONNECTED);
                        reconnect = false;
                    }
                    return;
                } catch (Throwable t) {
                    LOGGER.error(t, "Create channel to {} fail.", server);
                }
            }
            LOGGER.debug(
                "Selected collector grpc service is not available. Wait {}
seconds to retry",
                Config.Collector.GRPC_CHANNEL_CHECK_INTERVAL
            );
        }
    }
    ...
}
```

2. 在 Skywalking 平台中，完成 gRPC 服务器端的初始化

Skywalking 用 CoreModuleProvider 类的 prepare()方法完成 gRPC 服务器端的初始化，具体代码如下所示：

```
public class CoreModuleProvider extends ModuleProvider {
    @Override
    public void prepar() throws ServiceNotProvidedException,
ModuleStartException {
        ...
        //①加载 GRPCServer 类
        if (moduleConfig.isGRPCSslEnabled()) {
            grpcServer = new GRPCServer(moduleConfig.getGRPCHost(),
moduleConfig.getGRPCPort(), moduleConfig.getGRPCSslCertChainPath(),
moduleConfig.getGRPCSslKeyPath()
            );
        } else {
            grpcServer = new GRPCServer(moduleConfig.getGRPCHost(),
moduleConfig.getGRPCPort());
        }
        //②设置 GRPCServer 的连接参数
        if (moduleConfig.getMaxConcurrentCallsPerConnection() > 0) {
```

```
    grpcServer.setMaxConcurrentCallsPerConnection(moduleConfig.getMaxConcurr
entCallsPerConnection());
            }
            if (moduleConfig.getMaxMessageSize() > 0) {
    grpcServer.setMaxMessageSize(moduleConfig.getMaxMessageSize());
            }
            if (moduleConfig.getGRPCThreadPoolQueueSize() > 0) {
    grpcServer.setThreadPoolQueueSize(moduleConfig.getGRPCThreadPoolQueueSiz
e());
            }
            if (moduleConfig.getGRPCThreadPoolSize() > 0) {

grpcServer.setThreadPoolSize(moduleConfig.getGRPCThreadPoolSize());
            }
            //③执行 GRPCServer 类的 initialize()方法初始化 gRPC 通信渠道
            grpcServer.initialize();
            //④注册 GRPCHandlerRegister 类，用于处理 gRPC 请求
            this.registerServiceImplementation(GRPCHandlerRegister.class, new
GRPCHandlerRegisterImpl(grpcServer));
        }
    }
```

10.5.3 “基于 Kafka 实现通信渠道”的原理

Skywalking 用插件 kafka-reporter-plugin 在应用中初始化 Kafka 的生产者，用于异步地向 Kafka 发送异步消息。

1. 用插件 kafka-reporter-plugin 初始化 Kafka 的生产者

插件 kafka-reporter-plugin 在 Skywalking 中的存储路径如图 10-16 所示。Skywalking 通过 Java Agent 技术将插件 kafka-reporter-plugin 侵入应用，插件和应用一起启动。在启动的过程中，会去读取 Java Agent 中 Kafka 的配置信息，初始化 Kafka 的生产者（org.apache.kafka.clients.producer.KafkaProducer），KafkaProducer 类是框架 kafka-clients 提供的 API。

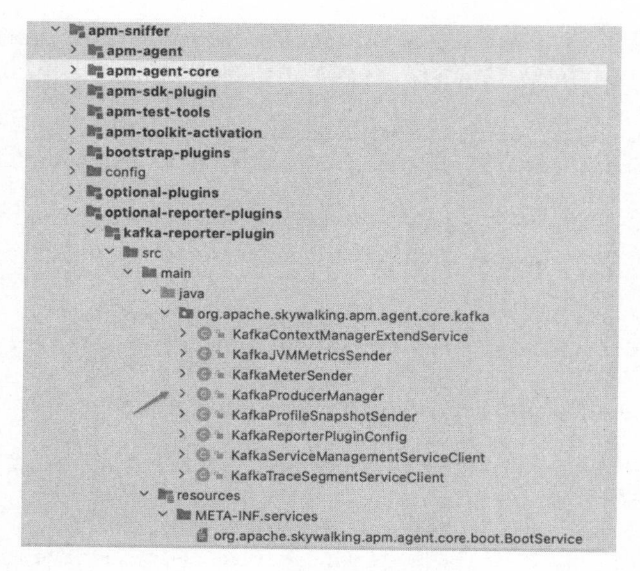

图 10-16

（1）Skywalking 用配置信息类 KafkaReporterPluginConfig 读取 Kafka 的配置信息，具体代码如下所示：

```
public class KafkaReporterPluginConfig {
    public static class Plugin {
        @PluginConfig(root = KafkaReporterPluginConfig.class)
        public static class Kafka {
            //①默认的 Kafka 服务器端的 IP 地址
            public static String BOOTSTRAP_SERVERS = "localhost:9092";
            //②度量消息的主题
            public static String TOPIC_METRICS = "skywalking-metrics";
            //③快照任务消息的主题
            public static String TOPIC_PROFILING = "skywalking-profilings";
            //④segments 消息的主题
            public static String TOPIC_SEGMENT = "skywalking-segments";
            //⑤managements 消息的主题
            public static String TOPIC_MANAGEMENT = "skywalking-managements";
            //⑥meters 消息的主题
            public static String TOPIC_METER = "skywalking-meters";
            public static Map<String, String> PRODUCER_CONFIG = new 
HashMap<>();
            //⑦获取消息主题的超时时间
            public static int GET_TOPIC_TIMEOUT = 10;
        }
    }
}
```

开发人员在配置文件 agent.config 中添加 Kafka 相关的配置信息，就可以覆盖 KafkaReporterPluginConfig 类中默认的配置信息，具体配置如下所示：

```
plugin.kafka.bootstrap_servers=${SW_KAFKA_BOOTSTRAP_SERVERS:localhost:90
92}
plugin.kafka.topic_metrics=${SW_KAFKA_TOPIC_METRICS:skywalking-metrics}
plugin.kafka.topic_profiling=${SW_KAFKA_TOPIC_PROFILING:
skywalking-profilings}
plugin.kafka.topic_segment=${SW_KAFKA_TOPIC_SEGMENT:skywalking-segments}
plugin.kafka.topic_management=${SW_KAFKA_TOPIC_MANAGEMENT:
skywalking-managements}
plugin.kafka.topic_meter=${SW_KAFKA_TOPIC_METER:skywalking-meters}
plugin.kafka.get_topic_timeout=${SW_KAFKA_GET_TOPIC_TIMEOUT:10}
```

（2）Skywalking 用 KafkaProducerManager 类初始化 Kafka 的生产者 KafkaProducer 类，具体代码如下所示：

```
@DefaultImplementor
public class KafkaProducerManager implements BootService, Runnable {
    ...
    @Override
    public void prepare() throws Throwable {
        Properties properties = new Properties();
        properties.setProperty(
            ProducerConfig.BOOTSTRAP_SERVERS_CONFIG,
KafkaReporterPluginConfig.Plugin.Kafka.BOOTSTRAP_SERVERS);
        KafkaReporterPluginConfig.Plugin.Kafka.PRODUCER_CONFIG.forEach((k,
v) -> properties.setProperty(k, v));
        AdminClient adminClient = AdminClient.create(properties);
        //①订阅主题消息
        DescribeTopicsResult topicsResult =
adminClient.describeTopics(Arrays.asList(
            KafkaReporterPluginConfig.Plugin.Kafka.TOPIC_MANAGEMENT,
            KafkaReporterPluginConfig.Plugin.Kafka.TOPIC_METRICS,
            KafkaReporterPluginConfig.Plugin.Kafka.TOPIC_PROFILING,
            KafkaReporterPluginConfig.Plugin.Kafka.TOPIC_SEGMENT,
            KafkaReporterPluginConfig.Plugin.Kafka.TOPIC_METER
        ));
        Set<String> topics = topicsResult.values().entrySet().stream()
            .map(entry -> {
                try {
                    entry.getValue().get(
                        KafkaReporterPluginConfig.Plugin.Kafka.
GET_TOPIC_TIMEOUT,TimeUnit.SECONDS);
                    return null;
```

```
                    } catch (InterruptedException | ExecutionException |
TimeoutException e) {
                        LOGGER.error(e, "Get KAFKA topic:{} error.",
entry.getKey());
                    }
                        return entry.getKey();
                }).filter(Objects::nonNull).collect(Collectors.toSet());
            if (!topics.isEmpty()) {
                throw new Exception("These topics" + topics + " don't exist.");
            }
            //②初始化生产者客户端 KafkaProducer
            producer = new KafkaProducer<>(properties, new StringSerializer(),
new BytesSerializer());
        }
    }
```

（3）Skywalking 通过 KafkaProducerManager 类获取 Kafka 的生产者 KafkaProducer 类，并生产不同主题消息，具体代码如下所示：

```
    KafkaProducer<String, Bytes> producer=ServiceManager.INSTANCE.findService
    (KafkaProducerManager.class).getProducer();
```

Skywalking 用不同的 API 生产主题消息，其中，主题“skywalking-segments”是用 KafkaTraceSegmentServiceClient 来发送的，具体代码如下所示：

```
    @OverrideImplementor(TraceSegmentServiceClient.class)
    public class KafkaTraceSegmentServiceClient implements BootService,
IConsumer<TraceSegment>, TracingContextListener {
        ...
        @Override
        public void boot() {
            carrier = new DataCarrier<>(CHANNEL_SIZE, BUFFER_SIZE,
BufferStrategy.IF_POSSIBLE);
            carrier.consume(this, 1);
            //①加载 KafkaProducer 类
            producer = ServiceManager.INSTANCE.
findService(KafkaProducerManager.class).getProducer();
        }
        @Override
    public void consume(final List<TraceSegment> data) {
        //②Kafka 作为应用的客户端，消费应用中的链路数据，并将链路数据推送到 Kafka 的服务器端
            data.forEach(traceSegment -> {
                SegmentObject upstreamSegment = traceSegment.transform();
                ProducerRecord<String, Bytes> record = new ProducerRecord<>(
                    topic,
                    upstreamSegment.getTraceSegmentId(),
```

```
            Bytes.wrap(upstreamSegment.toByteArray())
        );
        producer.send(record, (m, e) -> {
            if (Objects.nonNull(e)) {
                LOGGER.error("Failed to report TraceSegment.", e);
            }
        });
    });
}
```

2. 用插件 kafka-fetcher-plugin 初始化 Kafka 的消费者

在 Skywalking 中，插件 kafka-fetcher-plugin 会和服务平台一起启动，如图 10-17 所示。

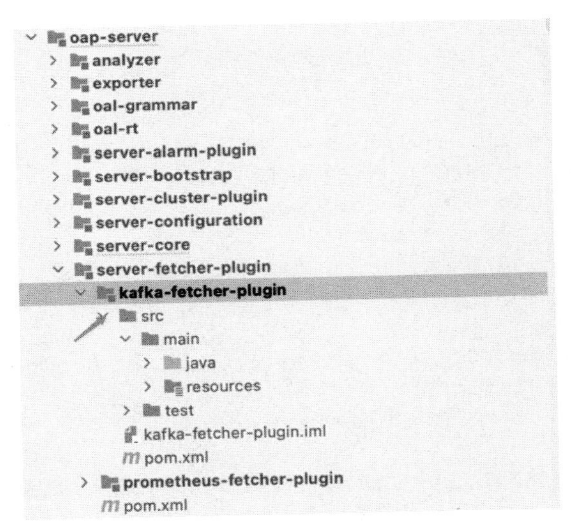

图 10-17

（1）Skywalking 用 KafkaFetcherHandlerRegister 类集成 Kafka 的客户端，并初始化 KafkaConsumer 类，具体代码如下所示：

```
@Slf4j
public class KafkaFetcherHandlerRegister implements Runnable {
    ...
    public KafkaFetcherHandlerRegister(KafkaFetcherConfig config) throws
ModuleStartException {
        this.config = config;
        //①读取 Kafka 的配置信息
        Properties properties = new Properties();
        properties.putAll(config.getKafkaConsumerConfig());
        properties.setProperty(ConsumerConfig.GROUP_ID_CONFIG,
config.getGroupId());
```

```
        properties.setProperty(ConsumerConfig.BOOTSTRAP_SERVERS_CONFIG,
config.getBootstrapServers());
        AdminClient adminClient = AdminClient.create(properties);
        Set<String> missedTopics =
adminClient.describeTopics(Lists.newArrayList(
    config.getTopicNameOfManagements(),config.getTopicNameOfMetrics(),
    config.getTopicNameOfProfiling(),config.getTopicNameOfTracingSegments(),
            config.getTopicNameOfMeters()
        )).values().entrySet().stream()
        .map(
            entry -> {
                try {
                    entry.getValue().get();
                    return null;
                } catch (InterruptedException
                    | ExecutionException e) {
                    }
                    return entry.getKey();
                }).filter(Objects::nonNull)
                .collect(Collectors.toSet());
        if (!missedTopics.isEmpty()) {
            log.info("Topics" + missedTopics.toString() + " not exist.");
            List<NewTopic> newTopicList = missedTopics.stream()
                .map(topic -> new NewTopic(
                    topic,config.getPartitions(),(short)
config.getReplicationFactor()
                )).collect(Collectors.toList());
            try {
                //②如果是 Skywalking 先启动，并且 Kafka 的服务器端没有主题（消息的生
产者没有启动），则 Skywalking 会自动创建消息主题，不需要开发人员手动创建主题
                adminClient.createTopics(newTopicList).all().get();
            } catch (Exception e) {
                throw new ModuleStartException("Failed to create Kafka Topics"
+ missedTopics + ".", e);
            }
        }
        if (config.isSharding() &&
StringUtil.isNotEmpty(config.getConsumePartitions())) {
            isSharding = true;
        } else {
            isSharding = false;
        }
        if (config.getKafkaHandlerThreadPoolSize() > 0) {
            threadPoolSize = config.getKafkaHandlerThreadPoolSize();
        }
```

```
            if (config.getKafkaHandlerThreadPoolQueueSize() > 0) {
                threadPoolQueueSize =
config.getKafkaHandlerThreadPoolQueueSize();
            }
            enableKafkaMessageAutoCommit = (boolean) properties.getOrDefault(
                ConsumerConfig.ENABLE_AUTO_COMMIT_CONFIG, true);
            //③初始化 Kafka 的消息生产者 KafkaConsumer
            consumer = new KafkaConsumer<>(properties, new StringDeserializer(),
new BytesDeserializer());
            executor = new ThreadPoolExecutor(threadPoolSize, threadPoolSize,
                60, TimeUnit.SECONDS,new
ArrayBlockingQueue(threadPoolQueueSize),
                new CustomThreadFactory("KafkaConsumer"),new
ThreadPoolExecutor.CallerRunsPolicy()
            );
        }
    }
```

（2）开发人员在文件 application.yml 中配置 Kafka 相关的配置信息，具体配置如下：

```
kafka-fetcher:
  selector: ${SW_KAFKA_FETCHER:-}
  default:
###①配置 Kafka 的 IP 地址
bootstrapServers: ${SW_KAFKA_FETCHER_SERVERS:localhost:9092}
###②配置 Kafka 的副本个数
partitions: ${SW_KAFKA_FETCHER_PARTITIONS:3}
###③配置 Kafka 的副本因子
replicationFactor: ${SW_KAFKA_FETCHER_PARTITIONS_FACTOR:2}
###④配置是否开启 Meter 系统
enableMeterSystem: ${SW_KAFKA_FETCHER_ENABLE_METER_SYSTEM:false}
###⑤配置是否开启日志功能
enableLog: ${SW_KAFKA_FETCHER_ENABLE_LOG:false}
###⑥配置是否共享消息主题
isSharding: ${SW_KAFKA_FETCHER_IS_SHARDING:false}
###⑦配置消费者分区数
consumePartitions: ${SW_KAFKA_FETCHER_CONSUME_PARTITIONS:""}
###⑧配置 Kafka 的处理器的线程池中的线程数
kafkaHandlerThreadPoolSize: ${SW_KAFKA_HANDLER_THREAD_POOL_SIZE:-1}
###⑨配置 Kafka 处理器的线程池中缓冲队列的初始容量大小
    kafkaHandlerThreadPoolQueueSize:
${SW_KAFKA_HANDLER_THREAD_POOL_QUEUE_SIZE:-1}
```

（3）Skywalking 用于消费链路主题消息的服务是“高内聚，低耦合”的。Skywalking 用不同的服务去消费主题消息，用服务 TraceSegmentHandler 类去消费其中的消息主题“skywalking-segments”，具体代码如下所示：

```
@Slf4j
public class TraceSegmentHandler implements KafkaHandler {
    private final KafkaFetcherConfig config;
    private final ISegmentParserService segmentParserService;
    private HistogramMetrics histogram;
private CounterMetrics errorCounter;
//①初始化 TraceSegmentHandler 类，它主要用于处理 Trace 数据
    public TraceSegmentHandler(ModuleManager moduleManager,
                              KafkaFetcherConfig config) {
        this.config = config;
        this.segmentParserService = moduleManager.find(AnalyzerModule.NAME)
            .provider().getService(ISegmentParserService.class);
        MetricsCreator metricsCreator
= moduleManager.find(TelemetryModule.NAME)
        .provider().getService(MetricsCreator.class);
        histogram = metricsCreator.createHistogramMetric(
            "trace_in_latency",
            "The process latency of trace data",
            new MetricsTag.Keys("protocol"),
            new MetricsTag.Values("kafka-fetcher")
        );
        errorCounter = metricsCreator.createCounter(
            "trace_analysis_error_count",
            "The error number of trace analysis",
            new MetricsTag.Keys("protocol"),
            new MetricsTag.Values("kafka-fetcher")
        );
}
//②消费 Kafka 中的链路消息
    @Override
    public void handle(final ConsumerRecord<String, Bytes> record) {
        try {
            SegmentObject segment =
SegmentObject.parseFrom(record.value().get());
            if (log.isDebugEnabled()) {
                log.debug(
                    "Fetched a tracing segment[{}] from service instance[{}].",
                    segment.getTraceSegmentId(),
                    segment.getServiceInstance()
                );
            }
            HistogramMetrics.Timer timer = histogram.createTimer();
            try {
                segmentParserService.send(segment);
            } catch (Exception e) {
```

```
                errorCounter.inc();
                log.error(e.getMessage(), e);
            } finally {
                timer.finish();
            }
        } catch (InvalidProtocolBufferException e) {
            log.error("handle record failed", e);
        }
    }
    ...
}
```

10.5.4 【实例】搭建 Kafka 环境，并用异步通信渠道 Kafka 收集 Spring Cloud Alibaba 应用的运行链路指标数据

本实例的源码在本书配套资源的“skywalking-kafka-consumer/”和“skywalking-kafka-provider/”目录下。

Skywalking 默认使用 gRPC 作为通信渠道，但是 gRPC 通信是同步的。如果开发人员想通过异步通信来减小对应用性能的影响，则可以考虑使用 Kafka 作为应用和 Skywalking 后端之间的通信渠道。

1. 搭建 Kafka 环境

本书不是一本专门介绍 Kafka 的书籍，这里只是演示单机版的 Kafka 环境的搭建。

（1）从官网下载 ZooKeeper 的安装包 apache-zookeeper-3.7.0-bin.tar.gz，并解压缩。将安装文件中的“conf/zoo_sample.cfg”文件重命名为“conf/zoo.cfg”，并执行命令“nohup bash bin/zkServer.sh start &”启动 ZooKeeper。

（2）从 Kafka 的官网下载安装包 kafka_2.13-2.8.0.tgz，并解压缩安装包。Kafka 在启动的过程中，会读取“/config/server.properties”中的配置信息。如果是单机环境，则开发人员可以不更改配置信息，Kafka 默认会连接 ZooKeeper 的 IP 地址“localhost:2181”。

开发人员执行命令“nohup bash bin/kafka-server-start.sh config/server.properties &”启动 Kafka 服务。

2. 配置 Skywalking 后端的 Kafka 信息

在 Skywalking 中的配置文件 application.yaml 中，配置 Kafka 的信息如下：

```
kafka-fetcher:
  selector: ${SW_KAFKA_FETCHER:default }
```

```
    default:
      bootstrapServers: ${SW_KAFKA_FETCHER_SERVERS:localhost:9092}
      partitions: ${SW_KAFKA_FETCHER_PARTITIONS:3}
      replicationFactor: ${SW_KAFKA_FETCHER_PARTITIONS_FACTOR:2}
      enableMeterSystem: ${SW_KAFKA_FETCHER_ENABLE_METER_SYSTEM:false}
      enableLog: ${SW_KAFKA_FETCHER_ENABLE_LOG:false}
      isSharding: ${SW_KAFKA_FETCHER_IS_SHARDING:false}
      consumePartitions: ${SW_KAFKA_FETCHER_CONSUME_PARTITIONS:""}
      kafkaHandlerThreadPoolSize: ${SW_KAFKA_HANDLER_THREAD_POOL_SIZE:-1}
      kafkaHandlerThreadPoolQueueSize:
${SW_KAFKA_HANDLER_THREAD_POOL_QUEUE_SIZE:-1}
```

3.用 Spring Cloud Alibaba 框架初始化两个应用

使用 Spring Cloud Alibaba 框架初始化两个应用：skywalking-kafka-consumer 和 skywalking-kafka-provider，详细的源码可以参考本书配套资源中的代码。

4. 配置应用中的 Kafka 信息

在应用的 Java Agent 中，开发人员可以修改配置文件“/agent/config/agent.config”，配置 Kafka 的信息如下：

```
plugin.kafka.bootstrap_servers=${SW_KAFKA_BOOTSTRAP_SERVERS:localhost:9092}
```

5. 启动 Skywalking

执行命令“bin/startup.sh”启动 Skywalking。

6. 模拟接口请求

执行命令“curl 127.0.0.1:8078/skywalking/kafka”，并观察 Skywalking 的 UI 控制台。在图 10-18 中已经生成了调用链路关系。

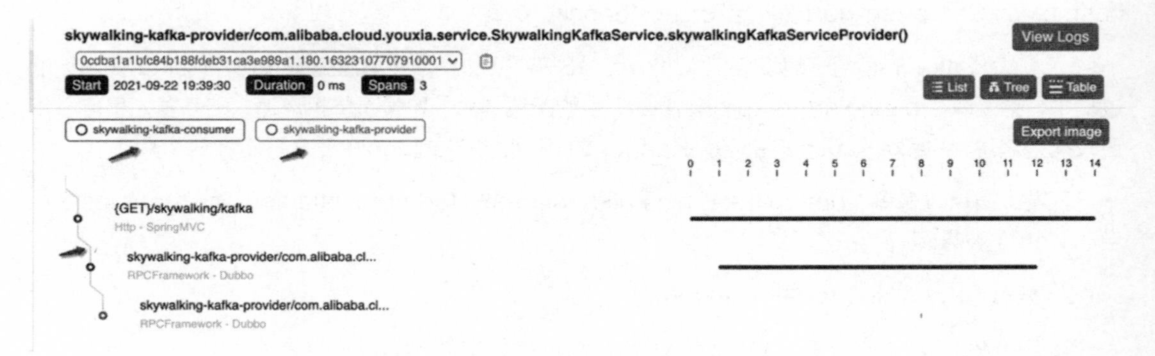

图 10-18

7. 用 Kafka 的后台命令控制台，查询 Kafka 上的链路消息主题

在 Kafka 的后台执行命令“bin/kafka-topics.sh --list --zookeeper localhost:2181”。在图 10-19 中，已经在 ZooKeeper 上生成了链路消息主题。

```
huxian@huxians-mbp kafka_2.13-2.8.0 % bin/kafka-topics.sh --list --zookeeper localhost:2181

__consumer_offsets
skywalking-managements
skywalking-meters
skywalking-metrics
skywalking-profilings
skywalking-segments
huxian@huxians-mbp kafka_2.13-2.8.0 %
```

图 10-19

10.6　用“注册中心”保证集群的高可用

Skywalking 是一个高可用的分布式链路追踪系统。既然是高可用的，那它需要通过技术手段来确保链路平台后端节点的可用性。

节点不可用的原因主要包括：硬件故障、系统资源耗尽等。Skywalking 引入了注册中心，并集成了注册中心的服务治理功能，这样开发人员可以实时地收到故障告警，并快速响应故障（软件开发人员可以利用注册中心和 Skywalking 进行二次开发，具体可以参考本书第 18 章）。

10.6.1　为什么需要注册中心

注册中心用于管理实例节点和相关服务的元数据，并提供基于元数据的服务治理功能（比如上下线服务、服务健康检查等）。服务订阅者可以直接从注册中心获取服务提供者的元数据，并完成 RPC 调用。

Skywalking 使用注册中心主要是为了管理它的实例节点，并且利用注册中心的服务治理功能保障 Skywaking 集群中实例节点的高可用性，从而提高 Java Agent 收集业务链路数据的可用性。

基于 Java Agent 的应用在接入 Skywalking 之后，会通过 gRPC 通信渠道将链路相关的数据发送到 Skwalking 集群中的节点。在引入注册中心之后，gRPC 通信渠道具备负载均衡的能力，所以按照负载均衡策略，Skywalking 集群中的节点会收到同一服务相同接口的调用链路数据。

在 Skywalking 中，为了管理后端节点处理数据的功能，后端节点被划分为 3 类：Receiver（接收）、Aggregator（聚合）和 Mixed（混合，包括接收和聚合）。开发人员可以在 CoreModuleConfig 中配置后端节点的角色，默认为 Mixed。

Skywalking 用基于注册中心的服务治理功能去管理 Skywalking 集群的 gRPC 通信渠道列表，

并为 Java Agent 提供高可用的 gRPC 通信渠道。用 Java Agent 采集业务链路数据的请求会被转发到 Skywalking 集群。图 10-20（a）和 10-20（b）描述了不同角色之间链路数据转发的过程。

- 如图 10-20（a）所示，当管理后端节点为 Aggregator 角色时，该节点不会从 Java Agent 收集链路数据。节点 A 为 Receiver 角色、节点 B 为 Mixed 角色、节点 C 和节点 D 为 Aggregator 角色，所以，节点 C 和节点 D 会相互转发数据，并且接收节点 B 的数据，不做任何处理直接持久化到数据库中。
- 如图 10-20（b）所示，当管理后端节点为 Receiver 角色时，该节点会利用 gRPC 或者 Kafka 通信渠道从 Java Agent 收集链路数据，在进行分析和聚合之后，会将处理结果转发给角色为 Mixed 和 Aggregator 的后端节点。节点 E 为 Receiver 角色，所以节点 E 向节点 F 和节点 G 转发数据。

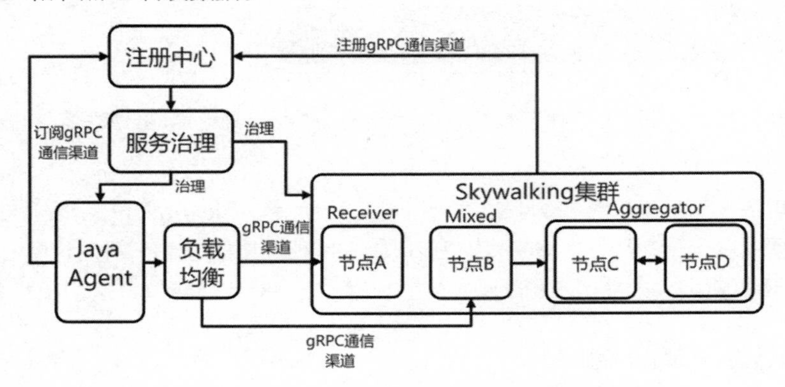

（a）在 Aggregator 角色下节点数据的传输

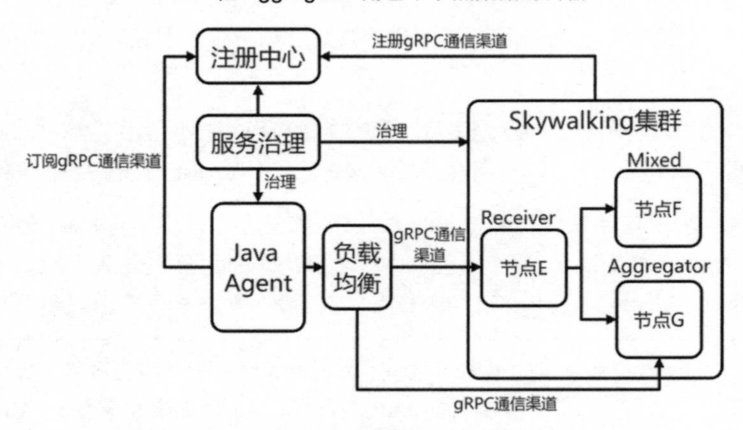

（b）在 Receiver 角色下节点数据的传输

图 10-20

如图 10-21 所示，如果数据请求转发到了角色为 Mixed（混合）和 Aggregator（聚合）的后端节点，则 Skywalking 会在后端节点的内存中进行数据聚合的操作。所以，需要在后端节点之间保持通信，并在节点之间进行增量的链路数据同步，确保节点之间数据的最终一致性，也确保最终聚合的数据的准确性。

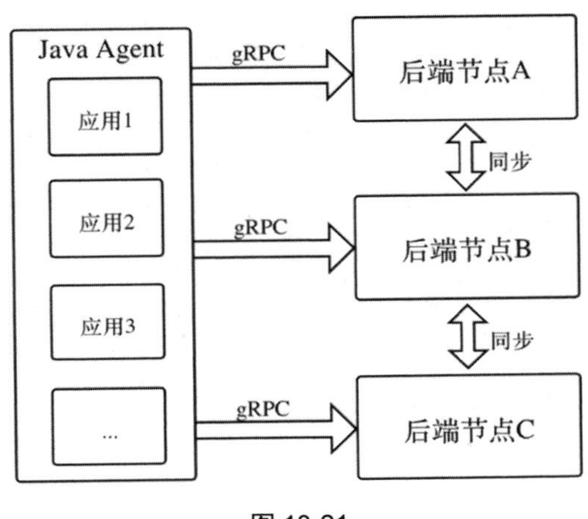

图 10-21

10.6.2 “用注册中心保证集群高可用”的原理

Skywalking 支持 Nacos、Consul、Eureka、ZooKeeper 等注册中心。但是，本书是一本关于 Spring Cloud Alibaba 的实战书籍，这里就结合 Nacos 分析 Skywalking “用注册中心保证集群高可用”的原理。

1. 注册 gRPC 节点到注册中心

Skywalking 支持用 gRPC、HTTP 和 Kafka 作为通信渠道，但是在启动的过程中，只会将 gRPC 节点注册到注册中心中。

（1）启动核心模块，在完成通信渠道的初始化后，调用 Nacos 的 NameService 注册实例节点，具体代码如下所示：

```
    public class CoreModuleProvider extends ModuleProvider {
        @Override
        public void start() throws ModuleStartException {
            ...
            //①gRPC 服务在启动完成后，会将 IP 地址封装为 Address 对象
            Address gRPCServerInstanceAddress = new
Address(moduleConfig.getGRPCHost(),
                moduleConfig.getGRPCPort(), true);
```

```
            TelemetryRelatedContext.INSTANCE.setId
(gRPCServerInstanceAddress.toString());
            //②如果后端节点的角色是混合或者聚合
            if (CoreModuleConfig.Role.Mixed.name().equalsIgnoreCase
(moduleConfig.getRole())
                || CoreModuleConfig.Role.Aggregator.name()
                    .equalsIgnoreCase(moduleConfig.getRole())) {
                //③则将 Address 对象转换成 RemoteInstance 对象
                RemoteInstance gRPCServerInstance = new
RemoteInstance(gRPCServerInstanceAddress);
                //④从模块管理器中获取 JVM 中注册中心模块的具体实现，如果配置的是 Nacos，则
使用模块的 ClusterModuleNacosProvider 类获取 Nacos 的集群管理类 NacosCoordinator，
并调用 NacosCoordinator 类的 registerRemote()方法完成注册
                this.getManager().find(ClusterModule.NAME).provider()
                    .getService(ClusterRegister.class).registerRemote(gRPCServe
rInstance);
            }
        }
    }
```

（2）Skywalking 初始化模块的 ClusterModuleNacosProvider 类，具体代码如下所示：

```
    public class ClusterModuleNacosProvider extends ModuleProvider {
        private final ClusterModuleNacosConfig config;
        private NamingService namingService;
        public ClusterModuleNacosProvider() {
            super();
            //①初始化 Nacos 的配置信息类 ClusterModuleNacosConfig
            this.config = new ClusterModuleNacosConfig();
        }
        @Override
        public String name() {
            return "nacos";
        }
        @Override
        public Class<? extends ModuleDefine> module() {
            return ClusterModule.class;
        }
        @Override
        public ModuleConfig createConfigBeanIfAbsent() {
            return config;
        }
        @Override
        public void prepare() throws ServiceNotProvidedException,
ModuleStartException {
            try {
```

```
            //②读取与 Nacos 相关的系统属性
            Properties properties = new Properties();
            properties.put(PropertyKeyConst.SERVER_ADDR,
config.getHostPort());
            properties.put(PropertyKeyConst.NAMESPACE,
config.getNamespace());
            if (StringUtil.isNotEmpty(config.getUsername()) &&
StringUtil.isNotEmpty(config.getAccessKey())) {
                throw new ModuleStartException("Nacos Auth method should
choose either username or accessKey, not both");
            }
            if (StringUtil.isNotEmpty(config.getUsername())) {
                properties.put(PropertyKeyConst.USERNAME,
config.getUsername());
                properties.put(PropertyKeyConst.PASSWORD,
config.getPassword());
            } else if (StringUtil.isNotEmpty(config.getAccessKey())) {
                properties.put(PropertyKeyConst.ACCESS_KEY,
config.getAccessKey());
                properties.put(PropertyKeyConst.SECRET_KEY,
config.getSecretKey());
            }
            //③用 Nacos Client 的 API NamingFactory 类的 createNamingService()
方法创建 NamingService 类
            namingService = NamingFactory.createNamingService(properties);
        } catch (Exception e) {
            throw new ModuleStartException(e.getMessage(), e);
        }
        //④初始化 Nacos 的协调者 NacosCoordinator
        NacosCoordinator coordinator = new NacosCoordinator(getManager(),
namingService, config);
        //⑤注册 ClusterRegister 类到 Nacos 的模块类 ClusterModuleNacosProvider 中
        this.registerServiceImplementation(ClusterRegister.class,
coordinator);
        //⑥注册 ClusterNodesQuery 类到模块类 ClusterModuleNacosProvider 中
        this.registerServiceImplementation(ClusterNodesQuery.class,
coordinator);
    }
    @Override
    public void start() throws ServiceNotProvidedException {
    }
    @Override
    public void notifyAfterCompleted() throws ServiceNotProvidedException {

    }
```

```
        //⑦Nacos 模块强依赖 Core 模块
        @Override
        public String[] requiredModules() {
            return new String[] {CoreModule.NAME};
        }
    }
```

（3）Skywalking 用协调者 NacosCoordinator 的 registerRemote()方法完成服务注册，具体代码如下所示：

```
    //①Nacos 集群协调者
    public class NacosCoordinator implements ClusterRegister, ClusterNodesQuery {
        private final ModuleDefineHolder manager;
        private final NamingService namingService;
        private final ClusterModuleNacosConfig config;
        private volatile Address selfAddress;
        private HealthCheckMetrics healthChecker;
        //②向 Nacos 注册中心注册 gRPC 实例节点
        @Override
        public void registerRemote(RemoteInstance remoteInstance) throws
ServiceRegisterException {
            if (needUsingInternalAddr()) {
                remoteInstance = new RemoteInstance(new
Address(config.getInternalComHost(), config.getInternalComPort(), true));
            }
            String host = remoteInstance.getAddress().getHost();
            int port = remoteInstance.getAddress().getPort();
            try {
                //③开启实例节点的健康检查
                initHealthChecker();
                //④调用 NamingService 的 registerInstance()方法注册实例节点
                namingService.registerInstance(config.getServiceName(), host,
port);
                //⑤设置实例节点为健康状态
                healthChecker.health();
            } catch (Throwable e) {
                //⑥如果出现异常，则设置实例节点为不健康状态
                healthChecker.unHealth(e);
                throw new ServiceRegisterException(e.getMessage());
            }
            this.selfAddress = remoteInstance.getAddress();
        }
        private void initHealthChecker() {
            if (healthChecker == null) {
                //⑦从模块管理器中获取度量创建器对象 MetricsCreator
```

```
            MetricsCreator metricCreator = manager.find(TelemetryModule.
NAME).provider().getService(MetricsCreator.class);
            //⑧创建名称为 cluster_nacos 的健康检查器
            healthChecker = metricCreator.createHealthCheckerGauge
("cluster_nacos", MetricsTag.EMPTY_KEY, MetricsTag.EMPTY_VALUE);
        }
    }
    ...
}
```

2. 获取注册中心的 gRPC 节点

Skywalking 用 RemoteClientManager 类从注册中心获取实例节点，并维护 gRPC 客户端，具体代码如下所示：

```
public class RemoteClientManager implements Service {
    private ClusterNodesQuery clusterNodesQuery;
    private volatile List<RemoteClient> usingClients;
    private GaugeMetrics gauge;
    private int remoteTimeout;
    //①启动 RemoteClientManager 类
    public void start() {
        Optional.ofNullable(sslContext).
ifPresent(DynamicSslContext::start);
        //②启动定时器，每隔 5s 执行 refresh()方法从注册中心获取最新的实例节点信息
Executors.newSingleThreadScheduledExecutor().scheduleWithFixedDelay(this::re
fresh, 1, 5, TimeUnit.SECONDS);
    }
    void refresh() {
        if (gauge == null) {
            gauge = moduleDefineHolder.find(TelemetryModule.NAME)
                .provider().getService(MetricsCreator.class)
                .createGauge("cluster_size", "Cluster size of current oap
node", MetricsTag.EMPTY_KEY,
                                    MetricsTag.EMPTY_VALUE);
        }
        try {
            if (Objects.isNull(clusterNodesQuery)) {
                synchronized (RemoteClientManager.class) {
                    if (Objects.isNull(clusterNodesQuery)) {
            //③从模块管理器中获取 Nacos 的集群管理类 ClusterModuleNacosProvider
                        this.clusterNodesQuery =
moduleDefineHolder.find(ClusterModule.NAME)
                            .provider().getService(ClusterNodesQuery.class);
                    }
                }
```

```
                }
                if (LOGGER.isDebugEnabled()) {
                    LOGGER.debug("Refresh remote nodes collection.");
                }
                //④从 Nacos 注册中心获取最新的实例节点列表
                List<RemoteInstance> instanceList =
clusterNodesQuery.queryRemoteNodes();
                instanceList = distinct(instanceList);
                Collections.sort(instanceList);
                gauge.setValue(instanceList.size());
                if (LOGGER.isDebugEnabled()) {
                    instanceList.forEach(instance -> LOGGER.debug("Cluster
instance: {}", instance.toString()));
                }
                if (!compare(instanceList)) {
                    if (LOGGER.isDebugEnabled()) {
                        LOGGER.debug("ReBuilding remote clients.");
                    }
                    //⑤按照最新的实例节点重新初始化 gRPC 客户端
                    reBuildRemoteClients(instanceList);
                }
                printRemoteClientList();
            } catch (Throwable t) {
                LOGGER.error(t.getMessage(), t);
            }
        }
        ...
    }
```

10.7 用“分布式配置中心”动态加载集群的配置信息

在 Skywalking 中，告警模块和核心模块的部分功能支持配置信息的动态加载，目前支持 Nacos、Consul、Apollo、ZooKeeper、Etcd 和 K8s 等配置中心。

10.7.1 为什么需要分布式配置中心

如图 10-22 所示，Skywalking 后端在启动的过程中会加载配置信息(比如 application.yaml、alarm-settings.yaml、gateways.yaml 等) 来初始化功能模块。

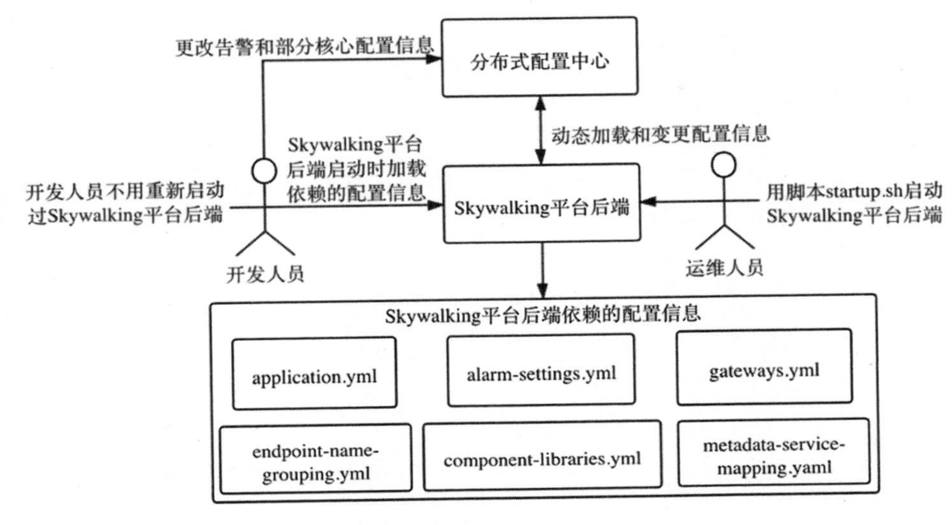

图 10-22

引入配置中心，可以提高 Skywalking 的可用性：

- 在 Skywalking 集成了配置中心（Nacos、Apollo 等）后，可以封装一个公共的配置模块，然后将不同配置中心的技术细节下沉到配置模块中，并统一暴露 SDK 给 Skywalking 的其他模块。
- 在 Skywalking 中，各个模块可以直接使用配置模块的 SDK 来动态地加载配置信息。
- 在 Skywaling 中，开发人员和运维人员可以在线直接修改配置中心中的配置信息，并动态覆盖正在运行的实例中的配置信息，实现配置信息的动态加载。

10.7.2 “用配置中心动态加载集群配置信息”的原理

Skywalking 支持多种配置中心，比如 Nacos、Apollo、Consul 等。本书是一本关于 Spring Cloud Alibaba 的书籍，所以下面只详细介绍 Skywaking 是如何集成 Nacos 配置中心，并动态加载集群配置信息的。

1. 用模块类 NacosConfigurationProvider 初始化 Nacos 配置中心

Skywalking 用模块类 NacosConfigurationProvider 来初始化 Nacos 配置中心，并通过模块管理器暴露 Nacos 配置中心。

（1）在 NacosConfigurationProvider 类中，用 initConfigReader()方法初始化 Nacos 配置中心的配置信息变更监听类 NacosConfigWatcherRegister，其中，initConfigReader()方法会在 Skywalking 模块管理器的初始化过程中同步地执行。

```
    public class NacosConfigurationProvider extends
AbstractConfigurationProvider {
        private NacosServerSettings settings;
        public NacosConfigurationProvider() {
            settings = new NacosServerSettings();
        }
        @Override
        public String name() {
            return "nacos";
        }
        @Override
        public ModuleConfig createConfigBeanIfAbsent() {
            return settings;
        }
        @Override
    protected ConfigWatcherRegister initConfigReader() throws
ModuleStartException {
        ...
            try {
                //返回Nacos配置中心配置信息变更监听类NacosConfigWatcherRegister
                return new NacosConfigWatcherRegister(settings);
            } catch (NacosException e) {
                throw new ModuleStartException(e.getMessage(), e);
            }
        }
    }
```

（2）Skywalking 用 NacosConfigWatcherRegister 类来初始化 Nacos 客户端的 ConfigService 类，具体代码如下所示：

```
    public class NacosConfigWatcherRegister extends ConfigWatcherRegister {
        private final NacosServerSettings settings;
        private final ConfigService configService;
         public NacosConfigWatcherRegister(NacosServerSettings settings) throws
NacosException {
            super(settings.getPeriod());
            this.settings = settings;
            this.configItemKeyedByName = new ConcurrentHashMap<>();
            this.listenersByKey = new ConcurrentHashMap<>();
            final int port = this.settings.getPort();
            final String serverAddr = this.settings.getServerAddr();
            final Properties properties = new Properties();
            properties.put(PropertyKeyConst.SERVER_ADDR, serverAddr + ":" +
port);
            properties.put(PropertyKeyConst.NAMESPACE,
settings.getNamespace());
```

```
        if (StringUtil.isNotEmpty(settings.getUsername())) {
            properties.put(PropertyKeyConst.USERNAME,
settings.getUsername());
            properties.put(PropertyKeyConst.PASSWORD,
settings.getPassword());
        } else if (StringUtil.isNotEmpty(settings.getAccessKey())) {
            properties.put(PropertyKeyConst.ACCESS_KEY,
settings.getAccessKey());
            properties.put(PropertyKeyConst.SECRET_KEY,
settings.getSecretKey());
        }
        //利用 Nacos 的配置信息初始化 Nacos 客户端的 ConfigService 类
        this.configService = NacosFactory.createConfigService(properties);
    }
    ...
}
```

2. 用抽象类 ConfigWatcherRegister 监听指定的 dataID，并动态更新应用的配置信息

Skywalking 在模块 configuration-api 中，定制了一套能够动态加载 Nacos 配置信息到应用中的基础框架。

（1）暴露抽象类 ConfigWatcherRegister 的 registerConfigChangeWatcher() 方法给 Skywalking 的其他模块（需要使用配置中心的模块），具体代码如下所示：

```
public abstract class ConfigWatcherRegister implements
DynamicConfigurationService {
    private Register register = new Register();
    private volatile boolean isStarted = false;
    @Override
  synchronized public void registerConfigChangeWatcher(ConfigChangeWatcher
watcher) {
      //①如果 Nacos 配置中心模块没有启动，则直接报错
        if (isStarted) {
            throw new IllegalStateException("Config Register has been started.
Can't register new watcher.");
        }
        //②将需要注册的监听器 ConfigChangeWatcher 包装在 WatcherHolder 类中
        WatcherHolder holder = new WatcherHolder(watcher);
        if (register.containsKey(holder.getKey())) {
            throw new IllegalStateException("Duplicate register, watcher=" +
watcher);
        }
        //③将包装器缓存在本地变量 register 中
        register.put(holder.getKey(), holder);
    }
```

```
    ...
}
```

（2）在 Skywalking 中，告警模块已经接入配置中心（比如 Nacos、Apollo 等），并通过抽象类 ConfigWatcherRegister 的 registerConfigChangeWatcher()方法注册自己的监听器，具体代码如下所示：

```
public class AlarmModuleProvider extends ModuleProvider {
    @Override
public void start()
throws ServiceNotProvidedException, ModuleStartException {
    //①用模块化管理器获取 Nacos 的模块类 NacosConfigurationProvider
        DynamicConfigurationService dynamicConfigurationService
 = getManager().find(ConfigurationModule.NAME)
            .provider().getService(DynamicConfigurationService.class);
        //②注册告警模块的配置变更监听器类 AlarmRulesWatcher
        dynamicConfigurationService.registerConfigChangeWatcher
            (alarmRulesWatcher);
    }
    ...
}
```

（3）在 Skywalking 中，Nacos 的配置模块主要通过抽象类 ConfigWatcherRegister 的 start()方法来启动，具体代码如下所示：

```
public abstract class ConfigWatcherRegister implements
DynamicConfigurationService {
    public void start() {
        isStarted = true;
        configSync();
        LOGGER.info("Current configurations after the bootstrap sync." +
LINE_SEPARATOR + register.toString());
        //①启动定时线程，定时执行 configSync()方法从配置中心同步最新的配置信息
        Executors.newSingleThreadScheduledExecutor()
                .scheduleAtFixedRate(
                    new RunnableWithExceptionProtection(
                        this::configSync,
                        t -> LOGGER.error("Sync config center error.", t)
                    ), syncPeriod, syncPeriod, TimeUnit.SECONDS);
    }
void configSync() {
    //②调用抽象类 ConfigWatcherRegister 的实现类的 readConfig()方法读取配置信息，
这里主要指 NacosConfigWatcherRegister 类的 readConfig()方法
        Optional<ConfigTable> configTable = readConfig(register.keys());
        configTable.ifPresent(config -> {
            config.getItems().forEach(item -> {
```

```
                String itemName = item.getName();
                WatcherHolder holder = register.get(itemName);
                if (holder != null) {
                    //③获取配置信息的监听器
                    ConfigChangeWatcher watcher = holder.getWatcher();
                    String newItemValue = item.getValue();
                    if (newItemValue == null) {
                        if (watcher.value() != null) {
                            //④如果没有新的配置信息，则执行配置删除事件
ConfigChangeWatcher.EventType.DELETE
                            watcher.notify(
                                new ConfigChangeWatcher.ConfigChangeEvent(null,
ConfigChangeWatcher.EventType.DELETE));
                        } else {
                        }
                    } else {
                        //⑤如果新的配置信息和监听器中的配置信息一致，则需更新应用的配
置信息，执行更新事件 ConfigChangeWatcher.EventType.MODIFY
                        if (!newItemValue.equals(watcher.value())) {
                            watcher.notify(new
ConfigChangeWatcher.ConfigChangeEvent(
                                newItemValue,
                                ConfigChangeWatcher.EventType.MODIFY
                            ));
                        } else {
                        }
                    }
                } else {
        });
    }
    ...
}
```

（4）本书用 Skywalking 告警模块的监听器类 AlarmRulesWatcher，来分析 Skywalking 的动态配置信息的通知机制，具体代码如下所示：

```
@Slf4j
public class AlarmRulesWatcher extends ConfigChangeWatcher {
    @Getter
    private volatile Map<String, List<RunningRule>> runningContext;
    private volatile Map<AlarmRule, RunningRule> alarmRuleRunningRuleMap;
    private volatile Rules rules;
    private volatile String settingsString;
    @Getter
    private final CompositeRuleEvaluator compositeRuleEvaluator;
    //①加载本地的告警配置文件中的配置信息，文件路径为"/config/alarm-settings.yml"
```

```
        public AlarmRulesWatcher(Rules defaultRules, ModuleProvider provider) {
            super(AlarmModule.NAME, provider, "alarm-settings");
            this.runningContext = new HashMap<>();
            this.alarmRuleRunningRuleMap = new HashMap<>();
            this.settingsString = Const.EMPTY_STRING;
            Expression expression = new Expression(new ExpressionContext());
            this.compositeRuleEvaluator = new
CompositeRuleEvaluator(expression);
            //②应用在启动过程中，将文件中的配置信息通知给告警模块
            notify(defaultRules);
        }
        //③暴露 notify()方法给 Nacos 模块的监听器，监听器通过该方法回调告警模块，并将最新
的配置信息更新到应用中
        @Override
        public void notify(ConfigChangeEvent value) {
            //④如果是删除事件，则调用本地的 notify()方法删除应用的配置信息
            if (value.getEventType().equals(EventType.DELETE)) {
                settingsString = Const.EMPTY_STRING;
                notify(new Rules());
            } else {
                //⑤如果是更新事件，则调用本地的 notify()方法更新应用的配置信息
                settingsString = value.getNewValue();
                RulesReader rulesReader = new RulesReader(new
StringReader(value.getNewValue()));
                Rules rules = rulesReader.readRules();
                notify(rules);
            }
    }
    //⑥用本地方法 notify()将最新的配置信息通知给应用，并执行更新和删除操作
        void notify(Rules newRules) {
            Map<AlarmRule, RunningRule> newAlarmRuleRunningRuleMap = new
HashMap<>();
            Map<String, List<RunningRule>> newRunningContext = new HashMap<>();
            newRules.getRules().forEach(rule -> {
                RunningRule runningRule =
alarmRuleRunningRuleMap.getOrDefault(rule, new RunningRule(rule));
                newAlarmRuleRunningRuleMap.put(rule, runningRule);
                String metricsName = rule.getMetricsName();
                List<RunningRule> runningRules =
newRunningContext.computeIfAbsent(metricsName, key -> new ArrayList<>());
                runningRules.add(runningRule);
            });
            this.rules = newRules;
            this.runningContext = newRunningContext;
            this.alarmRuleRunningRuleMap = newAlarmRuleRunningRuleMap;
```

```
        log.info("Update alarm rules to {}", rules);
    }
    ...
}
```

10.7.3 【实例】用配置中心动态地修改告警规则

本实例的源码在本书配套资源的“chapterten/dubbo-alarm-provider“和“chapterten/dubbo-alarm-consumer/”目录下。

Skywalking 支持用配置中心动态地修改告警规则，这样开发人员可以快速地变更线上业务的预警阈值，非常灵活。本实例将演示利用 Nacos 作为配置中心来动态地修改告警规则。

1. 在 Nacos 配置中心中添加配置文件 alarm.default.alarm-settings

在 Nacos 配置中心中添加配置文件 alarm.default.alarm-settings，如图 10-23 所示。

图 10-23

在配置文件 alarm.default.alarm-settings 中添加告警规则，具体规则如下所示：

```
###这条规则的语义是，在 2 分钟之内，如果有一次 RT（请求耗时）超过阈值 3s，则触发一条告警
rules:
  service_resp_time_rule:
    metrics-name: service_resp_time
    op: ">"
```

```
        threshold: 3000
        period: 2
        count: 1
        silence-period: 5
        message: Response time of service {name} is more than 3000ms in 1 minutes
of last 2 minutes.
```

2. 启动 Skywaking 的后端，动态加载告警规则

启动 Skywalking 的后端之后，观察启动日志。从图 10-24 可以看出，告警规则已经加载到后端实例中，并且 Skywalking 会用配置中心的告警规则，完全替换本地配置文件“config/alarm-settings.yml”中的配置信息。

```
2021-06-11 11:53:28,400 - org.apache.skywalking.oap.server.configuration.nacos.NacosConfigWatcherRegister - 138 [main] INFO  [] - Nacos config changed: alarm.default.alarm-settings: rules:
   service_resp_time_rule:
     metrics-name: service_resp_time
     op: ">"
     threshold: 3000
     period: 2
     count: 1
     silence-period: 5
     message: Response time of service {name} is more than 3000ms in 1 minutes of last 2 minutes.
2021-06-11 11:53:28,415 - org.apache.skywalking.oap.server.analyzer.provider.trace.UninstrumentedGatewaysConfig - 83 [main] INFO  [] - Updating uninstrumented gateways with: null
2021-06-11 11:53:28,428 - org.apache.skywalking.oap.server.core.alarm.provider.AlarmRulesWatcher - 102 [main] INFO  [] - Update alarm rules to Rules(rules=[AlarmRule(alarmRuleName=service_resp_time_rule, metricsName=service_resp_time, includeNames=[], includeNamesRegex=, excludeNames=[], excludeNamesRegex=, includeLabels=[], includeLabelsRegex=, excludeLabels=[], excludeLabelsRegex=, threshold=3000, op=>, period=2, count=1, silencePeriod=5, message=Response time of service {name} is more than 3000ms in 1 minutes of last 2 minutes., onlyAsCondition=false)], webhooks=[], grpchookSetting=null, slacks=null, wecchats=null, compositeRules=[], dingtalks=null, feishus=null)
2021-06-11 11:53:28,430 - org.apache.skywalking.oap.server.configuration.api.ConfigWatcherRegister - 68 [main] INFO  [] - Current configurations after the bootstrap sync.
Following dynamic config items are available.
---------------------------------------------
key:agent-analyzer.default.uninstrumentedGateways    module:agent-analyzer    provider:default    value(current):
key:configuration-discovery.default.agentConfigurations    module:configuration-discovery    provider:default    value(current):
key:core.default.endpoint-name-grouping    module:core    provider:default    value(current):
key:core.default.apdexThreshold    module:core    provider:default    value(current):
key:agent-analyzer.default.slowDBAccessThreshold    module:agent-analyzer    provider:default    value(current):
key:agent-analyzer.default.sampleRate    module:agent-analyzer    provider:default    value(current):10000
key:alarm.default.alarm-settings    module:alarm    provider:default    value(current):rules:
   service_resp_time_rule:
     metrics-name: service_resp_time
     op: ">"
     threshold: 3000
     period: 2
     count: 1
     silence-period: 5
     message: Response time of service {name} is more than 3000ms in 1 minutes of last 2 minutes.
```

图 10-24

3. 初始化本实例中的项目 dubbo-alarm-consumer 和 dubbo-alarm-provider

下面用项目 dubbo-alarm-provider 暴露 Dubbo 接口给服务订阅者 dubbo-alarm-consumer，具体代码如下所示：

```
//①暴露 Dubbo 接口
@DubboService(version = "1.0.0",group = "dubbo-alarm-provider")
public class DubboAlarmServiceImpl implements DubboAlarmService {
    @Autowired
    private AlarmManager alarmManager;
    @Override
    public String alarm() {
        return alarmManager.alarm();
    }
}
@Component
@RefreshScope
public class AlarmManager {
    @Resource
```

```
        private NacosConfig nacosConfig;
    public String alarm() {
        //②通过动态开关开启延迟执行的开关，默认关闭
            if(nacosConfig.getOpen().equals("true")){
                //③从 Nacos 配置中心动态读取延迟时间
                String time=nacosConfig.getTime();
                try {
                    Thread.sleep(Long.valueOf(time));
                }catch (InterruptedException interruptedException){
                    System.out.println(interruptedException.getMessage());
                }
                return "error!";
            }else{
                return "sucdess!";
            }
        }
    }
    @RestController
    @RequestMapping(value = "/alarm")
    public class AlarmController {
        //④订阅 Dubbo 接口 DubboAlarmService，超时时间为 2s，重试次数为 3 次
        @DubboReference(version = "1.0.0",group =
"dubbo-alarm-provider",timeout = 2000,retries = 3)
        private DubboAlarmService dubboAlarmService;
        //⑤暴露 RESTful API 接口
        @RequestMapping(value = "/query")
        public String alarm(){
            return dubboAlarmService.alarm();
        }
    }
```

4. 启动项目，模拟延迟故障

（1）在项目启动的过程中，延迟执行的开关是关闭的。在项目启动完成后，执行命令“curl 127.0.0.1:8045/alarm/query”，在 Skywalking 的 UI 控制台中可以看到正常执行的链路，如图 10-25 所示。

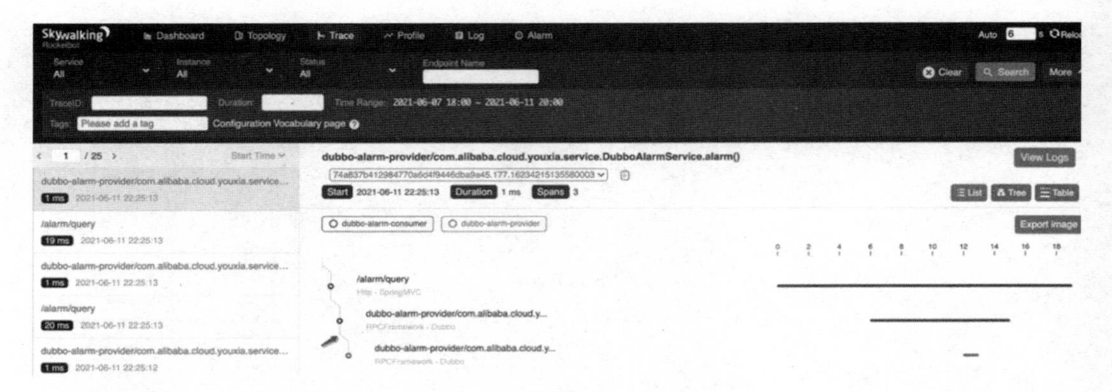

图 10-25

（2）在 Nacos 配置中心的配置文件 dubbo-alarm-provider-test.yaml 中，开启延迟执行的开关，并设置延迟时间为 3s，具体配置如下：

```
spring:
   youxia:
      sleep:
         time: 3000
         open: true
```

服务订阅者 dubbo-alarm-consumer 设置的调用超时时间为 2s，当服务提供者 dubbo-alarm-provider 被延迟 3s 执行之后，服务订阅者肯定会超时。

图 10-26 所示为执行命令“curl 127.0.0.1:8045/alarm/query”后 Skywalking 采集到的链路信息。从中可以看出，超时导致服务订阅者 dubbo-alarm-consumer 连续调用了 4 次 RPC 请求（设置了其重试 3 次）。

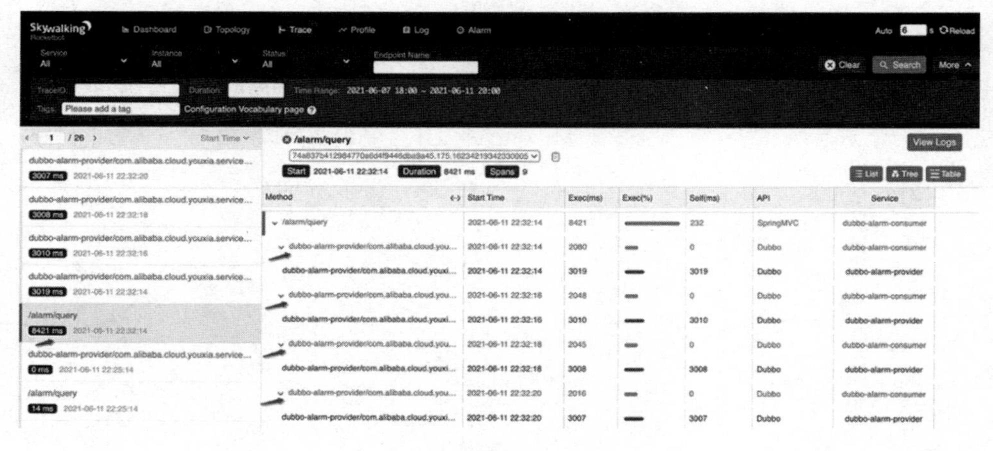

图 10-26

在 Skywalking 的告警管理中生成了一条告警信息，如图 10-27 所示。

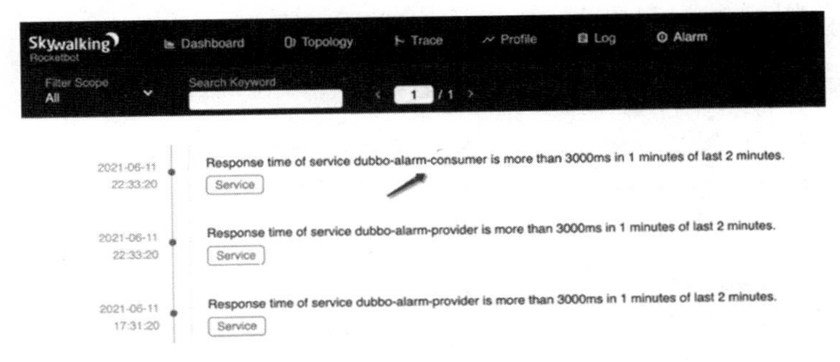

图 10-27

（3）动态调整告警规则，将阈值调整到 10s。

在 Nacos 配置中心的文件 alarm.default.alarm-settings 中，将属性字段“threshold”调整为 10s。再次执行命令“curl 127.0.0.1:8045/alarm/query”，Skywalking 并没有为超时的链路生成告警。

10.8　用探针采集链路追踪数据

Skywalking 主要用于从应用中采集链路追踪数据。既然是采集数据，就需要侵入应用的代码，以采集接口调用的详细信息。如果采用耦合式侵入，则需要更改应用的业务代码，以及应用所依赖的中间件客户端源码。Skywalking 采用“探针+Java Agent”的模式零侵入应用，应用不需要更改代码。

10.8.1　什么是探针

在 Skywalking 中，探针会在应用启动的过程中侵入应用。

如图 10-28 所示，Skywaking 为了能提升探针的覆盖率，支持很多中间件，比如 Dubbo、Redis、gRPC 等。Skywalking 采用 Java Agent 技术将这些探针零侵入应用，应用无感知。

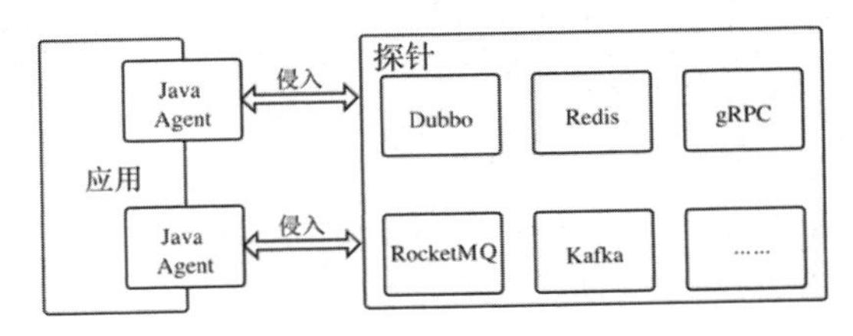

图 10-28

10.8.2 Dubbo 探针的原理

Skywalking 用插件模块 dubbo-2.7.x-plugin 来实现 Dubbo 探针。

1. 用 DubboInstrumentation 类实现 Dubbo 探针

Skywalking 在插件模块 dubbo-2.7.x-plugin 中，用 DubboInstrumentation 类去继承 ClassInstanceMethodsEnhancePluginDefine 类（字节码增强的公共类），来实现 Dubbo 探针，具体代码如下所示：

```
public class DubboInstrumentation
            extends ClassInstanceMethodsEnhancePluginDefine {
private static final String ENHANCE_CLASS
 = "org.apache.dubbo.monitor.support.MonitorFilter";
private static final String INTERCEPT_CLASS
 = "org.apache.skywalking.apm.plugin.asf.dubbo.DubboInterceptor";
    //①定义 Skywalking 拦截和增强的类为 MonitorFilter
    @Override
    protected ClassMatch enhanceClass() {
        return NameMatch.byName(ENHANCE_CLASS);
    }
    //②在 Dubbo 探针中不拦截构造方法
    @Override
    public ConstructorInterceptPoint[] getConstructorsInterceptPoints() {
        return null;
    }
    @Override
public InstanceMethodsInterceptPoint[]
getInstanceMethodsInterceptPoints() {
        return new InstanceMethodsInterceptPoint[] {
            new InstanceMethodsInterceptPoint() {
     //③定义 Skywalking 拦截的实例方法的名称是“invoke”，即拦截 MonitorFilter 类的
invoke()方法
                @Override
                public ElementMatcher<MethodDescription> getMethodsMatcher() {
                    return named("invoke");
                }
     //④定义在 Skywalking 的拦截切面生效后的拦截器，Dubbo 探针的拦截器是 Skywalking
自定义的 DubboInterceptor 类
                @Override
                public String getMethodsInterceptor() {
                    return INTERCEPT_CLASS;
                }
                @Override
                public boolean isOverrideArgs() {
```

```
                return false;
            }
        }
    };
  }
}
```

2. 用拦截器 DubboInterceptor 拦截 Dubbo 的 RPC 请求（包括服务提供者和服务消费者）

Skywalking 定义了中间件拦截器的 3 个切入点：在方法执行前、在方法执行后和在接口出现异常时，分别对应的方法是：beforeMethod()、afterMethod()和 handleMethodException()。

（1）在 Skywalking 中，用拦截器 DubboInterceptor 的 beforeMethod()方法拦截 MonitorFilter 类的 invoke()方法，拦截的触发点为“在 invoke()方法执行之前”，具体代码如下所示：

```
public class DubboInterceptor implements InstanceMethodsAroundInterceptor {
    public static final String ARGUMENTS = "arguments";
    //①在执行被拦截的 invoke()方法之前执行 beforeMethod()方法
    @Override
    public void beforeMethod(EnhancedInstance objInst, Method method,
Object[] allArguments, Class<?>[] argumentsTypes,
                            MethodInterceptResult result) throws Throwable {
        //②获取 Dubbo 的 RPC 对象 Invoker，以获取 Dubbo 请求的 URL
        Invoker invoker = (Invoker) allArguments[0];
        //③获取 Dubbo 的 RPC 对象 Invocation，以获取 Duboo 请求的方法参数等
        Invocation invocation = (Invocation) allArguments[1];
        //④获取 Dubbo 的 RPC 请求上下文对象 RpcContext
        RpcContext rpcContext = RpcContext.getContext();
        boolean isConsumer = rpcContext.isConsumerSide();
        URL requestURL = invoker.getUrl();
        AbstractSpan span;
        final String host = requestURL.getHost();
        final int port = requestURL.getPort();
        boolean needCollectArguments;
        int argumentsLengthThreshold;
        //⑤如果 RPC 请求的类型是消费者，则拦截消息者请求
        if (isConsumer) {
            final ContextCarrier contextCarrier = new ContextCarrier();
            span =
ContextManager.createExitSpan(generateOperationName(requestURL, invocation),
contextCarrier, host + ":" + port);
            CarrierItem next = contextCarrier.items();
            while (next.hasNext()) {
                next = next.next();
                rpcContext.getAttachments().put(next.getHeadKey(),
next.getHeadValue());
```

```
                if
(invocation.getAttachments().containsKey(next.getHeadKey())) {
                    invocation.getAttachments().remove(next.getHeadKey());
                }
            }
            needCollectArguments =
DubboPluginConfig.Plugin.Dubbo.COLLECT_CONSUMER_ARGUMENTS;
            argumentsLengthThreshold =
DubboPluginConfig.Plugin.Dubbo.CONSUMER_ARGUMENTS_LENGTH_THRESHOLD;
        } else {
            //⑥如果 RPC 请求的类型是生产者，则拦截生产者请求
            ContextCarrier contextCarrier = new ContextCarrier();
            CarrierItem next = contextCarrier.items();
            while (next.hasNext()) {
                next = next.next();
                next.setHeadValue(rpcContext.getAttachment
(next.getHeadKey()));
            }
            span = ContextManager.createEntrySpan
(generateOperationName(requestURL, invocation), contextCarrier);
            span.setPeer(rpcContext.getRemoteAddressString());
            needCollectArguments =
DubboPluginConfig.Plugin.Dubbo.COLLECT_PROVIDER_ARGUMENTS;
            argumentsLengthThreshold =
DubboPluginConfig.Plugin.Dubbo.PROVIDER_ARGUMENTS_LENGTH_THRESHOLD;
        }

        Tags.URL.set(span, generateRequestURL(requestURL, invocation));
        //⑦将 Dubbo 的 RPC 请求的信息收集到 Skywalking 的 Span 对象中
        collectArguments(needCollectArguments, argumentsLengthThreshold,
span, invocation);
        //⑧设置 Span 的类型为 ComponentsDefine.DUBBO
        span.setComponent(ComponentsDefine.DUBBO);
        //⑨用 SpanLayer 类的 asRPCFramework()方法异步地收集 Span，用 gRPC 通信渠道
发送数据给 Skywalking 的后端
        SpanLayer.asRPCFramework(span);
    }
    ...
}
```

（2）在 Skywalking 中，用拦截器 DubboInterceptor 的 afterMethod()方法拦截 MonitorFilter 类的 invoke()方法，拦截器的触发点为“在 invoke()方法执行之后”，具体代码如下所示：

```
@Override
public Object afterMethod(EnhancedInstance objInst, Method method, Object[]
allArguments, Class<?>[] argumentsTypes,
```

```
                                    Object ret) throws Throwable {
    //①解析 Dubbo 接口执行的结果，并将其转换为 Result 对象
    Result result = (Result) ret;
  try {
        //②获取 Result 对象中的异常信息
         if (result != null && result.getException() != null) {
             //③处理异常信息，并持久化 Dubbo 接口的业务异常信息
             dealException(result.getException());
         }
  } catch (RpcException e) {
    //④如果处理异常失败，则持久化 RpcException 异常信息
         dealException(e);
  }
  //⑤结束 Span 的生命周期
  ContextManager.stopSpan();
  //⑥返回 Dubbo 接口的处理结果给调用方
     return ret;
  }

  private void dealException(Throwable throwable) {
     //⑦获取上下文中活跃的 Span
  AbstractSpan span = ContextManager.activeSpan();
     //⑧用 Span 的 log()方法持久化异常信息
     span.log(throwable);
  }
```

（3）Skywalking 用拦截器 DubboInterceptor 的 handleMethodException()方法拦截 MonitorFilter 类的 invoke()方法，拦截的触发点为“在 Dubbo 接口出现异常时”，具体代码如下所示：

```
@Override
public void handleMethodException(EnhancedInstance objInst,
 Method method, Object[] allArguments,Class<?>[] argumentsTypes, Throwable t) {
    dealException(t);
}
```

10.8.3　“Skywalking 用探针来增强应用代码”的原理

在 Skywalking 中，所有的探针都会继承抽象类 ClassEnhancePluginDefine 和抽象类 AbstractClassEnhancePluginDefine。Skywalking 的 Java Agent 类 SkyWalkingAgent 在启动的过程中，会循环遍历抽象类 AbstractClassEnhancePluginDefine 的所有实现类（即探针），并执行 define()方法增强代码。

（1）在 Skywalking 中，用 Transformer 类的 transform()方法加载探针的具体代码如下所示：

```
    @Override
    public DynamicType.Builder<?> transform(final DynamicType.Builder<?> 
builder,
        final TypeDescription typeDescription,final ClassLoader classLoader,
        final JavaModule module) {
        List<AbstractClassEnhancePluginDefine> pluginDefines = 
pluginFinder.find(typeDescription);
        if (pluginDefines.size() > 0) {
            DynamicType.Builder<?> newBuilder = builder;
            EnhanceContext context = new EnhanceContext();
            //①循环遍历抽象模板类 AbstractClassEnhancePluginDefine 的所有实现类
            for (AbstractClassEnhancePluginDefine define : pluginDefines) {
                //②调用 AbstractClassEnhancePluginDefine 类的 define()方法增强代码
                DynamicType.Builder<?> possibleNewBuilder = define.define(
                    typeDescription, newBuilder, classLoader, context);
                    if (possibleNewBuilder != null) {
                        newBuilder = possibleNewBuilder;
                    }
                }
                if (context.isEnhanced()) {
                    LOGGER.debug("Finish the prepare stage for {}.", 
typeDescription.getName());
                }
                return newBuilder;
            }
        ...
        return builder;
    }
```

（2）用 AbstractClassEnhancePluginDefine 类的 define() 方法增强代码（AbstractClassEnhancePluginDefine 类是一个抽象模板类，它封装了增强代码的功能），具体代码如下所示：

```
    public abstract class AbstractClassEnhancePluginDefine {
        public DynamicType.Builder<?> define(TypeDescription typeDescription, 
DynamicType.Builder<?> builder,
            ClassLoader classLoader, EnhanceContext context) throws 
PluginException {
            String interceptorDefineClassName = this.getClass().getName();
            String transformClassName = typeDescription.getTypeName();
            if (StringUtil.isEmpty(transformClassName)) {
                LOGGER.warn("classname of being intercepted is not defined by {}.", 
interceptorDefineClassName);
                return null;
            }
```

```
        LOGGER.debug("prepare to enhance class {} by {}.", transformClassName,
interceptorDefineClassName);
        WitnessFinder finder = WitnessFinder.INSTANCE;
        String[] witnessClasses = witnessClasses();
        //①过滤白名单的类（不需要增强的类）
        if (witnessClasses != null) {
            for (String witnessClass : witnessClasses) {
                if (!finder.exist(witnessClass, classLoader)) {
                    LOGGER.warn("enhance class {} by plugin {} is not working.
Because witness class {} is not existed.", transformClassName,
interceptorDefineClassName, witnessClass);
                    return null;
                }
            }
        }
        //②过滤白名单的方法（不需要增强的方法）
        List<WitnessMethod> witnessMethods = witnessMethods();
        if (!CollectionUtil.isEmpty(witnessMethods)) {
            for (WitnessMethod witnessMethod : witnessMethods) {
                if (!finder.exist(witnessMethod, classLoader)) {
                    LOGGER.warn("enhance class {} by plugin {} is not working.
Because witness method {} is not existed.", transformClassName,
interceptorDefineClassName, witnessMethod);
                    return null;
                }
            }
        }
        //③调用抽象模板类 AbstractClassEnhancePluginDefine 的实现类
ClassEnhancePluginDefine 的 enhance()方法增强代码
        DynamicType.Builder<?> newClassBuilder =
this.enhance(typeDescription, builder, classLoader, context);
        context.initializationStageCompleted();
        LOGGER.debug("enhance class {} by {} completely.", transformClassName,
interceptorDefineClassName);
        //④将增强后的代码返给 JVM 的类加载器
        return newClassBuilder;
    }
    ...
}
```

（3）在 Skywalking 中，用 ClassEnhancePluginDefine 类的 enhance()方法增强代码，具体代码如下所示：

```
public abstract class ClassEnhancePluginDefine
                    extends AbstractClassEnhancePluginDefine {
    @Override
```

```
        protected DynamicType.Builder<?> enhance(TypeDescription
typeDescription, DynamicType.Builder<?> newClassBuilder,ClassLoader
classLoader,
            EnhanceContext context) throws PluginException {
            //①增加静态方法
            newClassBuilder = this.enhanceClass(typeDescription,
newClassBuilder, classLoader);
            //②增强构造方法和实例方法
            newClassBuilder = this.enhanceInstance(typeDescription,
newClassBuilder, classLoader, context);
            return newClassBuilder;
        }
        ...
    }
```

（4）在 Skywalking 中，用 ClassEnhancePluginDefine 类的 enhanceClass()方法增强静态方法的具体代码如下所示：

```
    private DynamicType.Builder<?> enhanceClass(TypeDescription
typeDescription, DynamicType.Builder<?> newClassBuilder,
    ClassLoader classLoader) throws PluginException {
    //①获取在探针（比如 DubboInstrumentation）中定义的静态方法的拦截切面列表
    StaticMethodsInterceptPoint[] staticMethodsInterceptPoints =
getStaticMethodsInterceptPoints();
    //②获取需要增强的类的名称
        String enhanceOriginClassName = typeDescription.getTypeName();
        if (staticMethodsInterceptPoints == null ||
staticMethodsInterceptPoints.length == 0) {
            return newClassBuilder;
    }
    //③遍历静态方法的拦截切面列表
    for (StaticMethodsInterceptPoint staticMethodsInterceptPoint :
staticMethodsInterceptPoints) {
        //④获取切面中的拦截器，比如 DubboInterceptor
            String interceptor =
staticMethodsInterceptPoint.getMethodsInterceptor();
            if (StringUtil.isEmpty(interceptor)) {
                throw new EnhanceException("no StaticMethodsAroundInterceptor
define to enhance class " + enhanceOriginClassName);
            }
            //⑤如果要覆盖静态方法的参数，则执行拦截
            if (staticMethodsInterceptPoint.isOverrideArgs()) {
                if (isBootstrapInstrumentation()) {
                    //⑥如果是优先级最高的 Bootstrap 基础探针，则在采用 byte-body 框架进行
字节码增强时，用 BootstrapInstrumentBoost 类作为字节码的抽取器
```

```
                newClassBuilder =
newClassBuilder.method(isStatic().and(staticMethodsInterceptPoint.getMethods
Matcher()))
                        .intercept(MethodDelegation.withDefaultConfiguration()
                        .withBinders(Morph.Binder.install(OverrideCallable.class))
                        .to(BootstrapInstrumentBoost.forInternalDelegateClass(in
terceptor)));
            } else {
                //⑦如果是普通的中间件探针，则在采用 byte-body 框架进行字节码增强时，用
StaticMethodsInterWithOverrideArgs 类作为字节码的抽取器
                newClassBuilder =
newClassBuilder.method(isStatic().and(staticMethodsInterceptPoint.getMethods
Matcher()))
                        .intercept(MethodDelegation.withDefaultConfiguration()
                        .withBinders(Morph.Binder.install(OverrideCallable.class))
                        .to(new
StaticMethodsInterWithOverrideArgs(interceptor)));
            }
        } else {
        //⑧如果不需要覆盖静态方法的参数，则执行拦截
        if (isBootstrapInstrumentation()) {
            //⑨如果是优先级最高的 Bootstrap 基础探针，则在采用 byte-body 框架进行字节
码增强时，用 BootstrapInstrumentBoost 类作为字节码的抽取器
            newClassBuilder =
newClassBuilder.method(isStatic().and(staticMethodsInterceptPoint.getMethods
Matcher()))
                    .intercept(MethodDelegation.withDefaultConfiguration()
                    .to(BootstrapInstrumentBoost.forInternalDelegateClass(inter
ceptor)));
        } else {
        //⑩如果是普通的中间件探针，则在采用 byte-body 框架进行字节码增强时，用
StaticMethodsInter 类作为字节码的抽取器
            newClassBuilder =
newClassBuilder.method(isStatic().and(staticMethodsInterceptPoint.getMethods
Matcher()))
                    .intercept(MethodDelegation.withDefaultConfiguration()
                    .to(new StaticMethodsInter(interceptor)));
            }
        }
    }
    return newClassBuilder;
}
```

（5）在 Skywalking 中，用 ClassEnhancePluginDefine 类的 enhanceInstance()方法增强类的实例方法的具体代码如下所示：

```
    private DynamicType.Builder<?>enhanceInstance(TypeDescription
typeDescription,
        DynamicType.Builder<?> newClassBuilder, ClassLoader classLoader,
        EnhanceContext context) throws PluginException {
        //①获取在探针（比如 DubboInstrumentation）中构造方法和实例方法的拦截切面
        ConstructorInterceptPoint[] constructorInterceptPoints =
getConstructorsInterceptPoints();
        InstanceMethodsInterceptPoint[] instanceMethodsInterceptPoints =
getInstanceMethodsInterceptPoints();
        String enhanceOriginClassName = typeDescription.getTypeName();
        boolean existedConstructorInterceptPoint = false;
        if (constructorInterceptPoints != null &&
constructorInterceptPoints.length > 0) {
            existedConstructorInterceptPoint = true;
        }
        boolean existedMethodsInterceptPoints = false;
        if (instanceMethodsInterceptPoints != null &&
instanceMethodsInterceptPoints.length > 0) {
            existedMethodsInterceptPoints = true;
        }
        //②如果不存在构造方法和实例方法的拦截切面，则直接返回原来的字节码给 JVM
        if (!existedConstructorInterceptPoint && !existedMethodsInterceptPoints) {
            return newClassBuilder;
        }
        if (!typeDescription.isAssignableTo(EnhancedInstance.class)) {
            if (!context.isObjectExtended()) {
                newClassBuilder = newClassBuilder.defineField(
                    CONTEXT_ATTR_NAME, Object.class, ACC_PRIVATE | ACC_VOLATILE)
                        .implement(EnhancedInstance.class)
                        .intercept(FieldAccessor.ofField(CONTEXT_ATTR_NAME));
                context.extendObjectCompleted();
                }
        }
        //③如果存在构造方法拦截切面，则执行增强构造方法
        if (existedConstructorInterceptPoint) {
            for (ConstructorInterceptPoint constructorInterceptPoint :
constructorInterceptPoints) {
                if (isBootstrapInstrumentation()) {
                    //④如果是优先级最高的 Bootstrap 基础探针，则在采用 byte-body 框架进行
字节码增强时，用 BootstrapInstrumentBoost 类作为字节码的抽取器
                    newClassBuilder = newClassBuilder.constructor
(constructorInterceptPoint.getConstructorMatcher())
                        .intercept(SuperMethodCall.INSTANCE.andThen(MethodDelega
tion.withDefaultConfiguration()
```

```
                    .to(BootstrapInstrumentBoost.forInternalDelegateClass(co
nstructorInterceptPoint
                    .getConstructorInterceptor()))));
            } else {
                //⑤如果是普通的中间件探针，则在采用 byte-body 框架进行字节码增强时，用
ConstructorInter 类作为字节码的抽取器
                newClassBuilder = newClassBuilder.constructor
(constructorInterceptPoint.getConstructorMatcher())
                    .intercept(SuperMethodCall.INSTANCE.andThen(MethodDelega
tion.withDefaultConfiguration()
                    .to(new ConstructorInter(constructorInterceptPoint
                    .getConstructorInterceptor(), classLoader))));
            }
        }
    }
    //⑥如果存在实例方法拦截切面，则执行增强实例方法
    if (existedMethodsInterceptPoints) {
        for (InstanceMethodsInterceptPoint instanceMethodsInterceptPoint :
instanceMethodsInterceptPoints) {
            String interceptor =
instanceMethodsInterceptPoint.getMethodsInterceptor();
            if (StringUtil.isEmpty(interceptor)) {
                throw new EnhanceException("no
InstanceMethodsAroundInterceptor define to enhance class " +
enhanceOriginClassName);
            }
            ElementMatcher.Junction<MethodDescription> junction =
not(isStatic()).and(instanceMethodsInterceptPoint.getMethodsMatcher());
            if (instanceMethodsInterceptPoint instanceof
DeclaredInstanceMethodsInterceptPoint) {
                junction = junction.and(ElementMatchers.<MethodDescription>
isDeclaredBy(typeDescription));
            }
            //⑦如果要覆盖实例方法的参数，则执行拦截
            if (instanceMethodsInterceptPoint.isOverrideArgs()) {
                if (isBootstrapInstrumentation()) {
                    //⑧如果是优先级最高的 Bootstrap 基础探针，则在采用 byte-body 框架
进行字节码增强时，用 BootstrapInstrumentBoost 类作为字节码的抽取器
                    newClassBuilder = newClassBuilder.method(junction)
                        .intercept(MethodDelegation.withDefaultConfiguration()
                        .withBinders(Morph.Binder.install(OverrideCallable.c
lass))
                        .to(BootstrapInstrumentBoost.forInternalDelegateClas
s(interceptor)));
                } else {
```

```
                        //⑨如果是普通的中间件探针，则在采用 byte-body 框架进行字节码增强
时，用 InstMethodsInterWithOverrideArgs 类作为字节码的抽取器。
                        newClassBuilder = newClassBuilder.method(junction)
                            .intercept(MethodDelegation.withDefaultConfiguration()
                            .withBinders(Morph.Binder.install(OverrideCallable.c
lass))
                            .to(new InstMethodsInterWithOverrideArgs(interceptor,
classLoader)));
                        }
                } else {
                    if (isBootstrapInstrumentation()) {
                        newClassBuilder = newClassBuilder.method(junction)
                            .intercept(MethodDelegation.withDefaultConfiguration()
                            .to(BootstrapInstrumentBoost.forInternalDelegateClas
s(interceptor)));
                    } else {
                        //⑩如果是普通的中间件探针，则在采用 byte-body 框架进行字节码增强
时，用 InstMethodsInter 类作为字节码的抽取器。
                        newClassBuilder = newClassBuilder.method(junction)
                            .intercept(MethodDelegation.withDefaultConfiguration()
                            .to(new InstMethodsInter(interceptor, classLoader)));
                    }
                }
            }
        }
        return newClassBuilder;
    }
```

10.8.4 【实例】模拟 Dubbo 服务故障，用 Dubbo 探针采集链路追踪数据

本实例的源码在本书配套资源的“chapterten/dubbo-error-consumer”和“chapterten/dubbo-error-provider/”目录下。

本实例主要包括两个服务：dubbo-error-consumer 和 dubbo-error-provider。dubbo-error-provider 暴露一个 Dubbo 接口给 dubbo-error-consumer。

1. 初始化 dubbo-error-consumer 和 dubbo-error-provider 服务

使用 Spring Cloud Alibaba 框架快速地初始化 dubbo-error-consumer 和 dubbo-error-provider 服务，在 dubbo-error-consumer 服务中定义一个 RESTful API 去消费 dubbo-error-provider 服务的接口 DubboErrorService，具体代码如下所示：

```
@DubboService(version = "1.0.0",group = "dubbo-error-provider")
public class DubboErrorServiceImpl implements DubboErrorService{
    private AtomicLong count=new AtomicLong();
    @Override
    public String simulationError() {
        Long result=count.incrementAndGet();
        //①使用计数器 AtomicLong，如果计数器的数字是 3 的倍数，则休眠 1s，模拟超时故障
        if(result%3==0){
            try {
                Thread.sleep(1000);
            }catch (InterruptedException io){
                System.out.println(io.getMessage());
            }
            return "出现超时故障了！";
        }
        return "成功！";
    }
}
@RestController
@RequestMapping("/dubbo")
public class ErrorController {
    //②设置消费端的 Dubbo 调用超时时间为 1s
    @DubboReference(version = "1.0.0",group =
"dubbo-error-provider",timeout = 1000)
    private DubboErrorService dubboErrorService;
    @RequestMapping(value = "/error")
    public String simulationError(){
        return dubboErrorService.simulationError();
    }
}
```

2. 启动 dubbo-error-consumer 和 dubbo-error-provider 服务

（1）在启动服务之后，在 Skywalking 的拓扑管理中可以看到 dubbo-error-consumer 和 dubbo-error-provider 服务的拓扑关系，如图 10-29 所示。dubbo-error-provider 服务提供的 DubboErrorService 接口会按照“计数器的数字是 3 的倍数就延迟”的规则添加 1s 的延迟，以模拟服务接口的超时异常。

（2）设置 dubbo-error-consumer 服务调用 DubboErrorService 接口的超时时间为 1s。这样，如果调用超时，则 dubbo-error-consumer 服务会重试一次（Dubbo 默认重试一次），并重新发起一次 RPC 调用。

（3）从 Skywalking 的链路关系中，可以看出 dubbo-error-consumer 服务确实调用了两次 dubbo-error-provider 服务，如图 10-30 所示。

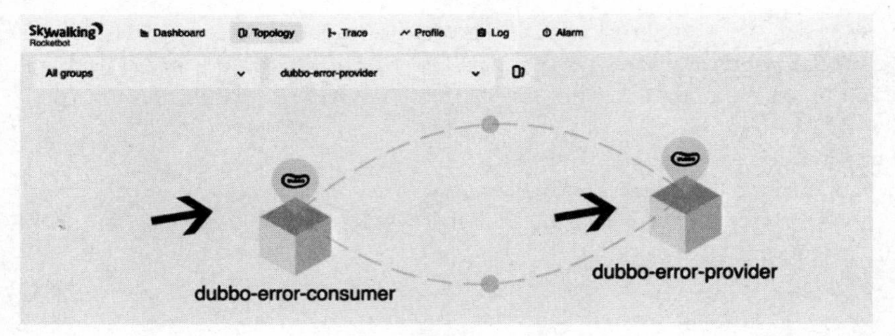

图 10-29

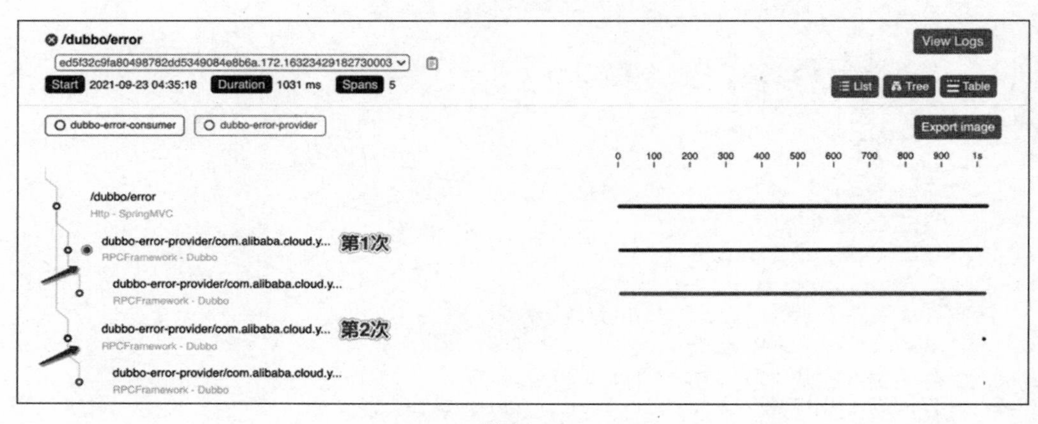

图 10-30

（4）在 Skywalking 中单击错误的链路节点，可以查看链路节点的错误信息，如图 10-31 所示。

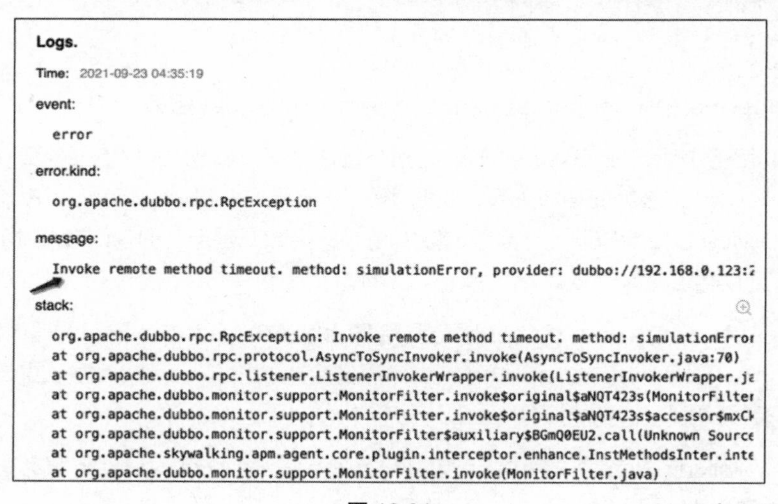

图 10-31

10.9 用 Elasticsearch 存储链路追踪数据

Skywalking 支持多种存储类型，比如 H2、MySQL、Elasticsearch 等。如果是线上部署，并且接入的应用的流量峰值比较大，则推荐采用 Elasticsearch 进行存储。

10.9.1 什么是 Elasticsearch

Elasticsearch 是一个分布式的免费的开源搜索和分析引擎，适用于存储包括文本、数字、地理空间、结构化和非结构化数据等类型的数据。Elasticsearch 在 Apache Lucene 的基础上开发而成，由 Elasticsearch N.V.（即现在的 Elastic）于 2010 年首次发布。

Elasticsearch 以其简单的 REST 风格 API、具有分布式特性、速度快和可扩展而闻名，是 Elastic Stack 的核心组件。Elastic Stack 是一套适用于数据采集、扩充、存储、分析和可视化的免费开源工具。

人们通常将 Elastic Stack 称为 ELK Stack（即 Elasticsearch + Logstash + Kibana）。目前 Elastic Stack 包括一系列丰富的轻量型数据采集代理。这些代理被统称为 Beats，可用来向 Elasticsearch 发送数据。

Elasticsearch 的主要特性如下。

1. 速度快

由于 Elasticsearch 是在 Lucene 基础上构建而成的，所以其在全文本搜索方面表现得十分出色。Elasticsearch 还是一个近实时的搜索平台，延时很短，一般只有 1s。因此，Elasticsearch 非常适用于对时间有严苛要求的用例，例如安全分析和基础设施监控。

2. 具有分布式的本质特征

Elasticsearch 中存储的文档分布在不同的容器中，这些容器被称为分片。可以复制这些分片以提供数据冗余副本，以防发生硬件故障时的数据丢失。

Elasticsearch 的分布式特性使得它可以扩展至数百台（甚至数千台）服务器，并处理 PB 量级的数据。

3. 提供的数据功能非常多

Elasticsearch 有大量强大的内置功能（例如数据汇总和索引生命周期管理），可以方便用户更高效地存储和搜索数据。

4. 组成 Elastic Stack 后，可简化数据采集、可视化和报告的过程

通过将 Elasticsearch 与 Beats 和 Logstash 进行集成，用户能够在向 Elasticsearch 中索引数据之前轻松地处理数据。

Kibana 不仅可以为 Elasticsearch 数据提供实时的可视化结果，还可以提供了用户接口，以便用户快速访问应用程序性能监测（APM）、日志和基础设施指标等数据。

10.9.2 存储链路追踪指标数据的原理

Skywalking 用模块 server-storage-plugin 来封装 Skywalking 的存储插件（包括 Elasticsearch、Influxdb、MySQL、PostgreSQL 及 Zipkin 等）。其中，Elasticsearch 存储是用 storage-elasticsearch7-plugin 和 storage-elasticsearch-plugin 这两个插件来实现的。

storage-elasticsearch7-plugin 主要依赖 Elasticsearch 7.x，storage-elasticsearch-plugin 主要依赖 Elasticsearch 6.x。

1. 读取 Elasticsearch 客户端的配置信息

（1）Skywalking 用模块类 StorageModuleElasticsearchProvider 来加载 Elasticsearch 存储组件，并通过 application.yml 中的 selector 标签“elasticsearch7”来配置 Elasticsearch 存储组件的配置信息，具体配置如下：

```
storage:
  selector: ${SW_STORAGE:elasticsearch7}
  elasticsearch7:
###①命名空间
nameSpace: ${SW_NAMESPACE:""}
###②IP 地址
    clusterNodes: ${SW_STORAGE_ES_CLUSTER_NODES:127.0.0.1:9200}
    protocol: ${SW_STORAGE_ES_HTTP_PROTOCOL:"http"}
    trustStorePath: ${SW_STORAGE_ES_SSL_JKS_PATH:""}
    trustStorePass: ${SW_STORAGE_ES_SSL_JKS_PASS:""}
dayStep: ${SW_STORAGE_DAY_STEP:1}
###③索引分片个数
indexShardsNumber: ${SW_STORAGE_ES_INDEX_SHARDS_NUMBER:1}
###④索引副本个数
    indexReplicasNumber: ${SW_STORAGE_ES_INDEX_REPLICAS_NUMBER:1}
    superDatasetDayStep: ${SW_SUPERDATASET_STORAGE_DAY_STEP:-1}
    superDatasetIndexShardsFactor:
${SW_STORAGE_ES_SUPER_DATASET_INDEX_SHARDS_FACTOR:5}
    superDatasetIndexReplicasNumber:
${SW_STORAGE_ES_SUPER_DATASET_INDEX_REPLICAS_NUMBER:0}
    user: ${SW_ES_USER:""}
```

```
      password: ${SW_ES_PASSWORD:""}
      secretsManagementFile: ${SW_ES_SECRETS_MANAGEMENT_FILE:""}
   bulkActions: ${SW_STORAGE_ES_BULK_ACTIONS:1000}
   ###⑤刷盘周期
      flushInterval: ${SW_STORAGE_ES_FLUSH_INTERVAL:10
      concurrentRequests: ${SW_STORAGE_ES_CONCURRENT_REQUESTS:2}
      resultWindowMaxSize: ${SW_STORAGE_ES_QUERY_MAX_WINDOW_SIZE:10000}
      metadataQueryMaxSize: ${SW_STORAGE_ES_QUERY_MAX_SIZE:5000}
      segmentQueryMaxSize: ${SW_STORAGE_ES_QUERY_SEGMENT_SIZE:200}
      profileTaskQueryMaxSize: ${SW_STORAGE_ES_QUERY_PROFILE_TASK_SIZE:200}
      oapAnalyzer: ${SW_STORAGE_ES_OAP_ANALYZER:"{\"analyzer\":
{\"oap_analyzer\":{\"type\":\"stop\"}}}"}
      oapLogAnalyzer: ${SW_STORAGE_ES_OAP_LOG_ANALYZER:"{\"analyzer\":
{\"oap_log_analyzer\":{\"type\":\"standard\"}}}"}
      advanced: ${SW_STORAGE_ES_ADVANCED:""}
```

（2）在 Skywalking 中，将配置类 StorageModuleElasticsearchConfig 中的变量与 Elasticsearch 的配置信息中的属性名称一一映射，具体代码如下所示：

```
@Getter
@Setter
public class StorageModuleElasticsearchConfig extends ModuleConfig {
    private String nameSpace;
    private String clusterNodes;
    String protocol = "http";
    private int dayStep = 1;
    private int indexReplicasNumber = 0;
    private int indexShardsNumber = 1;
    private int superDatasetDayStep = -1;
    private int superDatasetIndexReplicasNumber = 0;
    private String trustStorePass;
    private int resultWindowMaxSize = 10000;
    private int metadataQueryMaxSize = 5000;
    private int segmentQueryMaxSize = 200;
    private int profileTaskQueryMaxSize = 200;
    ...
}
```

2. 自建 Elasticsearch 索引

在 Skywalking 平台后端的启动过程中，会自建链路数据的 Elasticsearch 索引，具体索引信息见表 10-1。

表 10-1　Skywalking 链路数据的 Elasticsearch 索引

索引名称	功能描述
alarm_record	存储 Skywalking 的告警数据

续表

索引名称	功能描述
Segment	存储 Skywalking 的 segment（分段）数据
top_n_database_statement	存储 Skywalking 的 top_n 排名数据
profile_task	存储 Skywalking 的定时任务数据
browser_error_log	存储 Skywalking 收集的浏览器错误日志数据
Log	存储 Skywalking 的日志数据
profile_task_segment_snapshot	存储 profile 任务的 segment 快照数据
profile_task_log	存储 profile 任务的日志数据
zipkin_span	存储从 Zipkin 收集的链路数据
service_relation_client_side	存储客户端与 service 之间的调用关系数据
service_relation_server_side	存储服务器端与 service 之间的调用关系数据
endpoint_traffic	存储服务运行的端点流量数据
instance_traffic	存储服务运行的实例流量数据
service_traffic	存储服务运行的 service 流量数据
Events	存储事件的数据
ui_template	存储 UI 模板的数据
network_address_alias	存储网络地址别名的数据
service_instance_relation_server_side	存储服务器端的 service 与实例之间依赖关系的数据
service_instance_relation_client_side	存储客户端的 service 与实例之间依赖关系的数据
endpoint_relation_server_side	存储服务器端点之间关系的数据

3. 自建 Elasticsearch 索引的过程

Elasticsearch 不需要提前执行建表语句，这一点和关系型数据库完全不一样。下面来分析 Skywalking 自建 Elasticsearch 索引的过程。

在 Skywalking 中，制定了一套标准的自建 Elasticsearch 索引的规范，其扩展性非常好。

下面以索引“segment”为例子，分析新增一个 Elasticsearch 索引的流程。

（1）用注解@Stream 标注 SegmentRecord 类。

在 Skywalking 中，如果要用 SegmentRecord 类来自建“segment”索引，则需要用注解@Stream 标注 SegmentRecord 类：①设置索引的名称为“segment”；②设置索引的作用域为 DefaultScopeDefine.SEGMENT；③设置索引的对象构造器为 SegmentRecord.Builder.class；④设置索引的持久化处理器为 RecordStreamProcessor。

具体代码如下所示：

```
    @SuperDataset
    @Stream(name = SegmentRecord.INDEX_NAME, scopeId =
DefaultScopeDefine.SEGMENT, builder = SegmentRecord.Builder.class, processor =
RecordStreamProcessor.class)
    public class SegmentRecord extends Record {
    //①定义索引名称
    public static final String INDEX_NAME = "segment";
    //②定义 segment_id
    public static final String SEGMENT_ID = "segment_id";
    //③定义 trace_id
    public static final String TRACE_ID = "trace_id";
    //④定义 service_id
    public static final String SERVICE_ID = "service_id";
    //⑤定义 service_instance_id
    public static final String SERVICE_INSTANCE_ID = "service_instance_id";
    //⑥定义 endpoint_name
    public static final String ENDPOINT_NAME = "endpoint_name";
    //⑦定义 endpoint_id
        public static final String ENDPOINT_ID = "endpoint_id";
    ...
    //⑧用注解@Column 定义 Java 的 POJO 类和 Elasticsearch 索引的属性字段的映射关系，包括
SEGMENT_ID、TRACE_ID、SERVICE_ID 等
        @Setter
        @Getter
        @Column(columnName = SEGMENT_ID, length = 150)
        private String segmentId;
        @Setter
        @Getter
        @Column(columnName = TRACE_ID, length = 150)
        private String traceId;
        @Setter
        @Getter
        @Column(columnName = TopN.STATEMENT)
        private String statement;
        @Setter
        @Getter
        @Column(columnName = SERVICE_ID)
        private String serviceId;
        ...
    }
```

（2）用注解扫描器 AnnotationScan 和监听器 StreamAnnotationListener 扫描被注解@Stream 标注的类，比如 SegmentRecord 类。

在 Skywalking 中，在核心模块的 CoreModuleProvider 类的 prepare()方法被调用的过程中，

会注册监听器的 StreamAnnotationListener 类，并在调用 start()方法的过程中开启注解扫描器 AnnotationScan，具体代码如下所示：

```
public class CoreModuleProvider extends ModuleProvider {
    private final AnnotationScan annotationScan;
    public CoreModuleProvider() {
        super();
        //①初始化注解扫描器
        this.annotationScan = new AnnotationScan();
    }
    @Override
public void prepare() throws ServiceNotProvidedException,
ModuleStartException {
    //②用注解扫描器去注册监听器的 StreamAnnotationListener 类
        annotationScan.registerListener(new
StreamAnnotationListener(getManager()));
        ...
    }
    @Override
    public void start() throws ModuleStartException {
     try {
            ...
            //③用注解扫描器去扫描被注解@Stream 标注的类
            annotationScan.scan();
        } catch (IOException | IllegalAccessException |
InstantiationException | StorageException e) {
            throw new ModuleStartException(e.getMessage(), e);
        }
    }
    ...
}
```

（3）用注解扫描器去注册监听器。具体代码如下所示：

```
public class AnnotationScan {
    private final List<AnnotationListenerCache> listeners;
    public AnnotationScan() {
        this.listeners = new LinkedList<>();
}
//①注册监听器，并将其存储在本地缓存变量 listeners 中
public void registerListener(AnnotationListener listener) {
    //②将监听器转换为 AnnotationListenerCache 类
        listeners.add(new AnnotationListenerCache(listener));
    }
    ...
}
```

（4）注解扫描器扫描被注解@Stream 标注的类，并通知监听器 StreamAnnotationListener。具体代码如下所示：

```
public class AnnotationScan {
    public void scan() throws IOException, StorageException {
        ClassPath classpath =
ClassPath.from(this.getClass().getClassLoader());
        //①扫描包路径“org.apache.skywalking”下所有的 class 文件
        ImmutableSet<ClassPath.ClassInfo> classes =
classpath.getTopLevelClassesRecursive("org.apache.skywalking");
        for (ClassPath.ClassInfo classInfo : classes) {
            Class<?> aClass = classInfo.load();
            //②遍历监听器列表
            for (AnnotationListenerCache listener : listeners) {
                //③如果 aClass 的类型和监听器中监听的注解类型一样，则将 aClass 添加到
监听器中
                if (aClass.isAnnotationPresent(listener.annotation())) {
                    listener.addMatch(aClass);
                }
            }
        }
        //④扫描完成后遍历监听器列表，用监听器的 complete()方法通知索引的持久化处理器
去自建索引
        for (AnnotationListenerCache listener : listeners) {
            listener.complete();
        }
    }
    ...
}
```

（5）用监听器通知索引的持久化处理器（包括类 RecordStreamProcessor、类 MetricsStreamProcessor 和类 ManagementStreamProcessor 等）去自建索引。具体代码如下所示：

```
public class StreamAnnotationListener implements AnnotationListener {
    private final ModuleDefineHolder moduleDefineHolder;
    public StreamAnnotationListener(ModuleDefineHolder moduleDefineHolder) {
        this.moduleDefineHolder = moduleDefineHolder;
    }
    @Override
public Class<? extends Annotation> annotation() {
    //①定义监听器监听的注解为 Stream 类
        return Stream.class;
    }
    @SuppressWarnings("unchecked")
```

```
    @Override
    public void notify(Class aClass) throws StorageException {
        //②如果变量 aClass 和 Stream 类一致
            if (aClass.isAnnotationPresent(Stream.class)) {
                //③则获取 Stream 注解对象
                Stream stream = (Stream) aClass.getAnnotation(Stream.class);
                if (DisableRegister.INSTANCE.include(stream.name())) {
                    return;
                }
                //④解析注解@Stream 中的参数，比如索引的持久化处理器
                if (stream.processor().equals(RecordStreamProcessor.class)) {
            //⑤如果@Stream 中配置的索引的持久化处理器和 RecordStreamProcessor.class
一致，则调用 RecordStreamProcessor 类的 create()方法自建索引
                    RecordStreamProcessor.getInstance().
create(moduleDefineHolder, stream, aClass);
                } else if
(stream.processor().equals(MetricsStreamProcessor.class)) {
        //⑥如果在注解@Stream 中配置的索引的持久化处理器和 MetricsStreamProcessor.
class 一致，则调用 MetricsStreamProcessor 类的 create()方法自建索引
                    MetricsStreamProcessor.getInstance().
create(moduleDefineHolder, stream, aClass);
                } else if (stream.processor().equals(TopNStreamProcessor.class)) {
        //⑦如果在注解@Stream 中配置的索引的持久化处理器和 TopNStreamProcessor.class 一
致，则调用 TopNStreamProcessor 类的 create()方法自建索引
                    TopNStreamProcessor.getInstance().
    create(moduleDefineHolder, stream, aClass);
                } else if (stream.processor().equals(NoneStreamProcessor.class)) {
        //⑧如果在注解@Stream 中配置的索引的持久化处理器和 NoneStreamProcessor.class
一致，则调用 NoneStreamProcessor 类的 create()方法自建索引
                    //⑨匹配处理器 NoneStreamProcessor，调用 create()方法自建索引
                    NoneStreamProcessor.getInstance().create(moduleDefineHolder,
stream, aClass);
                } else
        //⑩如果在注解@Stream 中配置的索引的持久化处理器和 ManagementStreamProcessor.
class 一致，则调用 ManagementStreamProcessor 类的 create()方法自建索引
                if (stream.processor().equals(ManagementStreamProcessor.class)) {
                ManagementStreamProcessor.getInstance().
create(moduleDefineHolder, stream, aClass);
                } else {
                    throw new UnexpectedException("Unknown stream processor.");
                }
            } else {
                throw new UnexpectedException(
                        "Stream annotation listener could only parse the class
present stream annotation.");
```

```
            }
        }
    }
```

（6）用索引的持久化处理器去自建索引，比如索引“segment”的持久化处理器类RecordStreamProcessor。

Skywalking 用 RecordStreamProcessor 类的 create()方法自建索引“segment”的具体代码如下所示：

```
    public class RecordStreamProcessor implements StreamProcessor<Record> {
        private final static RecordStreamProcessor PROCESSOR = new
RecordStreamProcessor();
    private Map<Class<? extends Record>, RecordPersistentWorker> workers = new
HashMap<>();
        public static RecordStreamProcessor getInstance() {
            return PROCESSOR;
        }
        @Override
    public void in(Record record) {
        //①从缓存对象 workers 中，获取持久化工作类 RecordPersistentWorker
            RecordPersistentWorker worker = workers.get(record.getClass());
            if (worker != null) {
                //②执行获取持久化工作类 RecordPersistentWorker 的方法完成数据的持久化
                worker.in(record);
            }
        }
        @Override
        @SuppressWarnings("unchecked")
        public void create(ModuleDefineHolder moduleDefineHolder, Stream stream,
Class<? extends Record> recordClass) throws StorageException {
            //③从模块定义持有器中，获取模块 StorageModuleElasticsearch7Provider 类的
存储工厂类 StorageBuilderFactory
            final StorageBuilderFactory storageBuilderFactory =
moduleDefineHolder.find(StorageModule.NAME).provider()
                        .getService(StorageBuilderFactory.class);
            final Class<? extends StorageBuilder> builder =
storageBuilderFactory.builderOf(recordClass, stream.builder());
            //④从模块定义持有器中，获取模块 StorageModule 类的存储 DAO 类 StorageEs7DAO
            StorageDAO storageDAO = moduleDefineHolder.find(StorageModule.NAME).
provider().getService(StorageDAO.class);
            IRecordDAO recordDAO;
            try {
                //⑤实例化 RecordEsDAO 类
                recordDAO =
storageDAO.newRecordDao(builder.getDeclaredConstructor().newInstance());
```

```
            } catch (InstantiationException | IllegalAccessException |
NoSuchMethodException | InvocationTargetException e) {
                throw new UnexpectedException("Create " +
stream.builder().getSimpleName() + " record DAO failure.", e);
            }
            //⑥从模块定义持有器中，获取 StorageModels 类
            ModelCreator modelSetter =
moduleDefineHolder.find(CoreModule.NAME).provider().getService(ModelCreator.
class);
            //⑦向存储模型管理类 StorageModels 中添加新的模型
            Model model = modelSetter.add(
                recordClass, stream.scopeId(), new Storage(stream.name(),
DownSampling.Second), true);
            //⑧构造 RecordPersistentWorker 类
            RecordPersistentWorker persistentWorker = new
RecordPersistentWorker(moduleDefineHolder, model, recordDAO);
            //⑨缓存 RecordPersistentWorker 类
            workers.put(recordClass, persistentWorker);
        }
    }
```

（7）用 ModelInstalle 类的 whenCreating()方法创建索引。

Skywalking 用 ModelInstalle 类的 whenCreating()方法调用 ElasticSearchClient 类的 createIndex()方法，完成索引的创建，具体代码如下所示：

```
    public abstract class ModelInstaller implements
ModelCreator.CreatingListener {
        ...
        @Override
        public void whenCreating(Model model) throws StorageException {
            if (RunningMode.isNoInitMode()) {
            //①如果 Skywalking 的启动的模式是 noinit，且索引又没有初始化，则当前线程休眠 3s
                while (!isExists(model)) {
                    try {
                        log.info(
                            "table: {} does not exist. OAP is running in 'no-init'
mode, waiting... retry 3s later.",
                            model
                                .getName()
                        );
                        Thread.sleep(3000L);
                    } catch (InterruptedException e) {
                        log.error(e.getMessage());
                    }
                }
```

```
        } else {
        //②如果 Skywalking 的启动模式不是 noinit，则调用 createTable()方法创建索引
            if (!isExists(model)) {
                log.info("table: {} does not exist", model.getName());
                createTable(model);
            }
        }
    }
}
```

Skywalking 用 StorageEsInstaller 类的 createTable()方法创建索引，具体代码如下所示：

```
@Slf4j
public class StorageEsInstaller extends ModelInstaller {
    ...
    @Override
    protected void createTable(Model model) throws StorageException {
        if (model.isTimeSeries()) {
            createTimeSeriesTable(model);
        } else {
            createNormalTable(model);
        }
    }
//①创建普通的索引表
    private void createNormalTable(Model model) throws StorageException {
        ElasticSearchClient esClient = (ElasticSearchClient) client;
        String tableName = IndexController.INSTANCE.getTableName(model);
        try {
            if (!esClient.isExistsIndex(tableName)) {
                //②用 ElasticSearchClient 客户端创建索引
                boolean isAcknowledged = esClient.createIndex(tableName);
                log.info("create {} index finished, isAcknowledged: {}",
tableName, isAcknowledged);
                if (!isAcknowledged) {
                    throw new StorageException("create " + tableName + " time
series index failure, ");
                }
            }
        } catch (IOException e) {
            throw new StorageException("cannot create the normal index", e);
        }
    }
//③创建带有时间戳的表
    private void createTimeSeriesTable(Model model) throws StorageException {
        ElasticSearchClient esClient = (ElasticSearchClient) client;
        String tableName = IndexController.INSTANCE.getTableName(model);
```

```
            Map<String, Object> settings = createSetting(model);
            Map<String, Object> mapping = createMapping(model);
            //④获取带有时间戳的索引名称
            String indexName = TimeSeriesUtils.latestWriteIndexName(model);
            try {
             Boolean shouldUpdateTemplate = !esClient.isExistsTemplate(tableName);
                if (IndexController.INSTANCE.isMetricModel(model)) {
                  shouldUpdateTemplate = shouldUpdateTemplate
|| !structures.containsStructure(tableName, mapping);
                }
                if (shouldUpdateTemplate) {
                    structures.putStructure(tableName, mapping);
                    //⑤创建索引模板
                    boolean isAcknowledged = esClient.createOrUpdateTemplate(
                        tableName, settings, structures.getMapping(tableName));
                    log.info("create {} index template finished, isAcknowledged:
{}", tableName, isAcknowledged);
                    if (!isAcknowledged) {
                        throw new IOException("create " + tableName + " index
template failure, ");}
                    if (esClient.isExistsIndex(indexName)) {
                        Map<String, Object> historyMapping = (Map<String, Object>)
esClient.getIndex(indexName).get("mappings");
                        Map<String, Object> appendMapping =
structures.diffStructure(tableName, historyMapping);
                        if (!appendMapping.isEmpty()) {
                            //⑥更新索引模板
                            isAcknowledged = esClient.updateIndexMapping
(indexName, appendMapping);
                            log.info("update {} index finished, isAcknowledged: {},
append mappings: {}", indexName,
                                    isAcknowledged, appendMapping.toString()
                            );
                            if (!isAcknowledged) {
                                throw new StorageException("update " + indexName +
" time series index failure");
                            }
                        }
                    } else {
                        //⑦创建索引
                        isAcknowledged = esClient.createIndex(indexName);
                        log.info("create {} index finished, isAcknowledged: {}",
indexName, isAcknowledged);
                        if (!isAcknowledged) {
```

```
                        throw new StorageException("create " + indexName + " time
series index failure");
                  }
               }
            }
         } catch (IOException e) {
            throw new StorageException("cannot create " + tableName + " index
template", e);
         }
      }
      ...
   }
```

10.9.3 【实例】将 Skywalking 集群接入 Elasticsearch，并采集 Spring Cloud Alibaba 应用的链路追踪数据

本实例的源码在本书配套资源的"chapterten/use-es-skywalking-consumer"和"chapterten/skywalking-kafka-provider/"目录下。

本实例包含两个服务：use-es-skywalking-provider 和 use-es-skywalking-consumer。

1. 初始化 use-es-skywalking-provider 和 use-es-skywalking-consumer 服务

采用 Spring Cloud Alibaba 框架初始化 use-es-skywalking-provider 和 use-es-skywalking-consumer 服务，并在 use-es-skywalking-provider 服务中暴露 Dubbo 接口，用 use-es-skywalking-consumer 服务订阅 Dubbo 接口完成消费，具体代码如下所示：

```
   //①use-es-skywalking-provider 服务暴露 Dubbo 接口 AnchorService
   @DubboService(version = "1.0.0",group = "use-es-skywalking-provider")
   public class AnchorServiceImpl implements AnchorService{
      @Resource
      private AnchorManager anchorManager;
      @Override
   public Integer insertAnchor(AnchorBo anchorBo) {
      //②插入数据
         return anchorManager.insert(anchorBo);
      }
   }
   //③定义 AnchorManager 类，用于操作数据库类 AnchorMapper，进行数据的增加、删除、查询
和修改
   @Service
   public class AnchorManager {
      @Autowired
```

```
    private AnchorMapper anchorMapper;

    public Integer insert(AnchorBo anchorBo){
        AnchorEntity anchorEntity=new AnchorEntity();
        anchorEntity.setAnchorName(anchorBo.getAnchorName());
        anchorEntity.setAnchorLevel(anchorBo.getAnchorLevel());
        Integer integer=anchorMapper.insert(anchorEntity);
        return integer;
    }
}
//④ AnchorMapper 类继承 MyBatis-Plus 的数据库操作接口 BaseMapper
public interface AnchorMapper extends BaseMapper<AnchorEntity> {
}
//⑤定义 RESTful API 调用 Dubbo 接口 AnchorService
@RestController
@RequestMapping(value = "/consumer")
public class ConsumerController {
    @DubboReference(version = "1.0.0",group = "use-es-skywalking-provider")
    private AnchorService anchorService;
    @GetMapping(value = "/todo")
    public String consumer(){
        AnchorBo anchorBo=new AnchorBo();
        anchorBo.setAnchorLevel(RandomUtils.nextInt()+"");
        anchorBo.setAnchorName("test"+RandomUtils.nextLong());
        anchorService.insertAnchor(anchorBo);
        return "success";
    }
}
```

2. 启动服务并模拟用户请求，观察 Skywaking 的链路追踪，查看 Elasticsearch 中存储的数据

（1）在启动 use-es-skywalking-provider 和 use-es-skywalking-consumer 服务后，从 Nacos 注册中心中可以看到服务已经注册成功，如图 10-32 所示。

public | test | spring cloud alibaba practice | dev

Service List | spring cloud alibaba practice c7ba173f-29e5-4c58-ae78-b102be11c4f9

Service Name [Enter Service Name] Group Name [Enter Group Name] Hidden Empty Service: Search Create Service

Service Name	Group Name	Cluster Count	Instance Count	Healthy Instance Count	Trigger Protection Threshold	Operation
use-es-skywalking-consumer	use-es-skywalking-provider	1	1	1	false	Details | Code Example | Delete
use-es-skywalking-provider	use-es-skywalking-provider	1	1	1	false	Details | Code Example | Delete
SkyWalking_OAP_Cluster	DEFAULT_GROUP	1	1	1	false	Details | Code Example | Delete

图 10-32

（2）执行命令“curl 127.0.0.1:7864/consumer/todo”模拟用户请求，观察 Skywalking 的 UI 控制台。

为了让在 Elasticsearch 中存储的链路数据与 Skywalking UI 控制台中的链路数据对应，本实例在 Skywalking 的链路管理中选择了一个完整的链路，其 traceId 如下所示：

```
traceId=b626696a900647feac327ea0569964c1.96.16323438326830001
```

traceId 对应的完整链路关系如图 10-33 所示。

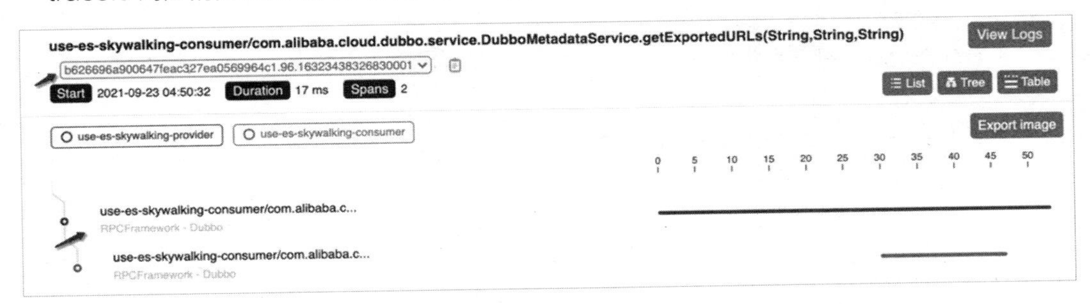

图 10-33

在 Elasticsearch UI 界面中，搜索对应 traceId 的链路数据，可以发现数据已经存储到 Elasticsearch 中了，如图 10-34 所示。

trace_id: b626696a900647feac327ea0569964c1.96.16323438326830001
data_binary: CjViNjI2Njk2YTkwMDY0N2ZlYWMzMjdlYTA1Njk5NjRjMS45Ni4xNjMyMzQzODMyNjgzMDAwMRI2NmQwNWVjZDYzYTA3NGQ2OGJkZDA1YTkzOWQ4ZDM1MWQuMjY2LjE2MzIzNDM4MzI3MTMwMDAwGooFEP///////////wEYivGY+sAvIJvxmPrALyrCAhI1YjYyNjY5NmE5MDA2NDdmZWFjMzI3ZWEwNTY5OTY0YzEuOTYuMTYzMjM0MzgzMjY4MzAwMDEaNWI2MjY2OTZhOTAwNjQ3ZmVhYzMyN2VhMDU2OTk2NGMxLjk2LjE2MzIzNDM4MzI2ODMwMDAwKhp1c2UtZXMtc2t5d2Fsa2luZy1wcm92aWRlcjIuZGIxZDYwNmZjNTU5NGM5ZDg3MWU2NDFiNzlmNDc4NmFAMTkyLjE2OC4wLjEyMzpxdXNlLWVzLXNreXdhbGtpbmctY29uc3VtZXIvY29tLmFsaWJhYmEuY2xvdWQuZHViYm8uc2VydmljZS5EdWJib01ldGFkYXRhU2VydmljZS4kaW52b2tlKF

trace_id: b626696a900647feac327ea0569964c1.96.16323438326830001
data_binary: CjViNjI2Njk2YTkwMDY0N2ZlYWMzMjdlYTA1Njk5NjRjMS45Ni4xNjMyMzQzODMyNjgzMDAwMRI1YjYyNjY5NmE5MDA2NDdmZWFjMzI3ZWEwNTY5OTY0YzEuOTYuMTYzMjM0MzgzMjY4MzAwMDAavwIQ////////////ARjs8Jj6wC8gofGY+sAvMnF1c2UtZXMtc2t5d2Fsa2luZy1jb25zdW1lci9jb20uYWxpYmFiYS5jbG91ZC5kdWJiby5zZXJ2aWNlLkR1YmJvTWV0YWRhdGFTZXJ2aWNlLiRpbnZva2UoU3RyaW5nLFN0cmluZ1tdLE9iamVjdFtdKToTMTkyLjE2OC4wLjEyMzoyNjc1MEABSAJQA2KVAQoDdXJsEo0BZHViYm86Ly8xOTIuMTY4LjAuMTIzOjI2NzUwL3VzZS1lcy1za3l3YWxraW5nLWNvbnN1bWVyL2NvbS5hbGliYWJhLmNsb3VkLmR1YmJvLnNlcnZpY2UuRHViYm9NZXRhZGF0YVNlcnZpY2UuJGludm9rZShTdHJpbmcsU3RyaW

图 10-34

第 11 章 分布式 Job——基于 Elastic Job

在微服务架构中，通常利用分布式 Job 去执行一些数据同步操作。在开源领域中有很多成熟的分布式 Job 中间件框架，Elastic Job 就是其中之一。

11.1 认识分布式 Job

分布式 Job，从字面上来理解是指，将分布式和 Job 组合在一起，从而形成分布式 Job；从功能上来讲，是将单节点中 Job 的规则及状态，同步到分布式系统中的其他节点，从而确保规则和状态在分布式系统中是一致性的（可以是强一致性，也可以是最终一致性）。

11.1.1 为什么需要分布式 Job

如图 11-1 所示，交易服务和支付服务被分别部署在节点 A 和节点 B 中。在交易服务和支付服务中，通常会存在如下业务场景：

（1）用户发起交易请求，调用交易服务。

（2）交付服务处理交易请求，存储交易记录及交易状态，并持久化到交易数据库中。

（3）交易服务发起支付请求，并调用支付服务。

（4）支付服务处理支付请求，存储支付记录及支付状态，异步调用第三方支付机构服务，返给交易服务一个发起支付请求成功的结果。

（5）支付服务通过定时轮询接口定时调用第三方支付机构提供的“支付结果轮询服务”查询支付结果，并将结果存储在支付数据库中。

（6）交易服务用定时轮询接口定时调用支付服务提供的“查询支付结果的接口”，从而获取交易请求的支付结果。

为了实现如上业务场景，需要在交易服务中开启一个本地 Job，以定时轮询支付服务；支付服务也需要开启一个本地 Job，以定时轮询第三方支付机构的服务。

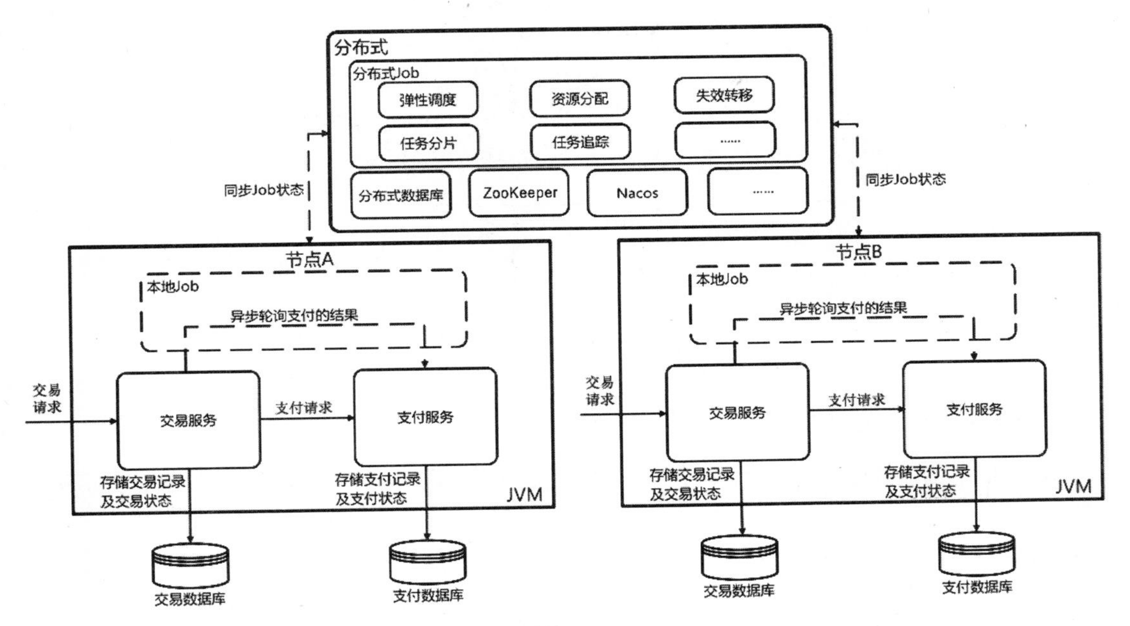

图 11-1

但是，本地 Job 不能解决如下技术问题：

（1）如果支付服务和交易服务采用集群部署，则本地 Job 会在不同的 JVM 进程中执行，但不能感知彼此的执行状态（即某个 JVM 进程中的 Job 不能感知其他 JVM 进程中 Job 的执行状态），从而不能保证 Job 对应的业务功能的数据一致性。

（2）本地 Job 不具备弹性调度、资源分配、失效转移、任务分片、任务追踪等分布式功能。如果只采用本地 Job，则不能充分利用硬件资源去解决一些 Job 性能问题（比如 Job 执行太慢），可以将 Job 分片，调度不同的机器去执行 Job。

1. 了解本地 Job

如果把业务功能放在一个普通的接口中，这个业务功能在服务中只会执行一次；如果把业务功能放在本地 Job 中，则在服务启动的过程中会加载本地 Job 和 Job 对应的定时执行规则。图 11-2 所示为业务功能在本地 Job 中的执行过程：

- 如果没有配置定时规则，则本地 Job 只执行一次；
- 如果配置了定时规则，则按照定时规则生成触发器，并定时触发执行业务功能。

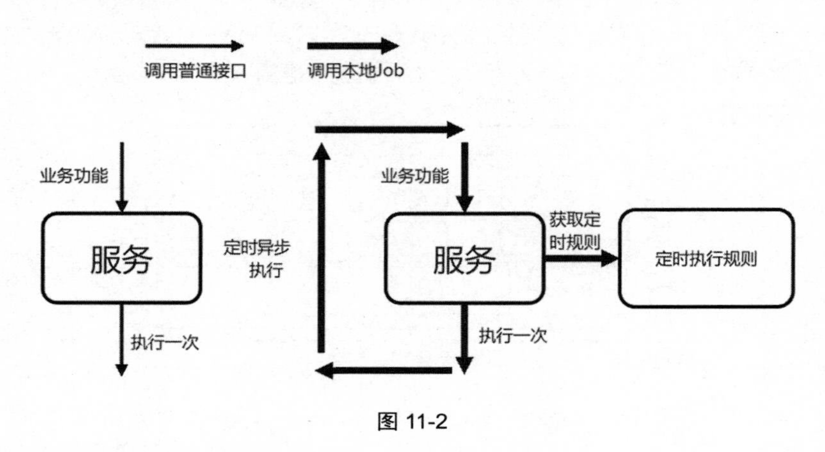

图 11-2

通常一个本地 Job 包含如下功能模块：

（1）触发器（Trigger）。

触发器是一个带有触发条件的执行器。当满足触发条件时，它就会调用执行器去执行指定的功能。

触发器大致分为如下两类。

① 基于时间的触发器：用时间作为触发器的触发条件，比如指定时间、定时时间等。Elastic Job 就是采用基于时间的触发器。

② 基于业务逻辑的触发器：用业务逻辑作为触发器的触发条件，比如账户余额小于 1000 时发送预警消息、商品库存低于 10.000 时发送预警消息等。通常软件开发人员自研的分布式 Job 中间件都会采用此类触发器。

（2）任务。

只要开发人员将需要定时执行的业务功能添加到任务中，触发器就能控制业务功能的执行。当满足触发条件时，本地 Job 会执行相应的业务功能。

（3）扫描器。

在本地 Job 中定义了触发器和任务后，本地 Job 需要一个扫描器去加载触发器，在 Java 的容器（通常指 Spring Framework 的 IOC 容器）中生效。

扫描器大致可以分为如下两类。

① 基于注解的扫描器：开发者需要定义一个注解，并将注解添加到需要执行的类或者方法上（包含业务功能）。扫描器会在服务启动的过程中，加载这些被注解标注的类或者方法，将其转换为触发器能够识别的 Job，然后按照触发规则执行业务功能。

② 基于 SPI 的扫描器：本地 Job 提供了一套统一的 SDK，开发者需要自定义一个 Job 类去实现 SDK 中的 Job 接口。扫描器采用 SPI 加载这些 Job 接口的实现类，然后将其和触发器绑定，从而能够按照触发规则去执行业务功能。

③ 基于切面的扫描器：本地 Job 提供了统一的切面规则，开发人员将需要定时执行的业务功能添加到指定的模块中（统一的包路径和方法），这样扫描器就可以选择性地加载类，从而将类转换为触发器能够识别的 Job，并按照触发规则执行业务功能。

2. 了解分布式 Job 系统

分布式 Job 系统需要具备如下功能，如图 11-3 所示。

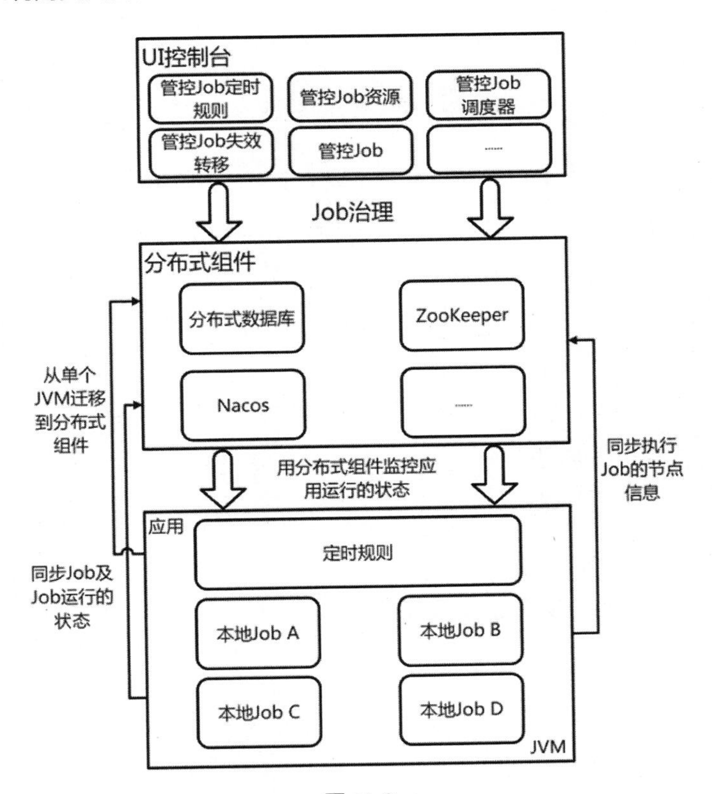

图 11-3

（1）统一的 UI 控制台（或者命令控制台）。

分布式 Job 的执行环境非常复杂，所以需要统一的 UI 控制台（或者命令控制台）去治理 Job（动态更改线上正在执行的 Job 的行为，比如开启、暂停、重新开启及关闭等）。

（2）注册中心。

分布式 Job 需要统一管理 Job 的元数据（以服务为维度），所以需要用注册中心去完成 Job 的注册/订阅。

（3）执行调度器。

为了保证分布式 Job 能够高效、高可用地执行，需要有一个统一的执行调度器去管控 Job 执行的路由规则（具体指执行 Job 的机器节点的路由规则）。

（4）分片。

为了确保分布式 Job 任务能够在作业机器组上能被均衡地执行，需要系统支持分片策略，比如平均分片策略、奇偶分片策略及轮询分片策略等。

（5）Job 追踪。

为了确保在分布式 Job 执行完后开发人员可以回放 Job 的执行过程，所以需要 Job 追踪功能。

11.1.2 认识 Elastic Job

Elastic Job 是一个分布式调度技术中间件框架，由两个相互独立的子项目 Elastic Job Lite 和 Elastic Job Cloud 组成。

- Elastic Job Lite：定位为轻量级无中心化解决方案，使用 Jar 包的形式提供分布式 Job 的协调服务。
- Elastic Job Cloud：利用 Mesos 来提供资源治理、应用分发及进程隔离等服务。

Elastic Job 的各个产品使用的是统一的 API，开发者可以将分布式 Job 直接部署在 ElasticJob-Lite 或者 ElasticJob-Cloud 中。

1. 了解 Elastic Job Lite

图 11-4 为 Elastic Job Lite 的整体架构。

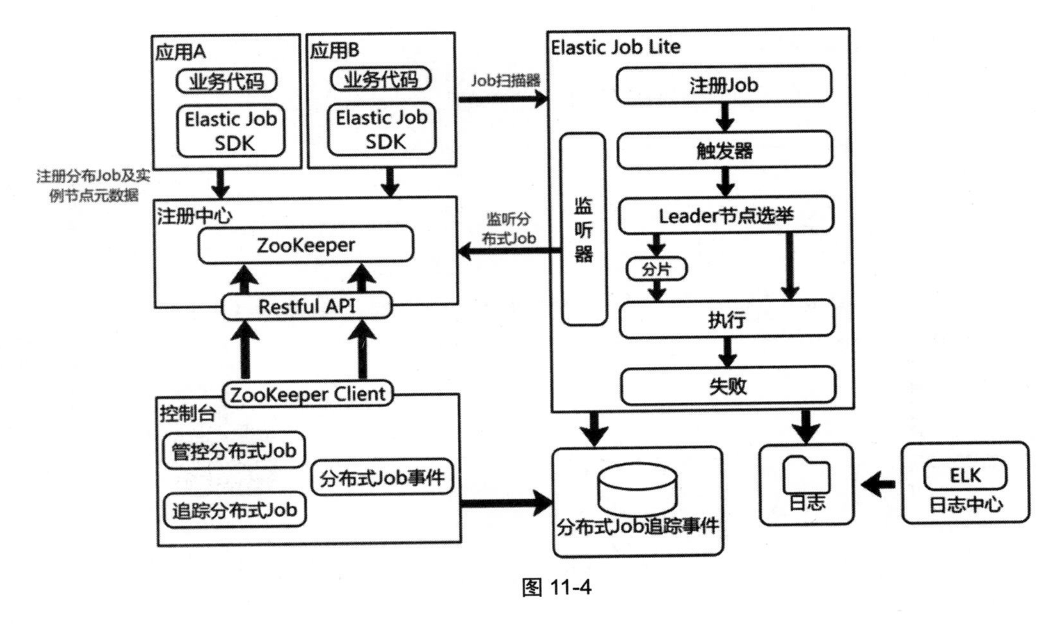

图 11-4

具体过程如下：

（1）应用在业务代码中依赖 Elastic Job SDK，并在配置文件中添加注册中心地址。

（2）应用在启动的过程中将分布式 Job 规则及实例节点元数据注册到注册中心中。

（3）在注册分布式 Job 的过程中，利用 Job 扫描器扫描并解析应用的分布式 Job 的实现类，并将它转换成基于 Quartz 的本地 Job 和触发器。

（4）利用 ZooKeeper 的分布式锁完成 Leader 节点的选举，并执行分片策略将本地 Job 转换成分布式 Job。

（5）利用监听器监听注册中心中分布式 Job 的更新，实时通知应用去执行定时任务。

（6）控制台在启动的过程中会连接 ZooKeeper 的注册中心，开发人员可以在控制台中管控分布式 Job，执行更新/终止分布式 Job 的操作。

（7）在应用本地的配置文件中，可以配置分布式 Job 追踪的数据源，并将分布式 Job 的状态和日志记录到数据源中。

（8）在控制台中，开发人员可以配置分布式 Job 追踪的数据源，这样就可以将分布式 Job 的状态和日志可视化，方便开发人员和运维人员实时查看分布式 Job 的执行情况。

（9）可以打通分布式 Job 的执行日志和 ELK（日志系统），方便运维人员和开发人员通过 ELK 的控制台访问 Job 的执行日志。

2. 了解 Elastic Job Cloud

图 11-5 为 Elastic Job Cloud 的整体架构。

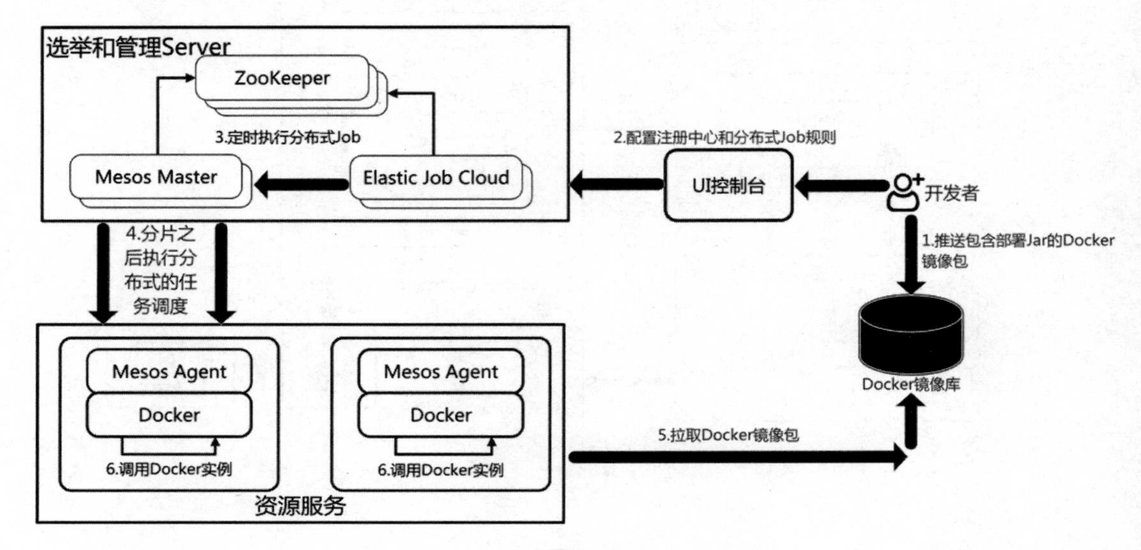

图 11-5

具体过程如下：

（1）开发人员将包含部署 Jar 包的 Docker 镜像包推送到 Docker 镜像库中。

（2）开发人员通过 UI 控制台配置注册中心和分布式 Job 规则。

（3）Elastic Job Cloud 通过定时器定时触发任务并执行任务分片。

（4）Mesos Master 调度一个 Mesos Agent 去执行任务分片。

（5）在资源服务中，Mesos Agent 在收到一个调度请求后，会将调度请求转发给 Docker 实例（从 Docker 镜像库中拉取镜像包，在部署镜像包后启动 Docker 实例），并在 Docker 实例上执行任务分片。

11.2 将应用接入 Elastic Job Lite

Elastic Job Lite 支持 3 种接入模式：Java API、Spring Boot Starter、Spring 命名空间。下面分别介绍。

11.2.1　将应用接入 Elastic Job Lite 的 3 种模式

开发人员可以利用业务服务所使用的基础框架将服务接入 Elastic Job Lite。如果基础框架为 Spring Framework，则可以利用 Spring 命名空间将服务接入 Elastic Job Lite。

1. 利用 Java API 接入

在服务中，开发人员可以使用 ZookeeperConfiguration 类来配置注册中心的属性，并连接注册中心（目前仅支持 ZooKeeper），具体代码如下所示：

```
@Getter
@Setter
@RequiredArgsConstructor
public final class ZookeeperConfiguration {
//①配置注册中心的地址
private final String serverLists;
//②配置分布式 Job 的命名空间
private final String namespace;
//③配置最小休眠时间
private int baseSleepTimeMilliseconds = 1000;
//④配置最大休眠时间
private int maxSleepTimeMilliseconds = 3000;
//⑤配置最大重试次数
private int maxRetries = 3;
//⑥配置 session 超时时间
private int sessionTimeoutMilliseconds;
//⑦配置连接超时时间
private int connectionTimeoutMilliseconds;
//⑧配置 ZooKeeper 的数字签名
   private String digest;
}
```

开发人员可以在服务中使用 JobConfiguration 类来配置分布式 Job 的规则，具体代码如下所示：

```
//用 JobConfiguration 类来配置分布式 Job 的规则
@Getter
@AllArgsConstructor(access = AccessLevel.PRIVATE)
public final class JobConfiguration {
    private final String jobName;
    private final String cron;
    private final int shardingTotalCount;
    private final String shardingItemParameters;
    private final String jobParameter;
    private final boolean monitorExecution;
    private final boolean failover;
```

```
    private final boolean misfire;
    private final int maxTimeDiffSeconds;
    private final int reconcileIntervalMinutes;
    private final String jobShardingStrategyType;
    private final String jobExecutorServiceHandlerType;
    private final String jobErrorHandlerType;
    private final Collection<String> jobListenerTypes;
    private final Collection<JobExtraConfiguration> extraConfigurations;
    private final String description;
    private final Properties props;
    private final boolean disabled;
    private final boolean overwrite;
    private final String label;
    private final boolean staticSharding;
    ...
}
```

2. 利用 Spring Boot Starter 接入

开发人员可以按照如下步骤，将服务接入 Elastic Job Lite。

（1）在应用的 pom.xml 中添加如下依赖：

```
<dependency>
    <groupId>org.apache.shardingsphere.elasticjob</groupId>
    <artifactId>elasticjob-lite-spring-boot-starter</artifactId>
    <version>3.0.0</version>
</dependency>
```

（2）在服务的配置文件（比如 bootstrap.yaml）中添加如下配置属性：

```
elasticjob:
  regCenter:
    serverLists: 127.0.0.1:2181
    namespace: spring-cloud-alibaba-job
  jobs:
    accessElasticMessageJob:
        elasticJobClass: com.alibaba.cloud.youxia.elasticjob.AccessElasticMessageJob
        cron: 0/20 * * * * ?
        shardingTotalCount: 1
        overwrite: true
```

3. 利用 Spring 命名空间接入

（1）在应用的 pom.xml 文件中添加如下依赖：

```
<dependency>
    <groupId>org.apache.shardingsphere.elasticjob</groupId>
```

```
        <artifactId>elasticjob-lite-spring-namespace</artifactId>
        <version>3.0.0</version>
    </dependency>
```

（2）在 Elastic Job Lite 中，首先用 elasticjob.xsd 来定义连接注册中心和配置分布式 Job 的语法规则，然后用 spring.schemas 将语法规则“elasticjob.xsd”集成到 Spring Framework 的规则引擎中，最后用 spring.handlers 将解析的配置信息转换为 POJO 对象，并注册到 Spring Framework 的 IOC 容器中，如图 11-6 所示。

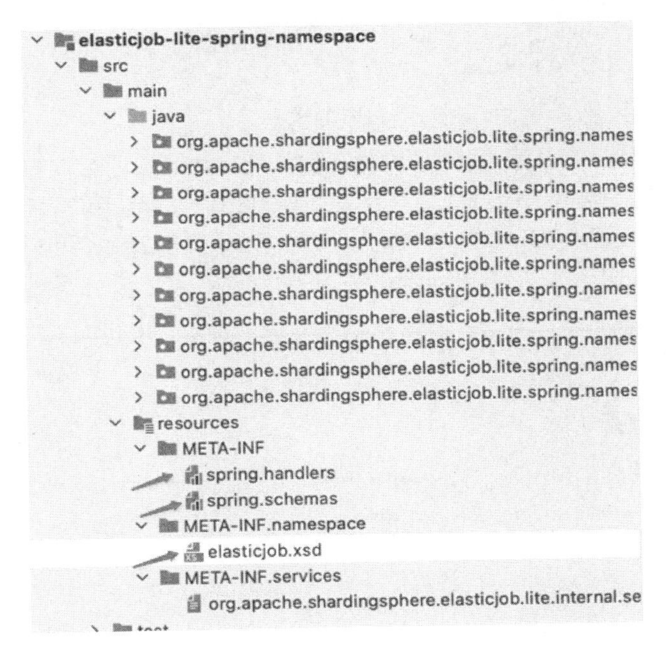

图 11-6

11.2.2 搭建 Elastic Job Lite 的分布式环境

Elastic Job Lite 的基础环境包括：ZooKeeper、数据库（MySQL、PostgreSQL 等）、UI 控制台（Elastic Job UI）。搭建过程如下。

（1）搭建 ZooKeeper 环境：可以参考本书第 10 章的搭建 ZooKeeper 环境来实现。

（2）搭建数据库：本书统一采用 MySQL，具体搭建过程读者可以参考 MySQL 的官网。

（3）从 Elastic Job 的官网下载安装包 apache-shardingsphere-elasticjob-3.0.0-lite-ui-bin.tar.gz，并解压缩，如图 11-7 所示。

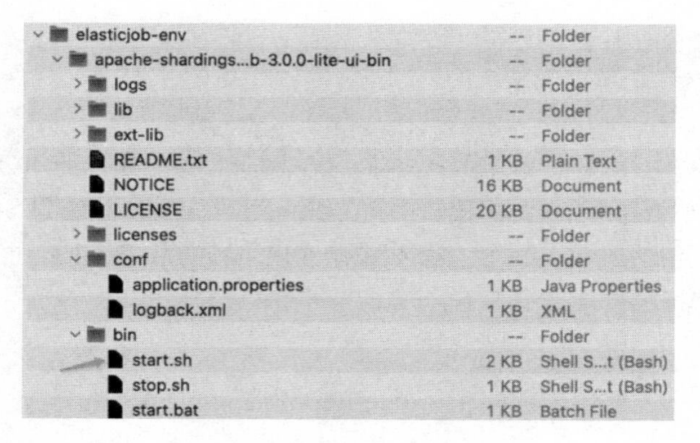

图 11-7

（4）使用 UI 控制台的默认配置信息，并执行命令“sh start.sh”启动 UI 控制台。在控制台启动成功后，软件开发人员访问 URL 地址“127.0.0.1:8088/”（默认用户名和密码都为 root），登录 UI 控制台，就可以使用“ADD”按钮添加一个注册中心（基于 ZooKeeper），并在“Operate”列表中单击“连接”按钮，如图 11-8 所示。

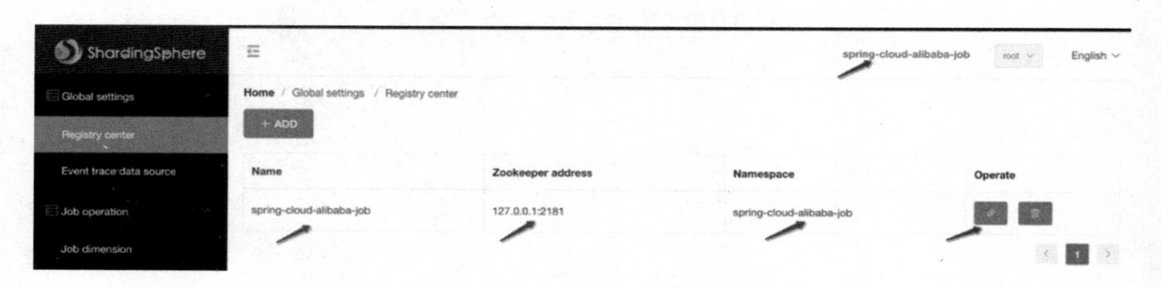

图 11-8

11.2.3 【实例】用 Spring Boot Starter 将 Spring Cloud Alibaba 应用接入 Elastic Job Lite

本实例的源码在本书配套资源的“/spring-cloud-alibaba-practice/chaptereleven/elasticjob-spring-cloud-alibaba/”目录下。

本实例将一个 Spring Cloud Alibaba 应用接入 Elastic Job Lite 中，并实时地修改 Job 的规则，观察 Job 执行的效果。

1. 初始化服务 elasticjob-spring-cloud-alibaba

用 Spring Cloud Alibaba 初始化一个服务 elasticjob-spring-cloud-alibaba，并添加相关的

POM 依赖，具体如下所示：

```
<dependency>
    <groupId>org.apache.shardingsphere.elasticjob</groupId>
    <artifactId>elasticjob-lite-spring-boot-starter</artifactId>
    <version>3.0.0</version>
</dependency>
```

2. 在服务中新建一个定时 Job，并在配置文件中添加注册中心和 Job 规则

（1）在服务中新建一个定时 Job 类 ElasticMessageJob，具体代码如下所示：

```
    //①新建一个定时 Job 类，实现 Elastic Job 的 SDK 接口 SimpleJob
@Component
public class ElasticMessageJob implements SimpleJob {
    @Override
public void execute(ShardingContext shardingContext) {
    //②定时打印一条消息
        System.out.println("发送定时消息！");
    }
}
```

（2）在配置文件 bootstrap.yaml 中，添加注册中心和 Job 规则，具体代码如下所示：

```
elasticjob:
  regCenter:
    //①配置基于 ZooKeeper 的注册中心地址
    serverLists: 127.0.0.1:2181
    //②配置 Job 所在命名空间的名称
    namespace: spring-cloud-alibaba-job
  jobs:
    elasticMessageJob:
        //③配置定时 Job 类的路径
        elasticJobClass: com.alibaba.cloud.youxia.elasticjob.ElasticMessageJob
        //④配置触发器中的 Cron 表达式，含义为每隔 20s 执行一次 Job
        cron: 0/20 * * * * ?
        //⑤配置分片数为 1
        shardingTotalCount: 1
        //⑥配置支持动态覆盖 Job 规则（默认不支持）
        overwrite: true
```

（3）启动服务，并观察 Job 执行的日志。如图 11-9 所示，定时器会每隔 19s 或者 20s 执行一次。定时器会出现 1s 偏差的主要原因是：Elastic Job 依赖分布式框架 ZooKeeper，所以在分布式环境下会出现一定的调度时延。当 Job 触发时，Elastic Job 需要调度分片去执行定时 Job，由于调度存在一定的时延，所以导致 Elastic Job 实际执行的时间周期和触发器的触发规则可能会存在一点偏差（采用 Job 执行业务功能是异步设计的一种技术手段，通常可以接受一定范围内的时延）。

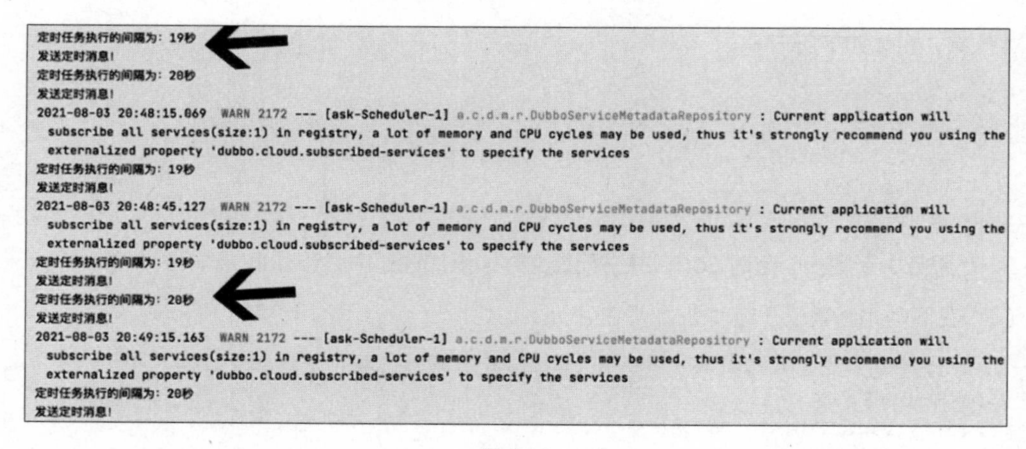

图 11-9

（4）在 UI 控制台中修改 Job 的配置信息，将定时间隔从 20s 更改为 15s，如图 11-10 所示。

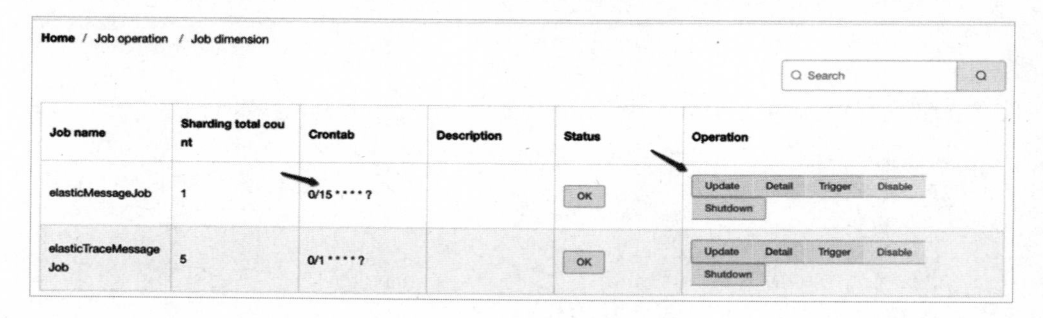

图 11-10

（5）在 IDEA 控制台可以观察到，定时器会每隔 14s 或者 15s 执行一次，如图 11-11 所示。

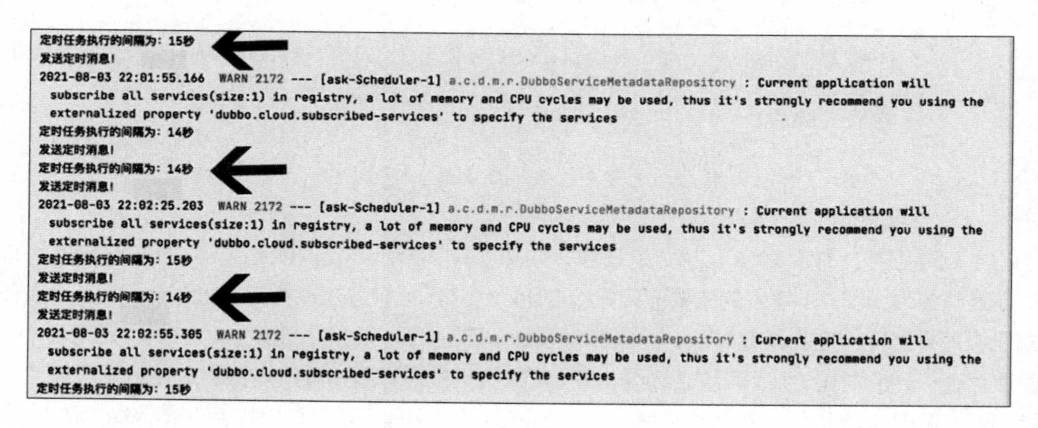

图 11-11

11.3 “实现 Elastic Job Lite 的本地 Job 和分布式 Job”的原理

Elastic Job Lite 主要用 Quartz 框架来实现本地 Job，用 ZooKeeper 框架来实现分布式 Job。

11.3.1 用 Quartz 框架实现本地 Job

下面从 Elastic Job 的接入模式（Spring Boot Starter）来分析“用 Quartz 框架实现本地 Job”的原理。

1. 用元数据加载连接注册中心和 Job 规则的配置信息

在 Elastic Job Lite 中，用元数据加载连接注册中心、Job 规则的配置信息，具体过程如下。

（1）在服务中添加如下 POM 依赖。

```
<dependency>
    <groupId>org.apache.shardingsphere.elasticjob</groupId>
    <artifactId>elasticjob-lite-spring-boot-starter</artifactId>
    <version>3.0.0-alpha</version>
</dependency>
```

（2）在配置文件 bootstrap.yaml 中添加如下配置信息。

```
elasticjob:
  jobs:
    nacosJob:
      elasticJobClass: com.alibaba.cloud.alarm.api.job.NacosAlarmJob
      cron: 0/5 * * * * ?
      shardingTotalCount: 1
```

（3）用 ElasticJobLiteAutoConfiguration 类加载配置信息，具体代码如下所示：

```
    @Configuration(proxyBeanMethods = false)
    @AutoConfigureAfter(DataSourceAutoConfiguration.class)
    //①用开关“elasticjob.enabled.true”控制 Elastic Job 的开启和关闭，默认为开启
    @ConditionalOnProperty(name = "elasticjob.enabled", havingValue = "true",
matchIfMissing = true)
    //②导入注册中心相关配置及初始化注册中心对象 ZookeeperRegistryCenter、Job 追踪相关配
置及追踪配置对象 TracingConfiguration、Job 快照存储相关配置及快照对象 SnapshotService
    @Import({ElasticJobRegistryCenterConfiguration.class,
ElasticJobTracingConfiguration.class,
ElasticJobSnapshotServiceConfiguration.class})
    //③加载连接注册中心、Job 规则的配置信息
    @EnableConfigurationProperties(ElasticJobProperties.class)
    public class ElasticJobLiteAutoConfiguration {
```

```
    @Configuration(proxyBeanMethods = false)
    //④初始化 Job 配置对象 ElasticJobBootstrapConfiguration 和 Job 启动类
ScheduleJobBootstrapStartupRunner
        @Import({ElasticJobBootstrapConfiguration.class,
ScheduleJobBootstrapStartupRunner.class})
        protected static class ElasticJobConfiguration {
        }
    }
```

2. 创建 Job 的启动实例对象，并将其注册到 Spring Framework 容器中

下面在 Elastic Job Lite 中，用 ElasticJobBootstrapConfiguration 类创建 Job 启动实例对象，并将实例对象注册到 Spring Framework 容器中。其中，Job 启动实例对象主要包括 OneOffJobBootstrap 类和 ScheduleJobBootstrap 类。

（1）在 Spring Framework 的 IOC 容器启动的过程中，用 Spring Framework 的注解 @PostConstruct 触发 createJobBootstrapBeans()方法加载 Job 启动实例对象所依赖的 Bean 对象，具体代码如下所示：

```
    public class ElasticJobBootstrapConfiguration
    implements ApplicationContextAware, BeanPostProcessor {
        @PostConstruct
        public void createJobBootstrapBeans() {
            //①从 Spring 的 IOC 容器中获取配置信息对象 ElasticJobProperties
            ElasticJobProperties elasticJobProperties =
applicationContext.getBean(ElasticJobProperties.class);
            //②获取在 IOC 容器中注册 Bean 的对象 SingletonBeanRegistry
            SingletonBeanRegistry singletonBeanRegistry =
((ConfigurableApplicationContext) applicationContext).getBeanFactory();
            //③从 Spring 的 IOC 容器中获取注册中心对象 CoordinatorRegistryCenter
            CoordinatorRegistryCenter registryCenter =
applicationContext.getBean(CoordinatorRegistryCenter.class);
            //④获取追踪 Job 的配置对象 TracingConfiguration
            TracingConfiguration<?> tracingConfig = getTracingConfiguration();
            //⑤创建 Job 的启动实例对象
            constructJobBootstraps(elasticJobProperties, singletonBeanRegistry,
registryCenter, tracingConfig);
        }
        ...
    }
```

（2）用 constructJobBootstraps()方法创建 Job 的启动实例对象，具体代码如下所示：

```
    private void constructJobBootstraps(final ElasticJobProperties
elasticJobProperties, final SingletonBeanRegistry singletonBeanRegistry,
```

```
                                   final CoordinatorRegistryCenter
registryCenter, final TracingConfiguration<?> tracingConfig) {
        //①遍历配置信息中的 Job 列表
        for (Map.Entry<String, ElasticJobConfigurationProperties> entry :
elasticJobProperties.getJobs().entrySet()) {
            ElasticJobConfigurationProperties jobConfigurationProperties =
entry.getValue();
            Preconditions.checkArgument(null !=
jobConfigurationProperties.getElasticJobClass()
                                || !Strings.isNullOrEmpty
(jobConfigurationProperties.getElasticJobType()),
                         "Please specific [elasticJobClass] or [elasticJobType]
under job configuration.");
            Preconditions.checkArgument(null ==
jobConfigurationProperties.getElasticJobClass()
                                || Strings.isNullOrEmpty
(jobConfigurationProperties.getElasticJobType()),
                         "[elasticJobClass] and [elasticJobType] are mutually
exclusive.");
            if (null != jobConfigurationProperties.getElasticJobClass()) {
             //②如果在配置信息中添加了 Elastic Job 的类文件,则调用 registerClassedJob()
方法去注册 Job 的启动实例对象
                registerClassedJob(entry.getKey(),
entry.getValue().getJobBootstrapBeanName(), singletonBeanRegistry,
registryCenter, tracingConfig, jobConfigurationProperties);
            } else if
(!Strings.isNullOrEmpty(jobConfigurationProperties.getElasticJobType())) {
            //③如果在配置信息中添加了 Elastic Job 类型，则调用 registerTypedJob()方法
去注册 Job 的启动实例对象
            registerTypedJob(entry.getKey(),
entry.getValue().getJobBootstrapBeanName(), singletonBeanRegistry,
registryCenter, tracingConfig, jobConfigurationProperties);
            }
        }
    }
    private void registerClassedJob(final String jobName, final String
jobBootstrapBeanName, final SingletonBeanRegistry singletonBeanRegistry, final
CoordinatorRegistryCenter registryCenter,
                                    final TracingConfiguration<?> tracingConfig,
final ElasticJobConfigurationProperties jobConfigurationProperties) {
        JobConfiguration jobConfig =
jobConfigurationProperties.toJobConfiguration(jobName);
        Optional.ofNullable(tracingConfig).
ifPresent(jobConfig.getExtraConfigurations()::add);
```

```
        //④从 Spring Framework 的 IOC 容器中，获取 SimpleJob 对象（在应用启动的过程中注
入的接口 SimpleJob 的实现类）
        ElasticJob elasticJob =
applicationContext.getBean(jobConfigurationProperties.getElasticJobClass());
        if (Strings.isNullOrEmpty(jobConfig.getCron())) {
            Preconditions.checkArgument
(!Strings.isNullOrEmpty(jobBootstrapBeanName), "The property
[jobBootstrapBeanName] is required for One-off job.");
            //⑤如果 Cron 表达式为空，则初始化 OneOffJobBootstrap 对象，并将其注册到 IOC
容器中
            singletonBeanRegistry.registerSingleton(jobBootstrapBeanName, new
OneOffJobBootstrap(registryCenter, elasticJob, jobConfig));
        } else {
            String beanName = !Strings.isNullOrEmpty(jobBootstrapBeanName) ?
jobBootstrapBeanName : jobConfig.getJobName() + "ScheduleJobBootstrap";
            //⑥如果 Cron 表达式不为空，则初始化 ScheduleJobBootstrap 对象，并将其注册到
IOC 容器中
            singletonBeanRegistry.registerSingleton(beanName, new
ScheduleJobBootstrap(registryCenter, elasticJob, jobConfig));
        }
    }
    private void registerTypedJob(final String jobName, final String
jobBootstrapBeanName, final SingletonBeanRegistry singletonBeanRegistry, final
CoordinatorRegistryCenter registryCenter,
                                  final TracingConfiguration<?> tracingConfig,
final ElasticJobConfigurationProperties jobConfigurationProperties) {
        JobConfiguration jobConfig =
jobConfigurationProperties.toJobConfiguration(jobName);
        Optional.ofNullable(tracingConfig).
ifPresent(jobConfig.getExtraConfigurations()::add);
        if (Strings.isNullOrEmpty(jobConfig.getCron())) {
    Preconditions.checkArgument(!Strings.isNullOrEmpty(jobBootstrapBeanName),
"The property [jobBootstrapBeanName] is required for One-off job.");
        //⑦如果如果 Cron 表达式为空，则按照 Elastic Job 类型初始化 OneOffJobBootstrap
对象，并将其注册到 IOC 容器中
            singletonBeanRegistry.registerSingleton(jobBootstrapBeanName, new
OneOffJobBootstrap(registryCenter,
jobConfigurationProperties.getElasticJobType(), jobConfig));
        } else {
            String beanName = !Strings.isNullOrEmpty(jobBootstrapBeanName) ?
jobBootstrapBeanName : jobConfig.getJobName() + "ScheduleJobBootstrap";
         //⑧如果如果 Cron 表达式为空,则按照 Elastic Job 类型初始化 ScheduleJobBootstrap
对象，并将其注册到 IOC 容器中
```

```
            singletonBeanRegistry.registerSingleton(beanName, new
ScheduleJobBootstrap(registryCenter,
jobConfigurationProperties.getElasticJobType(), jobConfig));
        }
    }
```

（3）初始化 OneOffJobBootstrap 和 ScheduleJobBootstrap 对象。

在 Elastic Job Lite 中，初始化 OneOffJobBootstrap 和 ScheduleJobBootstrap 对象的具体代码如下所示：

```
    //①初始化 OneOffJobBootstrap 对象
    public final class OneOffJobBootstrap implements JobBootstrap {
        private final JobScheduler jobScheduler;
    private final InstanceService instanceService;
        public OneOffJobBootstrap(final CoordinatorRegistryCenter regCenter,
final ElasticJob elasticJob, final JobConfiguration jobConfig) {
           Preconditions.checkArgument(Strings.isNullOrEmpty
(jobConfig.getCron()), "Cron should be empty.");
            //②在构造函数中初始化 JobScheduler 对象（Job 定时器）
            jobScheduler = new JobScheduler(regCenter, elasticJob, jobConfig);
            //③在构造函数中初始化 InstanceService 对象（存储执行 Job 的实例节点信息）
            instanceService = new InstanceService(regCenter,
jobConfig.getJobName());
    }
        public OneOffJobBootstrap(final CoordinatorRegistryCenter regCenter,
final String elasticJobType, final JobConfiguration jobConfig) {
            Preconditions.checkArgument(Strings.isNullOrEmpty
(jobConfig.getCron()), "Cron should be empty.");
            jobScheduler = new JobScheduler(regCenter, elasticJobType,
jobConfig);
            instanceService = new InstanceService(regCenter,
jobConfig.getJobName());
    }
        public void execute() {
            instanceService.triggerAllInstances();
        }
        @Override
        public void shutdown() {
            jobScheduler.shutdown();
        }
    }
    //④初始化 ScheduleJobBootstrap 对象
    public final class ScheduleJobBootstrap implements JobBootstrap {
        private final JobScheduler jobScheduler;
```

```
        public ScheduleJobBootstrap(final CoordinatorRegistryCenter regCenter,
final ElasticJob elasticJob, final JobConfiguration jobConfig) {
            jobScheduler = new JobScheduler(regCenter, elasticJob, jobConfig);
        }
        public ScheduleJobBootstrap(final CoordinatorRegistryCenter regCenter,
final String elasticJobType, final JobConfiguration jobConfig) {
            jobScheduler = new JobScheduler(regCenter, elasticJobType,
jobConfig);
        }
        public void schedule() {
            Preconditions.checkArgument(!Strings.isNullOrEmpty
(jobScheduler.getJobConfig().getCron()), "Cron can not be empty.");
        //⑤触发 Job 定时器定时执行 Job
    jobScheduler.getJobScheduleController().scheduleJob(jobScheduler
    .getJobConfig().getCron());
    }

        @Override
        public void shutdown() {
            jobScheduler.shutdown();
        }
    }
```

3. 初始化一个基于 Quartz 的定时器类 JobScheduler

在 Elastic Job Lite 中，Job 类 OneOffJobBootstrap 和 ScheduleJobBootstrap 在初始化的过程中，也会初始化一个基于 Quartz 的定时器类 JobScheduler。

（1）用 JobScheduler 类的构造函数 JobScheduler()触发定时器类 JobScheduler 的初始化，具体代码如下所示：

```
    //①在已经实例化一个 Job 对象 ElasticJob 后，调用这个构造函数去初始化一个定时器类
JobScheduler
    public JobScheduler(final CoordinatorRegistryCenter regCenter, final
ElasticJob elasticJob, final JobConfiguration jobConfig) {
    Preconditions.checkArgument(null != elasticJob, "Elastic job cannot be
null.");
    //②初始化注册中心对象 CoordinatorRegistryCenter
    this.regCenter = regCenter;
    //③获取在应用中配置的监听器列表
    Collection<ElasticJobListener> jobListeners =
getElasticJobListeners(jobConfig);
    //④初始化 SetUpFacade 对象
        setUpFacade = new SetUpFacade(regCenter, jobConfig.getJobName(),
jobListeners);
```

```
        String jobClassName =
JobClassNameProviderFactory.getProvider().getJobClassName(elasticJob);
    this.jobConfig = setUpFacade.setUpJobConfiguration(jobClassName,
jobConfig);
    //⑤初始化 SchedulerFacade 对象
        schedulerFacade = new SchedulerFacade(regCenter,
jobConfig.getJobName());
        //⑥初始化 LiteJobFacade 对象
    jobFacade = new LiteJobFacade(regCenter, jobConfig.getJobName(),
jobListeners, findTracingConfiguration().orElse(null));
    //⑦校验 Job 的配置属性值
    validateJobProperties();
    //⑧初始化一个 Elastic Job 执行器
    jobExecutor = new ElasticJobExecutor(elasticJob, this.jobConfig,
jobFacade);
    //⑨将注册中心对象设置到监听器中
    setGuaranteeServiceForElasticJobListeners(regCenter, jobListeners);
    //⑩创建一个 Job 控制器
        jobScheduleController = createJobScheduleController();
    }
```

（2）用 JobScheduler 类的 createJobScheduleController()方法创建一个 Job 控制器，具体代码如下所示：

```
    //①创建一个 Job 控制器
    private JobScheduleController createJobScheduleController() {
        //②用基于框架 Quartz 的定时器 Scheduler 对象和 Job 详情的对象，去初始化 Job 控制器
类 JobScheduleController
        JobScheduleController result = new
JobScheduleController(createScheduler(), createJobDetail(),
getJobConfig().getJobName());
        JobRegistry.getInstance().registerJob(getJobConfig().getJobName(),
result);
        registerStartUpInfo();
        return result;
    }
    //③创建一个基于 Quartz 框架的定时器类 Scheduler
    private Scheduler createScheduler() {
        Scheduler result;
    try {
        //④初始化一个基于 Quartz 框架的工厂类 StdSchedulerFactory
            StdSchedulerFactory factory = new StdSchedulerFactory();
            //⑤设置 Job 配置属性值到工厂类 StdSchedulerFactory 中
            factory.initialize(getQuartzProps());
            result = factory.getScheduler();
            //⑥在基于 Quartz 框架的定时器类 Scheduler 中设置触发器监听器
```

```
            result.getListenerManager().addTriggerListener(
                schedulerFacade.newJobTriggerListener());
        } catch (final SchedulerException ex) {
            throw new JobSystemException(ex);
        }
        return result;
    }
    //⑦创建一个 Job 详情的对象
    private JobDetail createJobDetail() {
        JobDetail result =
JobBuilder.newJob(LiteJob.class).withIdentity(getJobConfig().getJobName()).b
uild();
        result.getJobDataMap().put(JOB_EXECUTOR_DATA_MAP_KEY, jobExecutor);
        return result;
    }
```

4. 开启在服务中配置的分布式 Job 执行器

在 Elastic Job Lite 中，用 ScheduleJobBootstrapStartupRunner 类开启在服务中配置的分布式 Job 执行器，具体代码如下所示：

```
    @Setter
    @Slf4j
    public class ScheduleJobBootstrapStartupRunner implements CommandLineRunner,
ApplicationContextAware {
        private ApplicationContext applicationContext;
        @Override
        public void run(final String... args) {
            log.info("Starting ElasticJob Bootstrap.");
            //从 Spring Framework 的 IOC 容器中获取 ScheduleJobBootstrap 对象，并调用
schedule()方法开启在服务中配置的分布式 Job 执行器
            applicationContext.getBeansOfType(ScheduleJobBootstrap.class).
            values().forEach(ScheduleJobBootstrap::schedule);
            log.info("ElasticJob Bootstrap started.");
        }
    }
```

如果在应用的 Job 配置信息中没有定义 Job 的 Cron 表达式，则会用 OneOffJobBootstrap 类的 execute()方法执行 Job，并且只执行一次；如果在应用的 Job 配置信息中定义了 Job 的 Cron 表达式，则会用 ScheduleJobBootstrap 类的 schedule()方法定时执行 Job。

5. 用 JobScheduleController 类去操作定时器

在 Elastic Job Lite 中，用 JobScheduleController 类去操作定时器，包括：执行定时器、重新执行定时器、暂定定时器、暂停后再开启定时器（保留暂停之前执行的状态）。

（1）用 JobScheduleController 类的 scheduleJob()方法执行定时器，具体代码如下所示：

```
@RequiredArgsConstructor
public final class JobScheduleController {
    //①基于 Quartz 框架的定时器类 Scheduler
    private final Scheduler scheduler;
     public void scheduleJob(final String cron) {
        try {
            //②如果在 Quartz 框架中不存在需要执行的定时器
            if (!scheduler.checkExists(jobDetail.getKey())) {
                //③则调用调用定时器类 Scheduler 的 scheduleJob()方法创建一个可以执
行的 Job
                scheduler.scheduleJob(jobDetail, createCronTrigger(cron));
            }
            //④调用 Quartz 框架的定时器类 Scheduler 的 start()方法执行定时器中的 Job
            scheduler.start();
        } catch (final SchedulerException ex) {
            throw new JobSystemException(ex);
        }
    }
    ...
}
```

（2）用 JobScheduleController 类的 rescheduleJob ()方法重新执行定时器，具体代码如下所示：

```
//①重新执行 ScheduleJobBootstrap 类型的 Job
public synchronized void rescheduleJob(final String cron) {
try {
    //②获取 Cron 表达式触发器对象 CronTrigger
        CronTrigger trigger = (CronTrigger)
scheduler.getTrigger(TriggerKey.triggerKey(triggerIdentity));
        if (!scheduler.isShutdown() && null != trigger
&& !cron.equals(trigger.getCronExpression())) {
            //③调用 Quartz 框架的定时器类 Scheduler 的 rescheduleJob()方法，去重新
执行 ScheduleJobBootstrap 类型的定时器
            scheduler.rescheduleJob(TriggerKey.triggerKey(triggerIdentity),
createCronTrigger(cron));
        }
    } catch (final SchedulerException ex) {
        throw new JobSystemException(ex);
    }
}
//④重新执行 OneOffJobBootstrap 类型的 Job
public synchronized void rescheduleJob() {
try {
```

```
        //⑤获取 Simple 触发器对象 SimpleTrigger
            SimpleTrigger trigger = (SimpleTrigger)
scheduler.getTrigger(TriggerKey.triggerKey(triggerIdentity));
            if (!scheduler.isShutdown() && null != trigger) {
                //⑥调用 Quartz 框架的定时器类 Scheduler 的 rescheduleJob()方法，去
重新执行 OneOffJobBootstrap 类型的定时器
                scheduler.rescheduleJob(TriggerKey.
triggerKey(triggerIdentity), createOneOffTrigger());
            }
        } catch (final SchedulerException ex) {
            throw new JobSystemException(ex);
        }
    }
```

（3）用 JobScheduleController 类的 pauseJob ()方法去暂定定时器，具体代码如下所示：

```
    //①暂停一个正在运行的定时器
    public synchronized void pauseJob() {
    try {
        //②如果定时器没有被关闭
            if (!scheduler.isShutdown()) {
                //③则暂定定时器中所有的触发器
                scheduler.pauseAll();
            }
        } catch (final SchedulerException ex) {
            throw new JobSystemException(ex);
        }
    }
```

（4）用 JobScheduleController 类的 resumeJob ()方法开启一个已经暂停的定时器（保留暂停之前执行的状态），具体代码如下所示：

```
    //①开启一个已经暂停的定时器
    public synchronized void resumeJob() {
    try {
        //②如果定时器没有被关闭
            if (!scheduler.isShutdown()) {
                //③则开启已经暂停的定时器中的所有的触发器
                scheduler.resumeAll();
            }
        } catch (final SchedulerException ex) {
            throw new JobSystemException(ex);
        }
    }
```

11.3.2　用 ZooKeeper 框架实现分布式 Job

下面从 Elastic Job 的接入模式（Spring Boot Starter）来分析“用 ZooKeeper 框架实现分布式 Job”的原理。

1. 初始化注册中心对象 ZookeeperRegistryCenter

在 Elastic Job 中，用 ElasticJobRegistryCenterConfiguration 类去初始化注册中心对象 ZookeeperRegistryCenter，具体代码如下所示：

```
//①加载配置文件中 ZooKeeper 的配置属性
@EnableConfigurationProperties(ZookeeperProperties.class)
public class ElasticJobRegistryCenterConfiguration {
    //②用 init()方法初始化基于 ZooKeeper 的注册中心对象 ZookeeperRegistryCenter
    @Bean(initMethod = "init")
    public ZookeeperRegistryCenter zookeeperRegistryCenter(final
ZookeeperProperties zookeeperProperties) {
        return new
ZookeeperRegistryCenter(zookeeperProperties.toZookeeperConfiguration());
    }
}
```

在 Elastic Job 中，用注册中心对象 ZookeeperRegistryCenter 类的 init()方法，去初始化 ZooKeeper 的客户端 CuratorFramework 类（基于 Apache 的客户端中间件框架 Curator），具体代码如下所示：

```
@Slf4j
public final class ZookeeperRegistryCenter implements
CoordinatorRegistryCenter {
    @Getter(AccessLevel.PROTECTED)
    private final ZookeeperConfiguration zkConfig;
    private final Map<String, CuratorCache> caches = new
ConcurrentHashMap<>();
    @Getter
    private CuratorFramework client;

    public ZookeeperRegistryCenter(final ZookeeperConfiguration zkConfig) {
        this.zkConfig = zkConfig;
    }
    //①初始化 CuratorFramework 对象
    @Override
    public void init() {
        log.debug("Elastic job: zookeeper registry center init, server lists
is: {}.", zkConfig.getServerLists());
        CuratorFrameworkFactory.Builder builder =
CuratorFrameworkFactory.builder()
```

```
                .connectString(zkConfig.getServerLists())
                .retryPolicy(new
ExponentialBackoffRetry(zkConfig.getBaseSleepTimeMilliseconds(),
zkConfig.getMaxRetries(), zkConfig.getMaxSleepTimeMilliseconds()))
                .namespace(zkConfig.getNamespace());
        if (0 != zkConfig.getSessionTimeoutMilliseconds()) {
            builder.sessionTimeoutMs(zkConfig.
getSessionTimeoutMilliseconds());
        }
        if (0 != zkConfig.getConnectionTimeoutMilliseconds()) {
            builder.connectionTimeoutMs(zkConfig.
getConnectionTimeoutMilliseconds());
        }
        if (!Strings.isNullOrEmpty(zkConfig.getDigest())) {
            builder.authorization("digest",
zkConfig.getDigest().getBytes(StandardCharsets.UTF_8))
                    .aclProvider(new ACLProvider() {
                        @Override
                        public List<ACL> getDefaultAcl() {
                            return ZooDefs.Ids.CREATOR_ALL_ACL;
                        }
                        @Override
                        public List<ACL> getAclForPath(final String path) {
                            return ZooDefs.Ids.CREATOR_ALL_ACL;
                        }
                    });
        }
        client = builder.build();
        //②开启 ZooKeeper 客户端和服务器端的 RPC 连接
        client.start();
        try {
            if (!client.blockUntilConnected(zkConfig.
getMaxSleepTimeMilliseconds() * zkConfig.getMaxRetries(),
TimeUnit.MILLISECONDS)) {
                client.close();
                throw new KeeperException.OperationTimeoutException();
            }
        } catch (final Exception ex) {
            RegExceptionHandler.handleException(ex);
        }
    }
    ...
}
```

2. 用 ZookeeperRegistryCenter 类操控 ZooKeeper 中 Job 实例节点的信息

在 Elastic Job 中，用 ZookeeperRegistryCenter 类操控 ZooKeeper 中 Job 实例节点的信息。

（1）用 ZookeeperRegistryCenter 类的 persist()方法持久化一个 Job 实例节点，具体代码如下所示：

```
//①创建一个类型为“PERSISTENT”的持久化的 Job 实例节点
@Override
public void persist(final String key, final String value) {
try {
    //②如果路径节点不存在，则在 ZooKeeper 中新建一个路径节点
        if (!isExisted(key)) {
    client.create().creatingParentsIfNeeded().
withMode(CreateMode.PERSISTENT)
    .forPath(key, value.getBytes(StandardCharsets.UTF_8));
        } else {
            //③否则更新路径节点的值
            update(key, value);
        }
    } catch (final Exception ex) {
        RegExceptionHandler.handleException(ex);
    }
}
@Override
public void update(final String key, final String value) {
try {
    //④获取 ZooKeeper 的事务操控对象 TransactionOp
        TransactionOp transactionOp = client.transactionOp();
        //⑤更新路径节点的值（更新操作是线程安全的）
        client.transaction().forOperations(transactionOp.check().
forPath(key), transactionOp.setData().forPath(key,
value.getBytes(StandardCharsets.UTF_8)));
    } catch (final Exception ex) {
        RegExceptionHandler.handleException(ex);
    }
}
```

（2）用 ZookeeperRegistryCenter 类的 persistEphemeral()方法创建一个临时的 Job 实例节点，具体代码如下所示：

```
//在 ZooKeeper 中创建一个类型为“EPHEMERAL”的临时 Job 实例节点
@Override
public void persistEphemeral(final String key, final String value) {
    try {
```

```
            if (isExisted(key)) {
                client.delete().deletingChildrenIfNeeded().forPath(key);
            }
client.create().creatingParentsIfNeeded().withMode(CreateMode.EPHEMERAL).for
Path(key, value.getBytes(StandardCharsets.UTF_8));
            } catch (final Exception ex) {
            RegExceptionHandler.handleException(ex);
        }
    }
```

（3）用 ZookeeperRegistryCenter 类的 persistSequential()方法，去创建一个持久化的、顺序的 Job 实例节点，具体代码如下所示：

```
    //在 ZooKeeper 中创建一个类型为“PERSISTENT_SEQUENTIAL”、持久化的、顺序的 Job 实例
节点
    @Override
    public String persistSequential(final String key, final String value) {
        try {
            return
client.create().creatingParentsIfNeeded().withMode(CreateMode.PERSISTENT_SEQ
UENTIAL).forPath(key, value.getBytes(StandardCharsets.UTF_8));
        } catch (final Exception ex) {
            RegExceptionHandler.handleException(ex);
        }
        return null;
    }
```

（4）用 ZookeeperRegistryCenter 类的 persistEphemeralSequential ()方法，去创建一个临时的、顺序的 Job 实例节点，具体代码如下所示：

```
    //在 ZooKeeper 中创建一个类型为“EPHEMERAL_SEQUENTIAL”、临时的、顺序的 Job 实例节点
    @Override
    public void persistEphemeralSequential(final String key) {
        try {
    client.create().creatingParentsIfNeeded().withMode(CreateMode.EPHEMERAL_
SEQUENTIAL).forPath(key);
        } catch (final Exception ex) {
            RegExceptionHandler.handleException(ex);
        }
    }
```

11.3.3 【实例】在 Elastic Job 控制台中操控分布式 Job

本实例的源码在本书配套资源的“/spring-cloud-alibaba-practice/chaptereleven/elasticjob-operate-spring-cloud-alibaba/”目录下。

在 Elastic Job 控制台操控分布式 Job 的操作包括触发、失效、终止和删除等。

1. 初始化一个 Spring Cloud Alibaba 服务

用 Spring Cloud Alibaba 初始化一个服务，并添加一个分布式 Job 类 ElasticOperateMessageJob。配置的 Job 规则如下所示：

```
elasticjob:
  regCenter:
    serverLists: 127.0.0.1:2181
    namespace: spring-cloud-alibaba-job
  jobs:
    elasticOperateMessageJob:
        elasticJobClass:
com.alibaba.cloud.youxia.elasticjob.ElasticOperateMessageJob
        //每 1 分钟执行一次
        cron: 0 0/1 * * * ?
        shardingTotalCount: 1
        overwrite: true
```

2. 提前触发定时任务

单击“Trigger”按钮提前触发定时任务，如图 11-12 所示。

图 11-12

在 IDEA 控制台可以观察到定时任务已经执行，如图 11-13 所示。

```
当前Job执行的时间: Wed Aug 04 04:41:10 CST 2021
定时任务执行的间隔为: 59秒
发送定时消息!
2021-08-04 04:41:14.755  WARN 6719 --- [ask-Scheduler-1] a.c.d.m.r.DubboServiceMetadataRepository : Current application will
 subscribe all services(size:1) in registry, a lot of memory and CPU cycles may be used, thus it's strongly recommend you using the
 externalized property 'dubbo.cloud.subscribed-services' to specify the services
当前Job执行的时间: Wed Aug 04 04:41:29 CST 2021
定时任务执行的间隔为: 19秒
发送定时消息!
```

图 11-13

3. 失效定时任务

如图 11-14 所示单击“Disable”按钮，失效定时任务。

图 11-14

在定时任务失效后，在控制台中看到的“Enable”按钮如图 11-15 所示。可以再次单击该按钮去恢复定时任务。

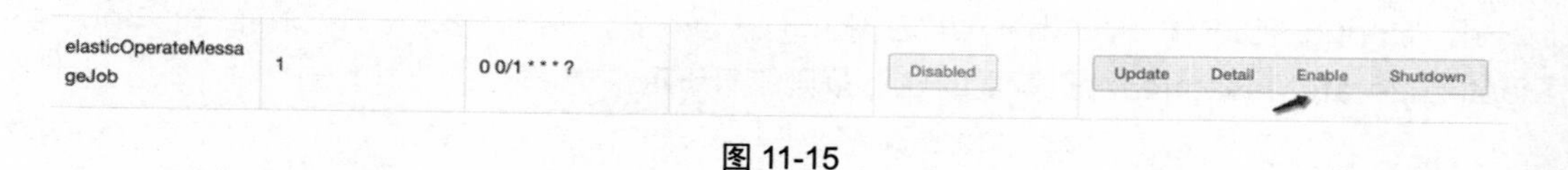

图 11-15

4. 终止定时任务

如图 11-16 所示单击“Shutdown”按钮终止定时任务。

图 11-16

终止后的定时任务的状态如图 11-17 所示。定时任务被终止后，其状态变为“Crashed”，这样就直接关闭它了，不能在线恢复。

图 11-17

定时任务终止后的服务运行日志如图 11-18 所示。如果需要恢复定时任务，则只能重启服务。

```
当前Job执行的时间: Wed Aug 04 04:51:10 CST 2021
定时任务执行的间隔为: 59秒
发送定时消息!
2021-08-04 04:51:15.756  WARN 6719 --- [ask-Scheduler-1] a.c.d.m.r.DubboServiceMetadataRepository : Current application will
 subscribe all services(size:1) in registry, a lot of memory and CPU cycles may be used, thus it's strongly recommend you using the
 externalized property 'dubbo.cloud.subscribed-services' to specify the services
2021-08-04 04:51:45.804  WARN 6719 --- [ask-Scheduler-1] a.c.d.m.r.DubboServiceMetadataRepository : Current application will
 subscribe all services(size:1) in registry, a lot of memory and CPU cycles may be used, thus it's strongly recommend you using the
 externalized property 'dubbo.cloud.subscribed-services' to specify the services
2021-08-04 04:51:54.935  INFO 6719 --- [NotifyService-0] org.quartz.core.QuartzScheduler          : Scheduler
 elasticOperateMessageJob_$_NON_CLUSTERED shutting down.
2021-08-04 04:51:54.936  INFO 6719 --- [NotifyService-0] org.quartz.core.QuartzScheduler          : Scheduler
 elasticOperateMessageJob_$_NON_CLUSTERED paused.
2021-08-04 04:51:54.942  INFO 6719 --- [NotifyService-0] org.quartz.core.QuartzScheduler          : Scheduler
 elasticOperateMessageJob_$_NON_CLUSTERED shutdown complete.
2021-08-04 04:52:15.854  WARN 6719 --- [ask-Scheduler-1] a.c.d.m.r.DubboServiceMetadataRepository : Current application will
 subscribe all services(size:1) in registry, a lot of memory and CPU cycles may be used, thus it's strongly recommend you using the
 externalized property 'dubbo.cloud.subscribed-services' to specify the services
2021-08-04 04:52:45.893  WARN 6719 --- [ask-Scheduler-1] a.c.d.m.r.DubboServiceMetadataRepository : Current application will
 subscribe all services(size:1) in registry, a lot of memory and CPU cycles may be used, thus it's strongly recommend you using the
 externalized property 'dubbo.cloud.subscribed-services' to specify the services
2021-08-04 04:53:15.915  WARN 6719 --- [ask-Scheduler-1] a.c.d.m.r.DubboServiceMetadataRepository : Current application will
```

图 11-18

5. 删除定时任务

如图 11–19 所示单击“Remove”按钮，删除定时任务。如果需要恢复定时任务，则只能重启服务。

elasticOperateMessageJob	1	0 0/1 * * * ?		Crashed	Update　Remove

图 11-19

删除后定时任务的状态如图 11–20 所示，在控制台中不会显示定时任务“elasticOperateMessageJob”。

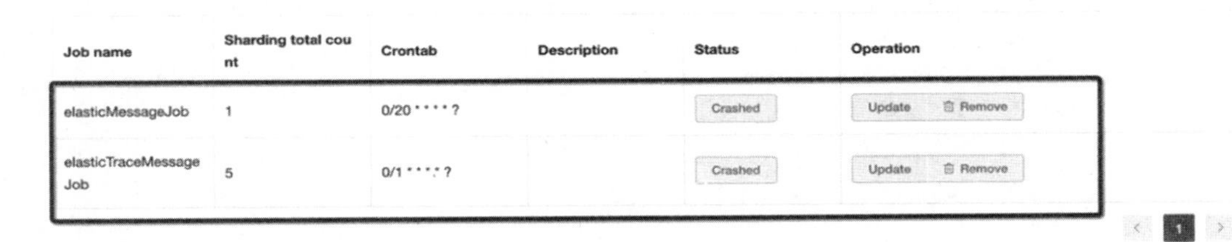

Job name	Sharding total count	Crontab	Description	Status	Operation
elasticMessageJob	1	0/20 * * * * ?		Crashed	Update　Remove
elasticTraceMessageJob	5	0/1 * * * * ?		Crashed	Update　Remove

图 11-20

11.4 “用 SPI 将 Job 分片策略插件化”的原理

Elastic Job 默认支持以下 3 种分片策略。

（1）平均分片策略：根据分片项平均分片，如果作业服务器的数量与分片总数无法整除，则多余的分片会被顺序地分配至每一个作业服务器。

（2）奇偶分片策略：根据作业名称哈希值的奇偶数，来决定按照作业服务器 IP 地址升序或降序的方式进行分片。

- 如果作业名称哈希值是偶数，则按照 IP 地址进行升序分片。
- 如果作业名称哈希值是奇数，则按照 IP 地址进行降序分片。在多个作业共同运行时，服务器负载均衡器可以让作业执行得更加均匀。

（3）轮询分片策略。根据作业名称轮询分片。

在 Elastic Job 中，Elastic Job 采用 SPI 机制实现分片策略的插件化。

11.4.1 用 SPI 工厂类 JobShardingStrategyFactory 加载分片策略

在 Elastic Job 中，用 SPI 工厂类 JobShardingStrategyFactory 来加载分片策略。

（1）初始化 SPI 工厂类 JobShardingStrategyFactory。

在 Elastic Job 中，SPI 工厂类 JobShardingStrategyFactory 主要是通过静态代码块来初始化并加载分片策略的，具体代码如下所示：

```
    @NoArgsConstructor(access = AccessLevel.PRIVATE)
    public final class JobShardingStrategyFactory {
        //①默认为平均分片策略
        private static final String DEFAULT_STRATEGY = "AVG_ALLOCATION";
        static {
            //②使用 ElasticJobServiceLoader 类注册 JobShardingStrategy 接口的所有实
现类，比如轮询策略实现类 RoundRobinByNameJobShardingStrategy
            ElasticJobServiceLoader.registerTypedService
            (JobShardingStrategy.class);
    }
    ...
    }
```

（2）在 Elastic Job 中，用工厂类 JobShardingStrategyFactory 的 getStrategy()方法暴露其支持的分片策略，具体代码如下所示：

```
    //①按照类型去获取分片策略
    public static JobShardingStrategy getStrategy(final String type) {
        //②如果类型为空，则直接返回平均分片策略
        if (Strings.isNullOrEmpty(type)) {
            return ElasticJobServiceLoader.getCachedTypedServiceInstance
(JobShardingStrategy.class, DEFAULT_STRATEGY).get();
        }
        //③用 ElasticJobServiceLoader 类的 getCachedTypedServiceInstance()方法获
取指定类型的分片策略的实现类，比如轮询策略实现类 RoundRobinByNameJobShardingStrategy
        return
ElasticJobServiceLoader.getCachedTypedServiceInstance(JobShardingStrategy.cl
ass, type).orElseThrow(() -> new JobConfigurationException("Cannot find
sharding strategy using type '%s'.", type));
    }
```

（3）在 Elastic Job 中，用 RoundRobinByNameJobShardingStrategy 类实现轮询分片策略，具体代码如下所示：

```
    public final class RoundRobinByNameJobShardingStrategy implements
JobShardingStrategy {
```

```
    private final AverageAllocationJobShardingStrategy
averageAllocationJobShardingStrategy = new
AverageAllocationJobShardingStrategy();
    //①执行轮询分片
        @Override
        public Map<JobInstance, List<Integer>> sharding(final List<JobInstance>
jobInstances, final String jobName, final int shardingTotalCount) {
            return
averageAllocationJobShardingStrategy.sharding(rotateServerList(jobInstances,
jobName), jobName, shardingTotalCount);
    }
    //②重新排序作业服务实例的 IP 地址列表
        private List<JobInstance> rotateServerList(final List<JobInstance>
shardingUnits, final String jobName) {
            int shardingUnitsSize = shardingUnits.size();
            int offset = Math.abs(jobName.hashCode()) % shardingUnitsSize;
            if (0 == offset) {
                return shardingUnits;
            }
            List<JobInstance> result = new ArrayList<>(shardingUnitsSize);
            for (int i = 0; i < shardingUnitsSize; i++) {
                int index = (i + offset) % shardingUnitsSize;
                result.add(shardingUnits.get(index));
            }
            return result;
    }
    //③返回轮询分片策略的类型“ROUND_ROBIN”
        @Override
        public String getType() {
            return "ROUND_ROBIN";
        }
    }
```

（4）在 Elastic Job 中，用 OdevitySortByNameJobShardingStrategy 类实现奇偶分片策略，具体代码如下所示：

```
    public final class OdevitySortByNameJobShardingStrategy implements
JobShardingStrategy {
        private final AverageAllocationJobShardingStrategy
averageAllocationJobShardingStrategy = new
AverageAllocationJobShardingStrategy();
        //①执行奇偶分片
        @Override
        public Map<JobInstance, List<Integer>> sharding(final List<JobInstance>
jobInstances, final String jobName, final int shardingTotalCount) {
            long jobNameHash = jobName.hashCode();
```

```
            if (0 == jobNameHash % 2) {
                Collections.reverse(jobInstances);
            }
            return averageAllocationJobShardingStrategy.sharding(jobInstances, 
jobName, shardingTotalCount);
        }
        //②返回奇偶分片类型“ODEVITY”
        @Override
        public String getType() {
            return "ODEVITY";
        }
    }
```

（5）在 Elastic Job 中，用 AverageAllocationJobShardingStrategy 类实现平均分片策略，具体代码如下所示：

```
    public final class AverageAllocationJobShardingStrategy implements 
JobShardingStrategy {
        //①执行平均分片
        @Override
        public Map<JobInstance, List<Integer>> sharding(final List<JobInstance> 
jobInstances, final String jobName, final int shardingTotalCount) {
            if (jobInstances.isEmpty()) {
                return Collections.emptyMap();
            }
            Map<JobInstance, List<Integer>> result = 
shardingAliquot(jobInstances, shardingTotalCount);
            addAliquant(jobInstances, shardingTotalCount, result);
            return result;
        }
        //②按照平均算法，重新排序作业服务实例的 IP 地址列表
        private Map<JobInstance, List<Integer>> shardingAliquot(final 
List<JobInstance> shardingUnits, final int shardingTotalCount) {
            Map<JobInstance, List<Integer>> result = new 
LinkedHashMap<>(shardingUnits.size(), 1);
            int itemCountPerSharding = shardingTotalCount / 
shardingUnits.size();
            int count = 0;
            for (JobInstance each : shardingUnits) {
                List<Integer> shardingItems = new 
ArrayList<>(itemCountPerSharding + 1);
                for (int i = count * itemCountPerSharding; i < (count + 1) * 
itemCountPerSharding; i++) {
                    shardingItems.add(i);
                }
                result.put(each, shardingItems);
```

```
            count++;
        }
        return result;
    }
    //③添加平均分片策略的算法种子
    private void addAliquant(final List<JobInstance> shardingUnits, final
int shardingTotalCount, final Map<JobInstance, List<Integer>> shardingResults) {
        int aliquant = shardingTotalCount % shardingUnits.size();
        int count = 0;
        for (Map.Entry<JobInstance, List<Integer>> entry :
shardingResults.entrySet()) {
            if (count < aliquant) {
                entry.getValue().add(shardingTotalCount /
shardingUnits.size() * shardingUnits.size() + count);
            }
            count++;
        }
    }
    //④返回平均分片策略的类型名称“AVG_ALLOCATION”
    @Override
    public String getType() {
        return "AVG_ALLOCATION";
    }
}
```

11.4.2　用 ShardingService 类触发 Job 去执行分片策略

在 Elastic Job 中，用 ShardingService 类触发 Job 去执行分片策略。

（1）在 Elastic Job 中，用构造函数来初始化类 ShardingService，具体代码如下所示：

```
@Slf4j
public final class ShardingService {
    private final String jobName;
    private final JobNodeStorage jobNodeStorage;
    private final LeaderService leaderService;
    private final ConfigurationService configService;
    private final InstanceService instanceService;
    private final InstanceNode instanceNode;
    private final ServerService serverService;
    private final ExecutionService executionService;
    private final JobNodePath jobNodePath;

    public ShardingService(final CoordinatorRegistryCenter regCenter, final
String jobName) {
        //①初始化 Job 名称
```

```
        this.jobName = jobName;
        //②初始化 Job 节点存储对象 JobNodeStorage，它主要是用于向注册中心（基于
ZooKeeper）持久化和获取 Job 节点数据
        jobNodeStorage = new JobNodeStorage(regCenter, jobName);
        //③初始化选举对象 LeaderService，它主要用于 Job 节点中的主从节点选举
        leaderService = new LeaderService(regCenter, jobName);
        //④初始化配置对象 ConfigurationService，它主要用于向配置中心（基于
ZooKeeper）持久化和获取配置数据
        configService = new ConfigurationService(regCenter, jobName);
        //⑤初始化实例对象 InstanceService，它主要用于维护 Job 实例数据
        instanceService = new InstanceService(regCenter, jobName);
        instanceNode = new InstanceNode(jobName);
        //⑥初始化服务对象 ServerService，它主要用于维护 Job 服务数据
        serverService = new ServerService(regCenter, jobName);
        //⑦初始化 Job 执行器对象 ExecutionService，它主要用于执行 Job
        executionService = new ExecutionService(regCenter, jobName);
        jobNodePath = new JobNodePath(jobName);
    }
    ...
}
```

（2）在 Elastic Job 中，用 ShardingService 类的 shardingIfNecessary()方法来判断是否要进行 Job 分片，具体代码如下所示：

```
public void shardingIfNecessary() {
    //①获取可用的 Job 实例节点列表，如果实例节点列表为空或者当前 Job 不需要分片，则直接
返回
    List<JobInstance> availableJobInstances =
instanceService.getAvailableJobInstances();
    if (!isNeedSharding() || availableJobInstances.isEmpty()) {
        return;
    }
    //②如果当前节点的角色是 Leader 节点，则执行分片，代码继续执行
    if (!leaderService.isLeaderUntilBlock()) {
        //③如果不是 Leader 节点，则休眠一段时间，直到 Leader 节点分片结束，否则直接返回
        blockUntilShardingCompleted();
        return;
    }
    //④如果还有其他已经分片的任务 Job 在执行中，则等待一段时间，默认休眠 100ms
    waitingOtherShardingItemCompleted();
    JobConfiguration jobConfig = configService.load(false);
    //⑤获取配置文件中的分片数（通过参数 shardingTotalCount 来配置）
    int shardingTotalCount = jobConfig.getShardingTotalCount();
    log.debug("Job '{}' sharding begin.", jobName);
    //⑥用 JobNodeStorage 类的 fillEphemeralJobNode()方法，调用注册中心对象
ZookeeperRegistryCenter 的 persistEphemeral()方法创建临时的 Job 实例节点
```

```
        jobNodeStorage.fillEphemeralJobNode(ShardingNode.PROCESSING, "");
        //⑦重置 ZooKeeper 中的分片节点实例信息
        resetShardingInfo(shardingTotalCount);
        //⑧获取分片策略
        JobShardingStrategy jobShardingStrategy = JobShardingStrategyFactory.
getStrategy(jobConfig.getJobShardingStrategyType());
        //⑨执行分片策略
        jobNodeStorage.executeInTransaction(new
PersistShardingInfoTransactionExecutionCallback(jobShardingStrategy.sharding
(availableJobInstances, jobName, shardingTotalCount)));
        log.debug("Job '{}' sharding complete.", jobName);
    }
```

11.4.3 【实例】给 Spring Cloud Alibaba 应用接入带有分片功能的分布式 Job

本实例的源码在本书配套资源的“/spring-cloud-alibaba-practice/chaptereleven/elasticjob-shard-spring-cloud-alibaba/”目录下。

本实例给 Spring Cloud Alibaba 应用接入带有分片功能的分布式 Job，并验证分布式 Job 的分片功能。

1. 初始化一个 Spring Cloud Alibaba 应用

使用 Spring Cloud Alibaba 初始化一个服务 elasticjob-shard-spring- cloud-alibaba，并添加一个分布式 Job 类 ElasticOperateMessageJob。配置的 Job 规则如下所示：

```
  elasticjob:
    regCenter:
      serverLists: 127.0.0.1:2181
      namespace: spring-cloud-alibaba-job
    jobs:
      elasticShardMessageJob:
          elasticJobClass:
com.alibaba.cloud.youxia.elasticjob.ElasticShardMessageJob
          cron: 0 0/5 * * * ?
          shardingTotalCount: 15
          overwrite: true
          ##配置平均分片策略
          jobShardingStrategyType: AVG_ALLOCATION
```

2. 搭建分片环境

下面使用 15 个分片、5 个分布式 Job 的作业节点实例，在单机环境下模拟分布式 Job 的集群分片场景。

（1）用“mvn clean install”命令打包服务 elasticjob-shard-spring-cloud-alibaba，生成 Jar 包 elasticjob-shard-spring-cloud-alibaba-1.0.0.release.jar。

（2）在本地新建 5 个文件（本书统一采用 Mac 环境），将 Jar 包 elasticjob-shard-spring-cloud-alibaba-1.0.0.release.jar 分别放入文件中，如图 11-21 所示。

elasticjob-shard

Name	Date Modified	Size	Kind
127.0.0.1	Today at 07:20	--	Folder
elasticjob-shard-sp...ba-1.0.0.release.jar	Today at 07:20	57.1 MB	Java JAR file
nohup.out	Today at 10:30	1.9 MB	Document
127.0.0.2	Today at 07:21	--	Folder
elasticjob-shard-sp...ba-1.0.0.release.jar	Today at 07:20	57.1 MB	Java JAR file
nohup.out	Today at 10:30	994 KB	Document
127.0.0.3	Today at 07:21	--	Folder
elasticjob-shard-sp...ba-1.0.0.release.jar	Today at 07:20	57.1 MB	Java JAR file
nohup.out	Today at 10:30	1 MB	Document
127.0.0.4	Today at 07:21	--	Folder
elasticjob-shard-sp...ba-1.0.0.release.jar	Today at 07:20	57.1 MB	Java JAR file
nohup.out	Today at 10:30	991 KB	Document
127.0.0.5	Today at 07:21	--	Folder
elasticjob-shard-sp...ba-1.0.0.release.jar	Today at 07:20	57.1 MB	Java JAR file
nohup.out	Today at 10:30	1.1 MB	Document

图 11-21

（3）在文件目录中分别执行启动命令启动服务，具体命令如下所示：

```
//在 127.0.0.1 文件目录中执行如下命令
nohup java -jar elasticjob-shard-spring-cloud-alibaba-1.0.0.release.jar
--server.port=8001 --dubbo.protocol.port=26000 &
//在 127.0.0.2 文件目录中执行如下命令
nohup java -jar elasticjob-shard-spring-cloud-alibaba-1.0.0.release.jar
--server.port=8002 --dubbo.protocol.port=26001 &
//在 127.0.0.3 文件目录中执行如下命令
nohup java -jar elasticjob-shard-spring-cloud-alibaba-1.0.0.release.jar
--server.port=8003 --dubbo.protocol.port=26002 &
//在 127.0.0.4 文件目录中执行如下命令
nohup java -jar elasticjob-shard-spring-cloud-alibaba-1.0.0.release.jar
--server.port=8004 --dubbo.protocol.port=26003 &
//在 127.0.0.5 文件目录中执行如下命令
nohup java -jar elasticjob-shard-spring-cloud-alibaba-1.0.0.release.jar
--server.port=8005 --dubbo.protocol.port=26004 &
```

这样在单机环境中，用 5 个进程去模拟集群环境，其中，“127.0.0.1:8001”“127.0.0.1:8002”“127.0.0.1:8003”“127.0.0.1:8004”“127.0.0.1:8005”为各个进程的 IP 地址。

（4）在服务启动后，在 Elastic Job 的 UI 控制台能够看到当前分布式 Job“elasticShardMessageJob”有 15 个分片，状态为“OK”，如图 11-22 所示。

图 11-22

（5）单击“Trigger”按钮调度 15 个分片对应的执行节点实例，观察进程中分布式 Job 执行的日志：

图 11–23 所示为 IP 地址“127.0.0.1:8001”运行的日志。在平均分片策略下，该节点执行了 3 个分片任务。

图 11–24 所示为 IP 地址“127.0.0.1:8002”运行的日志。在平均分片策略下，该节点执行了 3 个分片任务。

```
执行分布式Job的IP地址为： 127.0.0.1:8001
执行分布式Job的IP地址为： 127.0.0.1:8001
执行分布式Job的IP地址为： 127.0.0.1:8001
当前Job执行的时间： Wed Aug 04 13:18:30 CST 2021
定时任务执行的间隔为： 60秒
发送定时消息！
当前Job执行的时间： Wed Aug 04 13:18:30 CST 2021
定时任务执行的间隔为： 0秒
当前Job执行的时间： Wed Aug 04 13:18:30 CST 2021
定时任务执行的间隔为： 0秒
发送定时消息！
发送定时消息！
```

图 11-23

```
执行分布式Job的IP地址为： 127.0.0.1:8002
执行分布式Job的IP地址为： 127.0.0.1:8002
执行分布式Job的IP地址为： 127.0.0.1:8002
当前Job执行的时间： Wed Aug 04 13:18:30 CST 2021
当前Job执行的时间： Wed Aug 04 13:18:30 CST 2021
定时任务执行的间隔为： 60秒
发送定时消息！
当前Job执行的时间： Wed Aug 04 13:18:30 CST 2021
定时任务执行的间隔为： 0秒
发送定时消息！
定时任务执行的间隔为： 60秒
发送定时消息！
```

图 11-24

图 11–25 所示为 IP 地址“127.0.0.1:8003”运行的日志。在平均分片策略下，该节点执行了 3 个分片任务。

图 11–26 所示为 IP 地址“127.0.0.1:8004”运行的日志。在平均分片策略下，该节点执行了 3 个分片任务。

```
执行分布式Job的IP地址为： 127.0.0.1:8003
执行分布式Job的IP地址为： 127.0.0.1:8003
执行分布式Job的IP地址为： 127.0.0.1:8003
当前Job执行的时间： Wed Aug 04 13:18:30 CST 2021
定时任务执行的间隔为： 60秒
当前Job执行的时间： Wed Aug 04 13:18:30 CST 2021
定时任务执行的间隔为： 0秒
发送定时消息！
当前Job执行的时间： Wed Aug 04 13:18:30 CST 2021
定时任务执行的间隔为： 0秒
发送定时消息！
发送定时消息！
```

图 11-25

```
执行分布式Job的IP地址为： 127.0.0.1:8004
执行分布式Job的IP地址为： 127.0.0.1:8004
执行分布式Job的IP地址为： 127.0.0.1:8004
当前Job执行的时间： Wed Aug 04 13:18:30 CST 2021
当前Job执行的时间： Wed Aug 04 13:18:30 CST 2021
定时任务执行的间隔为： 59秒
发送定时消息！
定时任务执行的间隔为： 59秒
发送定时消息！
当前Job执行的时间： Wed Aug 04 13:18:30 CST 2021
定时任务执行的间隔为： 0秒
发送定时消息！
```

图 11-26

图 11–27 所示为 IP 地址“127.0.0.1:8005”运行的日志。在平均分片策略下，该节点执行了 3 个分片任务。

```
执行分布式Job的IP地址为：127.0.0.1:8005
执行分布式Job的IP地址为：127.0.0.1:8005
执行分布式Job的IP地址为：127.0.0.1:8005
当前Job执行的时间：Wed Aug 04 13:18:30 CST 2021
当前Job执行的时间：Wed Aug 04 13:18:30 CST 2021
当前Job执行的时间：Wed Aug 04 13:18:30 CST 2021
定时任务执行的间隔为：60秒
发送定时消息！
定时任务执行的间隔为：60秒
发送定时消息！
定时任务执行的间隔为：60秒
发送定时消息！
```

图 11-27

11.5 “实现分布式 Job 的事件追踪”的原理

Elastic Job 提供了事件追踪功能，可以查询、统计和监控 Job 的执行的过程。目前 Elastic Job 支持基于关系型数据库的事件追踪方式。Elastic Job 也支持 SPI 扩展，应用可以自定义基于 SPI 的个性化扩展插件，去替换默认的事件追踪方式。

在 Elastic Job 中，用 Guava 的事件机制和数据库来实现分布式 Job 的事件追踪，并用数据库持久化分布式 Job 的运行状态和日志。

11.5.1 用基于 Guava 的事件机制实现分布式 Job 的事件追踪

在 Elastic Job 中，Job 的事件追踪需要应用主动去开启。开启方式如下所示：

```
//在配置文件中配置数据源（基于 MySQL）和 Trace 的类型（基于 RDB）
spring:
  datasource:
url:jdbc:mysql://127.0.0.1:3306/job_trace?serverTimezone=UTC&useSSL=false&useUnicode=true&characterEncoding=UTF-8
    driver-class-name: com.mysql.cj.jdbc.Driver
    username: root
    password: root
elasticjob:
  tracing:
    type: RDB
```

在 Elastic Job 中，用 ElasticJobTracingConfiguration 类读取配置文件数据源，并将其加载到 Spring Framework 的 IOC 容器中。

```
public class ElasticJobTracingConfiguration {
    @Bean
    @ConditionalOnBean(DataSource.class)
    @ConditionalOnProperty(name = "elasticjob.tracing.type", havingValue = "RDB")
```

```
    public TracingConfiguration<DataSource> tracingConfiguration(final
DataSource dataSource) {
        //返回 Job 事件追踪的 trace 配置对象 TracingConfiguration
            return new TracingConfiguration<>("RDB", dataSource);
        }
    }
```

1. 加载 Elastic Job 的事件类 JobTracingEventBus

在 Elastic Job 中，在 LiteJobFacade 类初始化的过程中，会触发事件类 JobTracingEventBus 的初始化。

（1）用 LiteJobFacade 类的构造函数加载事件类 JobTracingEventBus，具体代码如下所示：

```
    private final JobTracingEventBus jobTracingEventBus;
    public LiteJobFacade(final CoordinatorRegistryCenter regCenter, final
String jobName, final Collection<ElasticJobListener> elasticJobListeners, final
TracingConfiguration<?> tracingConfig) {
    ...
    //如果在应用的配置文件中配置了 Job 事件追踪的配置信息，则使用对象
TracingConfiguration 去初始化 JobTracingEventBus 对象，否则使用默认配置对象
        this.jobTracingEventBus = null == tracingConfig ? new
JobTracingEventBus() : new JobTracingEventBus(tracingConfig);
    }
```

（2）初始化 Job 追踪事件总线类 JobTracingEventBus，具体代码如下所示：

```
    public final class JobTracingEventBus {
        private final ExecutorService executorService;
        private final EventBus eventBus;
    private volatile boolean isRegistered;
        public JobTracingEventBus() {
            executorService = null;
            eventBus = null;
        }

    public JobTracingEventBus(final TracingConfiguration<?> tracingConfig) {
        //①创建线程池，核心线程数为“CPU 处理器个数 ×2”
            executorService =
createExecutorService(Runtime.getRuntime().availableProcessors() * 2);
            //②创建事件总线对象 AsyncEventBus
            eventBus = new AsyncEventBus(executorService);
            //③配置对象 TracingConfiguration
            register(tracingConfig);
        }
    private ExecutorService createExecutorService(final int threadSize) {
```

```
            ThreadPoolExecutor threadPoolExecutor = new
ThreadPoolExecutor(threadSize, threadSize, 5L, TimeUnit.MINUTES,
                    new LinkedBlockingQueue<>(), new
BasicThreadFactory.Builder().namingPattern(String.join("-", "job-event",
"%s")).build());
            threadPoolExecutor.allowCoreThreadTimeOut(true);
            return
MoreExecutors.listeningDecorator(MoreExecutors.getExitingExecutorService(thr
eadPoolExecutor));
        }
        private void register(final TracingConfiguration<?> tracingConfig) {
            try {
                //④向事件总线对象中注册配置对象 TracingConfiguration 中的监听器
                eventBus.register(TracingListenerFactory.
                getListener(tracingConfig));
                //⑤标记配置对象已经注册成功
                isRegistered = true;
            } catch (final TracingConfigurationException ex) {
                log.error("Elastic job: create tracing listener failure, error is:
", ex);
            }
        }
        public void post(final JobEvent event) {
            //⑥如果已经注册成功，则通过事件总线发送 Job 事件 JobEvent
            if (isRegistered && !executorService.isShutdown()) {
                eventBus.post(event);
            }
        }
    }
```

（3）用 TracingListenerFactory 类加载应用所有的监听器接口 TracingListener 的实现类，具体代码如下所示：

```
    @NoArgsConstructor(access = AccessLevel.PRIVATE)
    public final class TracingListenerFactory {
        private static final Map<String, TracingListenerConfiguration>
LISTENER_CONFIGS = new HashMap<>();
        static {
            //①利用静态方法块及类加载器对象 ServiceLoader，去加载应用中所有接口
TracingListenerConfiguration 的实现类，并将其存储在本地缓存对象 LISTENER_CONFIGS 中
            for (TracingListenerConfiguration each :
ServiceLoader.load(TracingListenerConfiguration.class)) {
                LISTENER_CONFIGS.put(each.getType(), each);
            }
        }
        @SuppressWarnings("unchecked")
```

```
    public static TracingListener getListener(final TracingConfiguration
tracingConfig) throws TracingConfigurationException {
        //②从缓存对象 LISTENER_CONFIGS 中获取指定配置对象的监听器,比如,类型为“RDB”,
则对应的追踪监听器对象为 RDBTracingListenerConfiguration
        if (null == tracingConfig.getTracingStorageConfiguration() ||
Strings.isNullOrEmpty(tracingConfig.getType())
|| !LISTENER_CONFIGS.containsKey(tracingConfig.getType())) {
            throw new TracingConfigurationException(String.format("Can not
find executor service handler type '%s'.", tracingConfig.getType()));
        }
        //③返回追踪监听器对象
        return
LISTENER_CONFIGS.get(tracingConfig.getType()).createTracingListener(tracingC
onfig.getTracingStorageConfiguration().getStorage());
    }
}
```

用 RDBTracingListenerConfiguration 类去创建分布式 Job 追踪的事件监听器类 RDBTracingListener，具体代码如下所示：

```
public final class RDBTracingListenerConfiguration implements
TracingListenerConfiguration<DataSource> {
    @Override
    public TracingListener createTracingListener(final DataSource storage)
throws TracingConfigurationException {
        try {
            //①返回 RDBTracingListener 监听器对象，其中，storage 为在配置文件中配置
的数据库数据源
            return new RDBTracingListener(storage);
        } catch (final SQLException ex) {
            throw new TracingConfigurationException(ex);
        }
    }
    //②返回跟踪监听器的类型“RDB”
    @Override
    public String getType() {
        return "RDB";
    }
}
```

经过步骤（1）、步骤（2）和步骤（3）后，Elastic Job 的事件类 JobTracingEventBus、监听器就和 Guava 的事件机制绑定在一块了，这样就可以完成对 Job 的追踪。

2. 在分布式 Job 执行的过程中，发送分布式 Job 的运行状态和日志

在分布式 Job 执行的过程中，会用 LiteJobFacade 类的 postJobStatusTraceEvent()方法发

送 Job 运行的状态，具体代码如下所示：

```
@Slf4j
public final class ElasticJobExecutor {
    //①初始化 LiteJobFacade 对象
    private final JobFacade jobFacade;
    public void execute(){
        ...
        //②记录分布式 Job 运行的状态
        jobFacade.postJobStatusTraceEvent(shardingContexts.getTaskId(),
State.TASK_STAGING, String.format("Job '%s' execute begin.",
jobConfig.getJobName()));
        if (jobFacade.misfireIfRunning(shardingContexts.
getShardingItemParameters().keySet())) {
            //③如果分布式 Job 对应的分片已经在运行，则记录分布式 Job 的运行日志，并标记
为任务结束
            jobFacade.postJobStatusTraceEvent(shardingContexts.getTaskId(),
State.TASK_FINISHED, String.format(
                    "Previous job '%s' - shardingItems '%s' is still running,
misfired job will start after previous job completed.", jobConfig.getJobName(),
            shardingContexts.getShardingItemParameters().keySet()));
            return;
        }
        //④执行分布式 Job
        execute(jobConfig, shardingContexts,
ExecutionSource.NORMAL_TRIGGER);
        ...
    }
    private void execute(final JobConfiguration jobConfig, final
ShardingContexts shardingContexts, final ExecutionSource executionSource) {
        if (shardingContexts.getShardingItemParameters().isEmpty()) {
            //⑤如果分片参数为空，则记录分布式 Job 的运行日志，并标记为任务结束
            jobFacade.postJobStatusTraceEvent(shardingContexts.getTaskId(),
State.TASK_FINISHED, String.format("Sharding item for job '%s' is empty.",
jobConfig.getJobName()));
            return;
        }
        ...
        jobFacade.postJobStatusTraceEvent(taskId, State.TASK_RUNNING, "");
        try {
            process(jobConfig, shardingContexts, executionSource);
        } finally {
            jobFacade.registerJobCompleted(shardingContexts);
            if (itemErrorMessages.isEmpty()) {
                //⑥如果没有错误信息，则记录分布式 Job 的运行日志，并标记为任务结束
```

```
                jobFacade.postJobStatusTraceEvent(taskId,
State.TASK_FINISHED, "");
            } else {
                //⑦否则记录分布式 Job 的运行的日志，并标记为任务失败
                jobFacade.postJobStatusTraceEvent(taskId, State.TASK_ERROR,
itemErrorMessages.toString());
            }
        }
        ...
    }
    ...
}
```

在 Elastic Job 中，在分布式 Job 执行的过程中，会用 JobTracingEventBus 类的 postJobExecutionEvent()方法发送 Job 执行的日志，具体代码如下所示：

```
    //①采用事件溯源机制来追踪分布式 Job 的运行状态
    private void process(final JobConfiguration jobConfig, final
ShardingContexts shardingContexts, final int item, final JobExecutionEvent
startEvent) {
        //②用开始事件 startEvent 去记录分布式 Job 开始执行的日志
        jobFacade.postJobExecutionEvent(startEvent);
        log.trace("Job '{}' executing, item is: '{}'.", jobConfig.getJobName(),
item);
        JobExecutionEvent completeEvent;
        try {
            jobItemExecutor.process(elasticJob, jobConfig, jobFacade,
shardingContexts.createShardingContext(item));
            //③用结束事件 JobExecutionEvent（执行结果为成功）去记录分布式 Job 结束执行的
日志
            completeEvent = startEvent.executionSuccess();
            log.trace("Job '{}' executed, item is: '{}'.", jobConfig.getJobName(),
item);
            jobFacade.postJobExecutionEvent(completeEvent);
    } catch (final Throwable cause) {
            //④用结束事件 JobExecutionEvent（执行结果为失败）去记录分布式 Job 结束执行的
日志
            completeEvent =
startEvent.executionFailure(ExceptionUtils.transform(cause));
            jobFacade.postJobExecutionEvent(completeEvent);
            itemErrorMessages.put(item, ExceptionUtils.transform(cause));
            JobErrorHandler jobErrorHandler =
executorContext.get(JobErrorHandler.class);
            jobErrorHandler.handleException(jobConfig.getJobName(), cause);
        }
    }
```

3. 将追踪分布式 Job 运行状态的事件发送给事件的监听器类

在 Elastic Job 中，用 Guava 的事件机制，将追踪分布式 Job 运行状态的事件发送给事件的监听器类 RDBTracingListener，具体实现方式是，用 LiteJobFacade 类的 postJobStatusTraceEvent()方法发送 Job 运行状态的事件。具体代码如下所示：

```
    @Override
    public void postJobStatusTraceEvent(final String taskId, final State state,
final String message) {
        //①初始化任务上下文对象 TaskContext
        TaskContext taskContext = TaskContext.from(taskId);
        //②调用 JobTracingEventBus 类的 post()方法发送追踪 Job 运行状态的事件
        jobTracingEventBus.post(new JobStatusTraceEvent(taskContext.
getMetaInfo().getJobName(), taskContext.getId(),
            taskContext.getSlaveId(), Source.LITE_EXECUTOR,
taskContext.getType().name(),
taskContext.getMetaInfo().getShardingItems().toString(), state, message));
        if (!Strings.isNullOrEmpty(message)) {
            log.trace(message);
        }
    }
```

用 JobTracingEventBus 类的 post()方法发送分布式 Job 运行的追踪事件，具体代码如下所示：

```
    private final EventBus eventBus;
    public void post(final JobEvent event) {
    if (isRegistered && !executorService.isShutdown()) {
        //用框架 Guava 中 AsyncEventBus 类的 post()方法向监听器类 RDBTracingListener
发送事件
            eventBus.post(event);
        }
    }
```

在 Elastic Job 中，用 LiteJobFacade 类的 postJobExecutionEvent()方法发送分布式 Job 运行的日志溯源事件，具体代码如下所示：

```
    @Override
    public void postJobExecutionEvent(final JobExecutionEvent jobExecutionEvent) {
    //用框架 Guava 中 AsyncEventBus 类的 post()方法向监听器类 RDBTracingListener 发送
Job 运行的日志溯源事件
        jobTracingEventBus.post(jobExecutionEvent);
    }
```

4. 用监听器类 RDBTracingListener 监听分布式 Job 的运行状态和日志

在 Elastic Job 中，用监听器类 RDBTracingListener 监听分布式 Job 的运行状态和日志，具体代码如下所示：

```
public final class RDBTracingListener implements TracingListener {
    private final RDBJobEventStorage repository;
public RDBTracingListener(final DataSource dataSource)
throws SQLException {
    //①初始化持久化事件到数据库的对象 RDBJobEventStorage 中
        repository = new RDBJobEventStorage(dataSource);
    }
    @Override
public void listen(final JobExecutionEvent executionEvent) {
    //②监听事件对象 JobExecutionEvent（分布式 Job 执行的日志），并用
RDBJobEventStorage 对象的 addJobExecutionEvent()方法完成持久化
        repository.addJobExecutionEvent(executionEvent);
    }
    @Override
public void listen(final JobStatusTraceEvent jobStatusTraceEvent) {
    //③监听事件对象 JobStatusTraceEvent（分布式 Job 执行的状态），并用对象
RDBJobEventStorage 的 addJobStatusTraceEvent()方法完成持久化
        repository.addJobStatusTraceEvent(jobStatusTraceEvent);
    }
}
```

11.5.2 用数据库持久化分布式 Job 的运行状态和日志

在 Elastic Job 中，在分布式 Job 的运行状态和日志的事件被监听器监听到后，需要将它们持久化到数据库中。这样，开发人员就可以在 UI 控制台中实时地看到分布式 Job 执行过程的链路追踪。

1. 初始化数据库持久化类 RDBJobEventStorage

（1）在 RDBJobEventStorage 类中，使用 SPI 来加载 Elastic Job 支持的数据库类型（Elastic Job 默认支持 MySQL、Oracle、PostgreSQL 等 7 种数据库），具体代码如下所示：

```
static {
    //①使用 SPI 类 ServiceLoader 的 load()方法加载应用中 DatabaseType 接口的实现类
    for (DatabaseType each : ServiceLoader.load(DatabaseType.class)) {
        DATABASE_TYPES.put(each.getType(), each);
    }
}
//②PostgreSQL 类型的数据库
public final class PostgreSQLDatabaseType implements DatabaseType {
    @Override
public String getType() {
    //③返回数据库类型的名称为“PostgreSQL”
```

```
        return "PostgreSQL";
    }
    @Override
public int getDuplicateRecordErrorCode() {
    //④返回重复后的错误码为 0
        return 0;
    }
}
public final class SQLServerDatabaseType implements DatabaseType {
    @Override
public String getType() {
    //⑤返回数据库类型的名称为“SQLServer”
        return "SQLServer";
    }
    @Override
public String getDatabaseProductName() {
    //⑥返回数据库的产品名称为“Microsoft SQL Server”
        return "Microsoft SQL Server";
    }
    @Override
public int getDuplicateRecordErrorCode() {
    //⑦返回重复后的错误码为 1
        return 1;
    }
}
public final class OracleDatabaseType implements DatabaseType {
    @Override
public String getType() {
    //⑧返回数据库类型的名称为“Oracle”
        return "Oracle";
    }
    @Override
public int getDuplicateRecordErrorCode() {
    //⑨返回重复后的错误码为 1
        return 1;
    }
}
...
```

（2）在 RDBJobEventStorage 类中，从数据源类 DataSource 中解析出数据库的类型，具体代码如下所示：

```
private DatabaseType getDatabaseType(final DataSource dataSource) throws SQLException {
    //①获取数据库的数据库连接对象 Connection
try (Connection connection = dataSource.getConnection()) {
```

```
        //②获取数据库连接元数据中的数据库产品名称
            String databaseProductName =
connection.getMetaData().getDatabaseProductName();
            //③如果在本地数据库类型的缓存中存在匹配数据库产品名称的数据库类型
            for (DatabaseType each : DATABASE_TYPES.values()) {
                //④则直接返回该数据库类型
                if (each.getDatabaseProductName().equals(databaseProductName)) {
                    return each;
                }
            }
    }
    //⑤否则返回默认的数据库类型（默认的数据库类型是 SQL92）
        return new DefaultDatabaseType();
    }
```

（3）在 RDBJobEventStorage 类中，初始化数据库操作类 RDBStorageSQLMapper，具体代码如下所示：

```
    @Getter
    public final class RDBStorageSQLMapper {
        private final String createTableForJobExecutionLog;
        private final String createTableForJobStatusTraceLog;
        private final String createIndexForTaskIdStateIndex;
        private final String insertForJobExecutionLog;
        private final String insertForJobExecutionLogForComplete;
        private final String insertForJobExecutionLogForFailure;
        private final String updateForJobExecutionLog;
        private final String updateForJobExecutionLogForFailure;
        private final String insertForJobStatusTraceLog;
        private final String selectForJobStatusTraceLog;
        private final String selectOriginalTaskIdForJobStatusTraceLog;
        public RDBStorageSQLMapper(final String sqlPropertiesFileName) {
            Properties props = loadProps(sqlPropertiesFileName);
            //①加载持久化分布式 Job 执行日志的建表语句，表名称为“JOB_EXECUTION_LOG”
            createTableForJobExecutionLog =
props.getProperty("JOB_EXECUTION_LOG.TABLE.CREATE");
            //②加载持久化分布式 Job 执行状态的建表语句，表名称为“JOB_STATUS_TRACE_LOG”
            createTableForJobStatusTraceLog =
props.getProperty("JOB_STATUS_TRACE_LOG.TABLE.CREATE");
            //③加载“在表 JOB_STATUS_TRACE_LOG 中创建索引”的 SQL 语句
            createIndexForTaskIdStateIndex =
props.getProperty("TASK_ID_STATE_INDEX.INDEX.CREATE");
            //④加载“表 JOB_EXECUTION_LOG 数据更新”的插入和更新的 SQL 语句
            insertForJobExecutionLog =
props.getProperty("JOB_EXECUTION_LOG.INSERT");
```

```
            insertForJobExecutionLogForComplete =
props.getProperty("JOB_EXECUTION_LOG.INSERT_COMPLETE");
            insertForJobExecutionLogForFailure =
props.getProperty("JOB_EXECUTION_LOG.INSERT_FAILURE");
            updateForJobExecutionLog =
props.getProperty("JOB_EXECUTION_LOG.UPDATE");
            updateForJobExecutionLogForFailure =
props.getProperty("JOB_EXECUTION_LOG.UPDATE_FAILURE");
            insertForJobStatusTraceLog =
props.getProperty("JOB_STATUS_TRACE_LOG.INSERT");
            //⑤加载“查询表 JOB_STATUS_TRACE_LOG 的数据”的 SQL 语句
            selectForJobStatusTraceLog =
props.getProperty("JOB_STATUS_TRACE_LOG.SELECT");
            //⑥加载“查询表 JOB_STATUS_TRACE_LOG 中的原始任务 ID”的 SQL 语句
            selectOriginalTaskIdForJobStatusTraceLog =
props.getProperty("JOB_STATUS_TRACE_LOG.SELECT_ORIGINAL_TASK_ID");
        }
        @SneakyThrows
        private Properties loadProps(final String sqlPropertiesFileName) {
            Properties result = new Properties();
            //⑦从指定类型数据库的属性文件中加载属性，比如 MySQL 的属性文件路径为
“META-INF/sql/MySQL.properties”
            result.load(getPropertiesInputStream(sqlPropertiesFileName));
            return result;
        }
        private InputStream getPropertiesInputStream(final String
sqlPropertiesFileName) {
            //⑧获取属性文件流，如果应用没有配置数据库，则默认使用配置文件
SQL92.properties
            InputStream sqlPropertiesFile =
RDBJobEventStorage.class.getClassLoader().getResourceAsStream(String.format(
"META-INF/sql/%s", sqlPropertiesFileName));
            return null == sqlPropertiesFile ?
RDBJobEventStorage.class.getClassLoader().getResourceAsStream("META-INF/sql/
SQL92.properties") : sqlPropertiesFile;
        }
    }
```

（4）自动创建分布式 Job 运行日志，以及状态的表和索引。

在 Elastic Job 中，不需要手动创建分布式 Job 运行日志和分布式 Job 运行状态的表和索引。在应用启动的过程中会自动创建它们。具体代码如下所示：

```
    //①用 initTablesAndIndexes()方法初始化分布式 Job 的表和索引
    private void initTablesAndIndexes() throws SQLException {
        try (Connection connection = dataSource.getConnection()) {
```

```
            createJobExecutionTableAndIndexIfNeeded(connection);
            createJobStatusTraceTableAndIndexIfNeeded(connection);
        }
    }
    //②用 createJobExecutionTableAndIndexIfNeeded()方法创建分布式 Job 运行日志的表
和索引
    private void createJobExecutionTableAndIndexIfNeeded(final Connection
connection) throws SQLException {
        if (existsTable(connection, TABLE_JOB_EXECUTION_LOG) ||
existsTable(connection, TABLE_JOB_EXECUTION_LOG.toLowerCase())) {
            return;
        }
        createJobExecutionTable(connection);
    }
    //③用 createJobStatusTraceTableAndIndexIfNeeded()方法创建分布式 Job 运行状态的
表和索引
    private void createJobStatusTraceTableAndIndexIfNeeded(final Connection
connection) throws SQLException {
        if (existsTable(connection, TABLE_JOB_STATUS_TRACE_LOG) ||
existsTable(connection, TABLE_JOB_STATUS_TRACE_LOG.toLowerCase())) {
            return;
        }
        createJobStatusTraceTable(connection);
        createTaskIdIndexIfNeeded(connection);
    }
    //④使用数据库连接对象 Connection，去数据库的元数据中心中验证表是否已经存在
    private boolean existsTable(final Connection connection, final String
tableName) throws SQLException {
        DatabaseMetaData dbMetaData = connection.getMetaData();
        try (ResultSet resultSet = dbMetaData.getTables(connection.getCatalog(),
null, tableName, new String[]{"TABLE"})) {
            return resultSet.next();
        }
    }
    //⑤创建分布式 Job 的索引
    private void createTaskIdIndexIfNeeded(final Connection connection) throws
SQLException {
        if (existsIndex(connection, TABLE_JOB_STATUS_TRACE_LOG,
TASK_ID_STATE_INDEX) || existsIndex(connection,
TABLE_JOB_STATUS_TRACE_LOG.toLowerCase(), TASK_ID_STATE_INDEX.toLowerCase())) {
            return;
        }
        createTaskIdAndStateIndex(connection);
    }
    //⑥使用数据库连接对象 Connection，去数据库的元数据中心中验证索引是否已经存在
```

```
    private boolean existsIndex(final Connection connection, final String
tableName, final String indexName) throws SQLException {
        DatabaseMetaData dbMetaData = connection.getMetaData();
        try (ResultSet resultSet =
dbMetaData.getIndexInfo(connection.getCatalog(), null, tableName, false,
false)) {
            while (resultSet.next()) {
                if (indexName.equals(resultSet.getString("INDEX_NAME"))) {
                    return true;
                }
            }
        }
        return false;
    }
    //⑦使用 SQL 预处理对象 PreparedStatement，在数据库中创建分布式 Job 执行日志的表
    private void createJobExecutionTable(final Connection connection) throws
SQLException {
        try (PreparedStatement preparedStatement =
connection.prepareStatement(sqlMapper.getCreateTableForJobExecutionLog())) {
            preparedStatement.execute();
        }
    }
    //⑧使用 SQL 预处理对象 PreparedStatement，在数据库中创建分布式 Job 执行状态的表
    private void createJobStatusTraceTable(final Connection connection) throws
SQLException {
        try (PreparedStatement preparedStatement =
connection.prepareStatement(sqlMapper.getCreateTableForJobStatusTraceLog())) {
            preparedStatement.execute();
        }
    }
    //⑨使用 SQL 预处理对象 PreparedStatement，在数据库中创建分布式 Job 的索引
    private void createTaskIdAndStateIndex(final Connection connection) throws
SQLException {
        try (PreparedStatement preparedStatement =
connection.prepareStatement(sqlMapper.getCreateIndexForTaskIdStateIndex())) {
            preparedStatement.execute();
        }
    }
```

2. 添加分布式 Job 运行日志事件

在 Elastic Job 中，使用数据库持久化类 RDBJobEventStorage 添加分布式 Job 运行日志事件，具体代码如下所示：

```
    //①添加分布式 Job 运行日志事件到数据库中
    public boolean addJobExecutionEvent(final JobExecutionEvent
```

```
jobExecutionEvent) {
        //②如果事件还没完成，则插入一条新的数据
        if (null == jobExecutionEvent.getCompleteTime()) {
            return insertJobExecutionEvent(jobExecutionEvent);
        } else {
            if (jobExecutionEvent.isSuccess()) {
                return updateJobExecutionEventWhenSuccess(jobExecutionEvent);
            } else {
                return updateJobExecutionEventFailure(jobExecutionEvent);
            }
        }
    }
    //③使用 PreparedStatement 对象插入 JobExecutionEvent 对象
    private boolean insertJobExecutionEvent(final JobExecutionEvent
jobExecutionEvent) {
        boolean result = false;
        try (
            Connection connection = dataSource.getConnection();
            PreparedStatement preparedStatement =
connection.prepareStatement(sqlMapper.getInsertForJobExecutionLog())) {
                preparedStatement.setString(1, jobExecutionEvent.getId());
                preparedStatement.setString(2, jobExecutionEvent.getJobName());
                preparedStatement.setString(3, jobExecutionEvent.getTaskId());
                preparedStatement.setString(4,
jobExecutionEvent.getHostname());
                preparedStatement.setString(5, jobExecutionEvent.getIp());
                preparedStatement.setInt(6,
jobExecutionEvent.getShardingItem());
                preparedStatement.setString(7,
jobExecutionEvent.getSource().toString());
                preparedStatement.setBoolean(8, jobExecutionEvent.isSuccess());
                preparedStatement.setTimestamp(9, new
Timestamp(jobExecutionEvent.getStartTime().getTime()));
                preparedStatement.execute();
                result = true;
        } catch (final SQLException ex) {
            if (!isDuplicateRecord(ex)) {
                log.error(ex.getMessage());
            }
        }
        return result;
    }
    //④如果分布式 Job 运行日志事件执行成功，则更新数据库中对应事件的 ID 数据
    private boolean updateJobExecutionEventWhenSuccess(final JobExecutionEvent
jobExecutionEvent) {
```

```
        boolean result = false;
        try (
            Connection connection = dataSource.getConnection();
            PreparedStatement preparedStatement =
connection.prepareStatement(sqlMapper.getUpdateForJobExecutionLog())) {
                preparedStatement.setBoolean(1, jobExecutionEvent.isSuccess());
                //⑤插入事件完成的时间
                preparedStatement.setTimestamp(2, new
Timestamp(jobExecutionEvent.getCompleteTime().getTime()));
                preparedStatement.setString(3, jobExecutionEvent.getId());
                //⑥如果更新失败，则重新插入一条新的数据
                if (0 == preparedStatement.executeUpdate()) {
                return insertJobExecutionEventWhenSuccess(jobExecutionEvent);
            }
            result = true;
        } catch (final SQLException ex) {
            log.error(ex.getMessage());
        }
        return result;
    }
    //⑦如果分布式 Job 运行日志事件执行失败，则更新数据库中对应事件 ID 的数据
    private boolean updateJobExecutionEventFailure(final JobExecutionEvent
jobExecutionEvent) {
        boolean result = false;
        try (
            Connection connection = dataSource.getConnection();
            PreparedStatement preparedStatement = connection.prepareStatement
(sqlMapper.getUpdateForJobExecutionLogForFailure())) {
                preparedStatement.setBoolean(1, jobExecutionEvent.isSuccess());
                preparedStatement.setTimestamp(2, new
Timestamp(jobExecutionEvent.getCompleteTime().getTime()));
                //⑧插入事件执行失败的原因
                preparedStatement.setString(3,
truncateString(jobExecutionEvent.getFailureCause()));
                preparedStatement.setString(4, jobExecutionEvent.getId());
                //⑨如果更新失败，则重新插入一条新的数据
                if (0 == preparedStatement.executeUpdate()) {
                    return insertJobExecutionEventWhenFailure(jobExecutionEvent);
                }
                result = true;
        } catch (final SQLException ex) {
            log.error(ex.getMessage());
        }
        return result;
    }
```

3. 添加分布式 Job 运行状态事件

在 Elastic Job 中，使用数据库持久化类 RDBJobEventStorage 添加分布式 Job 运行状态事件，具体代码如下所示：

```
    //①使用数据库持久化类 RDBJobEventStorage 添加分布式 Job 运行状态事件
    public boolean addJobStatusTraceEvent(final JobStatusTraceEvent
jobStatusTraceEvent) {
        String originalTaskId = jobStatusTraceEvent.getOriginalTaskId();
        if (State.TASK_STAGING != jobStatusTraceEvent.getState()) {
            originalTaskId = getOriginalTaskId(jobStatusTraceEvent.getTaskId());
        }
        boolean result = false;
        try (
            Connection connection = dataSource.getConnection();
            //②使用数据库对象 PreparedStatement，持久化添加分布式 Job 运行状态事件
            PreparedStatement preparedStatement =
connection.prepareStatement(sqlMapper.getInsertForJobStatusTraceLog())) {
                preparedStatement.setString(1, UUID.randomUUID().toString());
                preparedStatement.setString(2,
jobStatusTraceEvent.getJobName());
                preparedStatement.setString(3, originalTaskId);
                preparedStatement.setString(4,
jobStatusTraceEvent.getTaskId());
                preparedStatement.setString(5,
jobStatusTraceEvent.getSlaveId());
                preparedStatement.setString(6,
jobStatusTraceEvent.getSource().toString());
                preparedStatement.setString(7,
jobStatusTraceEvent.getExecutionType());
                preparedStatement.setString(8,
jobStatusTraceEvent.getShardingItems());
                preparedStatement.setString(9,
jobStatusTraceEvent.getState().toString());
                preparedStatement.setString(10,
truncateString(jobStatusTraceEvent.getMessage()));
                preparedStatement.setTimestamp(11, new
Timestamp(jobStatusTraceEvent.getCreationTime().getTime()));
                preparedStatement.execute();
                result = true;
        } catch (final SQLException ex) {
            log.error(ex.getMessage());
        }
        return result;
    }
```

11.5.3 【实例】将 Spring Cloud Alibaba 应用接入 Elastic Job，并开启分布式 Job 的事件追踪

本实例的源码在本书配套资源的“/spring-cloud-alibaba-practice/chaptereleven/elasticjob-Job-trace-spring-cloud-alibaba/”目录下。

在 Elastic Job 模式不会自动开启分布式 Job 的事件追踪，需要手动添加配置信息和数据源。

1. 初始化一个 Spring Cloud Alibaba 应用

使用 Spring Cloud Alibaba 初始化一个服务 elasticjob-Job-trace-spring- cloud-alibaba，并在服务中添加如下配置信息：

```
###①配置分布式 Job 事件追踪的数据源
spring:
   datasource:
   username: root
   password: 123456huxian
   url: jdbc:mysql://127.0.0.1:3306/elasticjob-job?characterEncoding=
utf8&connectTimeout=1000&socketTimeout=3000&autoReconnect=true&useUnicode=tr
ue&useSSL=false&serverTimezone=UTC
driver-class-name: com.mysql.cj.jdbc.Driver
###②配置注册中心及分布式 Job 规则
elasticjob:
  regCenter:
    serverLists: 127.0.0.1:2181
    namespace: spring-cloud-alibaba-job
  jobs:
    elasticTraceMessageJob:
       elasticJobClass:
com.alibaba.cloud.youxia.elasticjob.ElasticTraceMessageJob
       cron: 0/1 * * * * ?
       shardingTotalCount: 1
       overwrite: true
       failover: true
  tracing:
    type: RDB
```

2. 添加 MySQL 的驱动程序包

如图 11-28 所示，在 UI 控制台的安装包的文件“ext-lib”中添加 MySQL 的驱动程序包 mysql-connector-java-8.0.23.jar。

elasticjob-env
elasticjob-shard
apache-shardings...b-3.0.0-lite-ui-bin
logs
lib
ext-lib
mysql-connector-java-8.0.23.jar 2.4 MB
README.txt 1 KB
NOTICE 16 KB
LICENSE 20 KB
licenses
conf
application.properties 1 KB
logback.xml 1 KB

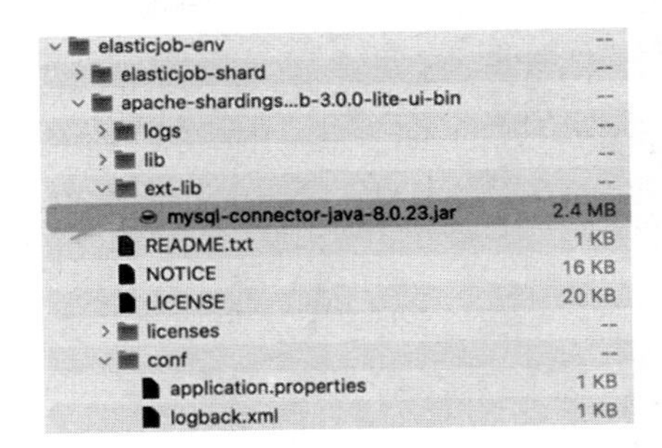

图 11-28

在启动 UI 控制台后，在控制台中添加分布式 Job 的事件追踪的数据源，如图 11-29 所示。

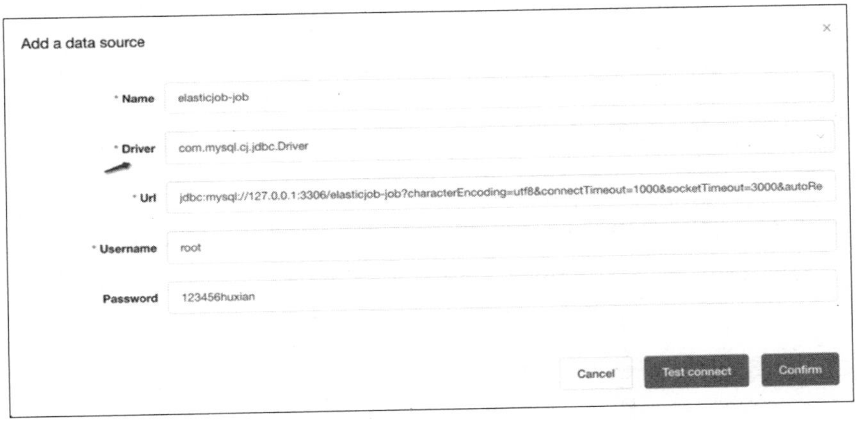

Add a data source
* Name
elasticjob-job
* Driver
com.mysql.cj.jdbc.Driver
* Url
jdbc:mysql://127.0.0.1:3306/elasticjob-job?characterEncoding=utf8&connectTimeout=1000&socketTimeout=3000&autoRe
* Username
root
Password
123456huxian
Cancel
Test connect
Confirm

图 11-29

在添加数据源后，单击数据源连接按钮，UI 控制台就可以从数据源中获取事件追踪的数据，如图 11-30 所示。

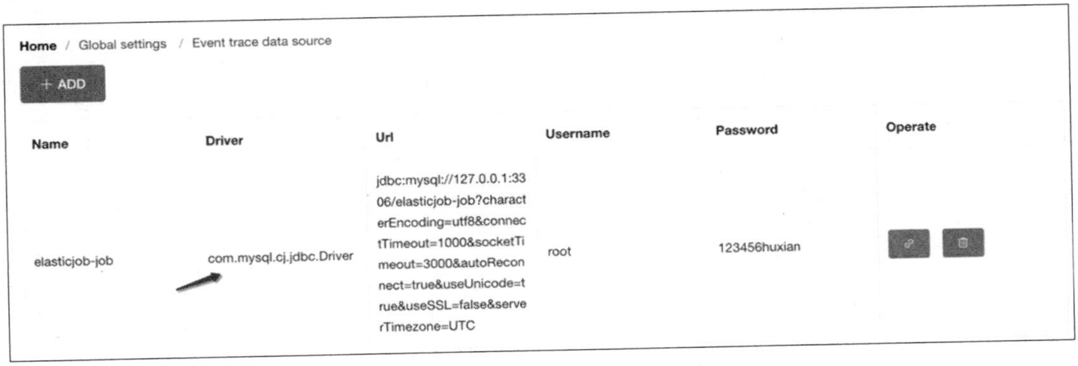

Home / Global settings / Event trace data source
+ ADD
Name
Driver
Url
Username
Password
Operate
elasticjob-job
com.mysql.cj.jdbc.Driver
jdbc:mysql://127.0.0.1:3306/elasticjob-job?characterEncoding=utf8&connectTimeout=1000&socketTimeout=3000&autoReconnect=true&useUnicode=true&useSSL=false&serverTimezone=UTC
root
123456huxian

图 11-30

3. 观察分布式 Job 的追踪事件数据

在启动服务后，定时任务开始执行。在 UI 控制台中单击“Job Trace”可以查看分布式 Job 的执行结果（如图 11-31 所示），可以看到名称为“elasticTraceMessageJob”的分布式 Job 的执行过程。

elasticTraceMessage Job	10.0.24.214	0	false		2021-8-4 17:32:12	1970-1-1 8:0:0
elasticTraceMessage Job	10.0.24.214	0	true		2021-8-4 17:31:46	2021-8-4 17:31:56
elasticTraceMessage Job	10.0.24.214	0	true		2021-8-4 17:31:20	2021-8-4 17:31:30
elasticTraceMessage Job	10.0.24.214	0	true		2021-8-4 17:30:54	2021-8-4 17:31:4
elasticTraceMessage Job	10.0.24.214	0	true		2021-8-4 17:30:28	2021-8-4 17:30:38
elasticTraceMessage Job	10.0.24.214	0	true		2021-8-4 17:30:2	2021-8-4 17:30:12

图 11-31

在 UI 控制台中单击“History status”可以查看分布式 Job 的历史状态（如图 11-32 所示）。在 UI 控制台中可以观察到名称为“elasticTraceMessageJob”的分布式 Job 的历史状态。

elasticTraceMessageJob	[0]	TASK_STAGING	2021-8-4 17:8:34	Job 'elasticTraceMessageJob' execute begin.
elasticTraceMessageJob	[0]	TASK_RUNNING	2021-8-4 17:8:34	

图 11-32

第 12 章

分库分表和读写分离

——基于 ShardingSphere

在微服务架构中，为了确保服务的“高内聚，低耦合”，通常会引入如下技术和架构方案：

- 采用领域驱动模型，从模型层面抽象出不同领域的业务对象。
- 采用 E-R 模型，将领域中的业务对象细化为数据库模型中的实体对象。
- 采用 E-R 模型，将服务拆分成微服务，用 RPC 完成微服务之间的通信。
- 将服务拆分为微服务后，如果需要一些非功能性设计（比如高性能），则可以对数据表进行数据分片（分库分表）。
- 为了提高服务的 TPS，对数据进行读写分离，从而隔离读和写的流量。

在开源社区中有很多成熟的数据分片的中间件框架，ShardingSphere 就是其中之一。

12.1 认识 ShardingSphere

在微服务架构领域，ShardingSphere 并不是分布式数据库，但是开发人员可以把 ShardingSphere 当作分布式数据库。

12.1.1 什么是分布式数据库

在了解“分布式数据库”的概念之前，需要先理解“内存数据库”和“本地数据库”。

1. 什么是内存数据库

当应用利用自身进程中的一个数据容器去存储数据时，那么这个数据容器就是内存数据库。在 Java 语言中，Java 提供的容器类（比如 HashMap、ArrayList 等）都可以作为内存数据库。

如图 12-1 所示，在同一个 JVM 中，Java 程序可以访问内存数据库，完成数据的增加、删除、修改和查询。

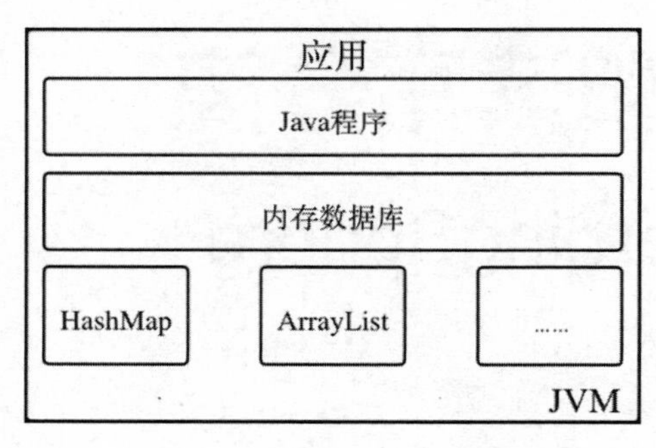

图 12-1

内存数据库的特性：

- 对数据的增加、删除、修改和查询都是基于内存的，所以内存数据库的性能非常好。
- 数据不是持久化的，所以内存数据库不是强一致性的。
- 如果不同的 Java 程序的内存数据库可以同步数据，则可以确保数据的最终一致性。

2. 什么是本地数据库

当应用利用自身部署的节点实例去存储数据时，那么这个节点实例就是本地数据库，比如本地文件。

如图 12-2 所示，在同一个节点实例中，Java 程序通过访问本地数据库，完成数据的增加、删除、修改和查询，并将数据持久化到本地节点实例的文件中。

本地数据库的特性：

- 数据是持久化的。在 Java 程序重新启动后，还可以读取已经持久化的数据。
- 数据的存储是基于单节点实例的，所以本地数据库不是强一致性的。
- 如果在不同的节点实例之间可以同步数据，则可以确保数据的最终一致性。

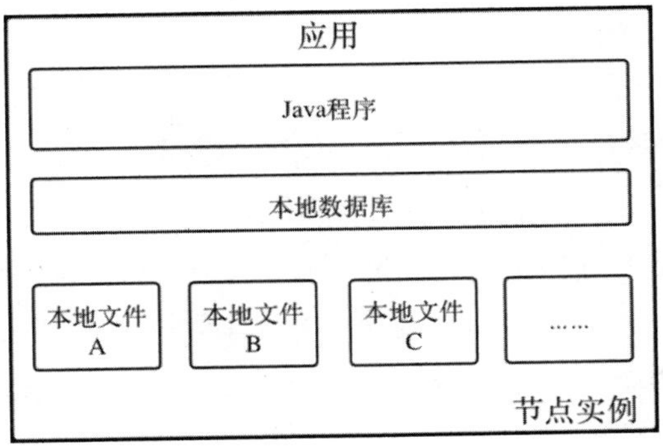

应用

Java程序

本地数据库

本地文件A　本地文件B　本地文件C　……

节点实例

图 12-2

3. 什么是分布式数据库

在应用利用节点实例（本 Java 程序部署节点实例除外）去存储数据时，那么这个节点实例就是一个分布式数据库，所有的应用都可以通过网络访问这个节点实例，并完成数据的增加、删除、修改和查询。

如图 12-3 所示，应用 A、应用 B 和应用 C 访问分布式数据库。在分布式数据库中，分布式数据库是一个包含 3 个节点实例（数据库 D、数据库 E 及数据 F）的集群。在节点之间的数据同步过程中，节点集群会采用分布式选举算法（比如 Raft）选出 Master 节点和 Slave 节点，数据会从 Master 节点往 Slave 节点同步，确保集群数据的一致性。

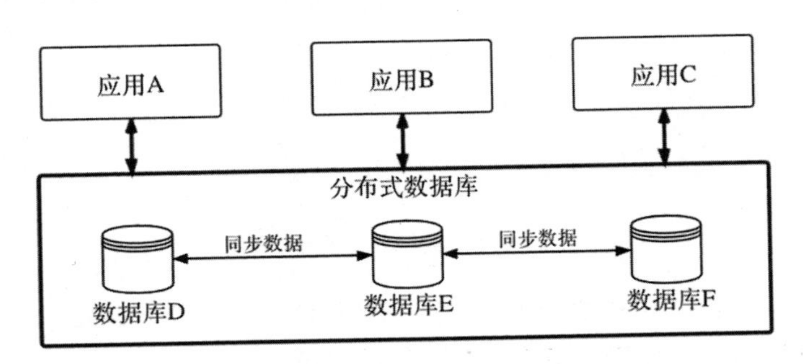

图 12-3

分布式数据库的特性：

- 数据可以存储在网络互通的任意节点实例中，数据通过网络完成传递。
- 数据通常都具备 CAP 理论中的 CP 或者 AP 特性。
- 数据是可以水平扩展的。
- 通常都能满足本地事务和分布式事务的场景。
- 在数据实例节点之间会进行数据同步，以确保数据的最终一致性。

12.1.2　什么是 ShardingSphere

能够同时具备数据水平扩展、分布式事务及分布式治理等功能的中间件并不多。ShardingSphere 是一套开源的分布式数据库解决方案组成的生态圈，它由 ShardingSphere JDBC、ShardingSphere Proxy 和 ShardingSphere Sidecar（规划中）这 3 款既支持独立部署，又支持混合部署配合使用的产品组成。它们均提供标准化的数据水平扩展、分布式事务和分布式治理等功能，可适用于 Java 同构、异构语言、云原生等各种多样化的应用场景。

ShardingSphere 的整体架构如图 12-4 所示，核心功能主要包括数据分片、分布式事务、数据治理、多模式连接及管控界面。

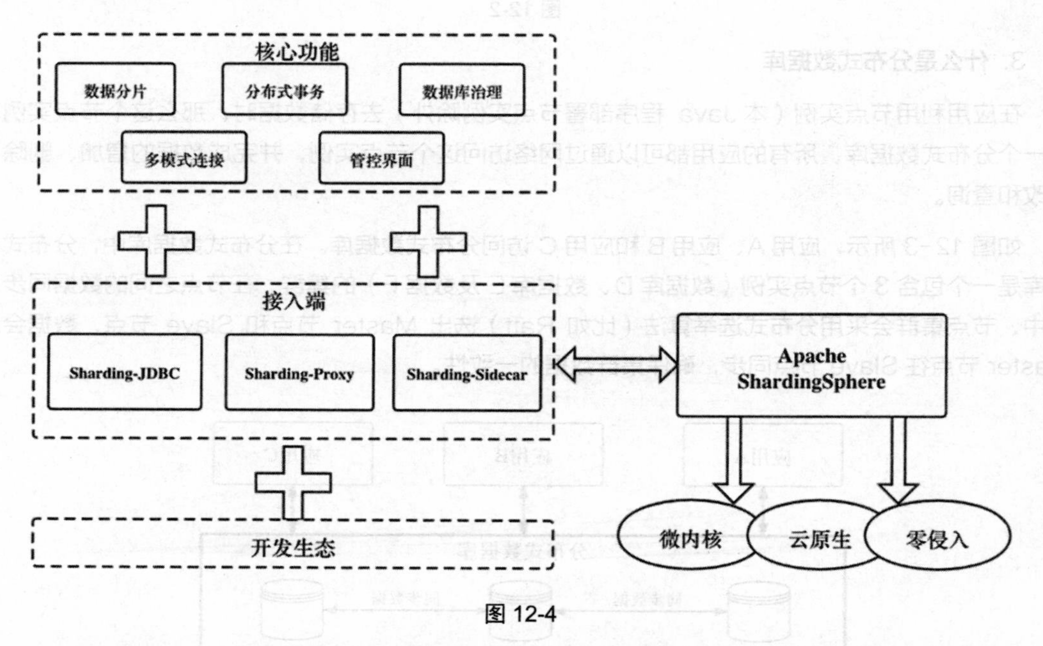

图 12-4

ShardingSphere 旨在分布式的场景下充分、合理地利用关系型数据库的计算和存储能力，而并非实现一个全新的关系型数据库。关系型数据库目前依然占有巨大市场份额，是企业核心系统的基石，未来较长一段时间内也难被撼动。ShardingSphere 更加注重在原有基础上提供一些增量的技术解决方案，而不是取代关系型数据库。

ShardingSphere 从 5.x 版本开始致力于可插拔架构，项目的功能组件能够以灵活的、可以插拔的方式进行扩展，目前支持数据分片、读写分离、数据加密、影子库压测等功能。SharedingSphere 支持 MySQL、PostgreSQL、SQLServer、Oracle 等 SQL 协议，它们均可以通过插件的方式植入项目。开发者能够像搭积木那样定制属于自己的独特系统。ShardingSphere 目前已提供数十个 SPI 作为系统的扩展点，并仍在不断增加中。ShardingSphere 已于 2020 年 4 月 16 日成为

Apache 软件基金会的顶级项目。

1. 认识 ShardingSphere JDBC

ShardingSphere JDBC 是一个轻量级的 Java 框架，它提供了一些 Java 的 JDBC 没有的功能（比如数据水平扩展、分布式事务和分布式治理等）。从功能的角度，可以将 ShardingSphere JDBC 理解为增强版的 JDBC，它能够完全兼容 JDBC 和各种 ORM 框架。

ShardingSphere JDBC 的整体架构如图 12-5 所示。

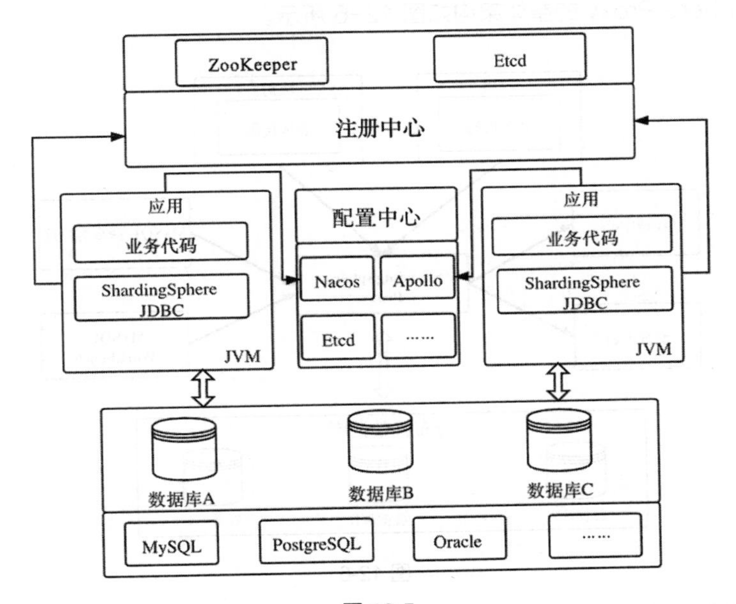

图 12-5

（1）应用要接入 ShardingSphere，需要依赖 ShardingSphere JDBC 的 SDK，并侵入业务代码。ShardingSphere JDBC 的 SDK 与应用的业务代码在同一个 JVM 进程中。

（2）ShardingSphere JDBC 会使用数据库客户端去连接分布式数据库（比如 MySQL、PostgreSQL 等）。

（3）ShardingSphere JDBC 会用注册中心（比如 ZooKeeper 和 Etcd 等）来完成分布式治理功能。

（4）ShardingSphere JDBC 会用配置中心（比如 Nacos、Apollo 及 Etcd 等）动态地加载分库分表中的配置信息。

（5）应用会强依赖 ShardingSphere JDBC 的 SDK。应用在升级 SDK 的版本时，需要考虑 SDK 的版本的兼容性。

2. 认识 ShardingSphere Proxy

ShardingSphere Proxy 定位为透明化的数据库代理。ShardingSphere Proxy 封装了一个数据库二进制协议的服务器端版本，用于完成对异构语言的支持（比如 Java、Go 等）。

目前 ShardingSphere Proxy 提供了 MySQL 和 PostgreSQL 的版本，可以使用任何兼容 MySQL/PostgreSQL 协议的访问客户端（如 MySQL Command Client、MySQL Workbench, Navicat 等）操作数据，并对开发人员和 DBA 更加友好。

ShardingSphere Proxy 的整体架构如图 12-6 所示。

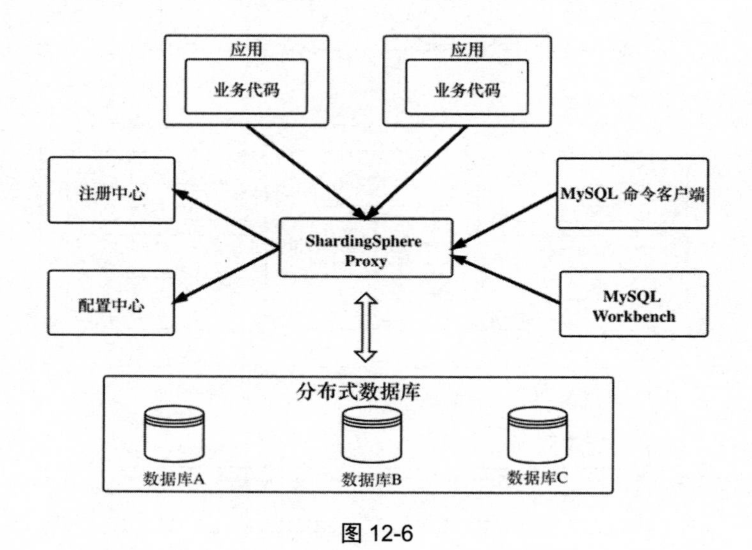

图 12-6

（1）应用（业务代码）不会直接连接分布式数据库，而是连接 ShardingSphere Proxy。

（2）用 ShardingSphere Proxy 去连接分布式数据库（比如数据库 A、数据库 B 及数据库 C），则需要将要执行的 SQL 语句通过数据客户端发送给分布式数据库。

（3）软件开发人员可以通过 MySQL 命令客户端或者第三方可视化控制台工具 MySQL Workbench 连接 ShardingSphere Proxy，从而维护分布式数据库。

（4）ShardingSphere Proxy 支持用注册中心来完成分布式治理功能，同时也支持用配置中心来动态地加载分库分表的配置信息。

3. 认识 ShardingSphere Sidecar

ShardingSphere Sidecar 被定位为 Kubernetes 的云原生数据库的代理，以 Sidecar（边车）的形式代理所有对数据库的访问。

ShardingSphere Sidecar 提供无中心和零侵入的接入模式，并提供应用与数据库交互的啮合层（即 Database Mesh，又被称为数据库网格）。ShardingSphere Sidecar 正在规划中，官方还没有落地。图 12-7 所示为 ShardingSphere Sidecar 的整体架构。

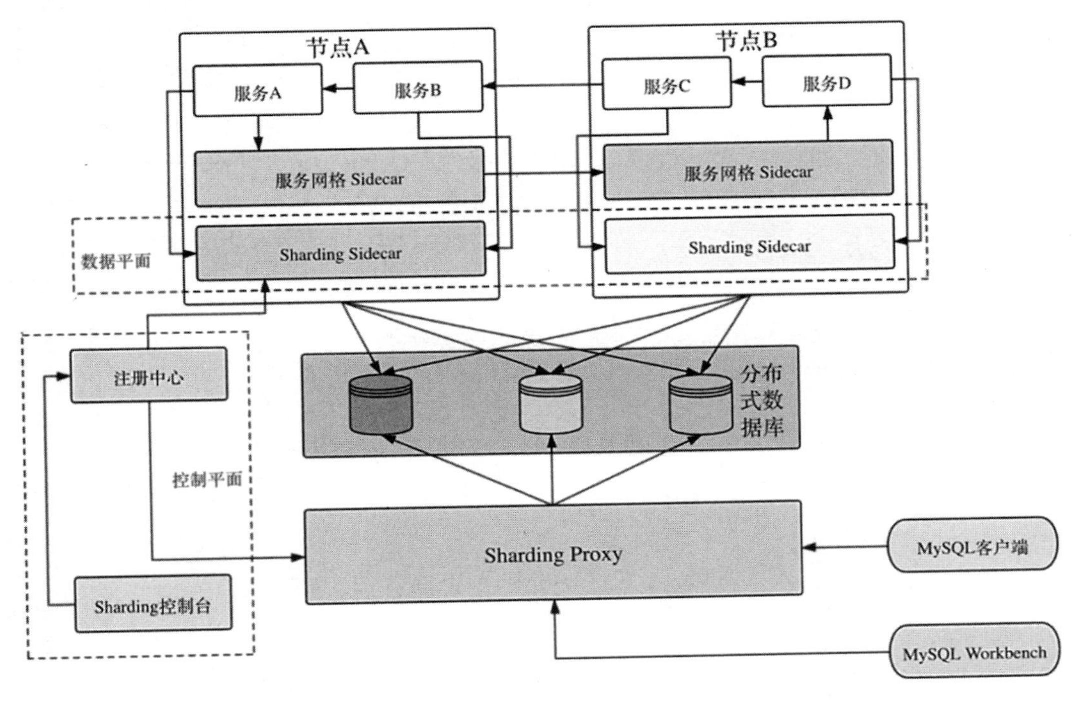

图 12-7

（1）在节点 A 中，包含服务 A 和服务 B，在节点 B 中包含服务 C 和服务 D。在服务 A、服务 B、服务 C、服务 D 之间通过服务网格 Sidecar 来完成 RPC 通信。

（2）在节点 A 中，服务 A 和服务 B 接入 ShardingSphere Sidecar；在节点 B 中，服务 C 和服务 D 接入 ShardingSphere Sidecar。

（3）节点 A 和节点 B 中的 ShardingSphere Sidecar 一起组成服务网格的数据平面，用于统一地调节和控制微服务之间的 RPC 通信（比如服务 A、服务 B、服务 C 及服务 D）

（4）在节点 A 和节点 B 中，通过 ShardingSphere Sidecar 连接分布式数据库。

（5）ShardingSphere Sidecar 和 Sharding Proxy 用注册中心完成分布式服务治理。

（6）在 ShardingSphere Sidecar 中，用注册中心和 Sharding 控制台组成服务网格的控制平面，用于管理服务注册/发现、服务负载均衡、请求路由及故障处理等。

4. 对比 ShardingSphere JDBC、ShardingSphere Proxy 和 ShardingSphere Sidecar

三者的对比见表 12-1。

表 12-1　ShardingSphere JDBC、ShardingSphere Proxy 和 ShardingSphere Sidecar 的对比

对比项	ShardingSphere JDBC	ShardingSphere Proxy	ShardingSphere Sidecar
数据库	任意数据库	MySQL/PostgreSQL	MySQL/PostgreSQL
连接消耗数	高	低	高
性能	损耗低	损耗略高	损耗低
无中心化	是	否	是
静态入口	无	有	无
异构语言	仅 Java	任意语言	任意语言

12.2　将应用接入 ShardingSphere JDBC

ShardingSphere JDBC 是一个轻量级的 Java 框架，需要采用技术手段将应用接入其中。目前 ShardingSphere JDBC 支持四种接入模式：Java API、Yaml、Spring Boot 及 Spring Framework 的命令空间。

12.2.1　用四种模式将应用接入 Shardingsphere JDBC

下面用 Java API、Yaml、Spring Boot 及 Spring Framework 的命名空间这四种模式将应用接入 Shardingsphere JDBC。

1. Java API

如果用 Java API 将应用接入 ShardingSphere JDBC 中，则开发人员需要在应用中定义数据分片配置对象 org.apache.shardingsphere.sharding.api.config.ShardingRuleConfiguration，具体代码如下所示：

```
   @Getter
   @Setter
   public final class ShardingRuleConfiguration implements RuleConfiguration {
   //①分片表规则列表
   private Collection<ShardingTableRuleConfiguration> tables = new
LinkedList<>();
   //②自动化分片表规则列表
   private Collection<ShardingAutoTableRuleConfiguration> autoTables = new
LinkedList<>();
   //③绑定表规则列表
   private Collection<String> bindingTableGroups = new LinkedList<>();
```

```
    //④广播表规则列表
    private Collection<String> broadcastTables = new LinkedList<>();
    //⑤默认分库策略，默认不分片
    private ShardingStrategyConfiguration defaultDatabaseShardingStrategy;
    //⑥默认分表策略，默认不分片
    private ShardingStrategyConfiguration defaultTableShardingStrategy;
    //⑦默认自增列生成器配置，默认值是雪花算法
    private KeyGenerateStrategyConfiguration defaultKeyGenerateStrategy;
    //⑧分片算法的名称和配置信息
    private Map<String, ShardingSphereAlgorithmConfiguration>
shardingAlgorithms = new LinkedHashMap<>();
    //⑨自增列生成算法的名称和配置信息
    private Map<String, ShardingSphereAlgorithmConfiguration> keyGenerators =
new LinkedHashMap<>();
    }
```

2. Yaml

如果用 Yaml 将应用接入 ShardingSphere JDBC，则开发人员需要在应用中定义数据分片配置对象 org.apache.shardingsphere.sharding.yaml.config.YamlShardingRuleConfiguration，具体代码如下所示：

```
    @Getter
    @Setter
    public final class YamlShardingRuleConfiguration implements
YamlRuleConfiguration {
        //①分片表规则列表
        private Map<String, YamlTableRuleConfiguration> tables = new
LinkedHashMap<>();
        //②自动化分片表规则列表
        private Map<String, YamlShardingAutoTableRuleConfiguration> autoTables
= new LinkedHashMap<>();
        //③绑定表规则列表
        private Collection<String> bindingTables = new ArrayList<>();
        //④广播表规则列表
        private Collection<String> broadcastTables = new ArrayList<>();
        //⑤默认分库策略，默认不分片
        private YamlShardingStrategyConfiguration defaultDatabaseStrategy;
        //⑥默认分表策略，默认不分片
        private YamlShardingStrategyConfiguration defaultTableStrategy;
        //⑦默认自增列生成器配置，默认值是雪花算法
        private YamlKeyGenerateStrategyConfiguration
defaultKeyGenerateStrategy;
        //⑧分片算法的名称和配置信息
        private Map<String, YamlShardingSphereAlgorithmConfiguration>
shardingAlgorithms = new LinkedHashMap<>();
```

```
        //⑨自增列生成算法的名称和配置信息
        private Map<String, YamlShardingSphereAlgorithmConfiguration>
keyGenerators = new LinkedHashMap<>();
        //⑩返回规则配置类型对象 ShardingRuleConfiguration
        @Override
        public Class<ShardingRuleConfiguration> getRuleConfigurationType() {
            return ShardingRuleConfiguration.class;
        }
    }
```

3. Spring Boot

如果用 Spring Boot 将应用接入 ShardingSphere JDBC，则需要利用 Spring Boot 的自动装配的原理去加载数据分片的配置信息，具体代码如下所示：

```
    @Configuration
    //①用注解@ComponentScan 扫描指定包路径下的 Bean
    @ComponentScan("org.apache.shardingsphere.spring.boot.converter")
    //②加载 SpringBootPropertiesConfiguration 对象中的属性值
    @EnableConfigurationProperties(SpringBootPropertiesConfiguration.class)
    //③用开关 spring.shardingsphere.enabled 控制是否开启 ShardingSphere JDBC，默认为 true
    @ConditionalOnProperty(prefix = "spring.shardingsphere", name = "enabled",
havingValue = "true", matchIfMissing = true)
    //④用条件注解@AutoConfigureBefore 控制 IOC 容器加载 SpringBootConfiguration 类的优先级， SpringBootConfiguration 类加载的优先级要高于 DataSourceAutoConfiguration 类
    @AutoConfigureBefore(DataSourceAutoConfiguration.class)
    @RequiredArgsConstr
    public class SpringBootConfiguration implements EnvironmentAware {
        ...
    }
```

4. Spring Framework 的命名空间

如果用 Spring Framework 的命名空间将应用接入 ShardingSphere JDBC，则需要利用 Spring Framework 的 spring.schemas 和 spring.handlers 文件去加载数据分片的配置信息。

如图 12-8 所示，在 ShardingSphere JDBC 的工程“shardingsphere-sharding-spring-namespace”中，定义了一个分片的规格文件 sharding.xsd。

如图 12-9 所示，在 ShardingSphere JDBC 的工程“shardingsphere-jdbc-core-spring-namespace”中，定义了一个数据源的规格文件 datasource.xsd。

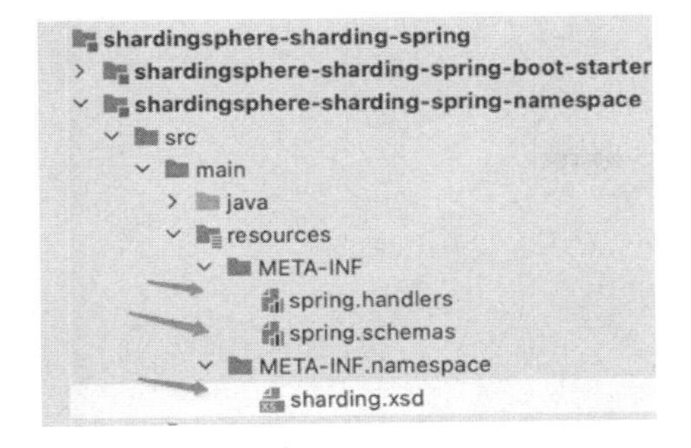

图 12-8

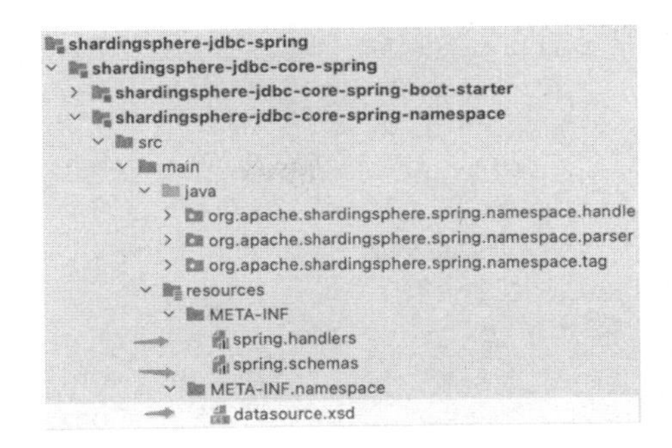

图 12-9

如果读者想了解规格文件“sharding.xsd”和“datasource.xsd”的具体配置信息，可以分别参考工程“shardingsphere-sharding-spring-namespace”中的文件“resources/META/namespace/sharding.xsd”和工程“shardingsphere-jdbc-core-spring-namespace”中的文件“resources/META/namespace/ datasource.xsd”。

12.2.2 【实例】用 Spring Boot 将应用接入 Shardingsphere JDBC 并完成分库分表

本实例的源码在本书配套资源的“/spring-cloud-alibaba-practice/chaptertwelve/shardingsphere-jdbc-use-spring-boot/”和“/spring-cloud-alibaba-practice/chaptertwelve/shardingsphere-spring-boot-mybatis-plus/”目录下。

本实例使用 Spring Boot 作为基础框架将应用接入 Shardingsphere JDBC。其中包含两个服务：shardingsphere-jdbc-use-spring-boot 和 shardingsphere-spring-boot -mybatis-plus。

1. 准备分库分表的数据库和数据库表

可以利用 Shardingsphere JDBC 完成对数据的分库、分库分表及分表。

（1）准备 SQL 脚本，具体脚本信息如下：

```
//①初始化收货地址表
CREATE TABLE `t_address` (
  `id` bigint NOT NULL AUTO_INCREMENT,
  `address_name` varchar(50) DEFAULT NULL,
  `address_id` bigint DEFAULT NULL,
  `is_deleted` int DEFAULT NULL,
```

```
  `gmt_create` datetime DEFAULT NULL,
  `gmt_modified` datetime DEFAULT NULL,
  PRIMARY KEY (`id`)
) ENGINE=InnoDB AUTO_INCREMENT=701971395235167233 DEFAULT CHARSET=utf8mb4
COLLATE=utf8mb4_0900_ai_ci
//②初始化订单表
CREATE TABLE `t_order` (
  `id` bigint DEFAULT NULL,
  `user_id` bigint DEFAULT NULL,
  `order_name` varchar(30) DEFAULT NULL,
  `address_id` bigint DEFAULT NULL,
  `status` int DEFAULT NULL,
  `is_deleted` int DEFAULT NULL,
  `gmt_create` datetime DEFAULT NULL,
  `gmt_modified` datetime DEFAULT NULL,
  `order_id` bigint DEFAULT NULL
) ENGINE=InnoDB DEFAULT CHARSET=utf8mb4 COLLATE=utf8mb4_0900_ai_ci
//③初始化订单子表
CREATE TABLE `t_order_item` (
  `id` bigint NOT NULL AUTO_INCREMENT,
  `order_id` bigint DEFAULT NULL,
  `user_id` bigint NOT NULL,
  `status` int NOT NULL DEFAULT '0',
  `good_id` bigint DEFAULT NULL,
  `order_item_id` bigint DEFAULT NULL,
  `is_deleted` int DEFAULT NULL,
  `gmt_create` datetime DEFAULT NULL,
  `gmt_modified` datetime DEFAULT NULL,
  PRIMARY KEY (`id`)
) ENGINE=InnoDB AUTO_INCREMENT=910927533569899393 DEFAULT CHARSET=utf8mb4
COLLATE=utf8mb4_0900_ai_ci
```

（2）准备分库、分库分表和分表的数据源。

分库和分库分表的数据源的脚本信息如下：

```
CREATE DATABASE `datasource_0`
CREATE DATABASE `datasource_1`
CREATE DATABASE `datasource_2`
```

分表的数据源的脚本信息如下：

```
CREATE DATABASE `ds`
```

在数据源 datasource_0、datasource_1 和 datasource_2 中创建分库和分库分表的数据表结构，如图 12-10 所示。在数据源 ds 中创建分表的数据表结构，如图 12-11 所示。

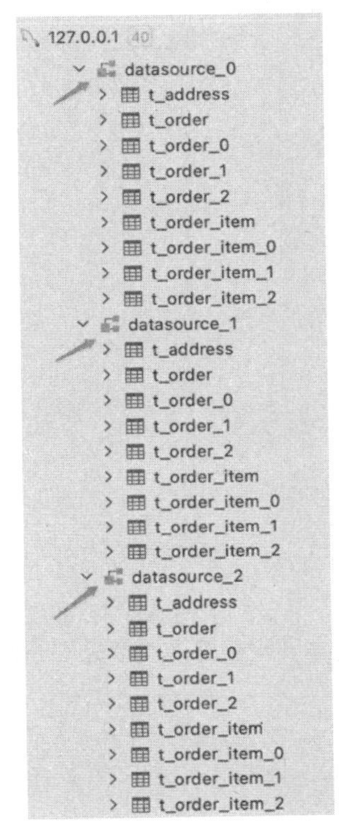

图 12-10

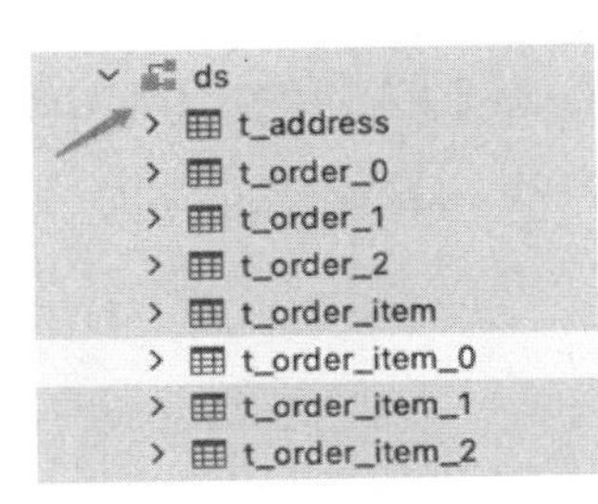

图 12-11

2. 初始化服务 shardingsphere-spring-boot-mybatis-plus

用 Spring Boot 基础框架快速初始化服务，并添加如下 POM 依赖：

```
<!—省略部分 POM 依赖-->
<dependency>
   <groupId>com.baomidou</groupId>
   <artifactId>mybatis-plus-boot-starter</artifactId>
   <version>3.4.2</version>
</dependency>
<dependency>
   <groupId>org.mybatis</groupId>
   <artifactId>mybatis</artifactId>
   <version>3.5.6</version>
</dependency>
<dependency>
   <groupId>org.mybatis</groupId>
   <artifactId>mybatis-spring</artifactId>
   <version>2.0.6</version>
</dependency>
```

在服务中，添加一个操作数据库的接口 AddressMapper 接口和配置文件 AddressMapper.xml，具体代码如下所示：

```
    //①基于 MyBatis-Plus 的数据库操作接口
    @Mapper
    public interface AddressMapper extends BaseMapper<Address> {
    }
    //②数据库操作接口对应的配置文件 AddressMapper.xml
    <?xml version="1.0" encoding="UTF-8" ?>
    <!DOCTYPE mapper PUBLIC "-//mybatis.org//DTD Mapper 3.0//EN"
"http://mybatis.org/dtd/mybatis-3-mapper.dtd">
    <mapper namespace="com.alibaba.cloud.youxia.mapper.AddressMapper">
    </mapper>
```

3. 初始化服务 shardingsphere-jdbc-use-spring-boot

用 Spring Boot 快速初始化服务 shardingsphere-spring-boot-mybatis-plus 并添加 POM 依赖，详细的配置信息可以参考本书配套资源中的代码。

在服务中添加 OrderManager 和 OrderController 类，主要用于插入需要分库或者分表的数据，具体代码如下所示：

```
    @Component
    public class OrderManager {
        @Resource
        private SnowFlakeService snowFlakeService;
        @Autowired
        private OrderMapper orderMapper;
        @Autowired
        private OrderItemMapper orderItemMapper;
        @Autowired
        private AddressMapper addressMapper;
    public synchronized Integer insertOrderInfo() {
        //①创建一个订单对象
            final Order orderEntity = new Order();
            orderEntity.setOrderName("test"+ snowFlakeService.nextId());
            orderEntity.setId(snowFlakeService.nextId()+RandomUtils.nextInt
(1,3));
            orderEntity.setGmtCreate(new Date());
            orderEntity.setGmtModified(new Date());
            orderEntity.setOrderId(snowFlakeService.nextId()+RandomUtils.
nextInt(1,3));
            orderEntity.setIsDeleted(0);
            orderEntity.setAddressId(snowFlakeService.nextId()+RandomUtils.
nextInt(1,3));
```

```
        orderEntity.setUserId(snowFlakeService.nextId()+RandomUtils.
nextInt(1,3));
        orderEntity.setStatus(0);
        //②向已经完成分库、分库分表及分表的数据库中插入订单对象数据
        Integer orderResult = orderMapper.insert(orderEntity);
        //③创建一个订单子对象
        final OrderItem orderItem = new OrderItem();
        orderItem.setId(snowFlakeService.nextId()+RandomUtils.nextInt
(1,3));
        orderItem.setGmtCreate(new Date());
        orderItem.setGmtModified(new Date());
        orderItem.setOrderItemId(snowFlakeService.nextId()+RandomUtils.
nextInt(1,3));
        orderItem.setUserId(orderEntity.getUserId());
        orderItem.setIsDeleted(0);
        orderItem.setOrderId(orderEntity.getOrderId());
        orderItem.setGoodId(snowFlakeService.nextId()+RandomUtils.
nextInt(1,3));
        orderItem.setStatus(0);
        //④向已经完成分库、分库分表及分表的数据库中插入订单子对象数据
        Integer orderItemResult = orderItemMapper.insert(orderItem);
        //⑤创建一个收货地址对象
        final Address address=new Address();
        address.setId(snowFlakeService.nextId()+RandomUtils.nextInt(1,3));
        address.setAddressId(orderEntity.getAddressId());
        address.setAddressName("test-address"+snowFlakeService.nextId());
        address.setGmtCreate(new Date());
        address.setGmtModified(new Date());
        address.setIsDeleted(0);
        //⑥向已经完成分库、分库分表及分表的数据库中插入收货地址对象数据
        Integer addressInsertResult=addressMapper.insert(address);
        return orderResult+orderItemResult+addressInsertResult;
    }
    //⑦查询订单对象
      public synchronized List<Order> selectOrderInfo() {
        List<Order> itemList=orderMapper.selectList(null);
        return itemList;
      }
    }
```

在服务中添加分库、分库分表及分表的配置信息，具体如下所示：

（1）在文件 application-sharding-databases.properties 中添加分库的配置信息，具体配置信息可以参考本书配套资源中的代码。

（2）在文件 application-sharding-databases-tables.properties 中添加分库分表的配置信息，具体配置信息可以参考本书配套资源中的代码。

（3）在文件 application-sharding-tables.properties 中添加分库分表的配置信息，具体配置信息可以参考本书配套资源中的代码。

4. 插入数据，验证分库策略

在服务 shardingsphere-jdbc-use-spring-boot 中，如果数据库的分库策略生效，则分片算法“datasource_$->{order_id % 3}”也会被执行。

（1）在文件 application.properties 中，修改如下配置信息：

```
#将服务的环境配置信息切换到分库策略的配置信息
spring.profiles.active=sharding-databases
```

（2）启动服务 shardingsphere-jdbc-use-spring-boot。在启动后，执行命令“curl 127.0.0.1:34998/order/insert”4 次。如果分库策略生效，则会在分库后的 t_order、t_order_item 表及 t_address 表中插入 4 条数据记录。

（3）通过观察数据库中数据的存储，可以发现分库策略已经生效。

①观察数据源 datasource_0 中的 t_order 表、t_order_item 表及 t_address 表的数据。

- 在 datasource_0.t_order 表中插入了 1 条数据，如图 12-12 所示，其中“order_id”的值为“611687225444691969”，满足分库策略“datasource_$->{order_id % 3}”（611687225444691969 %3 等于 0），所以数据被存储在 datasource_0.t_order 表中。

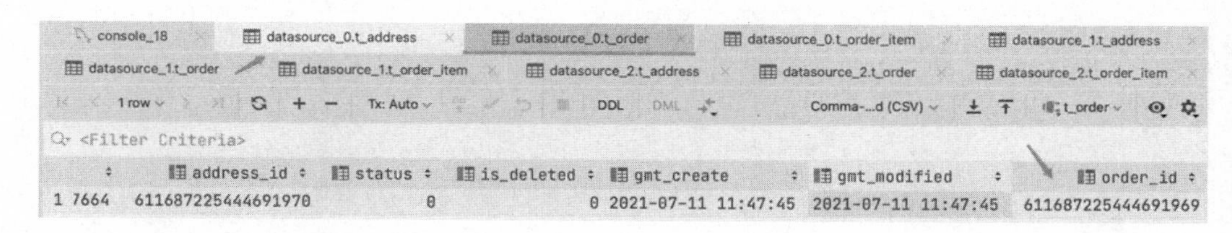

图 12-12

- 在 datasource_0.t_order_item 表中插入了 1 条数据，如图 12-13 所示，其中“order_id”的值为“611687225444691969”，满足分库策略“datasource_$->{order_id % 3}”（611687225444691969 %3 等于 0），所以数据被存储在 datasource_0.t_order_item 表中。

	id	order_id	user_id	status	good_id	order_item_id
1	611687225662795778	611687225444691969	611687225444691972	0	611687225662795780	611687225662795779

图 12-13

- 在 datasource_0.t_address 表中插入了 4 条收货地址对象数据(t_address 表是一个广播表，如图 12-14 所示。如果执行了 4 次命令，则会在 datasource_0.t_address 表中插入 4 条数据）。

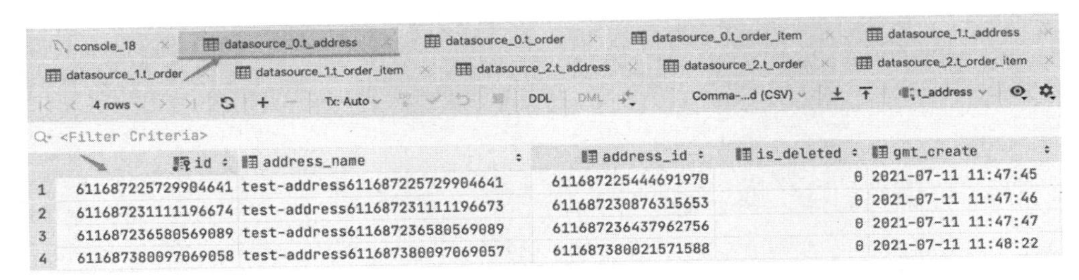

	id	address_name	address_id	is_deleted	gmt_create
1	611687225729904641	test-address611687225729904641	611687225444691970	0	2021-07-11 11:47:45
2	611687231111196674	test-address611687231111196673	611687230876315653	0	2021-07-11 11:47:46
3	611687236580569089	test-address611687236580569089	611687236437962756	0	2021-07-11 11:47:47
4	611687380097069058	test-address611687380097069057	611687380021571588	0	2021-07-11 11:48:22

图 12-14

②观察数据源 datasource_1 中的 t_order 表、t_order_item 表及 t_address 表的数据。

- 在 datasource_1.t_order 表中插入了 2 条数据，如图 12-15 所示，其中“order_id”的值为“611687236437962755”和“611687230876315651”，满足分库策略（611687236437962755%3 等于 1，611687230876315651%3 等于 1），所以数据被存储在 datasource_1.t_order 表中。

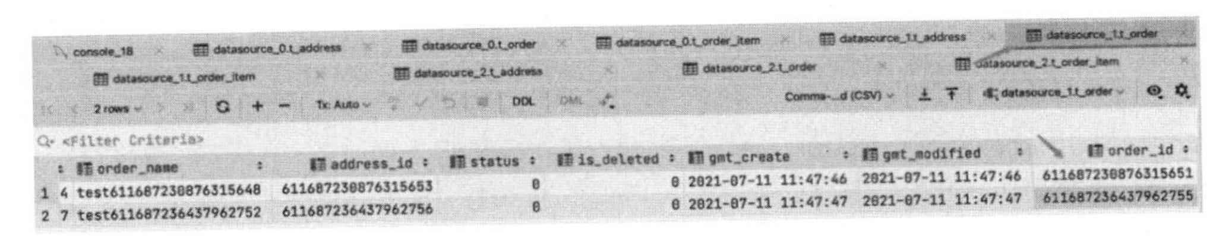

		order_name	address_id	status	is_deleted	gmt_create	gmt_modified	order_id
1	4	test611687230876315648	611687230876315653	0	0	2021-07-11 11:47:46	2021-07-11 11:47:46	611687230876315651
2	7	test611687236437962752	611687236437962756	0	0	2021-07-11 11:47:47	2021-07-11 11:47:47	611687236437962755

图 12-15

- 在 datasource_1.t_order_item 表中插入了 2 条数据，如图 12-16 所示，其中“order_id”的值为“611687236437962755”和“611687230876315651”，满足分库策略（611687236437962755%3 等于 1，611687230876315651%3 等于 1），所以数据被存储在 datasource_1. t_order_item 表中。

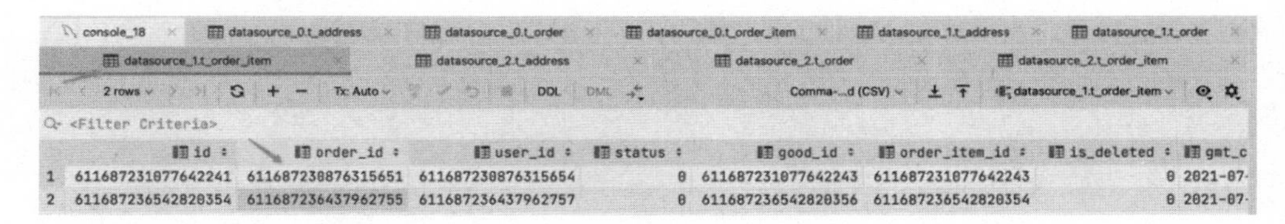

	id	order_id	user_id	status	good_id	order_item_id	is_deleted	gmt_c
1	611687231077642241	611687230876315651	611687230876315654	0	611687231077642243	611687231077642243	0	2021-07-
2	611687236542820354	611687236437962755	611687236437962757	0	611687236542820356	611687236542820354	0	2021-07-

图 12-16

- 在 datasource_1.t_address 表中插入了 4 条收货地址对象数据(t_address 表是一个广播表，如图 12-17 所示。如果执行了 4 次命令，则会在 datasource_1.t_address 表中插入 4 条数据）。

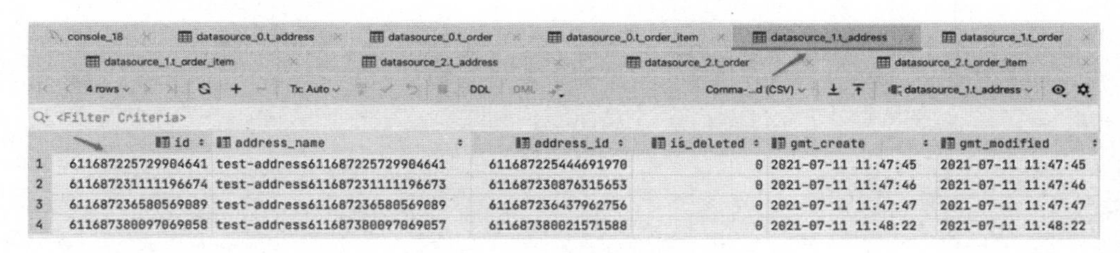

	id	address_name	address_id	is_deleted	gmt_create	gmt_modified
1	611687225729904641	test-address611687225729904641	611687225444691970	0	2021-07-11 11:47:45	2021-07-11 11:47:45
2	611687231111196674	test-address611687231111196673	611687230876315653	0	2021-07-11 11:47:46	2021-07-11 11:47:46
3	611687236580569089	test-address611687236580569089	611687236437962756	0	2021-07-11 11:47:47	2021-07-11 11:47:47
4	611687380097069058	test-address611687380097069057	611687380021571588	0	2021-07-11 11:48:22	2021-07-11 11:48:22

图 12-17

③观察数据源 datasource_2 中的 t_order 表、t_order_item 表及 t_address 表的数据。

- 在 datasource_2.t_order 表中插入了 2 条数据，如图 12-18 所示，其中“order_id”的值为“611687380021571588”，满足分库策略(611687380021571588%3 等于 2)，所以数据被存储在 datasource_2.t_order 表中。

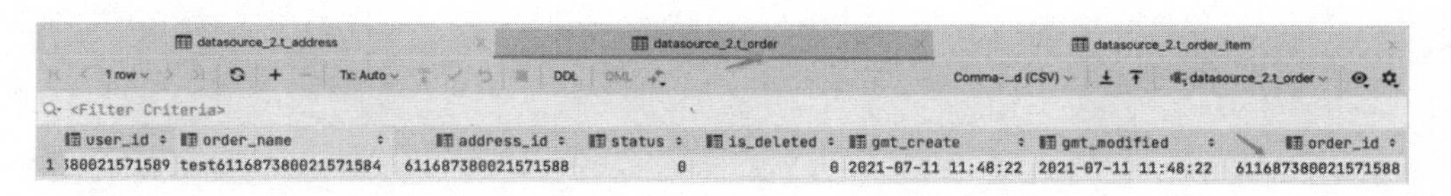

	user_id	order_name	address_id	status	is_deleted	gmt_create	gmt_modified	order_id
1	380021571589	test611687380021571584	611687380021571588	0	0	2021-07-11 11:48:22	2021-07-11 11:48:22	611687380021571588

图 12-18

- 在 datasource_2.t_order_item 表中插入了 1 条数据，如图 12-19 所示，其中“order_id”的值为“611687380021571588”，满足分库策略“datasource_$->{order_id % 3}”(611687380021571588%3 等于 2)，所以数据被存储在 datasource_2.t_order_item 表中。

	id	order_id	user_id	status	good_id	order_item_id	is_deleted	gmt_create	
1	611687380071903234	611687380021571588	611687380021571589	0	611687380071903235	611687380071903234	0	2021-07-11 11:48:22	20

图 12-19

- 在 datasource_2.t_address 表中插入了 4 条收货地址对象数据(t_address 表是一个广播表，如图 12-20 所示。如果执行了 4 次命令，则会在 datasource_2.t_address 表中插入 4 条数据）。

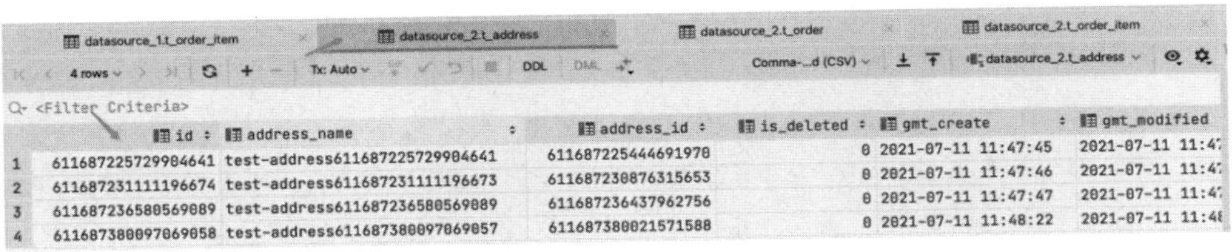

datasource_1.t_order_item　datasource_2.t_address　datasource_2.t_order　datasource_2.t_order_item

4 rows　Tx: Auto　DDL　DML　Comma-...d (CSV)　datasource_2.t_address

<Filter Criteria>

	id	address_name	address_id	is_deleted	gmt_create	gmt_modified
1	611687225729904641	test-address611687225729904641	611687225444691970	0	2021-07-11 11:47:45	2021-07-11 11:4
2	611687231111196674	test-address611687231111196673	611687230876315653	0	2021-07-11 11:47:46	2021-07-11 11:4
3	611687236580569089	test-address611687236580569089	611687236437962756	0	2021-07-11 11:47:47	2021-07-11 11:4
4	611687380097069058	test-address611687380097069057	611687380021571588	0	2021-07-11 11:48:22	2021-07-11 11:4

图 12-20

（4）综上所述，执行 4 次后的分库结果见表 12-2。

表 12-2　分库策略执行后的结果

执行结果	t_order	t_order_item	t_address
datasource_0	向 t_order 表中，插入 1 条数据，分库键 order_id 的值为 611687225444691969，其中，611687225444691969%3=0	t_order 和 t_order_item 是绑定表关系，向 t_order_item 表中，插入 1 条数据，分库键 order_id 的值为 611687225444691969，其中，611687225444691969%3=0	t_address 是广播表，向 t_address 表中插入 4 条数据
datasource_1	向 t_order 表中，插入 2 条数据，分库键 order_id 的值为 611687236437962755 和 611687230876315651，其中，61168723643796275%3=1，611687230876315651%3=1	t_order 和 t_order_item 是绑定表关系，向 t_order_item 表中，插入 2 条数据，分库键 order_id 的值为 611687236437962755 和 611687230876315651，其中，61168723643796275%3=1，611687230876315651%3=1	t_address 是广播表，向 t_address 表中插入 4 条数据
datasource_2	向 t_order 表中，插入 1 条数据，分库键 order_id 的值为 611687380021571588，其中，611687380021571588%3=2	t_order 和 t_order_item 是绑定表关系，向 t_order_item 表中，插入 1 条数据，分库键 order_id 的值为 611687380021571588，其中，611687380021571588%3=2	t_address 是广播表，向 t_address 表中插入 4 条数据
数据源中执行的总次数	总共执行 4 次	总共执行 4 次	总共执行 12 次

5. 插入数据，验证分库分表策略

在服务 shardingsphere-jdbc-use-spring-boot 中，如果数据库的分库分表策略生效，则分片算法 “datasource_$->{order_id % 3}” “t_order_$->{order_id % 3}” 和 “t_order_item_$-> {order_id % 3}” 也会被执行。

（1）在文件 application.properties 中，修改如下配置信息：

```
#从“服务的环境配置信息”切换到“分库分表策略的配置信息”
spring.profiles.active=sharding-databases-tables
```

（2）启动服务 shardingsphere-jdbc-use-spring-boot。在启动后，执行命令“curl 127.0.0.1:34998/order/insert”4 次。

（3）通过观察数据库中数据的存储，可以发现分库分表策略已经生效。

①观察数据源 datasource_0 中的 t_order_0 表、t_order_1 表、t_order_2 表、t_order_item_0 表、t_order_item_1 表、t_order_item_2 表，以及 t_address 表的数据。

- 在 datasource_0.t_order_0 表中插入了 2 条数据，如图 12-21 所示，其中“order_id”的值为“611743754529173505”和“611743768475234307”，满足分库分表策略（611743754529173505%3 等于 0，611743768475234307%3 等于 0），所以数据被存储在 datasource_0.t_order_0 表中。

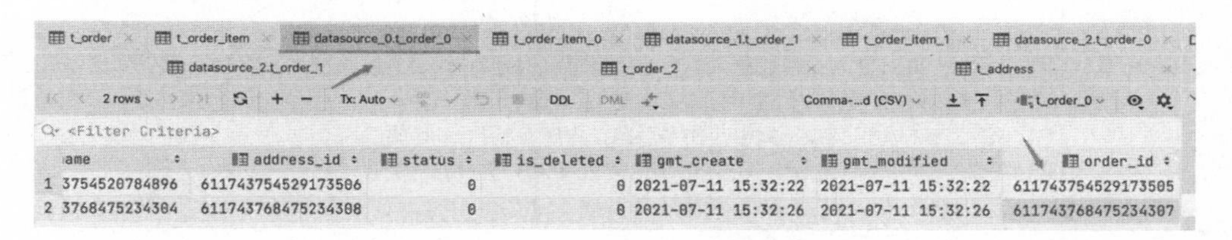

	ame	address_id	status	is_deleted	gmt_create	gmt_modified	order_id
1	3754520784896	611743754529173506	0	0	2021-07-11 15:32:22	2021-07-11 15:32:22	611743754529173505
2	3768475234304	611743768475234308	0	0	2021-07-11 15:32:26	2021-07-11 15:32:26	611743768475234307

图 12-21

- 在 datasource_0.t_order_item_0 表中插入了 2 条数据，如图 12-22 所示，其中“order_id”的值为“611743754529173505”和“611743768475234307”，满足分库分表策略（611743754529173505%3 等于 0，611743768475234307%3 等于 0），所以数据被存储在 datasource_0. t_order_item_0 表中。

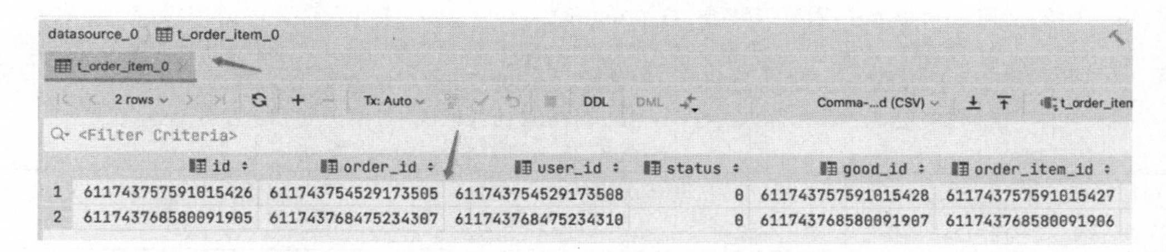

	id	order_id	user_id	status	good_id	order_item_id
1	611743757591015426	611743754529173505	611743754529173508	0	611743757591015428	611743757591015427
2	611743768580091905	611743768475234307	611743768475234310	0	611743768580091907	611743768580091906

图 12-22

- 在 datasource_0.t_address 表中插入了 4 条收货地址对象数据（t_address 表是一个广播表，如图 12-23 所示。如果执行了 4 次命令，则会在 datasource_0.t_address 表中插入 4 条数据）。

	id	address_name	address_id	is_deleted	gmt_create
1	611743757704261634	test-address611743757704261633	611743754529173506	0	2021-07-11 15:32:23
2	611743763374960641	test-address611743763374960641	611743763140079620	0	2021-07-11 15:32:24
3	611743768643006466	test-address611743768643006465	611743768475234308	0	2021-07-11 15:32:26
4	611743890906968066	test-address611743890906968065	611743890865025029	0	2021-07-11 15:32:55

图 12-23

②观察数据源 datasource_1 中的 t_order_0 表、t_order_1 表、t_order_2 表、t_order_item_0 表、t_order_item_表 1、t_order_item_2 表，以及 t_address 表的数据。

- 在 datasource_1.t_order_1 表中插入了 1 条数据，如图 12-24 所示，其中“order_id”的值为“611743763140079620”，满足分库分表策略（611743763140079620%3 等于 1），所以数据被存储在 datasource_1.t_order_1 表中。

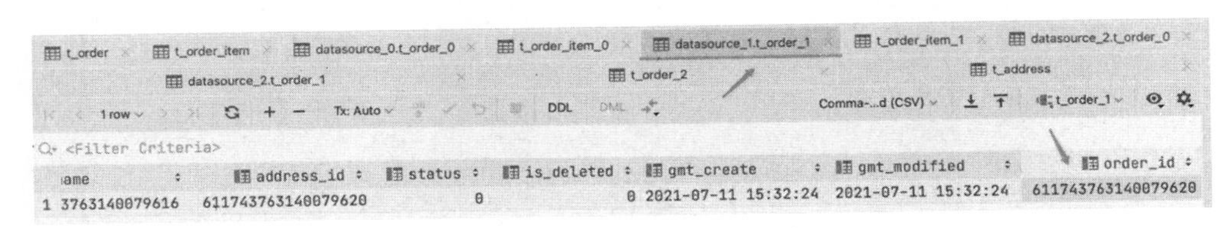

	ame	address_id	status	is_deleted	gmt_create	gmt_modified	order_id
1	3763140079616	611743763140079620	0	0	2021-07-11 15:32:24	2021-07-11 15:32:24	611743763140079620

图 12-24

- 在 datasource_1.t_order_item_1 表中插入了 1 条数据，如图 12-25 所示，其中“order_id”的值为“611743763140079620”，满足分库分表策略（611743763140079620%3 等于 1），所以数据被存储在 datasource_1.t_order_item_1 表中。

	id	order_id	user_id	status	good_id	order_item_id
1	611743763328823298	611743763140079620	611743763140079622	0	611743763328823300	611743763328823298

图 12-25

- 在 datasource_1.t_address 表中插入了 4 条收货地址对象数据（t_address 表是一个广播表，如图 12-26 所示。如果执行了 4 次命令，则会在 datasource_1.t_address 表中插入 4 条数据）。

	id	address_name	address_id	is_deleted	gmt_create
1	611743757704261634	test-address611743757704261633	611743754529173506	0	2021-07-11 15:32:23
2	611743763374960641	test-address611743763374960641	611743763140079620	0	2021-07-11 15:32:24
3	611743768643006466	test-address611743768643006465	611743768475234308	0	2021-07-11 15:32:26
4	611743890906968066	test-address611743890906968065	611743890865025029	0	2021-07-11 15:32:55

图 12-26

③ 观察数据源 datasource_2 中的 t_order_0 表、t_order_1 表、t_order_2 表、t_order_item_0 表、t_order_item_1 表、t_order_item_2 表及 t_address 表的数据。

- 在 datasource_2.t_order_2 表中插入了 1 条数据，如图 12-27 所示，其中“order_id”的值为“611743890865025027”，满足分库分表策略（611743890865025027%3 等于 2），所以数据被存储在 datasource_2.t_order_2 表中。

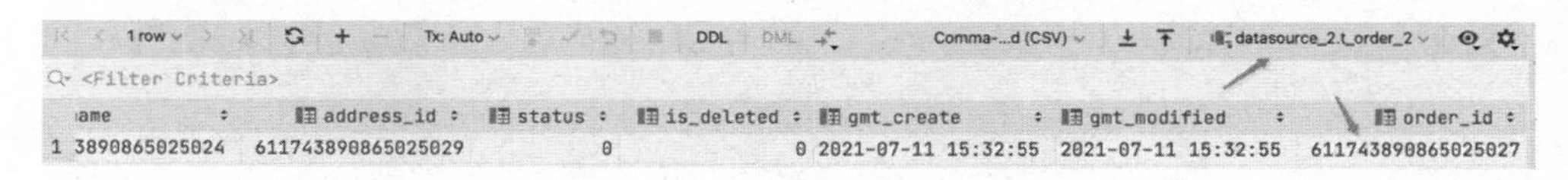

	ame	address_id	status	is_deleted	gmt_create	gmt_modified	order_id
1	3890865025024	611743890865025029	0	0	2021-07-11 15:32:55	2021-07-11 15:32:55	611743890865025027

图 12-27

- 在 datasource_2.t_order_item_2 表中插入了 1 条数据，如图 12-28 所示，其中“order_id”的值为“611743890865025027”，满足分库分表策略（611743890865025027%3 等于 1），所以数据被存储在 datasource_1.t_order_item_2 表中。

	id	order_id	user_id	status	good_id	order_item_id
1	611743890890190850	611743890865025027	611743890865025029	0	611743890890190851	611743890890190851

图 12-28

- 在 datasource_2.t_address 表中插入了 4 条收货地址对象数据(t_address 表是一个广播表，如图 12-29 所示。如果执行了 4 次命令，则会在 datasource_2.t_address 表中插入 4 条数据）。

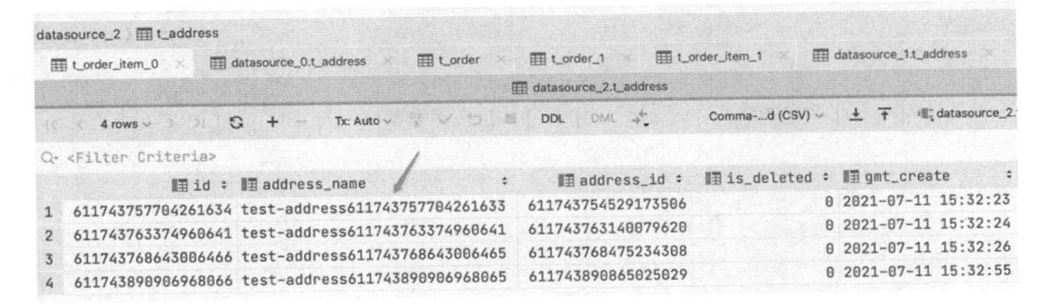

	id	address_name	address_id	is_deleted	gmt_create
1	611743757704261634	test-address611743757704261633	611743754529173506	0	2021-07-11 15:32:23
2	611743763374960641	test-address611743763374960641	611743763140079620	0	2021-07-11 15:32:24
3	611743768643006466	test-address611743768643006465	611743768475234308	0	2021-07-11 15:32:26
4	611743890906968066	test-address611743890906968065	611743890865025029	0	2021-07-11 15:32:55

图 12-29

（4）综上所述，执行 4 次命令后的分库分表的结果见表 12-2。

表 12-3　分库分表策略执行后的结果

执行结果	datasource_0	datasource_1	datasource_2
t_order_0	向 t_order_0 表中插入 2 条数据，分库键 order_id 的值为 611743754529173505 和 611743768475234307，其中，611743754529173505%3=0，611743768475234307%3=0	无	无
t_order_1	无	向 t_order_1 表中插入 1 条数据，分库键 order_id 的值为 611743763140079620，其中，611743763140079620%3=1	无
t_order_2	无	无	向 t_order_2 表中插入 1 条数据，分库键 order_id 的值为 611743890865025027，其中，611743890865025027%3=2
t_order_item_0	t_order_0 和 t_order_item_0 是绑定表关系，向 t_order_item_0 表中插入 2 条数据，分库键 order_id 的值为 611743754529173505 和 611743768475234307，其中，611743754529173505%3=1，611743768475234307%3=1	无	无
t_order_item_1	无	t_order_1 和 t_order_item_1 是绑定表关系，插入 1 条数据分库键 order_id 的值为 611743763140079620，其中，611743763140079620%3=1	无

续表

执行结果	datasource_0	datasource_1	datasource_2
t_order_item_2	无	无	向 t_order_item_2 表中插入 1 条数据，分库键 order_id 的值为 611743890865025027，其中，611743890865025027%3=2
t_address	t_address 是广播表，向 t_address 表中插入 4 条数据	t_address 是广播表，向 t_address 表中插入 4 条数据	t_address 是广播表，向 t_address 表中插入 4 条数据
数据源中执行的总次数	datasource_0.t_order_0 执行 2 次，datasource_0.t_order_item_0 执行 2 次，datasource_0.t_address 执行 4 次	datasource_1.t_order_1 执行 1 次，datasource_1.t_order_item_1 执行 1 次，datasource_1.t_address 执行 4 次	datasource_2.t_order_2 执行 1 次，datasource_2.t_order_item_2 执行 1 次，datasource_2.t_address 执行 4 次

6. 插入数据，验证分表策略

在服务 shardingsphere-jdbc-use-spring-boot 中，如果数据库的分表策略生效，则分片算法“t_order_$->{order_id % 3}”“和 t_order_item_$->{order_id % 3}”也会被执行。

（1）在文件 application.properties 中，修改如下配置信息：

```
#从“服务的环境配置信息”切换到“分库策略的配置信息”
spring.profiles.active=sharding-tables
```

（2）启动服务 shardingsphere-jdbc-use-spring-boot。在启动后，执行 4 次命令“curl 127.0.0.1:34998/order/insert”。

（3）通过观察数据库中数据的存储，可以发现分表策略已经生效。

①观察数据源 ds 中的 t_order_0 表和 t_order_item_0 表的数据。

- 在 ds.t_order_0 表中插入了 1 条数据，如图 12-30 所示，其中“order_id”的值为“611769891502321668”，满足分库分表策略（611769891502321668%3 等于 0），所以数据被存储在 ds.t_order_0 表中。

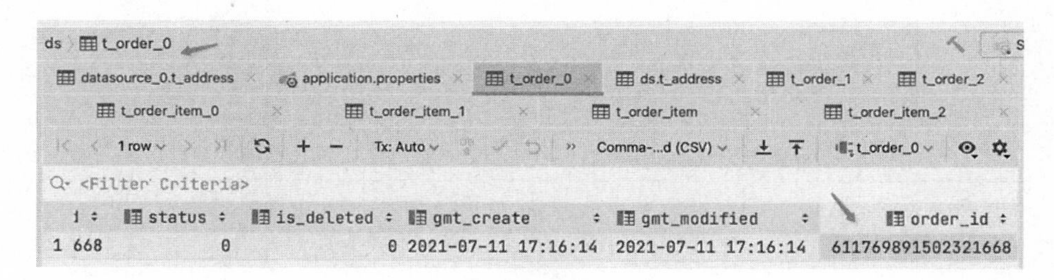

图 12-30

- 在 ds.t_order_item_0 表中插入了 1 条数据，如图 12-31 所示，其中“order_id”的值为“611769891502321668”，满足分库分表策略（611769891502321668%3 等于 0），所以数据被存储在 ds.t_order_item_0 表中。

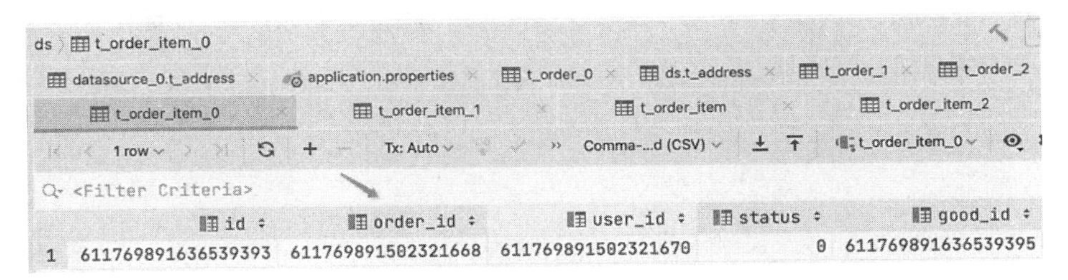

图 12-31

②观察数据源 ds 中的 t_order_1 表和 t_order_item_1 表的数据。

- 在 ds.t_order_1 表中插入了 1 条数据，如图 12-32 所示，其中“order_id”的值为“611769895797288964”，满足分库分表策略（611769895797288964%3 等于 1），所以数据被存储在 ds.t_order_1 表中。

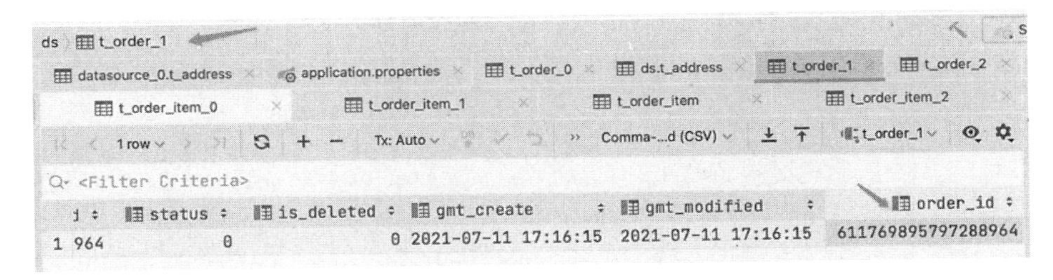

图 12-32

- 在 ds.t_order_item_1 表中插入了 1 条数据，如图 12-33 所示，其中“order_id”的值为“611769895797288964”，满足分库分表策略（611769895797288964%3 等于 1），所以数据被存储在 ds. t_order_item_1 表中。

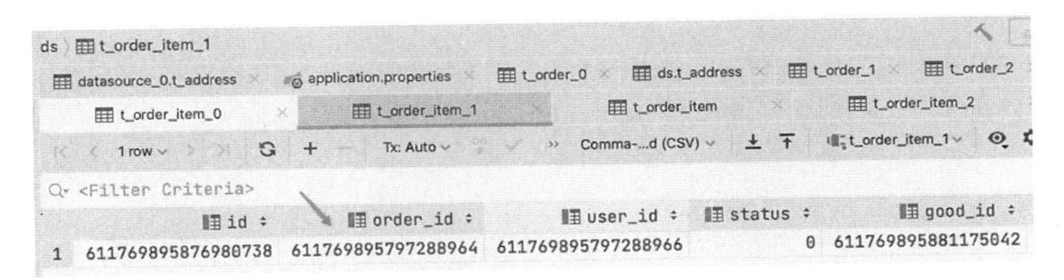

图 12-33

③观察数据源 ds 中的 t_order_2 表和 t_order_item_2 表的数据。

- 在 ds. t_order_2 表中插入了 2 条数据，如图 12-34 所示，其中“order_id”的值为“611769887123468291”和“611769900889174019”，满足分库分表策略（611769887123468291%3 等于 2，611769900889174019%3 等于 2），所以数据被存储在 ds. t_order_2 表中。

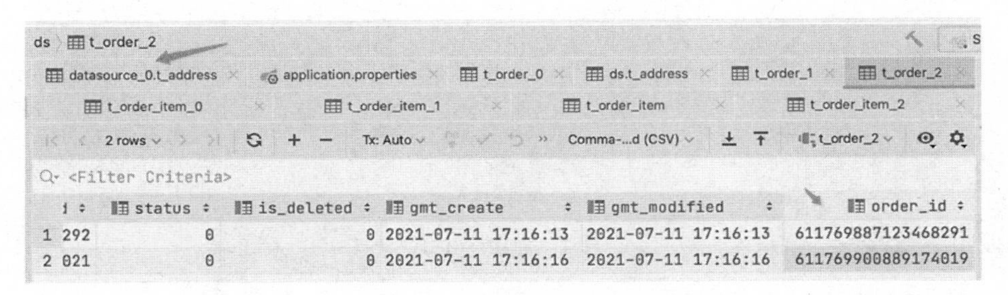

图 12-34

- 在 ds.t_order_item_2 表中插入 2 条数据，如图 12-35 所示，其中“order_id”的值为“611769887123468291”和“611769900889174019”，满足分库分表策略（611769887123468291%3 等于 2，611769900889174019%3 等于 2），所以数据被存储在 ds.t_order_item_2 表中。

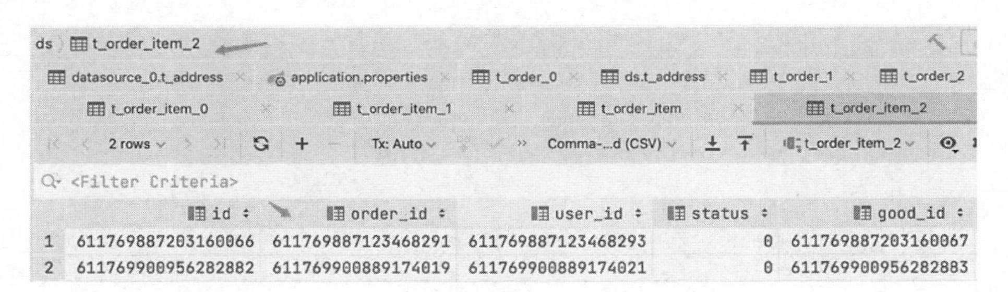

图 12-35

④观察数据源 ds 中的 t_address 表的数据。

如图 12-36 所示，在 ds.t_address 表中插入了 4 条收货地址对象数据（t_address 表是一个广播表。如果执行了 4 次命令，则会在 ds.t_address 表中插入 4 条数据）。

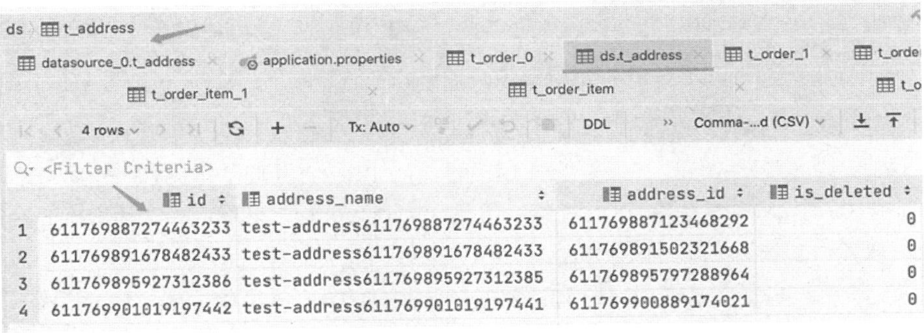

	id	address_name	address_id	is_deleted
1	611769887274463233	test-address611769887274463233	611769887123468292	0
2	611769891678482433	test-address611769891678482433	611769891502321668	0
3	611769895927312386	test-address611769895927312385	611769895797288964	0
4	611769901019197442	test-address611769901019197441	611769900889174021	0

图 12-36

（4）综上所述，执行 4 次后的分表的结果见表 12-4。

表 12-4　分表策略执行后的结果

执行结果	说明	备　注
t_order_0	向 t_order_0 表中插入 1 条数据，分表键 order_id 的值为 611769891502321668，其中，611769891502321668%3 等于 0	ds.t_order_0 表执行 1 次
t_order_1	向 t_order_1 表中插入 1 条数据，分表键 order_id 的值为 611769895797288964，其中，611769895797288964%3 等于 1	ds.t_order_1 表执行 1 次
t_order_2	向 t_order_2 表中插入 2 条数据，分表键 order_id 的值为 611769887123468291 和 611769900889174019，其中，611769887123468291%3 等于 2，611769900889174019%3 等于 2	ds.t_order_2 表执行 2 次
t_order_item_0	t_order_0 和 t_order_item_0 是绑定表的关系，向 t_order_item_0 表中插入 1 条数据，分表键 order_id 的值为 611769891502321668，其中，611769891502321668%3 等于 0	ds.t_order_item_0 表执行 1 次
t_order_item_1	t_order_1 和 t_order_item_1 是绑定表的关系，向 t_order_item_1 表中插入 1 条数据，分表键 order_id 的值为 611769895797288964，其中，611769895797288964%3 等于 1	ds.t_order_item_1 表执行 1 次
t_order_item_2	t_order_2 和 t_order_item_2 是绑定表的关系，向 t_order_item_2 表中插入 2 条数据，分表键 order_id 的值为 611769887123468291 和 611769900889174019，其中，611769887123468291%3 等于 2，611769900889174019%3 等于 2	ds.t_order_item_2 表执行 2 次
t_address	t_address 是广播表，向 t_address 表中插入 4 条数据	ds.t_address 执行 4 次

12.3　“用路由引擎实现分库分表”的原理

在 Shardingsphere JDBC 中，用路由引擎实现其分库分表。开发人员可以用 Java API、Spring Framework 的命令空间等不同的接入方式，去集成 Shardingsphere JDBC 的分库和分表到应用中。但是无论哪种接入方式，都会调用 Shardingsphere JDBC 的路由引擎。

本书会从“Shardingsphere JDBC 的 Spring Boot 接入模式”的角度去分析路由引擎的原理。

在 Shardingsphere JDBC 的路由引擎中，最核心的概念如下。

- 规则：主要指数据分片规则、读写分离规则、数据加密规则，以及影子库规则。
- 数据库数据源：主要指 Shardingsphere JDBC 提供的数据源代理 ShardingSphereDataSource 类。在 Shardingsphere JDBC 中，用数据源代理去代理源数据源 javax.sql.DataSource。
- 数据库连接：主要指 Shardingsphere JDBC 提供的代理连接 ShardingSphereConnection 类。在 Shardingsphere JDBC 中，用代理数据库连接代理源数据库连接 java.sql.Connection。
- 数据库连接声明：主要指 Shardingsphere JDBC 提供的连接声明 ShardingSphereStatement 类和 ShardingSpherePreparedStatement 类。在 Shardingsphere JDBC 中，用连接声明代理源数据库连接声明 java.sql.PreparedStatement 和 java.sql.Statement。
- 数据源规格上下文：主要指 Shardingsphere JDBC 提供的 SchemaContexts 类。在 Shardingsphere JDBC 中，用数据源规格上下文存储规格对象 ShardingSphereSchema，其中规格对象的内容包括规则、数据源、元数据（数据源元数据、逻辑规格元数据、表地址元数据和数据库缓存元数据）。
- 数据源事务上下文：主要指 Shardingsphere JDBC 提供的数据源事务 TransactionContexts 接口，它的主要实现类是 StandardTransactionContexts 类和 GovernanceTransactionContexts 类，Shardingsphere JDBC 用数据源事务上下文来管理事务管理引擎。

12.3.1 绑定分库分表规则和数据库数据源，并初始化路由引擎

Shardingsphere JDBC 绑定分库分表规则和数据库数据源，并初始化路由引擎的具体过程如下。

1. 用 SpringBootConfiguration 类加载分库分表规则和数据库数据源

开发人员可以在应用的属性文件 application.properties 中，添加如下配置信息：

```
spring.shardingsphere.datasource.names=datasource_0, datasource_1
//①配置数据源 datasource_0
spring.shardingsphere.datasource.datasource_0.type=com.zaxxer.hikari.HikariDataSource
spring.shardingsphere.datasource.datasource_0.driver-class-name=com.mysql.cj.jdbc.Driver
spring.shardingsphere.datasource.datasource_0.jdbc-url=jdbc:mysql://127.0.0.1:3306/datasource_0?serverTimezone=UTC&useSSL=false&useUnicode=true&characterEncoding=UTF-8
```

```
spring.shardingsphere.datasource.datasource_0.username=root
spring.shardingsphere.datasource.datasource_0.password=root
//②省略配置数据源 datasource_1
...
//③配置表规则
spring.shardingsphere.rules.sharding.tables.t_order.actual-data-nodes=da
tasource_$->{0..1}.t_order
//④配置分库策略
spring.shardingsphere.rules.sharding.tables.t_order.database-strategy.st
andard.sharding-column=order_id
spring.shardingsphere.rules.sharding.tables.t_order.database-strategy.st
andard.sharding-algorithm-name=database_inline
//⑤配置分库的分片算法
spring.shardingsphere.rules.sharding.sharding-algorithms.database_inline
.type=INLINE
spring.shardingsphere.rules.sharding.sharding-algorithms.database_inline
.props.algorithm-expression=datasource_$->{id % 2}
```

应用启动的过程中会读取上面添加的配置信息，并用 SpringBootConfiguration 类加载分库分表规则和数据库数据源，具体代码如下所示：

```
@Configuration
@ComponentScan("org.apache.shardingsphere.spring.boot.converter")
@EnableConfigurationProperties(SpringBootPropertiesConfiguration.class)
//①定义路由引擎开关"spring.shardingsphere.enabled"，默认为"true"
@ConditionalOnProperty(prefix = "spring.shardingsphere", name = "enabled",
havingValue = "true", matchIfMissing = true)
@AutoConfigureBefore(DataSourceAutoConfiguration.class)
@RequiredArgsConstructor
public class SpringBootConfiguration implements EnvironmentAware {
private final SpringBootPropertiesConfiguration props;
//②定义一个数据库数据源的本地缓存
    private final Map<String, DataSource> dataSourceMap = new
LinkedHashMap<>();
    @Bean
    @Autowired(required = false)
public DataSource shardingSphereDataSource(final
ObjectProvider<List<RuleConfiguration>> rules) throws SQLException {
    //③读取 application.properties 文件中前缀为
"spring.shardingsphere.rules.*"的路由规则信息，如果在应用中没有配置，则返回一个空的
规则列表对象
        Collection<RuleConfiguration> ruleConfigurations = Optional.
ofNullable(rules.getIfAvailable()).orElse(Collections.emptyList());
        //④在创建数据源代理 ShardingSphereDataSource 类的过程，将规则和数据源进行绑定
        return ShardingSphereDataSourceFactory.createDataSource
(dataSourceMap, ruleConfigurations, props.getProps());
```

```
    }
    //⑤定义一个分库分表的事务扫描器
        @Bean
        public ShardingTransactionTypeScanner shardingTransactionTypeScanner() {
            return new ShardingTransactionTypeScanner();
    }
    //⑥在应用启动的过程中，Shardingsphere-JDBC 读取配置文件中配置的数据源信息，生成
javax.sql.DataSource 对象，并将其存储在本地缓存变量 dataSourceMap 中
        @Override
        public final void setEnvironment(final Environment environment) {
            dataSourceMap.putAll(DataSourceMapSetter
            .getDataSourceMap(environment));
        }
    }
```

本书为了验证“用 SpringBootConfiguration 类加载规则和数据源的过程”，利用 Java 调试工具，在应用启动的过程中调试规则和数据源的加载过程。

通过 Java 断点调试规则的加载过程如图 12-37 所示。

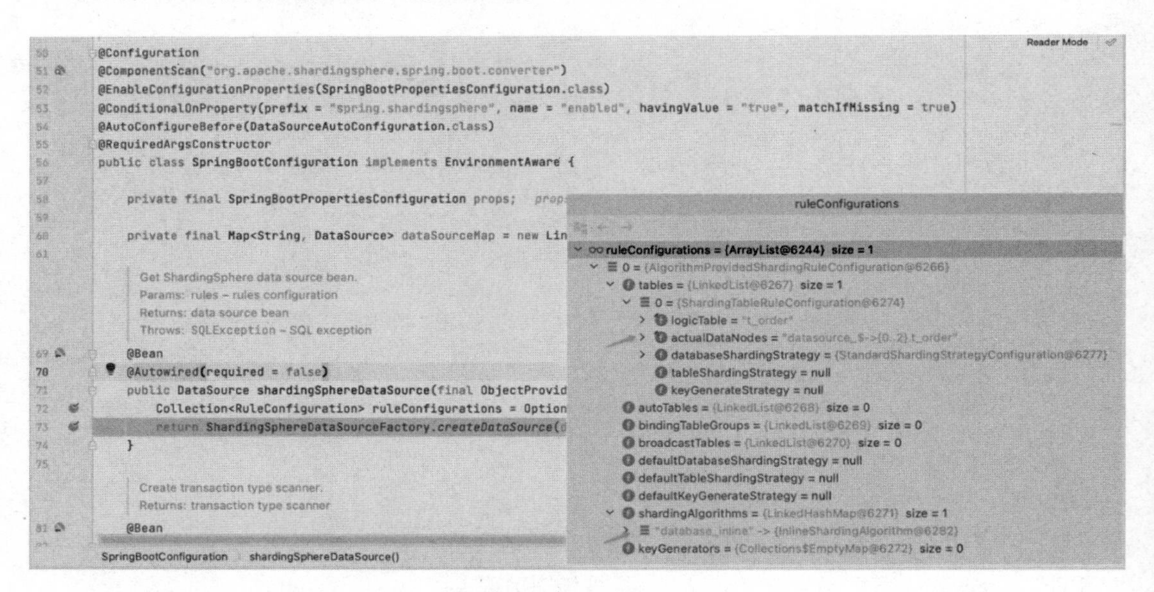

图 12-37

通过 Java 断点调试数据源加载过程如图 12-38 所示。

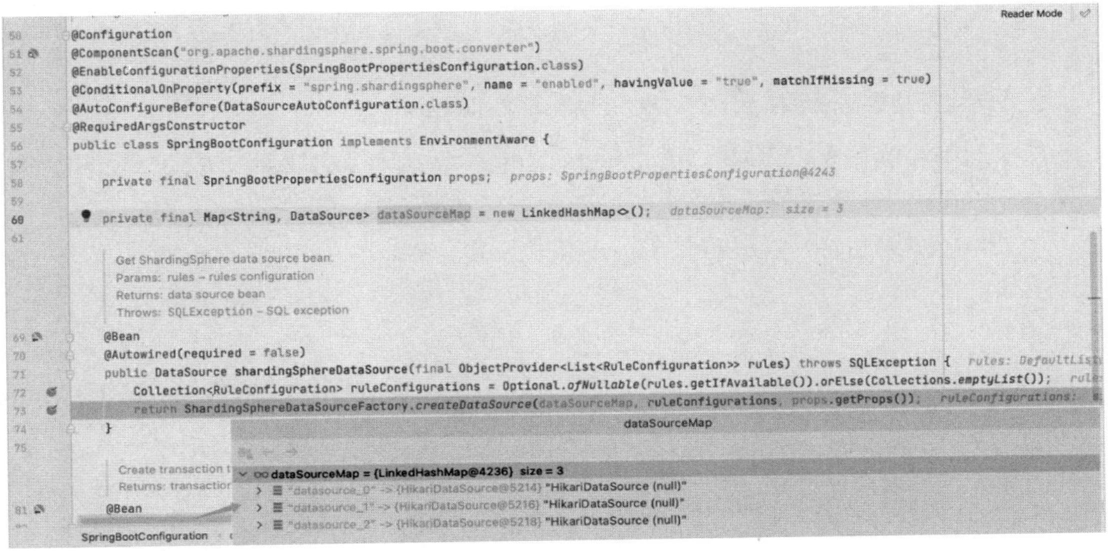

```
@Configuration
@ComponentScan("org.apache.shardingsphere.spring.boot.converter")
@EnableConfigurationProperties(SpringBootPropertiesConfiguration.class)
@ConditionalOnProperty(prefix = "spring.shardingsphere", name = "enabled", havingValue = "true", matchIfMissing = true)
@AutoConfigureBefore(DataSourceAutoConfiguration.class)
@RequiredArgsConstructor
public class SpringBootConfiguration implements EnvironmentAware {

    private final SpringBootPropertiesConfiguration props;   props: SpringBootPropertiesConfiguration@4243

    private final Map<String, DataSource> dataSourceMap = new LinkedHashMap<>();   dataSourceMap:  size = 3

    Get ShardingSphere data source bean.
    Params: rules – rules configuration
    Returns: data source bean
    Throws: SQLException – SQL exception
    @Bean
    @Autowired(required = false)
    public DataSource shardingSphereDataSource(final ObjectProvider<List<RuleConfiguration>> rules) throws SQLException {   rules: DefaultList
        Collection<RuleConfiguration> ruleConfigurations = Optional.ofNullable(rules.getIfAvailable()).orElse(Collections.emptyList());   rule
        return ShardingSphereDataSourceFactory.createDataSource(dataSourceMap, ruleConfigurations, props.getProps());   ruleConfigurations:
    }
                                dataSourceMap
    Create transaction
    Returns: transaction    dataSourceMap = {LinkedHashMap@4236}  size = 3
                              > "datasource_0" -> {HikariDataSource@5214} "HikariDataSource (null)"
    @Bean                     > "datasource_1" -> {HikariDataSource@5216} "HikariDataSource (null)"
                              > "datasource_2" -> {HikariDataSource@5218} "HikariDataSource (null)"
SpringBootConfiguration
```

图 12-38

2. 初始化数据源代理 ShardingSphereDataSource 的对象

在 Shardingsphere JDBC 中，应用在启动时会初始化数据源代理 ShardingSphereDataSource 的对象，具体代码如下所示：

```
@RequiredArgsConstructor
@Getter
public final class ShardingSphereDataSource
    extends AbstractUnsupportedOperationDataSource implements AutoCloseable {
    private final SchemaContexts schemaContexts;
    private final TransactionContexts transactionContexts;
public ShardingSphereDataSource(final Map<String, DataSource> dataSourceMap,
final Collection<RuleConfiguration> configurations, final Properties props)
throws SQLException {
      //①解析本地缓存变量 dataSourceMap 中的 javax.sql.DataSource 对象
        DatabaseType databaseType = createDatabaseType(dataSourceMap);
        //②创建全局的数据源规格上下文对象 SchemaContexts
        schemaContexts = new SchemaContextsBuilder(
                databaseType,
Collections.singletonMap(DefaultSchema.LOGIC_NAME, dataSourceMap),
        Collections.singletonMap(DefaultSchema.LOGIC_NAME, configurations),
props).build();
        //③创建全局的事务上下文对象 TransactionContexts
        transactionContexts = createTransactionContexts(databaseType,
dataSourceMap);
    }
```

```
    ...
}
```

3. 创建全局的数据源规格上下文对象 SchemaContexts

在 Shardingsphere JDBC 中，用 SchemaContextsBuilder 类创建全局的数据源规格上下文对象 SchemaContexts，具体步骤如下。

（1）用构造函数初始化 SchemaContextsBuilder 类，具体代码如下所示：

```
@Slf4j
public final class SchemaContextsBuilder {
    private final DatabaseType databaseType;
    private final Map<String, Map<String, DataSource>> dataSources;
    private final Map<String, Collection<RuleConfiguration>> ruleConfigs;
    private final Authentication authentication;
    private final ConfigurationProperties props;
    private final ExecutorKernel executorKernel;
    public SchemaContextsBuilder(final DatabaseType databaseType, final
Map<String, Map<String, DataSource>> dataSources,
                                final Map<String,
Collection<RuleConfiguration>> ruleConfigs, final Properties props) {
        this(databaseType, dataSources, ruleConfigs, new Authentication(),
props);
    }
    public SchemaContextsBuilder(final DatabaseType databaseType, final
Map<String, Map<String, DataSource>> dataSources,
                                final Map<String,
Collection<RuleConfiguration>> ruleConfigs, final Authentication
authentication, final Properties props) {
        //①初始化 DatabaseType 对象
        this.databaseType = databaseType;
        //②初始化数据源 dataSources 对象
        this.dataSources = dataSources;
        //③初始化规则
        this.ruleConfigs = ruleConfigs;
        //④初始化认证信息
        this.authentication = authentication;
        //⑤初始化属性信息
        this.props = new ConfigurationProperties(null == props ? new
Properties() : props);
        //⑥初始化线程池 ExecutorKernel
        executorKernel = new ExecutorKernel(this.props.
<Integer>getValue(ConfigurationPropertyKey.EXECUTOR_SIZE));
    }
    ...
}
```

（2）用 SchemaContextsBuilder 类的 build()方法创建 StandardSchemaContexts 对象，具体代码如下所示：

```
public SchemaContexts build() throws SQLException {
    Map <String,
ShardingSphereSchema> schemas = new LinkedHashMap <>(ruleConfigs.size(), 1);
//①如果应用中的规则不为空
for (String each: ruleConfigs.keySet()) {
    //②则解析规则，并创建 ShardingSphereSchema 对象
        schemas.put(each, createSchema(each));
    }
    return new StandardSchemaContexts(schemas, executorKernel,
authentication, props, databaseType);
}
private ShardingSphereSchema createSchema(final String schemaName) throws
SQLException {
    Map <String,
    DataSource> dataSources = this.dataSources.get(schemaName);
Collection <RuleConfiguration> ruleConfigs =
this.ruleConfigs.get(schemaName);
//③用 ShardingSphereRulesBuilder 类构造规则引擎中的规则
Collection <ShardingSphereRule> rules =
ShardingSphereRulesBuilder.build(ruleConfigs, dataSources.keySet());
//④创建元数据，并返回 ShardingSphereSchema 对象
    return new ShardingSphereSchema(schemaName, ruleConfigs, rules,
dataSources, createMetaData(schemaName, dataSources, rules));
}

private ShardingSphereMetaData createMetaData(final String schemaName,
final Map <String, DataSource> dataSourceMap, final Collection
<ShardingSphereRule> rules) throws SQLException {
long start = System.currentTimeMillis();
//⑤创建数据源元数据对象
DataSourcesMetaData dataSourceMetas = new DataSourcesMetaData(databaseType,
getDatabaseAccessConfigurationMap(dataSourceMap));
//⑥创建逻辑规格元数据对象
LogicSchemaMetaData logicSchemaMetaData = new
LogicSchemaMetaDataLoader(rules).load(databaseType, dataSourceMap, props);
//⑦创建表地址元数据对象
    TableAddressingMetaData tableAddressingMetaData =
TableAddressingMetaDataLoader.load(databaseType, dataSourceMap, rules);

//⑧创建数据库缓存元数据对象
CachedDatabaseMetaData cachedDatabaseMetaData =
createCachedDatabaseMetaData(dataSources.get(schemaName)).orElse(null);
```

```
        ShardingSphereMetaData result = new
ShardingSphereMetaData(dataSourceMetas, logicSchemaMetaData,
tableAddressingMetaData, cachedDatabaseMetaData);
        log.info("Load meta data for schema {} finished, cost {} milliseconds.",
schemaName, System.currentTimeMillis() - start);
        return result;
    }
    //⑨获取数据库访问配置对象 DatabaseAccessConfiguration
    private Map <String,
    DatabaseAccessConfiguration> getDatabaseAccessConfigurationMap(
        final Map < String, DataSource> dataSourceMap) throws SQLException {
        Map <String,DatabaseAccessConfiguration> result = new
    LinkedHashMap <>(dataSourceMap.size(), 1);
        for (Entry <String, DataSource> entry: dataSourceMap.entrySet()) {
            DataSource dataSource = entry.getValue();
            try (Connection connection = dataSource.getConnection()) {
                DatabaseMetaData metaData = connection.getMetaData();
                result.put(entry.getKey(),new
DatabaseAccessConfiguration(metaData.getURL(),metaData.getUserName()));
            }
        }
        return result;
    }
    private Optional <CachedDatabaseMetaData>
createCachedDatabaseMetaData(final Map <String, DataSource> dataSources) throws
SQLException {
        if (dataSources.isEmpty()) {
            return Optional.empty();
    }
    //⑩设置数据库连接的元数据的本地缓存
    try (Connection connection =
         dataSources.values().iterator().next().getConnection()) {
             return Optional.of(new CachedDatabaseMetaData
            (connection.getMetaData()));
        }
    }
```

4. 构造代理数据库连接类 ShardingSphereConnection

在 Shardingsphere JDBC 中，主要是用 ShardingSphereDataSource 类的 getConnection()方法构造代理数据库连接 ShardingSphereConnection 类，具体代码如下所示：

```
    private final SchemaContexts schemaContexts;
    private final TransactionContexts transactionContexts;
    @Override
    //返回代理数据库连接（包含规则、全局的数据源规格上下文对象等）
```

```
    public ShardingSphereConnection getConnection() {
        return new ShardingSphereConnection(getDataSourceMap(), schemaContexts,
transactionContexts, TransactionTypeHolder.get());
    }
    @Override
    public ShardingSphereConnection getConnection(final String username, final
String password) {
        return getConnection();
    }
    public Map<String, DataSource> getDataSourceMap() {
        return schemaContexts.getDefaultSchema().getDataSources();
    }
```

5. 初始化代理数据库连接声明类

在 Shardingsphere JDBC 中，用代理数据库连接类 ShardingSphereConnection 初始化代理数据库连接声明类 ShardingSphereStatement 和 ShardingSpherePreparedStatement，具体代码如下所示：

```
    @Getter
    public final class ShardingSphereConnection extends
AbstractConnectionAdapter {
        //①返回 ShardingSpherePreparedStatement 对象
        @Override
        public PreparedStatement prepareStatement(final String sql, final int
resultSetType, final int resultSetConcurrency) throws SQLException {
            return new ShardingSpherePreparedStatement(this, sql, resultSetType,
resultSetConcurrency);
    }
        //②返回 ShardingSphereStatement 对象
        @Override
        public Statement createStatement(final int resultSetType, final int
resultSetConcurrency) {
            return new ShardingSphereStatement(this, resultSetType,
resultSetConcurrency);
        }
        ...
    }
```

12.3.2　拦截 SQL 语句，并启动路由引擎

在 Shardingsphere JDBC 中，利用 Spring Boot 将数据库的数据源 ShardingSphereDataSource 加载到应用环境中，并绑定规则和数据库数据源。这样应用在 ORM 层（比如 MyBatis 或者 MyBatis-Plus）执行 SQL 语句（比如 INSERT、UPDATE、DELETE、SELECT 和 FLUSH 等）时，MyBatis 或者 MyBatis-Plus 会用这个已经初始化的数据源去创建

SqlSession 对象，从而完成对 SQL 的拦截。

1. 用 MyBatis 加载数据源类 ShardingSphereDataSource

在执行 SQL 的过程中，用 MyBatis 加载数据源 ShardingSphereDataSource，具体代码如下所示：

```
public class DefaultSqlSessionFactory implements SqlSessionFactory {
    @Override
    public SqlSession openSession() {
        returnopenSessionFromDataSource(configuration.
        getDefaultExecutorType(),null, false);
    }
    private SqlSession openSessionFromDataSource(ExecutorType execType,
TransactionIsolationLevel level, boolean autoCommit) {
        Transaction tx = null;
        try {
            //①从应用环境中获取环境对象 Environment
            final Environment environment = configuration.getEnvironment();
            //②从应用环境中获取事务工厂
            final TransactionFactory transactionFactory =
getTransactionFactoryFromEnvironment(environment);
            //③用事务工厂创建数据源 ShardingSphereDataSource 的事务对象 Transaction
            tx = transactionFactory.newTransaction(environment.
getDataSource(), level, autoCommit);
            //④创建 SQL 执行器（比如 SimpleExecutor 类、CachingExecutor 类等）
            final Executor executor = configuration.newExecutor(tx,
execType);
            //⑤返回 SqlSession 对象
            return new DefaultSqlSession(configuration, executor,
autoCommit);
        } catch (Exception e) {
           closeTransaction(tx);
           throw ExceptionFactory.wrapException("Error opening session.
Cause: " + e, e);
        } finally {
           ErrorContext.instance().reset();
        }
      }
      ...
}
```

用 IDEA 断点调试 MyBatis 加载数据源 ShardingSphereDataSource 的过程如图 12-39 所示：将断点设置在 DefaultSqlSessionFactory 类的 openSessionFromDataSource()方法中，在事务对象 Transaction 中会存储应用中配置的多数据源。

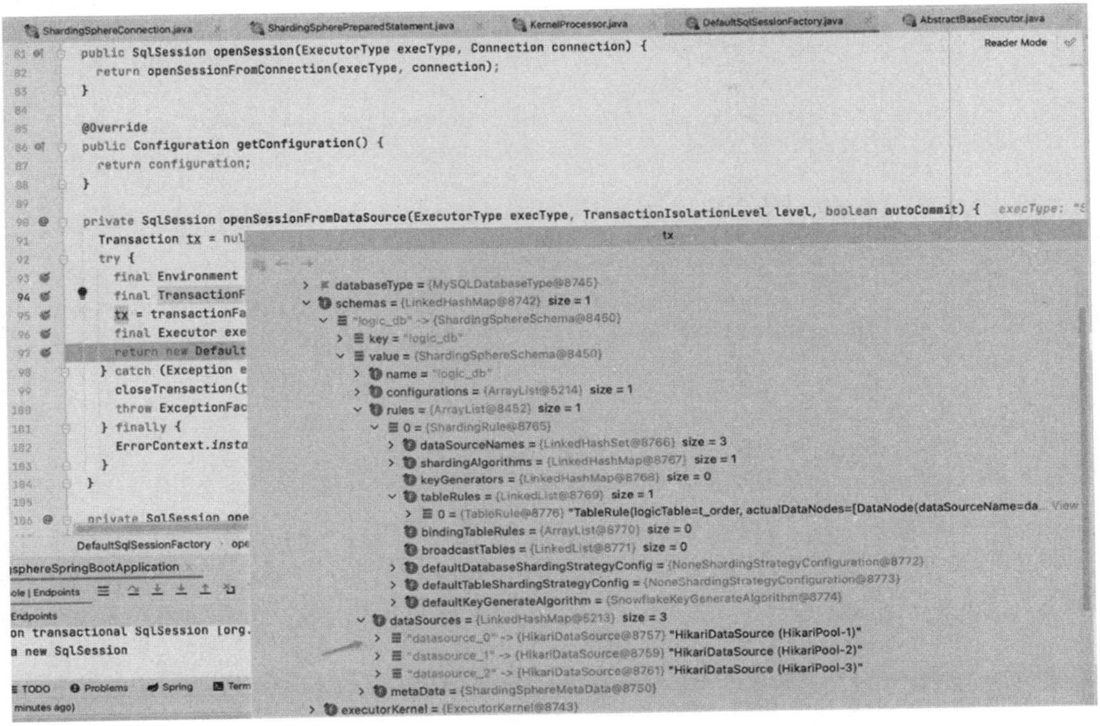

```
ShardingSphereConnection.java   ShardingSpherePreparedStatement.java   KernelProcessor.java   DefaultSqlSessionFactory.java   AbstractBaseExecutor.java
81    public SqlSession openSession(ExecutorType execType, Connection connection) {
82      return openSessionFromConnection(execType, connection);
83    }
84
85    @Override
86    public Configuration getConfiguration() {
87      return configuration;
88    }
89
90    private SqlSession openSessionFromDataSource(ExecutorType execType, TransactionIsolationLevel level, boolean autoCommit) {
91      Transaction tx = nul
92      try {
93        final Environment
94        final TransactionF
95        tx = transactionFa
96        final Executor exe
97        return new Default
98      } catch (Exception e
99        closeTransaction(t
100       throw ExceptionFac
101     } finally {
102       ErrorContext.insta
103     }
104   }
105
106   private SqlSession ope

tx
> databaseType = {MySQLDatabaseType@8745}
v schemas = {LinkedHashMap@8742} size = 1
  v "logic_db" -> {ShardingSphereSchema@8450}
    > key = "logic_db"
    v value = {ShardingSphereSchema@8450}
      > name = "logic_db"
      > configurations = {ArrayList@5214} size = 1
      v rules = {ArrayList@8452} size = 1
        v 0 = {ShardingRule@8765}
          > dataSourceNames = {LinkedHashSet@8766} size = 3
          > shardingAlgorithms = {LinkedHashMap@8767} size = 1
            keyGenerators = {LinkedHashMap@8768} size = 0
          v tableRules = {LinkedList@8769} size = 1
            > 0 = {TableRule@8776} "TableRule(logicTable=t_order, actualDataNodes=[DataNode(dataSourceName=da
            bindingTableRules = {ArrayList@8770} size = 0
            broadcastTables = {LinkedList@8771} size = 0
          > defaultDatabaseShardingStrategyConfig = {NoneShardingStrategyConfiguration@8772}
          > defaultTableShardingStrategyConfig = {NoneShardingStrategyConfiguration@8773}
          > defaultKeyGenerateAlgorithm = {SnowflakeKeyGenerateAlgorithm@8774}
      v dataSources = {LinkedHashMap@5213} size = 3
        > "datasource_0" -> {HikariDataSource@8757} "HikariDataSource (HikariPool-1)"
        > "datasource_1" -> {HikariDataSource@8759} "HikariDataSource (HikariPool-2)"
        > "datasource_2" -> {HikariDataSource@8761} "HikariDataSource (HikariPool-3)"
      > metaData = {ShardingSphereMetaData@8750}
> executorKernel = {ExecutorKernel@8743}
```

图 12-39

2. 获取代理数据库连接对象 ShardingSphereConnection

在执行 SQL 的过程中，用 Mybatis 的 MybatisSimpleExecutor 类的 doUpdate()方法获取代理数据库连接 ShardingSphereConnection 对象，具体代码如下所示：

```
public class MybatisSimpleExecutor extends AbstractBaseExecutor {
    @Override
    public int doUpdate(MappedStatement ms, Object parameter) throws
SQLException {
        Statement stmt = null;
        try {
            Configuration configuration = ms.getConfiguration();
            //①获取 MyBatis 的 RoutingStatementHandler 对象，用于处理 SQL 语句
            StatementHandler handler = configuration.newStatementHandler
(this, ms, parameter, RowBounds.DEFAULT, null, null);
            stmt = prepareStatement(handler, ms.getStatementLog(), false);
      //②在获取代理数据库连接声明对象后，调用 MyBatis 的 RoutingStatementHandler 类
的 update()方法执行 SQL 语句的更新
            return stmt == null ? 0 : handler.update(stmt);
        } finally {
```

```
            closeStatement(stmt);
        }
    }
    private Statement prepareStatement(StatementHandler handler, Log
statementLog, boolean isCursor) throws SQLException {
        Statement stmt;
        //③获取应用中的代理连接对象 ShardingSphereConnection
        Connection connection = getConnection(statementLog);
        //④获取应用中的代理数据库连接声明对象 ShardingSpherePreparedStatement
        stmt = handler.prepare(connection, transaction.getTimeout());
        if (stmt == null && !isCursor) {
            return null;
        } else {
            handler.parameterize(stmt);
            return stmt;
        }
    }
    ...
}
```

3. 执行 SQL 语句的更新

在 SQL 语句执行的过程中，用 MyBatis 的路由声明处理 RoutingStatementHandler 类的 update()方法执行 SQL 语句更新，具体代码如下所示：

```
public class RoutingStatementHandler implements StatementHandler {
    private final StatementHandler delegate;
    public RoutingStatementHandler(Executor executor, MappedStatement ms,
Object parameter, RowBounds rowBounds, ResultHandler resultHandler, BoundSql
boundSql) {
        switch (ms.getStatementType()) {
//①如果当前 SQL 的数据库连接的声明类型为 STATEMENT，则使用 SimpleStatementHandler
类的 update()方法执行 SQL 语句的更新
            case STATEMENT:
                delegate = new SimpleStatementHandler(executor, ms, parameter,
rowBounds, resultHandler, boundSql);
                break;
//②如果当前 SQL 的数据库连接的声明类型为 PREPARED,则使用 PreparedStatementHandler
类的 update()方法执行 SQL 语句的更新
            case PREPARED:
                delegate = new PreparedStatementHandler(executor, ms,
parameter, rowBounds, resultHandler, boundSql);
                break;
//③如果当前 SQL 的数据库连接的声明类型为 CALLABLE,则使用 CallableStatementHandler
类的 update()方法执行 SQL 语句的更新
            case CALLABLE:
```

```
                delegate = new CallableStatementHandler(executor, ms,
parameter, rowBounds, resultHandler, boundSql);
                break;
            default:
    //④如果没有匹配到 MyBatis 支持的数据库连接声明类型，则直接抛出异常
                throw new ExecutorException("Unknown statement type: " +
ms.getStatementType());
        }
    }
    @Override
    public int update(Statement statement) throws SQLException {
    //⑤执行 SQL 语句更新
        return delegate.update(statement);
    }
    ...
}
public class PreparedStatementHandler extends BaseStatementHandler {
    @Override
    public int update(Statement statement) throws SQLException {
        PreparedStatement ps = (PreparedStatement) statement;
    //⑥调用 ShardingSphere-JDBC 的数据库连接声明 ShardingSpherePreparedStatement
类的 execute()方法执行 SQL 路由
        ps.execute();
        int rows = ps.getUpdateCount();
        Object parameterObject = boundSql.getParameterObject();
        KeyGenerator keyGenerator = mappedStatement.getKeyGenerator();
        keyGenerator.processAfter(executor, mappedStatement, ps,
parameterObject);
    //⑦在返回 SQL 语句执行成功后影响的数据条数
        return rows;
    }
    ...
}
```

4. 预处理 SQL 语句

在 SQL 执行的过程中，用 Shardingsphere-JDBC 的 ShardingSpherePreparedStatement 类的 execute()方法预处理 SQL 语句，具体代码如下所示：

```
public final class ShardingSpherePreparedStatement extends
AbstractPreparedStatementAdapter {
    @Override
    public boolean execute() throws SQLException {
        try {
            //①清除数据库连接声明缓存
            clearPrevious();
```

```
                //②创建 SQL 语句执行的上下文对象 ExecutionContext
                executionContext = createExecutionContext();
                if (ExecutorConstant.MANAGED_RESOURCE) {
                    Collection<InputGroup<StatementExecuteUnit>> inputGroups =
getInputGroups();
                    cacheStatements(inputGroups);
                    reply();
                    return preparedStatementExecutor.execute(inputGroups,
executionContext.getSqlStatementContext().getSqlStatement(),
executionContext.getRouteContext().getRouteUnits());
                } else {
                    return rawExecutor.execute(getRawInputGroups(),
                        new RawSQLExecutorCallback());
                }
            } finally {
                clearBatch();
            }
        }
        //③获取数据库连接声明的执行单元对象的集合
        private Collection<InputGroup<StatementExecuteUnit>> getInputGroups()
throws SQLException {
            int maxConnectionsSizePerQuery =schemaContexts.
            getProps().<Integer>getValue
            (ConfigurationPropertyKey.MAX_CONNECTIONS_SIZE_PER_QUERY);
            return new PreparedStatementExecuteGroupEngine
             (maxConnectionsSizePerQuery, connection,
             statementOption,schemaContexts.getDefaultSchema().
             getRules()).generate(executionContext.
             getRouteContext(),executionContext.getExecutionUnits());
    }
    private ExecutionContext createExecutionContext() {
        //④创建逻辑 SQL 对象
        LogicSQL logicSQL = createLogicSQL();
        //⑤调用 KernelProcessor 类的 generateExecutionContext()方法绑定逻辑 SQL 对象
和数据库数据源规格对象，并创建 SQL 语句执行的上下文对象 ExecutionContext
        ExecutionContext result =
        kernelProcessor.generateExecutionContext(logicSQL,
schemaContexts.getDefaultSchema(), schemaContexts.getProps());
        findGeneratedKey(result).ifPresent(generatedKey ->
generatedValues.addAll(generatedKey.getGeneratedValues()));
        logSQL(logicSQL, result);
        return result;
    }
    private LogicSQL createLogicSQL() {
         List<Object> parameters = new ArrayList<>(getParameters());
```

```
        //⑥创建物理规格元数据对象 PhysicalSchemaMetaData
        PhysicalSchemaMetaData schemaMetaData =
        schemaContexts.getDefaultSchema().getMetaData()
        .getSchemaMetaData().getSchemaMetaData();
    //⑦用工厂类 SQLStatementContextFactory 创建 SQL 声明上下文对象
SQLStatementContext
        SQLStatementContext<?> sqlStatementContext =
        SQLStatementContextFactory.newInstance(schemaMetaData, parameters,
sqlStatement);
        return new LogicSQL(sqlStatementContext, sql, parameters);
    }
      ...
    }
```

5. 用路由引擎去拦截 SQL 语句

在 SQL 语句执行的过程中，创建路由引擎 SQLRouteEngine 类，用路由引擎去拦截 SQL 语句，具体代码如下所示：

```
    public final class KernelProcessor {
      //①在创建 SQL 语句执行的上下文对象 ExecutionContext 的过程中，调用
generateExecutionContext()方法创建路由引擎 SQLRouteEngine 类
      public ExecutionContext generateExecutionContext(final LogicSQL
logicSQL, final ShardingSphereSchema schema,
          final ConfigurationProperties props) {
          //②获取规则
          Collection<ShardingSphereRule> rules = schema.getRules();
          //③初始化路由引擎
          SQLRouteEngine sqlRouteEngine = new SQLRouteEngine(rules, props);
          //④获取逻辑 SQL 语句的声明上下文对象
          SQLStatementContext<?> sqlStatementContext =
logicSQL.getSqlStatementContext();
          //⑤获取路由上下文对象
          RouteContext routeContext = sqlRouteEngine.route(logicSQL, schema);
          //⑥用 SQLRewriteEntry 初始化一个实体对象 SQLRewriteEntry
          SQLRewriteEntry rewriteEntry =
          New SQLRewriteEntry(schema.getMetaData().getSchemaMetaData()
          .getConfiguredSchemaMetaData(), props, rules);
          //⑦将执行完成的结果进行重写，并保存在 SQLRewriteResult 对象中
          SQLRewriteResult rewriteResult =
          rewriteEntry.rewrite(logicSQL.getSql(), logicSQL.getParameters(),
sqlStatementContext, routeContext);
          //⑧构建一个执行 SQL 语句的容器单元
          Collection<ExecutionUnit> executionUnits =
          ExecutionContextBuilder.build(schema.getMetaData(),
              rewriteResult, sqlStatementContext);
```

```
        //⑨返回一个包含 SQL 语句执行单元的执行上下文对象
          return new ExecutionContext(sqlStatementContext, executionUnits,
routeContext);
      }
   }
```

6. 获取路由上下文对象 RouteContext

在 SQL 语句执行的过程中，用路由引擎 SQLRouteEngine 类获取路由上下文对象 RouteContext，具体代码如下所示：

```
   @RequiredArgsConstructor
   public final class SQLRouteEngine {
      private final Collection<ShardingSphereRule> rules;
      private final ConfigurationProperties props;
   private final SPIRoutingHook routingHook = new SPIRoutingHook();

      public RouteContext route(final LogicSQL logicSQL, final
ShardingSphereSchema schema) {
          routingHook.start(logicSQL.getSql());
          try {
   //①如果逻辑 SQL 对象被设置为"需要依赖所有的数据库规格",则初始化 AllSQLRouteExecutor
对象，否则初始化 PartialSQLRouteExecutor 对象
             SQLRouteExecutor executor =
isNeedAllSchemas(logicSQL.getSqlStatementContext().getSqlStatement())
    ? new AllSQLRouteExecutor() : new PartialSQLRouteExecutor(rules, props);
             //②调用路由执行器的 route()方法,将应用中的规则应用到需要执行的 SQL 语句中
             RouteContext result = executor.route(logicSQL, schema);
             routingHook.finishSuccess(result,
schema.getMetaData().getSchemaMetaData().getConfiguredSchemaMetaData());
             //③返回路由上下文对象 RouteContext
             return result;
          } catch (final Exception ex) {
             routingHook.finishFailure(ex);
             throw ex;
          }
   }
   //④如果 SQL 语句声明对象匹配 MySQLShowTablesStatement 类，则需要获取所有的数据库规
格列表
      private boolean isNeedAllSchemas(final SQLStatement sqlStatement) {
         return sqlStatement instanceof MySQLShowTablesStatement;
      }
      ...
   }
```

7. 处理 SQL 关联 LogicSQL 对象和 ShardingSphereSchema 对象

在 SQL 语句执行的过程中，用 PartialSQLRouteExecutor 类的 route()方法，去处理 SQL 关联 LogicSQL 对象和 ShardingSphereSchema 对象，具体代码如下所示：

```
public final class PartialSQLRouteExecutor implements SQLRouteExecutor {
static {
    //①用 SPI 接口 ShardingSphereServiceLoader 注册应用中所有 SQLRouter 接口的实现类，比如 ShardingSQLRouter 类
        ShardingSphereServiceLoader.register(SQLRouter.class);
    }
    private final ConfigurationProperties props;
    @SuppressWarnings("rawtypes")
    private final Map<ShardingSphereRule, SQLRouter> routers;
    public PartialSQLRouteExecutor(final Collection<ShardingSphereRule> rules, final ConfigurationProperties props) {
        this.props = props;
        //②用 SPI 接口 OrderedSPIRegistry 的 getRegisteredServices()方法加载 SQL 路由缓存对象 routers，Key 值为“规则”，Value 为“SQLRouter 对象”
        routers = OrderedSPIRegistry.getRegisteredServices(rules, SQLRouter.class);
    }
    @Override
    @SuppressWarnings({"unchecked", "rawtypes"})
    public RouteContext route(final LogicSQL logicSQL, final ShardingSphereSchema schema) {
        RouteContext result = new RouteContext();
        //③遍历 SQL 路由列表，并创建路由上下文对象
        for (Entry<ShardingSphereRule, SQLRouter> entry : routers.entrySet()) {
        //④如果路由执行单元为空，则调用 createRouteContext()方法创建一个新的路由上下文对象
            if (result.getRouteUnits().isEmpty()) {
                result = entry.getValue().createRouteContext(logicSQL, schema, entry.getKey(), props);
            } else {
        //⑤如果路由执行单元不为空，则调用 decorateRouteContext()方法更新原有的路由上下文对象
                entry.getValue().decorateRouteContext(result, logicSQL, schema, entry.getKey(), props);
            }
        }
        return result;
    }
    ...
}
```

8. 创建路由上下文对象，以拦截 SQL 语句和重写 SQL 语句

在 SQL 语句执行的过程中，用 ShardingSQLRouter 类的 createRouteContext()方法创建路由上下文对象，以拦截 SQL 语句和重写 SQL 语句，具体代码如下所示：

```
public final class ShardingSQLRouter implements SQLRouter<ShardingRule> {
    @Override
    public RouteContext createRouteContext(final LogicSQL logicSQL, final
ShardingSphereSchema schema, final ShardingRule rule, final
ConfigurationProperties props) {
        RouteContext result = new RouteContext();
        SQLStatement sqlStatement =
logicSQL.getSqlStatementContext().getSqlStatement();
        //①按照 SQLStatement 的类型，创建不同的数据库声明校验对象，比如
ShardingInsertStatementValidator 类
        Optional<ShardingStatementValidator> validator =
ShardingStatementValidatorFactory.newInstance(sqlStatement);
        //②校验数据库连接声明上下文、逻辑 SQL 的参数，以及数据库规则的元数据
        validator.ifPresent(v -> v.preValidate(rule,
logicSQL.getSqlStatementContext(), logicSQL.getParameters(),
schema.getMetaData()));
        //③将逻辑 SQL 对象、数据库数据源规格及规则组装在对象 ShardingConditions 中
        ShardingConditions shardingConditions =
createShardingConditions(logicSQL, schema, rule);
        boolean needMergeShardingValues =
isNeedMergeShardingValues(logicSQL.getSqlStatementContext(), rule);
        if (sqlStatement instanceof DMLStatement && needMergeShardingValues) {
            checkSubqueryShardingValues(logicSQL.getSqlStatementContext(),
rule, shardingConditions);
            mergeShardingConditions(shardingConditions);
        }
        //④用工厂类 ShardingRouteEngineFactory 构造一个路由引擎，比如
ShardingStandardRoutingEngine 类，然后用路由引擎构造一个路由上下文对象（包含规则、规格
元数据、分片条件对象 ShardingConditions 等）
        ShardingRouteEngineFactory.newInstance(rule, schema.getMetaData(),
logicSQL.getSqlStatementContext(), shardingConditions, props).route(result,
rule);
        validator.ifPresent(v -> v.postValidate(sqlStatement, result));
        //⑤返回路由上下文对象
        return result;
    }
    ...
}
```

9. 执行分库分表，并返回路由上下文对象

在执行 SQL 的过程中，用 ShardingStandardRoutingEngine 类的 route()方法执行分库分表，并返回路由上下文对象，具体代码如下所示：

```
    @RequiredArgsConstructor
    public final class ShardingStandardRoutingEngine implements
ShardingRouteEngine {
        private final String logicTableName;
        private final ShardingConditions shardingConditions;
        private final ConfigurationProperties properties;
        private final Collection<Collection<DataNode>> originalDataNodes = new
LinkedList<>();
        //①调用路由引擎的路由方法 route()执行分库分表
        @Override
        public void route(final RouteContext routeContext, final ShardingRule
shardingRule) {
            Collection<DataNode> dataNodes = getDataNodes(shardingRule,
shardingRule.getTableRule(logicTableName));
            routeContext.getOriginalDataNodes().addAll(originalDataNodes);
            for (DataNode each : dataNodes) {
                //②遍历数据源节点，将路由信息封装在路由单元对象 RouteUnit 中
                routeContext.getRouteUnits().add(
                        new RouteUnit(new RouteMapper(each.getDataSourceName(),
each.getDataSourceName()), Collections.singletonList(new
RouteMapper(logicTableName, each.getTableName()))));
            }
        }
        //③获取应用中配置的数据源节点列表
        private Collection<DataNode> getDataNodes(final ShardingRule
shardingRule, final TableRule tableRule) {
            ShardingStrategy databaseShardingStrategy =
createShardingStrategy(shardingRule.getDatabaseShardingStrategyConfiguration
(tableRule), shardingRule.getShardingAlgorithms());
            ShardingStrategy tableShardingStrategy =
createShardingStrategy(shardingRule.getTableShardingStrategyConfiguration(ta
bleRule), shardingRule.getShardingAlgorithms());
            if (isRoutingByHint(shardingRule, tableRule)) {
                return routeByHint(tableRule, databaseShardingStrategy,
tableShardingStrategy);
            }
            if (isRoutingByShardingConditions(shardingRule, tableRule)) {
                return routeByShardingConditions(shardingRule, tableRule,
databaseShardingStrategy, tableShardingStrategy);
            }
```

```
        return routeByMixedConditions(shardingRule, tableRule,
databaseShardingStrategy, tableShardingStrategy);
    }
    //④组合表的路由规则、数据库的分库策略、表的分表策略等，完成分库分表路由，并将路由信息封装在数据源节点 DataNode 中
    private Collection<DataNode> route0(final TableRule tableRule,
                                         final ShardingStrategy
databaseShardingStrategy, final List<ShardingConditionValue>
databaseShardingValues,
                                         final ShardingStrategy
tableShardingStrategy, final List<ShardingConditionValue> tableShardingValues) {
        Collection<String> routedDataSources = routeDataSources(tableRule,
databaseShardingStrategy, databaseShardingValues);
        Collection<DataNode> result = new LinkedList<>();
        for (String each : routedDataSources) {
            result.addAll(routeTables(tableRule, each,
tableShardingStrategy, tableShardingValues));
        }
        return result;
    }
    //⑤分库
    private Collection<String> routeDataSources(final TableRule tableRule,
final ShardingStrategy databaseShardingStrategy, final
List<ShardingConditionValue> databaseShardingValues) {
        if (databaseShardingValues.isEmpty()) {
            return tableRule.getActualDatasourceNames();
        }
        Collection<String> result = new
LinkedHashSet<>(databaseShardingStrategy.doSharding(tableRule.getActualDatas
ourceNames(), databaseShardingValues, properties));
        Preconditions.checkState(!result.isEmpty(), "no database route
info");
        Preconditions.checkState
(tableRule.getActualDatasourceNames().containsAll(result),
                "Some routed data sources do not belong to configured data
sources. routed data sources: `%s`, configured data sources: `%s`", result,
tableRule.getActualDatasourceNames());
        return result;
    }
    //⑥分表
    private Collection<DataNode> routeTables(final TableRule tableRule,
final String routedDataSource,
                                              final ShardingStrategy
tableShardingStrategy, final List<ShardingConditionValue> tableShardingValues) {
```

```
        Collection<String> availableTargetTables =
tableRule.getActualTableNames(routedDataSource);
        Collection<String> routedTables = new
LinkedHashSet<>(tableShardingValues.isEmpty()
                ? availableTargetTables :
tableShardingStrategy.doSharding(availableTargetTables, tableShardingValues,
properties));
        Collection<DataNode> result = new LinkedList<>();
        for (String each : routedTables) {
            result.add(new DataNode(routedDataSource, each));
        }
        return result;
    }
    //⑦创建分库分表的策略对象
    private ShardingStrategy createShardingStrategy(final
ShardingStrategyConfiguration shardingStrategyConfig, final Map<String,
ShardingAlgorithm> shardingAlgorithms) {
        return null == shardingStrategyConfig ? new NoneShardingStrategy()
                : ShardingStrategyFactory.newInstance (shardingStrategyConfig,
shardingAlgorithms.get(shardingStrategyConfig.getShardingAlgorithmName()));
    }
    ...
}
```

10. 重写 SQL 语句，并将分库分表的路由规则添加到 SQL 语句中

在执行 SQL 语句的过程中，用 SQLRewriteEntry 类重写 SQL 语句，并将分库分表的路由规则添加到 SQL 语句中，具体代码如下所示：

```
    //在 KernelProcessor 类的 generateExecutionContext()方法中获取路由上下文对象后，
开始重写 SQL 语句，并生成新的 SQL 执行单元对象 ExecutionUnit
    SQLRewriteEntry rewriteEntry = new
SQLRewriteEntry(schema.getMetaData().getSchemaMetaData().getConfiguredSchema
MetaData(), props, rules);
    SQLRewriteResult rewriteResult = rewriteEntry.rewrite(logicSQL.getSql(),
logicSQL.getParameters(), sqlStatementContext, routeContext);
    Collection<ExecutionUnit> executionUnits =
ExecutionContextBuilder.build(schema.getMetaData(), rewriteResult,
sqlStatementContext);
```

通过 Java 断点调试 SQLRewriteEntry 重写 SQL 语句的过程如图 12-40 所示。可以看出，在路由上下文对象中已经将数据源路由到“datasource_0”，将表路由到“t_order”。

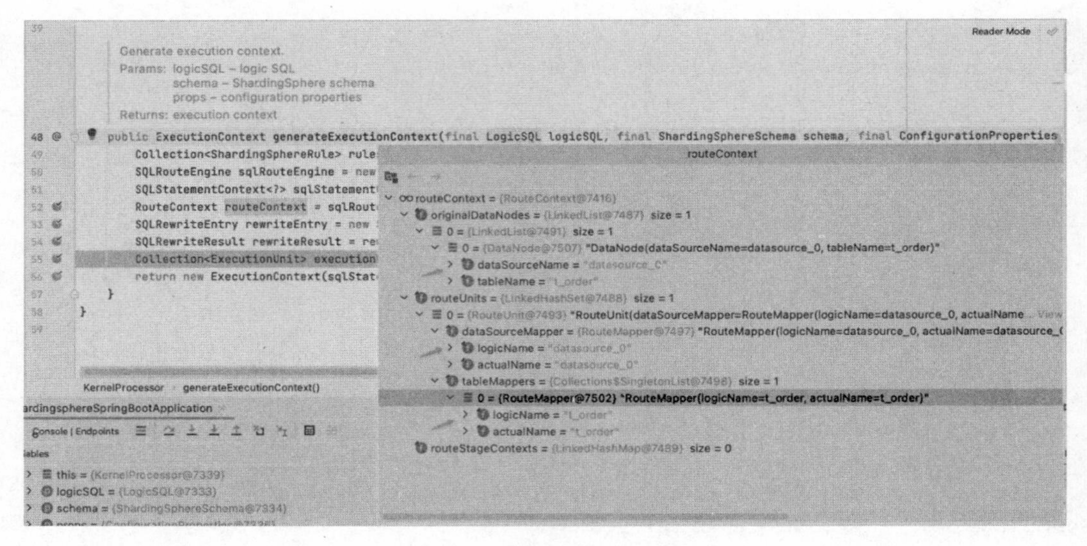

图 12-40

从图 12-41 中可以看出，在 SQL 语句重写后，在对象 SQLRewriteResult 中，已经将路由规则和执行的 SQL 语句绑定在一块。这样 SQL 语句就可以被路由到指定的数据库的数据表上，路由的整个过程，应用完全无感知，非常轻量级。

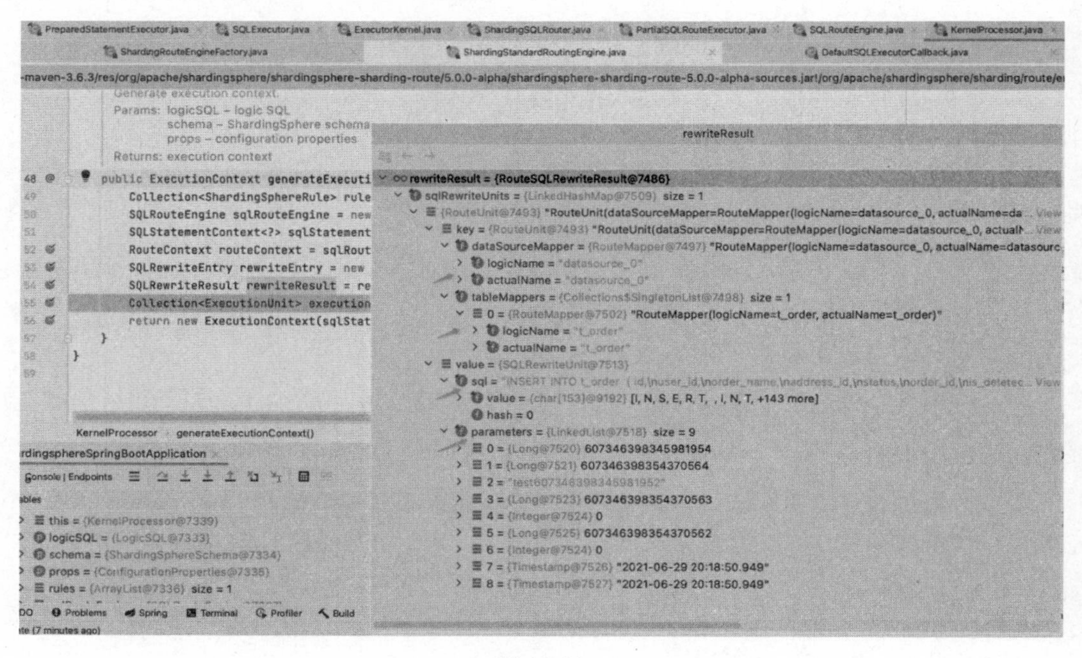

图 12-41

12.4 "读写分离"的原理

Shardingsphere JDBC 不仅支持分库分表，也支持读写分离。

12.4.1 读取应用配置文件中的数据库据源及读写分离规则

如果要实现读写分离，则 Shardingsphere JDBC 需要先读取应用配置文件中的数据库据源及读写分离规则，具体过程如下：

（1）在应用的配置文件（比如 application.properties）中添加如下配置信息：

```
//①配置读写数据源名称
spring.shardingsphere.datasource.names=primary_ds,replica_ds_0,replica_ds_1
//②配置数据源的公共信息
spring.shardingsphere.datasource.common.type=com.zaxxer.hikari.HikariDat
aSource
spring.shardingsphere.datasource.common.driver-class-name=com.mysql.jdbc.
Driver
spring.shardingsphere.datasource.common.username=root
spring.shardingsphere.datasource.common.password=123456huxian
//③配置写数据源 primary_ds
spring.shardingsphere.datasource.primary_ds.jdbc-url=jdbc:mysql://127.0.
0.1:3306/primary_ds?serverTimezone=UTC&useSSL=false&useUnicode=true&characte
rEncoding=UTF-8
//④配置读数据源 replica_ds_0
spring.shardingsphere.datasource.replica_ds_0.jdbc-url=jdbc:mysql://127.
0.0.1:3306/replica_ds_0?serverTimezone=UTC&useSSL=false&useUnicode=true&char
acterEncoding=UTF-8
//⑤配置读数据源 replica_ds_1
spring.shardingsphere.datasource.replica_ds_1.jdbc-url=jdbc:mysql://127.
0.0.1:3306/replica_ds_1?serverTimezone=UTC&useSSL=false&useUnicode=true&char
acterEncoding=UTF-8
spring.shardingsphere.rules.replica-query.load-balancers.round_robin.typ
e=ROUND_ROBIN
//⑥配置 Shardingsphere-JDBC 中读写分离中的写数据源
spring.shardingsphere.rules.replica-query.data-sources.ds_0.primary-data
-source-name=primary_ds
//⑦配置 Shardingsphere-JDBC 中读写分离中的读数据源
spring.shardingsphere.rules.replica-query.data-sources.ds_0.replica-data
-source-names=replica_ds_0,replica_ds_1
//⑧配置两个读数据源（replica_ds_0 和 replica_ds_1）之间的负载均衡策略为轮询
(round_robin)
```

```
    spring.shardingsphere.rules.replica-query.data-sources.ds_0.load-balance
r-name=round_robin
```

（2）当在应用中添加了数据库数据源和读写分离规则后，Shardingsphere JDBC 用 ReplicaQueryDataSourceRuleConfiguration 类读取数据源数据源及用 AlgorithmProvidedReplicaQueryRuleConfiguration 类，以读取读写分离的规则，具体代码如下所示：

```
    @NoArgsConstructor
    @AllArgsConstructor
    @Getter
    @Setter
    public final class AlgorithmProvidedReplicaQueryRuleConfiguration
implements RuleConfiguration {
    //①解析通配符“spring.shardingsphere.rules.replica-query. data-sources.*”
中的 data-sources 字段
    private Collection<ReplicaQueryDataSourceRuleConfiguration> dataSources =
new LinkedList<>();
    ...
    }
    @RequiredArgsConstructor
    @Getter
    public final class ReplicaQueryDataSourceRuleConfiguration {
        private final String name;
        //②解析通配符“spring.shardingsphere.rules.replica-query.*.
primary-data-source-name”中写数据源的名称 private final String
primaryDataSourceName;
    //③解析通配符“spring.shardingsphere.rules.replica-query.*.replica-data-
source-names”中读数据源的名称
        private final List<String> replicaDataSourceNames;
        //④解析通配符“spring.shardingsphere.rules.replica-query.load-
balancers.*”中的负载均衡的策略
        private final String loadBalancerName;
    }
```

12.4.2 使用 ReplicaQuerySQLRouter 类的 createRouteContext()方法创建读写分离的路由上下文对象 RouteContext

在 Shardingsphere JDBC 中，如果应用已经加载完成读写规则，则需要将读写规则组合在路由上下文对象中，这样才能将读写规则和 SQL 语句绑定，并重写 SQL 语句，具体代码如下所示：

```
   public final class ReplicaQuerySQLRouter
       implements SQLRouter<ReplicaQueryRule> {
       @Override
```

```
    public RouteContext createRouteContext(final LogicSQL logicSQL, final
ShardingSphereSchema schema, final ReplicaQueryRule rule, final
ConfigurationProperties props) {
        RouteContext result = new RouteContext();
       //①在调用 ReplicaQueryDataSourceRouter 类的 route()方法后，获取到数据源的名称
        String dataSourceName = new
ReplicaQueryDataSourceRouter(rule.getSingleDataSourceRule()).route(logicSQL.
getSqlStatementContext().getSqlStatement());
        result.getRouteUnits().add(new RouteUnit(new
RouteMapper(DefaultSchema.LOGIC_NAME, dataSourceName),
Collections.emptyList()));
        return result;
    }
    ...
}
@RequiredArgsConstructor
public final class ReplicaQueryDataSourceRouter {
    private final ReplicaQueryDataSourceRule rule;
    public String route(final SQLStatement sqlStatement) {
        //②如果当前 SQL 操作是写请求，则直接返回写数据源名称
        if (isPrimaryRoute(sqlStatement)) {
            PrimaryVisitedManager.setPrimaryVisited();
            return rule.getPrimaryDataSourceName();
        }
        //③从读写分离规则中获取到负载均衡器，从读数据源列表中按照负载均衡算法（默认是
轮询算法）选出一个读数据源的名称
        return rule.getLoadBalancer().getDataSource(rule.getName(),
rule.getPrimaryDataSourceName(), rule.getReplicaDataSourceNames());
    }
    //④判断 SQL 语句是否是写操作，如果是写操作，则返回 true，否则返回 false
    private boolean isPrimaryRoute(final SQLStatement sqlStatement) {
        return containsLockSegment(sqlStatement) || !(sqlStatement
instanceof SelectStatement) || PrimaryVisitedManager.getPrimaryVisited() ||
HintManager.isPrimaryRouteOnly();
    }
    private boolean containsLockSegment(final SQLStatement sqlStatement) {
        return sqlStatement instanceof SelectStatement &&
SelectStatementHandler.getLockSegment((SelectStatement)
sqlStatement).isPresent();
    }
}
```

12.4.3 使用 ReplicaQueryRuleSpringbootConfiguration 类加载应用的负载均衡器 ReplicaLoadBalanceAlgorithm 对象

在 Shardingsphere JDBC 中，应用可以用基于 Spring Boot 的 ReplicaQueryRuleSpringbootConfiguration 类加载应用的负载均衡器 ReplicaLoadBalanceAlgorithm 对象。

（1）实例化读写负载均衡算法注册器对象 ReplicaQueryAlgorithmProvidedBeanRegistry，具体代码如下所示：

```
    @Configuration
    @EnableConfigurationProperties(YamlReplicaQueryRuleSpringBootConfigurati
on.class)
    @ConditionalOnClass(YamlReplicaQueryRuleConfiguration.class)
    @Conditional(ReplicaQuerySpringBootCondition.class)
    @RequiredArgsConstructor
    public class ReplicaQueryRuleSpringbootConfiguration {
        @Bean
    public static ReplicaQueryAlgorithmProvidedBeanRegistry
replicaQueryAlgorithmProvidedBeanRegistry(final Environment environment) {
        //将对象 ReplicaQueryAlgorithmProvidedBeanRegistry 添加到 IOC 容器中
            return new ReplicaQueryAlgorithmProvidedBeanRegistry(environment);
        }
        ...
    }
```

（2）利用 Spring Framwork 的 BeanDefinitionRegistryPostProcessor 类的 postProcessBeanDefinitionRegistry() 方法注册负载均衡算法对象 ReplicaLoadBalanceAlgorithm，具体代码如下所示：

```
    public final class ReplicaQueryAlgorithmProvidedBeanRegistry extends
AbstractAlgorithmProvidedBeanRegistry<ReplicaLoadBalanceAlgorithm> {

        private static final String ALGORITHMS =
"spring.shardingsphere.rules.replica-query.load-balancers.";
        public ReplicaQueryAlgorithmProvidedBeanRegistry(final Environment
environment) {
            super(environment);
    }
    //①注册 ReplicaLoadBalanceAlgorithm 类，并调用抽象类
AbstractAlgorithmProvidedBeanRegistry 的 registerBean()方法
        @Override
        public void postProcessBeanDefinitionRegistry(final
BeanDefinitionRegistry registry) {
```

```
            registerBean(ALGORITHMS, ReplicaLoadBalanceAlgorithm.class,
registry);
        }
    }
    @RequiredArgsConstructor(access = AccessLevel.PROTECTED)
    public abstract class AbstractAlgorithmProvidedBeanRegistry<T extends
ShardingSphereAlgorithm> implements BeanDefinitionRegistryPostProcessor,
BeanPostProcessor {

        private final Environment environment;
        //②注册读写分离的负载均衡器，比如轮询算法
        @SuppressWarnings("all")
        protected void registerBean(final String preFix, final Class<T>
algorithmClass, final BeanDefinitionRegistry registry) {
            Map<String, Object> paramMap = PropertyUtil.handle(environment,
preFix, Map.class);
            Set<String> keys = paramMap.keySet().stream().map(key -> {
                return key.contains(".") ? key.substring(0, key.indexOf(".")) :
key;
            }).collect(Collectors.toSet());
            Map<String, YamlShardingSphereAlgorithmConfiguration>
shardingAlgorithmMap = new LinkedHashMap<>();
            keys.forEach(each -> {
                String type = environment.getProperty(preFix + each + ".type");
                Map<String, Object> propsMap = PropertyUtil.handle(environment,
preFix + each + ".props", Map.class);
                YamlShardingSphereAlgorithmConfiguration config = new
YamlShardingSphereAlgorithmConfiguration();
                config.setType(type);
                config.getProps().putAll(propsMap);
                shardingAlgorithmMap.put(each, config);
            });
            ShardingSphereServiceLoader.register(algorithmClass);
            shardingAlgorithmMap.forEach((k, v) -> {
                ShardingSphereAlgorithm algorithm = TypedSPIRegistry.
getRegisteredService(algorithmClass, v.getType(), v.getProps());
                BeanDefinitionBuilder builder =
BeanDefinitionBuilder.genericBeanDefinition(algorithm.getClass());
                builder.addPropertyValue("props", v.getProps());
                registry.registerBeanDefinition(k,
builder.getBeanDefinition());
            });
        }
        //③在负载均衡器注册完成后执行初始化
        @Override
```

```
    public Object postProcessAfterInitialization(final Object bean, final
String beanName) {
        if (bean instanceof ShardingSphereAlgorithmPostProcessor) {
            ((ShardingSphereAlgorithmPostProcessor) bean).init();
        }
        return bean;
    }
    ...
    }
```

12.5 用 Netty 实现 Shardingsphere Proxy 的通信渠道

在 Shardingsphere Proxy 中，主要是用 Netty 来实现其通信渠道。

12.5.1 “Shardingsphere Proxy 通信渠道”的原理

Shardingsphere Proxy 通信渠道主要用于完成客户端（Java 应用或者命令控制台）和 Shardingsphere Proxy 之间的 RPC 通信。下面来看看通信渠道初始化的过程。

1. 初始化通信渠道

在 Shardingsphere Proxy 的通信渠道中，用 ShardingSphereProxy 类的 start()方法初始化通信渠道的具体代码如下所示：

```
public final class ShardingSphereProxy {
    private EventLoopGroup bossGroup;
    private EventLoopGroup workerGroup;
    @SneakyThrows(InterruptedException.class)
    public void start(final int port) {
        try {
            //①创建事件对象组 EventLoopGroup
            createEventLoopGroup();
            ServerBootstrap bootstrap = new ServerBootstrap();
            //②初始化 ServerBootstrap 对象
            initServerBootstrap(bootstrap);
            //③绑定指定的端口号，并开启 Netty 的 NIO 通信渠道
            ChannelFuture future = bootstrap.bind(port).sync();
            log.info("ShardingSphere-Proxy start success.");
            future.channel().closeFuture().sync();
        } finally {
            //④如果通信渠道开启失败，则释放线程资源和后台执行器上下文资源
            workerGroup.shutdownGracefully();
            bossGroup.shutdownGracefully();
```

```
            BackendExecutorContext.getInstance().getExecutorKernel().close();
        }
    }
    ...
}
```

2. 创建 EventLoopGroup 对象

在 Shardingsphere Proxy 的通信渠道中，将接收和处理客户端的 RPC 通信请求的两个事件组进行隔离，具体代码如下所示：

```
public final class ShardingSphereProxy {
    private EventLoopGroup bossGroup;
    private EventLoopGroup workerGroup;
private void createEventLoopGroup() {
    //①判断操作系统是否开启了“基于 Epoll 的 NIO 通信“，如果开启了，则用
EpollEventLoopGroup 类初始化接收事件的 bossGroup 对象，否则用 NioEventLoopGroup 类初
始化
        bossGroup = Epoll.isAvailable() ? new EpollEventLoopGroup(1) : new
NioEventLoopGroup(1);
    //②判断操作系统是否开启了“基于 Epoll 的 NIO 通信”，如果开启了，则用
EpollEventLoopGroup 类初始化处理事件的 workerGroup 对象，否则用 NioEventLoopGroup 类
初始化
        workerGroup = Epoll.isAvailable() ? new EpollEventLoopGroup() : new
NioEventLoopGroup();
    }
    ...
}
```

3. 初始化 ServerBootstrap 对象

在 Shardingsphere Proxy 的通信渠道中，用 Netty 的 ServerBootstrap 类来开启通信渠道，具体代码如下所示：

```
public final class ShardingSphereProxy {
    //①初始化 Netty 的 ServerBootstrap 类
private void initServerBootstrap(final ServerBootstrap bootstrap) {
    //②绑定接收和处理客户端的 RPC 通信请求的事件组对象 EventLoopGroup
        bootstrap.group(bossGroup, workerGroup)
                //③绑定 NIO 通道
                .channel(Epoll.isAvailable() ?
EpollServerSocketChannel.class : NioServerSocketChannel.class)
                //④标识当服务器请求处理线程全满时，用于临时存放已完成三次握手的请求的
队列的最大长度
                .option(ChannelOption.SO_BACKLOG, 128)
                //⑤设置写缓存的水位对象 WriteBufferWaterMark
```

```
                .option(ChannelOption.WRITE_BUFFER_WATER_MARK, new
WriteBufferWaterMark(8 * 1024 * 1024, 16 * 1024 * 1024))
                //⑥设置缓冲区为 PooledByteBufAllocator
                .option(ChannelOption.ALLOCATOR,
PooledByteBufAllocator.DEFAULT)
                //⑦设置子 ChannelOption 对象的缓冲区为 PooledByteBufAllocator
                .childOption(ChannelOption.ALLOCATOR,
PooledByteBufAllocator.DEFAULT)
                //⑧设置 TCP_NODELAY 的值为 true
                .childOption(ChannelOption.TCP_NODELAY, true)
                //⑨添加事件处理器对象 LoggingHandler
                .handler(new LoggingHandler(LogLevel.INFO))
                //⑩添加子事件处理器对象 ServerHandlerInitializer
                .childHandler(new ServerHandlerInitializer());
        }
        ...
    }
```

4. 添加渠道管道流 FrontendChannelInboundHandler 类

在 Shardingsphere Proxy 的通信渠道中，用事件处理器 ServerHandlerInitializer 类添加渠道管道流 FrontendChannelInboundHandler 类，具体代码如下所示：

```
    @RequiredArgsConstructor
    public final class ServerHandlerInitializer
           extends ChannelInitializer<SocketChannel> {
        @Override
    protected void initChannel(final SocketChannel socketChannel) {
        //①初始化数据库协议引擎
            DatabaseProtocolFrontendEngine databaseProtocolFrontendEngine =
DatabaseProtocolFrontendEngineFactory.newInstance(getDatabaseType());
            ChannelPipeline pipeline = socketChannel.pipeline();
            //②添加编码器引擎对象 DatabasePacketCodecEngine
            pipeline.addLast(new
PacketCodec(databaseProtocolFrontendEngine.getCodecEngine()));
            //③添加前置通道处理器 FrontendChannelInboundHandler
            pipeline.addLast(new
FrontendChannelInboundHandler(databaseProtocolFrontendEngine));
        }
        //④获取数据库的数据源类型，默认为 MySQL
        private DatabaseType getDatabaseType() {
            return ProxyContext.getInstance().getSchemaContexts()
            .getSchemas().isEmpty() ? new MySQLDatabaseType()
            : ProxyContext.getInstance().getSchemaContexts().getDatabaseType();
        }
    }
```

12.5.2　【实例】搭建通信渠道环境，将 Spring Cloud Alibaba 应用接入 Shardingsphere Proxy

本实例的源码在本书配套资源的“/spring-cloud-alibaba-practice/chaptertwelve/shardingsphere-proxy-use-spring-cloud-alibaba/”目录下。

在 ShardingSphere 中，应用强依赖 ShardingSphere JDBC 的 SDK 去完成分库分表。如果应用接入了 Shardingsphere Proxy，则不用考虑 SDK 版本的兼容性问题，非常轻量级。

1. 搭建 Shardingsphere Proxy 的运行环境

搭建 Shardingsphere Proxy 的运行环境的步骤如下：

（1）从官网下载安装包 apache-shardingsphere-5.0.0-alpha-shardingsphere-proxy-bin.tar.gz，并解压缩，如图 12-42 所示。

Name	Size	Kind	Date Modified
apache-shardingsp...ingsphere-proxy-bin	--	Folder	Jun 13, 2021 at 15:09
logs	--	Folder	Jun 15, 2021 at 15:27
conf	--	Folder	Jun 13, 2021 at 15:09
README.txt	2 KB	Plain Text	Jun 13, 2021 at 15:09
lib	--	Folder	Jun 13, 2021 at 15:09
NOTICE	12 KB	Document	Jun 13, 2021 at 15:09
licenses	--	Folder	Jun 13, 2021 at 15:09
bin	--	Folder	Jun 13, 2021 at 15:09
start.sh	2 KB	Shell S...t (Bash)	Jun 13, 2021 at 15:09
stop.sh	1 KB	Shell S...t (Bash)	Jun 13, 2021 at 15:09
start.bat	2 KB	Batch File	Jun 13, 2021 at 15:09
LICENSE	16 KB	Document	Jun 13, 2021 at 15:09

图 12-42

（2）在文件“/conf/server.yaml”中配置代理连接的认证信息，server.yaml 的具体配置信息可以参考本书配套资源中的代码。

（3）在文件“/conf/config-sharding.yaml”中配置分库分表的规则，config-sharding.yaml 的具体配置信息可以参考本书配套资源中的代码。

（4）用脚本“/bin/start.sh”启动 Shardingsphere Proxy。

2. 初始化服务 shardingsphere-proxy-use-spring-cloud-alibaba

使用 Spring Cloud Alibaba 作为基础框架快速初始化服务 shardingsphere-proxy-use-spring-cloud-alibaba，具体代码可以参考本书配套资源中的代码。

如果要将服务接入 Shardingsphere Proxy，则需要在服务中添加一个配置文件 bootstrap.yaml，并在配置文件中配置数据源代理的连接信息，具体配置信息可以参考本书配套资源中的代码。

3. 启动服务，并验证分库分表策略

在 IDEA 中，在启动服务 shardingsphere-proxy-use-spring-cloud-alibaba 后，执行 1 次“curl 127.0.0.1:6789/order/insert”命令，向数据库中执行 1 次插入语句。服务中的执行日志如图 12-43 所示，总共产生了 3 条 SQL 语句（作用于 3 张表 t_order、t_order_item 及 t_address）。

```
Creating a new SqlSession
SqlSession [org.apache.ibatis.session.defaults.DefaultSqlSession@3815deed] was not registered for synchronization because synchronization is not active
2021-07-14 08:57:20.807  INFO 18195 --- [nio-6789-exec-1] com.zaxxer.hikari.HikariDataSource       : HikariPool-1 - Starting...
2021-07-14 08:57:21.016  INFO 18195 --- [nio-6789-exec-1] com.zaxxer.hikari.HikariDataSource       : HikariPool-1 - Start completed.
JDBC Connection [HikariProxyConnection@740711323 wrapping com.mysql.cj.jdbc.ConnectionImpl@4adc5b10] will not be managed by Spring
==>  Preparing: INSERT INTO t_order ( id, user_id, order_name, address_id, status, order_id, is_deleted, gmt_create, gmt_modified ) VALUES ( ?, ?, ?, ?, ?, ?, ?, ?, ? )
==> Parameters: 612610710526390275(Long), 612610710534778884(Long), test612610710526390272(String), 612610710534778882(Long), 0(Integer), 612610710534778882(Long), 0(Integer), 2021-07-14 08:57:20.779(Timestamp), 2021-07-14 08:57:20.779(Timestamp)
<==    Updates: 1
Closing non transactional SqlSession [org.apache.ibatis.session.defaults.DefaultSqlSession@3815deed]
Creating a new SqlSession
SqlSession [org.apache.ibatis.session.defaults.DefaultSqlSession@eba45e] was not registered for synchronization because synchronization is not active
JDBC Connection [HikariProxyConnection@1047380972 wrapping com.mysql.cj.jdbc.ConnectionImpl@4adc5b10] will not be managed by Spring
==>  Preparing: INSERT INTO t_order_item ( id, order_id, order_item_id, user_id, status, good_id, is_deleted, gmt_create, gmt_modified ) VALUES ( ?, ?, ?, ?, ?, ?, ?, ?, ? )
==> Parameters: 612610714502590465(L[illegible]12610710534778882(Long), 612610714502590467(Long), 612610710534778884(Long), 0(Integer), 612610714502590468(Long), 0(Integer), 2021-07-14 08:57:21.725(Timestamp), 2021-07-14 08:57:21.725(Timestamp)
<==    Updates: 1
Closing non transactional SqlSession [org.apache.ibatis.session.defaults.DefaultSqlSession@eba45e]
Creating a new SqlSession
SqlSession [org.apache.ibatis.session.defaults.DefaultSqlSession@741a6df7] was not registered for synchronization because synchronization is not active
JDBC Connection [HikariProxyConnection@832451507 wrapping com.mysql.cj.jdbc.ConnectionImpl@4adc5b10] will not be managed by Spring
==>  Preparing: INSERT INTO t_address ( id, address_id, address_name, is_deleted, gmt_create, gmt_modified ) VALUES ( ?, ?, ?, ?, ?, ? )
==> Parameters: 612610714787803138(Long[illegible]710534778882(Long), test-address612610714787803137(String), 0(Integer), 2021-07-14 08:57:21.793(Timestamp), 2021-07-14 08:57:21.793(Timestamp)
<==    Updates: 1
Closing non transactional SqlSession [org.apache.ibatis.session.defaults.DefaultSqlSession@741a6df7]
```

图 12-43

在 Shardingsphere Proxy 中执行 SQL 语句的过程如下：

（1）图 12-44 所示为 Shardingsphere Proxy 执行表“t_order”的逻辑 SQL 语句的日志：Shardingsphere Proxy 将表“t_order”的逻辑 SQL 语句从物理表“datasource_0.t_order”路由到物理表“datasource_1.t_order_1”。这样 Shardingsphere Proxy 会向物理表“datasource_1.t_order_1”中插入 1 条数据。

（2）图 12-45 所示为 Shardingsphere Proxy 执行表“t_order_item”的逻辑 SQL 语句的日志：Shardingsphere Proxy 将表“t_order_item”的逻辑 SQL 语句从物理表“datasource_2.t_order_item”路由到物理表“datasource_1.t_order_item_1”。这样 Shardingsphere Proxy 会向物理表“datasource_1. t_order_item_1”中插入 1 条数据。

（3）图 12-46 所示为 Shardingsphere Proxy 执行表“t_address”的逻辑 SQL 语句的日志：Shardingsphere Proxy 将表“t_address”的逻辑 SQL 语句从物理表“datasource_2.t_address”路由到物理表“datasource_0.t_address”、物理表“datasource_1.t_address”和物理表“datasource_2.t_address”。这样 Shardingsphere Proxy 会向这 3 张物理表中插入 3 条数据

（t_address 是广播表，所以会在分库后的数据源中都插入 1 条数据，主要是用来冗余表数据）。

```
[INFO ] 08:57:21.646 [ShardingSphere-Command-11] ShardingSphere-SQL - Logic SQL: SELECT @@session.transaction_read_only
[INFO ] 08:57:21.646 [ShardingSphere-Command-11] ShardingSphere-SQL - SQLStatement: MySQLSelectStatement(limit=Optional.empty, lock=Optional.empty)
[INFO ] 08:57:21.646 [ShardingSphere-Command-11] ShardingSphere-SQL - Actual SQL: datasource_0 ::: SELECT @@session.transaction_read_only
[INFO ] 08:57:21.672 [ShardingSphere-Command-8] ShardingSphere-SQL - Logic SQL: INSERT INTO t_order  ( id,
user_id,
order_name,
address_id,
status,
order_id,
is_deleted,
gmt_create,
gmt_modified )  VALUES  ( ?,
?,
?,
?,
?,
?,
?,
?,
? )
[INFO ] 08:57:21.672 [ShardingSphere-Command-8] ShardingSphere-SQL - SQLStatement: MySQLInsertStatement(setAssignment=Optional.empty, onDuplicateKeyColumns=Optional.empty)
[INFO ] 08:57:21.672 [ShardingSphere-Command-8] ShardingSphere-SQL - Actual SQL: datasource_1 ::: INSERT INTO t_order_1  ( id,
user_id,
order_name,
address_id,
status,
order_id,
is_deleted,
gmt_create,
gmt_modified )  VALUES  (?, ?, ?, ?, ?, ?, ?, ?, ?) ::: [612610710526390275, 612610710534778884, test612610710526390272, 612610710534778882, 0, 612610710534778882, 0, 2021-07-13 19:57:20.000779, 2021
-07-13 19:57:20.000779]
```

图 12-44

```
[INFO ] 08:57:21.744 [ShardingSphere-Command-14] ShardingSphere-SQL - Logic SQL: SELECT @@session.transaction_read_only
[INFO ] 08:57:21.744 [ShardingSphere-Command-14] ShardingSphere-SQL - SQLStatement: MySQLSelectStatement(limit=Optional.empty, lock=Optional.empty)
[INFO ] 08:57:21.744 [ShardingSphere-Command-14] ShardingSphere-SQL - Actual SQL: datasource_2 ::: SELECT @@session.transaction_read_only
[INFO ] 08:57:21.772 [ShardingSphere-Command-15] ShardingSphere-SQL - Logic SQL: INSERT INTO t_order_item  ( id,
order_id,
order_item_id,
user_id,
status,
good_id,
is_deleted,
gmt_create,
gmt_modified )  VALUES  ( ?,
?,
?,
?,
?,
?,
?,
?,
? )
[INFO ] 08:57:21.772 [ShardingSphere-Command-15] ShardingSphere-SQL - SQLStatement: MySQLInsertStatement(setAssignment=Optional.empty, onDuplicateKeyColumns=Optional.empty)
[INFO ] 08:57:21.772 [ShardingSphere-Command-15] ShardingSphere-SQL - Actual SQL: datasource_1 ::: INSERT INTO t_order_item_1  ( id,
order_id,
order_item_id,
user_id,
status,
good_id,
is_deleted,
gmt_create,
gmt_modified )  VALUES  (?, ?, ?, ?, ?, ?, ?, ?, ?) ::: [612610714502590465, 612610710534778882, 612610714502590467, 612610710534778884, 0, 612610714502590468, 0, 2021-07-13 19:57:21.000725, 2021-07-
13 19:57:21.000725]
```

图 12-45

```
[INFO ] 08:57:21.802 [ShardingSphere-Command-1] ShardingSphere-SQL - Logic SQL: SELECT @@session.transaction_read_only
[INFO ] 08:57:21.804 [ShardingSphere-Command-1] ShardingSphere-SQL - SQLStatement: MySQLSelectStatement(limit=Optional.empty, lock=Optional.empty)
[INFO ] 08:57:21.805 [ShardingSphere-Command-1] ShardingSphere-SQL - Actual SQL: datasource_2 ::: SELECT @@session.transaction_read_only
[INFO ] 08:57:21.808 [ShardingSphere-Command-2] ShardingSphere-SQL - Logic SQL: INSERT INTO t_address  ( id,
address_id,
address_name,
is_deleted,
gmt_create,
gmt_modified )  VALUES  ( ?,
?,
?,
?,
?,
? )
[INFO ] 08:57:21.808 [ShardingSphere-Command-2] ShardingSphere-SQL - SQLStatement: MySQLInsertStatement(setAssignment=Optional.empty, onDuplicateKeyColumns=Optional.empty)
[INFO ] 08:57:21.808 [ShardingSphere-Command-2] ShardingSphere-SQL - Actual SQL: datasource_0 ::: INSERT INTO t_address  ( id,
address_id,
address_name,
is_deleted,
gmt_create,
gmt_modified )  VALUES  (?, ?, ?, ?, ?, ?) ::: [612610714787803138, 612610710534778882, test-address612610714787803137, 0, 2021-07-13 19:57:21.000793, 2021-07-13 19:57:21.000793]
[INFO ] 08:57:21.808 [ShardingSphere-Command-2] ShardingSphere-SQL - Actual SQL: datasource_1 ::: INSERT INTO t_address  ( id,
address_id,
address_name,
is_deleted,
gmt_create,
gmt_modified )  VALUES  (?, ?, ?, ?, ?, ?) ::: [612610714787803138, 612610710534778882, test-address612610714787803137, 0, 2021-07-13 19:57:21.000793, 2021-07-13 19:57:21.000793]
[INFO ] 08:57:21.808 [ShardingSphere-Command-2] ShardingSphere-SQL - Actual SQL: datasource_2 ::: INSERT INTO t_address  ( id,
address_id,
address_name,
is_deleted,
gmt_create,
gmt_modified )  VALUES  (?, ?, ?, ?, ?, ?) ::: [612610714787803138, 612610710534778882, test-address612610714787803137, 0, 2021-07-13 19:57:21.000793, 2021-07-13 19:57:21.000793]
```

图 12-46

4. 验证数据库中的数据

查看物理表“datasource_1.t_order_1”中的数据，与分库分表的策略一致，如图 12-47 所示。

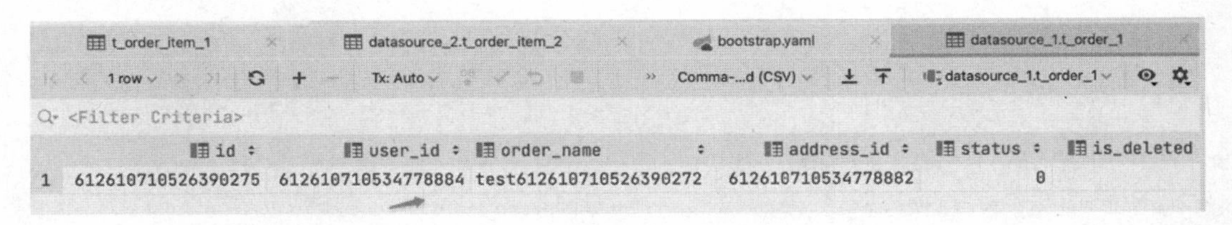

	id	user_id	order_name	address_id	status	is_deleted
1	612610710526390275	612610710534778884	test612610710526390272	612610710534778882	0	

图 12-47

查看物理表“datasource_1. t_order_item_1”中的数据，与分库分表的策略一致，如图 12-48 所示。

	id	order_id	user_id	status	good_id	order_item_id
1	612610714502590465	612610710534778882	612610710534778884	0	612610714502590468	61261071450259046

图 12-48

查看物理表“datasource_0.t_address”“datasource_1.t_address”和“datasource_2.t_address”中的数据，与分库分表的策略一致，如图 12-49、12-50 和 12-51 所示。

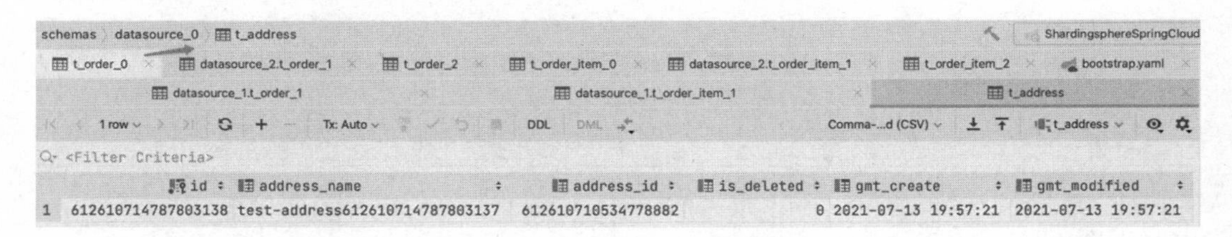

	id	address_name	address_id	is_deleted	gmt_create	gmt_modified
1	612610714787803138	test-address612610714787803137	612610710534778882	0	2021-07-13 19:57:21	2021-07-13 19:57:21

图 12-49

	id	address_name	address_id	is_deleted	gmt_create	gmt_modified
1	612610714787803138	test-address612610714787803137	612610710534778882	0	2021-07-13 19:57:21	2021-07-13 19:57:2

图 12-50

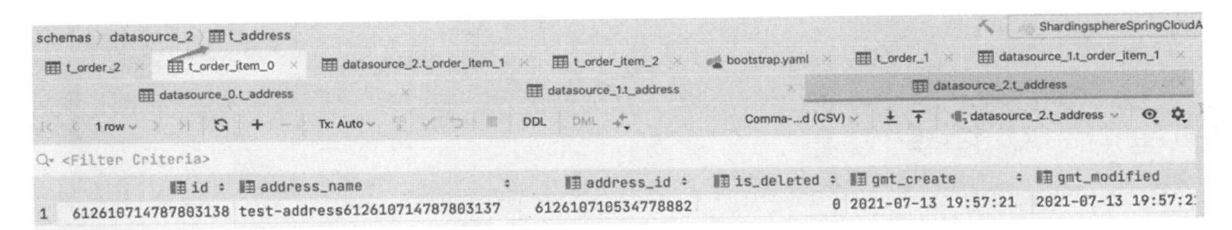

图 12-51

12.6　“使用 SQL 解析引擎实现 Shardingsphere Proxy 分库分表”的原理

在 Shardingsphere Proxy 中，Netty 作为通信渠道在客户端和 Sharding Proxy 之间进行 RPC 通信。Shardingsphere Proxy 在接收到 RPC 请求后，需要用 SQL 解析引擎处理与 SQL 相关的 RPC 请求，并模拟数据库的客户端去调用数据库服务器。

12.6.1　为什么需要 SQL 解析引擎

图 12-52 描述了应用接入 Shardingsphere Proxy 后 SQL 语句的执行过程：

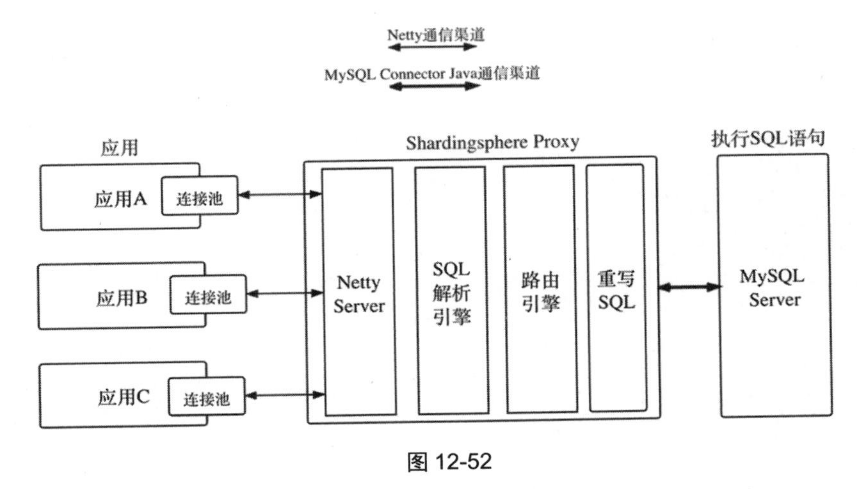

图 12-52

（1）在应用中，通过连接池（比如 Druid）去连接 Shardingsphere Proxy 的 Netty Server。

（2）Netty Server 在接收到应用执行 SQL 语句的 RPC 请求后，将其转发给 SQL 解析引擎，由 SQL 解析引擎模拟 MySQL 客户端去解析 SQL 语句。

（3）在路由引擎中，Shardingsphere Proxy 会拦截 SQL 语句，将数据源和规则绑定，并重写 SQL 语句。

（4）用通信渠道 MySQL Connector Java 调用 MySQL Server，去执行重写后的 SQL 语句，并返回执行完 SQL 语句后的结果。

在 Shardingsphere Proxy 中，SQL 解析引擎的主要功能如下：

- 在解析 SQL 语句的过程中，SQL 解析引擎校验 SQL 的语法（Shardingsphere Proxy 目前能够支持的数据库包括 MySQL、Oracle、PostgreSQL 等）；将 SQL 语句转换为数据库数据源声明；调用路由引擎将数据源和规则绑定，并重写 SQL 语句；将 SQL 语句转换为数据库数据源声明，并调用数据库数据源的声明对象访问数据库。
- 在执行完成 SQL 语句后，SQL 解析引擎处理 SQL 语句执行后的结果，并返给客户端。

12.6.2 使用命令设计模式实现 SQL 语句的路由

在 Shardingsphere Proxy 中，Netty Server 在接收并处理客户端的 RPC 请求后，会将 RPC 请求转发到 SQL 解析引擎。Shardingsphere Proxy 在解析不同类型的 SQL 语句后，生成不同的命令对象，用不同的类去处理 SQL 语句，从而实现按照 SQL 语句的类型执行命令的路由。总之，Shardingsphere Proxy 是用命令设计模式实现 SQL 语句的路由的。

1. 处理 RPC 请求

在 FrontendChannelInboundHandler 类中，用 channelActive()方法去激活通信渠道，具体代码如下所示：

```
    @Slf4j
    public final class FrontendChannelInboundHandler extends
ChannelInboundHandlerAdapter {
        @Override
    public void channelActive(final ChannelHandlerContext context) {
        //①注册渠道 ID
            ChannelThreadExecutorGroup.getInstance().
            register(context.channel().id());
            //②调用数据库协议引擎的认证引擎类 AuthenticationEngine 的 handshake()方
法，完成渠道处理器上下文的“通信握手”
            backendConnection.setConnectionId(databaseProtocolFrontendEngine.
            getAuthEngine().handshake(context));
        }
        ...
    }
```

在 FrontendChannelInboundHandler 类中，用 channelRead()方法处理 RPC 请求，具体代码如下所示：

```
@Slf4j
public final class FrontendChannelInboundHandler
    extends ChannelInboundHandlerAdapter {
    @Override
public void channelRead(final ChannelHandlerContext context, final Object
message) {
     //①校验 RPC 请求（执行 SQL 请求）的权限
        if (!authorized) {
            authorized = auth(context, (ByteBuf) message);
            return;
        }
        //②判定 RPC 请求（执行 SQL 请求）是否支持 Hint
        boolean supportHint =
ProxyContext.getInstance().getSchemaContexts().getProps().<Boolean>getValue(
ConfigurationPropertyKey.PROXY_HINT_ENABLED);
        //③判定 RPC 请求的链接（执行 SQL 请求）是否支持“线程私有”（即一个链接配对一个
线程，对其他连接不开放）
        boolean isOccupyThreadForPerConnection =
databaseProtocolFrontendEngine.getFrontendContext().isOccupyThreadForPerConn
ection();
        //④构造一个异步线程池对象 ExecutorService
        ExecutorService executorService =
CommandExecutorSelector.getExecutorService(
                isOccupyThreadForPerConnection, supportHint,
backendConnection.getTransactionStatus().getTransactionType(),
context.channel().id());
        //⑤执行异步线程池的 execute()方法处理 RPC 请求（执行 SQL 请求）
        executorService.execute(new
CommandExecutorTask(databaseProtocolFrontendEngine, backendConnection,
context, message));
    }
    //⑥校验通信渠道的合法性
    private boolean auth(final ChannelHandlerContext context, final ByteBuf
message) {
        try (
        PacketPayload payload
         = databaseProtocolFrontendEngine.getCodecEngine().
           createPacketPayload(message)) {
            AuthenticationResult authResult =
databaseProtocolFrontendEngine.getAuthEngine().auth(context, payload);
            if (authResult.isFinished()) {
                backendConnection.setUsername(authResult.getUsername());
                backendConnection.setCurrentSchema(authResult.getDatabase());
            }
            return authResult.isFinished();
```

```
            } catch (final Exception ex) {
                log.error("Exception occur: ", ex);
                context.write(databaseProtocolFrontendEngine.
getCommandExecuteEngine().getErrorPacket(ex));
            }
            return false;
        }
        ...
    }
```

2. 异步地处理 RPC 请求（执行 SQL 请求）

在 Shardingsphere Proxy 中，用 CommandExecutorTask 类的 run()方法异步地处理 RPC 请求，具体代码如下所示：

```
    @RequiredArgsConstructor
    @Slf4j
    public final class CommandExecutorTask implements Runnable {
        private final DatabaseProtocolFrontendEngine
databaseProtocolFrontendEngine;
        private final BackendConnection backendConnection;
        private final ChannelHandlerContext context;
        private final Object message;

        @Override
        public void run() {
            RootInvokeHook rootInvokeHook = new SPIRootInvokeHook();
            //①启动 Shardingsphere Proxy 的 hook 机制，在异步 RPC 请求处理完后，能够优雅
地释放与连接相关的资源
            rootInvokeHook.start();
            boolean isNeedFlush = false;
            int connectionSize = 0;
            try (PacketPayload payload = databaseProtocolFrontendEngine.
getCodecEngine().createPacketPayload((ByteBuf) message)) {
                ConnectionStatus connectionStatus =
backendConnection.getConnectionStatus();
                //②判断通信渠道连接的状态
                if
(!backendConnection.getTransactionStatus().isInConnectionHeldTransaction()) {
                    //③如果状态不是“事务处理中”，且连接并没有持有任何事务，则将通信渠道
连接加锁，加锁期间其他事务不能使用该连接
                    connectionStatus.waitUntilConnectionRelease();
                    //④如果状态不是“事务处理中”，且连接并没有持有任何事务，则设置通信渠
道连接的状态标识 isUsing 为 true，表示其已经被占用，在事务执行期间不能被复用
                    connectionStatus.switchToUsing();
                }
```

```
            //⑤异步地处理 SQL 请求
            isNeedFlush = executeCommand(context, payload, backendConnection);
            connectionSize = backendConnection.getConnectionSize();
        } catch (final Exception ex) {
            processException(ex);
        } finally {
            Collection<SQLException> exceptions = closeExecutionResources();
            //⑥如果 isNeedFlush 为 true，则调用 flush()方法刷新通道处理上下文类
            if (isNeedFlush) {
                context.flush();
            }
            if (!backendConnection.getTransactionStatus().
isInConnectionHeldTransaction()) {
                exceptions.addAll(backendConnection.closeConnections(false));
            }
            processClosedExceptions(exceptions);
            rootInvokeHook.finish(connectionSize);
        }
    }
```

3. 执行 SQL 请求的命令

在 Shardingsphere Proxy 中，用 CommandExecutorTask 类的 executeCommand()方法执行 SQL 请求的命令，具体代码如下所示：

```
    private boolean executeCommand(final ChannelHandlerContext context, final
PacketPayload payload, final BackendConnection backendConnection) throws
SQLException {
        //①从数据库协议引擎中，获取 SQL 解析引擎 CommandExecuteEngine 对象，比如
MySQLCommandExecuteEngine
        CommandExecuteEngine commandExecuteEngine =
databaseProtocolFrontendEngine.getCommandExecuteEngine();
        //②解析 SQL 请求的类型，比如 COM_QUERY 类型。如果使用的是 MySQL 数据源，则是
MySQL 的 MySQLCommandPacketType。如果使用的是 PostgreSQL 数据源，则是 PostgreSQL 的
PostgreSQLCommandPacketType
        //③解析 SQL 语句执行后响应的数据包的类型，比如 COM_CREATE_DB 类。如果使用的是
MySQL 数据源，则是 MySQL 的 MySQLCommandPacketType。如果使用的是 PostgreSQL 数据源，则
是 PostgreSQL 的 PostgreSQLCommandPacketType
        CommandPacketType type =
commandExecuteEngine.getCommandPacketType(payload);
        //④获取 SQL 请求的数据包对象 CommandPacket
        CommandPacket commandPacket =
commandExecuteEngine.getCommandPacket(payload, type, backendConnection);
        //⑤获取能够处理 SQL 请求的命令处理器对象 CommandExecutor，比如
MySQLComStmtPrepareExecutor、MySQLComStmtExecuteExecutor 等
```

```
        CommandExecutor commandExecutor = commandExecuteEngine.
getCommandExecutor(type, commandPacket, backendConnection);
        //⑥如果是执行 SQL 语句（增加、删除、查询、修改），则用
MySQLComStmtExecuteExecutor 类的 execute()方法执行 SQL 语句
        Collection<DatabasePacket<?>> responsePackets =
commandExecutor.execute();
        if (responsePackets.isEmpty()) {
            return false;
        }
        responsePackets.forEach(context::write);
        if (commandExecutor instanceof QueryCommandExecutor) {
            //⑦如果 SQL 请求的类型是 QueryCommandExecutor，则调用
CommandExecuteEngine 类的 writeQueryData()方法写入查询语句执行后返回的数据
            commandExecuteEngine.writeQueryData(context, backendConnection,
(QueryCommandExecutor) commandExecutor, responsePackets.size());
            return true;
        }
        return databaseProtocolFrontendEngine.getFrontendContext().
isFlushForPerCommandPacket();
    }
```

4. 写入查询语句执行后返回的数据

在Shardingsphere Proxy中，用MySQLCommandExecuteEngine类的writeQueryData()方法写入查询语句执行后返回的数据，具体代码如下所示：

```
    public final class MySQLCommandExecuteEngine implements
CommandExecuteEngine {
        @Override
        public void writeQueryData(final ChannelHandlerContext context,
                                final BackendConnection backendConnection, final
QueryCommandExecutor queryCommandExecutor, final int headerPackagesCount)
throws SQLException {
            if (ResponseType.QUERY != queryCommandExecutor.getResponseType()
|| !context.channel().isActive()) {
                return;
            }
            int count = 0;
            int flushThreshold =
ProxyContext.getInstance().getSchemaContexts().getProps().<Integer>getValue(
ConfigurationPropertyKey.PROXY_FRONTEND_FLUSH_THRESHOLD);
            int currentSequenceId = 0;
            //①循环遍历 QueryCommandExecutor 类
            while (queryCommandExecutor.next()) {
                count++;
                //②判断渠道处理上下文的模式和状态
```

```
            while (!context.channel().isWritable() &&
context.channel().isActive()) {
                //③如果渠道处理上下文是读模式，且渠道已经被激活，则调用
ChannelHandlerContext 的 flush()方法重新刷新渠道处理上下文的环境
                context.flush();
                //④如果渠道处理上下文是读模式，且渠道已经被激活，则调用 doAwait()方法
锁住资源
                backendConnection.getResourceLock().doAwait();
            }
            //⑤从命令执行器中获取查询后返回的数据，并写入渠道处理上下文中
            DatabasePacket<?> dataValue = queryCommandExecutor.getQueryData();
            context.write(dataValue);
            if (flushThreshold == count) {
                context.flush();
                count = 0;
            }
            currentSequenceId++;
        }
        context.write(new MySQLEofPacket(++currentSequenceId +
headerPackagesCount));
    }
    ...
}
```

5. 确定命令执行器的类型

在 Shardingsphere Proxy 中，用 MySQLCommandExecutorFactory 类的 newInstance() 方法确定命令执行器的类型，具体代码如下所示：

```
@NoArgsConstructor(access = AccessLevel.PRIVATE)
@Slf4j
public final class MySQLCommandExecutorFactory {
    public static CommandExecutor newInstance(final MySQLCommandPacketType
commandPacketType, final CommandPacket commandPacket, final BackendConnection
backendConnection) {
        log.debug("Execute packet type: {}, value: {}", commandPacketType,
commandPacket);
        switch (commandPacketType) {
            //①如果命令类型为 COM_QUIT，则返回 MySQLComQuitExecutor 对象
            case COM_QUIT:
                return new MySQLComQuitExecutor();
            //②如果命令类型为 COM_INIT_DB，则返回 MySQLComInitDbExecutor 对象
            case COM_INIT_DB:
                return new MySQLComInitDbExecutor((MySQLComInitDbPacket)
commandPacket, backendConnection);
            //③如果命令类型为 COM_FIELD_LIST，则返回 MySQLComFieldListPacketExecutor 对象
```

```
            case COM_FIELD_LIST:
                return new MySQLComFieldListPacketExecutor
((MySQLComFieldListPacket) commandPacket, backendConnection);
            //④如果命令类型为 COM_QUERY，则返回 MySQLComQueryPacketExecutor 对象
            case COM_QUERY:
                return new MySQLComQueryPacketExecutor((MySQLComQueryPacket)
commandPacket, backendConnection);
            //⑤如果命令类型为 COM_PING，则返回 MySQLComPingExecutor 对象
            case COM_PING:
                return new MySQLComPingExecutor();
            //⑥如果命令类型为 COM_STMT_PREPARE，则返回 MySQLComStmtPrepareExecutor 对象
            case COM_STMT_PREPARE:
                return new MySQLComStmtPrepareExecutor
((MySQLComStmtPreparePacket) commandPacket);
            //⑦如果命令类型为 COM_STMT_EXECUTE，则返回 MySQLComStmtExecuteExecutor 对象
            case COM_STMT_EXECUTE:
                return new MySQLComStmtExecuteExecutor
((MySQLComStmtExecutePacket) commandPacket, backendConnection);
            //⑧如果命令类型为 COM_STMT_RESET，则返回 MySQLComStmtResetExecutor 对象
            case COM_STMT_RESET:
                return new MySQLComStmtResetExecutor
((MySQLComStmtResetPacket) commandPacket);
            //⑨如果命令类型为 COM_STMT_CLOSE，则返回 MySQLComStmtCloseExecutor 对象
            case COM_STMT_CLOSE:
                return new MySQLComStmtCloseExecutor
((MySQLComStmtClosePacket) commandPacket);
            //⑩如果没有匹配到命令类型，则直接抛出异常
            default:
                return new MySQLUnsupportedCommandExecutor(commandPacketType);
        }
    }
    ...
}
```

12.6.3 “使用 MySQLComStmtPrepareExecutor 类处理 SQL 请求”的原理

在 Shardingsphere Proxy 中，如果使用 MySQL 作为数据源，则用 MySQL 的命令执行 MySQLComStmtPrepareExecutor 类来预处理 SQL 请求。

1. 预处理 SQL 请求

在 Shardingsphere Proxy 中，SQL 解析引擎能够处理不同类型数据库的多种类型 SQL 请求。SQL 解析引擎利用“高内聚，低耦合”的思想，将处理 SQL 请求的功能封装在不同的命令执行类中。其中，MySQLComStmtPrepareExecutor 类用来处理客户端调用 Shardingsphere Proxy

的增加、删除、修改和查询的 SQL 请求，具体代码如下所示：

```
@RequiredArgsConstructor
public final class MySQLComStmtPrepareExecutor implements CommandExecutor {
    private static final MySQLBinaryStatementRegistry
PREPARED_STATEMENT_REGISTRY = MySQLBinaryStatementRegistry.getInstance();
    private final MySQLComStmtPreparePacket packet;
    private int currentSequenceId;
    @Override
    public Collection<DatabasePacket<?>> execute() {
        //①初始化 SQL 解析引擎对象 ShardingSphereSQLParserEngine
        ShardingSphereSQLParserEngine sqlStatementParserEngine = new
ShardingSphereSQLParserEngine(
    DatabaseTypeRegistry.getTrunkDatabaseTypeName(ProxyContext.getInstance()
.getSchemaContexts().getDatabaseType()));
        //②调用解析引擎的 parse()方法解析和执行 SQL 语句，返回数据库处理后的结果对象
SQLStatement
        SQLStatement sqlStatement =
sqlStatementParserEngine.parse(packet.getSql(), true);
        if (!MySQLComStmtPrepareChecker.isStatementAllowed(sqlStatement)) {
            throw new UnsupportedPreparedStatementException();
        }
        int parameterCount = sqlStatement.getParameterCount();
        int projectionCount = getProjectionCount(sqlStatement);
        int statementId =
PREPARED_STATEMENT_REGISTRY.register(packet.getSql(), parameterCount);
        //③将 SQL 语句执行的结果封装在数据对象 DatabasePacket 中，并返回对象列表
        return createPackets(statementId, projectionCount, parameterCount);
    }
    //④创建 DatabasePacket 对象列表
    private Collection<DatabasePacket<?>> createPackets(final int
statementId, final int projectionCount, final int parameterCount) {
        Collection<DatabasePacket<?>> result = new LinkedList<>();
        result.add(new MySQLComStmtPrepareOKPacket(++currentSequenceId,
statementId, projectionCount, parameterCount, 0));
        if (parameterCount > 0) {
            result.addAll(createParameterColumnDefinition41Packets
(parameterCount));
        }
        if (projectionCount > 0) {
            result.addAll(createProjectionColumnDefinition41Packets
(projectionCount));
        }
        return result;
    }
```

```
    ...
}
```

2. 解析 SQL 语句

Shardingsphere Proxy 统一用 ShardingSphereSQLParserEngine 类处理 MySQL 数据源的 MySQLComStmtExecuteExecutor 命令和 PostgreSQL 数据源的 PostgreSQLComBindExecutor 命令。其中，初始化 ShardingSphereSQLParserEngine 类的具体代码如下所示：

```
public final class ShardingSphereSQLParserEngine {
    private final SQLStatementParserEngine sqlStatementParserEngine;
    private final DistSQLStatementParserEngine
distSQLStatementParserEngine;
    private final ParsingHookRegistry parsingHookRegistry;
public ShardingSphereSQLParserEngine(final String databaseTypeName) {
    //①初始化 SQL 声明的解析引擎
        sqlStatementParserEngine =
SQLStatementParserEngineFactory.getSQLStatementParserEngine(databaseTypeName);
    //②初始化 distSQL 声明的解析引擎
        distSQLStatementParserEngine = new DistSQLStatementParserEngine();
        parsingHookRegistry = ParsingHookRegistry.getInstance();
    }
    ...
}
```

在 Shardingsphere Proxy 中，用 ShardingSphereSQLParserEngine 类的 parse()方法解析 SQL 语句，具体代码如下所示：

```
public SQLStatement parse(final String sql, final boolean useCache) {
    //①在开始解析 SQL 语句后，开启 OpenTracingParsingHook 的链路追踪
    parsingHookRegistry.start(sql);
try {
    //②调用 parse0()方法解析和执行 SQL 请求
        SQLStatement result = parse0(sql, useCache);
        parsingHookRegistry.finishSuccess(result);
        return result;
} catch (final Exception ex) {
    //③如果出现异常，则在链路追踪中记录异常信息
        parsingHookRegistry.finishFailure(ex);
        throw ex;
    }
}
private SQLStatement parse0(final String sql, final boolean useCache) {
    try {
        //④解析 SQL 语句
```

```
            return sqlStatementParserEngine.parse(sql, useCache);
        } catch (final SQLParsingException originalEx) {
            try {
                //⑤返回解析的结果对象 SQLStatement
                return distSQLStatementParserEngine.parse(sql);
            } catch (final SQLParsingException ignored) {
                throw originalEx;
            }
        }
    }
```

3. 解析和执行 SQL 请求

在 Shardingsphere Proxy 中，用 SQLStatementParserEngine 类的 parse()方法解析和执行 SQL 请求的具体代码如下所示：

```
    public final class SQLStatementParserEngine {
        //①初始化 SQL 解析引擎
    private final SQLParserEngine parserEngine;
        //②初始化 SQL 访问引擎
        private final SQLVisitorEngine visitorEngine;
        private final Cache<String, SQLStatement> cache =
CacheBuilder.newBuilder().softValues().initialCapacity(2000).maximumSize(655
35).build();
        public SQLStatementParserEngine(final String databaseTypeName) {
            parserEngine = new SQLParserEngine(databaseTypeName);
            visitorEngine = new SQLVisitorEngine(databaseTypeName, "STATEMENT");
        }
    public SQLStatement parse(final String sql, final boolean useCache) {
        //③如果不使用本地缓存，则调用 parse()方法解析 SQL 语句
            if (!useCache) {
                return parse(sql);
            }
            //④从本地缓存中获取 SQLStatement 对象
            Optional<SQLStatement> statement =
Optional.ofNullable(cache.getIfPresent(sql));
            if (statement.isPresent()) {
                return statement.get();
            }
            //⑤将最新的结果存储在本地缓存中
            SQLStatement result = parse(sql);
            cache.put(sql, result);
            return result;
        }
        private SQLStatement parse(final String sql) {
            //⑥用 SQL 访问引擎访问解析树 ParseTree
```

```
        return visitorEngine.visit(parserEngine.parse(sql, false));
    }
}
```

4. 访问解析树

在 Shardingsphere Proxy 中，用 SQLVisitorEngine 类的方法 visit()访问解析树。下面来分析下具体访问的过程。

（1）在 Shardingsphere Proxy 中，用 SQLVisitorRule 类定义了访问解析树的规则，具体代码如下所示：

```
@RequiredArgsConstructor
public enum SQLVisitorRule {
//①将以“Select”开头的 SQL 语句定义为 DML 类型
SELECT("Select", SQLStatementType.DML),
//②将以“Insert”开头的 SQL 语句定义为 DML 类型
INSERT("Insert", SQLStatementType.DML),
//③将以“Update”开头的 SQL 语句定义为 DML 类型
UPDATE("Update", SQLStatementType.DML),
//④将以“Delete”开头的 SQL 语句定义为 DML 类型
DELETE("Delete", SQLStatementType.DML),
//⑤将以“Replace”开头的 SQL 语句定义为 DML 类型
REPLACE("Replace", SQLStatementType.DML),
//⑥将以“CreateTable”开头的 SQL 语句定义为 DDL 类型
CREATE_TABLE("CreateTable", SQLStatementType.DDL),
//⑦将以“AlterTable”开头的 SQL 语句定义为 DDL 类型
ALTER_TABLE("AlterTable", SQLStatementType.DDL),
//⑧将以“DropTable”开头的 SQL 语句定义为 DDL 类型
DROP_TABLE("DropTable", SQLStatementType.DDL),
//⑨将以“TruncateTable”开头的 SQL 语句定义为 DDL 类型
TRUNCATE_TABLE("TruncateTable", SQLStatementType.DDL),
//⑩将以“CreateTable”开头的 SQL 语句定义为 DDL 类型
    CREATE_TABLE("CreateTable", SQLStatementType.DDL),
    ...
}
```

（2）在 Shardingsphere Proxy 中，用 SQLVisitorFactory 类的 newInstance()方法初始化解析树访问器 ParseTreeVisitor 对象，具体代码如下所示：

```
@RequiredArgsConstructor
public final class SQLVisitorEngine {
    private final String databaseType;
    private final String visitorType;
public <T> T visit(final ParseTree parseTree) {
    //利用工厂类 SQLVisitorFactory 初始化解析树访问器对象 ParseTreeVisitor
```

```
        ParseTreeVisitor<T> visitor = SQLVisitorFactory.newInstance
(databaseType, visitorType, SQLVisitorRule.valueOf(parseTree.getClass()));
        return parseTree.accept(visitor);
    }
    ...
}
```

（3）在 Shardingsphere Proxy 中，用 SQLVisitorFactory 类的方法初始化解析树访问器对象 ParseTreeVisitor，具体代码如下所示：

```
@NoArgsConstructor(access = AccessLevel.PRIVATE)
public final class SQLVisitorFactory {

public static <T> ParseTreeVisitor<T> newInstance(
final String databaseType, final String visitorType, final SQLVisitorRule
visitorRule) {
    //①用 SQL 访问器的工厂类初始化一个 SQLVisitorFacade 对象，如果是 MySQL 数据库，则
加载 MySQLStatementSQLVisitorFacade 对象
        SQLVisitorFacade facade =
SQLVisitorFacadeRegistry.getInstance().getSQLVisitorFacade(databaseType,
visitorType);
        //②创建解析树访问器
        return createParseTreeVisitor(facade, visitorRule.getType());
}

    @SneakyThrows(ReflectiveOperationException.class)
    private static <T> ParseTreeVisitor<T> createParseTreeVisitor(final
SQLVisitorFacade visitorFacade, final SQLStatementType type) {
        switch (type) {
            case DML:
                //③如果 SQL 语句是 DML 类型，则返回 MySQLDMLStatementSQLVisitor 对
象（数据库为 MySQL）
                return (ParseTreeVisitor)
visitorFacade.getDMLVisitorClass().getConstructor().newInstance();
            case DDL:
                //④如果 SQL 语句是 DDL 类型，则返回 MySQLDDLStatementSQLVisitor 对
象（数据库为 MySQL）
                return (ParseTreeVisitor)
visitorFacade.getDDLVisitorClass().getConstructor().newInstance();
            case TCL:
            //⑤如果 SQL 语句是 TCL 类型，则返回 MySQLTCLStatementSQLVisitor 对象（数
据库为 MySQL）
                return (ParseTreeVisitor)
visitorFacade.getTCLVisitorClass().getConstructor().newInstance();
            case DCL:
```

```
            //⑥如果 SQL 语句是 DCL 类型，则返回 MySQLDCLStatementSQLVisitor 对象（数据库为 MySQL）
                return (ParseTreeVisitor)
visitorFacade.getDCLVisitorClass().getConstructor().newInstance();
            case DAL:
            //⑦如果 SQL 语句是 DAL 类型，则返回 MySQLDALStatementSQLVisitor 对象（数据库为 MySQL）
                return (ParseTreeVisitor)
visitorFacade.getDALVisitorClass().getConstructor().newInstance();
            case RL:
            //⑧如果 SQL 语句是 RL 类型，则返回 MySQLRLStatementSQLVisitor 对象（数据库为 MySQL）
                return (ParseTreeVisitor)
visitorFacade.getRLVisitorClass().getConstructor().newInstance();
            default:
            //⑨如果没有匹配到 SQL 语句类型，则抛出异常
                throw new SQLParsingException("Can not support SQL statement type: `%s`", type);
        }
    }
    ...
}
```

5. 用 SQLParserEngine 类的 parse()方法解析 SQL 语句

在 Shardingsphere Proxy 中，用 SQLParserEngine 类的 parse()方法调用 MySQL Server 解析 SQL 语句，具体代码如下所示：

```
@RequiredArgsConstructor
public final class SQLParserEngine {

    private static final Map<String, SQLParserExecutor> EXECUTORS = new ConcurrentHashMap<>();
    private final String databaseType;
public ParseTree parse(final String sql, final boolean useCache) {
    //①初始化执行 SQL 语句的执行器对象 SQLParserExecutor
        SQLParserExecutor executor = EXECUTORS.containsKey(databaseType) ? EXECUTORS.get(databaseType) : EXECUTORS.computeIfAbsent(databaseType, SQLParserExecutor::new);
        //②调用执行器的 parse()方法解析 SQL 语句
        return executor.parse(sql, useCache);
    }
}
```

6. 用 SQLParserExecutor 类的 parse()方法解析 SQL 语句

在 Shardingsphere Proxy 中，用 SQLParserExecutor 类的 parse()方法解析 SQL 语句，具体代码如下所示：

```
@RequiredArgsConstructor
public final class SQLParserExecutor {
    private final String databaseType;
    private final Cache<String, ParseTree> cache =
CacheBuilder.newBuilder().softValues().initialCapacity(2000).maximumSize(655
35).build();

    public ParseTree parse(final String sql, final boolean useCache) {
    //①如果不使用缓存，则重新解析 SQL 语句，否则读取本地缓存
        if (!useCache) {
            return parse(sql);
        }
        return parseAndCacheParseTree(sql);
    }
    //②解析 SQL 语句
    private ParseTree parse(final String sql) {
        ParseASTNode result = twoPhaseParse(sql);
        if (result.getRootNode() instanceof ErrorNode) {
            throw new SQLParsingException("Unsupported SQL of `%s`", sql);
        }
        return result.getRootNode();
    }
    //③解析和缓存解析树
    private ParseTree parseAndCacheParseTree(final String sql) {
        Optional<ParseTree> parseTree =
Optional.ofNullable(cache.getIfPresent(sql));
        if (parseTree.isPresent()) {
            return parseTree.get();
        }
        ParseTree result = parse(sql);
        cache.put(sql, result);
        return result;
    }

    private ParseASTNode twoPhaseParse(final String sql) {
        SQLParser sqlParser = SQLParserFactory.newInstance(databaseType, sql);
        try {
            //④设置断言模式为 PredictionMode.SLL
            setPredictionMode((Parser) sqlParser, PredictionMode.SLL);
            //⑤返回解析的结果对象 ParseASTNode
```

```
            return (ParseASTNode) sqlParser.parse();
        } catch (final ParseCancellationException ex) {
            //⑥如果出现异常，则重置 SQLParser 对象
            ((Parser) sqlParser).reset();
            //⑦如果出现异常，则设置断言模式为 PredictionMode.LL
            setPredictionMode((Parser) sqlParser, PredictionMode.LL);
            return (ParseASTNode) sqlParser.parse();
        }
    }
    //⑧设置 SQL 语句的断言模式
    private void setPredictionMode(final Parser sqlParser, final
PredictionMode mode) {
        sqlParser.setErrorHandler(new BailErrorStrategy());
        sqlParser.getInterpreter().setPredictionMode(mode);
    }
    ...
}
```

12.6.4 “使用 MySQLComStmtExecuteExecutor 类处理 SQL 请求”的原理

如果使用 MySQL 作为数据源，命令类型为 COM_STMT_EXECUTE，则用 MySQL 的命令执行类 MySQLComStmtExecuteExecutor 来处理 SQL 请求。

1. 用 MySQLComStmtExecuteExecutor 类的 execute()方法处理 SQL 请求

在 ShardingSphereSQLParserEngine 中，用 MySQLComStmtExecuteExecutor 类的 execute()方法处理 SQL 请求，具体代码如下所示：

```
public final class MySQLComStmtExecuteExecutor implements
QueryCommandExecutor {
    @Override
    public Collection<DatabasePacket<?>> execute() throws SQLException {
        //①如果触发熔断的阈值，则返回异常对象 CircuitBreakException
        if (ProxyContext.getInstance().
getSchemaContexts().isCircuitBreak()) {
            throw new CircuitBreakException();
        }
        //②用 JDBCDatabaseCommunicationEngine 类的 execute()方法执行 SQL 语句
        BackendResponse backendResponse =
databaseCommunicationEngine.execute();
        //③处理 MySQL Server 返回的数据，并返回包装后的数据包 DatabasePacket
        return backendResponse instanceof QueryResponse ?
processQuery((QueryResponse) backendResponse) : processUpdate((UpdateResponse)
backendResponse);
    }
    ...
```

```
    }
```

2. 用 JDBCDatabaseCommunicationEngine 类的 execute()方法处理 SQL 请求

Shardingsphere Proxy 用 JDBCDatabaseCommunicationEngine 类处理 MySQL 数据源的命令 MySQLComStmtExecuteExecutor 和 PostgreSQL 数据源的命令 PostgreSQLComBindExecutor。其中，初始化 JDBCDatabaseCommunicationEngine 类的 execute()方法的具体代码如下所示：

```
    @RequiredArgsConstructor
    public final class JDBCDatabaseCommunicationEngine implements
DatabaseCommunicationEngine {
        @Override
        public BackendResponse execute() throws SQLException {
            //①调用 KernelProcessor 类的 generateExecutionContext()方法，生成执行上
下文对象（包括数据源、路由规则等）
            ExecutionContext executionContext =
kernelProcessor.generateExecutionContext(logicSQL, schema,
ProxyContext.getInstance().getSchemaContexts().getProps());
            //②记录 SQL 语句执行的日志
            logSQL(executionContext);
            //③调用 doExecute()方法执行 SQL 语句
            return doExecute(executionContext);
        }
        private BackendResponse doExecute(final ExecutionContext
executionContext) throws SQLException {
            if (executionContext.getExecutionUnits().isEmpty()) {
                return new UpdateResponse();
            }
            //④判断在执行上下文对象 ExecutionContext 中是否存在 XA 模式事务，如果存在，则
直接抛出 TableModifyInTransactionException 异常
            sqlExecuteEngine.checkExecutePrerequisites(executionContext);
            //⑤调用 JDBCExecuteEngine 类的 execute()方法执行 SQL 语句
            response = sqlExecuteEngine.execute(executionContext);
            Collection<String> routeDataSourceNames =
executionContext.getRouteContext().getRouteUnits().stream()
                    .map(RouteUnit::getDataSourceMapper).map(RouteMapper::getLo
gicName).collect(Collectors.toList());
refreshTableMetaData(executionContext.getSqlStatementContext().getSqlStateme
nt(), routeDataSourceNames);
            return merge(executionContext.getSqlStatementContext());
        }
        ...
    }
```

3. 用 JDBCExecuteEngine 类的 execute()方法处理 SQL 请求

在 Shardingsphere Proxy 中，用 JDBCExecuteEngine 类的 execute()方法处理 SQL 请求，具体代码如下所示：

```
    public final class JDBCExecuteEngine implements SQLExecuteEngine {
        @Override
        public BackendResponse execute(final ExecutionContext executionContext)
throws SQLException {
            Collection<ExecuteResult> executeResults =
execute(executionContext,
                    executionContext.getSqlStatementContext().getSqlStatement()
instanceof InsertStatement, ExecutorExceptionHandler.isExceptionThrown());
            ExecuteResult executeResult = executeResults.iterator().next();
            //①如果 SQL 语句执行后的结果对象是 ExecuteQueryResult，则调用
getExecuteQueryResponse()方法处理 SQL 语句执行的结果
            if (executeResult instanceof ExecuteQueryResult) {
                return getExecuteQueryResponse(((ExecuteQueryResult)
executeResult).getQueryHeaders(), executeResults);
            } else {
                //②根据 SQL 语句的数据库声明对象,设置返回结果类型。如果是 InsertStatement
对象，则设置结果类型为“INSERT”
                UpdateResponse result = new UpdateResponse(executeResults);
                if (executionContext.getSqlStatementContext().getSqlStatement()
instanceof InsertStatement) {
                    result.setType("INSERT");
                } else if
(executionContext.getSqlStatementContext().getSqlStatement() instanceof
DeleteStatement) {
                    result.setType("DELETE");
                } else if
(executionContext.getSqlStatementContext().getSqlStatement() instanceof
UpdateStatement) {
                    result.setType("UPDATE");
                }
                return result;
            }
        }

        private Collection<ExecuteResult> execute(final ExecutionContext
executionContext, final boolean isReturnGeneratedKeys, final boolean
isExceptionThrown) throws SQLException {
            int maxConnectionsSizePerQuery =
ProxyContext.getInstance().getSchemaContexts().getProps().<Integer>getValue(
ConfigurationPropertyKey.MAX_CONNECTIONS_SIZE_PER_QUERY);
```

```
        // ③ExecutorConstant.MANAGED_RESOURCE 默认为 true，调用
executeWithManagedResource()方法处理 SQL 请求
        return ExecutorConstant.MANAGED_RESOURCE ?
executeWithManagedResource(executionContext, maxConnectionsSizePerQuery,
isReturnGeneratedKeys, isExceptionThrown)
                : executeWithUnmanagedResource(executionContext,
maxConnectionsSizePerQuery);
    }

    private Collection<ExecuteResult> executeWithManagedResource(final
ExecutionContext executionContext,final int maxConnectionsSizePerQuery,
  final boolean isReturnGeneratedKeys,final boolean isExceptionThrown) throws
SQLException {
        //④解析数据库类型，如果是 MySQL 类型，则是 MySQLDatabaseType 对象
        DatabaseType databaseType =
ProxyContext.getInstance().getSchemaContexts().getDatabaseType();
        //⑤调用 SQLExecutor 类的 execute()方法处理 SQL 请求
        return
sqlExecutor.execute(generateInputGroups(executionContext.getExecutionUnits(),
maxConnectionsSizePerQuery, isReturnGeneratedKeys,
executionContext.getRouteContext()),
                new ProxySQLExecutorCallback(databaseType,
executionContext.getSqlStatementContext(), backendConnection, accessor,
isExceptionThrown, isReturnGeneratedKeys, true),
                new ProxySQLExecutorCallback(databaseType,
executionContext.getSqlStatementContext(), backendConnection, accessor,
isExceptionThrown, isReturnGeneratedKeys, false));
    }
    //⑥生成 InputGroup 对象（主要包括数据库声明执行单元）
    @SuppressWarnings({"unchecked", "rawtypes"})
    private Collection<InputGroup<StatementExecuteUnit>>
generateInputGroups(final Collection<ExecutionUnit> executionUnits, final int
maxConnectionsSizePerQuery, final boolean isReturnGeneratedKeys,
                                                                 final
RouteContext routeContext) throws SQLException {
        Collection<ShardingSphereRule> rules =
ProxyContext.getInstance().getSchema(backendConnection.getSchemaName()).getR
ules();
        ExecuteGroupEngine executeGroupEngine =
accessor.getExecuteGroupEngine(backendConnection, maxConnectionsSizePerQuery,
new StatementOption(isReturnGeneratedKeys), rules);
        return (Collection<InputGroup<StatementExecuteUnit>>)
executeGroupEngine.generate(routeContext, executionUnits);
    }
    //⑦遍历执行 SQL 语句后返回的数据，并添加到对象 QueryResponse 中
```

```
    private BackendResponse getExecuteQueryResponse(final List<QueryHeader>
queryHeaders, final Collection<ExecuteResult> executeResults) {
        QueryResponse result = new QueryResponse(queryHeaders);
        for (ExecuteResult each : executeResults) {
            result.getQueryResults().add(((ExecuteQueryResult)
each).getQueryResult());
        }
        return result;
    }
    ...
}
```

4. 用 SQLExecutor 类的 execute()方法处理 SQL 请求

在 Shardingsphere Proxy 中，用 SQLExecutor 类的 execute()方法处理 SQL 请求，具体代码如下所示：

```
@RequiredArgsConstructor
public final class SQLExecutor {
    private final ExecutorKernel executorKernel;
    private final boolean serial;
    public <T> List<T> execute(final
Collection<InputGroup<StatementExecuteUnit>> inputGroups, final
SQLExecutorCallback<T> callback) throws SQLException {
        return execute(inputGroups, null, callback);
    }
    public <T> List<T> execute(final
Collection<InputGroup<StatementExecuteUnit>> inputGroups, final
SQLExecutorCallback<T> firstCallback, final SQLExecutorCallback<T> callback)
throws SQLException {
        try {
            //调用 ExecutorKernel 类的 execute()方法处理 SQL 请求
            return executorKernel.execute(inputGroups, firstCallback,
callback, serial);
        } catch (final SQLException ex) {
            ExecutorExceptionHandler.handleException(ex);
            return Collections.emptyList();
        }
    }
}
```

5. 用 ExecutorKernel 类的 execute()方法处理 SQL 请求

在 Shardingsphere Proxy 中，用 ExecutorKernel 类的 execute()方法处理 SQL 请求，具体代码如下所示：

```
@Getter
```

```
public final class ExecutorKernel implements AutoCloseable {
    private final ShardingSphereExecutorService executorService;
  public ExecutorKernel(final int executorSize) {
    //①初始化异步线程池对象 ShardingSphereExecutorService
      executorService = new ShardingSphereExecutorService(executorSize);
    }

    public <I, O> List<O> execute(final Collection<InputGroup<I>>
inputGroups, final ExecutorCallback<I, O> callback) throws SQLException {
        return execute(inputGroups, null, callback, false);
    }

    public <I, O> List<O> execute(final Collection<InputGroup<I>>
inputGroups,
                              final ExecutorCallback<I, O> firstCallback,
final ExecutorCallback<I, O> callback, final boolean serial) throws SQLException {
        if (inputGroups.isEmpty()) {
            return Collections.emptyList();
        }
        //②如果 serial 为"true"，则串行执行，调用 serialExecute()方法；如果 serial
为"false"，则并行执行，调用 parallelExecute()方法
        return serial ? serialExecute(inputGroups, firstCallback, callback) :
parallelExecute(inputGroups, firstCallback, callback);
    }
    //③串行执行 SQL 请求
    private <I, O> List<O> serialExecute(final Collection<InputGroup<I>>
inputGroups, final ExecutorCallback<I, O> firstCallback, final
ExecutorCallback<I, O> callback) throws SQLException {
        Iterator<InputGroup<I>> inputGroupsIterator = inputGroups.iterator();
        InputGroup<I> firstInputs = inputGroupsIterator.next();
        //④同步地执行 SQL 请求
        List<O> result = new LinkedList<>(syncExecute(firstInputs, null ==
firstCallback ? callback : firstCallback));
        for (InputGroup<I> each : Lists.newArrayList(inputGroupsIterator)) {
            result.addAll(syncExecute(each, callback));
        }
        return result;
    }
    //⑤并行地执行 SQL 请求
    private <I, O> List<O> parallelExecute(final Collection<InputGroup<I>>
inputGroups, final ExecutorCallback<I, O> firstCallback, final
ExecutorCallback<I, O> callback) throws SQLException {
        Iterator<InputGroup<I>> inputGroupsIterator = inputGroups.iterator();
        InputGroup<I> firstInputs = inputGroupsIterator.next();
```

```
        Collection<ListenableFuture<Collection<O>>> restResultFutures =
asyncExecute(Lists.newArrayList(inputGroupsIterator), callback);
        //⑥异步地执行 SQL 请求
        return getGroupResults(syncExecute(firstInputs, null ==
firstCallback ? callback : firstCallback), restResultFutures);
    }

    private <I, O> Collection<O> syncExecute(final InputGroup<I> inputGroup,
final ExecutorCallback<I, O> callback) throws SQLException {
        return callback.execute(inputGroup.getInputs(), true,
ExecutorDataMap.getValue());
    }

    private <I, O> Collection<ListenableFuture<Collection<O>>>
asyncExecute(final List<InputGroup<I>> inputGroups, final ExecutorCallback<I,
O> callback) {
        Collection<ListenableFuture<Collection<O>>> result = new
LinkedList<>();
        for (InputGroup<I> each : inputGroups) {
            result.add(asyncExecute(each, callback));
        }
        return result;
    }
    //⑦用线程池对象 ShardingSphereExecutorService，异步地执行 SQL 请求
    private <I, O> ListenableFuture<Collection<O>> asyncExecute(final
InputGroup<I> inputGroup, final ExecutorCallback<I, O> callback) {
        Map<String, Object> dataMap = ExecutorDataMap.getValue();
        return executorService.getExecutorService().submit(() ->
callback.execute(inputGroup.getInputs(), false, dataMap));
    }
    ...
}
```

6. 用 DefaultSQLExecutorCallback 类的 execute()方法处理 SQL 请求

在 Shardingsphere Proxy 中，用 DefaultSQLExecutorCallback 类的 execute()方法处理 SQL 请求。

```
@RequiredArgsConstructor
public abstract class DefaultSQLExecutorCallback<T> implements
SQLExecutorCallback<T> {
    //①处理 SQL 请求
    @Override
    public final Collection<T> execute(final
Collection<StatementExecuteUnit> statementExecuteUnits, final boolean
isTrunkThread, final Map<String, Object> dataMap) throws SQLException {
```

```
        Collection<T> result = new LinkedList<>();
        //②遍历 SQL 语句的执行单元列表，调用 execute0()方法处理 SQL 请求
        for (StatementExecuteUnit each : statementExecuteUnits) {
            result.add(execute0(each, isTrunkThread, dataMap));
        }
        return result;
    }
    private T execute0(final StatementExecuteUnit statementExecuteUnit,
final boolean isTrunkThread, final Map<String, Object> dataMap) throws
SQLException {
        ExecutorExceptionHandler.setExceptionThrown(isExceptionThrown);
        DataSourceMetaData dataSourceMetaData =
getDataSourceMetaData(statementExecuteUnit.getStorageResource().getConnectio
n().getMetaData());
        SQLExecutionHook sqlExecutionHook = new SPISQLExecutionHook();
        try {
            ExecutionUnit executionUnit =
statementExecuteUnit.getExecutionUnit();
            //③开启处理 SQL 请求的链路追踪
            sqlExecutionHook.start(executionUnit.getDataSourceName(),
executionUnit.getSqlUnit().getSql(), executionUnit.getSqlUnit().getParameters(),
dataSourceMetaData, isTrunkThread, dataMap);
    // ④DefaultSQLExecutorCallback 类是一个抽象模板类，需要调用实现类
ProxySQLExecutorCallback 的 executeSQL()方法处理 SQL 请求
            T result = executeSQL(executionUnit.getSqlUnit().getSql(),
statementExecuteUnit.getStorageResource(),
statementExecuteUnit.getConnectionMode());
            //⑤结束对执行 SQL 请求的链路追踪
            sqlExecutionHook.finishSuccess();
            return result;
        } catch (final SQLException ex) {
            //⑥将异常信息添加到链路追踪中
            sqlExecutionHook.finishFailure(ex);
            ExecutorExceptionHandler.handleException(ex);
            return null;
        }
    }
    ...
}
```

7. 用 ProxySQLExecutorCallback 类的 executeSQL ()方法处理 SQL 请求

在 Shardingsphere Proxy 中，用 ProxySQLExecutorCallback 类的 executeSQL()方法处理 SQL 请求，具体代码如下所示：

```
    public final class ProxySQLExecutorCallback extends
DefaultSQLExecutorCallback<ExecuteResult> {
        @Override
        public ExecuteResult executeSQL(final String sql, final Statement
statement, final ConnectionMode connectionMode) throws SQLException {
            if (fetchMetaData && !hasMetaData) {
                hasMetaData = true;
                return executeSQL(statement, sql, connectionMode, true);
            }
            return executeSQL(statement, sql, connectionMode, false);
        }

    private ExecuteResult executeSQL(final Statement statement, final String sql,
final ConnectionMode connectionMode, final boolean withMetadata) throws
SQLException {
        //①将数据源的数据连接的声明对象 Statement 添加到 BackendConnection 对象中
（BackendConnection 是与 MySQL Server 的连接）
            backendConnection.add(statement);
            //②调用 StatementAccessor 类的 execute()方法处理 SQL 语句
            if (accessor.execute(statement, sql, isReturnGeneratedKeys)) {
                //③获取 SQL 语句执行的结果
                ResultSet resultSet = statement.getResultSet();
                backendConnection.add(resultSet);
                //④返回结果给客户端
                return new ExecuteQueryResult(withMetadata ?
getQueryHeaders(sqlStatementContext, resultSet.getMetaData()) : null,
createQueryResult(resultSet, connectionMode));
            }
            return new ExecuteUpdateResult(statement.getUpdateCount(),
isReturnGeneratedKeys ? getGeneratedKey(statement) : 0L);
        }
        ...
    }
```

8. 用 StatementAccessor 类或者 PreparedStatementAccessor 类的 execute ()方法处理 SQL 请求

在 Shardingsphere Proxy 中，用 StatementAccessor 或者 PreparedStatementAccessor 类的 execute ()方法处理 SQL 请求，具体代码如下所示：

```
    //①在 DatabaseCommunicationEngineFactory 类中初始化 StatementAccessor 类或者
PreparedStatementAccessor 类
    public DatabaseCommunicationEngine newTextProtocolInstance(final
SQLStatement sqlStatement, final String sql, final BackendConnection
backendConnection) {
```

```
    LogicSQL logicSQL = createLogicSQL(sqlStatement, sql,
Collections.emptyList(), backendConnection);
    //②如果使用 newTextProtocolInstance()方法来实例化一个
JDBCDatabaseCommunicationEngine 对象，则初始化 StatementAccessor 对象
        return new JDBCDatabaseCommunicationEngine(
                logicSQL,
ProxyContext.getInstance().getSchema(backendConnection.getSchemaName()), new
JDBCExecuteEngine(backendConnection, new StatementAccessor()));
    }

    public DatabaseCommunicationEngine newBinaryProtocolInstance(final
SQLStatement sqlStatement, final String sql, final List<Object> parameters,
final BackendConnection backendConnection) {
    LogicSQL logicSQL = createLogicSQL(sqlStatement, sql, new
ArrayList<>(parameters), backendConnection);
    //③如果使用 newBinaryProtocolInstance ()方法来实例化一个
JDBCDatabaseCommunicationEngine 对象，则初始化 PreparedStatementAccessor 对象
        return new JDBCDatabaseCommunicationEngine(
                logicSQL,
ProxyContext.getInstance().getSchema(backendConnection.getSchemaName()), new
JDBCExecuteEngine(backendConnection, new PreparedStatementAccessor()));
    }

    public final class StatementAccessor implements JDBCAccessor {
        @Override
        public StatementExecuteGroupEngine getExecuteGroupEngine(final
BackendConnection backendConnection,
                                                         final int
maxConnectionsSizePerQuery, final StatementOption option, final
Collection<ShardingSphereRule> rules) {
            return new StatementExecuteGroupEngine(maxConnectionsSizePerQuery,
backendConnection, option, rules);
        }
        @Override
    public boolean execute(final Statement statement, final String sql, final
boolean isReturnGeneratedKeys) throws SQLException {
        //④直接调用 MySQL Server 执行 SQL 语句，并将结果保存在 Statement 对象中
            return statement.execute(sql, isReturnGeneratedKeys ?
Statement.RETURN_GENERATED_KEYS : Statement.NO_GENERATED_KEYS);
        }
    }

    public final class PreparedStatementAccessor implements JDBCAccessor {

        @Override
```

```
    public ExecuteGroupEngine<?> getExecuteGroupEngine(final
BackendConnection backendConnection,
                                                    final int
maxConnectionsSizePerQuery, final StatementOption option, final
Collection<ShardingSphereRule> rules) {
        return new PreparedStatementExecuteGroupEngine
(maxConnectionsSizePerQuery, backendConnection, option, rules);
    }

    @Override
  public boolean execute(final Statement statement, final String sql, final
boolean isReturnGeneratedKeys) throws SQLException {
    //⑤直接调用 MySQL Server 处理 SQL 语句，并将结果保存在 PreparedStatement 对象中
        return ((PreparedStatement) statement).execute();
    }
  }
```

第 13 章 分布式缓存——基于 Redis

在微服务架构中，针对“读多写少”的业务场景，架构师和技术专家会引入分布式缓存技术来提升服务的整体 QPS 处理能力。

在开源领域中有很多成熟的分布式缓存技术，Redis 是其中被广泛使用的分布式缓存技术之一。

13.1 认识缓存

从是否跨进程的角度，可以将缓存分为本地缓存和分布式缓存。在开发过程中使用得最多的是本地缓存。

13.1.1 什么是本地缓存

本地缓存可以分为基于内存的本地缓存和基于文件的本地缓存。

1. 基于内存的本地缓存

图 13-1 所示为基于内存的本地缓存。在请求服务 A 和服务 B 时，并没有调用数据库和文件系统去进行数据的读和写，而是直接从 JVM 中获取数据（即从内存中获取数据），并返回给调用方。这个在 JVM 中提供数据的容器就是本地缓存。

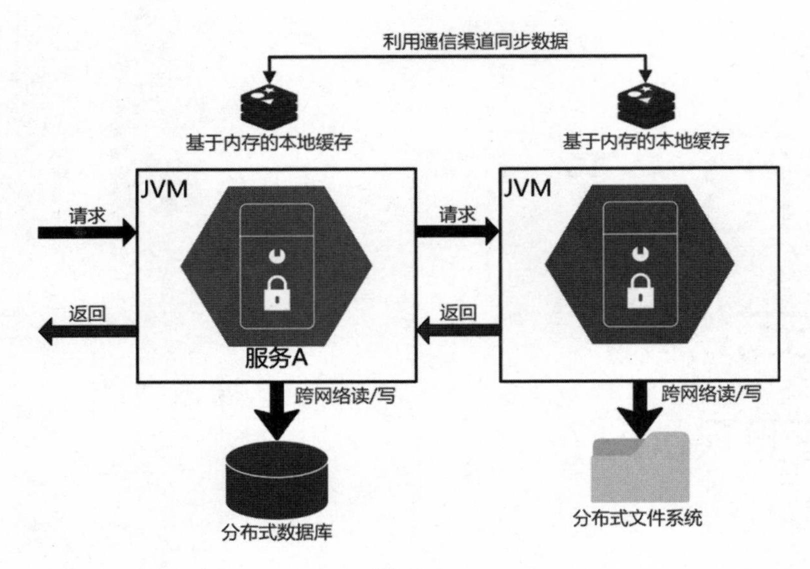

图 13-1

基于内存的本地缓存具有如下特性：

（1）作用域为 JVM 级别。服务 A 和服务 B 只能在 JVM 的内存中操作缓存，比如增加、删除、更新缓存数据。

（2）是非持久化的。当服务 A 和服务 B 重启后，JVM 中的缓存数据会丢失，并且不能恢复。

（3）在服务 A 的缓存和服务 B 的缓存之间可以互相同步数据，但是同步缓存数据具备一定的延迟性（即并不能确保数据的强一致性）。

2. 基于文件的本地缓存

图 13-2 所示为基于文件的本地缓存。在请求服务 A 时，服务 A 需要将缓存数据持久化到基于文件的本地缓存中（通常是指硬盘），并利用通信渠道将增量变更的数据同步到服务 B 的本地缓存中。

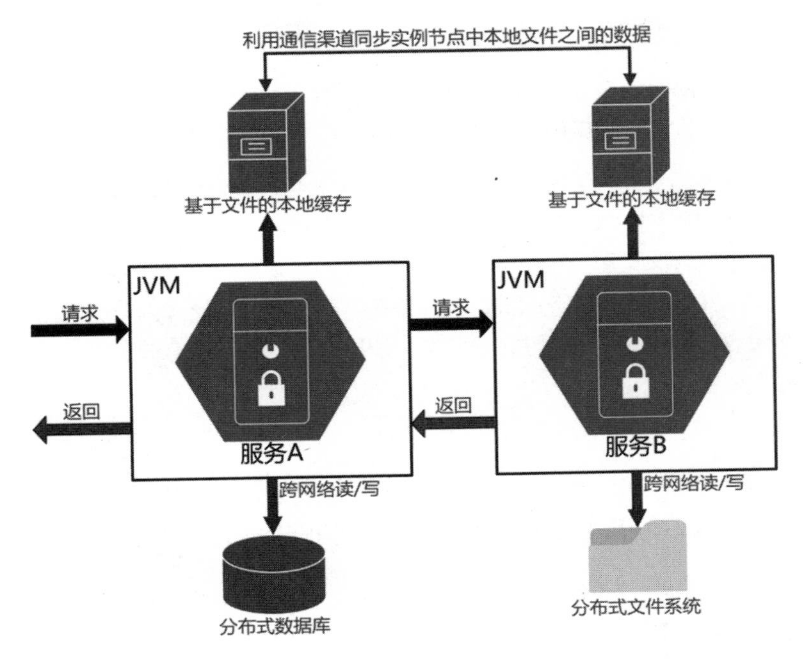

图 13-2

基于文件的本地缓存具备如下特性：

（1）作用域为 JVM 级别。如果调用方请求服务 A 与服务 B，则服务 A 和服务 B 会读/写各自本机的本地缓存数据，并返回各自本机最新的缓存数据给调用方。

（2）是持久化的。将缓存数据持久化到本地硬盘中。

（3）服务 A 和服务 B 的本地缓存会利用通信渠道进行双向同步。

13.1.2　什么是分布式缓存

从字面的角度来分析，分布式缓存是指具备分布式特性的缓存。从功能的角度来分析，分布式缓存主要包括如下功能。

（1）高性能的通信渠道。

跨进程或者实例节点的 RPC 调用，需要可靠性、高性能和稳定性，所以，高性能的通信渠道是分布式缓存最关键的技术之一。在分布式缓存中，不同实例节点之间的缓存，需要利用通信渠道去同步缓存数据，以确保各节点缓存数据的一致性（比如在 Master 节点和 Slave 节点之间的数据同步）。

（2）高可用的存储设备。

存储设备是分布式缓存中最重要的组件，所以，在保证高性能的前提下，存储设备的高可用性是非常关键的技术指标。通常会采用如下技术策略。

- 基于事件的快照文件：分布式缓存可以利用事件快速地恢复增量的数据。
- 全量的数据文件：在节点实例重启的过程中，分布式缓存可以利用全量的数据文件来恢复已经持久化的数据
- 内存：如果开启了基于内存的数据存储策略，则在数据存储到内存后直接返回存储结果。为了确保数据存储的高可用性，通常会采取异步事件机制将存储到内存中的数据异步地持久化到文件中。

（3）多语言性。

分布式缓存必须支持多种语言，比如支持 Java、Go 及 Java Script 等。当分布式缓存在业务中大量使用时，不同语言的应用都充分利用分布式缓存来提升服务的 QPS。

（4）客户端。

分布式缓存通常用于实现高可用、高性能的服务器端，但是需要一个客户端去连接服务器端，从而能够高效地操控分布式缓存中的缓存对象。

（5）高效的选举算法。

为了保证高并发和高吞吐量，分布式缓存的服务器端通常都采用集群部署或者主从部署，所以需要有一个高效的选举算法来维护集群中 Master 节点和 Salve 节点的角色。

（6）命令控制台。

分布式缓存需要一个命令控制台，这样运维人员和开发人员可以管理和维护线上的分布式缓存中的数据。

（7）UI 控制台。

为了确保数据的安全性，并不是所有的开发人员都具有线上机器的访问权限，所以分布式缓存需要一个可视化的 UI 控制台，来实时地管理和维护缓存数据。

如图 13-3 所示，服务 A 和服务 B 在处理调用方的数据请求时，如果是写请求，则会调用分布式缓存将数据写入分布式缓存中；如果是读请求，则会读取分布式缓存中的数据。分布式缓存通常都是一个集群，在集群的存储节点之间会同步缓存数据。对于应用来说，从分布式缓存集群中读/写数据和从本地缓存中读/写数据是一样的效果。在分布式缓存中，将比较复杂的技术细节下沉到底层中间件中了。

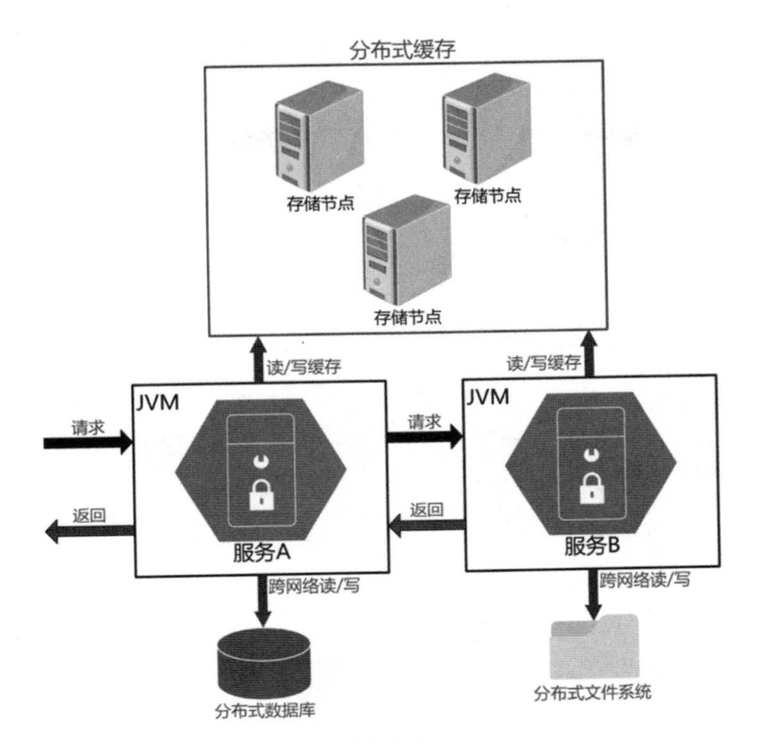

图 13-3

13.1.3　什么是 Redis

Redis 是一个基于 C 语言的高性能的 Key-Value 数据库，开发人员可以将它作为分布式数据库、分布式缓存及消息中间件引入应用中。在应用中，通常是将 Redis 作为分布式缓存，来解决大流量业务场景下的高性能、高并发和高可用等非功能性技术问题。

Redis 作为一个高性能的数据库，具有以下特性。

- 高性能：支持基于内存的数据存储，读/写速度非常快，支持 10 万以上的 QPS。
- 非常丰富的数据类型：支持字符串、散列、列表、集合、有序集合等数据类型。
- 单进程单线程：Redis 所有的命令都是“原子”的。
- 支持数据持久化：Redis 可以将数据持久化到磁盘中，在 Redis 所在的节点重启后可以恢复全量的缓存数据。
- 支持多种集环境：支持主从环境、哨兵集群环境、Redis Cluster 集群环境、Codis 集群环境。
- 支持分布式锁：比如 RedissonLock 等。

- 支持消息中间件的部分功能：Redis 将 JDK 中的容器类的作用域（比如 HashMap、ArrayList 等）从单一的服务级别扩大到服务之外的分布式环境中（独立的数据节点集群环境），并且增加了很多分布式环境下特有的数据结构，这样开发人员可以利用这些数据结构来作为消息中间件的载体，存储消息和消费消息。

13.1.4 Redis 的整体架构

图 13-4 为 Redis 的整体架构。

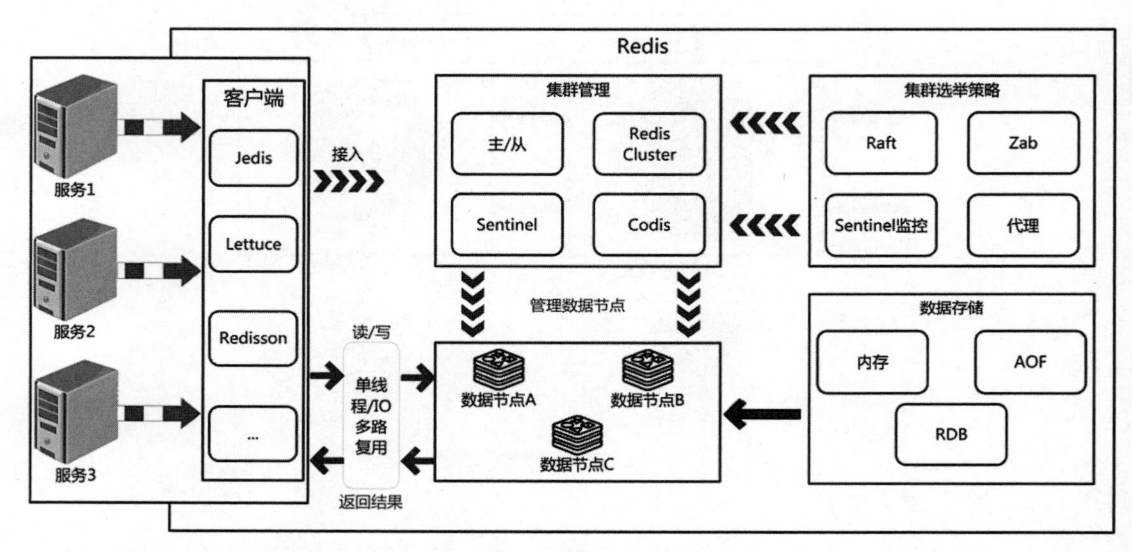

图 13-4

架构中包括如下几个部分。

1. 客户端

客户端是应用服务和 Redis 数据节点之间的代理，应用服务利用客户端提供的 SDK（目前比较成熟的客户端有 Jedis、Lettuce 及 Redisson 等。），可以建立应用服务和 Redis 数据节点之间的通信渠道，这样应用服务就可以使用 Redis 的功能了（比如分布式数据库、分布式缓存及消息中间件）。

2. 集群管理

集群管理是 Redis 中最重要的功能之一。目前 Redis 支持主从、Redis Cluster、Sentinel 及 Codis 集群管理。主从集群、Redis Cluster、Sentinel 是 Redis 自带的集群管理功能。Codis 是豌豆荚开源的 Redis 集群管理框架。

3. 集群选举策略

在启动集群的过程中，集群中的数据节点需要选举产生 Master 节点和 Salve 节点。Redis 目前支持多种集群选举策略，包括 Zab、Raft、主从集群等。

4. 数据存储

Redis 目前支持 3 种数据存储模式：内存、AOF 及 RDB。其中，AOF 和 RDB 是持久化的存储模式。

13.2　搭建 Redis 集群环境

本书采用“单节点，多实例”的方式来搭建 Redis 的集群环境，包括主从环境、Sentinel 集群环境、Codis 集群环境，以及 Redis Cluster 集群环境。本书统一采用 Redis 6.x 版本。

从 Redis 官网下载安装包“redis-6.2.1.tar.gz”，解压缩安装包后，在安装包的路径下执行编译命令“make”，可以得到 src 目录。在 src 目录下有“redis-server”“redis-cli”和“redis-sentinel”等脚本。

13.2.1　搭建主从环境

这里在单节点中启动 3 个进程来模拟实例节点（用端口号来区分实例），其中包括 1 个 Master 节点和 2 个 Slave 节点。

1. 准备主从环境

准备主从环境的过程如下。

（1）在本地（本书统一采用 Mac 环境）新建 3 个文件夹，分别如下。

- 文件夹“/redis-env/redis-master-slave/6470-master”，用于启动一个 Master 节点，端口号为 6470。
- 文件夹“/redis-env/redis-master-slave/6471-slave”，用于启动一个 Slave 节点，端口号为 6471。
- 文件夹“/redis-env/redis-master-slave/6472-salve”，用于启动一个 Slave 节点，端口号为 6472。

将编译好的 Redis 安装包“redis-6.2.1”分别复制到 3 个文件夹中，如图 13-5 所示。

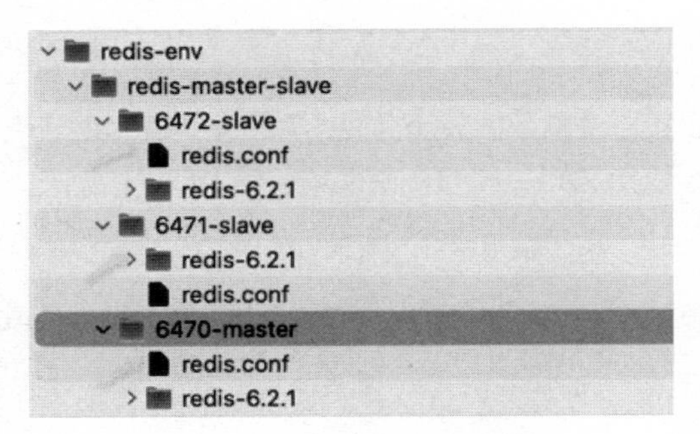

图 13-5

（2）在 3 个文件夹中新增配置文件“redis.conf”，具体配置信息如下。

- Master 节点所在文件夹“6470-master”中的“redis.conf”的配置信息如下所示：

```
#①配置端口
port 6470
#②绑定 IP 地址
bind 127.0.0.1
#③配置 Redis 后台执行
daemonize yes
#④关闭保护模式
protected-mode no
#⑤配置 Redis 的密码
requirepass 12345678
#⑥配置 Slave 节点访问 Master 节点的密码
masterauth 12345678
#⑦配置 Master 节点运行的进程号
pidfile /var/run/redis_6470.pid
```

- Slave 节点所在文件夹“6471-slave”中的“redis.conf”的配置信息如下所示：

```
#①配置 Slave 节点的端口
port 6471
#②绑定 IP 地址
bind 127.0.0.1
#③配置 Slave 节点的密码
requirepass 12345678
#④关闭保护模式
protected-mode no
#⑤配置访问 Master 节点的密码
masterauth 12345678
#⑥配置后台执行
daemonize yes
#⑦指定 Master 节点的 IP 地址
slaveof 127.0.0.1 6470
```

```
#⑧配置 Slave 节点运行的进程号
pidfile /var/run/redis_6471.pid
```

- Slave 节点所在文件夹“6472-slave”中的“redis.conf”的配置信息如下所示：

```
#①配置 Slave 节点的端口
port 6472
#②绑定 IP 地址
bind 127.0.0.1
#③配置 Slave 节点的密码
requirepass 12345678
#④关闭保护模式
protected-mode no
#⑤配置访问 Master 节点的密码
masterauth 12345678
#⑥配置后台执行
daemonize yes
#⑦指定 Master 节点的 IP 地址
slaveof 127.0.0.1 6470
#⑧配置 Slave 节点运行的进程号
pidfile /var/run/redis_6472.pid
```

2. 启动和验证主从环境

启动和验证主从环境的过程如下。

（1）在文件夹“6470-master”“6471-slave”及“6472-slave”中分别执行如下命令：

```
redis-6.2.1/src/redis-server redis.conf
```

（2）执行命令“ps -ef | grep redis”查看 Redis 进程，如果看到端口为 6470、6471 及 6472 的 Redis 进程（如图 13-6 所示），则表示已经启动成功。

```
501 39579     1   0  1:03PM ??         0:00.97 redis-6.2.1/src/redis-server 127.0.0.1:6470
501 39616     1   0  1:06PM ??         0:00.18 redis-6.2.1/src/redis-server 127.0.0.1:6471
501 39627     1   0  1:07PM ??         0:00.03 redis-6.2.1/src/redis-server 127.0.0.1:6472
```

图 13-6

（3）在文件夹“6470-master”中，执行命令“redis-6.2.1/src/redis-cli -h 127.0.0.1 - p 6470 -a 12345678”去连接 Master 节点。执行 Redis 的命令“INFO replication”后观察 Master 节点的状态，如果看到如图 13-7 所示效果，则 Master 节点运行正常。

（4）在文件夹“6471-slave”中，执行命令“redis-6.2.1/src/redis-cli -h 127.0.0.1 -p 6471 -a 12345678”去连接 Slave 节点。执行 Redis 的命令“INFO replication”后观察 Slave 节点的状态，如果看到如图 13-8 所示效果，则 Slave 节点运行正常。

```
huxian@huxians-MacBook-Pro 6470-master % redis-6.2.1/src/redis-cli -h 127.0.0.1 -p 6470 -a 12345678
Warning: Using a password with '-a' or '-u' option on the command line interface may not be safe.
127.0.0.1:6470> INFO replicaton
127.0.0.1:6470> INFO replication
# Replication
role:master
connected_slaves:2
slave0:ip=127.0.0.1,port=6471,state=online,offset=994,lag=0
slave1:ip=127.0.0.1,port=6472,state=online,offset=994,lag=0
master_failover_state:no-failover
master_replid:ba3dd50ad2abcc085881de143714da98861bec88
master_replid2:0000000000000000000000000000000000000000
master_repl_offset:994
second_repl_offset:-1
repl_backlog_active:1
repl_backlog_size:1048576
repl_backlog_first_byte_offset:1
repl_backlog_histlen:994
```

图 13-7

```
huxian@huxians-MacBook-Pro 6471-slave % redis-6.2.1/src/redis-cli -h 127.0.0.1 -p 6471 -a 12345678
Warning: Using a password with '-a' or '-u' option on the command line interface may not be safe.
127.0.0.1:6471> INFO replication
# Replication
role:slave
master_host:127.0.0.1
master_port:6470
master_link_status:up
master_last_io_seconds_ago:7
master_sync_in_progress:0
slave_repl_offset:1316
slave_priority:100
slave_read_only:1
connected_slaves:0
master_failover_state:no-failover
master_replid:ba3dd50ad2abcc085881de143714da98861bec88
master_replid2:0000000000000000000000000000000000000000
master_repl_offset:1316
second_repl_offset:-1
repl_backlog_active:1
repl_backlog_size:1048576
repl_backlog_first_byte_offset:1
repl_backlog_histlen:1316
```

图 13-8

（5）在文件夹“6472-slave”中，执行命令“redis-6.2.1/src/redis-cli -h 127.0.0.1 -p 6472 -a 12345678”去连接 Slave 节点。执行 Redis 的命令“INFO replication”后观察 Slave 节点的状态，如果看到如图 13-9 所示效果，则 Slave 节点运行正常。

```
huxian@huxians-MacBook-Pro 6472-slave % redis-6.2.1/src/redis-cli -h 127.0.0.1 -p 6472 -a 12345678
Warning: Using a password with '-a' or '-u' option on the command line interface may not be safe.
127.0.0.1:6472> INFO replication
# Replication
role:slave
master_host:127.0.0.1
master_port:6470
master_link_status:up
master_last_io_seconds_ago:3
master_sync_in_progress:0
slave_repl_offset:1470
slave_priority:100
slave_read_only:1
connected_slaves:0
master_failover_state:no-failover
master_replid:ba3dd50ad2abcc085881de143714da98861bec88
master_replid2:0000000000000000000000000000000000000000
master_repl_offset:1470
second_repl_offset:-1
repl_backlog_active:1
repl_backlog_size:1048576
repl_backlog_first_byte_offset:43
repl_backlog_histlen:1428
```

图 13-9

经过以上步骤后，Redis 的“一主二从”的主从环境就搭建完毕了。

13.2.2　搭建 Sentinel 集群环境

本书在单节点中启动 6 个进程来模拟实例节点（用端口号来区分实例），其中包括 1 个 Master 节点、2 个 Slave 节点和 3 个 Sentinel 节点，如图 13-10 所示。

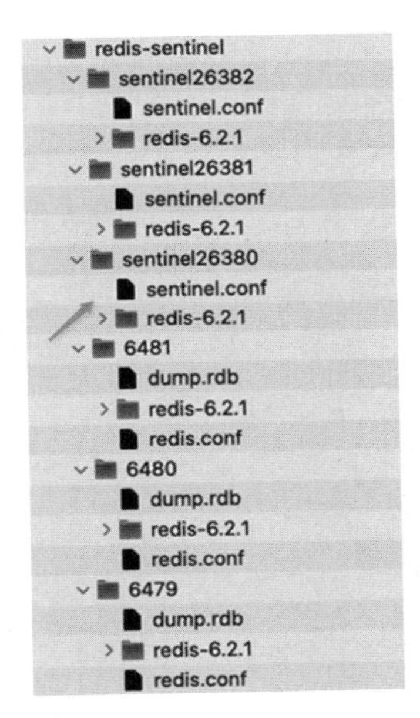

图 13-10

1. 准备 Sentinel 节点的基础环境

准备 Sentinel 节点的基础环境的过程如下。

（1）新建文件夹“redis-sentinel”，在其中创建 3 个子文件夹“6481”“6480”和“6479”用于搭建 Redis 主从环境，创建 3 个子文件夹“sentinel26382”“sentinel26381”和“sentinel26380”，用于搭建 3 个 Sentinel 节点。

（2）在文件夹“6481”“6480”和“6479”中添加主从环境的配置文件“redis.conf”。

- 文件夹“6481”中的配置文件“redis.conf”的具体配置信息如下所示：

```
###①配置 Slave 端口
port 6481
###②绑定 IP 地址
bind 127.0.0.1
###③配置密码
requirepass 12345678
```

```
###④关闭保护模式
protected-mode no
###⑤配置访问 Master 节点的密码
masterauth 12345678
###⑥配置后台执行
daemonize yes
###⑦配置 Master 节点的 IP 地址，将 Slave 节点和 Master 节点绑定
slaveof 127.0.0.1 6479
###⑧配置 Slave 节点运行的进程号
pidfile /var/run/redis_6481.pid
```

- 文件夹“6480”中的配置文件“redis.conf”的具体配置信息如下所示：

```
###①配置 Slave 端口
port 6480
###②绑定 IP 地址
bind 127.0.0.1
###③配置密码
requirepass 12345678
###④关闭保护模式
protected-mode no
###⑤配置后台执行
daemonize yes
###⑥配置访问 Master 节点的密码
masterauth 12345678
###⑦配置 Master 节点的 IP 地址，将 Slave 节点和 Master 节点绑定
slaveof 127.0.0.1 6479
###⑧配置 Slave 节点运行的进程号
pidfile /var/run/redis_6480.pid
```

- 文件夹“6479”中配置文件“redis.conf”的具体配置信息如下所示：

```
###①配置 Master 节点的端口
port 6479
###②绑定 IP 地址
bind 127.0.0.1
###③配置后台执行
daemonize yes
###④关闭保护模式
protected-mode no
###⑤配置密码
requirepass 12345678
###⑥配置 Slave 访问 Master 节点的密码
masterauth 12345678
###⑦配置 Master 节点运行的进程号
pidfile /var/run/redis_6479.pid
```

（3）在文件夹“sentinel26382”“sentinel26381”和“sentinel26380”中添加 Sentinel 节点的配置文件“sentinel.conf”。

- 文件夹“sentinel26382”中的配置文件“sentinel.conf”的具体配置信息如下所示：

```
###①绑定 IP 地址
bind 127.0.0.1
###②配置访问 Master 节点的密码
sentinel auth-pass mymaster 12345678
###③配置后台运行
daemonize yes
###④配置关闭保护模式
protected-mode no
###⑤配置哨兵的端口，客户端通过该端口来发现 Redis
port 26382
###⑥配置 Sentinel 节点的进程号，该文件会自动生成，进程号中的数字为哨兵的端口
pidfile "/var/run/redis-sentinel-26382.pid"
###⑦配置 Sentinelsentinel 监控的 Master 节点的名字为 mymaster，初始 Master 节点为
127.0.0.1 6479。“2”表示，如果 2 个及以上个数哨兵认定被监控的 Master 节点为宕机，则 Master
节点才被判定为宕机
sentinel monitor mymaster 127.0.0.1 6479 2
```

- 文件夹“sentinel26381”中的配置文件“sentinel.conf”的具体配置信息如下所示：

```
###①绑定 IP 地址
bind 127.0.0.1
###②配置访问 Master 节点的密码
sentinel auth-pass mymaster 12345678
###③配置后台运行
daemonize yes
###④配置关闭保护模式
protected-mode no
###⑤配置哨兵的端口，客户端通过该端口来发现 Redis
port 26381
###⑥配置 Sentinel 节点的进程号，这个文件会自动生成，进程号中的数字为哨兵的端口
pidfile "/var/run/redis-sentinel-26381.pid"
###⑦配置 Sentinelsentinel 监控的 Master 节点的名字为 mymaster，初始 Master 节点为
127.0.0.1 6479。“2”表示，如果 2 个及以上个数哨兵认定被监控的 Master 节点为宕机，则 Master
节点才被判定为宕机
sentinel monitor mymaster 127.0.0.1 6479 2
```

- 文件夹“sentinel26380”中的配置文件“sentinel.conf”的具体配置信息如下所示：

```
###①绑定 IP 地址
bind 127.0.0.1
###②配置访问 Master 节点的密码
sentinel auth-pass mymaster 12345678
```

```
###③配置后台运行
daemonize yes
###④配置关闭保护模式
protected-mode no
###⑤配置哨兵的端口，客户端通过该端口来发现 Redis
port 26380
###⑥配置 Sentinel 节点的进程号，这个文件会自动生成，进程号中的数字为哨兵的端口
pidfile "/var/run/redis-sentinel-26380.pid"
###⑦配置 Sentinelsentinel 监控的 Master 节点的名字为 mymaster，初始 Master 节点为
127.0.0.1 6479。“2”表示，如果 2 个及以上个数哨兵认定被监控的 Master 节点为宕机，则 Master
节点才被判定为宕机
sentinel monitor mymaster 127.0.0.1 6479 2
```

2. 启动主从环境和 Sentinel 集群

（1）启动主从环境。

在文件夹“6481”“6480”和“6479”中分别执行命令“redis-6.2.1/src/redis-server redis.conf”，执行后的结果如图 13-11 所示。

```
huxian@huxians-MacBook-Pro 6479 % ps -ef | grep redis
  501 44966     1   0  3:43AM ??        0:00.92 redis-6.2.1/src/redis-server 127.0.0.1:6481
  501 44997     1   0  3:45AM ??        0:00.25 redis-6.2.1/src/redis-server 127.0.0.1:6480
  501 45012     1   0  3:46AM ??        0:00.06 redis-6.2.1/src/redis-server 127.0.0.1:6479
  501 45019 41178   0  3:46AM ttys002   0:00.00 grep redis
```

图 13-11

（2）启动 Sentinel 环境。

在文件夹“sentinel26382”“sentinel26381”和“sentinel26380”中分别执行命令“redis-6.2.1/src/redis-sentinel sentinel.conf”，执行后的结果如图 13-12 所示。

```
huxian@huxians-MacBook-Pro sentinel26382 % ps -ef | grep redis
  501 44966     1   0  3:43AM ??        0:01.62 redis-6.2.1/src/redis-server 127.0.0.1:6481
  501 44997     1   0  3:45AM ??        0:01.03 redis-6.2.1/src/redis-server 127.0.0.1:6480
  501 45012     1   0  3:46AM ??        0:00.77 redis-6.2.1/src/redis-server 127.0.0.1:6479
  501 45042     1   0  3:48AM ??        0:00.58 redis-6.2.1/src/redis-sentinel 127.0.0.1:26380 [sentinel]
  501 45057     1   0  3:49AM ??        0:00.09 redis-6.2.1/src/redis-sentinel 127.0.0.1:26381 [sentinel]
  501 45062     1   0  3:49AM ??        0:00.04 redis-6.2.1/src/redis-sentinel 127.0.0.1:26382 [sentinel]
  501 45064 41178   0  3:49AM ttys002   0:00.00 grep redis
```

图 13-12

（3）验证“当 Master 节点宕机后，Sentinel 节点重新选举 Master 节点”的故障场景。

- 在主从集群中，节点“127.0.0.1:6479”为 Master 节点，图 13-13 所示为集群正常运行的状态。

```
huxian@huxians-MacBook-Pro 6479 % redis-6.2.1/src/redis-cli -h 127.0.0.1 -p 6479 -a 12345678
Warning: Using a password with '-a' or '-u' option on the command line interface may not be safe.
127.0.0.1:6479> INFO replication
# Replication
role:master
connected_slaves:2
slave0:ip=127.0.0.1,port=6480,state=online,offset=152000,lag=1
slave1:ip=127.0.0.1,port=6481,state=online,offset=152133,lag=0
master_failover_state:no-failover
master_replid:8ab04c529100d19c544d361350505a089652f215
master_replid2:0000000000000000000000000000000000000000
master_repl_offset:152133
second_repl_offset:-1
repl_backlog_active:1
repl_backlog_size:1048576
repl_backlog_first_byte_offset:1
repl_backlog_histlen:152133
127.0.0.1:6479> 
```

图 13-13

- 在文件夹“6479”中执行命令“redis-6.2.1/src/redis-cli -h 127.0.0.1 -p 6479 -a 12345678”连接到 Master 节点后，执行“shutdown”命令关闭 Master 节点“127.0.0.1:6479”，如图 13-14 所示。

```
huxian@huxians-MacBook-Pro 6479 % redis-6.2.1/src/redis-cli -h 127.0.0.1 -p 6479 -a 12345678
Warning: Using a password with '-a' or '-u' option on the command line interface may not be safe.
127.0.0.1:6479> shutdown
not connected> 
```

图 13-14

- Sentinel 的 3 个节点会投票选举，判定“127.0.0.1:6479”已经宕机，并在剩余的两个节点“127.0.0.1:6480”和“127.0.0.1:6481”中选出 1 个 Master 节点，选举后的结果如图 13-15 所示，节点“127.0.0.1:6480”为新的 Master 节点。

```
huxian@huxians-MacBook-Pro 6479 % redis-6.2.1/src/redis-cli -h 127.0.0.1 -p 6480 -a 12345678
Warning: Using a password with '-a' or '-u' option on the command line interface may not be safe.
127.0.0.1:6480> INFO replication
# Replication
role:master
connected_slaves:1
slave0:ip=127.0.0.1,port=6481,state=online,offset=356299,lag=1
master_failover_state:no-failover
master_replid:e555948356a9596820bee7b820ca65ce803192bf
master_replid2:8ab04c529100d19c544d361350505a089652f215
master_repl_offset:356299
second_repl_offset:218431
repl_backlog_active:1
repl_backlog_size:1048576
repl_backlog_first_byte_offset:1
repl_backlog_histlen:356299
```

图 13-15

- 在重新启动节点“127.0.0.1:6479”后，Sentinel 节点会重新选举，选举后的结果如图 13-16 所示，节点“127.0.0.1:6479”在加入主从环境中后，由原先的 Master 节点变成了 Slave 节点。

```
huxian@huxians-MacBook-Pro 6479 % redis-6.2.1/src/redis-cli -h 127.0.0.1 -p 6480 -a 12345678
Warning: Using a password with '-a' or '-u' option on the command line interface may not be safe.
127.0.0.1:6480> INFO replication
# Replication
role:master
connected_slaves:2
slave0:ip=127.0.0.1,port=6481,state=online,offset=405726,lag=0
slave1:ip=127.0.0.1,port=6479,state=online,offset=405726,lag=0
master_failover_state:no-failover
master_replid:e555948356a9596820bee7b820ca65ce803192bf
master_replid2:8ab04c529100d19c544d361350505a089652f215
master_repl_offset:405726
second_repl_offset:218431
repl_backlog_active:1
repl_backlog_size:1048576
repl_backlog_first_byte_offset:1
repl_backlog_histlen:405726
```

图 13-16

13.2.3 搭建 Codis 集群环境

本书统一采用 Mac 环境，使用“单节点，多进程”的方式来搭建 Codis 集群环境。

Codis 集群依赖的软件环境如下。

- Go 语言：Codis 是基于 Go 语言的中间件，所以需要在 Mac 环境下安装 Go 语言。
- 分布式选举框架：Codis 采用比较成熟的分布式选举框架（比如 ZooKeeper、Etcd 等）来管理 Redis 集群。本书统一采用 ZooKeeper。
- Redis：Codis 主要采用代理 Redis 的模式，来实现复用 Redis 的分布式缓存能力。本书统一使用 Codis 3.2.2，该版本主要依赖 Redis 3.2.11。
- Codis 安装包：Codis 官方提供了两种类型的安装包：源码包和二进制包。本书统一采用二进制包来部署 Codis 集群。

在 Codis 集群中主要包括以下组件。

- Codis Server：基于 Redis 源码开发开发的中间件框架（Codis 的版本是强依赖 Redis 的版本的），主要是增加了一些 Redis 没有的数据结构，以支持 Codis 中与 Slot 相关的操作及数据迁移指令。
- Codis Proxy：代理客户端去连接 Redis Server 的中间服务，并实现 Redis 协议。应用在使用 Codis Proxy 时，除不支持部分命令外，其他的体验和原生的 Redis 一样。对于同一个业务集群而言，可以同时部署多个 Codis Proxy 实例，在不同的 Codis Proxy 之间由 Codis Dashboard 保证状态同步。
- Codis Sentinel：主要监控 Codis Server 节点，以确保在 Codis Server 集群运行过程中 Master 节点和 Salve 节点的高可用。
- Codis Dashboard：Codis 的集群管理工具，支持在线添加/删除 Codis Proxy 和 Codis Server，以及数据迁移等操作。在集群状态发生改变时，Codis Dashboard 可以维护集群中所有 Codis Proxy 实例的分布式状态的一致性。对于同一个业务集群而言，在同一个时刻 Codis Dashboard 只能有 0 个或者 1 个。对集群中 Codis Proxy 实例状态的修改，都必须通过 Codis Dashboard 来完成。

- Codis Admin：Codis 的集群管理命令行工具，主要用来控制 Codis Proxy、Codis Dashboard 状态，以及访问外部存储。
- Codis FE：Codis 的集群管理界面。在 Codis 中，多个集群实例可以共享同一个前端展示页面。开发人员可以通过配置文件来管理后端 Codis Dashboard 列表，其中的配置文件可自动更新。
- Storage：为 Codis 中的集群实例提供外部存储，目前支持 ZooKeeper、Etcd 和 Fs 等。

1. 准备 Codis 的基础环境

准备工作主要包括：安装 Go 语言和 ZooKeeper，以及下载指定版本的 Codis 安装包。

（1）从 Codis 的官网下载二进制安装包“codis3.2.2-go1.9.2-osx.zip”，如图 13-17 所示，其中，Codis 3.2.2 版本依赖 Go 语言的最低版本为 1.9.2。

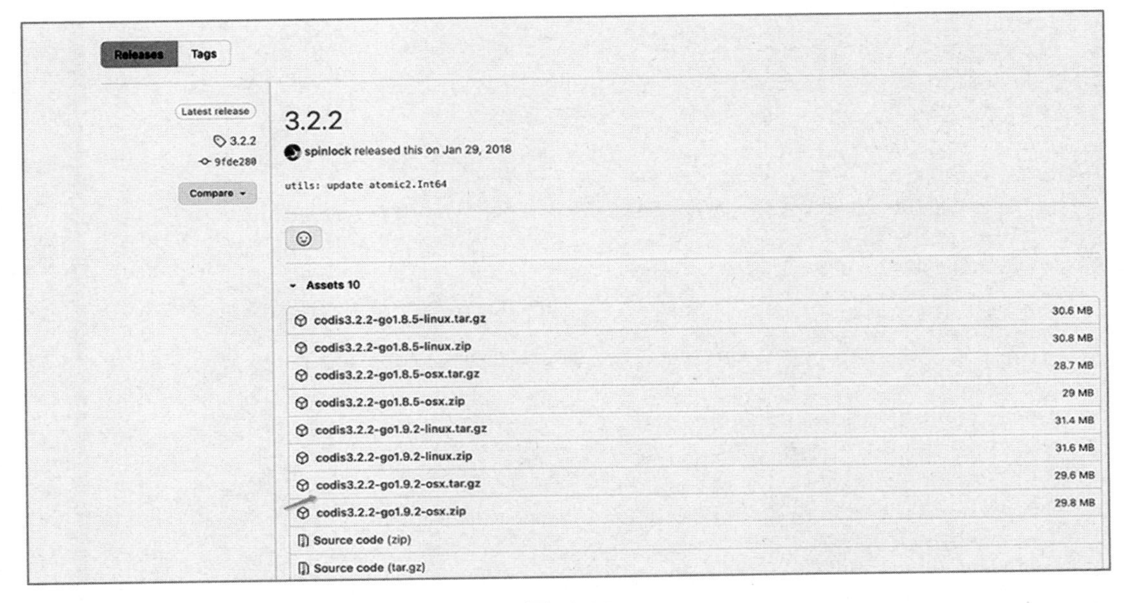

图 13-17

（2）从 Go 语言的官网下载 Go 语言的安装包“go1.9.2.darwin-amd64.pkg”，在安装完成后，执行命令“go version”查看 Go 语言的版本，如图 13-18 所示。

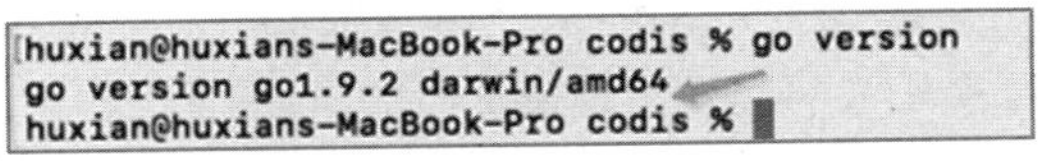

```
huxian@huxians-MacBook-Pro codis % go version
go version go1.9.2 darwin/amd64
huxian@huxians-MacBook-Pro codis %
```

图 13-18

（3）安装 ZooKeeper（可以参考本书的第 10 章），然后执行命令“ps -ef | grep java”查看 ZooKeeper 的进程信息。

（4）将 Codis 的安装包“codis3.2.2-go1.9.2-osx.zip”解压缩到文件夹“redis-env”中，重新命名解压缩后的安装包为“codis”。

在本书中，Codis 的组件共用一套二进制包，但是会隔离不同组件的配置和日志文件。

2. 搭建 Codis Server 环境

（1）在文件夹“redis-env/codis/data/redis/data/config”中添加 3 个配置文件：redis_master_6460.conf、redis_slave_6461.conf 和 redis_slave_6462.conf。

- Codis Server 节点“127.0.0.1:6460”的配置文件“redis_master_6460.conf”的具体配置信息如下所示：

```
###①允许后台运行
daemonize yes
###②设置端口，最好是非默认端口
port 6460
###③绑定登录 IP 地址，出于安全考虑，最好是内网
bind 127.0.0.1
###④命名并指定当前 Redis 的 PID 路径，用来区分多个 Redis
pidfile "/Users/huxian/Downloads/redis-env/codis/data/redis/data/config/
redis_6460.pid"
###⑤命名并指定当前 Redis 日志文件路径
logfile "/Users/huxian/Downloads/redis-env/codis/data/redis/data/logs/
redis_6460.log"
###⑥指定 RDB 文件名，用来备份数据到硬盘并区分不同 Redis，当使用内存超过可用内存的 45%
时触发快照功能
dbfilename "dump_6460.rdb"
###⑦指定当前 Redis 的根目录，用来存放 RDB/AOF 文件
dir "/Users/huxian/Downloads/redis-env/codis/data/codis_6460"
###⑧所有 codis-proxy 集群相关的 redis-server 认证密码必须全部一致
requirepass "12345678"
###⑨配置 Codis Server 的容量大小限制
maxmemory 100000kb
###⑩LRU 的策略有四种，根据具体情况选择
maxmemory-policy allkeys-lru
masterauth "12345678"
appendonly yes
appendfilename "appendonly_6460.aof"
```

- Codis Server 节点“127.0.0.1:6461”的配置文件“redis_master_6461.conf”的具体配置信息如下所示：

```
###①允许后台运行
daemonize yes
###②设置端口，最好是非默认端口
port 6461
###③绑定登录IP地址，出于安全考虑，最好是内网
bind 127.0.0.1
###④命名并指定当前Redis的PID路径，用来区分多个Redis
pidfile "/Users/huxian/Downloads/redis-env/codis/data/redis/data/config/
redis_6461.pid"
###⑤命名并指定当前Redis日志文件路径
logfile "/Users/huxian/Downloads/redis-env/codis/data/redis/data/logs/
redis_6461.log"
###⑥指定RDB文件名，用来备份数据到硬盘并区分不同Redis，当使用内存超过可用内存的45%
时触发快照功能
dbfilename "dump_6461.rdb"
###⑦指定当前Redis的根目录，用来存放RDB/AOF文件
dir "/Users/huxian/Downloads/redis-env/codis/data/codis_6461"
###⑧所有Codis Proxy集群的redis server认证密码必须全部一致
requirepass "12345678"
###⑨配置Codis Server的容量大小限制
maxmemory 100000kb
maxmemory-policy allkeys-lru
###⑩如果做故障切换，则不论是主节点还是从节点都要填写密码，且要保持一致
masterauth "12345678"
```

- Codis Server 节点"127.0.0.1:6462"的配置文件"redis_master_6462.conf"的具体配置信息如下所示：

```
###①允许后台运行
daemonize yes
###②设置端口，最好是非默认端口
port 6462
###③绑定登录IP地址，出于安全考虑，最好是内网
bind 127.0.0.1
###④命名并指定当前Redis的PID路径，用来区分多个Redis
pidfile "/Users/huxian/Downloads/redis-env/codis/data/redis/data/config/
redis_6462.pid"
###⑤命名并指定当前Redis日志文件路径
logfile "/Users/huxian/Downloads/redis-env/codis/data/redis/data/logs/
redis_6462.log"
###⑥指定RDB文件名，用来备份数据到硬盘并区分不同Redis，当使用内存超过可用内存的45%
时触发快照功能
dbfilename "dump_6462.rdb"
###⑦指定当前Redis的根目录，用来存放RDB/AOF文件
dir "/Users/huxian/Downloads/redis-env/codis/data/codis_6462"
###⑧所有Codis proxy集群的Redis Server认证密码必须全部一致
```

```
    requirepass "12345678"
    maxmemory 100000kb
    ###⑨LRU 的策略有四种，根据具体情况选择
    maxmemory-policy allkeys-lru
    ###⑩如果做故障切换，则主节点和从节点都要填写密码且要保持一致
    masterauth "12345678"
```

（2）执行如下命令启动 3 个 Codis Server 节点，如图 13-19 所示。

```
    ###启动 127.0.0.1:6460 节点
    sudo codis/codis-server
codis/data/redis/data/config/redis_master_6460.conf
    ###启动 127.0.0.1:6461 节点
    sudo codis/codis-server
codis/data/redis/data/config/redis_slave_6461.conf
    ###启动 127.0.0.1:6462 节点
    sudo codis/codis-server
codis/data/redis/data/config/redis_slave_6462.conf
```

```
huxian@huxians-MacBook-Pro redis-env % ps -ef | grep codis-server
    0 79680     1   0  1:21AM ??         0:01.04 codis/codis-server 127.0.0.1:6460
    0 79707     1   0  1:25AM ??         0:00.70 codis/codis-server 127.0.0.1:6461
    0 79715     1   0  1:26AM ??         0:00.70 codis/codis-server 127.0.0.1:6462
```

图 13-19

3. 搭建 Codis Sentinel 环境

（1）在文件夹“redis-env/codis/data/redis/data/config”中添加 3 个配置文件：redis_sentinel_26370.conf、redis_sentinel_26371.conf 和 redis_sentinel_26372.conf。

- Codis Sentinel 节点“127.0.0.1: 26370”的配置文件“redis_sentinel_26370.conf”的具体配置信息如下所示：

```
    ###①绑定 IP 地址
    bind 127.0.0.1
    protected-mode no
    ###②绑定端口号 26370
    port 26370
    dir "/Users/huxian/Downloads/redis-env/codis/data/sentinel_26370"
    pidfile "/Users/huxian/Downloads/redis-env/codis/data/redis/data/config/
sentinel_26370.pid"
    logfile "/Users/huxian/Downloads/redis-env/codis/data/redis/data/logs/
sentinel_26370.log"
    ###③配置后台启动
    daemonize yes
```

- Codis Sentinel 节点“127.0.0.1: 26371”的配置文件“redis_sentinel_26371.conf”的具体配置信息如下所示：

```
   bind 127.0.0.1
   protected-mode no
   port 26371
   dir "/Users/huxian/Downloads/redis-env/codis/data/sentinel_26371"
   pidfile "/Users/huxian/Downloads /redis-env/codis/data/redis/data/config/
sentinel_26371.pid"
   logfile "/Users/huxian/Downloads /redis-env/codis/data/redis/data/logs/
sentinel_26371.log"
   daemonize yes
```

- Codis Sentinel 节点“127.0.0.1: 26372”的配置文件“redis_sentinel_26372.conf”的具体配置信息如下所示：

```
   bind 127.0.0.1
   protected-mode no
   port 26372
   dir "/Users/huxian/Downloads/redis-env/codis/data/sentinel_26372"
   pidfile "/Users/huxian/redis-env /codis/data/redis/data/config/
sentinel_26372.pid"
   logfile "/Users/huxian/redis-env /codis/data/redis/data/logs/
sentinel_26372.log"
   daemonize yes
```

（2）执行如下命令启动 3 个 Codis Sentinel 节点，如图 13-20 所示。

```
   ###启动 127.0.0.1:26370 的 Sentinel 节点
   sudo codis/redis-sentinel codis/data/redis/data/config/
redis_sentinel_26370.conf
   ###启动 127.0.0.1:26371 的 Sentinel 节点
   sudo codis/redis-sentinel codis/data/redis/data/config/
redis_sentinel_26371.conf
   ###启动 127.0.0.1:26372 的 Sentinel 节点
   sudo codis/redis-sentinel codis/data/redis/data/config/
redis_sentinel_26372.conf
```

```
huxian@huxians-MacBook-Pro redis-env % ps -ef | grep redis-sentinel
    0 79829     1   0  1:49AM ??         0:00.90 codis/redis-sentinel 127.0.0.1:26370 [sentinel]
    0 79863     1   0  1:53AM ??         0:00.25 codis/redis-sentinel 127.0.0.1:26371 [sentinel]
    0 79868     1   0  1:54AM ??         0:00.23 codis/redis-sentinel 127.0.0.1:26372 [sentinel]
```

图 13-20

4. 搭建 Codis Proxy 环境

（1）在文件夹“redis-env/codis/data/redis/data/config”中添加一个配置文件 codis_proxy_19000.conf，具体配置信息如下所示：

```
###①项目名称
product_name = "codis-cluster"
###②设置登录 dashboard 的密码（与真实 Redis 中的 requirepass 一致）
product_auth = "12345678"
###③客户端（redis-cli）的登录密码（与真实 Redis 中的 requirepass 不一致），是登录
codis 的密码
session_auth = "12345678"
###④管理的端口，0.0.0.0 即对所有 IP 地址开放。出于安全考虑，可以限制内网
admin_addr = "127.0.0.1:11083"
proto_type = "tcp4"
###⑤客户端(redis-cli)访问代理的端口号。0.0.0.0 即对所有 IP 地址开放
proxy_addr = "127.0.0.1:19000"
###⑥外部配置存储类型，这里用的是 ZooKeeper。
jodis_name = "zookeeper"
###⑦配置 ZooKeeper 的连接地址，这里是 3 台就填 3 台的地址，本书采用单机版本
jodis_addr = "127.0.0.1:2181"
jodis_timeout="20s"
jodis_compatible=false
###⑧Codis 代理的最大连接数，默认是 1000。如果是高并发则要调大
proxy_max_clients = 100000
session_max_pipeline = 100000
backend_max_pipeline = 204800
session_recv_bufsize = "256kb"
session_recv_timeout = "0s"
```

（2）执行如下命令启动一个 Codis Proxy 节点，如图 13-21 所示

```
sudo codis/codis-proxy --ncpu=1 --config codis/data/redis/data/config/
codis_proxy_19000.conf --log /Users/huxian/redis-env/codis/data/ redis/
data/logs/proxy_19000.log &
```

```
huxian@huxians-MacBook-Pro redis-env % ps -ef | grep codis-proxy
    0 79914 79076   0  2:01AM ttys003    0:00.04 sudo codis/codis-proxy --ncpu=1 --config=/Users/huxian/Downloads/c
odis/data/redis/data/config/codis_proxy_19000.conf --log=/Users/huxian/Downloadscodis/data/redis/data/logs/proxy_19
000.log
  501 79921 79076   0  2:02AM ttys003    0:00.00 grep codis-proxy
```

图 13-21

Codis Proxy 节点在启动成功后，状态为等待上线，如图 13-22 所示。

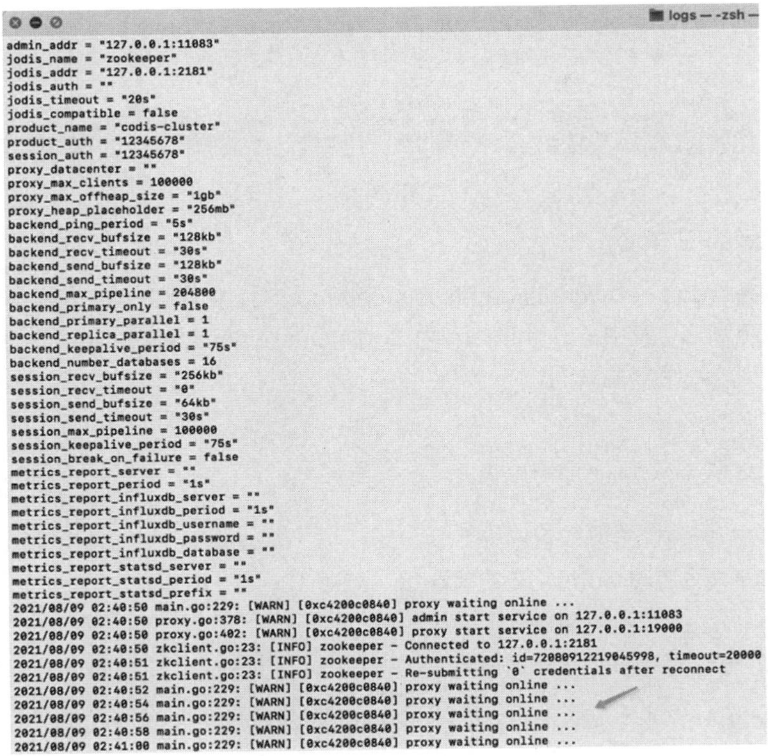

```
logs — -zsh —
admin_addr = "127.0.0.1:11083"
jodis_name = "zookeeper"
jodis_addr = "127.0.0.1:2181"
jodis_auth = ""
jodis_timeout = "20s"
jodis_compatible = false
product_name = "codis-cluster"
product_auth = "12345678"
session_auth = "12345678"
proxy_datacenter = ""
proxy_max_clients = 100000
proxy_max_offheap_size = "1gb"
proxy_heap_placeholder = "256mb"
backend_ping_period = "5s"
backend_recv_bufsize = "128kb"
backend_recv_timeout = "30s"
backend_send_bufsize = "128kb"
backend_send_timeout = "30s"
backend_max_pipeline = 204800
backend_primary_only = false
backend_primary_parallel = 1
backend_replica_parallel = 1
backend_keepalive_period = "75s"
backend_number_databases = 16
session_recv_bufsize = "256kb"
session_recv_timeout = "0"
session_send_bufsize = "64kb"
session_send_timeout = "30s"
session_max_pipeline = 100000
session_keepalive_period = "75s"
session_break_on_failure = false
metrics_report_server = ""
metrics_report_period = "1s"
metrics_report_influxdb_server = ""
metrics_report_influxdb_period = "1s"
metrics_report_influxdb_username = ""
metrics_report_influxdb_password = ""
metrics_report_influxdb_database = ""
metrics_report_statsd_server = ""
metrics_report_statsd_period = "1s"
metrics_report_statsd_prefix = ""
2021/08/09 02:40:50 main.go:229: [WARN] [0xc4200c0840] proxy waiting online ...
2021/08/09 02:40:50 proxy.go:378: [WARN] [0xc4200c0840] admin start service on 127.0.0.1:11083
2021/08/09 02:40:50 proxy.go:402: [WARN] [0xc4200c0840] proxy start service on 127.0.0.1:19000
2021/08/09 02:40:50 zkclient.go:23: [INFO] zookeeper - Connected to 127.0.0.1:2181
2021/08/09 02:40:51 zkclient.go:23: [INFO] zookeeper - Authenticated: id=72080912219045998, timeout=20000
2021/08/09 02:40:51 zkclient.go:23: [INFO] zookeeper - Re-submitting `0` credentials after reconnect
2021/08/09 02:40:52 main.go:229: [WARN] [0xc4200c0840] proxy waiting online ...
2021/08/09 02:40:54 main.go:229: [WARN] [0xc4200c0840] proxy waiting online ...
2021/08/09 02:40:56 main.go:229: [WARN] [0xc4200c0840] proxy waiting online ...
2021/08/09 02:40:58 main.go:229: [WARN] [0xc4200c0840] proxy waiting online ...
2021/08/09 02:41:00 main.go:229: [WARN] [0xc4200c0840] proxy waiting online ...
```

图 13-22

5. 搭建 Codis Dashboard 环境

（1）在文件夹“redis-env/codis/data/redis/data/config”中添加一个配置文件 codis_dashboard_18080.conf，具体配置信息如下所示：

```
###①外部配置存储类型，比如 ZooKeeper
coordinator_name = "zookeeper"
###②配置 ZooKeeper 的连接地址，这里是 3 台就填 3 台的地址
coordinator_addr = "127.0.0.1:2181"
###③项目名称
product_name = "codis-cluster"
###④所有 Redis 的登录密码（与真实 Redis 中的 requirepass 一致），因为要登录进去修改数据
product_auth = "12345678"
###⑤codis-dashboard 的通信端口。0.0.0.0 表示对所有开放，最好使用内网地址
admin_addr = "127.0.0.1:18080"
```

（2）执行如下命令启动一个 Codis Dashboard 节点，如图 13-23 所示。

```
sudo codis/codis-dashboard --ncpu=1
--config=/Users/huxian/Downloads/redis-env/codis/data/redis/data/config/codi
```

```
s_dashboard_18080.conf --log=/Users/huxian/Downloads/redis-env/codis/data/
redis/data/logs/codis_dashboard_18080.log --log-level=INFO &
```

```
huxian@huxians-MacBook-Pro redis-env % ps -ef | grep codis-dashboard
    0 79970 79076   0  2:09AM ttys003    0:00.03 sudo codis/codis-dashboard --ncpu=1 --config=/Users/huxian/Downloa
ds/codis/data/redis/data/config/codis_dashboard_18080.conf --log=/Users/huxian/Downloads/codis/data/redis/data/logs
/codis_dashboard_18080.log --log-level=INFO
```

图 13-23

6. 搭建 Codis FE 环境

（1）在文件夹“redis-env/codis/data/redis/data/config”中添加一个配置文件“codis.json”，具体配置信息如下所示。其中，dashboard 为 Codis Dashboard 暴露的后台 IP 地址。

```
[
    {
        "name": "codis-cluster",
        "dashboard": "127.0.0.1:18080"
    }
]
```

（2）执行如下命令启动 Codis FE 的前端 UI，如图 13-24 所示。

```
    sudo codis/codis-fe --ncpu=1
--log=/Users/huxian/Downloads/redis-env/codis/data/redis/data/logs/fe.log
--log-level=INFO --dashboard-list=/Users/huxian/Downloads/redis-env/codis/
data/redis/data/config/codis.json --listen=127.0.0.1:8078 &
```

```
huxian@huxians-MacBook-Pro redis-env % ps -ef | grep codis-fe
    0 80009 79076   0  2:16AM ttys003    0:00.04 sudo codis/codis-fe --ncpu=1 --log=/Users/huxian/Downloads/codis/d
ata/redis/data/logs/fe.log --log-level=INFO --dashboard-list=/Users/huxian/Downloads/codis/redis/data/config/codis.
json --listen=127.0.0.1:8078
  501 80013 79076   0  2:16AM ttys003    0:00.00 grep codis-fe
```

图 13-24

（3）在浏览器中输入“127.0.0.1:8078”，访问 Codis FE 的前端 UI，如图 13-25 所示。通过 Codis FE 的前端控制台可以管控 Codis 集群“codis-cluster-test”。

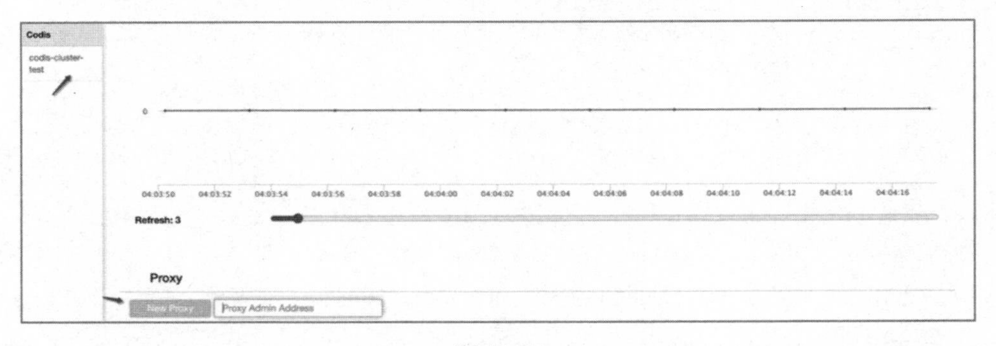

图 13-25

7. 通过 Codis FE 的前端 UI 控制台操控 Codis 集群“codis-cluster-test”

Codis 支持两种操控 Codis 集群“codis-cluster-test”的方式：

- 用后台命令操控 Codis Admin，比如执行命令“sudo codis/codis-admin--dashboard=127.0.0.1:18080 --create-proxy --addr=127.0.0.1:11083”。其中，“127.0.0.1:18080”为 Codis Dashboard 的后台 IP 地址，“127.0.0.1:11083”为 Codis Proxy 的后台 IP 地址。
- 用 Codis FE 的前端 UI 控制台去操控 Codis 集群“codis-cluster-test”。

本书统一采用前端 UI 控制台去操控 Codis 集群“codis-cluster-test”。

（1）在控制台中添加 Codis Proxy 和 Codis Dashboard 的关联关系。

在控制台中，添加 Codis Proxy 和 Codis Dashboard 的关联关系如图 13-26 所示。在添加完成后，观察 Codis Proxy 的运行日志，可以看到此时 Codis Proxy 已经上线，如图 13-27 所示。

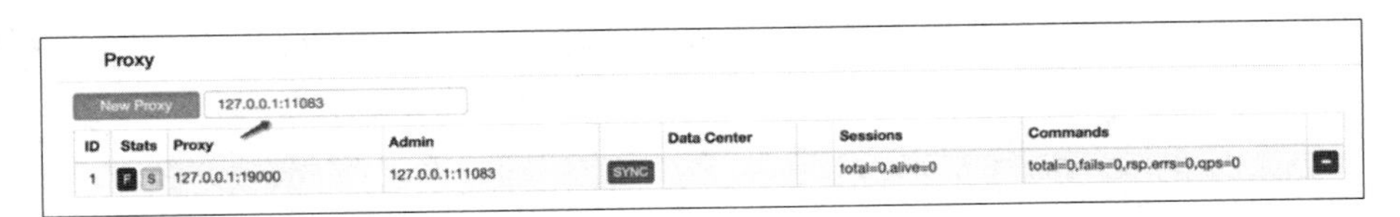

图 13-26

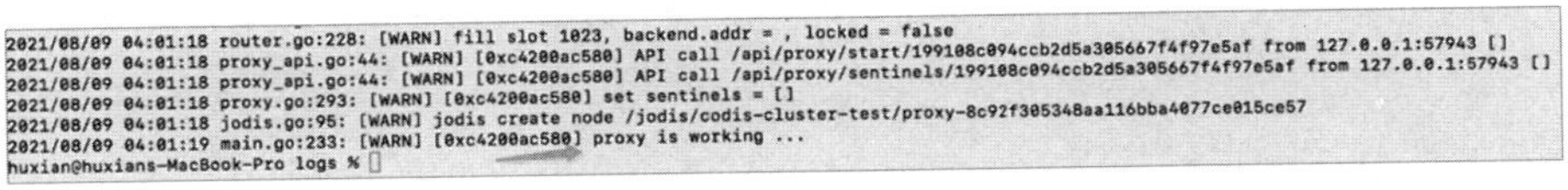

```
2021/08/09 04:01:18 router.go:228: [WARN] fill slot 1023, backend.addr = , locked = false
2021/08/09 04:01:18 proxy_api.go:44: [WARN] [0xc4200ac580] API call /api/proxy/start/199108c094ccb2d5a305667f4f97e5af from 127.0.0.1:57943 []
2021/08/09 04:01:18 proxy_api.go:44: [WARN] [0xc4200ac580] API call /api/proxy/sentinels/199108c094ccb2d5a305667f4f97e5af from 127.0.0.1:57943 []
2021/08/09 04:01:18 proxy.go:293: [WARN] [0xc4200ac580] set sentinels = []
2021/08/09 04:01:18 jodis.go:95: [WARN] jodis create node /jodis/codis-cluster-test/proxy-8c92f305348aa116bba4077ce015ce57
2021/08/09 04:01:19 main.go:233: [WARN] [0xc4200ac580] proxy is working ...
huxian@huxians-MacBook-Pro logs % 
```

图 13-27

（2）创建 Group。

Codis FE 中的 Group 的名称只能是 1 ~ 9999 的任意值（包含 1 和 9999），且不能重复，如图 13-28 所示。在控制台中输入“1”，单击“New Group”按钮去创建一个新的 Group，创建完成后如图 13-29 所示。

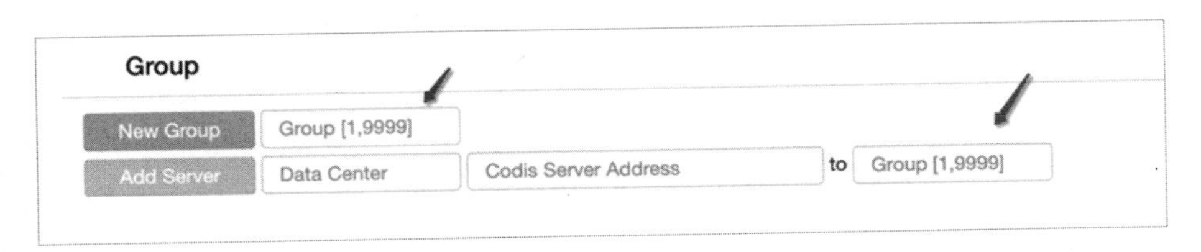

图 13-28

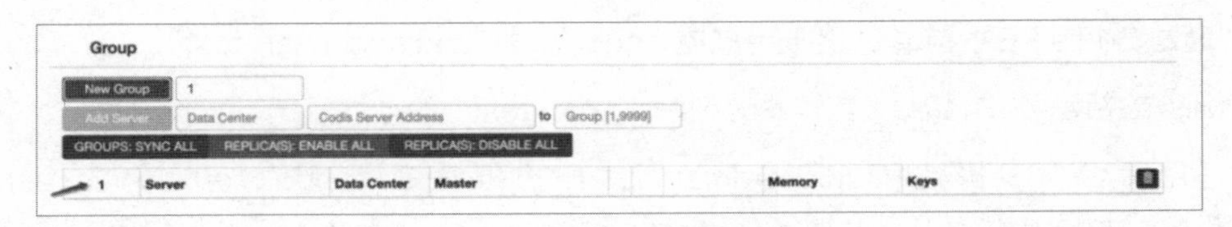

图 13-29

（3）添加 Codis Server。

在控制台中输入 Codis Server 的 IP 地址及 Group 的名称，如图 13-30 所示，然后再单击“Add Server” 按钮添加 Codis Server。

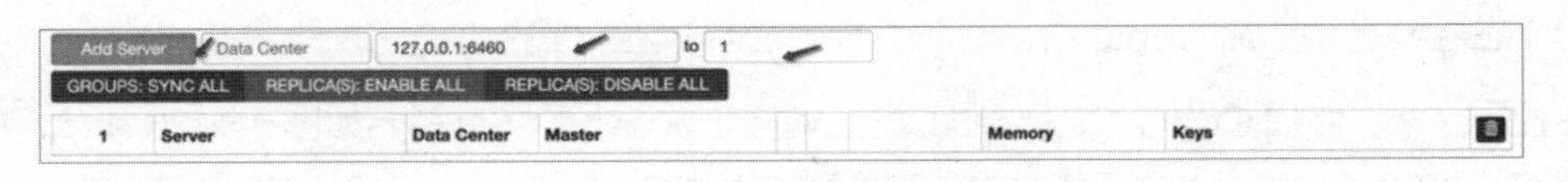

图 13-30

在添加完成后，在控制台中可以看到 Codis Server 节点“127.0.0.1:6460”，如图 13-31 所示。

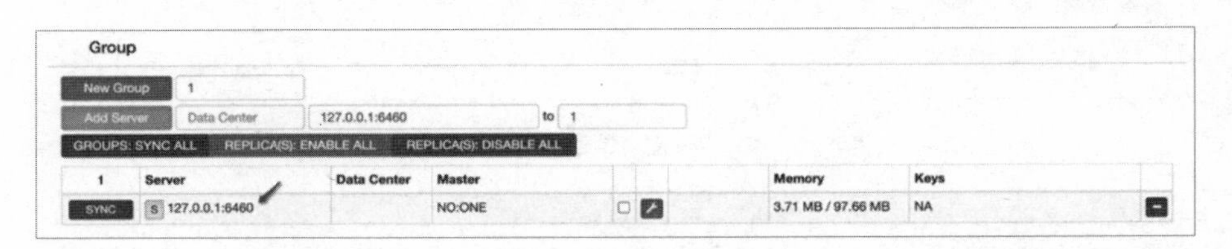

图 13-31

同理添加本书的其他两个 Codis Server 节点“127.0.0.1:6461”和“127.0.0.1:6462”，图 13-32 所示为添加完成后的效果。

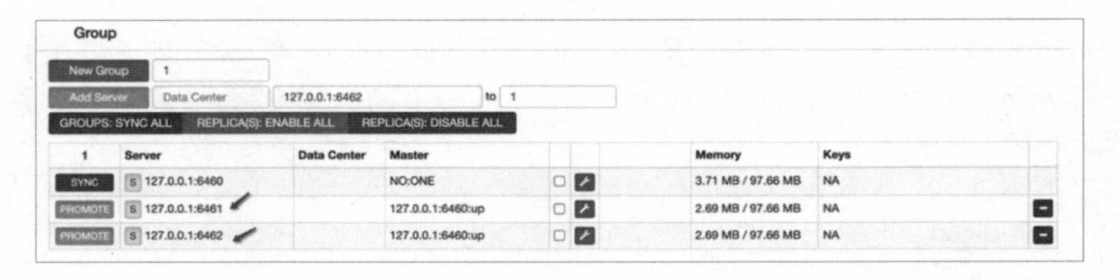

图 13-32

在 Codis Server 服务列表中单击如图 13-33 所示的按钮，将节点“127.0.0.1:6461”和“127.0.0.1:6462”设置为节点“127.0.0.1:6460”的 Slave 节点（在 Codis FE 中默认将第 1 个添加的节点当作 Master 节点）。设置完成后，3 个节点的运行状态如图 13-34 所示。

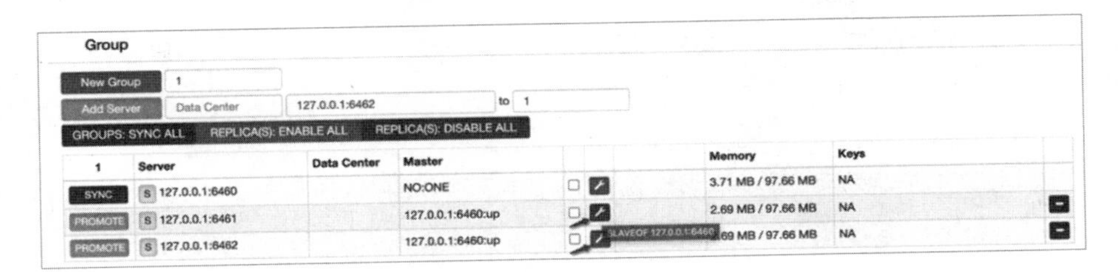

图 13-33

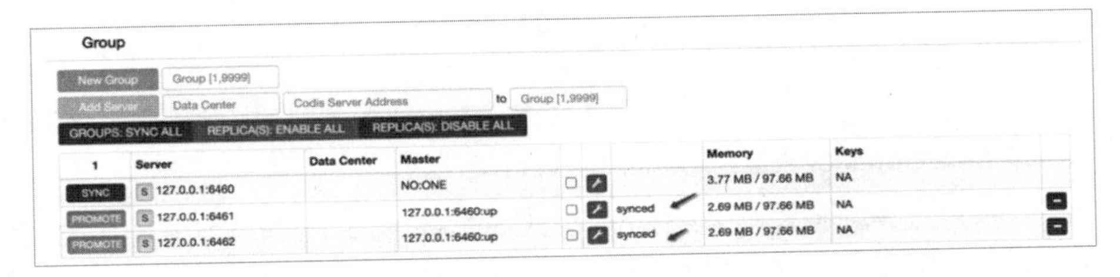

图 13-34

执行命令“sudo codis/redis-cli -a 12345678 -p 6460”连接到 Codis Server 节点“127.0.0.1:6460”，并输入 Codis 的命令 “INFO replication”。如图 13-35 所示，节点“127.0.0.1:6460”已经是 Master 节点，节点“127.0.0.1:6461”和“127.0.0.1:6462”是 Slave 节点。

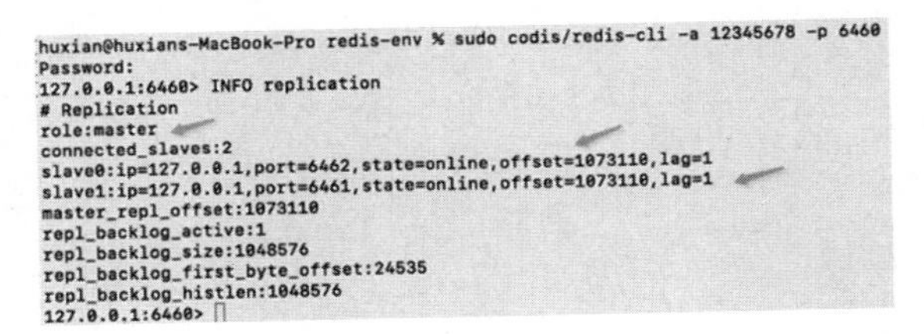

图 13-35

（4）将 Codis Server 添加到 Group 中后，需要给 Group 分配 Slot，这样 Codis 才会将缓存数据存储到指定 Codis Proxy 中。

如图 13-36 所示，Codis 支持将下标为 0～1023 的 Slot 分配到指定的 Group 中。

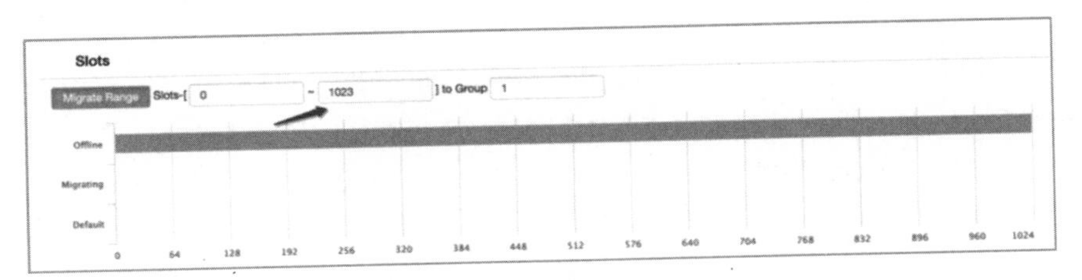

图 13-36

开发人员可以根据实际情况，将 Slot 平均地分配到不同的 Group 中。本书直接将 0～1023 的 Slot 全部分配到 Group 1 中，如图 13-37 所示。

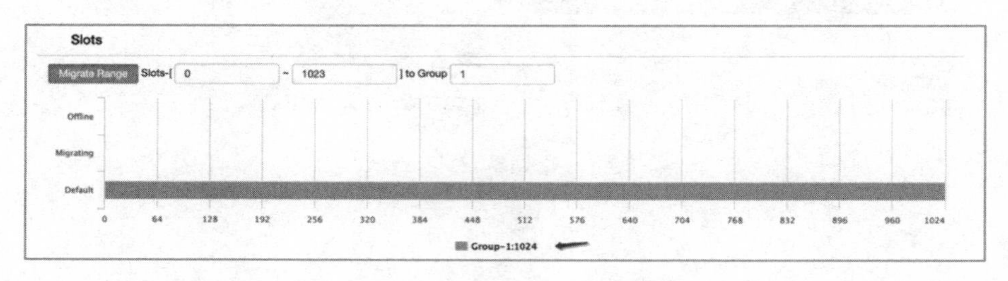

图 13-37

（5）添加 Codis Sentinel 去监控 Codis Server。

如图 13-38 在控制台中输入 Codis Sentinel 节点的 IP 地址，比如“127.0.0.1: 26370”，并单击“Add Sentinel”按钮，同理输入其他两个 Codis Sentinel 节点的 IP 地址“127.0.0.1: 26371”和“127.0.0.1: 26372”，添加完成后如图 13-39 所示。

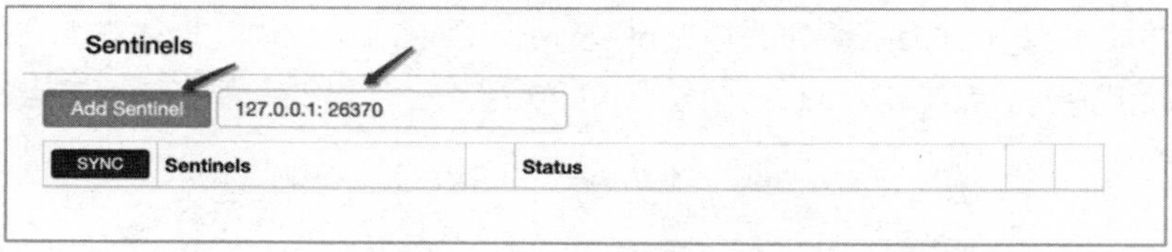

图 13-38

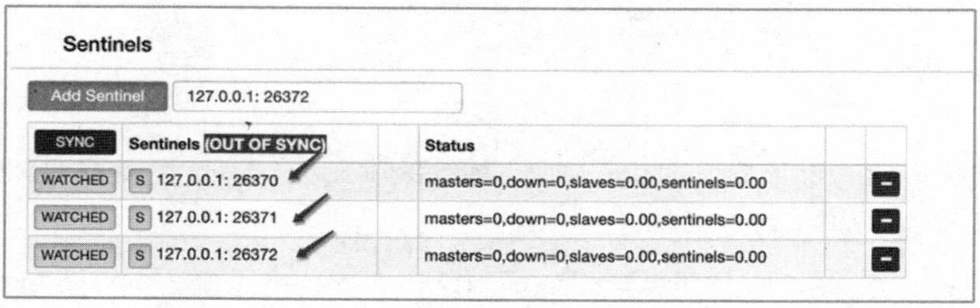

图 13-39

在控制台中单击“SYNC”按钮，从 Codis Dashboard 同步 Codis Server 节点及 Sentinel 节点元数据到 Codis Sentinel 中。如图 13-40 所示，可以看到已经同步成功，3 个 Sentinel 节点已经开始监控 Codis Server 中的 1 个 Master 节点和 2 个 Slave 节点，以及 3 个 Codis Sentinel 节点。

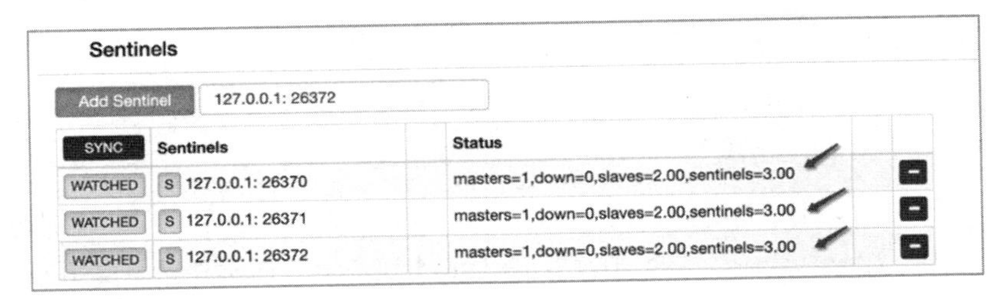

图 13-40

为了进一步验证 Codis Sentinel 节点的监控状态，采用命令“sudo codis/redis-cli -h 127.0.0.1 -p 26370 -a 12345678”去连接 Codis Sentinel 节点“127.0.0.1: 26370”，运行 Codis 的命令“info sentinel”去查看节点的状态。如图 13-41 所示，可以发现 Codis Sentinel 集群已经完成对 Codis Server 节点及自身节点的监控。

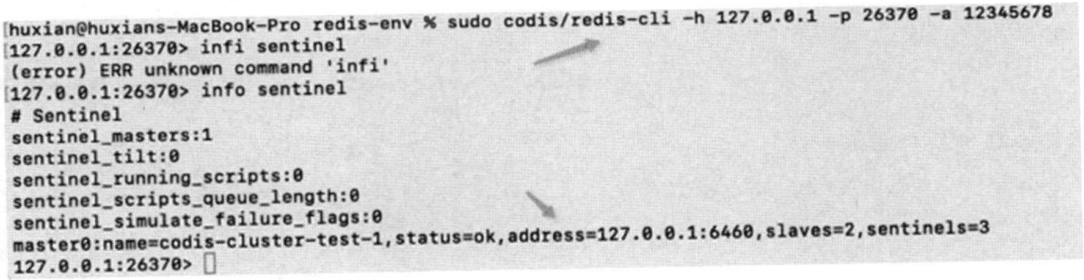

```
huxian@huxians-MacBook-Pro redis-env % sudo codis/redis-cli -h 127.0.0.1 -p 26370 -a 12345678
127.0.0.1:26370> infi sentinel
(error) ERR unknown command 'infi'
127.0.0.1:26370> info sentinel
# Sentinel
sentinel_masters:1
sentinel_tilt:0
sentinel_running_scripts:0
sentinel_scripts_queue_length:0
sentinel_simulate_failure_flags:0
master0:name=codis-cluster-test-1,status=ok,address=127.0.0.1:6460,slaves=2,sentinels=3
127.0.0.1:26370>
```

图 13-41

13.2.4　搭建 Redis Cluster 集群环境

本书在单节点中启动 6 个进程（用进程来模拟实例节点，用端口号来区分实例），其中包括 3 个 Master 和 3 个 Slave，如图 13-42 所示。

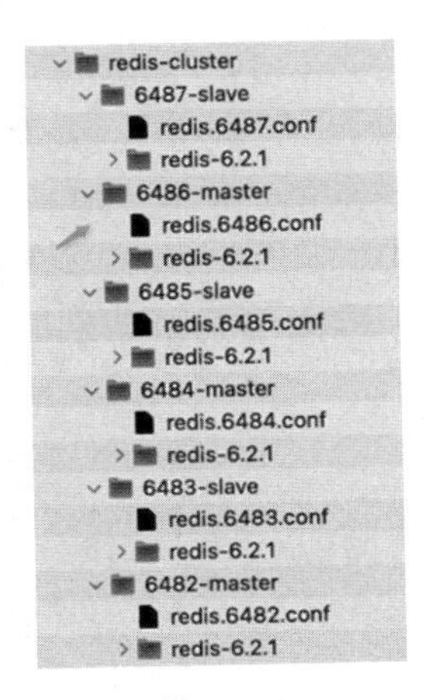

图 13-42

1. 准备 Cluster 集群环境

在搭建 Cluster 集群环境的过程中，需要新建 6 个文件夹“6482-master”“6484-master”“6486-master”“6483-slave”“6485-slave”和“6487-slave”，并在文件夹中添加集群配置文件。

（1）在“6482-master”中添加配置文件 redis.6482.conf，具体配置信息如下：

```
###①绑定 IP 地址
bind 127.0.0.1
###②配置端口号
port 6482
###③配置后台启动
daemonize yes
###④配置节点的进程号
pidfile /var/run/redis_6482.pid
###⑤配置日志路径
logfile "/Users/huxian/Downloads/redis-env/redis-cluster/6482-master/
redis-6.2.1/log/redis-6482.log"
###⑥配置数据存储路径
dir "/Users/huxian/Downloads/redis-env/redis-cluster/6482-master/
redis-6.2.1/data"
###⑦配置 RDB 文件名称
dbfilename "dump-6482.rdb"
###⑧配置 AOF 文件名称
appendfilename "appendonly-6482.aof"
###⑨开启 Redis Cluster
cluster-enabled yes
###⑩添加集群配置文件的路径
cluster-config-file "/Users/huxian/Downloads/redis-env/redis-cluster/
6482-master/redis-6.2.1/nodes-6482.conf"
requirepass 12345678
masterauth 12345678
```

（2）在“6483-master”中添加配置文件 redis.6483.conf，具体配置信息如下：

```
###①绑定 IP 地址
bind 127.0.0.1
###②配置端口号
port 6483
###③配置后台启动
daemonize yes
###④配置节点的进程号
pidfile /var/run/redis_6483.pid
###⑤配置日志路径
logfile "/Users/huxian/Downloads/redis-env/redis-cluster/6483-slave/
```

```
redis-6.2.1/log/redis-6483.log"
   ###⑥配置数据存储路径
   dir "/Users/huxian/Downloads/redis-env/redis-cluster/6483-slave/
redis-6.2.1/data"
   ###⑦配置 RDB 文件名称
   dbfilename "dump-6483.rdb"
   ###⑧配置 AOF 文件名称
   appendfilename "appendonly-6483.aof"
   ###⑨开启 Redis Cluster
   cluster-enabled yes
   ###⑩添加集群配置文件的路径
   cluster-config-file "/Users/huxian/Downloads/redis-env/redis-cluster/
6483-slave/redis-6.2.1/nodes-6483.conf"
   requirepass 12345678
   masterauth 12345678
```

（3）在“6484-master”中添加配置文件 redis.6484.conf，具体配置信息如下：

```
   ###①绑定 IP 地址
   bind 127.0.0.1
   ###②配置端口号
   port 6484
   ###③配置后台启动
   daemonize yes
   ###④配置节点的进程号
   pidfile /var/run/redis_6484.pid
   ###⑤配置日志路径
   logfile "/Users/huxian/Downloads/redis-env/redis-cluster/6484-master/
redis-6.2.1/log/redis-6484.log"
   ###⑥配置数据存储路径
   dir "/Users/huxian/Downloads/redis-env/redis-cluster/6484-master/
redis-6.2.1/data"
   ###⑦配置 RDB 文件名称
   dbfilename "dump-6484.rdb"
   ###⑧配置 AOF 文件名称
   appendfilename "appendonly-6484.aof"
   ###⑨开启 Redis Cluster
   cluster-enabled yes
   ###⑩添加集群配置文件的路径
   cluster-config-file "/Users/huxian/Downloads/redis-env/redis-cluster/
6484-master/redis-6.2.1/nodes-6484.conf"
   requirepass 12345678
   masterauth 12345678
```

（4）在“6485-slave”中添加配置文件 redis.6485.conf，具体配置信息如下：

```
   ###①绑定 IP 地址
```

```
bind 127.0.0.1
###②配置端口号
port 6485
###③配置后台启动
daemonize yes
###④配置节点的进程号
pidfile /var/run/redis_6485.pid
###⑤配置日志路径
logfile "/Users/huxian/Downloads/redis-env/redis-cluster/6485-slave/
redis-6.2.1/log/redis-6485.log"
###⑥配置数据存储路径
dir "/Users/huxian/Downloads/redis-env/redis-cluster/6485-slave/
redis-6.2.1/data"
###⑦配置 RDB 文件名称
dbfilename "dump-6485.rdb"
###⑧配置 AOF 文件名称
appendfilename "appendonly-6485.aof"
###⑨开启 Redis Cluster
cluster-enabled yes
###⑩添加集群配置文件的路径
cluster-config-file "/Users/huxian/Downloads/redis-env/redis-cluster/
6485-slave/redis-6.2.1/nodes-6485.conf"
requirepass 12345678
masterauth 12345678
```

（5）在“6486-master”中添加配置文件 redis.6486.conf，具体配置信息如下：

```
###①绑定 IP 地址
bind 127.0.0.1
###②配置端口号
port 6486
###③配置后台启动
daemonize yes
###④配置节点的进程号
pidfile /var/run/redis_6486.pid
###⑤配置日志路径
logfile "/Users/huxian/Downloads/redis-env/redis-cluster/6486-master/
redis-6.2.1/log/redis-6486.log"
###⑥配置数据存储路径
dir "/Users/huxian/Downloads/redis-env/redis-cluster/6486-master/
redis-6.2.1/data"
###⑦配置 RDB 文件名称
dbfilename "dump-6486.rdb"
###⑧配置 AOF 文件名称
appendfilename "appendonly-6486.aof"
###⑨开启 Redis Cluster
```

```
    cluster-enabled yes
    ###⑩添加集群配置文件的路径
    cluster-config-file "/Users/huxian/Downloads/redis-env/redis-cluster/
6486-master/redis-6.2.1/nodes-6486.conf"
    requirepass 12345678
    masterauth 12345678
```

（6）在“6487-slave”中添加配置文件 redis.6487.conf，具体配置信息如下：

```
    ###①绑定 IP 地址
    bind 127.0.0.1
    ###②配置端口号
    port 6487
    ###③配置后台启动
    daemonize yes
    ###④配置节点的进程号
    pidfile /var/run/redis_6487.pid
    ###⑤配置日志路径
    logfile "/Users/huxian/Downloads/redis-env/redis-cluster/6487-slave/
redis-6.2.1/log/redis-6487.log"
    ###⑥配置数据存储路径
    dir "/Users/huxian/Downloads/redis-env/redis-cluster/6487-slave/
redis-6.2.1/data"
    ###⑦配置 RDB 文件名称
    dbfilename "dump-6487.rdb"
    ###⑧配置 AOF 文件名称
    appendfilename "appendonly-6487.aof"
    ###⑨开启 Redis Cluster
    cluster-enabled yes
    ###⑩添加集群配置文件的路径
    cluster-config-file "/Users/huxian/Downloads/redis-env/redis-cluster/
6487-slave/redis-6.2.1/nodes-6487.conf"
    requirepass 12345678
    masterauth 12345678
```

2. 启动基础环境，验证 Redis Cluster 集群环境

启动基础环境，验证 Redis Cluster 集群环境的过程如下。

（1）启动 6 个 Redis 节点“127.0.0.1:6482”“127.0.0.1:6483”“127.0.0.1:6484”“127.0.0.1:6485”“127.0.0.1:6486”“127.0.0.1:6487”：

- 在文件夹“6482-master”中执行命令“redis-6.2.1/src/redis-server redis.6482.conf”。
- 在文件夹“6483-slave”中执行命令“redis-6.2.1/src/redis-server redis.6483.conf”。
- 在文件夹“6484-master”中执行命令“redis-6.2.1/src/redis-server

redis.6484.conf”。

- 在文件夹“6485-slave”中执行命令“redis-6.2.1/src/redis-server redis.6485.conf”。
- 在文件夹“6486-master”中执行命令“redis-6.2.1/src/redis-server redis.6486.conf”。
- 在文件夹“6487-slave”中执行命令“redis-6.2.1/src/redis-server redis.6487.conf”。

执行完成后，用命令“ps -ef | grep redis”查看 Redis 进程。如图 13-43 所示，6 个节点已经启动成功。

```
huxian@huxians-MacBook-Pro 6487-slave % ps -ef | grep redis
  501 46169     1   0  5:44AM ??         0:03.04 redis-6.2.1/src/redis-server 127.0.0.1:6482 [cluster]
  501 46210     1   0  5:47AM ??         0:00.94 redis-6.2.1/src/redis-server 127.0.0.1:6483 [cluster]
  501 46219     1   0  5:48AM ??         0:00.63 redis-6.2.1/src/redis-server 127.0.0.1:6484 [cluster]
  501 46231     1   0  5:48AM ??         0:00.39 redis-6.2.1/src/redis-server 127.0.0.1:6485 [cluster]
  501 46239     1   0  5:48AM ??         0:00.24 redis-6.2.1/src/redis-server 127.0.0.1:6486 [cluster]
  501 46248     1   0  5:49AM ??         0:00.04 redis-6.2.1/src/redis-server 127.0.0.1:6487 [cluster]
  501 46261 41178   0  5:49AM ttys002    0:00.00 grep redis
```

图 13-43

（2）执行命令“redis-6.2.1/src/redis-cli -a 12345678 --cluster create 127.0.0.1:6482 127.0.0.1:6484 127.0.0.1:6486 127.0.0.1:6483 127.0.0.1:6485 127.0.0.1:6487--cluster-replicas 1”创建 Redis Cluster，如图 13-44 所示。

```
huxian@huxians-MacBook-Pro 6487-slave % redis-6.2.1/src/redis-cli -a 12345678 --cluster create 127.0
:6485 127.0.0.1:6487 --cluster-replicas 1

Warning: Using a password with '-a' or '-u' option on the command line interface may not be safe.
>>> Performing hash slots allocation on 6 nodes...
Master[0] -> Slots 0 - 5460
Master[1] -> Slots 5461 - 10922
Master[2] -> Slots 10923 - 16383
Adding replica 127.0.0.1:6485 to 127.0.0.1:6482
Adding replica 127.0.0.1:6487 to 127.0.0.1:6484
Adding replica 127.0.0.1:6483 to 127.0.0.1:6486
>>> Trying to optimize slaves allocation for anti-affinity
[WARNING] Some slaves are in the same host as their master
M: 811924874962b6d62d570cc9eab3a7f85c09cc3b 127.0.0.1:6482
   slots:[0-5460] (5461 slots) master
M: e14eb0ba98f973183138781994f9e2526b046feb 127.0.0.1:6484
   slots:[5461-10922] (5462 slots) master
M: 5660ac280ac36638d58e38bfa559b5fcab8bf96b 127.0.0.1:6486
   slots:[10923-16383] (5461 slots) master
S: 9cbf5f7ca41b0077e4fedea011ed1be98db1b2a2 127.0.0.1:6483
   replicates 811924874962b6d62d570cc9eab3a7f85c09cc3b
S: c62ea2be90d599d405153616c5cce8e6ee19475e 127.0.0.1:6485
   replicates e14eb0ba98f973183138781994f9e2526b046feb
S: 71dabf50fdfd0d9486e445f3a0bf35c06bddaa63 127.0.0.1:6487
   replicates 5660ac280ac36638d58e38bfa559b5fcab8bf96b
Can I set the above configuration? (type 'yes' to accept): yes
>>> Nodes configuration updated
>>> Assign a different config epoch to each node
>>> Sending CLUSTER MEET messages to join the cluster
Waiting for the cluster to join
.
>>> Performing Cluster Check (using node 127.0.0.1:6482)
M: 811924874962b6d62d570cc9eab3a7f85c09cc3b 127.0.0.1:6482
   slots:[0-5460] (5461 slots) master
   1 additional replica(s)
M: 5660ac280ac36638d58e38bfa559b5fcab8bf96b 127.0.0.1:6486
   slots:[10923-16383] (5461 slots) master
   1 additional replica(s)
M: e14eb0ba98f973183138781994f9e2526b046feb 127.0.0.1:6484
   slots:[5461-10922] (5462 slots) master
   1 additional replica(s)
S: 9cbf5f7ca41b0077e4fedea011ed1be98db1b2a2 127.0.0.1:6483
   slots: (0 slots) slave
   replicates 811924874962b6d62d570cc9eab3a7f85c09cc3b
S: 71dabf50fdfd0d9486e445f3a0bf35c06bddaa63 127.0.0.1:6487
   slots: (0 slots) slave
   replicates 5660ac280ac36638d58e38bfa559b5fcab8bf96b
S: c62ea2be90d599d405153616c5cce8e6ee19475e 127.0.0.1:6485
   slots: (0 slots) slave
   replicates e14eb0ba98f973183138781994f9e2526b046feb
[OK] All nodes agree about slots configuration.
>>> Check for open slots...
>>> Check slots coverage...
[OK] All 16384 slots covered.
```

图 13-44

（3）执行命令“redis-6.2.1/src/redis-cli -a 12345678 -p 6483”去连接 Redis 节点，并输入命令“cluster nodes”。如图 13-45 所示，可以看到 Redis Cluster 已经启动成功。

- 执行命令的节点“127.0.0.1:6483”为 Slave 节点。
- 节点“127.0.0.1:6482”“127.0.0.1:6484”和“127.0.0.1:6486”为 Master 节点。
- 节点“127.0.0.1:6485”和“127.0.0.1:6487”为 Slave 节点。

```
huxian@huxians-MacBook-Pro 6487-slave % redis-6.2.1/src/redis-cli -a 12345678 -p 6483
Warning: Using a password with '-a' or '-u' option on the command line interface may not be safe.
127.0.0.1:6483> cluster nodes
c62ea2be90d599d405153616c5cce8e6ee19475e 127.0.0.1:6485@16485 slave e14eb0ba98f973183138781994f9e2526b046feb 0 1628287934000 2 connected
71dabf50fdfd0d9486e445f3a0bf35c06bddaa63 127.0.0.1:6487@16487 slave 5660ac280ac36638d58e38bfa559b5fcab8bf96b 0 1628287935575 3 connected
e14eb0ba98f973183138781994f9e2526b046feb 127.0.0.1:6484@16484 master - 0 1628287932000 2 connected 5461-10922
5660ac280ac36638d58e38bfa559b5fcab8bf96b 127.0.0.1:6486@16486 master - 0 1628287934567 3 connected 10923-16383
811924874962b6d62d570cc9eab3a7f85c09cc3b 127.0.0.1:6482@16482 master - 0 1628287933560 1 connected 0-5460
9cbf5f7ca41b0077e4fedea011ed1be98db1b2a2 127.0.0.1:6483@16483 myself,slave 811924874962b6d62d570cc9eab3a7f85c09cc3b 0 1628287933000 1 connected
```

图 13-45

13.3　将 Spring Cloud Alibaba 应用接入 Redis

可以使用以下几种方式将 Spring Cloud Alibaba 应用接入 Redis。

- spring-boot-starter-data-redis：Spring Boot 官方提供的一个“开箱即用”的 Starter 框架。
- Jedis：Redis 官方推荐的 Redis 客户端。
- redisson-spring-boot-starter：Redisson 官方提供的 Redis 客户端，它主要是将原有的 Redisson 客户端 Spring Boot 化。
- Lettuce：一个可扩展的线程安全的 Redis 客户端，支持同步、异步和响应式通信模式。

13.3.1　【实例】集成 spring-boot-starter-data-redis，将 Spring Cloud Alibaba 应用接入 Redis 主从环境

本实例用 Spring Boot 提供的 Starter 框架 spring-boot-starter-data-redis，将 Spring Cloud Alibaba 应用接入 Redis 主从环境。

1. 初始化一个 Spring Cloud Alibaba 服务

使用 Spring Cloud Alibaba 快速初始化一个服务“use-springbootdataredis-spring-cloud-alibaba”，并在 pom.xml 文件中添加 Redis 客户端相关的 Jar 包依赖，具体代码如下所示：

```
<dependency>
    <groupId>org.springframework.boot</groupId>
    <artifactId>spring-boot-starter-data-redis</artifactId>
```

```
        <version>2.3.2.RELEASE</version>
    </dependency>
```

在配置文件 bootstrap.yaml 中添加连接 Redis 的配置信息，具体如下所示：

```
spring:
  redis:
    password: 12345678
    host: 127.0.0.1
    port: 6470
```

2. 使用 RedisTemplate 类操作 Redis 中的缓存

（1）初始化一个 RedisTemplate 对象，具体代码如下所示：

```
@Configuration
public class SpringBootDataRedisConfig {
    @Bean
    @ConditionalOnMissingBean(name = "redisTemplate")
    public RedisTemplate<Object, Object>
redisTemplate(RedisConnectionFactory redisConnectionFactory) {
        RedisTemplate<Object, Object> template = new RedisTemplate<>();
        //①定义一个基于 Jackson 的序列化对象 fastJsonRedisSerializer
        Jackson2JsonRedisSerializer<Object> fastJsonRedisSerializer = new
Jackson2JsonRedisSerializer<Object>(Object.class);
        //②使用 fastJsonRedisSerializer 对象序列化 Redis 缓存对象中的 Value 值
        template.setValueSerializer(fastJsonRedisSerializer);
        template.setHashValueSerializer(fastJsonRedisSerializer);
        //③使用 StringRedisSerializer 对象序列化 Redis 缓存对象中的 Key 值
        template.setKeySerializer(new StringRedisSerializer());
        template.setHashKeySerializer(new StringRedisSerializer());
        template.setConnectionFactory(redisConnectionFactory);
        //④返回 RedisTemplate 对象
        return template;
    }
}
```

（2）使用 RedisTemplate 类连接 Redis 主从环境，并读取缓存对象，具体代码如下所示：

```
@Service
public class SpringDataRedisService {
    @Autowired
    private RedisTemplate redisTemplate;

    @PostConstruct
    public void redisInit(){
        ValueOperations<String, User> valueOperations =
redisTemplate.opsForValue();
```

```
        User item=new User();
        item.setUserId(12345L);
        item.setUserName("测试");
        //①塞入一个缓存对象 User
        valueOperations.set("12345",item);
        //②启动一个线程池，读取缓存对象
        ExecutorService executorService= Executors.newFixedThreadPool(1);
        executorService.execute(new RedisOperateThread());
    }
    class RedisOperateThread implements Runnable{
        @Override
        public void run() {
            while (true){
                try {
                    Thread.sleep(2000);
                }catch (InterruptedException e){
                    System.out.println(e.getMessage());
                }
                ValueOperations<String, User> valueOperations =
redisTemplate.opsForValue();
                //③从 Redis 主从环境中读取缓存对象
                Object value=valueOperations.get("12345");
                ObjectMapper objectMapper = new ObjectMapper();
                //④将缓存对象转化为 JSON 字符串
                String objStr="";
                try {
                    objStr = objectMapper.writeValueAsString(value);
                }catch (JsonProcessingException e){
                    e.printStackTrace();
                }
                User user = null;
                try {
                    //⑤将 JSON 字符串转化为 POJO 对象
                    user = objectMapper.readValue(objStr,User.class);
                } catch (Exception e) {
                    e.printStackTrace();
                }
                System.out.println("通过线程获取, user:"+user.toString());
            }
        }
    }
}
```

3. 启动服务，验证 RedisTemplate 类操作缓存的功能

启动本书的 Redis 主从环境，然后启动“use-springbootdataredis-spring-cloud-alibaba”

服务，在 IDEA 的控制台中，如果观察到如图 13-46 所示的服务运行的日志，则表示向 Redis 主从环境中塞入/读取缓存成功。

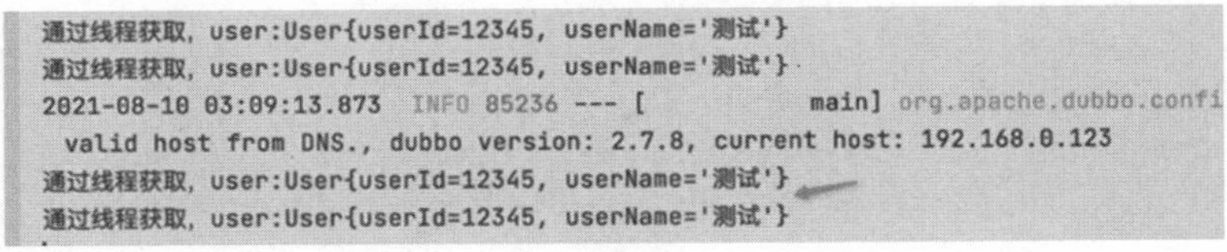

```
通过线程获取, user:User{userId=12345, userName='测试'}
通过线程获取, user:User{userId=12345, userName='测试'}
2021-08-10 03:09:13.873  INFO 85236 --- [           main] org.apache.dubbo.confi
 valid host from DNS., dubbo version: 2.7.8, current host: 192.168.0.123
通过线程获取, user:User{userId=12345, userName='测试'}
通过线程获取, user:User{userId=12345, userName='测试'}
```

图 13-46

4. 使用 Redis 的后台命令连接主从环境，查看缓存对象

在 Redis 主从环境中，执行命令“redis-6.2.1/src/redis-cli -h 127.0.0.1 -p 6470 -a 12345678 --raw”连接 Redis 主从环境中的 Master 节点。

图 13-47 所示为执行完“get 12345”命令后 Redis 返回的缓存对象。

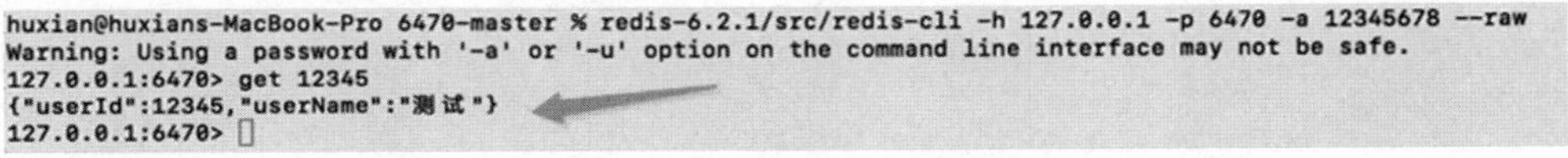

```
huxian@huxians-MacBook-Pro 6470-master % redis-6.2.1/src/redis-cli -h 127.0.0.1 -p 6470 -a 12345678 --raw
Warning: Using a password with '-a' or '-u' option on the command line interface may not be safe.
127.0.0.1:6470> get 12345
{"userId":12345,"userName":"测试"}
127.0.0.1:6470> 
```

图 13-47

13.3.2 【实例】集成 redisson-spring-boot-starter，将 Spring Cloud Alibaba 应用接入 Redis Sentinel 环境

本实例用 Redisson 提供的 redisson-spring-boot-starter，将 Spring Cloud Alibaba 应用接入 Redis Sentinel 环境。

1. 初始化一个 Spring Cloud Alibaba 服务

使用 Spring Cloud Alibaba 快速初始化一个服务“use-redisson-spring-cloud-alibaba”，并在 pom.xml 文件中添加 redisson-spring-boot-starter 的 Jar 包依赖，具体代码如下所示：

```
<dependency>
    <groupId>org.redisson</groupId>
    <artifactId>redisson-spring-boot-starter</artifactId>
    <version>3.16.1</version>
    <exclusions>
        <exclusion>
            <groupId>org.redisson</groupId>
            <artifactId>redisson-spring-data-25</artifactId>
        </exclusion>
    </exclusions>
</dependency>
```

```xml
<dependency>
    <groupId>org.redisson</groupId>
    <artifactId>redisson-spring-data-23</artifactId>
    <version>3.13.6</version>
</dependency>
<dependency>
    <groupId>org.springframework.boot</groupId>
    <artifactId>spring-boot-starter-data-redis</artifactId>
    <version>2.3.2.RELEASE</version>
    <exclusions>
        <exclusion>
            <groupId>io.lettuce</groupId>
            <artifactId>lettuce-core</artifactId>
        </exclusion>
    </exclusions>
</dependency>
```

在文件 application.yaml 中添加 Sentinel 环境中 Redis Server 的 Master 节点信息，具体代码如下所示：

```yaml
spring:
  cloud:
    redisson:
      config-file: classpath:redisson.yaml
  ###配置 Master 节点的 IP 地址
  redis:
    host: 127.0.0.1
    port: 6480
    password: 12345678
```

在文件 redisson.yaml 中添加 Redisson 中 Sentinel 节点的配置信息，具体代码如下所示：

```yaml
###①省略部分配置信息
sentinelServersConfig:
  ###②配置 Redis 主从环境的授权访问密码
  password: 12345678
  ###③配置哨兵的 IP 地址列表
  sentinelAddresses:
    - "redis://127.0.0.1:26382"
    - "redis://127.0.0.1:26381"
    - "redis://127.0.0.1:26380"
   masterName: "mymaster"
threads: 0
nettyThreads: 0
codec: !<org.redisson.codec.JsonJacksonCodec> {}
"transportMode":"NIO"
...
```

2. 用 Redisson 的 RedissonClient 类操控缓存对象

本实例用 Redisson 的 RedissonClient 类操控缓存对象，具体代码如下所示：

```
@Service
public class RedissonService {
    private String cacheMapName;
    private String cacheListName;
    private String cacheSetName;
    private RMap<String,String> rMap;
    private RSet<Good> rSet;
    private RList<Good> rList;
    @Autowired
    private RedissonClient redissonClient;
    @PostConstruct
    public void initRedis(){
        cacheMapName="test1234";
        cacheListName="testlist1234";
        cacheSetName="testSet1234";
        this.rList=redissonClient.getList(cacheListName);
        this.rMap = redissonClient.getMap(this.cacheMapName);
        this.rSet = redissonClient.getSet(this.cacheSetName);
        ExecutorService executorService= Executors.newFixedThreadPool(1);
        executorService.execute(new ReadThread());
    }
    public void put(String key,String value,Good item){
        rMap.put(key,value);
        rList.add(item);
        rSet.add(item);
    }
    class ReadThread implements Runnable {
        @Override
        public void run() {
            while (true) {
                try {
                    Thread.sleep(2000);
                } catch (InterruptedException e) {
                    System.out.println(e.getMessage());
                }
                Good item = new Good();
                String key = RandomUtils.nextLong() + "";
                item.setGoodId(Long.valueOf(key));
                item.setGoodName("商品名称");
                String objectJson = JSON.toJSONString(item);
                put(key, objectJson, item);
                String value = rMap.get(key);
```

```
                    System.out.println("缓存 Map 中的值为: " + value);
                    Iterator<Good> iteratorList = rList.iterator();
                    while (iteratorList.hasNext()) {
                        Good item1 = (Good) iteratorList.next();
                        System.out.println("缓存 List 中的值为: " +
item1.toString());
                    }
                    Iterator<Good> iteratorSet = rSet.iterator();
                    while (iteratorSet.hasNext()) {
                        Good item2 = (Good) iteratorSet.next();
                        System.out.println("缓存 Set 中的值为: "+item2.toString());
                    }
                }
            }
        }
    }
```

3. 启动服务，并验证 Redisson 操控缓存的效果

启动本书 Redis 的 Sentinel 集群环境和本实例服务，在 IDEA 控制台可以观察到如下所示的运行日志：

```
缓存 Set 中的值为: Good{goodId=4260218375227088896, goodName='商品名称'}
缓存 List 中的值为: Good{goodId=4953715990736233472, goodName='商品名称'}
缓存 Map 中的值为: {"goodId":1413231056868889600,"goodName":"商品名称"}
```

4. 使用 Redis 的“redis-cli”命令连接 Redis 的主节点，查看缓存对象

在执行 Redis 的连接命令“redis-6.2.1/src/redis-cli -h 127.0.0.1 -p 6479 -a 12345678 --raw”后，分别执行如下命令查看缓存对象。

（1）使用命令“hgetall test1234”查看名称为“test1234”的 Hash 缓存对象，如图 13-48 所示。

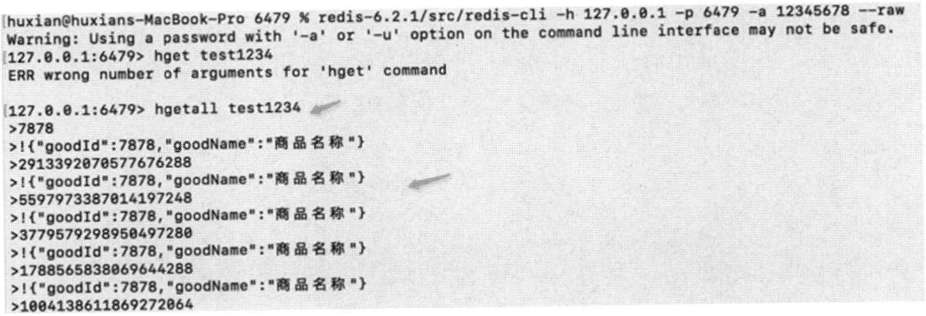

```
huxian@huxians-MacBook-Pro 6479 % redis-6.2.1/src/redis-cli -h 127.0.0.1 -p 6479 -a 12345678 --raw
Warning: Using a password with '-a' or '-u' option on the command line interface may not be safe.
127.0.0.1:6479> hget test1234
ERR wrong number of arguments for 'hget' command

127.0.0.1:6479> hgetall test1234
>7878
>!{"goodId":7878,"goodName":"商品名称"}
>2913392070577676288
>!{"goodId":7878,"goodName":"商品名称"}
>5597973387014197248
>!{"goodId":7878,"goodName":"商品名称"}
>3779579298950497280
>!{"goodId":7878,"goodName":"商品名称"}
>1788565838069644288
>!{"goodId":7878,"goodName":"商品名称"}
>1004138611869272064
```

图 13-48

（2）使用命令“hget testlist1234”查看名称为“testlist1234”的列表缓存对象，如图 13-49 所示。

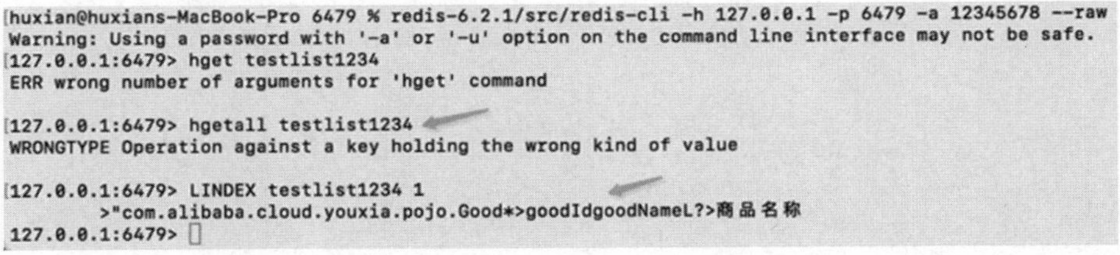

```
huxian@huxians-MacBook-Pro 6479 % redis-6.2.1/src/redis-cli -h 127.0.0.1 -p 6479 -a 12345678 --raw
Warning: Using a password with '-a' or '-u' option on the command line interface may not be safe.
127.0.0.1:6479> hget testlist1234
ERR wrong number of arguments for 'hget' command

127.0.0.1:6479> hgetall testlist1234
WRONGTYPE Operation against a key holding the wrong kind of value

127.0.0.1:6479> LINDEX testlist1234 1
        >"com.alibaba.cloud.youxia.pojo.Good*>goodIdgoodNameL?>商品名称
127.0.0.1:6479> 
```

图 13-49

（3）使用命令“SMEMBERS testSet1234”查看名称为“testSet1234”的有序集合缓存对象，如图 13-50 所示。

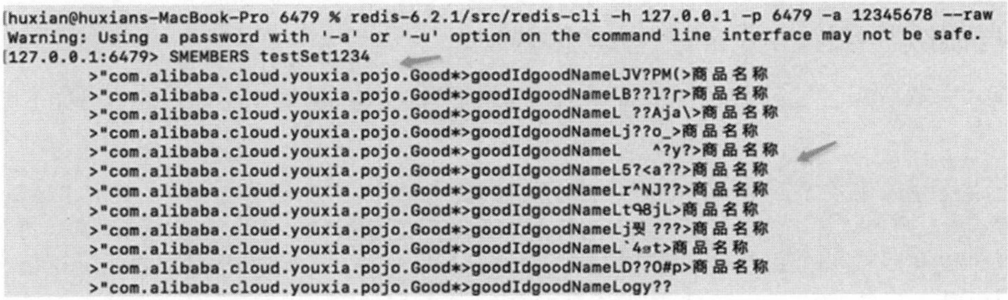

```
huxian@huxians-MacBook-Pro 6479 % redis-6.2.1/src/redis-cli -h 127.0.0.1 -p 6479 -a 12345678 --raw
Warning: Using a password with '-a' or '-u' option on the command line interface may not be safe.
127.0.0.1:6479> SMEMBERS testSet1234
        >"com.alibaba.cloud.youxia.pojo.Good*>goodIdgoodNameLJV?PM(>商品名称
        >"com.alibaba.cloud.youxia.pojo.Good*>goodIdgoodNameLB??l?┌>商品名称
        >"com.alibaba.cloud.youxia.pojo.Good*>goodIdgoodNameL ??Aja\>商品名称
        >"com.alibaba.cloud.youxia.pojo.Good*>goodIdgoodNameLj??o_>商品名称
        >"com.alibaba.cloud.youxia.pojo.Good*>goodIdgoodNameL   ^?y?>商品名称
        >"com.alibaba.cloud.youxia.pojo.Good*>goodIdgoodNameL5?<a??>商品名称
        >"com.alibaba.cloud.youxia.pojo.Good*>goodIdgoodNameLr^NJ??>商品名称
        >"com.alibaba.cloud.youxia.pojo.Good*>goodIdgoodNameLt98jL>商品名称
        >"com.alibaba.cloud.youxia.pojo.Good*>goodIdgoodNameLj梨 ???>商品名称
        >"com.alibaba.cloud.youxia.pojo.Good*>goodIdgoodNameL`4st>商品名称
        >"com.alibaba.cloud.youxia.pojo.Good*>goodIdgoodNameLD??O#p>商品名称
        >"com.alibaba.cloud.youxia.pojo.Good*>goodIdgoodNameLogy??
```

图 13-50

13.3.3 【实例】集成 Jedis，将 Spring Cloud Alibaba 应用接入 Redis Codis 集群环境

本实例用中间件框架 Jedis 将 Spring Cloud Alibaba 应用接入 Redis Codis 集群环境。

1. 初始化一个 Spring Cloud Alibaba 服务

使用 Spring Cloud Alibaba 快速初始化一个服务“use-jedis-spring-cloud-alibaba”，并在 pom.xml 文件中添加 Jedis 的 Jar 包依赖，具体代码如下所示：

```
<dependency>
    <groupId>org.springframework.boot</groupId>
    <artifactId>spring-boot-starter-data-redis</artifactId>
    <version>2.3.2.RELEASE</version>
    <exclusions>
        <exclusion>
            <groupId>io.lettuce</groupId>
            <artifactId>lettuce-core</artifactId>
```

```
            </exclusion>
        </exclusions>
    </dependency>
<dependency>
    <groupId>redis.clients</groupId>
    <artifactId>jedis</artifactId>
    <version>3.2.0</version>
</dependency>
```

在 bootstrap.yaml 中添加 Codis 的配置信息，具体如下所示：

```
###其中，127.0.0.1:19000 为本书的 Codis Proxy 节点
redis:
  node:
    max-total: 10
    host: 127.0.0.1
    port: 19000
    password: 12345678
```

2. 读取 Jedis 的配置信息，并初始化 Jedis 连接池

本实例主要是用 JedisConnectConfig 和 JedisConfig 类读取 Jedis 的配置信息，并初始化 Jedis 连接池。

（1）用 JedisConfig 类读取 Jedis 连接 Codis 的配置信息，具体代码如下所示：

```
//加载连接 Codis 集群的配置信息
@Data
@Configuration
@ConfigurationProperties(prefix = "redis.node")
public class JedisConfig {
    private Integer maxTotal;
    private String host;
    private Integer port;
    private String password;
}
```

（2）用 JedisConnectConfig 类初始化 Jedis 的连接池，具体代码如下所示：

```
@Configuration
public class JedisConnectConfig {
    @Autowired
    private JedisConfig jedisConfig;
    //①初始化连接池配置对象 JedisPoolConfig
    public JedisPoolConfig jedisPoolConfig(){
        JedisPoolConfig jedisPoolConfig = new JedisPoolConfig();
        jedisPoolConfig.setMaxTotal(jedisConfig.getMaxTotal());
        return jedisPoolConfig;
```

```
    }
    //②初始化连接池对象
        @Bean
        public JedisPool jedisPool(){
            JedisPoolConfig jedisPoolConfig = jedisPoolConfig();
            return new JedisPool(jedisPoolConfig,jedisConfig.getHost(),
jedisConfig.getPort(),3000,jedisConfig.getPassword());
        }
```

3. 用 Jedis 的连接池操控缓存对象

Jedis 通过 JedisPool 类去连接 Redis 服务器端，从而实现操控缓存的功能。本实例用 JedisService 类来调用 JedisPool 类的 API 去操控缓存对象，具体代码如下所示：

```
@Service
public class JedisService {

    @Autowired
    private JedisPool jedisPool;

    @PostConstruct
    public void initJedis(){
        //①获取 JedisPool 连接池中的客户端 Jedis 对象
        Jedis source=jedisPool.getResource();
        Order item=new Order();
        item.setOrderAmount(new BigDecimal(100));
        item.setOrderId(10000L);
        item.setOrderName("测试订单");
        String objectJson = JSON.toJSONString(item);
        //②用 Jedis 对象的 set()方法设置一个缓存
        source.set("10000",objectJson);
        ExecutorService executorService= Executors.newFixedThreadPool(1);
        executorService.execute(new ReadValueThread());
    }
    //③在线程中，用客户端对象 Jedis 的 get()方法读取缓存
    class ReadValueThread implements Runnable{
        @Override
        public void run() {
            while (true){
                try {
                    Thread.sleep(2000);
                }catch (InterruptedException e){
                    System.out.println(e.getMessage());
                }
                Jedis source=jedisPool.getResource();
                String value=source.get("10000");
```

```
                System.out.println("用 Jedis 客户端从 Codis 服务器端获取值为:
"+value);
            }
        }
    }
}
```

4. 启动服务，验证 Jedis 连接池操控缓存的效果

在启动本书的 Codis 集群环境中启动“use-jedis-spring-cloud-alibaba”服务，可以在 IDEA 控制台观察到如图 13-51 所示的运行日志。

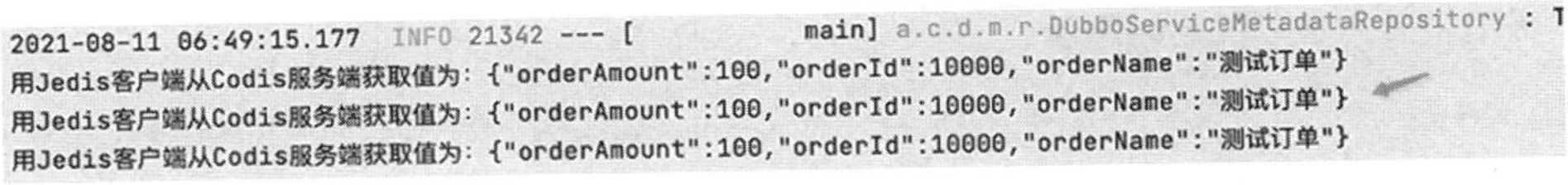

```
2021-08-11 06:49:15.177  INFO 21342 --- [           main] a.c.d.m.r.DubboServiceMetadataRepository : T
用Jedis客户端从Codis服务端获取值为: {"orderAmount":100,"orderId":10000,"orderName":"测试订单"}
用Jedis客户端从Codis服务端获取值为: {"orderAmount":100,"orderId":10000,"orderName":"测试订单"}
用Jedis客户端从Codis服务端获取值为: {"orderAmount":100,"orderId":10000,"orderName":"测试订单"}
```

图 13-51

5. 连接 Codis Proxy，查看缓存对象

使用命令“sudo codis/redis-cli -h 127.0.0.1 -p 19000 -a 12345678 --raw”连接 Codis Proxy 成功后，执行命令“get 10000”可以查询到缓存对象，如图 13-52 所示。

```
huxian@huxians-MacBook-Pro redis-env % sudo codis/redis-cli -h 127.0.0.1 -p 19000 -a 12345678 --raw
127.0.0.1:19000> get 10000
{"orderAmount":100,"orderId":10000,"orderName":"测试订单"}
127.0.0.1:19000>
```

图 13-52

13.3.4 【实例】集成 Lettuce，将 Spring Cloud Alibaba 应用接入 Redis Cluster 集群环境

本实例主要是使用中间件框架 Lettuce 将 Spring Cloud Alibaba 应用接入 Redis Cluster 集群环境。

1. 初始化一个 Spring Cloud Alibaba 服务

使用 Spring Cloud Alibaba 快速初始化一个服务“use-lettuce-spring-cloud-alibaba”，并在 pom.xml 文件中添加中间件框架 Lettuce 的 Jar 包依赖，具体代码如下所示：

```
<dependency>
    <groupId>io.lettuce</groupId>
    <artifactId>lettuce-core</artifactId>
    <version>6.1.4.RELEASE</version>
</dependency>
<dependency>
```

```
        <groupId>org.springframework.data</groupId>
        <artifactId>spring-data-redis</artifactId>
        <version>2.2.7.RELEASE</version>
    </dependency>
```

在文件 bootstrap.yaml 中添加连接 Redis Cluster 集群环境的配置信息，具体代码如下所示：

```
alibaba:
  lettuce:
    redisUrl: redis://127.0.0.1:6482,redis://127.0.0.1:6484,
redis://127.0.0.1:6486
    password: 12345678
```

2. 初始化 Lettuce 的基础环境

Lettuce 是一个连接 Redis 的中间框架。从 Redis 的角度来看，它是一个客户端，Redis 是服务器端，客户端连接服务器端需要先初始化基础环境。本实例采用 LoadLettuceResource 类和 LettuceConfig 类来初始化 Lettuce 的基础环境。

（1）用 LettuceConfig 类读取“Lettuce 连接 Redis Cluster 集群环境”的配置信息，具体代码如下所示：

```
@Configuration
@ConfigurationProperties(prefix = "alibaba.lettuce")
public class LettuceConfig {
    //①读取 Redis 的 Redis Cluster 的 URL
    private String redisUrl;
    //②读取集群节点的访问密码
    private String password;
    //③将 redisUrl 转换为链表
    private List<String> hostUrlList;
    public void setPassword(String password) {
        this.password = password;
    }
    public String getPassword() {
        return password;
    }
    public void setHostUrlList(List<String> hostUrlList) {
        this.hostUrlList = hostUrlList;
    }
    public void setRedisUrl(String redisUrl) {
        this.redisUrl = redisUrl;
    }
    public List<String> getHostUrlList() {
        String[] hostUrlArrays=null;
        if(StringUtils.isNotEmpty(redisUrl)){
```

```
            hostUrlArrays=redisUrl.split(",");
        }
        hostUrlList= Arrays.asList(hostUrlArrays);
        return hostUrlList;
    }
    public String getRedisUrl() {
        return redisUrl;
    }
}
```

（2）用 LoadLettuceResource 类初始化 Lettuce 与 Redis 之间的通信渠道，具体代码如下所示：

```
@Configuration
public class LoadLettuceResource {
    @Autowired
    private LettuceConfig lettuceConfig;
    //①初始化集群选项对象 ClusterClientOptions，自动重连，最多重定向 1 次
    @Bean
    ClusterClientOptions clusterClientOptions(){
        return ClusterClientOptions.builder().autoReconnect(true).
maxRedirects(1).build();
    }
    //②初始化集群连接的客户端对象 RedisClusterClient
    @Bean
    RedisClusterClient redisClusterClient(ClusterClientOptions options){
        List<RedisURI> arrayList=new ArrayList<>();
        List<String> hostUrlList=lettuceConfig.getHostUrlList();
        for(String s:hostUrlList){
            RedisURI redisURI=RedisURI.create(s);
            redisURI.setPassword(lettuceConfig.getPassword());
            arrayList.add(redisURI);
        }
        RedisClusterClient redisClusterClient =
RedisClusterClient.create(arrayList);
        redisClusterClient.setOptions(options);
        return redisClusterClient;
    }
    //③初始化集群连接对象 StatefulRedisClusterConnection
    @Bean(destroyMethod = "close")
    StatefulRedisClusterConnection<String,String>
statefulRedisClusterConnection(RedisClusterClient redisClusterClient){
        return redisClusterClient.connect();
    }
}
```

3. 操控缓存对象

下面用 Lettuce 的通信渠道类 StatefulRedisClusterConnection 操控缓存对象，具体代码如下所示：

```
@Service
public class LettuceService {
    //①引入 StatefulRedisClusterConnection 对象
    @Autowired
    private StatefulRedisClusterConnection statefulRedisClusterConnection ;
    @PostConstruct
public void initRedisResource(){
    //②调用 StatefulRedisClusterConnection 对象的 sync()方法,返回执行 Redis 命令
的集群命令对象 RedisAdvancedClusterCommands
        RedisAdvancedClusterCommands<String,String>
clusterCommands=statefulRedisClusterConnection.sync();
        //③执行 Redis 的 setex 命令，设置一个有效期为 100s 的缓存
        clusterCommands.setex("rediscluster",100000,"true");
        ExecutorService executorService= Executors.newFixedThreadPool(1);
        executorService.execute(new ReadRedisCluster());
}
//④在线程中，用 RedisAdvancedClusterCommands 的 get()方法读取缓存值
    class ReadRedisCluster implements Runnable{
        @Override
        public void run() {
            while (true){
                try {
                    Thread.sleep(2000);
                }catch (InterruptedException e){
                    System.out.println(e.getMessage());
                }
                RedisAdvancedClusterCommands<String,String>
clusterCommands=statefulRedisClusterConnection.sync();
                String value=clusterCommands.get("rediscluster");
                System.out.println("Redis Cluster 中的值为:"+value);
            }
        }
    }
}
```

4. 启动服务，验证 Lettuce 操控缓存的效果

在启动“use-lettuce-spring-cloud-alibaba”服务后，可以在 IDEA 控制台观察到如图 13-53 所示的日志。

```
2021-08-11 03:37:44.235  INFO 98346 --- [ing.beat.sender] com.alibaba.nacos.client
  service use-lettuce-spring-cloud-alibaba@@use-lettuce-spring-cloud-alibaba with i
  healthy=true, enabled=true, ephemeral=true, clusterName='DEFAULT', serviceName='u
  .register.source=SPRING_CLOUD}}
Redis Cluster中的值为:true
```

图 13-53

5. 连接 Redis Cluster 集群环境，查看缓存对象

使用“redis-cli”命令连接 Redis Cluster 集群环境，查看缓存对象，具体命令如下所示：

```
redis-6.2.1/src/redis-cli -h 127.0.0.1 -p 6486 -a 12345678
```

通过 Redis 的后台命令可以查看到缓存对象，如图 13-54 所示。

```
huxian@huxians-MacBook-Pro 6482-master % redis-6.2.1/src/redis-cli -h 127.0.0.1 -p 6486 -a 12345678
Warning: Using a password with '-a' or '-u' option on the command line interface may not be safe.
127.0.0.1:6486> get rediscluster
"true"
127.0.0.1:6486> 
```

图 13-54

13.4 “用分布式缓存 Redis 和 Redisson 框架实现分布式锁”的原理

Redis 不仅支持分布式缓存，还支持分布式锁。Redis 结合 Redisson 框架实现分布式锁，是 Redis 分布式锁最常见的技术方案。

13.4.1 什么是分布式锁

如图 13-55 所示，商品服务被部署在两个 JVM 进程中，当购买商品的流量请求同时到达两个商品服务后，商品服务 A 和商品服务 B 会同时处理购买商品的请求。在分布式环境中，对于这种同时处理商品购买业务流程的业务场景，如果在数据库层没有加锁（比如行锁、表锁、乐观锁等），则需要商品服务在业务代码中加锁；如果只是使用 JVM 级别的锁（比如 JDK 自带的乐观锁和悲观锁），则 JVM 进程不能感知锁的状态，这样并不能达到分布式环境下服务节点之间资源互斥的效果（还是会出现死锁的场景）。

比如商品服务处理 TPS 为 10K 的流量请求：10K 的流量请求被商品服务处理后（异步缓存+I/O 多路复用），最终到达分布式数据库的写请求可能只有 1K。流量请求到达分布式数据库后，会并发地操作分布式数据库中的一条记录。为了确保分布式数据库高并发处理数据后数据的强一致性，以及避免死锁问题，商品服务需要使用分布式锁来协调并发的写请求，将并发流量的资源竞争问题隔离在应用层。

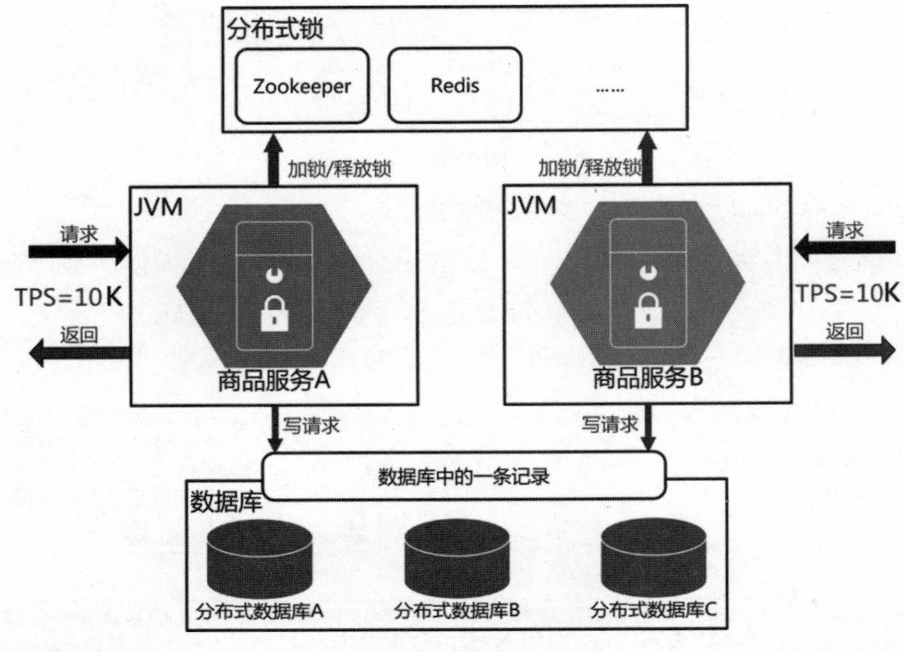

图 13-55

分布式锁通常包括如下功能特性。

（1）互斥性。

如果在服务中使用了分布式锁，则在任意时刻服务中所有运行的实例节点只能有一个能获取分布式锁（通常是指服务中的某一个接口，比如支付接口）。在分布式环境中，分布式锁的范围是整个集群中所有的服务节点实例。

（2）防止死锁。

如果持有锁的接口所在的服务崩溃了，接口所持有的分布式锁还没释放，那么其他服务的接口就不能获取锁。在多线程环境下，这样非常容易造成死锁。为了避免死锁问题，分布式锁通常都具备时效性（即存在过期时间）。

（3）由“持锁人”解锁。

为了确保分布式锁的安全性，分布式锁中的加锁和解锁必须是同一个接口。

（4）可重入。

如果一个订单支付接口加锁成功，则在锁失效之前，订单支付接口可以多次加锁，并且能够获取锁对象。

13.4.2　初始化 RedissonClient 并连接 Redis 的服务器端

如果要将“基于分布式缓存 Redis 和 Redisson 框架的分布式锁”引入应用，则需要先初始化 Redisson 的客户端 Redisson 类 Client，并连接 Redis 的服务器端。

本书以 Redisson 开箱即用的 Starter 框架“redisson-spring-boot-starter”来进行 RedissonClient 类的初始化。

1. 读取连接 Redis 的配置信息

用 Spring Boot 的自动装配 Redisson 类 AutoConfiguration 读取连接 Redis 的配置信息，具体代码如下所示：

```
    @Configuration
    @ConditionalOnClass({Redisson.class, RedisOperations.class})
    @AutoConfigureBefore(RedisAutoConfiguration.class)
    @EnableConfigurationProperties({RedissonProperties.class,
RedisProperties.class})
    public class RedissonAutoConfiguration {
        @Autowired
        private RedissonProperties redissonProperties;
        @Autowired
        private RedisProperties redisProperties;
        @Bean(destroyMethod = "shutdown")
        @ConditionalOnMissingBean(RedissonClient.class)
        public RedissonClient redisson() throws IOException {
            Config config = null;
            Method clusterMethod = ReflectionUtils.findMethod
(RedisProperties.class, "getCluster");
            Method timeoutMethod = ReflectionUtils.findMethod
(RedisProperties.class, "getTimeout");
            Object timeoutValue = ReflectionUtils.invokeMethod(timeoutMethod,
redisProperties);
            int timeout;
            //①从配置文件中读取超时时间，默认是 10s
            if(null == timeoutValue){
                timeout = 10000;
            }else if (!(timeoutValue instanceof Integer)) {
                Method millisMethod = ReflectionUtils.findMethod
(timeoutValue.getClass(), "toMillis");
                timeout = ((Long) ReflectionUtils.invokeMethod(millisMethod,
timeoutValue)).intValue();
            } else {
                timeout = (Integer)timeoutValue;
            }
```

```
        //②在 Redisson 的配置文件中，如果 config 字段不为空，则解析 config（config
字段数据支持 YAML 和 Key-Value 两种格式）
        if (redissonProperties.getConfig() != null) {
            try {
                //③读取 config 中的配置信息，并初始化配置信息对象 Config
                config = Config.fromYAML(redissonProperties.getConfig());
            } catch (IOException e) {
                try {
                    config = Config.fromJSON(redissonProperties.getConfig());
                } catch (IOException e1) {
                    throw new IllegalArgumentException("Can't parse config", e1);
                }
            }
        //④在 Redisson 的配置文件中，如果 file 字段不为空，则解析 file 的属性值（file
字段数据的格式支持 YAML 和 Key-Value）
        } else if (redissonProperties.getFile() != null) {
            try {
                InputStream is = getConfigStream();
                //⑤读取 file 字段中的配置信息，并初始化配置信息对象 Config
                config = Config.fromYAML(is);
            } catch (IOException e) {
                try {
                    InputStream is = getConfigStream();
                    config = Config.fromJSON(is);
                } catch (IOException e1) {
                    throw new IllegalArgumentException("Can't parse config", e1);
                }
            }
        //⑥在 Redis 的配置文件中，如果 sentinel 字段不为空，则解析 sentinel 的属性值
        } else if (redisProperties.getSentinel() != null) {
            Method nodesMethod = ReflectionUtils.findMethod(Sentinel.class,
"getNodes");
            Object nodesValue = ReflectionUtils.invokeMethod(nodesMethod,
redisProperties.getSentinel());
            String[] nodes;
            if (nodesValue instanceof String) {
                nodes = convert(Arrays.asList(((String)nodesValue).
split(",")));
            } else {
                nodes = convert((List<String>)nodesValue);
            }
            //⑦如果需要连接 Sentinel 环境，则初始化配置信息对象 Config
            config = new Config();
            config.useSentinelServers()
                .setMasterName(redisProperties.getSentinel().getMaster())
```

```
                .addSentinelAddress(nodes)
                .setDatabase(redisProperties.getDatabase())
                .setConnectTimeout(timeout)
                .setPassword(redisProperties.getPassword());
        //⑧在 RedisProperties 类中，如果 clusterMethod()方法的返回值不为空，则加载
Redis 配置的集群信息
        } else if (clusterMethod != null &&
ReflectionUtils.invokeMethod(clusterMethod, redisProperties) != null) {
            Object clusterObject =
ReflectionUtils.invokeMethod(clusterMethod, redisProperties);
            Method nodesMethod =
ReflectionUtils.findMethod(clusterObject.getClass(), "getNodes");
            List<String> nodesObject = (List)
ReflectionUtils.invokeMethod(nodesMethod, clusterObject);
            String[] nodes = convert(nodesObject);
            //⑨如果需要连接 Redis Cluster 环境，则初始化配置信息对象 Config
            config = new Config();
            config.useClusterServers()
                .addNodeAddress(nodes)
                .setConnectTimeout(timeout)
                .setPassword(redisProperties.getPassword());
        } else {
            //⑩如果以上条件都不满足，则加载单机版 Redis 的配置信息，并初始化配置信息对
象 Config
            config = new Config();
            String prefix = REDIS_PROTOCOL_PREFIX;
            Method method = ReflectionUtils.findMethod(RedisProperties.class,
"isSsl");
            if (method != null && (Boolean)ReflectionUtils.invokeMethod
(method, redisProperties)) {
                prefix = REDISS_PROTOCOL_PREFIX;
            }
            config.useSingleServer()
                .setAddress(prefix + redisProperties.getHost() + ":" +
redisProperties.getPort())
                .setConnectTimeout(timeout)
                .setDatabase(redisProperties.getDatabase())
                .setPassword(redisProperties.getPassword());
        }
        if (redissonAutoConfigurationCustomizers != null) {
            for (RedissonAutoConfigurationCustomizer customizer :
redissonAutoConfigurationCustomizers) {
                customizer.customize(config);
            }
        }
```

```
        ...

    }
    ...
}
```

2. 初始化 RedissonClient 类

初始化 RedissonClient 类的过程如下。

（1）在 Redisson 加载完配置信息后，初始化 RedissonClient 类，具体代码如下所示：

```
@Bean(destroyMethod = "shutdown")
@ConditionalOnMissingBean(RedissonClient.class)
public RedissonClient redisson() throws IOException {
    ...
    //①加载配置对象 Config
    Config config = null;
    //②用 Redisson 类的 create()方法创建 RedissonClient 对象
    return Redisson.create(config);
}
```

（2）加载RedissonClient类所依赖的资源，比如Redis连接管理器对象ConnectionManager。

```
public class Redisson implements RedissonClient {
    protected final ConnectionManager connectionManager;
    protected final CommandAsyncExecutor commandExecutor;
    protected final EvictionScheduler evictionScheduler;
    protected final WriteBehindService writeBehindService;
    //①用 create()方法创建一个 RedissonClient 对象
    public static RedissonClient create(Config config) {
        return new Redisson(config);
    }
    protected Redisson(Config config) {
        this.config = config;
        Config configCopy = new Config(config);
        //②创建 Reids 连接管理器对象 ConnectionManager
        connectionManager = ConfigSupport.createConnectionManager
(configCopy);
        RedissonObjectBuilder objectBuilder = null;
        if (config.isReferenceEnabled()) {
            objectBuilder = new RedissonObjectBuilder(this);
        }
        //③创建执行 Redis 命令的对象 CommandSyncService
        commandExecutor = new CommandSyncService(connectionManager,
objectBuilder);
        //④初始化定时清理任务
```

```
        evictionScheduler = new EvictionScheduler(commandExecutor);
        //⑤初始化异步写服务
        writeBehindService = new WriteBehindService(commandExecutor);
    }
    ...
}
```

3. 加载 Redis 连接管理器对象 ConnectionManager

Redisson 支持如下的 Redis 集群环境。

- 主从环境：Redisson 用 MasterSlaveConnectionManager 类来管理连接 Redis 主从环境的通信渠道。
- 单机环境：Redisson 用 SingleConnectionManager 类管理连接 Redis 单机环境的通信渠道。
- Sentinel 集群环境：Redisson 用 SentinelConnectionManager 类管理连接 Sentinel 集群环境的通信渠道。
- Redis Cluster 集群环境：Redisson 用 ClusterConnectionManager 类管理连接 Redis Cluster 集群环境的通信渠道。
- Redis 副本环境：Redisson 用 ReplicatedConnectionManager 类管理连接 Redis 副本环境的通信渠道。

在 Redisson 中，加载 Redis 连接管理器对象 ConnectionManager 的具体代码如下所示：

```
public class ConfigSupport {
    public static ConnectionManager createConnectionManager(Config
configCopy) {
        UUID id = UUID.randomUUID();
        //①如果是连接主从环境的配置信息，则初始化 MasterSlaveConnectionManager 对象
        if (configCopy.getMasterSlaveServersConfig() != null) {
            validate(configCopy.getMasterSlaveServersConfig());
            return new
MasterSlaveConnectionManager(configCopy.getMasterSlaveServersConfig(),
configCopy, id);
        //②如果是连接单机环境的配置信息，则初始化 SingleConnectionManager 对象
        } else if (configCopy.getSingleServerConfig() != null) {
            validate(configCopy.getSingleServerConfig());
            return new
SingleConnectionManager(configCopy.getSingleServerConfig(), configCopy, id);
        //③如果是连接 Sentinel 集群环境的配置信息，则初始化
SentinelConnectionManager 对象
        } else if (configCopy.getSentinelServersConfig() != null) {
            validate(configCopy.getSentinelServersConfig());
```

```
            return new SentinelConnectionManager(configCopy.
getSentinelServersConfig(), configCopy, id);
        //④如果是连接 Redis Cluster 集群环境的配置信息，则初始化
ClusterConnectionManager 对象
        } else if (configCopy.getClusterServersConfig() != null) {
            validate(configCopy.getClusterServersConfig());
            return new ClusterConnectionManager(configCopy.
getClusterServersConfig(), configCopy, id);
        //⑤如果是连接 Redis 副本环境的配置信息，则初始化
ReplicatedConnectionManager 对象
        } else if (configCopy.getReplicatedServersConfig() != null) {
            validate(configCopy.getReplicatedServersConfig());
            return new ReplicatedConnectionManager(configCopy.
getReplicatedServersConfig(), configCopy, id);
        //⑥如果以上条件否不满足，则初始化默认的连接管理器对象
        } else if (configCopy.getConnectionManager() != null) {
            return configCopy.getConnectionManager();
        }else {
            throw new IllegalArgumentException("server(s) address(es) not
defined!");
        }
    }
    ...
}
```

4. 加载执行 Redis 命令的 CommandSyncService 类

用 Redission 作为客户端调用 Redis 服务器端，除需要通信渠道外，还需要缓存数据的命令。CommandSyncService 类就是 Redis 命令的执行器。

（1）用构造函数初始化 CommandSyncService 类，具体代码如下所示：

```
public class CommandSyncService extends CommandAsyncService implements
CommandExecutor {
    final Logger log = LoggerFactory.getLogger(getClass());
    //①用构造函数加载 CommandSyncService 对象
    public CommandSyncService(ConnectionManager connectionManager,
RedissonObjectBuilder objectBuilder) {
        super(connectionManager, objectBuilder, RedissonObjectBuilder.
ReferenceType.DEFAULT);
    }
    //②异步处理读请求命令（从连接管理器中读取编码器和解码器对象 Codec）
    @Override
    public <T, R> R read(String key, RedisCommand<T> command, Object... params) {
        return read(key, connectionManager.getCodec(), command, params);
    }
```

```
    //③异步处理读请求命令（在调用的过程中定义编码器和解码器对象 Codec）
    @Override
    public <T, R> R read(String key, Codec codec, RedisCommand<T> command,
Object... params) {
        RFuture<R> res = readAsync(key, codec, command, params);
        return get(res);
    }
    //④异步处理 eval 脚本的读请求（从连接管理器中读取编码器和解码器对象 Codec）
    @Override
    public <T, R> R evalRead(String key, RedisCommand<T> evalCommandType,
String script, List<Object> keys, Object... params) {
        return evalRead(key, connectionManager.getCodec(), evalCommandType,
script, keys, params);
    }
    //⑤异步处理 eval 脚本的读请求（在调用的过程中定义编码器和解码器对象 Codec）
    @Override
    public <T, R> R evalRead(String key, Codec codec, RedisCommand<T>
evalCommandType, String script, List<Object> keys, Object... params) {
        RFuture<R> res = evalReadAsync(key, codec, evalCommandType, script,
keys, params);
        return get(res);
    }
    //⑥异步处理 eval 脚本的写请求（从连接管理器中读取编码器和解码器对象 Codec）
    @Override
    public <T, R> R evalWrite(String key, RedisCommand<T> evalCommandType,
String script, List<Object> keys, Object... params) {
        return evalWrite(key, connectionManager.getCodec(), evalCommandType,
script, keys, params);
    }
    //⑦异步处理 eval 脚本的写请求（在调用的过程中定义编码器和解码器对象 Codec）
    @Override
    public <T, R> R evalWrite(String key, Codec codec, RedisCommand<T>
evalCommandType, String script, List<Object> keys, Object... params) {
        RFuture<R> res = evalWriteAsync(key, codec, evalCommandType, script,
keys, params);
        return get(res);
    }
}
```

（2）用 RedisExecutor 类异步处理缓存命令

Redission 主要是用 RedisExecutor 类来异步处理缓存命令，具体代码如下所示：

```
public class CommandAsyncService implements CommandAsyncExecutor {
    //①异步处理缓存命令
    public <V, R> void async(boolean readOnlyMode, NodeSource source, Codec
```

```
codec,
            RedisCommand<V> command, Object[] params, RPromise<R>
mainPromise,
            boolean ignoreRedirect) {
        RedisExecutor<V, R> executor = new RedisExecutor<>(readOnlyMode,
source, codec, command, params, mainPromise,
                                                    ignoreRedirect,
connectionManager, objectBuilder, referenceType);
        //②调用 RedisExecutor 类的 execute()方法执行缓存命令
        executor.execute();
    }
    ...
}
```

13.4.3 "用 Redisson 框架的 RedissonLock 类实现分布式锁"的原理

Redisson 用 RedissonLock 类实现分布式锁，下面分析具体的原理。

1. 用 RedissonClient 接口提供获取分布式锁的 getLock()方法

在 Redisson 初始化完成 RedissonClient 类后，开发人员就可以利用它提供的 getLock()方法来获取分布式锁对象 RedissonLock，具体代码如下所示：

```
//Redisson 类是 RedissonClient 接口的实现类
public class Redisson implements RedissonClient {
    ...
    @Override
    public RLock getLock(String name) {
        //返回一个分布式锁对象
        return new RedissonLock(commandExecutor, name);
    }
}
```

2. 初始化分布式锁 RedissonLock

Redisson 用构造函数来初始化分布式锁 RedissonLock，具体代码如下所示：

```
public class RedissonLock extends RedissonBaseLock {
    protected long internalLockLeaseTime;
    protected final LockPubSub pubSub;
    final CommandAsyncExecutor commandExecutor;
    //用构造函数来加载分布式锁依赖的资源
    public RedissonLock(CommandAsyncExecutor commandExecutor, String name) {
        super(commandExecutor, name);
        this.commandExecutor = commandExecutor;
        this.internalLockLeaseTime = commandExecutor.getConnectionManager().
getCfg().getLockWatchdogTimeout();
```

```
        this.pubSub = commandExecutor.getConnectionManager().
getSubscribeService().getLockPubSub();
    }
    ...
}
```

3. 执行同步加锁

为了提高加锁的成功率，Redisson 支持同步加锁，并提供了同步加锁中的预加锁 tryLock()方法和加锁 lock()方法，下面分析具体的加锁过程。

（1）RedissonLock 用 tryLock()方法提供预加锁的 API。

在 RedissonLock 中封装了各种不同入参的 tryLock()方法来实现预加锁功能，具体代码如下所示：

```
//①用默认参数来预加锁
@Override
public boolean tryLock() {
    return get(tryLockAsync());
}
//②指定加锁周期来预加锁
@Override
public boolean tryLock(long waitTime, TimeUnit unit) throws
InterruptedException {
    return tryLock(waitTime, -1, unit);
}
//③执行预加锁，如果预加锁失败，则返回 false
@Override
public boolean tryLock(long waitTime, long leaseTime, TimeUnit unit) throws
InterruptedException {
    long time = unit.toMillis(waitTime);
    long current = System.currentTimeMillis();
    long threadId = Thread.currentThread().getId();
    Long ttl = tryAcquire(waitTime, leaseTime, unit, threadId);
    if (ttl == null) {
        return true;
    }
    time -= System.currentTimeMillis() - current;
    if (time <= 0) {
        acquireFailed(waitTime, unit, threadId);
        return false;
    }
    current = System.currentTimeMillis();
    RFuture<RedissonLockEntry> subscribeFuture = subscribe(threadId);
    if (!subscribeFuture.await(time, TimeUnit.MILLISECONDS)) {
```

```
            if (!subscribeFuture.cancel(false)) {
                subscribeFuture.onComplete((res, e) -> {
                    if (e == null) {
                        unsubscribe(subscribeFuture, threadId);
                    }
                });
            }
            acquireFailed(waitTime, unit, threadId);
            return false;
        }
        try {
            time -= System.currentTimeMillis() - current;
            if (time <= 0) {
                acquireFailed(waitTime, unit, threadId);
                return false;
            }
            while (true) {
                long currentTime = System.currentTimeMillis();
                ttl = tryAcquire(waitTime, leaseTime, unit, threadId);
                if (ttl == null) {
                    return true;
                }
                time -= System.currentTimeMillis() - currentTime;
                if (time <= 0) {
                    acquireFailed(waitTime, unit, threadId);
                    return false;
                }
                currentTime = System.currentTimeMillis();
                if (ttl >= 0 && ttl < time) {
                    subscribeFuture.getNow().getLatch().tryAcquire(ttl,
TimeUnit.MILLISECONDS);
                } else {
                    subscribeFuture.getNow().getLatch().tryAcquire(time,
TimeUnit.MILLISECONDS);
                }
                time -= System.currentTimeMillis() - currentTime;
                if (time <= 0) {
                    acquireFailed(waitTime, unit, threadId);
                    return false;
                }
            }
        } finally {
            unsubscribe(subscribeFuture, threadId);
        }
    }
```

在 RedissonLock 中，主要是利用预加锁来解决分布式锁的资源竞争的问题，具体代码如下所示：

```
//①用 tryLockInnerAsync()方法异步地执行 eval 脚本，并批量执行 Redis 的命令
<T> RFuture<T> tryLockInnerAsync(long waitTime, long leaseTime, TimeUnit
unit, long threadId, RedisStrictCommand<T> command) {
    //②异步执行脚本文件中的 Redis 命令
    return evalWriteAsync(getRawName(), LongCodec.INSTANCE, command,
            "if (redis.call('exists', KEYS[1]) == 0) then " +
                    "redis.call('hincrby', KEYS[1], ARGV[2], 1); " +
                    "redis.call('pexpire', KEYS[1], ARGV[1]); " +
                    "return nil; " +
                    "end; " +
                    "if (redis.call('hexists', KEYS[1], ARGV[2]) == 1) then " +
                    "redis.call('hincrby', KEYS[1], ARGV[2], 1); " +
                    "redis.call('pexpire', KEYS[1], ARGV[1]); " +
                    "return nil; " +
                    "end; " +
                    "return redis.call('pttl', KEYS[1]);",
    Collections.singletonList(getRawName()), unit.toMillis(leaseTime),
getLockName(threadId));
}
```

（2）RedissonLock 用 lock()方法提供同步加锁的 API。

在 RedissonLock 中封装了各种不同入参的 lock()方法来实现同步加锁，具体代码如下所示：

```
//①默认的分布式锁的构造函数，其中锁的周期为“-1”。在线程持有锁时其他线程不能中断
@Override
public void lock() {
    try {
        lock(-1, null, false);
    } catch (InterruptedException e) {
        throw new IllegalStateException();
    }
}
//②在调用过程中可以设置加锁周期
@Override
public void lock(long leaseTime, TimeUnit unit) {
    try {
        //③在线程持有锁的周期内，其他线程不能中断锁
        lock(leaseTime, unit, false);
    } catch (InterruptedException e) {
        throw new IllegalStateException();
    }
}
//④执行加锁
```

```
    private void lock(long leaseTime, TimeUnit unit, boolean interruptibly)
throws InterruptedException {
        long threadId = Thread.currentThread().getId();
        Long ttl = tryAcquire(-1, leaseTime, unit, threadId);
        if (ttl == null) {
            return;
        }
        //⑤订阅异步通信渠道（底层基于Netty）
        RFuture<RedissonLockEntry> future = subscribe(threadId);
        //⑥如果锁是可以中断的，则执行syncSubscriptionInterrupted()方法，并同步执行
RFuture类的sync()方法
        if (interruptibly) {
            commandExecutor.syncSubscriptionInterrupted(future);
        } else {
            //⑦否则执行CommandAsyncExecutor类的syncSubscription()方法
            commandExecutor.syncSubscription(future);
        }
        try {
            //⑧用无限循环获取分布式锁的TTL时间，如果TTL为空，则直接返回
            while (true) {
                ttl = tryAcquire(-1, leaseTime, unit, threadId);
                if (ttl == null) {
                    break;
                }
                //⑨如果TTL大于0，则尝试获取信号量。如果获取失败，则抛出异常，中断加锁过程
                if (ttl >= 0) {
                    try {
                        future.getNow().getLatch().tryAcquire(ttl,
TimeUnit.MILLISECONDS);
                    } catch (InterruptedException e) {
                        if (interruptibly) {
                            throw e;
                        }
                        future.getNow().getLatch().tryAcquire(ttl,
TimeUnit.MILLISECONDS);
                    }
                } else {
                    if (interruptibly) {
                        future.getNow().getLatch().acquire();
                    } else {
                        future.getNow().getLatch().acquireUninterruptibly();
                    }
                }
            }
        } finally {
```

```
        //⑩如果加锁异常，则取消异步通信渠道的订阅
            unsubscribe(future, threadId);
        }
    }
```

总的来说，Redisson 将实现分布式锁拆分为两个阶段：“校验是否可以加锁”和“执行加锁”。

在“校验是否可以加锁”的阶段，Redisson 执行 Redis 的相关脚本（与 Redis 进行 RPC 通信）；在“执行加锁”的阶段，利用 Java 本地的信号量来完成本地加锁，这样能有效降低分布式锁的性能损耗。

4. 执行异步加锁

为了提高加锁的效率，Redisson 支持异步加锁。下面分析具体的加锁过程。

（1）RedissonLock 用 tryLockAsync ()方法提供预加锁的 API。

在 RedissonLock 中封装了各种不同入参的 tryLockAsync ()方法来实现预加锁，具体代码如下所示：

```
    //①用默认参数异步地预加锁
    @Override
    public RFuture<Boolean> tryLockAsync() {
        return tryLockAsync(Thread.currentThread().getId());
    }
    //②用指定的线程 ID 异步地预加锁
    @Override
    public RFuture<Boolean> tryLockAsync(long threadId) {
        return tryAcquireOnceAsync(-1, -1, null, threadId);
    }
    //③用指定的锁超时等待时间异步地预加锁
    @Override
    public RFuture<Boolean> tryLockAsync(long waitTime, TimeUnit unit) {
        return tryLockAsync(waitTime, -1, unit);
    }
    @Override
    //④用指定的锁超时等待时间和锁的生命周期来异步地预加锁
    public RFuture<Boolean> tryLockAsync(long waitTime, long leaseTime, TimeUnit
unit) {
        long currentThreadId = Thread.currentThread().getId();
        return tryLockAsync(waitTime, leaseTime, unit, currentThreadId);
    }
    //⑤入参锁超时等待时间、锁的生命周期和线程 ID 异步地预加锁
    @Override
    public RFuture<Boolean> tryLockAsync(long waitTime, long leaseTime, TimeUnit
unit,
```

```
            long currentThreadId) {
        RPromise<Boolean> result = new RedissonPromise<Boolean>();
        AtomicLong time = new AtomicLong(unit.toMillis(waitTime));
        long currentTime = System.currentTimeMillis();
        RFuture<Long> ttlFuture = tryAcquireAsync(waitTime, leaseTime, unit,
currentThreadId);
        ttlFuture.onComplete((ttl, e) -> {
            if (e != null) {
                result.tryFailure(e);
                return;
            }
            if (ttl == null) {
                if (!result.trySuccess(true)) {
                    unlockAsync(currentThreadId);
                }
                return;
            }
            long el = System.currentTimeMillis() - currentTime;
            time.addAndGet(-el);
            if (time.get() <= 0) {
                trySuccessFalse(currentThreadId, result);
                return;
            }
            long current = System.currentTimeMillis();
            AtomicReference<Timeout> futureRef = new AtomicReference<Timeout>();
            RFuture<RedissonLockEntry> subscribeFuture =
subscribe(currentThreadId);
            subscribeFuture.onComplete((r, ex) -> {
                if (ex != null) {
                    result.tryFailure(ex);
                    return;
                }
                if (futureRef.get() != null) {
                    futureRef.get().cancel();
                }
                long elapsed = System.currentTimeMillis() - current;
                time.addAndGet(-elapsed);
                //异步地预加锁
                tryLockAsync(time, waitTime, leaseTime, unit, subscribeFuture,
result, currentThreadId);
                });
                //等待异步地预加锁的结果，如果超时，则取消等待，并返回加锁失败的结果
                if (!subscribeFuture.isDone()) {
                    Timeout scheduledFuture =
commandExecutor.getConnectionManager().newTimeout(new TimerTask() {
```

```
                @Override
                public void run(Timeout timeout) throws Exception {
                    if (!subscribeFuture.isDone()) {
                        subscribeFuture.cancel(false);
                        trySuccessFalse(currentThreadId, result);
                    }
                }
            }, time.get(), TimeUnit.MILLISECONDS);
            futureRef.set(scheduledFuture);
        }
    });

    return result;
}
//⑥执行预加锁，但不返回预加锁的结果
private void tryLockAsync(AtomicLong time, long waitTime, long leaseTime,
TimeUnit unit,
        RFuture<RedissonLockEntry> subscribeFuture, RPromise<Boolean>
result, long currentThreadId) {
    ...
}
```

（2）RedissonLock 用 lockAsync()方法提供同步加锁的 API。

在 RedissonLock 中封装了各种不同入参的 lockAsync()方法来实现同步加锁，具体代码如下所示：

```
//①用默认参数来异步地加锁
@Override
public RFuture<Void> lockAsync() {
    return lockAsync(-1, null);
}
//②用指定的锁的生命周期来异步地加锁
@Override
public RFuture<Void> lockAsync(long leaseTime, TimeUnit unit) {
    long currentThreadId = Thread.currentThread().getId();
    return lockAsync(leaseTime, unit, currentThreadId);
}
//③用指定的线程 ID 来异步地加锁
@Override
public RFuture<Void> lockAsync(long currentThreadId) {
    return lockAsync(-1, null, currentThreadId);
}
//④用指定的锁的生命周期和线程 ID 来异步地加锁，返回异步结果对象 RFuture
@Override
```

```
    public RFuture<Void> lockAsync(long leaseTime, TimeUnit unit, long
currentThreadId) {
        RPromise<Void> result = new RedissonPromise<Void>();
        RFuture<Long> ttlFuture = tryAcquireAsync(-1, leaseTime, unit,
currentThreadId);
        ttlFuture.onComplete((ttl, e) -> {
            if (e != null) {
                result.tryFailure(e);
                return;
            }
            if (ttl == null) {
                if (!result.trySuccess(null)) {
                    unlockAsync(currentThreadId);
                }
                return;
            }
            RFuture<RedissonLockEntry> subscribeFuture =
subscribe(currentThreadId);
            subscribeFuture.onComplete((res, ex) -> {
                if (ex != null) {
                    result.tryFailure(ex);
                    return;
                }
                lockAsync(leaseTime, unit, subscribeFuture, result,
currentThreadId);
            });
        });
        return result;
    }
    //⑤执行异步加锁，但不返回结果
    private void lockAsync(long leaseTime, TimeUnit unit,
            RFuture<RedissonLockEntry> subscribeFuture, RPromise<Void>
result, long currentThreadId) {
        ...
    }
```

13.4.4 【实例】在 Spring Cloud Alibaba 应用中验证分布式锁的功能

本实例主要是用 Redisson 提供的 redisson-spring-boot-starter，将 Spring Cloud Alibaba 应用接入 Redis 的 Sentinel 环境，并验证“用分布式缓存 Redis 和 Redisson 框架实现的分布式锁”功能。

1. 初始化一个 Spring Cloud Alibaba 服务

用 Spring Cloud Alibaba 快速初始化两个服务“use-redisson-distribute-lock-spring-cloud-alibaba" 和 " call-distribute-server-cloud-alibaba"，具体的 pom.xml 和 application.yaml 中的配置信息可以参考本书配套资源中的源码。

2. 在"use-redisson-distribute-lock-spring-cloud-alibaba"服务中添加分布式锁

（1）在"use-redisson-distribute-lock-spring-cloud-alibaba"服务中，初始化分布式锁 RedissonLock，具体代码如下所示：

```
@Configuration
public class RedissonLockConfig {
    //①加载 RedissonClient 对象
    @Autowired
    private RedissonClient redissonClient;
    @Bean(name ="goodRLock")
    public RLock initGoodRLock(){
        //②初始化一个全局分布式锁对象 RLock，分布式锁的名称为 goodRLock
        RLock rLock=redissonClient.getLock("goodRLock");
        return rLock;
    }
}
```

（2）在"use-redisson-distribute-lock-spring-cloud-alibaba"服务中定义一个 Dubbo 接口 GoodServiceImpl，具体代码如下所示：

```
@DubboService(version = "1.0.0",group =
"use-redisson-distribute-lock-spring-cloud-alibaba",timeout =
3000,executes=20,actives = 20,connections = 20)
public class GoodServiceImpl implements GoodService {
    @Autowired
    private Example2ProductMapper example2ProductMapper;
    //①从 Spring IOC 容器中加载分布式锁对象 RLock
    @Resource
    @Qualifier(value = "goodRLock")
    private RLock rlock;
    @Resource
    private SyncOrAsyncConfig syncOrAsyncConfig;

    @Override
    public DefaultResult<GoodDTO> updateGoodNum(GoodServiceRequest
goodServiceRequest) {
        DefaultResult<GoodDTO> result = new DefaultResult<>();
        final GoodDTO returnItem = new GoodDTO();
        final Example2ProductBo example2ProductBo = (Example2ProductBo)
goodServiceRequest.getRequestData();
        //②如果将 SyncOrAsyncConfig 中的 use 字段设置 1，则分布式锁生效
        if (syncOrAsyncConfig.getUse()==1) {
```

```
            Thread thread = Thread.currentThread();
            final long threadId = thread.getId();
            System.out.println("加同步分布式锁，开始扣减库存：" +
example2ProductBo.toString() + "扣减库处理请求的线程 ID 为：" + threadId + " 当前请
求的" + "uuid 为：" + goodServiceRequest.getUuid() + " 扣减库存请求的线程 ID 为：" +
goodServiceRequest.getThreadId() + " 执行扣减库存的时间：" + new
Date().toString());
            //③如果将 type 设置为 sync，则使用同步模式的分布式锁
            if (syncOrAsyncConfig.getType().equals("sync")) {
               try {

                     //④在同步模式下预加锁
                     boolean tryLockResult = rlock.tryLock(1L, 5L,
TimeUnit.SECONDS);
                    if (tryLockResult) {
                        //⑤同步加锁，锁的生命周期为 5s
                        rlock.lock(5L, TimeUnit.SECONDS);
                        execute(result, returnItem, example2ProductBo);
                    }              } catch (InterruptedException exception) {
                    result.setData(null);
                    result.setCode("500");
                    result.setMessage("加同步分布式锁，库存扣减失败！");
                    System.out.println(exception.getMessage());
               } catch (Exception e) {
                    result.setData(null);
                    result.setCode("500");
                    result.setMessage("加同步分布式锁，库存扣减失败！");
                    System.out.println(e.getMessage());
               } finally {
                    if (rlock.isLocked()) {
                       rlock.unlock();
                    }
               }
            //⑥如果将 type 设置为 async，则使用异步模式的分布式锁
            } else if (syncOrAsyncConfig.getType().equals("async")) {
                try {
                    System.out.println("加异步分布式锁，开始扣减库存：" +
example2ProductBo.toString() + "扣减库存处理请求的线程 ID 为：" + threadId + " 当前
请求的" + "uuid 为：" + goodServiceRequest.getUuid() + " 扣减库存请求的线程 ID 为："
+ goodServiceRequest.getThreadId() + " 执行扣减库存的时间：" + new
Date().toString());
                    //⑦在异步模式下预加锁
                    RFuture<Boolean> rFuture = rlock.tryLockAsync(1, 5,
TimeUnit.SECONDS, threadId);
                    if (rFuture.isSuccess()) {
```

```
                    //⑧异步加锁，锁的生命周期为 5s
                    rlock.lockAsync(5, TimeUnit.SECONDS, threadId);
                    execute(result, returnItem, example2ProductBo);
                }
            } catch (Exception e) {
                result.setData(null);
                result.setCode("500");
                result.setMessage("加异步分布式锁，库存扣减失败！");
                System.out.println(e.getMessage());
            } finally {
                if (rlock.isLocked()) {
                    rlock.unlockAsync();
                }
            }
        }
    } else {
        //⑨不加分布式锁，扣减库存
        try {
            Thread thread = Thread.currentThread();
            final long threadId = thread.getId();
            System.out.println("不加分布式锁，开始扣减库存：" +
example2ProductBo.toString() + "扣减库存处理请求的线程 ID 为：" + threadId + " 当前
请求的" + "uuid 为：" + goodServiceRequest.getUuid() + " 扣减库存请求的线程 ID 为："
+ goodServiceRequest.getThreadId() + " 执行扣减库存的时间：" + new
Date().toString());
            execute(result, returnItem, example2ProductBo);
        } catch (Exception e) {
            result.setData(null);
            result.setCode("500");
            result.setMessage("不加分布式锁库存扣减失败！");
            System.out.println(e.getMessage());
        }
    }
    return result;
}
//⑩扣减库存
private void execute(DefaultResult<GoodDTO> result, GoodDTO returnItem,
Example2ProductBo example2ProductBo) {
    List<Example2ProductEntity> querySyncResult1 =
example2ProductMapper.queryGoodInfoByGoodId(example2ProductBo);
    if (!CollectionUtils.isEmpty(querySyncResult1)) {
        Example2ProductEntity item = querySyncResult1.get(0);
        System.out.println("开始扣减库存，扣除之前的商品库存为：" +
item.getNum() + " 商品 ID 为：" + item.getGoodId());
    }
```

```
        example2ProductMapper.updateGoodNum(example2ProductBo);
        List<Example2ProductEntity> queryResult2 =
example2ProductMapper.queryGoodInfoByGoodId(example2ProductBo);
        if (!CollectionUtils.isEmpty(queryResult2)) {
            Example2ProductEntity item = queryResult2.get(0);
            System.out.println("开始扣减库存，扣除之后的商品库存为：" +
item.getNum() + " 商品 ID 为：" + item.getGoodId());
        }
        returnItem.setGoodId(queryResult2.get(0).getGoodId());
        returnItem.setNum(queryResult2.get(0).getNum());
        result.setData(returnItem);
        result.setCode("200");
        result.setMessage("库存扣减成功！");
    }
}
```

（3）在“call-distribute-server-cloud-alibaba”服务中定义一个 CallService 类去消费 Dubbo 接口 GoodServiceImpl，具体代码如下所示：

```
@Service
public class CallService {
    @Resource
    private CallConfig callConfig;
    //①设置 Dubbo 消费者属性
    @DubboReference(version = "1.0.0",group =
"use-redisson-distribute-lock-spring-cloud-alibaba",timeout = 3000,
            actives = 20,connections = 20)
    private GoodService goodService;
    @PostConstruct
public void initCall(){
    //②启动 50 个线程，模拟 50 个用户购买商品并扣减库存
        ExecutorService executorService=Executors.newFixedThreadPool(50);
        executorService.execute(new CallThread());
    }
    class CallThread implements Runnable{
        @Override
        public void run() {
            while (true){
                if(callConfig.isRemote()){
                    final Thread thread=Thread.currentThread();
                    GoodServiceRequest goodServiceRequest = new
GoodServiceRequest();
                    goodServiceRequest.setUuid(RandomUtils.nextLong() +"");
                    goodServiceRequest.setThreadId(thread.getId()+"");
                    Example2ProductBo example2ProductBo = new
Example2ProductBo();
```

```
                        //③扣减同一个商品的库存
                        example2ProductBo.setGoodId(7878L);
                        example2ProductBo.setId(3467374334L);
                        goodServiceRequest.setRequestData(example2ProductBo);
                        System.out.println("开始扣减库存: " +
example2ProductBo.toString()+" 扣减库存的源线程 ID 为:
"+goodServiceRequest.getThreadId()
                        +" 扣减库存的分布式发号器 ID 为:
"+goodServiceRequest.getUuid());
                        goodService.updateGoodNum(goodServiceRequest);
                    }
                }
            }
        }
    }
```

3. 搭建分布式环境

（1）使用 Maven 命令“mvn clean install -Dmaven.test.skip=true”打包服务，并生成部署服务的 Jar 包，如图 13-56 和 13-57 所示。

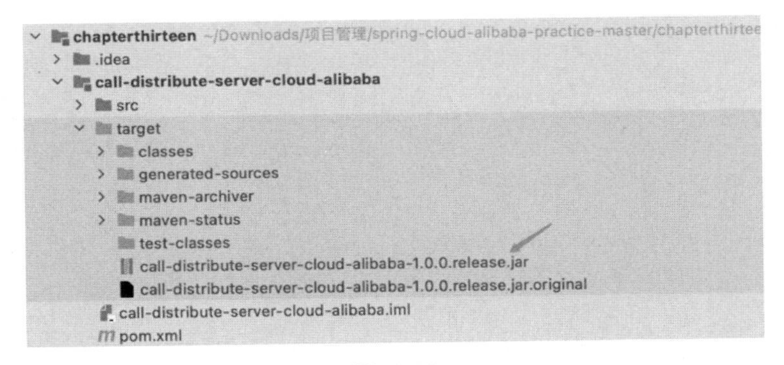

图 13-56

图 13-57

（2）将打包的 Jar 包复制到文件夹“chapterthirteen-env”中，如图 13-58 所示。

Name	Size	Kind	Date
chapterthirteen-env	--	Folder	Yesterday at 03:34
use-redisson-distribute-lock-spring-cloud-alibaba	--	Folder	Yesterday at 03:35
127.0.0.1/8032/26790	--	Folder	Yesterday at 03:41
nohup.out	Zero bytes	Document	Today at 03:10
use-redisson-distribute-lock-spring-cloud-alibaba-1.0.0.release.jar	74.1 MB	Java JAR file	Today at 03:10
127.0.0.1/8031/26789	--	Folder	Yesterday at 03:40
nohup.out	Zero bytes	Document	Today at 03:10
use-redisson-distribute-lock-spring-cloud-alibaba-1.0.0.release.jar	74.1 MB	Java JAR file	Today at 03:10
127.0.0.1/8030/26788	--	Folder	Yesterday at 03:40
nohup.out	Zero bytes	Document	Today at 03:10
use-redisson-distribute-lock-spring-cloud-alibaba-1.0.0.release.jar	74.1 MB	Java JAR file	Today at 03:10
call-distribute-server-cloud-alibaba	--	Folder	Yesterday at 03:35
127.0.0.1/8022/26414	--	Folder	Yesterday at 03:39
nohup.out	11.3 MB	Document	Today at 03:05
call-distribute-server-cloud-alibaba-1.0.0.release.jar	47.3 MB	Java JAR file	Today at 03:04
127.0.0.1/8021/26413	--	Folder	Yesterday at 03:39
nohup.out	11.3 MB	Document	Today at 03:05
call-distribute-server-cloud-alibaba-1.0.0.release.jar	47.3 MB	Java JAR file	Today at 03:04
127.0.0.1/8023/26412	--	Folder	Yesterday at 03:39
nohup.out	11.9 MB	Document	Today at 03:04
call-distribute-server-cloud-alibaba-1.0.0.release.jar	47.3 MB	Java JAR file	Today at 03:04

图 13-58

（3）执行命令后启动 6 个服务进程，“use-redisson-distribute-lock-spring-cloud-alibaba”服务和“call-distribute-server-cloud-alibaba”服务各 3 个进程，具体命令如下所示：

```
nohup java -jar use-redisson-distribute-lock-spring-cloud-alibaba-1.0.0.release.jar --server.port=8030 --dubbo.protocol.port=26788 &
nohup java -jar use-redisson-distribute-lock-spring-cloud-alibaba-1.0.0.release.jar --server.port=8031 --dubbo.protocol.port=26789 &
nohup java -jar use-redisson-distribute-lock-spring-cloud-alibaba-1.0.0.release.jar --server.port=8032 --dubbo.protocol.port=26790 &
nohup java -jar call-distribute-server-cloud-alibaba-1.0.0.release.jar --server.port=8021 --dubbo.protocol.port=26413 &
nohup java -jar call-distribute-server-cloud-alibaba-1.0.0.release.jar --server.port=8022 --dubbo.protocol.port=26414 &
nohup java -jar call-distribute-server-cloud-alibaba-1.0.0.release.jar --server.port=8023 --dubbo.protocol.port=26412 &
```

（4）在 Nacos 控制台可以看到 6 个进程已经启动成功，如图 13-59 所示。

public | test | spring cloud alibaba practice | dev

Service List | spring cloud alibaba practice c7ba173f-29e5-4c58-ae78-b102be11c4f9

Service Name: Enter Service Name　Group Name: Enter Group Name　Hidden Empty Service:　Search　Create Service

Service Name	Group Name	Cluster Count	Instance Count	Healthy Instance Count	Trigger Protection Threshold	Operation
call-distribute-server-cloud-alibaba	DEFAULT_GROUP	1	3	3	false	Details | Code Example | Delete
use-redisson-distribute-lock-spring-cloud-alibaba	DEFAULT_GROUP	1	3	3	false	Details | Code Example | Delete

图 13-59

4. 不开启分布式锁，观察“use-redisson-distribute-lock-spring-cloud-alibaba”服务的运行日志

（1）服务进程“127.0.0.1:8030”的运行日志如图 13-60 所示。在多线程环境下扣减库存的顺序是不一致的，会出现线程不安全和重复扣减的问题。

```
开始扣减库存，扣除之前的商品库存为：983848040427632 商品ID为：7878
开始扣减库存，扣除之前的商品库存为：983848040427632 商品ID为：7878
开始扣减库存，扣除之后的商品库存为：983848040427629 商品ID为：7878
开始扣减库存，扣除之后的商品库存为：983848040427629 商品ID为：7878
不加分布式锁，开始扣减库存：Example2ProductBo(id=3467374334, goodId=7878, goodName=null, num=null)扣减库处理请求的线程ID为：348 当前请求的uuid为：436138905553865341 扣减库存源请求的线程ID为：95 执行扣减
库存的时间：Sat Aug 14 04:36:18 CST 2021
开始扣减库存，扣除之前的商品库存为：983848040427626 商品ID为：7878
开始扣减库存，扣除之后的商品库存为：983848040427623 商品ID为：7878
不加分布式锁，开始扣减库存：Example2ProductBo(id=3467374334, goodId=7878, goodName=null, num=null)扣减库处理请求的线程ID为：348 当前请求的uuid为：7380011159366208909 扣减库存源请求的线程ID为：95 执行扣
减库存的时间：Sat Aug 14 04:36:18 CST 2021
开始扣减库存，扣除之前的商品库存为：983848040427623 商品ID为：7878
开始扣减库存，扣除之后的商品库存为：983848040427620 商品ID为：7878
不加分布式锁，开始扣减库存：Example2ProductBo(id=3467374334, goodId=7878, goodName=null, num=null)扣减库处理请求的线程ID为：348 当前请求的uuid为：7832488971878090881 扣减库存源请求的线程ID为：91 执行扣
减库存的时间：Sat Aug 14 04:36:18 CST 2021
开始扣减库存，扣除之前的商品库存为：983848040427620 商品ID为：7878
开始扣减库存，扣除之后的商品库存为：983848040427618 商品ID为：7878
不加分布式锁，开始扣减库存：Example2ProductBo(id=3467374334, goodId=7878, goodName=null, num=null)扣减库处理请求的线程ID为：348 当前请求的uuid为：8068890255268984342 扣减库存源请求的线程ID为：91 执行扣
减库存的时间：Sat Aug 14 04:36:18 CST 2021
开始扣减库存，扣除之前的商品库存为：983848040427615 商品ID为：7878
开始扣减库存，扣除之后的商品库存为：983848040427612 商品ID为：7878
不加分布式锁，开始扣减库存：Example2ProductBo(id=3467374334, goodId=7878, goodName=null, num=null)扣减库处理请求的线程ID为：348 当前请求的uuid为：6108496233517034321 扣减库存源请求的线程ID为：91 执行扣
减库存的时间：Sat Aug 14 04:36:18 CST 2021
开始扣减库存，扣除之前的商品库存为：983848040427608 商品ID为：7878
不加分布式锁，开始扣减库存：Example2ProductBo(id=3467374334, goodId=7878, goodName=null, num=null)扣减库处理请求的线程ID为：347 当前请求的uuid为：2713653309196438006 扣减库存源请求的线程ID为：92 执行扣
减库存的时间：Sat Aug 14 04:36:18 CST 2021
开始扣减库存，扣除之前的商品库存为：983848040427606 商品ID为：7878
开始扣减库存，扣除之后的商品库存为：983848040427606 商品ID为：7878
开始扣减库存，扣除之后的商品库存为：983848040427605 商品ID为：7878
huxian@huxians-MacBook-Pro 127.0.0.1:8030:26788 %
```

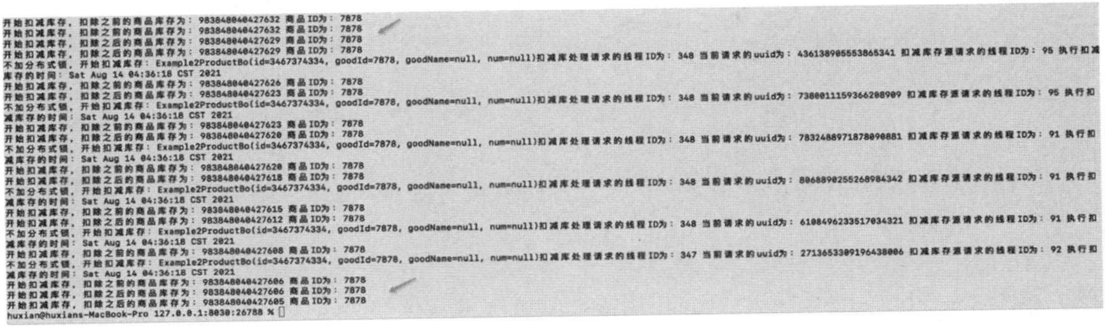

图 13-60

（2）服务进程“127.0.0.1:8031”的运行日志如图 13-61 所示。在多线程环境下扣减库存的顺序是不一致的，会出现线程不安全和重复扣减的问题。

```
开始扣减库存，扣除之前的商品库存为：983848040265478 商品ID为：7878
开始扣减库存，扣除之前的商品库存为：983848040265478 商品ID为：7878
开始扣减库存，扣除之前的商品库存为：983848040265478 商品ID为：7878
开始扣减库存，扣除之后的商品库存为：983848040265476 商品ID为：7878
开始扣减库存，扣除之后的商品库存为：983848040265475 商品ID为：7878
开始扣减库存，扣除之后的商品库存为：983848040265475 商品ID为：7878
不加分布式锁，开始扣减库存：Example2ProductBo(id=3467374334, goodId=7878, goodName=null, num=null)扣减库处理请求的线程ID为：347 当前请求的uuid为：8351459171903140232 扣减库存源请求的线程ID为：95 执行扣
减库存的时间：Sat Aug 14 04:38:57 CST 2021
不加分布式锁，开始扣减库存：Example2ProductBo(id=3467374334, goodId=7878, goodName=null, num=null)扣减库处理请求的线程ID为：348 当前请求的uuid为：8012063019390943173 扣减库存源请求的线程ID为：92 执行扣
减库存的时间：Sat Aug 14 04:38:57 CST 2021
开始扣减库存，扣除之前的商品库存为：983848040265472 商品ID为：7878
开始扣减库存，扣除之前的商品库存为：983848040265472 商品ID为：7878
开始扣减库存，扣除之后的商品库存为：983848040265470 商品ID为：7878
开始扣减库存，扣除之后的商品库存为：983848040265469 商品ID为：7878
不加分布式锁，开始扣减库存：Example2ProductBo(id=3467374334, goodId=7878, goodName=null, num=null)扣减库处理请求的线程ID为：347 当前请求的uuid为：7846728256464155064 扣减库存源请求的线程ID为：92 执行扣
减库存的时间：Sat Aug 14 04:38:57 CST 2021
开始扣减库存，扣除之前的商品库存为：983848040265466 商品ID为：7878
开始扣减库存，扣除之后的商品库存为：983848040265463 商品ID为：7878
不加分布式锁，开始扣减库存：Example2ProductBo(id=3467374334, goodId=7878, goodName=null, num=null)扣减库处理请求的线程ID为：347 当前请求的uuid为：1149199408672620221 扣减库存源请求的线程ID为：95 执行扣减
库存的时间：Sat Aug 14 04:38:57 CST 2021
不加分布式锁，开始扣减库存：Example2ProductBo(id=3467374334, goodId=7878, goodName=null, num=null)扣减库处理请求的线程ID为：348 当前请求的uuid为：6898073324609300615 扣减库存源请求的线程ID为：91 执行扣
减库存的时间：Sat Aug 14 04:38:57 CST 2021
开始扣减库存，扣除之前的商品库存为：983848040265457 商品ID为：7878
开始扣减库存，扣除之前的商品库存为：983848040265457 商品ID为：7878
开始扣减库存，扣除之后的商品库存为：983848040265455 商品ID为：7878
开始扣减库存，扣除之后的商品库存为：983848040265454 商品ID为：7878
不加分布式锁，开始扣减库存：Example2ProductBo(id=3467374334, goodId=7878, goodName=null, num=null)扣减库处理请求的线程ID为：348 当前请求的uuid为：2753734214768989012 扣减库存源请求的线程ID为：92 执行扣
减库存的时间：Sat Aug 14 04:38:57 CST 2021
开始扣减库存，扣除之前的商品库存为：983848040265454 商品ID为：7878
huxian@huxians-MacBook-Pro 127.0.0.1:8031:26789 %
```

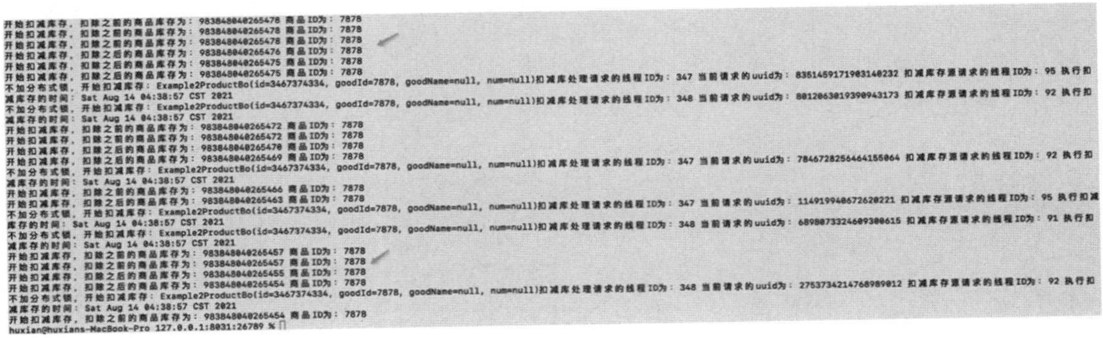

图 13-61

（3）服务进程“127.0.0.1:8032”的运行日志如图 13-62 所示。在多线程环境下扣减库存的顺序是不一致的，会出现线程不安全和重复扣减的问题。

```
开始扣减库存，扣除之前的商品库存为：983848040190882 商品ID为：7878
开始扣减库存，扣除之前的商品库存为：983848040190882 商品ID为：7878
开始扣减库存，扣除之后的商品库存为：983848040190880 商品ID为：7878
开始扣减库存，扣除之后的商品库存为：983848040190880 商品ID为：7878
不加分布式锁，开始扣减库存：Example2ProductBo(id=3467374334, goodId=7878, goodName=null, num=null)扣减库处理请求的线程ID为：343 当前请求的uuid为：4597817760860991241 扣减库存源请求的线程ID为：91 执行扣减库存的时间：Sat Aug 14 04:40:12 CST 2021
开始扣减库存，扣除之前的商品库存为：983848040190879 商品ID为：7878
开始扣减库存，扣除之后的商品库存为：983848040190877 商品ID为：7878
不加分布式锁，开始扣减库存：Example2ProductBo(id=3467374334, goodId=7878, goodName=null, num=null)扣减库处理请求的线程ID为：343 当前请求的uuid为：7886613660542376101 扣减库存源请求的线程ID为：92 执行扣减库存的时间：Sat Aug 14 04:40:12 CST 2021
开始扣减库存，扣除之前的商品库存为：983848040190870 商品ID为：7878
开始扣减库存，扣除之后的商品库存为：983848040190867 商品ID为：7878
不加分布式锁，开始扣减库存：Example2ProductBo(id=3467374334, goodId=7878, goodName=null, num=null)扣减库处理请求的线程ID为：343 当前请求的uuid为：4924482603296311047 扣减库存源请求的线程ID为：92 执行扣减库存的时间：Sat Aug 14 04:40:12 CST 2021
开始扣减库存，扣除之前的商品库存为：983848040190865 商品ID为：7878
开始扣减库存，扣除之后的商品库存为：983848040190864 商品ID为：7878
不加分布式锁，开始扣减库存：Example2ProductBo(id=3467374334, goodId=7878, goodName=null, num=null)扣减库处理请求的线程ID为：343 当前请求的uuid为：6444653223353808011 扣减库存源请求的线程ID为：95 执行扣减库存的时间：Sat Aug 14 04:40:12 CST 2021
开始扣减库存，扣除之前的商品库存为：983848040190862 商品ID为：7878
开始扣减库存，扣除之后的商品库存为：983848040190859 商品ID为：7878
不加分布式锁，开始扣减库存：Example2ProductBo(id=3467374334, goodId=7878, goodName=null, num=null)扣减库处理请求的线程ID为：347 当前请求的uuid为：61764784264522750 扣减库存源请求的线程ID为：95 执行扣减库存的时间：Sat Aug 14 04:40:12 CST 2021
不加分布式锁，开始扣减库存：Example2ProductBo(id=3467374334, goodId=7878, goodName=null, num=null)扣减库处理请求的线程ID为：343 当前请求的uuid为：2002515304305358143 扣减库存源请求的线程ID为：92 执行扣减库存的时间：Sat Aug 14 04:40:12 CST 2021
开始扣减库存，扣除之前的商品库存为：983848040190856 商品ID为：7878
开始扣减库存，扣除之前的商品库存为：983848040190856 商品ID为：7878
开始扣减库存，扣除之后的商品库存为：983848040190853 商品ID为：7878
开始扣减库存，扣除之后的商品库存为：983848040190853 商品ID为：7878
huxian@huxians-MacBook-Pro 127.0.0.1:8032:26790 %
```

图 13-62

5. 开启分布式锁，观察“use-redisson-distribute-lock-spring-cloud-alibaba”服务的运行日志

（1）服务进程“127.0.0.1:8030”的运行日志如图 13-63 所示。分布式锁会严格控制线程扣减库存的顺序，不会出现线程安全方面的问题。

```
加同步分布式锁，开始扣减库存：Example2ProductBo(id=3467374334, goodId=7878, goodName=null, num=null)扣减库处理请求的线程ID为：358 当前请求的uuid为：4564820018899272784 扣减库存源请求的线程ID为：90 执行扣减库存的时间：Sat Aug 14 03:16:25 CST 2021
开始扣减库存，扣除之前的商品库存为：983848041640425 商品ID为：7878
开始扣减库存，扣除之后的商品库存为：983848041640424 商品ID为：7878
加同步分布式锁，开始扣减库存：Example2ProductBo(id=3467374334, goodId=7878, goodName=null, num=null)扣减库处理请求的线程ID为：358 当前请求的uuid为：2522366302719524378 扣减库存源请求的线程ID为：90 执行扣减库存的时间：Sat Aug 14 03:16:25 CST 2021
开始扣减库存，扣除之前的商品库存为：983848041640424 商品ID为：7878
开始扣减库存，扣除之后的商品库存为：983848041640423 商品ID为：7878
加同步分布式锁，开始扣减库存：Example2ProductBo(id=3467374334, goodId=7878, goodName=null, num=null)扣减库处理请求的线程ID为：358 当前请求的uuid为：8487749400037965937 扣减库存源请求的线程ID为：90 执行扣减库存的时间：Sat Aug 14 03:16:25 CST 2021
开始扣减库存，扣除之前的商品库存为：983848041640422 商品ID为：7878
开始扣减库存，扣除之后的商品库存为：983848041640421 商品ID为：7878
加同步分布式锁，开始扣减库存：Example2ProductBo(id=3467374334, goodId=7878, goodName=null, num=null)扣减库处理请求的线程ID为：358 当前请求的uuid为：1329482563169614506 扣减库存源请求的线程ID为：90 执行扣减库存的时间：Sat Aug 14 03:16:25 CST 2021
开始扣减库存，扣除之前的商品库存为：983848041640421 商品ID为：7878
开始扣减库存，扣除之后的商品库存为：983848041640420 商品ID为：7878
huxian@huxians-MacBook-Pro 127.0.0.1:8030:26788 %
```

图 13-63

（2）服务进程“127.0.0.1:8031”的运行日志如图 13-64 所示。分布式锁会严格控制线程扣减库存的顺序，不会出现线程安全方面的问题。

```
加同步分布式锁，开始扣减库存：Example2ProductBo(id=3467374334, goodId=7878, goodName=null, num=null)扣减库处理请求的线程ID为：338 当前请求的uuid为：6113737857501182536 扣减库存源请求的线程ID为：90 执行扣减库存的时间：Sat Aug 14 03:17:21 CST 2021
开始扣减库存，扣除之前的商品库存为：983848041625615 商品ID为：7878
开始扣减库存，扣除之后的商品库存为：983848041625614 商品ID为：7878
加同步分布式锁，开始扣减库存：Example2ProductBo(id=3467374334, goodId=7878, goodName=null, num=null)扣减库处理请求的线程ID为：338 当前请求的uuid为：5412552248857309801 扣减库存源请求的线程ID为：90 执行扣减库存的时间：Sat Aug 14 03:17:21 CST 2021
开始扣减库存，扣除之前的商品库存为：983848041625612 商品ID为：7878
开始扣减库存，扣除之后的商品库存为：983848041625611 商品ID为：7878
加同步分布式锁，开始扣减库存：Example2ProductBo(id=3467374334, goodId=7878, goodName=null, num=null)扣减库处理请求的线程ID为：338 当前请求的uuid为：8768710224375984460 扣减库存源请求的线程ID为：90 执行扣减库存的时间：Sat Aug 14 03:17:21 CST 2021
开始扣减库存，扣除之前的商品库存为：983848041625611 商品ID为：7878
开始扣减库存，扣除之后的商品库存为：983848041625610 商品ID为：7878
加同步分布式锁，开始扣减库存：Example2ProductBo(id=3467374334, goodId=7878, goodName=null, num=null)扣减库处理请求的线程ID为：338 当前请求的uuid为：5908530739685165497 扣减库存源请求的线程ID为：90 执行扣减库存的时间：Sat Aug 14 03:17:21 CST 2021
开始扣减库存，扣除之前的商品库存为：983848041625610 商品ID为：7878
开始扣减库存，扣除之后的商品库存为：983848041625609 商品ID为：7878
huxian@huxians-MacBook-Pro 127.0.0.1:8031:26789 %
```

图 13-64

（3）服务进程“127.0.0.1:8032”的运行日志如图 13-65 所示。分布式锁会严格地控制线程扣减库存的顺序，不会出现线程安全方面问题。

```
加同步分布式锁，开始扣减库存：Example2ProductBo(id=3467374334, goodId=7878, goodName=null, num=null)扣减库处理请求的线程ID为：348 当前请求的uuid为：7950465041239575465 扣减库存源请求的线程ID为：90 执行扣减库存的时间：Sat Aug 14 03:17:54 CST 2021
开始扣减库存，扣除之前的商品库存为：983848041616464 商品ID为：7878
开始扣减库存，扣除之后的商品库存为：983848041616463 商品ID为：7878
加同步分布式锁，开始扣减库存：Example2ProductBo(id=3467374334, goodId=7878, goodName=null, num=null)扣减库处理请求的线程ID为：348 当前请求的uuid为：1305028946588304720 扣减库存源请求的线程ID为：90 执行扣减库存的时间：Sat Aug 14 03:17:54 CST 2021
开始扣减库存，扣除之前的商品库存为：983848041616462 商品ID为：7878
开始扣减库存，扣除之后的商品库存为：983848041616461 商品ID为：7878
加同步分布式锁，开始扣减库存：Example2ProductBo(id=3467374334, goodId=7878, goodName=null, num=null)扣减库处理请求的线程ID为：348 当前请求的uuid为：8001565042367058273 扣减库存源请求的线程ID为：90 执行扣减库存的时间：Sat Aug 14 03:17:54 CST 2021
开始扣减库存，扣除之前的商品库存为：983848041616455 商品ID为：7878
开始扣减库存，扣除之后的商品库存为：983848041616454 商品ID为：7878
加同步分布式锁，开始扣减库存：Example2ProductBo(id=3467374334, goodId=7878, goodName=null, num=null)扣减库处理请求的线程ID为：348 当前请求的uuid为：8702463474951060400 扣减库存源请求的线程ID为：90 执行扣减库存的时间：Sat Aug 14 03:17:54 CST 2021
开始扣减库存，扣除之前的商品库存为：983848041616451 商品ID为：7878
开始扣减库存，扣除之后的商品库存为：983848041616450 商品ID为：7878
huxian@huxians-MacBook-Pro 127.0.0.1:8032:26790 %
```

图 13-65

第 14 章 服务注册/订阅路由、全链路蓝绿发布和灰度发布——基于 Discovery

将单个服务拆分成多个服务，大大增加了业务功能发布上线的成本。为了确保业务功能的可用性，软件开发人员需要用真实的用户流量来验证业务功能逻辑的正确性。如果上线功能的流量非常大，则开发人员还需要用真实的用户流量去验证业务功能处理流量洪峰的能力，因此，软件开发人员需要解决如下问题：

（1）如何用最少的服务节点去验证业务功能逻辑的正确性，并将故障控制在最小范围内。

（2）如何将业务功能稳定地发布上线，并对业务调用方是无感知的。

（3）如何用最小的成本去解决业务功能发布失败后的回滚问题。

软件开发人员可以利用 Discovery 框架来解决上述问题。Discovery 框架具备服务注册/订阅路由、全链路蓝绿发布和灰度发布等功能。

14.1 认识服务注册/订阅路由、蓝绿发布和灰度发布

在分析 Discovery 框架的原理之前，先认识服务注册/订阅路由、蓝绿发布和灰度发布。

14.1.1 什么是服务注册路由、服务订阅路由

服务注册路由是指，在将服务注册到注册中心的过程中读取服务注册规则，并过滤服务注册的请求。如果过滤后没有服务实例需要注册，则直接返回。

服务订阅路由是指，在订阅服务的过程中读取服务订阅的规则，并过滤服务订阅的请求，然后将过滤后的服务列表更新到负载均衡器中。这样服务订阅者可以获取最新的可以订阅的服务实例。

服务注册/订阅路由有以下 6 步，如图 14-1 所示。

（1）用规则过滤需要注册的交易服务；

（2）注册交易服务；

（3）订单服务订阅交易服务；

（4）过滤需要订阅的交易服务；

（5）更新负载均衡器；

（6）订单服务从负载均衡器中获取可用的服务实例。

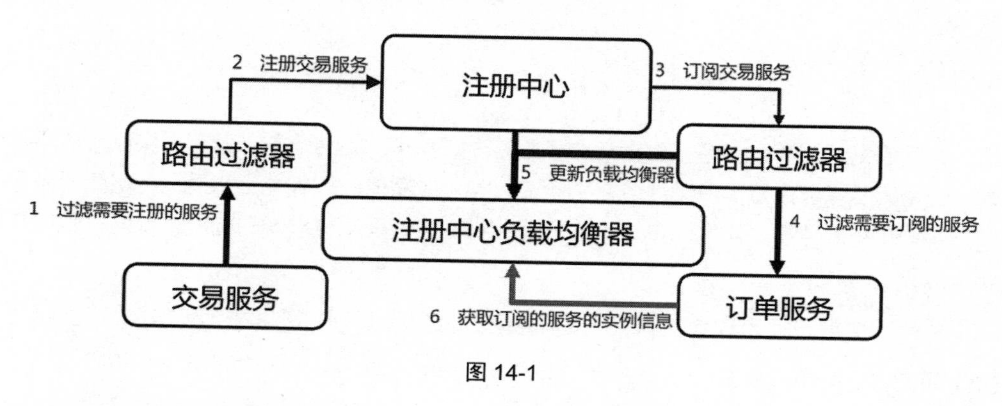

图 14-1

14.1.2 什么是蓝绿发布

蓝绿发布（Blue-Green Deployment）是指，软件开发人员在不下线服务的旧版本的情况下上线服务的新版本，并通过在用户请求流量中添加路由规则，将流量在服务的新旧版本之间切换。蓝绿发布属于无损发布。

1. 核心流程

图 14-2 所示为蓝绿发布的核心流程。

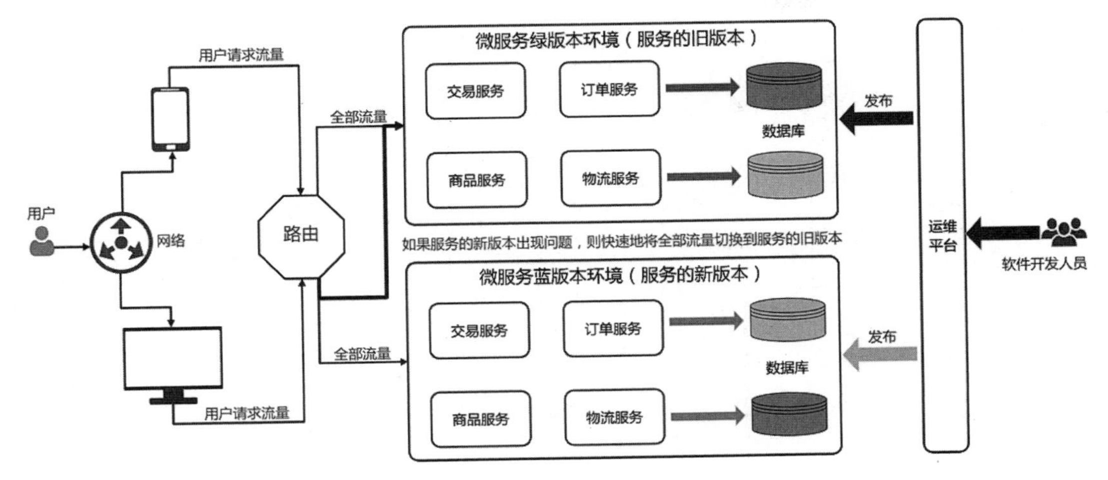

图 14-2

（1）软件开发人员在原有的绿版本环境（服务的旧版本）的基础上，搭建一个蓝版本的环境（服务的新版本），以发布蓝版本的微服务。

（2）在服务路由层添加用户标记，让用户请求流量从绿版本环境切换到蓝版本环境中（包括底层存储设备，比如数据库）。

（3）软件开发人员可以在蓝版本环境中验证需要发布的业务功能（绿版本环境不会下线）。

（4）如果需要发布的业务功能没有问题，则下线绿版本环境。如果有问题，则通过运维工具快速地将用户请求流量切换到绿版本环境中。

2. 优缺点

蓝绿发布的优缺点如下。

- 优点：在新版本升级的过程中，如果出现问题，可以快速地用线上的老版本进行回滚，软件开发人员可以非常灵活地控制请求流量的走向。
- 缺点：采用蓝绿发布，需要软件开发人员同时部署和维护两套环境，机器成本非常大，如果新版本出现问题，不能及时全部切换到旧版本上，则会造成大面积的技术故障。

14.1.3　什么是灰度发布

灰度发布（Gray Deployment）又名金丝雀发布（Canary Deployment），是指：在不下线服务的旧版本的情况下部署服务的新版本；将低比例流量（例如 20%）切换到新版本，高比例流量（例如 80%）仍走旧版本；通过监控观察新版本的流量是否存在异常，如果没有异常，则逐步扩大新版本的流量，最终将所有的流量从旧版本切换到新版本上，并下线旧版本。灰度发布也属于无损发布。

1. 核心流程

图 14–3 所示为灰度发布的核心流程。

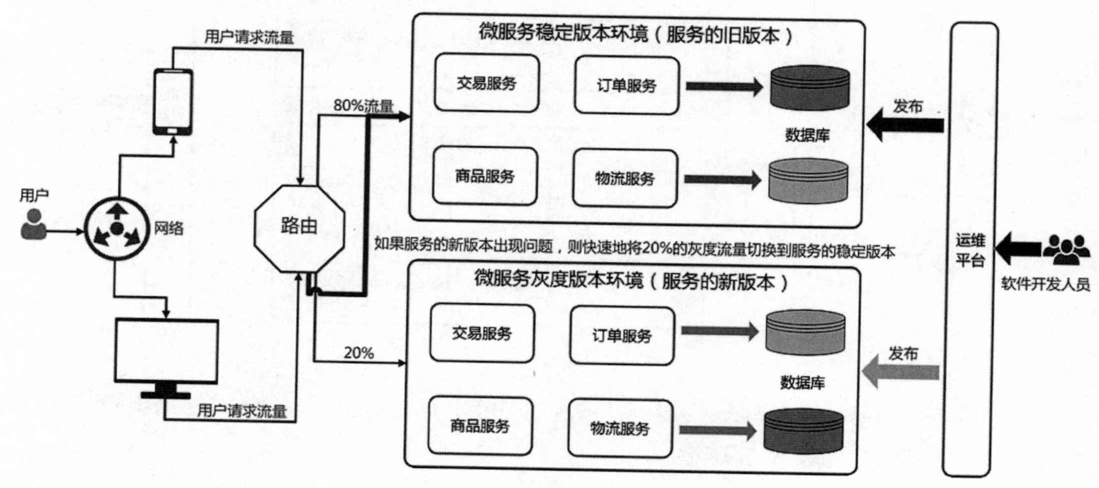

图 14-3

（1）软件开发人员在原有稳定版本环境的基础上，增加一个灰度版本环境（通常是用少量的服务实例搭建一套灰度环境），并将需要灰度发布的微服务发布到灰度环境中。

（2）将一部分用户请求流量（比如 20%的流量）切换到灰度环境，这样可以用 20%的流量去验证需要发布的业务功能。

（3）如果在灰度环境严重业务功能没问题，则逐步扩大灰度流量的比例，直到稳定版本的流量比例为 0（在切换流量的过程中，需要按照服务的 QPS 去扩容灰度环境中的服务的个数）。

（4）如果灰度环境严重业务功能存在问题，则直接将流量切换到稳定版本环境。

2. 优缺点

- 优点：灰度发布非常灵活，它不需要用户标记，线上部署非常简单。如果新版本出现问题，则技术故障率集中在较小的灰度流量上，事故的影响范围可以控制到最小，也能够快速地恢复故障。
- 缺点：软件开发人员和运维人员在线上实施灰度发布时，需要修改流量的配比，这会增加一些操作成本，并且在较低流量下去验证业务功能并不一定能够发现所有的问题。

总之，灰度发布是用少量的服务实例去验证需要发布的功能，能够提前发现问题，并确保线上业务的稳定性。

14.1.4 认识微服务治理框架 Discovery

Discovery 是开源的一款微服务治理框架，它底层主要基于 Spring Cloud 和 Spring Cloud Alibaba，并集成了 Sentinel、Skywalking 等主流的中间件框架。Discovery 包括非常多的服务治理功能，其中最核心的功能是服务注册/订阅路由、全链路蓝绿发布及灰度发布。

1. 整体架构

服务注册/订阅路由、全链路蓝绿发布及灰度发布的整体架构如图 14-4 所示。

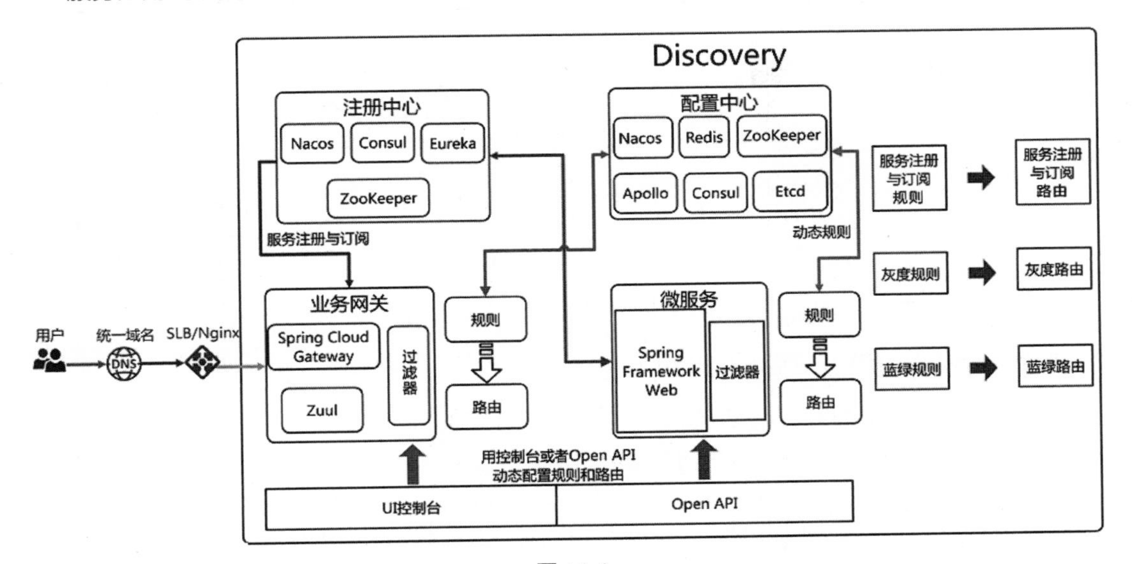

图 14-4

Discovery 是基于规则来驱动服务路由的，规则包括：服务注册/订阅规则、灰度规则和蓝绿规则。

Discovery 整合了主流的注册中心和配置中心，前者主要用于服务路由过程中的服务注册/订阅，后者主要用于动态配置规则。

Discovery 利用过滤器拦截 Spring Cloud Gateway、Zuul 及 RESTful API 的请求，并设置加载规则的标签。这样在全链路的服务请求中，Discovery 会按需加载规则并驱动服务路由。

Discovery 可以通过 UI 控制台和 Open API 动态地变更规则和路由信息。

2. Discovery 的服务注册/订阅路由

Discovery 的服务注册/订阅路由的流程如图 14-5 所示。

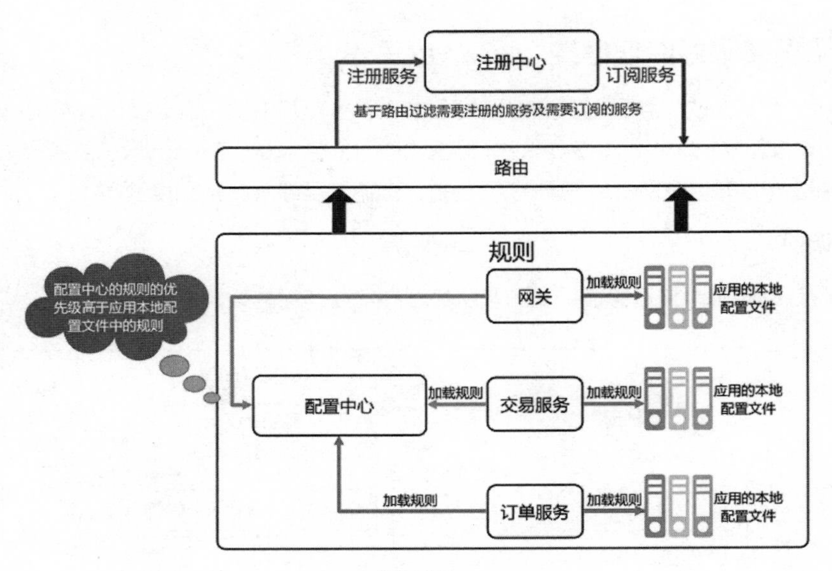

图 14-5

Discovery 会在服务注册/订阅过程中进行前置处理，触发规则监听器去加载规则。

Discovery 在加载了规则后，会将规则转换为服务路由，并更新到服务的负载均衡中。

- 如果是服务注册路由，则会按照规则生成的服务路由，直接过滤需要注册的服务，比如“按照 IP 地址前缀，过滤需要注册的服务的实例”。
- 如果是服务订阅注册，则会按照规则生成的服务路由，去过滤从注册中心获取的服务实例列表，从而生成最终的服务实例列表，并更新到服务路由负载均衡器中。

Discovery 的服务注册/订阅路由实际上是利用注册中心的服务路由机制来完成服务路由的。

3. Discovery 的全链路蓝绿发布

Discovery 的全链路蓝绿发布的流程如图 14-6 所示。

Discovery 从配置中心或者应用本地配置文件中读取规则，其中，配置中的规则的优先级要高于应用本地配置文件中的规则。

Discovery 利用规则监听器，将规则转换成服务路由（如果服务请求中设置的加载规则命中蓝路由的规则，则将规则转换为蓝路由）。

Discovery 基于路由中的服务实例列表过滤从注册中心获取的原始的服务实例列表，并生成最终的服务实例列表，更新到服务订阅的负载均衡器中。

Discovery 的全链路蓝绿发布是按照规则（比如版本号、区域等），将全量的流量从绿路由所在绿版本环境切换到蓝路由所在的蓝版本环境。

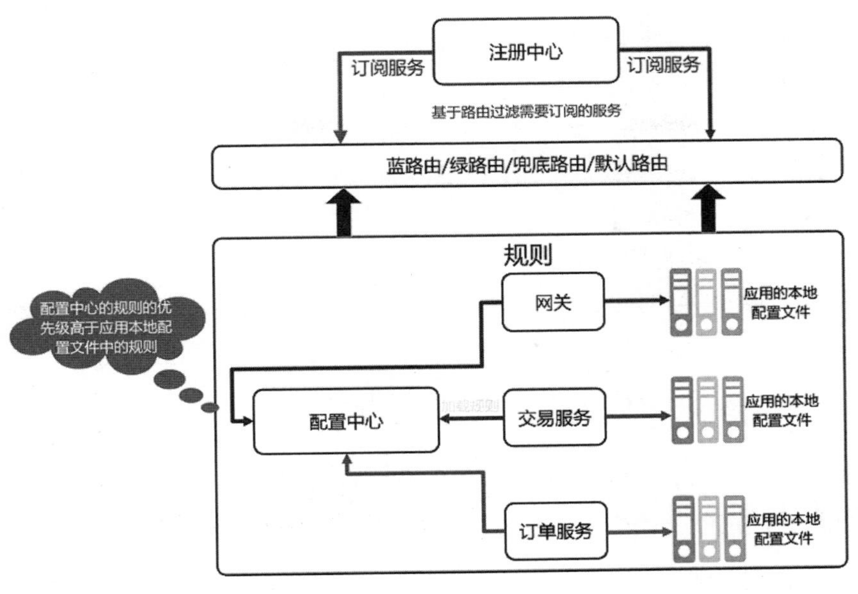

图 14-6

4. Discovery 的全链路灰度发布

Discovery 的全链路灰度发布的流程如图 14-7 所示。

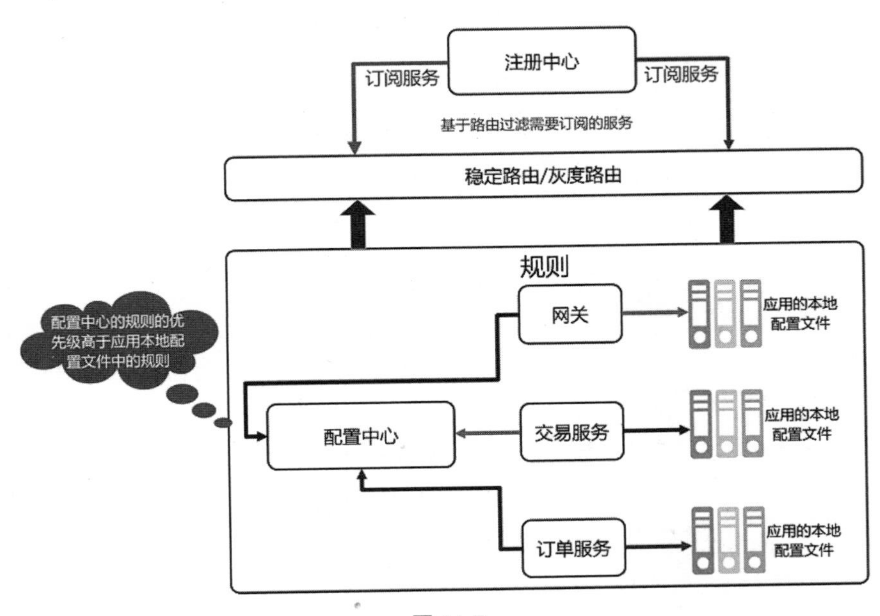

图 14-7

Discovery 在配置中心或者应用本地配置文件中读取规则，其中，配置中的规则的优先级要高于应用本地配置文件中的规则。

Discovery 利用规则监听器，将规则转换成服务路由（比如，将灰度规则和稳定规则，转换为灰度路由和稳定路由）。在全链路灰度发布中，Discovery 采用调用比例来分配灰度和稳定路由的流量，所以会同时触发灰度路由和稳定路由（如果用版本号来区分灰度流量和稳定流量，则需要启动指定版本的服务实例，同时触发指定的规则）。

Discovery 会基于路由中的服务实例列表，过滤从注册中心获取的原始的服务实例列表，并生成最终的服务实例列表，更新到服务订阅的负载均衡器中。

Discovery 的全链路灰度发布是按照灰度条件去控制灰度和稳定路由的调用比例的。

14.2 “用插件机制来集成主流的注册中心和配置中心”的原理

Discovery 采用插件化机制来集成主流的注册中心和配置中心。下面结合 Nacos 来分析 Discovery 的插件化机制。

14.2.1 集成主流的注册中心

Discovery 支持将 Nacos、ZooKeeper、Consul 及 Eureka 作为应用的注册中心。

1. 用包装器、适配器及上下文环境来实现插件机制

如图 14-8 所示，Discovery 主要采用包装器、适配器及上下文环境来实现插件机制。

（1）包装器：Discovery 在 Spring Cloud 和 Spring Cloud Alibaba 的基础上集成了主流的注册中心。

Spring Cloud 支持将 ZooKeeper、Consul 及 Eureka 作为应用的注册中心。比如，针对 ZooKeeper，Spring Cloud 封装了 ZookeeperServiceRegistry 类和 ZookeeperServerList 类。

Spring Cloud Alibaba 支持用 Nacos 作为应用的注册中心，Spring Cloud Alibaba 封装了 NacosServiceRegistry 类和 NacosServerList 类。

在主流注册中心暴露的 SDK 的基础之上，Discovery 通过包装器增强了原有注册中心的功能。比如，针对 Nacos，Discovery 用包装器 NacosServiceRegistryDecorator 类来增强 Spring Cloud Alibaba 的 NacosServiceRegistry 类的功能。

（2）适配器：Discovery 采用适配器屏蔽主流注册中心的技术实现的差异性，这样 Discovery 中的其他模块可以采用统一的 API 去集成注册中心。

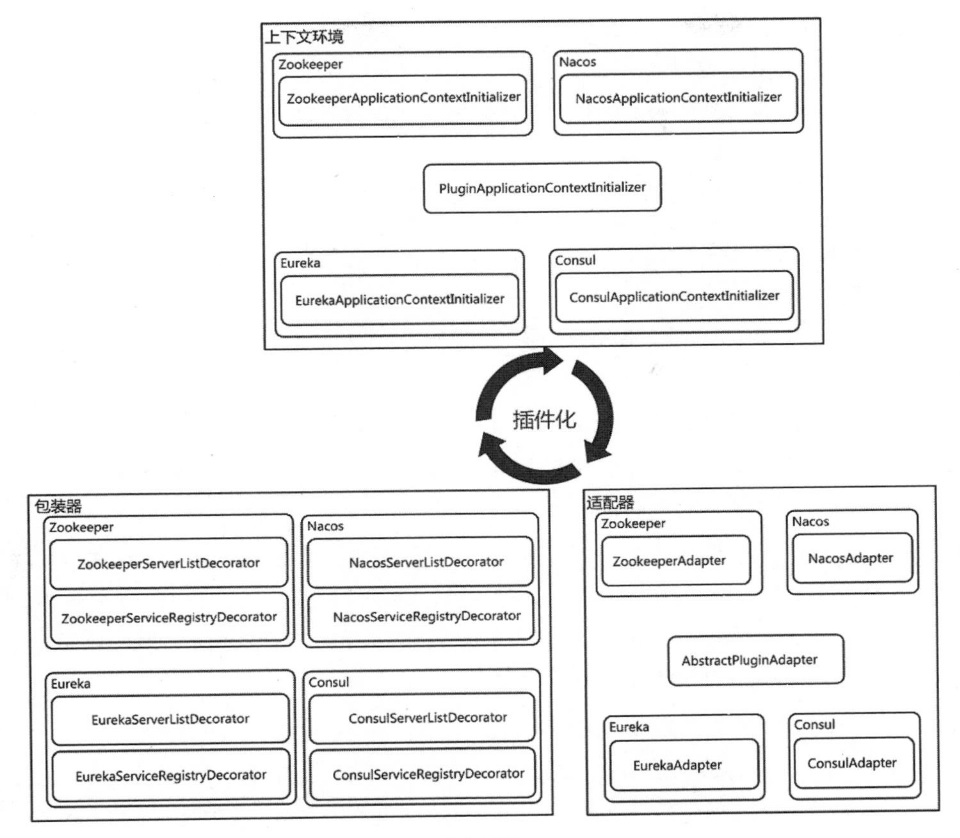

图 14-8

Discovery 将公共功能（比如与服务注册/订阅相关的缓存规则）下沉到抽象类 AbstractPluginAdapter 中。

针对不同的注册中心，Discovery 定义了个性化的类，封装了一些差异性的功能。比如 Discovery 在 NacosAdapter 类中增加了 Nacos 元数据及 Nacos 服务 ID 相关的功能。

Discovery 通过自动配置类，将不同注册中心的适配类加载到 Spring Framework 的 IOC 容器中。比如，加载 Nacos 注册中心适配类的具体代码如下所示：

```
@Configuration
@RibbonClients(defaultConfiguration = { PluginLoadBalanceConfiguration.
class, NacosLoadBalanceConfiguration.class })
public class NacosAutoConfiguration {
    ...
    @Bean
    public PluginAdapter pluginAdapter() {
        return new NacosAdapter();
```

```
        }
    }
```

软件开发人员在应用服务中添加如下 POM 依赖，即可激活基于 Nacos 的适配器。

```
    <dependency>
        <groupId>com.nepxion</groupId>
<artifactId>discovery-plugin-register-center-starter-nacos</artifactId>
        <version>6.11.0</version>
    </dependency>
```

（3）上下文环境：Discovery 在服务启动过程中，使用上下文环境来拦截应用的启动，并完成包装器的初始化。

Discovery 用抽象模板类 PluginApplicationContextInitializer 来实现 Spring Framework 的应用上下文加载器 ApplicationContextInitializer 类的 initialize()方法。这样在应用启动时，会优先执行 initialize()方法来初始化 Discovery 的包装器。

Discovery 用不同注册中心的上下文环境的实现类来继承抽象模板类 PluginApplicationContextInitializer，以实现抽象模板类的抽象方法 afterInitialization()。如果是采用 Nacos 作为注册中心，则用 NacosApplicationContextInitializer 类实现抽象模板类 PluginApplicationContextInitializer，其他注册中心以此类推。

Discovery 在实现类中完成包装器的初始化。

在采用基于插件机制的注册中心时，软件开发人员在应用服务中只能同时使用一种类型的注册中心。如果想使用 Discovery 不支持的注册中心类型，则需要利用插件机制自行扩展。

2. 拦截应用的启动

Discovery 用插件类 PluginApplicationContextInitializer 拦截应用的启动，具体代码如下所示：

```
    public abstract class PluginApplicationContextInitializer implements
ApplicationContextInitializer<ConfigurableApplicationContext>, Ordered {
        //①重写 Spring Framework 的 ApplicationContextInitializer 类的 initialize()
方法，这样 Discovery 可以拦截应用的启动
        @Override
        public void initialize(ConfigurableApplicationContext applicationContext) {
            //②如果applicationContext对象的类型不匹配AnnotationConfigApplicationContext
类，则初始化默认配置属性
            if (!(applicationContext instanceof
AnnotationConfigApplicationContext)) {
```

```
            LogoBanner logoBanner = new LogoBanner
(PluginApplicationContextInitializer.class, "/com/nepxion/discovery/
resource/logo.txt", "Welcome to Nepxion", 9, 5, new Color[] { Color.red,
Color.green, Color.cyan, Color.blue, Color.yellow, Color.magenta, Color.red,
Color.green, Color.cyan }, true);
            NepxionBanner.show(logoBanner, new Description(BannerConstant.
VERSION + ":", DiscoveryConstant.DISCOVERY_VERSION, 0, 1), new Description
(BannerConstant.GITHUB + ":", BannerConstant.NEPXION_GITHUB + "/Discovery", 0, 1));
            initializeDefaultProperties(applicationContext);
        }
        //③添加前置 Bean 处理器 InstantiationAwareBeanPostProcessorAdapter 类，
主要用于包装 Spring Framework 的 IOC 容器中的 Bean，这样可以增强原有 Bean 的功能
        applicationContext.getBeanFactory().addBeanPostProcessor(new
InstantiationAwareBeanPostProcessorAdapter() {
            @Override
            public Object postProcessAfterInitialization(Object bean, String
beanName) throws BeansException {
                //④如果 bean 的类型匹配 Spring Cloud 的服务发现类 DiscoveryClient，
则将 DiscoveryClient 类包装为 DiscoveryClientDecorator 类，并将其重新放入 IOC 容器中覆
盖原来的 DiscoveryClient 类
                if (bean instanceof DiscoveryClient) {
                    DiscoveryClient discoveryClient = (DiscoveryClient) bean;
                    return new DiscoveryClientDecorator(discoveryClient,
applicationContext);
                } else {
                    //⑤如果不匹配，则调用 afterInitialization()方法执行一些与注册中
心相关的个性功能
                    return afterInitialization(applicationContext, bean,
beanName);
                }
            }
        });
    }
    //⑥定义一个抽象方法 afterInitialization()，主要用于初始化一些与注册中心相关的个
性功能
    protected abstract Object afterInitialization
(ConfigurableApplicationContext applicationContext, Object bean, String
beanName) throws BeansException;
    ...
}
```

3. 处理 Nacos 的服务注册并重写 Nacos 元数据

Discovery 用 NacosApplicationContextInitializer 类处理 Nacos 的服务注册并重写 Nacos 元数据，具体代码如下所示：

```
    public class NacosApplicationContextInitializer extends
PluginApplicationContextInitializer {
        @Override
        protected Object afterInitialization(ConfigurableApplicationContext
applicationContext, Object bean, String beanName) throws BeansException {
            //①如果bean的类型匹配NacosServiceRegistry，则用Discovery的
NacosServiceRegistryDecorator类包装Spring Cloud Alibaba的NacosServiceRegistry类
            if (bean instanceof NacosServiceRegistry) {
                NacosServiceRegistry nacosServiceRegistry =
(NacosServiceRegistry) bean;
                NacosDiscoveryProperties nacosDiscoveryProperties =
applicationContext.getBean(NacosDiscoveryProperties.class);
                //②将包装类NacosServiceRegistryDecorator植入Spring Framework的
IOC容器中，并替换Spring Cloud Alibaba的NacosServiceRegistry类
                return new NacosServiceRegistryDecorator
(nacosDiscoveryProperties, nacosServiceRegistry, applicationContext);
            } else if (bean instanceof NacosDiscoveryProperties) {
                ConfigurableEnvironment environment =
applicationContext.getEnvironment();
                NacosDiscoveryProperties nacosDiscoveryProperties =
(NacosDiscoveryProperties) bean;
                Map<String, String> metadata = nacosDiscoveryProperties.
getMetadata();
                String groupKey = PluginContextAware.getGroupKey(environment);
                //③设置groupKey
                if (!metadata.containsKey(groupKey)) {
                    metadata.put(groupKey, DiscoveryConstant.DEFAULT);
                }
                //④设置版本号
                if (!metadata.containsKey(DiscoveryConstant.VERSION)) {
                    metadata.put(DiscoveryConstant.VERSION,
DiscoveryConstant.DEFAULT);
                }
                //⑤设置区域
                if (!metadata.containsKey(DiscoveryConstant.REGION)) {
                    metadata.put(DiscoveryConstant.REGION,
DiscoveryConstant.DEFAULT);
                }
                //⑥设置Group的前缀
                String prefixGroup = getPrefixGroup(applicationContext);
                if (StringUtils.isNotEmpty(prefixGroup)) {
                    metadata.put(groupKey, prefixGroup);
                }
                //⑦设置Git的版本号
                String gitVersion = getGitVersion(applicationContext);
                if (StringUtils.isNotEmpty(gitVersion)) {
                    metadata.put(DiscoveryConstant.VERSION, gitVersion);
```

```
            }
            //⑧重写 Nacos 元数据
            metadata.put(DiscoveryMetaDataConstant.SPRING_BOOT_VERSION,
SpringBootVersion.getVersion());
            metadata.put(DiscoveryMetaDataConstant.SPRING_APPLICATION_NAME,
PluginContextAware.getApplicationName(environment));
            ...
            //⑨过滤元数据
            MetadataUtil.filter(metadata, environment);
            return bean;
        } else {
            return bean;
        }
    }
}
```

14.2.2　集成主流的配置中心

Discovery 支持将 Nacos、ZooKeeper、Consul、Apollo、Etcd 及 Redis 作为应用的配置中心。如图 14-9 所示，Discovery 主要是采用配置中心操控器和适配器来实现插件机制。

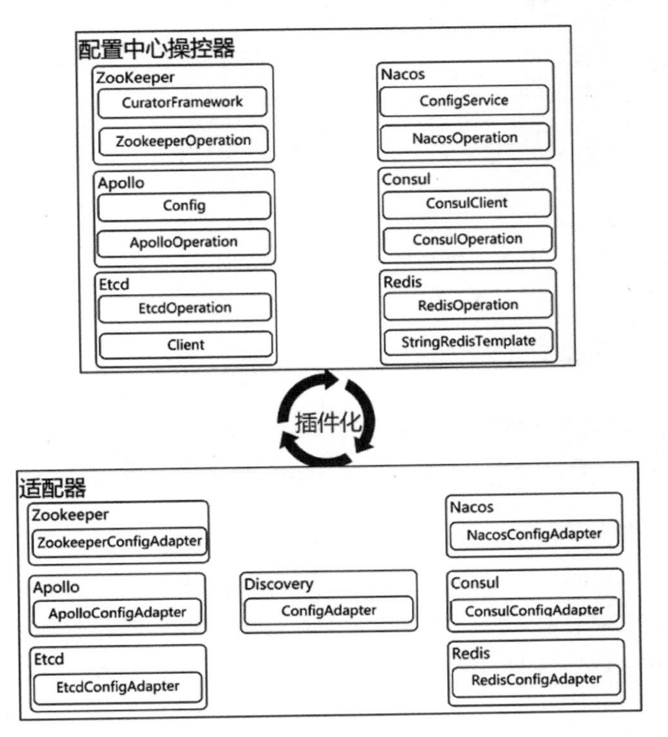

图 14-9

（1）配置中心操控器：Discovery 利用配置中心操控器来屏蔽不同配置中心的技术差异性。比如，Discovery 用 NacosOperation 类来封装 Nacos 配置中心的 SDK 类 ConfigService，这样 Discovery 的其他模块可以直接使用 NacosOperation 类来操控配置中心的配置属性。

（2）适配器：Discovery 利用适配器让使用者能够无感知地切换配置中心的类型。比如，针对 Nacos 配置中心，Discovery 使用 NacosConfigAdapter 类实现 ConfigAdapter 接口，这样使用者可以通过 POM 依赖快速地切换配置中心的类型，具体如下所示。

```
<dependency>
    <groupId>com.nepxion</groupId>
    <artifactId>discovery-plugin-config-center-starter-nacos</artifactId>
    <version>6.11.0</version>
</dependency>
```

Discovery 通过自动配置类 NacosConfigAutoConfiguration，将 NacosConfigAdapter 类注入 Spring Framework 的 IOC 容器中，这样应用可以直接通过 ConfigAdapter 接口来使用不同的配置中心，具体代码如下所示：

```
@Configuration
public class NacosConfigAutoConfiguration {
    @Bean
    public ConfigAdapter configAdapter() {
        return new NacosConfigAdapter();
    }
    ...
}
```

在采用基于插件机制的配置中心时，软件开发人员在应用服务中只能同时使用一种类型的配置中心。如果想使用 Discovery 不支持的配置中心类型，则需要利用插件机制自行扩展。

14.3 “用 Open API 和配置中心动态变修改规则”的原理

Discovery 支持两种动态修改路由规则的方式。

- Open API：软件开发人员可以利用 Open API 实时地修改路由规则。
- 配置中心：软件开发人员可以利用配置中心的控制台实时地修改路由规则。Discovery 支持的配置中心主要包括 Nacos、ZooKeeper、Apollo、Redis、Etcd 及 Consul。

14.3.1　用 Open API 动态修改规则

软件开发人员可以利用 Discovery 的 Open API 动态更新规则（也支持查看规则），修改操作主要包括更新和删除规则。下面从“更新操作”的角度去分析用 Open API 动态修改路由规则的原理。

1. 定义 Open API

Discovery 用 ConfigEndpoint 类定义 Open API，具体代码如下所示：

```
@RestController
@RequestMapping(path = "/config")
@Api(tags = { "配置接口" })
public class ConfigEndpoint {
    @Autowired
    private ConfigResource configResource;
    //①异步更新规则
    @RequestMapping(path = "/update-async", method = RequestMethod.POST)
    @ApiOperation(value = "异步更新规则配置", notes = "", response =
ResponseEntity.class, httpMethod = "POST")
    @ResponseBody
    public ResponseEntity<?> updateAsync(@RequestBody @ApiParam(value = "
规则配置内容", required = true) String config) {
        return doUpdate(config, true);
    }
    //②同步更新规则
    @RequestMapping(path = "/update-sync", method = RequestMethod.POST)
    @ApiOperation(value = "同步更新规则配置", notes = "", response =
ResponseEntity.class, httpMethod = "POST")
    @ResponseBody
    public ResponseEntity<?> updateSync(@RequestBody @ApiParam(value = "规
则配置内容", required = true) String config) {
        return doUpdate(config, false);
    }
    //③调用 ConfigResource 接口的 update()方法更新规则
    private ResponseEntity<?> doUpdate(String config, boolean async) {
        try {
            configResource.update(config, async);
            return ResponseUtil.getSuccessResponse(true);
        } catch (Exception e) {
            return ResponseUtil.getFailureResponse(e);
        }
    }
    ...
}
```

2. 用 ConfigResource 接口的实现类 ConfigResourceImpl 的 update()方法同步或者异步地更新规则

ConfigResourceImpl 类是 ConfigResource 接口的实现类，其 update()方法的具体代码如下所示：

```
public class ConfigResourceImpl implements ConfigResource {
    @Autowired
    private PluginContextAware pluginContextAware;
    @Autowired
    private PluginEventWapper pluginEventWapper;
    @Override
public void update(String config, boolean async) {
    //①获取配置信息“spring.application.discovery.control.enabled”，如果其值为 true，则动态更新规则，否则抛出异常
        Boolean discoveryControlEnabled = 
pluginContextAware.isDiscoveryControlEnabled();
        if (!discoveryControlEnabled) {
            throw new DiscoveryException("Discovery control is disabled");
        }
        //②获取配置信息“spring.application.config.rest.control.enabled”，如果其值为 true，则动态更新规则，否则抛出异常
        Boolean isConfigRestControlEnabled = 
pluginContextAware.isConfigRestControlEnabled();
        if (!isConfigRestControlEnabled) {
            throw new DiscoveryException("Config rest control is disabled");
        }
        //③调用 PluginEventWapper 类的 fireRuleUpdated()方法更新规则，如果参数async 为 true，则为异步调用，否则为同步调用
        pluginEventWapper.fireRuleUpdated(new 
RuleUpdatedEvent(SubscriptionType.PARTIAL, config), async);
    }
    ...
}
```

3. 用 PluginEventWapper 类的 fireRuleUpdated()方法同步或者异步地更新规则

用 PluginEventWapper 类的 fireRuleUpdated()方法更新规则，具体代码如下所示：

```
public class PluginEventWapper {
    @Autowired
    private PluginPublisher pluginPublisher;
    @Autowired
    private PluginSubscriber pluginSubscriber;

public void fireRuleUpdated(RuleUpdatedEvent ruleUpdatedEvent, boolean
```

```
async) {
    //①async 的默认值为 true
        if (async) {
            //②如果 async 为 true,则用 PluginPublisher 类的 asyncPublish()方法异步
地更新规则
            pluginPublisher.asyncPublish(ruleUpdatedEvent);
        } else {
            //③如果 async 为 false, 则用 PluginSubscriber 类的 onRuleUpdated()方
法同步地更新规则
            pluginSubscriber.onRuleUpdated(ruleUpdatedEvent);
        }
    }
    ...
}
```

4. 用 PluginPublisher 类的 asyncPublish()方法异步地更新规则

用 PluginPublisher 类的 asyncPublish()方法异步地更新规则，具体代码如下。

（1）用工厂类 EventControllerFactory 获取 Discovery 封装的前置事件控制器 EventControllerImpl，并利用异步事件框架 Event Bus 发送更新规则的异步事件 RuleUpdatedEvent，具体代码如下所示：

```
public class PluginPublisher {
    @Autowired
    private EventControllerFactory eventControllerFactory;
    //获取前置事件控制器 EventControllerImpl 对象,并用其 post()方法异步地发送更新规
则的事件 RuleUpdatedEvent
    public void asyncPublish(Object object) {
        eventControllerFactory.getAsyncController().post(object);
    }
    ...
}
```

Discovery 基于 Google 的 Guava 框架，封装了一个异步事件框架 Event Bus。

（2）熟悉 Guava 框架的开发人员都知道，Guava 框架需要用注解@Subscribe 去订阅事件。Discovery 用 PluginSubscriber 类的 onRuleUpdated()方法结合注解@Subscribe，去订阅发送更新规则的异步事件 RuleUpdatedEvent，具体代码如下所示：

```
//①开启 Guava 的事件通道
@EventBus
public class PluginSubscriber {
```

```
    @Autowired
    private PluginConfigParser pluginConfigParser;
    @Autowired
    private PluginAdapter pluginAdapter;
    @Autowired
    private PluginEventWapper pluginEventWapper;
    //②订阅事件 RuleUpdatedEvent
    @Subscribe
    public void onRuleUpdated(RuleUpdatedEvent ruleUpdatedEvent) {
        //③获取配置信息“spring.application.discovery.control.enabled”，如果
其值为 false，则直接返回
        Boolean discoveryControlEnabled =
pluginContextAware.isDiscoveryControlEnabled();
        if (!discoveryControlEnabled) {
            LOG.info("Discovery control is disabled, ignore to subscribe");
            return;
        }
        LOG.info("Rule updating has been triggered");
        if (ruleUpdatedEvent == null) {
            throw new DiscoveryException("RuleUpdatedEvent can't be null");
        }
        //④获取订阅事件的类型，比如全局事件 DiscoveryConstant.GLOBAL 或局部事件
DiscoveryConstant.PARTIAL
        SubscriptionType subscriptionType =
ruleUpdatedEvent.getSubscriptionType();
        String rule = ruleUpdatedEvent.getRule();
        try {
            RuleEntity ruleEntity = pluginConfigParser.parse(rule);
            switch (subscriptionType) {
                case GLOBAL:
                    //⑤如果是全局事件，则调用 PluginAdapter 接口的
setDynamicGlobalRule()方法动态更新规则
                    pluginAdapter.setDynamicGlobalRule(ruleEntity);
                    break;
                case PARTIAL:
                    //⑥如果是局部事件，则调用 PluginAdapter 接口的
setDynamicPartialRule()方法动态更新规则
                    pluginAdapter.setDynamicPartialRule(ruleEntity);
                    break;
            }
            //⑦发布参数变更事件 ParameterChangedEvent
            pluginEventWapper.fireParameterChanged();
        } catch (Exception e) {
            LOG.error("Parse rule xml failed", e);
            pluginEventWapper.fireRuleFailure(new
```

```
RuleFailureEvent(subscriptionType, rule, e));
                throw e;
            }
            //⑧刷新负载均衡的缓存
            refreshLoadBalancer();
        }

        // ⑨在规则或者版本更新后，强制刷新负载均衡缓存
        private void refreshLoadBalancer() {
            ZoneAwareLoadBalancer<?> loadBalancer =
loadBalanceListenerExecutor.getLoadBalancer();
            if (loadBalancer == null) {
                return;
            }
            //⑩用 ZoneAwareLoadBalancer 的 updateListOfServers()方法刷新负载均衡的缓存
            loadBalancer.updateListOfServers();
        }
        ...
    }
```

（3）用 AbstractPluginAdapter 类的 setDynamicPartialRule()方法动态地更新全局和局部规则，具体代码如下所示：

```
    public abstract class AbstractPluginAdapter implements PluginAdapter {
        @Autowired
        protected RuleCache ruleCache;
        //①将局部规则更新到缓存中，缓存的 Key 为
DiscoveryConstant.DYNAMIC_PARTIAL_RULE，具体名称为 dynamic-partial-rule
        @Override
        public void setDynamicPartialRule(RuleEntity ruleEntity) {
            ruleCache.put(DiscoveryConstant.DYNAMIC_PARTIAL_RULE, ruleEntity);
            assembleDynamicRule();
        }
        //②将全局规则更新到缓存中，缓存的 Key 为
DiscoveryConstant.DYNAMIC_GLOBAL_RULE，具体名称为 dynamic-global-rule
        @Override
        public void setDynamicGlobalRule(RuleEntity ruleEntity) {
            ruleCache.put(DiscoveryConstant.DYNAMIC_GLOBAL_RULE, ruleEntity);
            assembleDynamicRule();
        }
        //③利用动态的全局规则和动态的局部规则的缓存，组装出最终的动态规则，并更新到缓存中，
缓存 Key 的名称为 dynamic-rule
        private void assembleDynamicRule() {
            RuleEntity dynamicPartialRule = getDynamicPartialRule();
            RuleEntity dynamicGlobalRule = getDynamicGlobalRule();
```

```
            RuleEntity dynamicRule =
RuleEntityWrapper.assemble(dynamicPartialRule, dynamicGlobalRule);
            ruleCache.put(DiscoveryConstant.DYNAMIC_RULE, dynamicRule);
        }
        ...
    }
```

5. 用 PluginSubscriber 类的 onRuleUpdated()方法同步地更新规则

用 PluginSubscriber 类的 onRuleUpdated()方法同步地更新规则，同步地更新规则和异步地更新规则实际上调用的都是 PluginSubscriber 类的 onRuleUpdated()方法，只是同步方式是直接调用并等待结果，异步方式则是通过 Guava 的异步事件机制异步地调用。

14.3.2 用配置中心动态修改规则

软件开发人员可以利用配置中心去动态地修改规则，并实时同步到应用中。下面结合 Nacos 来分析“用配置中心动态修改规则”的原理。

1. 在应用启动的过程中从配置中心加载规则

应用在启动的过程中会从配置中心加载规则，即用 ConfigInitializer 类的 initialize()方法从配置中心加载规则，具体代码如下所示：

```
    public class ConfigInitializer {
        @Autowired
        private PluginAdapter pluginAdapter;
        @Autowired
        private PluginConfigParser pluginConfigParser;
        @PostConstruct
        public void initialize() {
            //①如果配置文件中的“spring.application.register.control.enabled”和
“spring.application.discovery.control.enabled=true”为 true，则从配置中心远程加载
规则和从本地配置文件中加载规则，否则直接返回
            Boolean registerControlEnabled =
pluginContextAware.isRegisterControlEnabled();
            Boolean discoveryControlEnabled =
pluginContextAware.isDiscoveryControlEnabled();
            if (!registerControlEnabled && !discoveryControlEnabled) {
                LOG.info("Register and Discovery controls are all disabled, ignore
to initialize");
                return;
            }
            LOG.info("------------- Load Discovery Config --------------");
            //②从配置中心远程加载规则
            String[] remoteConfigList = getRemoteConfigList();
```

```
        if (remoteConfigList != null) {
            //③获取配置中心的局部规则
            String partialRemoteConfig = remoteConfigList[0];
            if (StringUtils.isNotEmpty(partialRemoteConfig)) {
                LOG.info("Initialize partial remote config...");
                try {
                    RuleEntity ruleEntity = pluginConfigParser.parse
(partialRemoteConfig);
                    //④用 PluginAdapter 接口的 setDynamicPartialRule()方法将局部
规则更新到缓存中
                    pluginAdapter.setDynamicPartialRule(ruleEntity);
                } catch (Exception e) {
                    LOG.error("Initialize partial remote config failed", e);
                }
            }
            //⑤获取配置中心的全局规则
            String globalRemoteConfig = remoteConfigList[1];
            if (StringUtils.isNotEmpty(globalRemoteConfig)) {
                LOG.info("Initialize global remote config...");
                try {
                    RuleEntity ruleEntity = pluginConfigParser.parse
(globalRemoteConfig);
                    //⑥用 PluginAdapter 接口的 setDynamicGlobalRule()方法将全局规
则更新到缓存中
                    pluginAdapter.setDynamicGlobalRule(ruleEntity);
                } catch (Exception e) {
                    LOG.error("Initialize global remote config failed", e);
                }
            }
        }
        //⑦从本地配置文件中获取规则
        String[] localConfigList = getLocalConfigList();
        if (localConfigList != null) {
            String localConfig = localConfigList[0];
            if (StringUtils.isNotEmpty(localConfig)) {
                LOG.info("Initialize local config...");
                try {
                    //⑧用 PluginConfigParser 类的 parse()方法解析本地配置文件，应用
可以通过属性“spring.application.config.path”配置文件路径，例如
“spring.application.config.path=classpath:rule-weight.xml”
                    RuleEntity ruleEntity = pluginConfigParser.parse
(localConfig);
                    //⑨用 PluginAdapter 接口的 setLocalRule()方法将本地规则更新到
缓存中
                    pluginAdapter.setLocalRule(ruleEntity);
```

```
                } catch (Exception e) {
                    LOG.error("Initialize local config failed", e);
                }
            }
        }
        if (remoteConfigList == null && localConfigList == null) {
            LOG.info("No configs are found");
        }
        //⑩用属性“spring.application.parameter.event.onstart.enabled”来配置是否触发 fireParameterChanged 的 EventBus 事件，默认值为 true，并触发 fireParameterChanged 的 EventBus 事件
        if (parameterEventOnstartEnabled) {
            pluginEventWapper.fireParameterChanged();
        }
        LOG.info("--------------------------------------------------");
    }
    ...
}
```

2. 用监听器订阅规则，并实时地将规则同步到应用中

Discovery 用监听器订阅规则，并实时地将规则同步到应用中。

（1）在应用启动的过程中，添加监听器全局规则和局部规则的监听器，具体代码如下所示：

```
public class NacosConfigAdapter extends ConfigAdapter {
    @Autowired
    private NacosOperation nacosOperation;
    private Listener partialListener;
    private Listener globalListener;
    //①在应用启动后，执行 subscribeConfig()方法去订阅并监听配置中心的规则
    @PostConstruct
    @Override
    public void subscribeConfig() {
        //②初始化一个局部规则的监听器
        partialListener = subscribeConfig(false);
        //③初始化一个全局规则的监听器
        globalListener = subscribeConfig(true);
    }
    private Listener subscribeConfig(boolean globalConfig) {
        //④获取 Nacos 配置中心配置文件的 group
        String group = getGroup();
        //⑤获取 Nacos 配置中心配置文件的 dataId
        String dataId = getDataId(globalConfig);
        logSubscribeStarted(globalConfig);
        try {
```

```
            //⑥用 Discovery 封装的基于 Nacos 配置中心的 NacosOperation 类的
subscribeConfig()方法去订阅和监听规则
            return nacosOperation.subscribeConfig(group, dataId,
executorService, new NacosSubscribeCallback() {
                //⑦如果配置中心的规则有变更，则通过 callback()方法回调通知应用，并执
行 callbackConfig()方法去动态地更新应用中的规则
                @Override
                public void callback(String config) {
                    callbackConfig(config, globalConfig);
                }
            });
        } catch (Exception e) {
            logSubscribeFailed(e, globalConfig);
        }

        return null;
    }
    //⑧取消订阅规则
    @Override
    public void unsubscribeConfig() {
        unsubscribeConfig(partialListener, false);
        unsubscribeConfig(globalListener, true);
        executorService.shutdownNow();
    }
    //⑨用 Discovery 封装的基于 Nacos 配置中心的 NacosOperation 类的
unsubscribeConfig()方法去取消订阅规则
    private void unsubscribeConfig(Listener listener, boolean globalConfig) {
        if (listener == null) {
            return;
        }
        String group = getGroup();
        String dataId = getDataId(globalConfig);
        logUnsubscribeStarted(globalConfig);
        try {
            nacosOperation.unsubscribeConfig(group, dataId, listener);
        } catch (Exception e) {
            logUnsubscribeFailed(e, globalConfig);
        }
    }
    ...
}
```

（2）用 ConfigAdapter 类的 callbackConfig()方法将配置中心中变更后的最新规则同步到应用中。

```
public abstract class ConfigAdapter extends RemoteConfigLoader implements
PluginConfigAdapter {
```

```
        @Autowired
    private PluginAdapter pluginAdapter;
    @Autowired
        private PluginEventWapper pluginEventWapper;

        public void callbackConfig(String config, boolean globalConfig) {
            SubscriptionType subscriptionType =
getSubscriptionType(globalConfig);
            //①如果从配置中心获取的规则不为空，则执行回调
            if (StringUtils.isNotBlank(config)) {
                logUpdatedEvent(globalConfig);
                RuleEntity ruleEntity = null;
                if (globalConfig) {
                    //②从缓存中获取全局规则
                    ruleEntity = pluginAdapter.getDynamicGlobalRule();
                } else {
                    //③从缓存中获取局部规则
                    ruleEntity = pluginAdapter.getDynamicPartialRule();
                }
                String rule = null;
                if (ruleEntity != null) {
                    //④获取规则的内容
                    rule = ruleEntity.getContent();
                }
                //⑤如果配置中心最新的规则和缓存中的规则不一致
                if (!StringUtils.equals(rule, config)) {
                    //⑥则调用fireRuleUpdated()方法发送包含最新规则的事件RuleUpdatedEvent
                    fireRuleUpdated(new RuleUpdatedEvent(subscriptionType,
config), true);
                } else {
                    //⑦否则不做更新，并记录日志
                    logUpdatedSame(globalConfig);
                }
            } else {
                //⑧如果从配置中心获取的最新的配置规则为空，说明是要删除规则，则调用
fireRuleCleared()方法发送删除规则的事件RuleClearedEvent
                logClearedEvent(globalConfig);
                fireRuleCleared(new RuleClearedEvent(subscriptionType), true);
            }
    }
    //⑨调用PluginEventWapper类的fireRuleUpdated()方法发送变更规则的事件
RuleUpdatedEvent，这样订阅事件的方法即可异步地更新规则到本地应用中
    public void fireRuleUpdated(RuleUpdatedEvent ruleUpdatedEvent, boolean
async) {
        pluginEventWapper.fireRuleUpdated(ruleUpdatedEvent, async);
```

```
    }
    //⑩调用 PluginEventWapper 类的 fireRuleCleared()方法发送删除规则的事件
RuleClearedEvent，这样订阅事件的方法即可异步地在本地应用中删除规则
    public void fireRuleCleared(RuleClearedEvent ruleClearedEvent, boolean
async) {
        pluginEventWapper.fireRuleCleared(ruleClearedEvent, async);
    }
        ...
    }
```

14.3.3 【实例】在 Spring Cloud Alibaba 应用中用 Nacos 配置中心变更规则，并验证规则动态变更的效果

本实例的源码在本书配套资源的“chapterfourteen/dynamic-routing-cloud-alibaba”目录下。

1. 初始化服务

使用 Spring Cloud Alibaba 初始化一个服务 dynamic-routing-cloud-alibaba，并添加 POM 依赖，部分配置如下，详细的配置信息可以参考本书配套资源中的代码。

```
    <!—添加 Discovery 封装的 Nacos 注册中心的 POM 依赖-->
    <dependency>
        <groupId>com.nepxion</groupId>
        <artifactId>discovery-plugin-register-center-starter-nacos</artifactId>
        <version>6.11.0</version>
    </dependency>
    <!—添加 Discovery 封装的 Nacos 配置中心的 POM 依赖-->
    <dependency>
        <groupId>com.nepxion</groupId>
<artifactId>discovery-plugin-config-center-starter-nacos</artifactId>
        <version>6.11.0</version>
    </dependency>
```

2. 添加规则

为了验证规则动态更新的效果，本实例分别在本地文件和配置中心中添加规则。

（1）在本地文件“resources/rule-count.xml”中添加规则，具体配置如下所示：

```
    <?xml version="1.0" encoding="UTF-8"?>
    <rule>
        <register>
            <count filter-value="1">
```

```
                <!-- 表示服务 dynamic-routing-cloud-alibaba，最大实例注册数为 2，全局
配置值 1 将不起作用，以局部配置值为准 -->
                <service service-name="dynamic-routing-cloud-alibaba"
filter-value="2"/>
            </count>
        </register>
    </rule>
```

（2）在 Nacos 配置中心中，添加 dataId 名称为“global-cloud-alibaba”的全局规则，以及 dataId 名称为“dynamic-routing-cloud-alibaba”的局部规则。

①配置中心全局规则的配置信息如下所示：

```
    <?xml version="1.0" encoding="UTF-8"?>
    <rule>
        <register>
            <count filter-value="2">
                <!-- 表示服务 dynamic-routing-cloud-alibaba，最大实例注册数为 3，全局
配置值 2 将不起作用，以局部配置值为准 -->
                <service service-name="dynamic-routing-cloud-alibaba"
filter-value="3"/>
            </count>
        </register>
    </rule>
```

②配置中心全局规则配置信息如下所示：

```
    <?xml version="1.0" encoding="UTF-8"?>
    <rule>
        <register>
            <count filter-value="3">
                <!-- 表示服务 dynamic-routing-cloud-alibaba，最大实例注册数为 7，全局
配置值 3 将不起作用，以局部配置值为准 -->
                <service service-name="dynamic-routing-cloud-alibaba"
filter-value="7"/>
            </count>
        </register>
    </rule>
```

规则优先级依次递增，具体如下所示：

本地规则 → 配置中心全部规则 → 配置中心局部规则

3. 启动服务，实时地读取规则，验证规则加载的优先级

在服务中用一个线程实时地读取规则，具体代码如下所示：

```
    @Component
```

```
@Configuration
public class ReadConfigService {
    @Autowired
    private PluginAdapter pluginAdapter;
    @PostConstruct
    public void readConfig(){
        ExecutorService executorService= Executors.newFixedThreadPool(1);
        executorService.execute(new ReadConfigThread());
    }

    class ReadConfigThread implements Runnable{
        @Override
        public void run() {
            while (true){
                try {
                    Thread.sleep(2000);
                }catch (InterruptedException e){
                    System.out.println(e.getMessage());
                }
                RuleEntity ruleEntity=pluginAdapter.getRule();
                if(null!=ruleEntity) {
                    Map<String, Integer> result1 =
ruleEntity.getRegisterEntity().getCountFilterEntity().getFilterMap();
                    Integer result2 =
ruleEntity.getRegisterEntity().getCountFilterEntity().getFilterValue();
                    System.out.println("基于服务注册数量的规则(指定服务名称)为:" +
result1.toString());
                    System.out.println("基于服务注册数量的规则(不指定服务名称)为:"
+ result2);
                }
            }
        }
    }
}
```

启动服务后，在 IDEA 控制台输出日志，如图 14-10 所示，服务最终加载了配置中心的局部规则。

```
基于服务注册数量的规则（指定服务名称）为:{dynamic-routing-cloud-alibaba=7}
基于服务注册数量的规则（不指定服务名称）为:3
```

图 14-10

4. 在配置中心的控制台中变更局部规则，验证动态修改规则的效果

如图 14-11 所示修改局部规则，并单击“Publish”按钮更新规则。

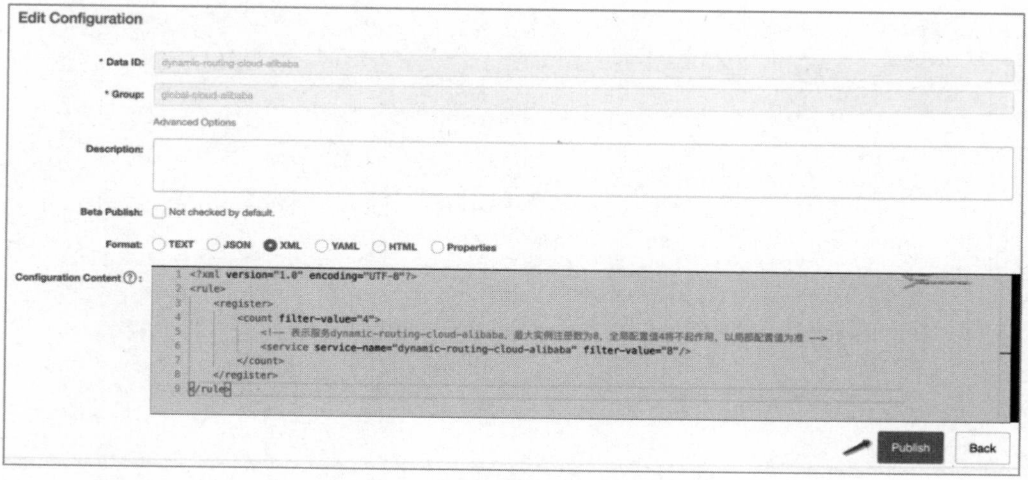

```
<?xml version="1.0" encoding="UTF-8"?>
<rule>
    <register>
        <count filter-value="4">
            <!-- 表示服务dynamic-routing-cloud-alibaba，最大实例注册数为8，全局配置值4将不起作用，以局部配置值为准 -->
            <service service-name="dynamic-routing-cloud-alibaba" filter-value="8"/>
        </count>
    </register>
</rule>
```

图 14-11

在 IDEA 控制台中实时地输出日志，如图 14-12 所示，说明规则已经动态地更新到服务中。

```
2021-08-28 13:50:43.214  INFO 91086 --- [12.176-thread-1] c.n.d.p.f.parser.xml.XmlConfigParser      : Rule content=
<?xml version="1.0" encoding="UTF-8"?>
<rule>
    <register>
        <count filter-value="4">
            <!-- 表示服务dynamic-routing-cloud-alibaba，最大实例注册数为8，全局配置值4将不起作用，以局部配置值为准 -->
            <service service-name="dynamic-routing-cloud-alibaba" filter-value="8"/>
        </count>
    </register>
</rule>
基于服务注册数量的规则（指定服务名称）为:{dynamic-routing-cloud-alibaba=8}
基于服务注册数量的规则（不指定服务名称）为:4
```

图 14-12

14.4 “用服务注册/订阅实现服务的路由”的原理

Discovery 支持服务在注册/订阅过程中完成服务的路由。服务的路由就是指“服务路由”，下面可能会不统一，是细微语境的不同。

“服务注册”支持以下的服务路由策略。

- 黑/白名单：在服务注册的过程中，可以基于 IP 地址进行过滤，从而实现服务的路由。
- 服务注册数量的限制：在服务注册的过程中，可以限制指定服务名称的实例注册的数量，从而实现服务的路由。

“服务订阅”支持以下的服务路由策略。

- 黑/白名单：在服务订阅的过程中，可以基于 IP 地址进行过滤，从而实现服务的路由。
- 版本号：在服务订阅的过程中，可以指定服务提供者和服务消费者的版本号，并通过版本号完成服务过滤，从而实现服务的路由。
- 区域：在服务订阅的过程中，可以指定服务提供者和服务消费者的区域，并通过区域完成服务过滤，从而实现服务的路由。
- 权重：在服务订阅的过程中，可以指定服务提供者和服务消费者的权重，并通过权重完成服务过滤，从而实现服务的路由。

14.4.1　用“服务注册的前置处理和注册监听器”实现基于服务注册的服务路由

1. 服务注册的前置处理

Discovery 为了实现基于服务注册的服务路由，采用 NacosServiceRegistryDecorator 类的 register()方法进行服务注册的前置处理，具体代码如下所示：

```
public class NacosServiceRegistryDecorator extends NacosServiceRegistry {
    private NacosServiceRegistry serviceRegistry;
    private ConfigurableApplicationContext applicationContext;
    private ConfigurableEnvironment environment;
    @Override
    public void register(Registration registration) {
        //①如果“spring.application.register.control.enabled”为 true，则开启
基于服务注册的灰度路由
        Boolean registerControlEnabled =
PluginContextAware.isRegisterControlEnabled(environment);
        if (registerControlEnabled) {
            try {
                //②从 Spring Framework 的 IOC 容器中获取 Discovery 的注册监听执行器
对象 RegisterListenerExecutor
                RegisterListenerExecutor registerListenerExecutor =
applicationContext.getBean(RegisterListenerExecutor.class);
                //③调用 RegisterListenerExecutor 类的 onRegister()方法执行服务注
册的前置处理
                registerListenerExecutor.onRegister(registration);
            } catch (BeansException e) {
            }
        }
        //④调用 Spring Cloud Alibaba 的 NacosServiceRegistry 类的 register()方
法实现服务的注册
        serviceRegistry.register(registration);
    }
    ...
}
```

2. 利用自动配置类 PluginAutoConfiguration，初始化前置处理的注册监听器

Discovery 利用自动配置类 PluginAutoConfiguration 初始化前置处理的注册监听器，具体代码如下所示：

```
@Configuration
@EnableEventBus
public class PluginAutoConfiguration {
    //①初始化注册监听执行器 RegisterListenerExecutor 类
    @Bean
    public RegisterListenerExecutor registerListenerExecutor() {
        return new RegisterListenerExecutor();
    }
    //②初始化限制服务注册数量的注册监听器 CountFilterRegisterListener 类
    @Bean
    @ConditionalOnMissingBean
    public CountFilterRegisterListener countFilterRegisterListener() {
        return new CountFilterRegisterListener();
    }
    //③初始化服务注册的黑/白名单注册监听器 HostFilterDiscoveryListener 类
    @Bean
    @ConditionalOnMissingBean
    public HostFilterDiscoveryListener hostFilterDiscoveryListener() {
        return new HostFilterDiscoveryListener();
    }
    ...
}
```

3. 绑定 Discovery 的注册监听器 RegisterListener 类和 Spring Cloud 的 Registration 类

用 RegisterListenerExecutor 类绑定 Discovery 的注册监听器 RegisterListener 和 Spring Cloud 的 Registration 类，具体代码如下所示：

```
public class RegisterListenerExecutor {
    //①加载 Spring Framework 中注册监听器 RegisterListener 接口的实现对象，比如
CountFilterRegisterListener 对象
    @Autowired
    private List<RegisterListener> registerListenerList;
    public void onRegister(Registration registration) {
    //②遍历注册监听器列表，调用监听器的 onRegister()方法去绑定 Spring Cloud 的
Registration 类
        for (RegisterListener registerListener : registerListenerList) {
            registerListener.onRegister(registration);
        }
    }
    ...
}
```

在 Discovery 中，默认存在以下两种注册监听器。

（1）CountFilterRegisterListener：限制服务注册数量的注册监听器。它在服务注册之前，从注册中心获取已经注册成功的服务实例数量，如果这个服务数量大于或者等于配置文件中配置的允许注册的服务实例的阈值，则直接抛出异常，终止本次服务注册，具体代码如下所示：

```
public class CountFilterRegisterListener extends AbstractRegisterListener {
    @Autowired
    @Lazy
    protected DiscoveryClientDecorator discoveryClient;
    @Override
    public void onRegister(Registration registration) {
        //①获取需要注册的服务 ID，通常都是应用中配置的服务名称
        String serviceId = pluginAdapter.getServiceId();
        //②获取需要注册服务的 IP 地址
        String host = pluginAdapter.getHost();
        //③获取需要注册服务的端口号
        int port = pluginAdapter.getPort();
        applyCountFilter(serviceId, host, port);
    }
    private void applyCountFilter(String serviceId, String host, int port) {
        //④用 PluginAdapter 接口的实现类的 getRule()方法获取灰度路由规则（如果采用
Nacos 作为注册中心，则实现类是 NacosAdapter）
        RuleEntity ruleEntity = pluginAdapter.getRule();
        if (ruleEntity == null) {
            return;
        }
        RegisterEntity registerEntity = ruleEntity.getRegisterEntity();
        if (registerEntity == null) {
            return;
        }
        //⑤获取包含服务注册数量阈值的对象 CountFilterEntity
        CountFilterEntity countFilterEntity =
registerEntity.getCountFilterEntity();
        if (countFilterEntity == null) {
            return;
        }
        Integer globalFilterValue = countFilterEntity.getFilterValue();
        Map<String, Integer> filterMap = countFilterEntity.getFilterMap();
        Integer filterValue = filterMap.get(serviceId);
        //⑥如果局部值存在，则取局部值，否则取全局值
        Integer maxCount = null;
        if (filterValue != null) {
            maxCount = filterValue;
        } else {
```

```
            maxCount = globalFilterValue;
        }
        if (maxCount == null) {
            return;
        }
        //⑦用 DiscoveryClientDecorator 类的 getRealInstances()方法从注册中心获取服务名称为 serviceId 的实例节点数
        int count = discoveryClient.getRealInstances(serviceId).size();
        //⑧如果注册中心的实例节点数大于或者等于配置文件中配置的实例节点的阈值，则直接抛出异常，终止本次服务注册
        if (count >= maxCount) {
            onRegisterFailure(maxCount, serviceId, host, port);
        }
    }
    //⑨抛出异常
    private void onRegisterFailure(int maxCount, String serviceId, String host, int port) {
        String description = serviceId + " for " + host + ":" + port + " is rejected to register to Register server, reach max limited count=" + maxCount;
        pluginEventWapper.fireRegisterFailure(new RegisterFailureEvent(DiscoveryConstant.REACH_MAX_LIMITED_COUNT, description, serviceId, host, port));
        throw new DiscoveryException(description);
    }
    @Override
    public int getOrder() {
        //⑩由于通过服务数来判断是否注册满，所以需要第一优先级执行“限制服务注册数量的注册监听器”，否则服务列表已经被其他注册监听器过滤过了，其数目就不准确了
        return HIGHEST_PRECEDENCE;
    }
    ...
}
```

（2）HostFilterRegisterListener：服务注册的黑/白名单注册监听器，如果指定前缀的 IP 地址被设置为黑名单，则终止服务注册，具体代码如下所示：

```
public class HostFilterRegisterListener extends AbstractRegisterListener {
    @Override
    public void onRegister(Registration registration) {
        //①获取需要注册服务的服务名称、IP 地址及端口号
        String serviceId = pluginAdapter.getServiceId();
        String host = pluginAdapter.getHost();
        int port = pluginAdapter.getPort();
        applyHostFilter(serviceId, host, port);
    }
    private void applyHostFilter(String serviceId, String host, int port) {
```

```
        //②用 NacosAdapter 类的 getRule()方法获取路由规则
        RuleEntity ruleEntity = pluginAdapter.getRule();
        if (ruleEntity == null) {
            return;
        }
        RegisterEntity registerEntity = ruleEntity.getRegisterEntity();
        if (registerEntity == null) {
            return;
        }
        //③获取包含黑/白名单地址的对象 HostFilterEntity
        HostFilterEntity hostFilterEntity =
registerEntity.getHostFilterEntity();
        if (hostFilterEntity == null) {
            return;
        }
        //④获取过滤类型，比如黑名单 BLACKLIST 或者白名单 WHITELIST
        FilterType filterType = hostFilterEntity.getFilterType();
        //⑤获取全局黑/白名单规则
        List<String> globalFilterValueList =
hostFilterEntity.getFilterValueList();
        //⑥获取指定服务名称的局部黑/白名单规则
        Map<String, List<String>> filterMap =
hostFilterEntity.getFilterMap();
        List<String> filterValueList = filterMap.get(serviceId);
        if (CollectionUtils.isEmpty(globalFilterValueList) &&
CollectionUtils.isEmpty(filterValueList)) {
            return;
        }
        List<String> allFilterValueList = new ArrayList<String>();
        if (CollectionUtils.isNotEmpty(globalFilterValueList)) {
            allFilterValueList.addAll(globalFilterValueList);
        }
        if (CollectionUtils.isNotEmpty(filterValueList)) {
            allFilterValueList.addAll(filterValueList);
        }
        switch (filterType) {
            //⑦如果过滤类型是黑名单，则校验黑名单
            case BLACKLIST:
                validateBlacklist(filterType, allFilterValueList, serviceId,
host, port);
                break;
            case WHITELIST:
                //⑧如果过滤类型是白名单，则校验白名单
                validateWhitelist(filterType, allFilterValueList, serviceId,
host, port);
```

```
                break;
            }
        }
        //⑨校验黑名单，如果需要注册的服务的 IP 地址的前缀能够匹配黑名单中的规则之一，则直接抛出异常，终止本次服务注册
        private void validateBlacklist(FilterType filterType, List<String> allFilterValueList, String serviceId, String host, int port) {
            for (String filterValue : allFilterValueList) {
                if (host.startsWith(filterValue)) {
                    onRegisterFailure(filterType, allFilterValueList, serviceId, host, port);
                }
            }
        }
        //⑩校验白名单，如果需要注册的服务的 IP 地址的前缀能够匹配白名单中的规则之一，则可以注册；如果所有的白名单规则都不匹配，则直接抛出异常，终止本次服务注册
        private void validateWhitelist(FilterType filterType, List<String> allFilterValueList, String serviceId, String host, int port) {
            boolean matched = true;
            for (String filterValue : allFilterValueList) {
                if (host.startsWith(filterValue)) {
                    matched = false;
                    break;
                }
            }
            if (matched) {
                onRegisterFailure(filterType, allFilterValueList, serviceId, host, port);
            }
        }
        //抛出异常
        private void onRegisterFailure(FilterType filterType, List<String> allFilterValueList, String serviceId, String host, int port) {
            String description = serviceId + " for " + host + ":" + port + " is rejected to register to Register server, not match host " + filterType + "=" + allFilterValueList;
            pluginEventWapper.fireRegisterFailure(new RegisterFailureEvent(filterType.toString(), description, serviceId, host, port));
            throw new DiscoveryException(description);
        }
    ...
    }
```

14.4.2 用“服务订阅前置处理 + 注册监听器”实现基于服务订阅的服务路由

1. 服务订阅的前置处理

Discovery 用包装器 DiscoveryClientDecorator 类来重写 Spring Cloud 的 DiscoveryClient 类的获取服务实例列表的方法 getInstances()和获取服务信息列表的方法 getServices()，并完成服务订阅的前置处理，具体代码如下所示：

```
    public class DiscoveryClientDecorator implements DiscoveryClient,
DiscoveryClientDelegate<DiscoveryClient> {
        ...
        @Override
        public List<ServiceInstance> getInstances(String serviceId) {
            //①使用 Spring Cloud 的服务订阅客户端 DiscoveryClient，获取服务 ID 的实例列表
            List<ServiceInstance> instances = getRealInstances(serviceId);
            //②获取配置信息“spring.application.discovery.control.enabled”的值，
如果为 true，则执行服务订阅的前置处理，否则直接返回从注册中心获取的实例列表
            Boolean discoveryControlEnabled =
PluginContextAware.isDiscoveryControlEnabled(environment);
            if (discoveryControlEnabled) {
                try {
                    //③从 Spring Framework 的 IOC 容器中获取服务订阅监听执行器的对象
DiscoveryListenerExecutor
                    DiscoveryListenerExecutor discoveryListenerExecutor =
applicationContext.getBean(DiscoveryListenerExecutor.class);
                    //④调用 DiscoveryListenerExecutor 类的 onGetInstances()方法，执
行基于服务订阅的灰度路由，过滤服务实例
                    discoveryListenerExecutor.onGetInstances(serviceId,
instances);
                } catch (BeansException e) {
                }
            }
            return instances;
        }
        //⑤从注册中心获取服务实例列表
        public List<ServiceInstance> getRealInstances(String serviceId) {
            return discoveryClient.getInstances(serviceId);
        }
        @Override
        public List<String> getServices() {
            //⑥使用 Spring Cloud 的服务订阅客户端 DiscoveryClient，获取所有的服务信息列表
            List<String> services = getRealServices();
            //⑦获取配置信息“spring.application.discovery.control.enabled”的值，
如果为 true，则执行服务订阅的前置处理，否则直接返回从注册中心获取的所有服务信息列表
            Boolean discoveryControlEnabled =
```

```
PluginContextAware.isDiscoveryControlEnabled(environment);
        if (discoveryControlEnabled) {
            try {
                //⑧从 Spring Framework 的 IOC 容器中获取服务订阅监听执行器的对象
DiscoveryListenerExecutor
                DiscoveryListenerExecutor discoveryListenerExecutor =
applicationContext.getBean(DiscoveryListenerExecutor.class);
                //⑨调用 DiscoveryListenerExecutor 类的 onGetServices()方法，执行
基于服务订阅的灰度路由，过滤服务信息
                discoveryListenerExecutor.onGetServices(services);
            } catch (BeansException e) {
            }
        }
        return services;
    }
    //⑩从注册中心获取服务信息列表
    public List<String> getRealServices() {
        return discoveryClient.getServices();
    }
    ...
}
```

2. 初始化前置处理的订阅监听器

Discovery 用自动配置类 PluginAutoConfiguration 初始化前置处理的订阅监听器，具体代码如下所示：

```
@Configuration
@EnableEventBus
public class PluginAutoConfiguration {
    //①初始化服务订阅监听执行器类 DiscoveryListenerExecutor
    @Bean
    public DiscoveryListenerExecutor discoveryListenerExecutor() {
        return new DiscoveryListenerExecutor();
    }
    //②初始化基于服务订阅的黑/白名单的过滤监听器
    @Bean
    @ConditionalOnMissingBean
    public HostFilterDiscoveryListener hostFilterDiscoveryListener() {
        return new HostFilterDiscoveryListener();
    }
    //③初始化基于服务订阅的版本号的过滤监听器
    @Bean
    @ConditionalOnMissingBean
    public VersionFilterDiscoveryListener versionFilterDiscoveryListener() {
        return new VersionFilterDiscoveryListener();
```

```
    }
    //④初始化基于服务订阅的区域的过滤监听器
    @Bean
    @ConditionalOnMissingBean
    public RegionFilterDiscoveryListener regionFilterDiscoveryListener() {
        return new RegionFilterDiscoveryListener();
    }
    ...
}
```

3. 过滤指定服务 ID 的服务实例和服务信息

Discovery 用服务订阅监听执行器类 DiscoveryListenerExecutor，过滤指定服务 ID 的服务实例和服务信息，具体代码如下所示：

```
public class DiscoveryListenerExecutor {
    //①加载 Spring Framework 中所有 DiscoveryListener 接口的实现类的对象，比如
HostFilterDiscoveryListener 对象
    @Autowired
    private List<DiscoveryListener> discoveryListenerList;
    //②执行监听器的 onGetInstances()方法，按照规则过滤服务实例
    public void onGetInstances(String serviceId, List<ServiceInstance>
instances) {
        for (DiscoveryListener discoveryListener : discoveryListenerList) {
            discoveryListener.onGetInstances(serviceId, instances);
        }
    }
    //③执行监听器的 onGetServices()方法，按照规则过滤服务信息
    public void onGetServices(List<String> services) {
        for (DiscoveryListener discoveryListener : discoveryListenerList) {
            discoveryListener.onGetServices(services);
        }
    }
}
```

在 Discovery 中，默认存在以下 3 种服务订阅监听器。

（1）HostFilterDiscoveryListener：订阅服务的黑/白名单监听器，具体代码如下所示：

```
public class HostFilterDiscoveryListener extends AbstractDiscoveryListener {
    @Override
    public void onGetInstances(String serviceId, List<ServiceInstance>
instances) {
        applyHostFilter(serviceId, instances);
    }

    private void applyHostFilter(String providerServiceId,
```

```
List<ServiceInstance> instances) {
          //①获取动态规则
          RuleEntity ruleEntity = pluginAdapter.getRule();
          if (ruleEntity == null) {
             return;
          }
          DiscoveryEntity discoveryEntity = ruleEntity.getDiscoveryEntity();
          if (discoveryEntity == null) {
             return;
          }
          //②获取黑/白名单规则
          HostFilterEntity hostFilterEntity =
discoveryEntity.getHostFilterEntity();
          if (hostFilterEntity == null) {
             return;
          }
          FilterType filterType = hostFilterEntity.getFilterType();
          //③获取全局规则
          List<String> globalFilterValueList =
hostFilterEntity.getFilterValueList();
          Map<String, List<String>> filterMap = hostFilterEntity.getFilterMap();
          //④获取局部规则
          List<String> filterValueList = filterMap.get(providerServiceId);
          if (CollectionUtils.isEmpty(globalFilterValueList) &&
CollectionUtils.isEmpty(filterValueList)) {
             return;
          }
          List<String> allFilterValueList = new ArrayList<String>();
          if (CollectionUtils.isNotEmpty(globalFilterValueList)) {
             allFilterValueList.addAll(globalFilterValueList);
          }
          if (CollectionUtils.isNotEmpty(filterValueList)) {
             allFilterValueList.addAll(filterValueList);
          }
          Iterator<ServiceInstance> iterator = instances.iterator();
          while (iterator.hasNext()) {
             ServiceInstance instance = iterator.next();
             String host = instance.getHost();
             switch (filterType) {
                //⑤校验黑名单规则
                case BLACKLIST:
                   if (validateBlacklist(allFilterValueList, host)) {
                      //⑥如果 IP 地址匹配黑名单规则，则删除服务实例
                      iterator.remove();
                   }
```

```
                    break;
                //⑦校验白名单规则
                case WHITELIST:
                    if (validateWhitelist(allFilterValueList, host)) {
                        //⑧如果 IP 地址不匹配白名单规则，则删除服务实例
                        iterator.remove();
                    }
                    break;
            }
        }
    }
    private boolean validateBlacklist(List<String> allFilterValueList,
String host) {
        //⑨如果黑名单规则中的IP地址前缀匹配需要订阅的服务的IP地址的前缀,则返回true,
否则返回 false
        for (String filterValue : allFilterValueList) {
            if (host.startsWith(filterValue)) {
                return true;
            }
        }
        return false;
    }
    private boolean validateWhitelist(List<String> allFilterValueList,
String host) {
        //⑩如果需要订阅的服务的IP 地址匹配白名单规则的 IP 地址前缀列表中的一个，则返回
false，否则返回 true
        boolean matched = true;
        for (String filterValue : allFilterValueList) {
            if (host.startsWith(filterValue)) {
                matched = false;
                break;
            }
        }
        return matched;
    }
    ...
}
```

（2）VersionFilterDiscoveryListener：基于版本号的服务订阅监听器，具体代码如下所示：

```
public class VersionFilterDiscoveryListener extends
AbstractDiscoveryListener {
    @Override
    public void onGetInstances(String serviceId, List<ServiceInstance>
instances) {
        //①获取当前服务消费者的服务 ID
```

```
        String consumerServiceId = pluginAdapter.getServiceId();
        //②获取当前服务消费者的版本号
        String consumerServiceVersion = pluginAdapter.getVersion();
        applyVersionFilter(consumerServiceId, consumerServiceVersion,
serviceId, instances);
    }

    private void applyVersionFilter(String consumerServiceId, String
consumerServiceVersion, String providerServiceId, List<ServiceInstance>
instances) {
        ...
        //③获取版本规则列表
        List<VersionEntity> versionEntityList =
versionEntityMap.get(consumerServiceId);
        if (CollectionUtils.isEmpty(versionEntityList)) {
            return;
        }
        List<String> allNoFilterValueList = null;
        boolean providerConditionDefined = false;
        //④遍历版本规则列表
        for (VersionEntity versionEntity : versionEntityList) {
            String providerServiceName =
versionEntity.getProviderServiceName();
            if (StringUtils.equalsIgnoreCase(providerServiceName,
providerServiceId)) {
                providerConditionDefined = true;
                //⑤获取规则中的服务消费者的版本号列表
                List<String> consumerVersionValueList =
versionEntity.getConsumerVersionValueList();
                //⑥获取规则中服务提供者的版本号列表
                List<String> providerVersionValueList =
versionEntity.getProviderVersionValueList();
                if (CollectionUtils.isNotEmpty(consumerVersionValueList)) {
                    //⑦如果规则中的服务消费者版本列表包含当前服务消费者的版本号，则向
allNoFilterValueList 列表中添加规则中服务提供者的版本号
                    if
(consumerVersionValueList.contains(consumerServiceVersion)) {
                        if (allNoFilterValueList == null) {
                            allNoFilterValueList = new ArrayList<String>();
                        }
                        if
(CollectionUtils.isNotEmpty(providerVersionValueList)) {

allNoFilterValueList.addAll(providerVersionValueList);
                        }
```

```
                    }
                } else {
                    if (allNoFilterValueList == null) {
                        allNoFilterValueList = new ArrayList<String>();
                    }
                    if (CollectionUtils.isNotEmpty(providerVersionValueList)) {
                        allNoFilterValueList.
addAll(providerVersionValueList);
                    }
                }
            }
        }
        //⑧如果 allNoFilterValueList 不为空，则遍历服务实例列表，按照规则进行过滤
        if (allNoFilterValueList != null) {
            if (allNoFilterValueList.isEmpty()) {
                return;
            } else {
                Iterator<ServiceInstance> iterator = instances.iterator();
                while (iterator.hasNext()) {
                    ServiceInstance instance = iterator.next();
                    String instanceVersion =
pluginAdapter.getInstanceVersion(instance);
                    //⑨如果服务实例的版本号不在规则的服务提供者的版本号列表中，则删除
该服务实例
                    if (!allNoFilterValueList.contains(instanceVersion)) {
                        iterator.remove();
                    }
                }
            }
        } else {
            //⑩如果 allNoFilterValueList 为 null，则意味着定义的版本关系都不匹配，
直接清空所有的服务实例
            if (providerConditionDefined) {
                instances.clear();
            }
        }
    }
    ...
}
```

（3）RegionFilterDiscoveryListener：基于区域的服务订阅监听器，具体代码如下所示：

```
public class RegionFilterDiscoveryListener extends
AbstractDiscoveryListener {
    ...
    @Override
```

```
    public void onGetInstances(String serviceId, List<ServiceInstance>
instances) {
        //①获取当前服务消费者 ID
        String consumerServiceId = pluginAdapter.getServiceId();
        //②获取当前服务消费者的区域
        String consumerServiceRegion = pluginAdapter.getRegion();
        applyRegionFilter(consumerServiceId, consumerServiceRegion,
serviceId, instances);
    }

    private void applyRegionFilter(String consumerServiceId, String
consumerServiceRegion, String providerServiceId, List<ServiceInstance>
instances) {
        ...
        RuleEntity ruleEntity = pluginAdapter.getRule();
        if (ruleEntity == null) {
            return;
        }
        ...
        //③获取区域规则
        List<RegionEntity> regionEntityList =
regionEntityMap.get(consumerServiceId);
        if (CollectionUtils.isEmpty(regionEntityList)) {
            return;
        }

        //④计算出当前区域的消费端所能调用提供端的区域号列表
        List<String> allNoFilterValueList = null;
        //⑤提供端规则未作任何定义
        boolean providerConditionDefined = false;
        for (RegionEntity regionEntity : regionEntityList) {
            String providerServiceName =
regionEntity.getProviderServiceName();
            if (StringUtils.equalsIgnoreCase(providerServiceName,
providerServiceId)) {
                providerConditionDefined = true;

                List<String> consumerRegionValueList =
regionEntity.getConsumerRegionValueList();
                List<String> providerRegionValueList =
regionEntity.getProviderRegionValueList();
                //⑥判断 consumer-region-value 的值是否包含当前消费端的区域号，如果
consumerRegionValueList 为空，则表示消费端区域列表未指定，那么任意消费端区域可以访问指定
区域提供端区域
                if (CollectionUtils.isNotEmpty(consumerRegionValueList)) {
```

```
                    if
(consumerRegionValueList.contains(consumerServiceRegion)) {
                        if (allNoFilterValueList == null) {
                            allNoFilterValueList = new ArrayList<String>();
                        }
                        if
(CollectionUtils.isNotEmpty(providerRegionValueList)) {
                            allNoFilterValueList.
addAll(providerRegionValueList);
                        }
                    }
                } else {
                    if (allNoFilterValueList == null) {
                        allNoFilterValueList = new ArrayList<String>();
                    }
                    if (CollectionUtils.isNotEmpty(providerRegionValueList)) {
                        allNoFilterValueList.
addAll(providerRegionValueList);
                    }
                }
            }
        }

        if (allNoFilterValueList != null) {
            //⑦如果 allNoFilterValueList 为空列表，则意味着区域对应关系未做任何定义（即所有的 providerRegionValueList 为空），不需要执行过滤，直接返回
            if (allNoFilterValueList.isEmpty()) {
                return;
            } else {
                Iterator<ServiceInstance> iterator = instances.iterator();
                while (iterator.hasNext()) {
                    ServiceInstance instance = iterator.next();
                    String instanceRegion =
pluginAdapter.getInstanceRegion(instance);
                    //⑧如果服务提供者的区域不在区域规则的列表中，则直接删除服务实例
                    if (!allNoFilterValueList.contains(instanceRegion)) {
                        iterator.remove();
                    }
                }
            }
        } else {
            if (providerConditionDefined) {
                //⑨如果 allNoFilterValueList 为 null，则意味着定义的区域关系都不匹配，直接清空所有实例
                instances.clear();
```

```
            }
        }
    }
    ...
}
```

14.4.3 【实例】在 Spring Cloud Alibaba 应用中配置服务注册的路由规则

本实例的源码在本书配套资源的“chapterfourteen/register-gray-route-alibaba”目录下。

本实例在 Spring Cloud Alibaba 应用中配置服务注册的路由规则，并演示服务路由的效果。

1. 初始化服务

使用 Spring Cloud Alibaba 初始化一个服务 register-gray-route-trade-server，并添加 POM 依赖，部分配置信息如下，详细的配置信息可以参考本书配套资源中的代码。

```
<!—添加 Discovery 封装的 Nacos 注册中心的 POM 依赖-->
<dependency>
    <groupId>com.nepxion</groupId>
    <artifactId>discovery-plugin-register-center-starter-nacos</artifactId>
        <version>6.11.0</version>
    </dependency>
<!—添加 Discovery 封装的 Spring Cloud Gateway 的 POM 依赖-->
<dependency>
    <groupId>com.nepxion</groupId>
    <artifactId>discovery-plugin-strategy-starter-gateway</artifactId>
    <version>6.11.0</version>
</dependency>
```

2. 添加服务配置信息和灰度路由规则

在服务 register-gray-route-trade-server 中添加服务配置信息和路由规则，部分配置信息和路由规则如下。

（1）基于注册服务数量的路由规则如下（文件路径为“resources/ rule-count.xml”）。

```
<?xml version="1.0" encoding="UTF-8"?>
<rule>
    <register>
        <count filter-value="1">
            <!-- 表示服务 register-gray-route-trade-server，最大实例注册数为 2，
全局配置值 1 将不起作用，以局部配置值为准 -->
```

```
            <service service-name=" register-gray-route-trade-server "
filter-value="2"/>
         </count>
      </register>
   </rule>
```

（2）基于黑名单的路由规则如下（文件路径为“resources/ rule-blacklist.xml”）。

```
   <?xml version="1.0" encoding="UTF-8"?>
   <rule>
      <register>
         <blacklist filter-value="192.168">
            <!-- 表示服务 register-gray-route-trade-server，不允许以“192.168”
为前缀的 IP 地址注册 -->
            <service service-name="register-gray-route-trade-server"
filter-value="192.168"/>
         </blacklist>
      </register>
   </rule>
```

（3）基于白名单的路由规则如下（文件路径为“resources/ rule-whitelist.xml”）。

```
   <?xml version="1.0" encoding="UTF-8"?>
   <rule>
      <register>
         <whitelist filter-value="192.167">
            <!-- 表示服务 register-gray-route-trade-server，只允许以“192.167”
为前缀的 IP 地址注册 -->
            <service service-name="register-gray-route-trade-server"
filter-value="192.167"/>
         </whitelist>
      </register>
   </rule>
```

3. 演示服务路由效果

下面演示“注册服务数量的路由规则”的路由效果。

（1）使用 Maven 的打包命令“mvn clean install -Dmaven.test.skip=true”，生成部署 Jar 包“register-gray-route-trade-server-1.0.0.release.jar”。

（2）执行启动服务的 3 条命令，如果命令都执行成功，则启动 3 个服务进程，具体命令如下。

```
   nohup java -jar register-gray-route-trade-server-1.0.0.release.jar
--server.port=8023 > nohup-8023.log &
   nohup java -jar register-gray-route-trade-server-1.0.0.release.jar
--server.port=8024 > nohup-8024.log &
```

```
nohup java -jar register-gray-route-trade-server-1.0.0.release.jar
--server.port=8025 > nohup-8025.log &
```

执行完前两个命令后，可以在 Nacos 控制台看到端口号为 8023 和 8024 的服务进程已经注册成功，如图 14-13 所示。

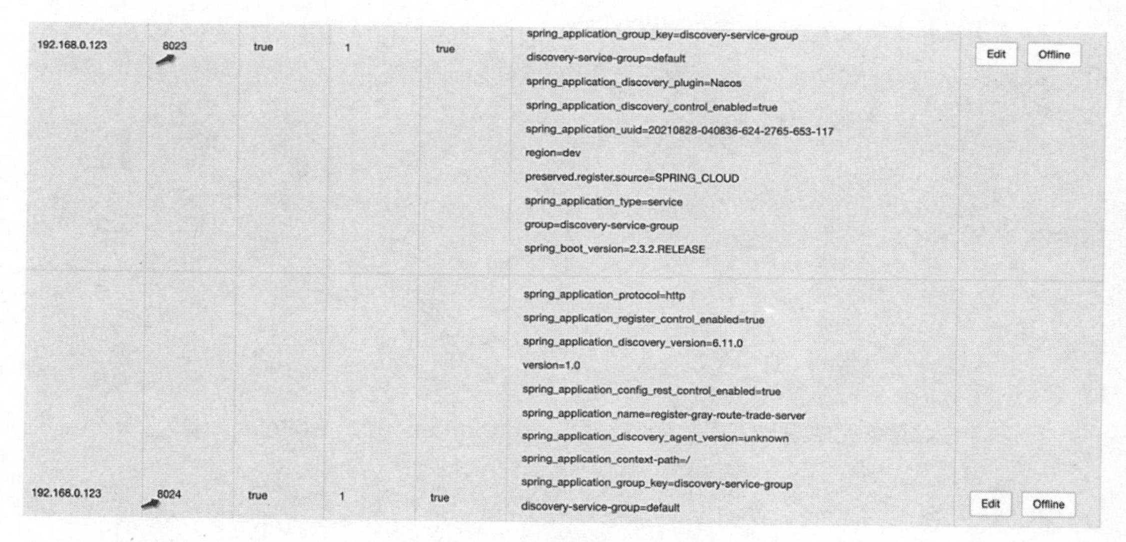

图 14-13

执行第 3 个命令注册端口号为 8025 的服务进程，服务启动日志如下：

```
    2021-08-28 04:21:08.970 ERROR 87799---[main] o.s.boot.SpringApplication:
    Application run failed
org.springframework.context.ApplicationContextException: Failed to start bean
    'webServerStartStop'; nested exception is
    com.nepxion.discovery.common.exception.DiscoveryException:
    register-gray-route-trade-server for 192.168.0.123:8025 is rejected to
    register to Register server, reach max limited count=2
```

从日志中可以看出，注册的实例数已经达到上限数，服务在启动的过程中报错，并打印了错误信息。

14.4.4 【实例】在 Spring Cloud Alibaba 应用中配置服务订阅的路由规则

> 本实例的源码在本书配套资源的“chapterfourteen/discovery-gray-route-alibaba”目录下。

本实例在 Spring Cloud Alibaba 应用中配置服务订阅的路由规则，并演示服务路由的效果。本实例包括如下 4 个服务。

- analog-gateway-traffic-server：模拟用户调用网关的流量，并计算服务路由的调用比例。
- discovery-gray-route-gateway-server：基于 Spring Cloud Gateway 的网关服务。
- discovery-gray-route-good-service-server：模拟电商业务中的商品服务。
- discovery-gray-route-trade-service-server：模拟电商业务中的订单服务。

1. 初始化服务

使用 Spring Cloud Alibaba 初始化 4 个服务，并添加 POM 依赖，部分配置信息如下，详细的配置信息可以参考本书配套资源中的代码。

```
<!—添加 Discovery 封装的 Nacos 注册中心的 POM 依赖-->
<dependency>
    <groupId>com.nepxion</groupId>
    <artifactId>discovery-plugin-register-center-starter-nacos</artifactId>
        <version>6.11.0</version>
    </dependency>
<!—添加 Discovery 封装的 Spring Cloud Gateway 的 POM 依赖-->
<dependency>
    <groupId>com.nepxion</groupId>
    <artifactId>discovery-plugin-strategy-starter-gateway</artifactId>
    <version>6.11.0</version>
</dependency>
```

2. 添加服务配置信息和路由规则

在 discovery-gray-route-gateway-server 、 discovery-gray-route-trade-service-server 和 discovery-gray-route-good-service-server 服务中添加服务配置信息和灰度路由策略。其中，discovery-gray-route-gateway-server 的部分配置信息和路由规则如下，其他服务可以参考本书配套资源中的代码。

（1）基于区域的服务路由规则（配置文件的路径为“resources/rule-region.xml”）。

```
<?xml version="1.0" encoding="UTF-8"?>
    <rule>
        <discovery
            <region>
        ①<!-- 表示 dev 区域的消费端服务 discovery-gray-route-gateway-server，允许访
问 dev 区域的提供端服务 discovery-gray-route-trade-service-server-->
    <service consumer-service-name="discovery-gray-route-gateway-server"
provider-service-name="discovery-gray-route-trade-service-server"
consumer-region-value="dev" provider-region-value="dev"/>
```

```
        ②<!-- 表示 qa 区域的消费端服务 discovery-gray-route-gateway-server，允许访问 qa 区域的提供端服务 discovery-gray-route-trade-service-server-->
        <service consumer-service-name="discovery-gray-route-gateway-server"
provider-service-name="discovery-gray-route-trade-service-server"
consumer-region-value="qa" provider-region-value="qa"/>
        ③<!-- 表示 pre 区域的消费端服务 discovery-gray-route-gateway-server，允许访问 pre 区域的提供端服务 discovery-gray-route-trade-service-server-->
                <service
consumer-service-name="discovery-gray-route-gateway-server"
provider-service-name="discovery-gray-route-trade-service-server"
consumer-region-value="pre" provider-region-value="pre"/>
        ④<!-- 表示 online 区域的消费端服务 discovery-gray-route-gateway-server，允许访问 online 区域的提供端服务 discovery-gray-route-trade-service-server-->
    <service consumer-service-name="discovery-gray-route-gateway-server"
provider-service-name="discovery-gray-route-trade-service-server"
consumer-region-value="online" provider-region-value="online"/>
            </region>
        </discovery>
    </rule>
```

（2）基于版本的服务路由规则（配置文件的路径为“resources/rule-version.xml”）。

```
    <?xml version="1.0" encoding="UTF-8"?>
        <rule>
            <discovery>
                <version>
        ①<!-- 表示网关 discovery-gray-route-gateway-server 的 1.0，允许访问提供端服务 discovery-gray-route-trade-service-server 的 1.0 版本 -->
    <service consumer-service-name="discovery-gray-route-gateway-server"
provider-service-name="discovery-gray-route-trade-service-server"
consumer-version-value="1.0" provider-version-value="1.0"/>
    ②<!-- 表示网关 discovery-gray-route-gateway-server 的 1.1，允许访问提供端服务 discovery-gray-route-trade-service-server 的 1.1 版本 -->
    <service consumer-service-name="discovery-gray-route-gateway-server"
provider-service-name="discovery-gray-route-trade-service-server"
consumer-version-value="1.1" provider-version-value="1.1"/>
            </version>
        </discovery>
    </rule>
```

（3）基于权重的服务路由规则（配置文件的路径为“resources/rule-weight.xml”）。

```
    <?xml version="1.0" encoding="UTF-8"?>
        <rule>
            <discovery>
                <weight>
        ①<!--基于版本的权重的服务路由规则-->
```

```
    ②<!-- 表示在消费端服务 discovery-gray-route-gateway-server 访问提供端服务
discovery-gray-route-trade-service-server 时，提供端服务
discovery-gray-route-trade-service-server 的 1.0 版本提供 90%的流量,1.1 版本提供 10%
的流量 -->
   <service consumer-service-name="discovery-gray-route-gateway-server"
provider-service-name="discovery-gray-route-trade-service-server"
provider-weight-value="1.0=90;1.1=10" type="version"/>
    ③<!--基于区域的权重的服务路由规则-->
    ④<!-- 表示在消费端服务 discovery-gray-route-gateway-server 访问提供端服务
discovery-gray-route-trade-service-server 时，提供端服务
discovery-gray-route-trade-service-server 的 dev 区域提供 90%的流量，qa 区域提供 10%
的流量 -->
      <service consumer-service-name="discovery-gray-route-gateway-server"
provider-service-name="discovery-gray-route-trade-service-server"
provider-weight-value="dev=90;qa=10" type="region"/>
         </weight>
      </discovery>
   </rule>
```

（4）基于黑/白名单的灰度路由策略（配置文件的路径为“resources/rule-weight. xml”）。

```
   <?xml version="1.0" encoding="UTF-8"?>
      <rule>
         <discovery>
   <!—IP 地址的前缀为“192.168”的机器不能订阅服务 discovery-gray-route-trade-
service-server -->
         <blacklist filter-value="192.168">
            <service service-name="discovery-gray-route-trade-service-
server" filter-value="192.168"/>
         </blacklist>
      </discovery>
   </rule>
```

3. 演示服务路由效果

（1）在服务中 analog-gateway-traffic-server 中模拟用户调用网关的流量，具体代码如下所示：

```
@Configuration
public class StarterConfig {
    @Bean
    public RestTemplate restTemplate(){
        return new RestTemplate();
    }
    @PostConstruct
    public void init() {
        execute();
```

```
    }
    @Resource
    private RestTemplate restTemplate;

    private AtomicInteger atomicInteger6756=new AtomicInteger();
    private AtomicInteger atomicInteger8030=new AtomicInteger();

    public void execute(){
        //①用 10000 次请求来模拟用户访问网关的流量
        for(int i=0; i<10000;i++){
            String result = restTemplate.getForObject("http://127.0.0.1:
28988/discovery-gray-route-trade-service-server/tradeManager/buygood",
                    String.class);
            System.out.println("结果为: " + result);
            //②记录服务提供者对应的实例被调用的次数，比如实例“127.0.0.1:6756”和
“127.0.0.1:8030”
            if(result.contains("6756")) {
                atomicInteger6756.incrementAndGet();
            }else{
                atomicInteger8030.incrementAndGet();
            }
            Integer totalCount=atomicInteger6756.get()+
atomicInteger8030.get();
            double ratio1=atomicInteger6756.get()*1.0d/totalCount;
            double ratio2=atomicInteger8030.get()*1.0d/totalCount;
            //③统计服务实例被调用比例
            System.out.println("抽样调用"+i+"次");
            System.out.println("127.0.0.1:6756 调用比例为: "+ratio1);
            System.out.println("127.0.0.1:8030 调用比例为: "+ratio2);
        }
    }
}
```

（2）分别启动 discovery-gray-route-gateway-server、discovery-gray-route-trade-service-server 和 discovery-gray-route-good-service-server 服务，其中，discovery-gray-route-trade-service-server 服务启动两个实例：127.0.0.1:6756 和 127.0.0.1:8030。

（3）修改网关 discovery-gray-route-gateway-server 服务中配置文件（文件路径为“resources/bootstrap.properties”）中的属性字段，开启权重灰度路由策略。具体如下：

```
spring.application.config.path=classpath:rule-weight.xml
```

在 Nacos 注册中心控制台上，修改服务节点“127.0.0.1:8030”的版本号为“1.1”，修改区域为“qa”，如图 14-14 所示。启动服务 analog-gateway-traffic-server，观察统计服务实例被调用的比例。部分统计结果如图 14-15 所示，在抽样 62 次后实例“127.0.0.1:6756”的流量比

例为“0.90476”，接近配置的灰度流量“0.9”；在抽样 62 次后实例“127.0.0.1:8030”的流量比例为“0.09523”，接近配置的灰度流量“0.1”。同理，可以开启版本和区域的路由策略，并验证实例“127.0.0.1:8030”和“127.0.0.1:8030”的调用比例，本书就不做统一的效果演示了。

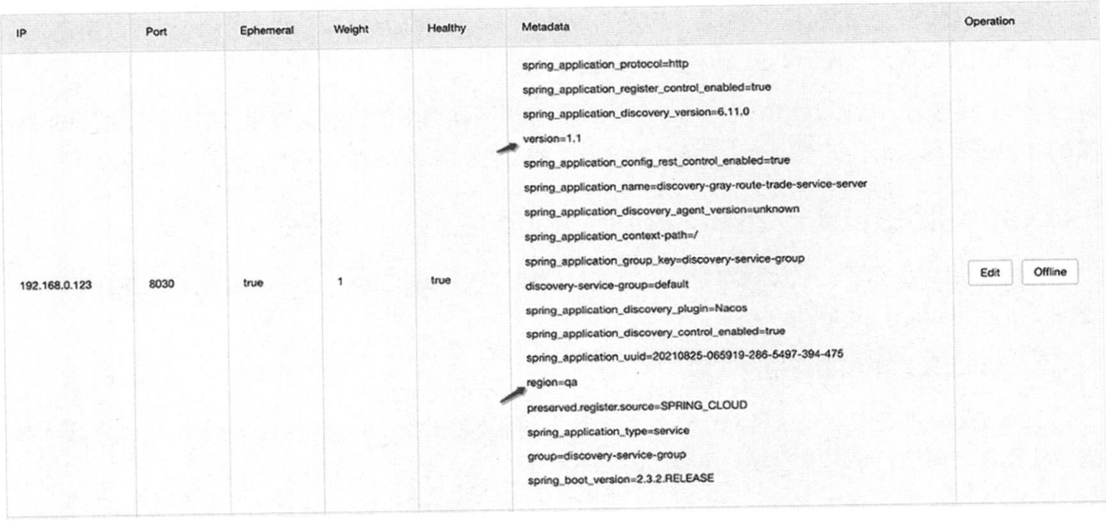

IP	Port	Ephemeral	Weight	Healthy	Metadata	Operation
192.168.0.123	8030	true	1	true	spring_application_protocol=http spring_application_register_control_enabled=true spring_application_discovery_version=6.11.0 version=1.1 spring_application_config_rest_control_enabled=true spring_application_name=discovery-gray-route-trade-service-server spring_application_discovery_agent_version=unknown spring_application_context-path=/ spring_application_group_key=discovery-service-group discovery-service-group=default spring_application_discovery_plugin=Nacos spring_application_discovery_control_enabled=true spring_application_uuid=20210825-065919-286-5497-394-475 region=qa preserved.register.source=SPRING_CLOUD spring_application_type=service group=discovery-service-group spring_boot_version=2.3.2.RELEASE	Edit Offline

图 14-14

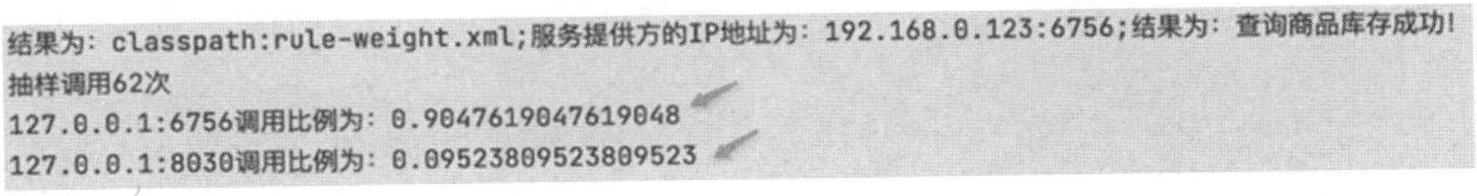

```
结果为: classpath:rule-weight.xml;服务提供方的IP地址为: 192.168.0.123:6756;结果为: 查询商品库存成功!
抽样调用62次
127.0.0.1:6756调用比例为: 0.9047619047619048
127.0.0.1:8030调用比例为: 0.09523809523809523
```

图 14-15

（4）修改网关 discovery-gray-route-gateway-server 服务中配置文件（文件路径为“resources/bootstrap.properties”）中的属性字段，开启黑/白名单灰度路由策略。具体如下：

```
spring.application.config.path=classpath:rule-white-black.xml
```

在重启网关 discovery-gray-route-gateway-server 服务后，服务 analog-gateway-traffic-server 在调用网关时如果出现如下日志，则说明黑/白名单配置已经生效（服务 discovery-gray-route-trade-service-server 不允许 IP 地址前缀为“192.168”的机器实例订阅）。

```
2021-08-25 07:42:38.962  WARN 60904 ---
[ctor-http-nio-2] .p.f.l.l.NotificationLoadBalanceListener : ********** No
server instances found for serviceId=discovery-gray-route-trade-service-server,
perhaps they are isolated, filtered or not registered **********
```

14.5 “用路由过滤器实现全链路的蓝绿发布和灰度发布”的原理

Discovery 支持 3 种路由过滤器：适配 Spring Cloud Gateway 网关的路由过滤器、适配 Zuul 网关的路由过滤器、适配 RESTful API 的路由过滤器。

这里使用 Spring Cloud Gateway 作为网关分析两种路由过滤器的原理，不分析适配 Zuul 网关的路由过滤器原理。如果读者对适配 Zuul 网关的路由过滤器感兴趣，可以查阅相关源码。

14.5.1 用路由过滤器适配 Spring Cloud Gateway 网关

Discovery 定义了一个负载均衡前置路由过滤器“DefaultGatewayStrategyRouteFilter”以适配 Spring Cloud Gateway 网关。

1. 启动负载均衡前置路由过滤器

Discovery 基于 Spring Boot 定义了一个自动配置类 GatewayStrategyAutoConfiguration 来启动负载均衡前置路由过滤器，具体代码如下所示：

```
@Configuration
@AutoConfigureBefore(RibbonClientConfiguration.class)
public class GatewayStrategyAutoConfiguration {
    @Bean
    @ConditionalOnMissingBean
    public GatewayStrategyRouteFilter gatewayStrategyRouteFilter() {
        //将 DefaultGatewayStrategyRouteFilter 对象注入 Spring Framework 的 IOC
容器中，启动路由过滤器，并拦截请求网关的服务请求
        return new DefaultGatewayStrategyRouteFilter();
    }
    ...
}
```

2. 处理 Header 参数

用路由过滤器 AbstractGatewayStrategyRouteFilter 类的 filter()方法处理 Header 参数，其中，AbstractGatewayStrategyRouteFilter 类继承 Spring Cloud Gateway 的全局过滤器接口 GlobalFilter，并实现它的 filter()方法。即利用 Spring Cloud Gateway 全局过滤器先拦截调用网关的请求，然后增加规则过滤的逻辑。具体实现过程如下。

（1）处理内部的 Header 参数的转发，具体代码如下所示：

```
public abstract class AbstractGatewayStrategyRouteFilter implements
GatewayStrategyRouteFilter {
```

```
    ...
    @Override
    public Mono<Void> filter(ServerWebExchange exchange, GatewayFilterChain 
chain) {
        //①把 ServerWebExchange 放入 ThreadLocal
        GatewayStrategyContext.getCurrentContext().setExchange(exchange);
        //②通过过滤器设置路由 Header 信息，并传递到链路中的各个服务器端
        ServerHttpRequest request = exchange.getRequest();
        ServerHttpRequest.Builder requestBuilder = request.mutate();
        //③处理内部 Header 参数的转发
        applyInnerHeader(request, requestBuilder);
        ...
    }
    private void applyInnerHeader(ServerHttpRequest request, 
ServerHttpRequest.Builder requestBuilder) {
        //④设置 Group，Key 为“n-d-service-group”
        GatewayStrategyFilterResolver.setHeader(request, requestBuilder, 
DiscoveryConstant.N_D_SERVICE_GROUP, pluginAdapter.getGroup(), 
gatewayHeaderPriority);
        //⑤设置服务的类型，Key 为“n-d-service-type”
        GatewayStrategyFilterResolver.setHeader(request, requestBuilder, 
DiscoveryConstant.N_D_SERVICE_TYPE, pluginAdapter.getServiceType(), false);
        String serviceAppId = pluginAdapter.getServiceAppId();
        if (StringUtils.isNotEmpty(serviceAppId)) {
            GatewayStrategyFilterResolver.setHeader(request, requestBuilder, 
DiscoveryConstant.N_D_SERVICE_APP_ID, serviceAppId, false);
        }
        //⑥设置服务 ID，Key 为“n-d-service-id”
        GatewayStrategyFilterResolver.setHeader(request, requestBuilder, 
DiscoveryConstant.N_D_SERVICE_ID, pluginAdapter.getServiceId(), false);
        //⑦设置服务 IP 地址，Key 为“n-d-service-address”
        GatewayStrategyFilterResolver.setHeader(request, requestBuilder, 
DiscoveryConstant.N_D_SERVICE_ADDRESS, pluginAdapter.getHost() + ":" + 
pluginAdapter.getPort(), false);
        //⑧设置服务版本，Key 为“n-d-service-version”
        GatewayStrategyFilterResolver.setHeader(request, requestBuilder, 
DiscoveryConstant.N_D_SERVICE_VERSION, pluginAdapter.getVersion(), false);
        //⑨设置服务范围，Key 为“n-d-service-region”
        GatewayStrategyFilterResolver.setHeader(request, requestBuilder, 
DiscoveryConstant.N_D_SERVICE_REGION, pluginAdapter.getRegion(), false);
        //⑩设置服务环境信息，Key 为“n-d-service-env”
        GatewayStrategyFilterResolver.setHeader(request, requestBuilder, 
DiscoveryConstant.N_D_SERVICE_ENVIRONMENT, pluginAdapter.getEnvironment(), 
false);
        //设置服务区域，Key 为“n-d-service-zone”
```

```
            GatewayStrategyFilterResolver.setHeader(request, requestBuilder,
DiscoveryConstant.N_D_SERVICE_ZONE, pluginAdapter.getZone(), false);
        }
        ...
    }
```

（2）处理外部的 Header 参数的转发，具体代码如下所示：

```
    //①处理调用网关的服务请求的 Header 的转发，将调用网关的服务请求传递过来的 Header 传递到下游服务
    private void applyOuterHeader(ServerHttpRequest request,
ServerHttpRequest.Builder requestBuilder) {
        String routeEnvironment = getRouteEnvironment();
        //②设置路由环境信息
        if (StringUtils.isNotEmpty(routeEnvironment)) {
            GatewayStrategyFilterResolver.setHeader(request, requestBuilder,
DiscoveryConstant.N_D_ENVIRONMENT, routeEnvironment, false);
        }
        //③如果设置参数“gatewayCoreHeaderTransmissionEnabled”为 true，则不属于蓝绿灰度范畴的 Header
        if (gatewayCoreHeaderTransmissionEnabled) {
            Map<String, String> headerMap = strategyWrapper.getHeaderMap();
            if (MapUtils.isNotEmpty(headerMap)) {
                for (Map.Entry<String, String> entry : headerMap.entrySet()) {
                    String key = entry.getKey();
                    String value = entry.getValue();
                    GatewayStrategyFilterResolver.setHeader(request,
requestBuilder, key, value, gatewayHeaderPriority);
                }
            }
            String routeVersion = getRouteVersion();
            ...
            //④将配置中心的路由版本号设置到请求链路中
            if (StringUtils.isNotEmpty(routeVersion)) {
                GatewayStrategyFilterResolver.setHeader(request, requestBuilder,
DiscoveryConstant.N_D_VERSION, routeVersion, gatewayHeaderPriority);
            }
            ...
        } else {
            //⑤当核心 Header 的传值开关关闭时，执行忽略 Header 设置的相关逻辑
            GatewayStrategyFilterResolver.ignoreHeader(requestBuilder,
DiscoveryConstant.N_D_VERSION);
            ...
        }
    }
```

14.5.2　用路由过滤器适配 RESTful API

Discovery 定义了一个负载均衡前置路由过滤器 DefaultServiceStrategyRouteFilter 来适配 RESTful API。

1. 启动负载均衡前置路由过滤器

用自动配置类 ServiceStrategyAutoConfiguration 启动负载均衡前置路由过滤器的具体代码如下所示：

```
@Configuration
@AutoConfigureBefore(RibbonClientConfiguration.class)
public class ServiceStrategyAutoConfiguration {
    @Bean
    @ConditionalOnMissingBean
    public ServiceStrategyRouteFilter serviceStrategyRouteFilter() {
        return new DefaultServiceStrategyRouteFilter();
    }
    ...
}
```

2. 处理 Header 参数

用路由过滤器 AbstractServiceStrategyRouteFilter 抽象类的 doFilterInternal ()方法处理 Header 参数，其中，AbstractServiceStrategyRouteFilter 抽象类继承 Spring Web 的抽象过滤器类 OncePerRequestFilter。即利用 Spring Web 自带的过滤器先拦截 RESTful API 请求，然后增加规则过滤的逻辑。具体实现过程如下。

（1）Discovery 处理内部 Header 的转发的具体代码如下所示：

```
public abstract class AbstractServiceStrategyRouteFilter extends
ServiceStrategyRouteFilter {
    @Override
    protected void doFilterInternal(HttpServletRequest request,
HttpServletResponse response, FilterChain filterChain) throws ServletException,
IOException {
        //①如果当前请求的 URL 在过滤拦截请求的白名单中，则退出过滤，直接返回
        boolean isExclusion =
serviceStrategyFilterExclusion.isExclusion(request, response);
        if (isExclusion) {
            filterChain.doFilter(request, response);
            return;
        }
        // ②通过过滤器设置路由 Header 头部信息，并传递到链路中的各个服务器端
```

```
        ServiceStrategyRouteFilterRequest
serviceStrategyRouteFilterRequest = new
ServiceStrategyRouteFilterRequest(request);
        //③处理内部 Header 的转发
        applyInnerHeader(serviceStrategyRouteFilterRequest);
        ...
    }
    //④处理内部 Header 的转发，即把本地服务的相关属性封装成 Header 转发到下游服务
    private void applyInnerHeader(ServiceStrategyRouteFilterRequest
serviceStrategyRouteFilterRequest) {
        //⑤通过 Spring OncePerRequestFilter 实现 Header 前置拦截外部 Header，对于
内部 Header 则不需要处理
    }
    ...
}
```

（2）Discovery 处理外部 Header 的转发的具体代码如下所示：

```
//①处理外部 Header 的转发，即将外部服务传递过来的 Header 参数转发到下游服务
private void applyOuterHeader(ServiceStrategyRouteFilterRequest
serviceStrategyRouteFilterRequest) {
    String routeEnvironment = getRouteEnvironment();
    if (StringUtils.isNotEmpty(routeEnvironment)) {
    //②设置环境信息
    ServiceStrategyFilterResolver.setHeader
(serviceStrategyRouteFilterRequest, DiscoveryConstant.N_D_ENVIRONMENT,
routeEnvironment, false);
    }
    if (feignCoreHeaderTransmissionEnabled ||
restTemplateCoreHeaderTransmissionEnabled) {
        Map<String, String> headerMap = strategyWrapper.getHeaderMap();
        if (MapUtils.isNotEmpty(headerMap)) {
            for (Map.Entry<String, String> entry : headerMap.entrySet()) {
                String key = entry.getKey();
                String value = entry.getValue();
                ServiceStrategyFilterResolver.setHeader
(serviceStrategyRouteFilterRequest, key, value, serviceHeaderPriority);
            }
        }
        //③获取路由版本号
        String routeVersion = getRouteVersion();
        ...
        if (StringUtils.isNotEmpty(routeVersion)) {
        //④将版本号设置到服务请求头中
  ServiceStrategyFilterResolver.setHeader(serviceStrategyRouteFilterReques
t, DiscoveryConstant.N_D_VERSION, routeVersion, serviceHeaderPriority);
```

```
        }
        ...
    }
}
```

14.5.3 【实例】在 Spring Cloud Alibaba 应用中配置全链路灰度发布的规则，并验证全链路灰度发布的效果

本实例的源码在本书配套资源的“chapterfourteen/ gray-publish-cloud-alibaba”目录下。

本实例总共包括如下 6 个服务。

- gray-publish-trade-server：模拟电商中交易服务。
- gray-publish-order-server：模拟电商中的订单服务。
- gray-publish-good-server：模拟电商中的商品服务。
- gray-publish-logistics-server：模拟电商中的物流服务。
- gray-publish-gateway-server：模拟业务网关。
- gray-publish-gateway-traffic-server：模拟用户调用业务网关的请求流量。

1. 初始化服务

使用 Spring Cloud Alibaba 初始化 6 个服务，并添加 POM 依赖，部分配置信息如下，详细的配置信息可以参考本书配套资源中的代码。

```
    <!–添加 Discovery 封装的 Nacos 注册中心的 POM 依赖-->
    <dependency>
        <groupId>com.nepxion</groupId>
        <artifactId>discovery-plugin-register-center-starter-nacos
</artifactId>
            <version>6.11.0</version>
        </dependency>
    <!–添加 Discovery 封装的 Spring Cloud Gateway 的 POM 依赖-->
    <dependency>
        <groupId>com.nepxion</groupId>
        <artifactId>discovery-plugin-strategy-starter-gateway</artifactId>
        <version>6.11.0</version>
    </dependency>
    <!–添加 Nacos 配置中心的 POM 依赖-->
    <dependency>
        <groupId>com.alibaba.cloud</groupId>
        <artifactId>spring-cloud-starter-alibaba-nacos-config</artifactId>
```

```
    <version>2.2.5.RELEASE</version>
</dependency>
```

2. 添加服务配置信息和灰度路由规则

在网关服务 gray-publish-gateway-server 中添加配置信息和灰度路由规则，其他服务可以参考本书配套资源中的代码。

（1）在文件 application.properties 和 bootstrap.properties 中的部分配置信息如下：

```
#规则文件的格式，支持 XML 和 JSON 格式，如缺失则默认为 XML 格式
spring.application.config.format=xml
#规则文件的路径
spring.application.config.path=classpath:rule-gray-publish-version.xml
#服务的版本号为 1.0
spring.cloud.nacos.discovery.metadata.version=1.0
...
```

（2）在文件 rule-gray-publish-version.xml 中添加灰度路由规则，部分配置信息如下：

```
<?xml version="1.0" encoding="UTF-8"?>
<rule>
    <strategy-release>
        <conditions type="gray">
            <!--灰度路由 1，用条件 expression 驱动灰度路由，其中，灰度流量为 10%，稳定
流量为 90%-->
             <condition id="gray-condition-1" expression="#H['a'] == '1'"
version-id="gray-route=10;stable-route=90"/>
            <!--灰度路由 2，用条件 expression 驱动灰度路由，其中，灰度流量为 20%，稳定
流量为 80%-->
            <condition id="gray-condition-2" expression="#H['a'] == '1' and
#H['b'] == '2'" version-id="gray-route=20;stable-route=80"/>
            <!--灰度路由 3，用条件 expression 驱动灰度路由，其中，灰度流量为 50%，稳定
流量为 50%-->
            <condition id="gray-condition-3" expression="#H['a'] == '1' and
#H['b'] == '3'" version-id="gray-route=50;stable-route=50"/>
            <!--灰度路由 4，用条件 expression 驱动灰度路由，其中，灰度流量为 80%，稳定
流量为 20%-->
            <condition id="gray-condition-4" expression="#H['a'] == '1' and
#H['b'] == '4'" version-id="gray-route=80;stable-route=20"/>
            <!--灰度路由 5，用条件 expression 驱动灰度路由，其中，灰度流量为 90%，稳定
流量为 10%-->
            <condition id="gray-condition-5" expression="#H['a'] == '1' and
#H['b'] == '5'" version-id="gray-route=90;stable-route=10"/>
            <!--灰度路由 6，用条件 expression 驱动灰度路由，其中，灰度流量为 100%，稳
定流量为 0% -->
            <condition id="gray-condition-6" expression="#H['a'] == '1' and
```

```
#H['b'] == '6'" version-id="gray-route=100;stable-route=0"/>
            <!-- 兜底路由，用条件 expression 驱动兜底路由，其中，稳定流量为 100%，灰度
流量为 0% -->
            <condition id="basic-condition"
version-id="gray-route=0;stable-route=100"/>
        </conditions>
        <routes>
            <!--将版本号为 1.1 的服务设置为灰度服务-->
            <route id="gray-route"
type="version">{"gray-publish-gateway-server":"1.1",
"gray-publish-trade-server":"1.1", "gray-publish-order-server":"1.1",
                "gray-publish-good-server":"1.1","gray-publish-logistics-
server":"1.1"}</route>
            <!--将版本号为 1.0 的服务设置为稳定服务-->
            <route id="stable-route"
type="version">{"gray-publish-gateway-server":"1.0",
"gray-publish-trade-server":"1.0", "gray-publish-order-server":"1.0",
                "gray-publish-good-server":"1.0","gray-publish-logistics-
server":"1.0"}</route>
        </routes>
    </strategy-release>
</rule>
```

3. 演示全链路灰度发布的效果

（1）为了演示全链路灰度发布的效果，这里统一采用动态开关切换“带有条件的灰度规则”，具体配置如下：

```
#表示统一使用灰度路由 1，其中，灰度流量为 10%，稳定流量为 90%
gray.publish.traffic.grayType=gray-condition-1
gray.publish.trade.grayType=gray-condition-1
gray.publish.order.grayType=gray-condition-1
gray.publish.good.grayType=gray-condition-1
```

（2）在服务 gray-publish-gateway-traffic-server 中统计交易服务 gray-publish-trade-server 的流量占比，具体代码如下所示：

```
private AtomicInteger atomicIntegerGray=new AtomicInteger();
private AtomicInteger atomicIntegerStable=new AtomicInteger();
public void execute() {
    try {
        HttpHeaders headers = new HttpHeaders();
        if (grayConfig.getLabel().equals("open")) {
            //①如果动态开关设置为“gray-condition-1”，则设置请求头中的“a=1”
            if (grayConfig.getGrayType().equals("gray-condition-1")) {
                headers.add("a", "1");
```

```
            } else if (grayConfig.getGrayType().equals("gray-condition-2")) {
            //②如果动态开关设置为"gray-condition-2",则设置请求头中的"a=1"和"b=2"
                headers.add("a", "1");
                headers.add("b", "2");
            } else if (grayConfig.getGrayType().equals("gray-condition-3")) {
            //③如果动态开关设置为"gray-condition-3",则设置请求头中的"a=1"和"b=3"
                headers.add("a", "1");
                headers.add("b", "3");
            } else if (grayConfig.getGrayType().equals("gray-condition-4")) {
            //④如果动态开关设置为"gray-condition-4",则设置请求头中的"a=1"和"b=4"
                headers.add("a", "1");
                headers.add("b", "4");
            } else if (grayConfig.getGrayType().equals("gray-condition-5")) {
            //⑤如果动态开关设置为"gray-condition-5",则设置请求头中的"a=1"和"b=5"
                headers.add("a", "1");
                headers.add("b", "5");
            } else if (grayConfig.getGrayType().equals("gray-condition-6")) {
            //⑥如果动态开关设置为"gray-condition-6",则设置请求头中的"a=1"和"b=6"
                headers.add("a", "1");
                headers.add("b", "6");
            } else if (grayConfig.getGrayType().equals("basic-condition")) {
                    //do nothing
            }
        }
        String url = "";
        //⑦轮询网关
        if (null != grayConfig.getServer()
&& !"".equals(grayConfig.getServer())) {
            String[] serverArrays = grayConfig.getServer().split(";");
            Integer size = serverArrays.length;
            Integer randomIndex = RandomUtils.nextInt() % size;
            url = serverArrays[randomIndex];
        }
        ResponseEntity<String> result = restTemplate.exchange(url,
HttpMethod.GET,
                new HttpEntity<String>(null, headers), String.class);
        System.out.println("结果为: " + result.getBody());
        String applicationName = "gray-publish-order-server";
        //⑧统计灰度路由流量和稳定路由流量
        if (null != result.getBody() && !"".equals(result.getBody())) {
            if (result.getBody().contains(applicationName + "," +
"GrayRoute")) {
                atomicIntegerGray.incrementAndGet();
            } else if (result.getBody().contains(applicationName + "," +
"StableRoute")) {
```

```
                atomicIntegerStable.incrementAndGet();
            }
        }
        //⑨换算出服务 gray-publish-gateway-traffic-server 调用交易服务
gray-publish-order-server 的流量比例
        Integer totalCount = atomicIntegerGray.get() +
atomicIntegerStable.get();
        double ratio1 = atomicIntegerGray.get() * 1.0d / totalCount;
        double ratio2 = atomicIntegerStable.get() * 1.0d / totalCount;
        System.out.println("抽样调用" + totalCount + "次");
        System.out.println("灰度路由调用比例为：" + ratio1);
        System.out.println("稳定路由调用比例为：" + ratio2);
    } catch (Exception e) {
        System.out.println(e.getMessage());
    }
}
```

在服务 gray-publish-trade-server 中，统计订单服务 gray-publish-order-server 的流量占比的实现可以参考本书配套资源中的代码。

在服务 gray-publish-order-server 中，统计商品服务 gray-publish-good-server 的流量占比的实现可以参考本书配套资源中的代码。

在服务 gray-publish-good-server 中，统计物流服务 gray-publish-logistics-server 的流量占比的实现可以参考本书配套资源中的代码。

（3）执行脚本命令去启动全链路灰度发布中的所有服务。

①启动服务 gray-publish-trade-server。

```
#服务 gray-publish-trade-server，启动一个灰度进程（服务版本号为 1.1）和两个稳定进程（服务版本号为 1.0）
nohup java -jar gray-publish-trade-server-1.0.0.release.jar
--server.port=8020  --spring.cloud.nacos.discovery.metadata.version=1.0 >
nohup-8020.log &
nohup java -jar gray-publish-trade-server-1.0.0.release.jar
--server.port=8021  --spring.cloud.nacos.discovery.metadata.version=1.0 >
nohup-8021.log &
nohup java -jar gray-publish-trade-server-1.0.0.release.jar
--server.port=8022  --spring.cloud.nacos.discovery.metadata.version=1.1 >
nohup-8022.log &
```

②启动服务 gray-publish-order-server。

```
#服务 gray-publish-order-server，启动一个灰度进程（服务版本号为 1.1）和两个稳定进程（服务版本号为 1.0）
```

```
    nohup java -jar gray-publish-order-server-1.0.0.release.jar
--server.port=8023  --spring.cloud.nacos.discovery.metadata.version=1.0 >
nohup-8023.log &
    nohup java -jar gray-publish-order-server-1.0.0.release.jar
--server.port=8024  --spring.cloud.nacos.discovery.metadata.version=1.0 >
nohup-8024.log &
    nohup java -jar gray-publish-order-server-1.0.0.release.jar
--server.port=8025  --spring.cloud.nacos.discovery.metadata.version=1.1 >
nohup-8025.log &
```

③启动服务 gray-publish-good-server。

```
    #服务 gray-publish-good-server，启动一个灰度进程（服务版本号为 1.1）和两个稳定进程
（服务版本号为 1.0）
    nohup java -jar gray-publish-good-server-1.0.0.release.jar
--server.port=8026  --spring.cloud.nacos.discovery.metadata.version=1.0 >
nohup-8026.log &
    nohup java -jar gray-publish-good-server-1.0.0.release.jar
--server.port=8027  --spring.cloud.nacos.discovery.metadata.version=1.0 >
nohup-8027.log &
    nohup java -jar gray-publish-good-server-1.0.0.release.jar
--server.port=8028  --spring.cloud.nacos.discovery.metadata.version=1.1 >
nohup-8028.log &
```

④启动服务 gray-publish-logistics-server。

```
    #服务 gray-publish-logistics-server，启动一个灰度进程（服务版本号为 1.1）和两个稳
定进程（服务版本号为 1.0）
    nohup java -jar gray-publish-logistics-server-1.0.0.release.jar
--server.port=8029  --spring.cloud.nacos.discovery.metadata.version=1.0 >
nohup-8029.log &
    nohup java -jar gray-publish-logistics-server-1.0.0.release.jar
--server.port=8030  --spring.cloud.nacos.discovery.metadata.version=1.0 >
nohup-8030.log &
    nohup java -jar gray-publish-logistics-server-1.0.0.release.jar
--server.port=8031  --spring.cloud.nacos.discovery.metadata.version=1.1 >
nohup-8031.log &
```

⑤启动服务 gray-publish-gateway-server。

```
    #服务 gray-publish-gateway-server，启动一个灰度进程（服务版本号为 1.1）和两个稳定进
程（服务版本号为 1.0）
    nohup java -jar gray-publish-gateway-server-1.0.0.release.jar
--server.port=8032  --spring.cloud.nacos.discovery.metadata.version=1.0 >
nohup-8032.log &
    nohup java -jar gray-publish-gateway-server-1.0.0.release.jar
--server.port=8033  --spring.cloud.nacos.discovery.metadata.version=1.0 >
nohup-8033.log &
```

```
    nohup java -jar gray-publish-gateway-server-1.0.0.release.jar
--server.port=8034  --spring.cloud.nacos.discovery.metadata.version=1.1 >
nohup-8034.log &
```

在 Nacos 注册中心控制台可以看到服务启动的结果，如图 14-16 所示。

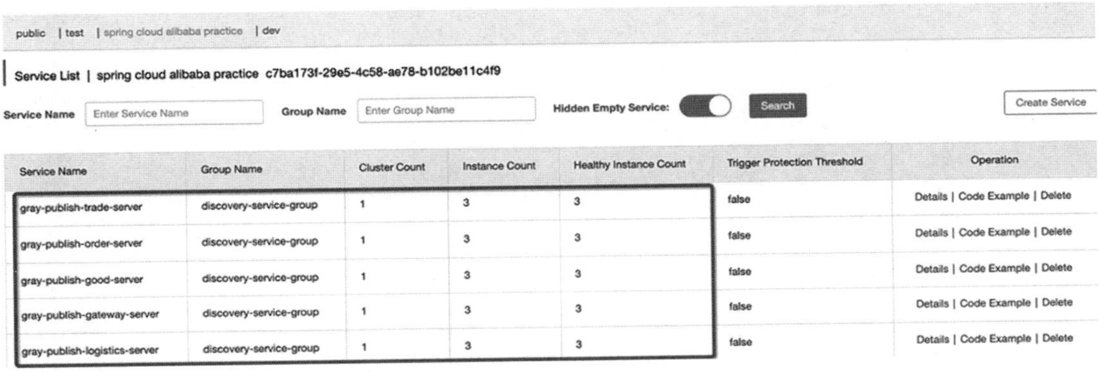

public | test | spring cloud alibaba practice | dev

Service List | spring cloud alibaba practice c7ba173f-29e5-4c58-ae78-b102be11c4f9

Service Name [Enter Service Name] Group Name [Enter Group Name] Hidden Empty Service: Search Create Service

Service Name	Group Name	Cluster Count	Instance Count	Healthy Instance Count	Trigger Protection Threshold	Operation
gray-publish-trade-server	discovery-service-group	1	3	3	false	Details | Code Example | Delete
gray-publish-order-server	discovery-service-group	1	3	3	false	Details | Code Example | Delete
gray-publish-good-server	discovery-service-group	1	3	3	false	Details | Code Example | Delete
gray-publish-gateway-server	discovery-service-group	1	3	3	false	Details | Code Example | Delete
gray-publish-logistics-server	discovery-service-group	1	3	3	false	Details | Code Example | Delete

图 14-16

（4）在 Nacos 控制台切换灰度路由条件，观察路由流量比例。本书只展示交易服务 gray-publish-trade-server 的流量占比，其他服务读者可以运行本书配套资源中的代码，观察流量占比。

①统一设置灰度路由条件为 gray-condition-1，其中，灰度流量为 10%，稳定流量为 90%。如图 14-17 所示，流量占比为：灰度路由调用比例为 0.1，稳定路由调用比例为 0.9，符合灰度规则。

```
结果为：交易服务路由类型：gray-publish-trade-server,StableRoute;classpath:rule-gray-publish-version.xml;
抽样调用10次
灰度路由调用比例为：0.1
稳定路由调用比例为：0.9
```

图 14-17

②统一设置灰度路由条件为 gray-condition-2，其中，灰度流量为 20%，稳定流量为 80%，如图 14-18 所示。

```
结果为：交易服务路由类型：gray-publish-trade-server,StableRoute;classpath:rule-gray-publish-version.xml;3
抽样调用10次
灰度路由调用比例为：0.2
稳定路由调用比例为：0.8
```

图 14-18

③统一设置灰度路由条件为 gray-condition-3，其中，灰度流量为 50%，稳定流量为 50%，

如图 14-19 所示。

```
结果为：交易服务路由类型：gray-publish-trade-server,GrayRoute;classpath:rule-gray-publish-version.xml;
抽样调用10次
灰度路由调用比例为：0.5
稳定路由调用比例为：0.5
```

图 14-19

④统一设置灰度路由条件为 gray-condition-4，其中，灰度流量为 80%，稳定流量为 20%，如图 14-20 所示。

```
结果为：交易服务路由类型：gray-publish-trade-server,GrayRoute;classpath:rule-gray-publish-version.xml;
抽样调用15次
灰度路由调用比例为：0.8
稳定路由调用比例为：0.2
```

图 14-20

⑤统一设置灰度路由条件为 gray-condition-5，其中，灰度流量为 90%，稳定流量为 10%，如图 14-21 所示。

```
结果为：交易服务路由类型：gray-publish-trade-server,GrayRoute;classpath:rule-gray-publish-version.xml;
抽样调用30次
灰度路由调用比例为：0.9
稳定路由调用比例为：0.1
```

图 14-21

⑥统一设置灰度路由条件为 gray-condition-6，其中，灰度流量为 100%，稳定流量为 0，如图 14-22 所示。

```
结果为：交易服务路由类型：gray-publish-trade-server,GrayRoute;classpath:rule-gray-publish-version.xml;
抽样调用46次
灰度路由调用比例为：1.0
稳定路由调用比例为：0.0
```

图 14-22

⑦统一设置灰度路由条件为 basic-condition，其中，灰度流量为 0，稳定流量为 100%，如图 14-23 所示。

```
结果为：交易服务路由类型：gray-publish-trade-server,StableRoute;classpath:rule-gray-publish-version.xml;
抽样调用21次
灰度路由调用比例为：0.0
稳定路由调用比例为：1.0
```

图 14-23

14.5.4 【实例】在 Spring Cloud Alibaba 应用中配置全链路蓝绿发布的规则，并验证全链路蓝绿发布的效果

本实例的源码在本书配套资源的“chapterfourteen/ blue-green-publish-cloud-alibaba”目录下。

本实例总共包括如下 6 个服务。

- blue-green-trade-server：模拟电商中交易服务。
- blue-green-order-server：模拟电商中的订单服务。
- blue-green-good-server：模拟电商中的商品服务。
- blue-green-logistics-server：模拟电商中的物流服务。
- blue-green-gateway-server：模拟业务网关。
- blue-green-gateway-traffic-server：模拟用户调用业务网关的请求流量。

1. 初始化服务

使用 Spring Cloud Alibaba 初始化 6 个服务，并添加 POM 依赖，部分配置信息如下，详细配置信息可以参考本书配套资源中的代码。

```
<!—添加 Discovery 封装的 Nacos 注册中心的 POM 依赖-->
<dependency>
    <groupId>com.nepxion</groupId>
    <artifactId>discovery-plugin-register-center-starter-nacos</artifactId>
        <version>6.11.0</version>
    </dependency>
<!—添加 Discovery 封装的 Spring Cloud Gateway 的 POM 依赖-->
<dependency>
    <groupId>com.nepxion</groupId>
    <artifactId>discovery-plugin-strategy-starter-gateway</artifactId>
    <version>6.11.0</version>
</dependency>
```

2. 添加服务配置信息，以及蓝路由、绿路由及兜底路由规则

下面在服务 blue-green-gateway-server 中添加配置信息，以及蓝路由、绿路由及兜底路由规则，其他服务可以参考本书配套资源中的代码。

（1）文件 application.properties 和 bootstrap.properties 中的部分配置信息如下：

```
#规则文件的格式，支持 XML 和 JSON 格式，如缺失则默认为 XML 格式
spring.application.config.format=xml
#规则文件的路径
spring.application.config.path=classpath:rule-blue-green-version.xml
#服务的版本号为 1.0
spring.cloud.nacos.discovery.metadata.version=1.0
...
```

（2）在文件 rule-blue-green-version.xml 中添加蓝路由、绿路由及兜底路由规则，部分配置信息如下：

```
<?xml version="1.0" encoding="UTF-8"?>
<rule>
    <strategy-release>
        <conditions type="blue-green">
            <!-- 蓝路由，条件驱动规则为："a=1" -->
            <condition id="blue-condition" expression="#H['a'] == '1'"
version-id="blue-route"/>
            <!-- 绿路由，条件驱动规则为："a=1"和"b=2" -->
            <condition id="green-condition" expression="#H['a'] == '1' and
#H['b'] == '2'" version-id="green-route"/>
            <!-- 兜底路由，条件驱动规则为："a=1"和"b=3" -->
            <condition id="basic-condition" expression="#H['a'] == '1' and
#H['b'] == '3'" version-id="basic-route"/>
        </conditions>
        <!—详细的路由规则-->
        <routes>
            <!—配置全链路蓝路由，统一路由到服务版本 1.0-->
            <route id="blue-route"
type="version">{"blue-green-gateway-server":"1.0",
"blue-green-trade-server":"1.0", "blue-green-order-server":"1.0",
        "blue-green-good-server":"1.0",
"blue-green-logistics-server":"1.0"}</route>
            <!—配置全链路绿路由，统一路由到服务版本 1.1-->
            <route id="green-route" type="version">
{"blue-green-gateway-server":"1.1", "blue-green-trade-server":"1.1",
"blue-green-order-server":"1.1",
            "blue-green-good-server":"1.1","blue-green-
logistics-server":"1.1"}</route>
           <!—配置全链路兜底路由，统一路由到服务版本 1.2-->
           <route id="basic-route" type="version">{"blue-green- gateway-
server":"1.2", "blue-green-trade-server":"1.2", "blue-green-order-server":
"1.2",
            "blue-green-good-server":"1.2","blue-green-logistics-
server":"1.2"}</route>
        </routes>
```

```
    </strategy-release>
</rule>
```

为了区分蓝路由、绿路由及兜底路由，这里将 3 种路由方式指向不同的服务版本。

3. 演示全链路蓝绿发布的效果

本书为了演示全链路蓝绿发布的效果，将需要进行蓝绿发布的服务启动，并且每个版本启动一个进程。例如交易服务 blue-green-trade-server，启动 4 个进程，版本号分别为：1.0、1.1、1.2、1.3（默认版本号为 1.0）。

（1）将需要进行蓝绿发布的服务版本号设置为 1.0，例如交易服务 blue-green-trade-server 如图 14-24 所示，全链路请求会统一走蓝路由。

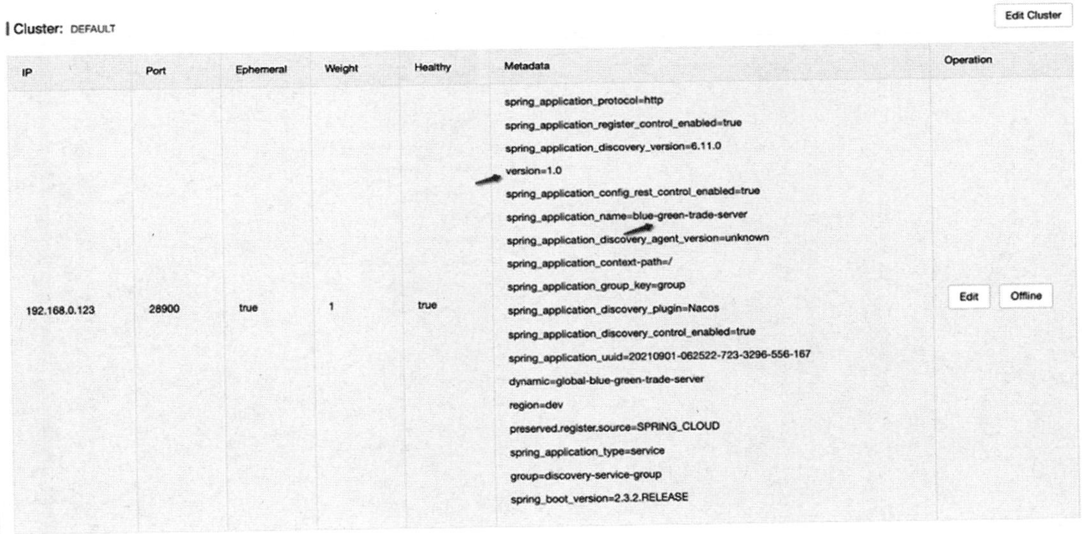

Edit Cluster

| Cluster: DEFAULT

IP	Port	Ephemeral	Weight	Healthy	Metadata	Operation
192.168.0.123	28900	true	1	true	spring_application_protocol=http spring_application_register_control_enabled=true spring_application_discovery_version=6.11.0 version=1.0 spring_application_config_rest_control_enabled=true spring_application_name=blue-green-trade-server spring_application_discovery_agent_version=unknown spring_application_context-path=/ spring_application_group_key=group spring_application_discovery_plugin=Nacos spring_application_discovery_control_enabled=true spring_application_uuid=20210901-062522-723-3296-556-167 dynamic=global-blue-green-trade-server region=dev preserved.register.source=SPRING_CLOUD spring_application_type=service group=discovery-service-group spring_boot_version=2.3.2.RELEASE	Edit Offline

图 14-24

在服务 blue-green-gateway-traffic-server 中，在模拟的用户请求中添加“a=1”的请求头，用于设置当前流量统一走蓝路由，部分代码如下所示。

```
    HttpHeaders headers = new HttpHeaders();
    //在 Header 请求头中，添加请求头参数“a=1”
    headers.add("a","1");
    ResponseEntity<String> result =
restTemplate.exchange("http://127.0.0.1:28920/blue-green-trade-server/trade/
buyGood", HttpMethod.GET,
                    new HttpEntity<String>(null,headers),String.class);
```

演示效果如图 14-25 所示。

```
结果为：交易服务路由类型：blueRoute;classpath:rule-blue-green-version.xml;交易服务提供方的IP地址为：192.168.0.123:28900订单服务路由类型：blueRoute;
抽样调用89次
蓝路由调用比例为：1.0
绿路由调用比例为：0.0
兜底路由调用比例为：0.0
```

图 14-25

（2）将需要进行蓝绿发布的服务版本号设置为 1.1，例如交易服务 blue-green-trade-server 如图 14-26 所示，全链路请求会统一走绿路由。

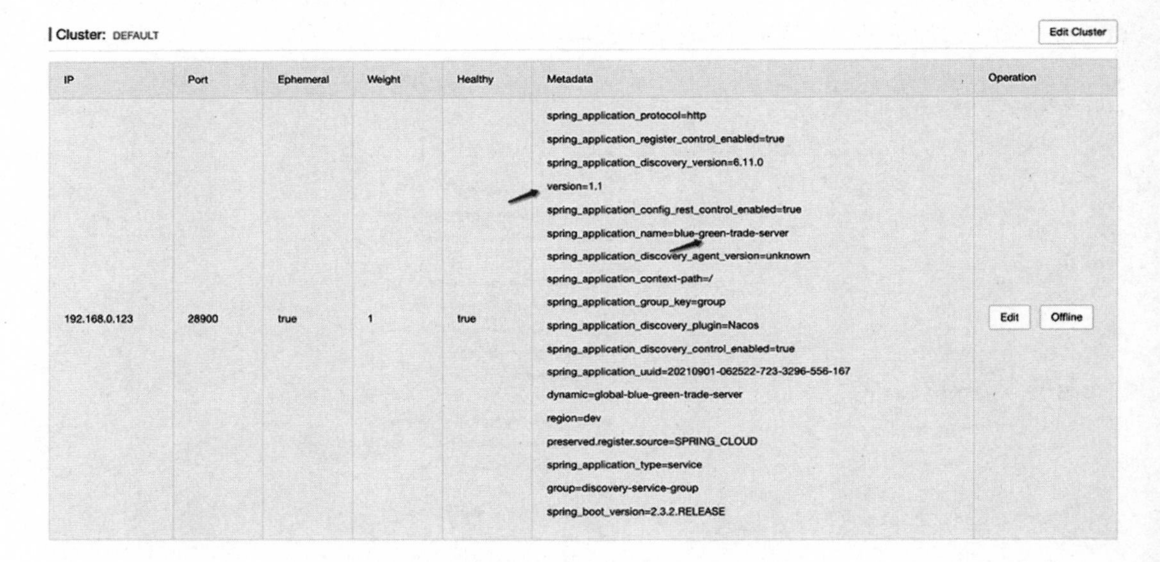

| Cluster: DEFAULT | Edit Cluster

IP	Port	Ephemeral	Weight	Healthy	Metadata	Operation
192.168.0.123	28900	true	1	true	spring_application_protocol=http spring_application_register_control_enabled=true spring_application_discovery_version=6.11.0 version=1.1 spring_application_config_rest_control_enabled=true spring_application_name=blue-green-trade-server spring_application_discovery_agent_version=unknown spring_application_context-path=/ spring_application_group_key=group spring_application_discovery_plugin=Nacos spring_application_discovery_control_enabled=true spring_application_uuid=20210901-062522-723-3296-556-167 dynamic=global-blue-green-trade-server region=dev preserved.register.source=SPRING_CLOUD spring_application_type=service group=discovery-service-group spring_boot_version=2.3.2.RELEASE	Edit Offline

图 14-26

在服务 blue-green-gateway-traffic-server 中，在模拟的用户请求中添加“a=1”和“b=2”的请求头，用于设置当前流量统一走绿路由，部分代码如下所示。

```
    HttpHeaders headers = new HttpHeaders();
    //在 Header 请求头中，添加请求头参数“a=1”和“b=2”
    headers.add("a","1");
    headers.add("b","2");
    ResponseEntity<String> result =
restTemplate.exchange("http://127.0.0.1:28920/blue-green-trade-server/trade/
buyGood", HttpMethod.GET,new HttpEntity<String>(null,headers),String.class);
```

演示效果如图 14-27 所示。

```
结果为：交易服务路由类型：greenRoute;classpath:rule-blue-green-version.xml;交易服务提供方的IP地址为：192.168.0.123:28900订单服务路由类型：
抽样调用9次
蓝路由调用比例为：0.0
绿路由调用比例为：1.0
兜底路由调用比例为：0.0
```

图 14-27

（3）将需要进行蓝绿发布的服务版本号设置为 1.2，如图 14-28 所示。例如交易服务 blue-green-trade-server 如图 14-29 所示，全链路请求会统一走兜底路由。

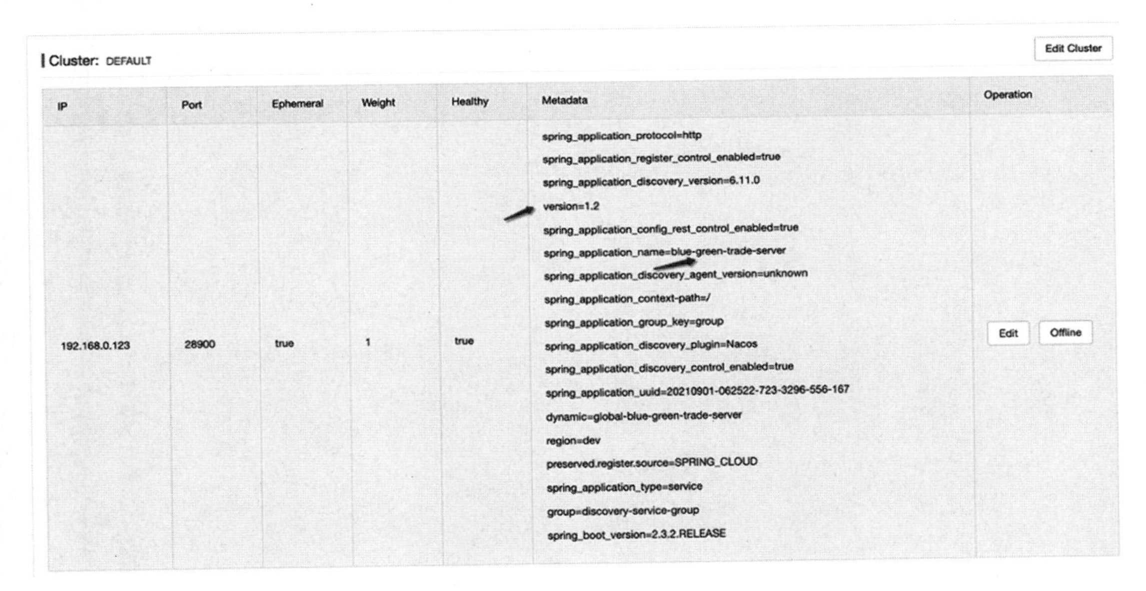

图 14-28

```
结果为：交易服务路由类型：defaultRoute;classpath:rule-blue-green-version.xml;交易服务提供方的IP地址为：192.168.0.123:28900订单服务路由类型：
抽样调用5次
蓝路由调用比例为：0.0
绿路由调用比例为：0.0
兜底路由调用比例为：1.0
```

14-29

在服务 blue-green-gateway-traffic-server 中，去掉模拟的用户请求中的请求头，部分代码如下所示。

```
HttpHeaders headers = new HttpHeaders();
//在 Header 请求头中，添加请求头参数“a=1”和“b=3”
headers.add("a","1");
headers.add("b","3");
```

```
    ResponseEntity<String> result = restTemplate.exchange ("http://127.0.0.1:
28920/blue-green-trade-server/trade/buyGood", HttpMethod.GET,
                        new HttpEntity<String>(null,headers),String.class);
```

项目实战篇

第 15 章

【项目】全链路日志平台

——基于 ELK、FileBeat、Kafka、Spring Cloud Alibaba 及 Skywalking

在微服务架构中，经常需要用日志信息快速地定位问题，所以需要一个高效的全链路日志平台。ELK 是一个比较成熟的、开源的日志收集和存储框架。本项目会基于 ELK、FileBeat、Kafka、Spring Cloud Alibaba 及 Skywalking，构建一个全链路日志平台。

15.1 全链路日志平台整体架构

图 15-1 所示为全链路日志平台整体架构，说明如下。

（1）微服务统一采用 Spring Cloud Alibaba 作为基础框架。

（2）将微服务接入 Skywalking 中，使得微服务具备分布式链路追踪的能力。

（3）使用 Skywalking 提供的插件“apm-toolkit-logback”，将分布式链路的 traceId 关联到微服务的日志文件中，关联的字段值为“TID”。

（4）使用中间件 FileBeat 收集日志文件中的日志，并异步地存储到 Kafka 中。

（5）使用中间件 Logstash 异步地消费 FileBeat 向 Kafka 中生产的日志消息，Logstash 会将日志消息转换为 Logstash 支持的数据结构的数据并存储在 Elasticsearch 中。

（6）软件开发人员使用 Kibana 提供的控制台访问日志。

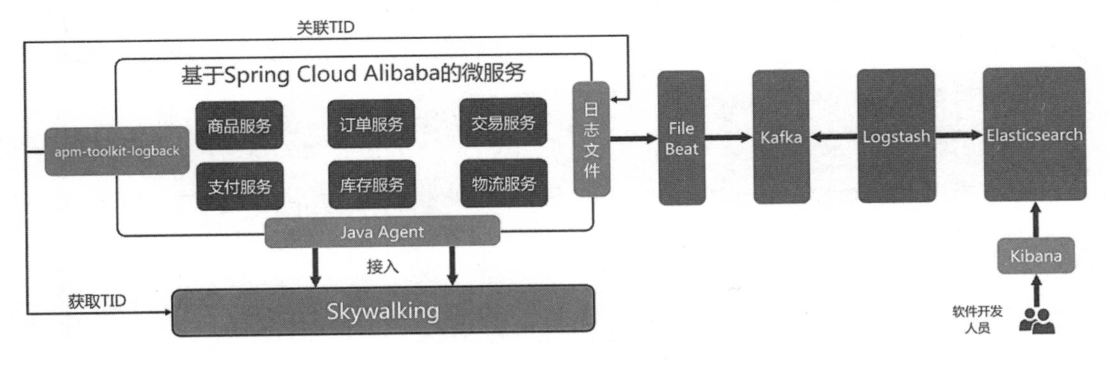

图 15-1

15.2 搭建环境

1. 初始化微服务

本实例的源码在本书配套资源的“chapterfiveteen/”目录下。

本项目用电商中比较常见的业务（比如“用户下单并完成订单支付”）作为微服务的业务场景，并通过微服务来模拟业务请求链路。

在电商微服务架构中，“用户下单并完成订单支付”的业务请求要经过“交易服务”“订单服务”“商品服务”等微服务。为了不增加业务流程的复杂度，本项目模拟从“交易服务”到“订单服务”的订单创建过程。项目中的微服务统一采用 Spring Cloud Alibaba 作为基础框架，并接入全链路日志平台。

使用 Spring Cloud Alibaba 快速初始化 3 个服务：elk-front-api-server（前置 API 服务）、elk-trade-dubbo-server（电商交易服务）和 elk-order-dubbo-server（电商订单服务）。

调用链路关系如下：

前置 API 服务 → 电商交易服务 → 电商订单服务

3 个服务的具体代码可以参考本书配套资源中代码，这里就不展示了。

2. 统一日志框架

本项目统一封装了两个模块：elk-common-logback 和 elk-common-log-format。前者用于拦截服务提供者的 Dubbo 请求，并统计 Dubbo 提供者的访问日志；后者用于统一日志格式。

（1）在微服务配置文件中添加如下配置信息，开启统计 Dubbo 访问日志：

```
dubbo.provider.accesslog=true
```

（2）在模块 elk-common-logback 中封装了一个过滤器 DubboLogFilter 类，用于拦截服务提供者的 Duboo 请求，部分代码如下所示：

```
    //①当 Dubbo 接口是服务提供者时，过滤器 DubboLogFilter 才会执行
    @Activate(group = {CommonConstants.PROVIDER}, value =
Constants.ACCESS_LOG_KEY, order = 90000)
    public class DubboLogFilter implements Filter {
       //②实现 Dubbo 过滤器的 invoke()方法，拦截 Dubbo 的请求
       @Override
        public Result invoke(Invoker<?> invoker, Invocation invocation) throws
RpcException {
              ...
        }
        ...
    }
```

统一日志框架的具体代码请参考本书配套资源中的代码。

3. 搭建 Kafka 环境

可以参考本书 10.5.4 中的步骤，这里就不展开了。

4. 搭建 ELK 环境

（1）搭建 Elasticsearch 环境。

可以参考本书 10.2.2 中的步骤，这里就不展开了。

（2）搭建 Kibana 环境。

可以参考本书 10.2.2 中的步骤，这里就不展开了。

（3）搭建 logstash 环境。

从官网下载 logstash 的安装包“logstash-7.13.4-darwin-x86_64.tar.gz”并解压缩，生成文件“logstash-7.13.4”，在文件夹“/logstash-7.13.4/config”中添加配置文件“kafka-logstash-ad-es.conf”，具体配置信息如下：

```
    #①从 Kafka 中读取日志
    input {
```

```
  kafka {
    bootstrap_servers => "127.0.0.1:9092"
    codec => "json"
    topics => "SpringCloudAlibabaPractice"
    group_id => "consumer-group-01"
    consumer_threads => 1
    decorate_events => false
    session_timeout_ms => "60000"
    request_timeout_ms => "90000"
  }
}
#②将日志输出到 Elasticsearch 中
output {
  elasticsearch {
    hosts => ["127.0.0.1:9200"]
    manage_template => true
    template_overwrite => true
  }
  stdout{codec => rubydebug}
}
```

执行 Shell 命令启动 logstash，具体命令如下：

```
sh bin/logstash -f config/kafka-logstash-ad-es.conf
--config.reload.automatic --path.data=../data/logstash
```

5. 搭建 FileBeat 环境

从官网下载 FileBeat 的安装包 filebeat-7.13.1-darwin-x86_64.tar.gz，解压缩安装包生成文件夹“filebeat-7.13.1-darwin-x86_64”，在文件夹中添加一个配置文件“filebeat.yml”，具体配置信息如下：

```
#配置微服务的日志路径
filebeat.inputs:
- type: log
  enabled: true
  paths:
    -../../application/elk-front-api-server/logs/*.log
- type: log
  enabled: true
  paths:
    -../../application/elk-order-dubbo-server/logs/*.log
- type: log
  enabled: true
  paths:
    -../../application/elk-trade-dubbo-server/logs/*.log
```

```
output.kafka:
  enabled: true
  #如果是集群，则用","分割
  hosts: ["127.0.0.1:9092"]
  topic: SpringCloudAlibabaPractice
  group_id: producer-group-01
```

执行 Shell 命令启动 FileBeat，具体命令如下：

```
filebeat -e -c filebeat.yml
```

6. 搭建 Skywalking 环境

可以参考本书 10.2.2 节中的步骤，这里就不展开了。

15.3 将 Spring Cloud Alibaba 应用接入全链路日志平台

15.3.1 将微服务接入全链路日志平台

使用 Maven 打包 elk-front-api-server（前置 API 服务）、elk-trade-dubbo-server（电商交易服务）和 elk-order-dubbo-server（电商订单服务）并生成部署包。将部署包复制到文件夹“/elk-env/application”中，如图 15-2 所示。

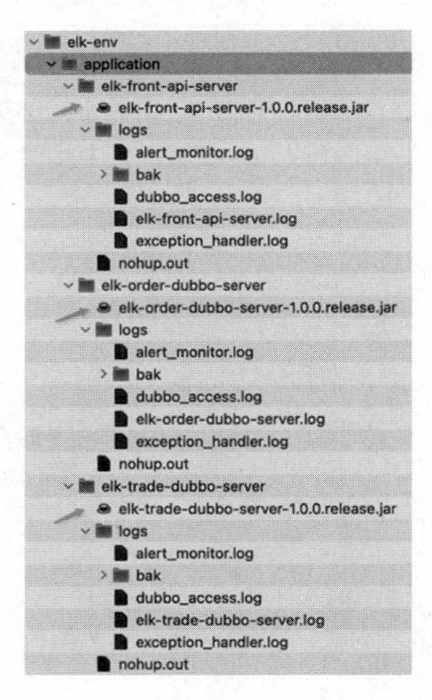

图 15-2

按顺序执行 Shell 命令，并启动微服务，具体命令如下：

```
#①启动微服务 Elk-Order-Dubbo-Server，并将其接入 Skywalking
nohup java -javaagent:/Users/huxian/Downloads/skywalking-env/agent/
skywalking-agent.jar -Dskywalking.agent.service_name=Elk-Order-Dubbo-Server
-jar elk-order-dubbo-server-1.0.0.release.jar --server.port=8021
--dubbo.protocol.port=26721
#②启动微服务 Elk-Trade-Dubbo-Server，并将其接入 Skywalking
nohup java -javaagent:/Users/huxian/Downloads/skywalking-env/agent/
skywalking-agent.jar -Dskywalking.agent.service_name=Elk-Trade-Dubbo-Server
-jar elk-trade-dubbo-server-1.0.0.release.jar --server.port=8022
--dubbo.protocol.port=26722 &
#③启动微服务 Elk-Front-Api-Server，并将其接入 Skywalking
nohup java -javaagent:/Users/huxian/Downloads/skywalking-env/agent/
skywalking-agent.jar -Dskywalking.agent.service_name=Elk-Front-Api-Server
-jar elk-front-api-server-1.0.0.release.jar --server.port=8020
--dubbo.protocol.port=26720 &
```

15.3.2 使用全链路日志平台查询业务日志

执行完 “curl 127.0.0.1:8020/order/createOrder” 命令后，就可以调用前置 API 服务 Elk-Front-Api-Server 去创建订单了。在 Skywalking 中会生成一条分布式调用链路，如图 15-3 所示。

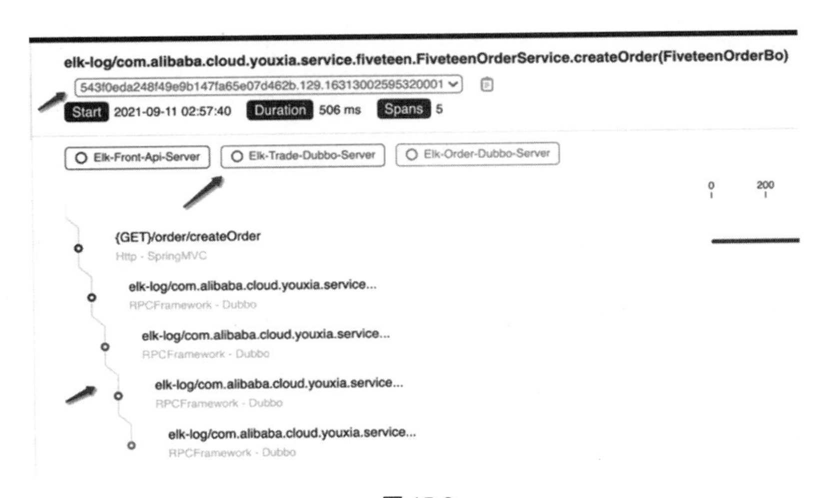

图 15-3

其中，分布式链路的 traceId 为 “543f0eda248f49e9b147fa65e07d462b.129.16313002595320001”。使用这个 traceId 在 Kibana 的控制台中可以快速地搜索到日志信息，如图 15-4 所示。

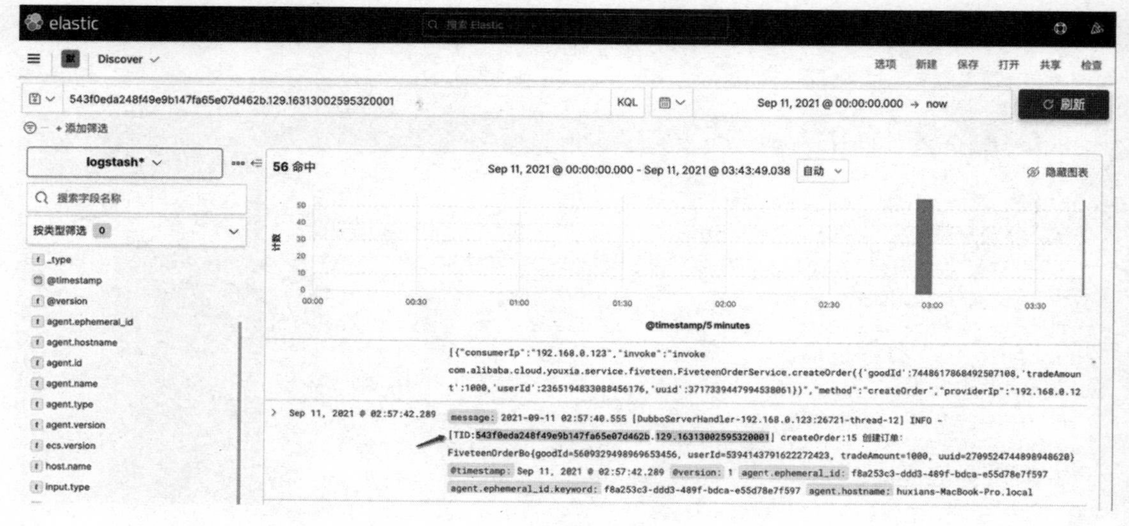
elastic
Discover
选项
新建
保存
打开
共享
检查
543f0eda248f49e9b147fa65e07d462b.129.16313002595320001
KQL
Sep 11, 2021 @ 00:00:00.000 → now
刷新
+ 添加筛选
logstash*
搜索字段名称
按类型筛选 0
_type
@timestamp
@version
agent.ephemeral_id
agent.hostname
agent.id
agent.name
agent.type
agent.version
ecs.version
host.name
input.type
56 命中
Sep 11, 2021 @ 00:00:00.000 - Sep 11, 2021 @ 03:43:49.038
自动
隐藏图表
计数
@timestamp/5 minutes
[{"consumerIp":"192.168.0.123","invoke":"invoke
com.alibaba.cloud.youxia.service.fiveteen.FiveteenOrderService.createOrder({'goodId':7448617868492507108,'tradeAmoun
t':1000,'userId':2365194833088456176,'uuid':3717339447994538061})","method":"createOrder","providerIp":"192.168.0.12
Sep 11, 2021 @ 02:57:42.289
message: 2021-09-11 02:57:40.555 [DubboServerHandler-192.168.0.123:26721-thread-12] INFO -
[TID:543f0eda248f49e9b147fa65e07d462b.129.16313002595320001] createOrder:15 创建订单:
FiveteenOrderBo{goodId=5609329498969653456, userId=5394143791622272423, tradeAmount=1000, uuid=2709524744898948620}
@timestamp: Sep 11, 2021 @ 02:57:42.289 @version: 1 agent.ephemeral_id: f8a253c3-ddd3-489f-bdca-e55d78e7f597
agent.ephemeral_id.keyword: f8a253c3-ddd3-489f-bdca-e55d78e7f597 agent.hostname: huxians-MacBook-Pro.local

图 15-4

16

第 16 章 【项目】在企业中落地中台架构

在企业数字化转型的过程中，需要搭建技术中台、业务中台及数据中台。在中台落地的过程中，通常都会以用户中台为业务抓手整合与用户相关的通用业务，形成不同的能力中心，比如账户中心、商品中心、交易中心等。

本章将介绍落地中台架构过程中的一些业务功能和技术架构的方法论。

16.1 某跨境支付公司中台架构

面对数字化转型大潮，某独角兽跨境支付公司开始尝试落地中台，这里就带读者鸟瞰一下跨境支付中台架构。

16.1.1 跨境支付中台架构

图 16-1 所示为某跨境公司的跨境支付中台架构。

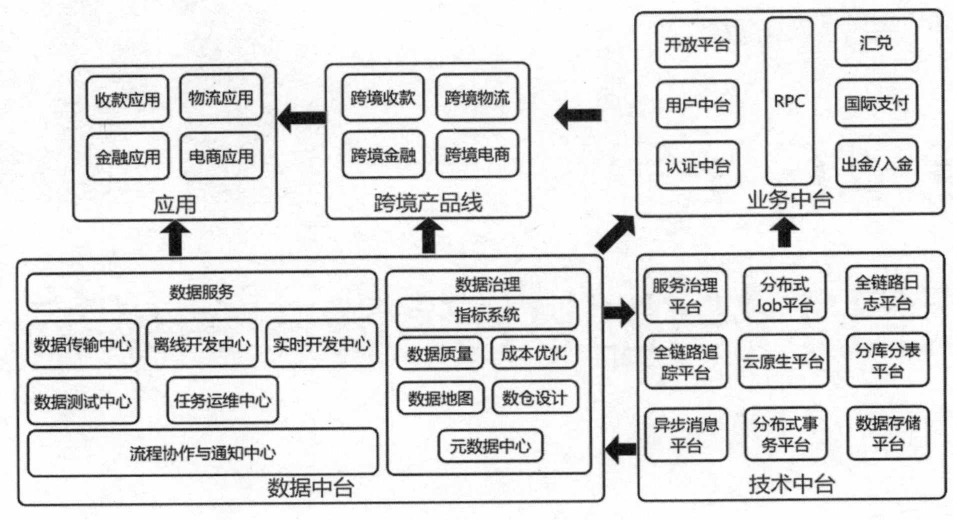

图 16-1

说明如下。

（1）业务中台：其中沉淀了用户中台、认证中台及开放平台等通用能力。在业务底层沉淀了汇兑、国际支付，以及出金/入金的“原子”业务能力。

（2）技术中台：其中沉淀了服务治理平台、分布式 Job 平台、全链路日志平台、全链路追踪平台、云原生平台、分库分表平台、异步消息平台、分布式事务平台，以及数据存储平台等。技术中台支撑业务中台及数据中台，数据中台支撑应用、跨境产品线及业务中台，业务中台主要支撑跨境产品线。

（3）数据中台：其中沉淀了与数据治理相关的数据能力，比如指标系统、数据质量、成本优化、数据地图、数仓设计，以及元数据中心等。在数据服务方面，主要沉淀了数据传输中心、离线开发中心、实时开发中心、数据测试中心和任务运维中心等。

（4）跨境产品线：主要包括跨境收款、跨境物流、跨境金融及跨境电商。

（5）应用：主要包括收款应用、物流应用、金融应用及电商应用等。

跨境支付中台利用开放平台，将比较核心的跨境支付能力，开放给第三方应用和具备一定资质的第三方合作供应商。

跨境支付中台对内是整合公司现有的跨境支付业务，形成可以复用的底层原子跨境支付能力。

在跨境支付中台落地的过程中，首先规划落地的是跨境的用户中台，利用用户中台作为抓手整合业务线的各种产品，并将用户服务和数据模型统一起来，这样才能将“长在用户上的业务逐步地统一起来”。

16.1.2 跨境支付用户中台架构

落地用户中台是一次快速梳理现有业务边界的过程：如果企业内部的产品能够统一用户服务及用户数据模型，那么各个产品在用户域可以互通；可以将成熟产品的流量快速引流到新产品，从而提升新产品的行业竞争力。

图 16-2 所示为某跨境公司的跨境支付用户中台架构。

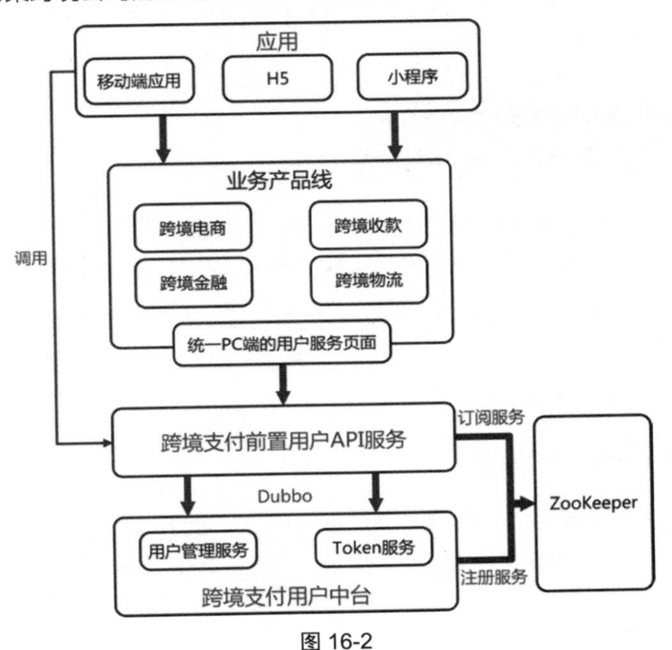

图 16-2

- 用户管理服务：主要是提供用户登录、用户注册、用户信息管理及用户安全等功能。
- Token 服务：主要是用于认证用户信息，并维护用户登录之后的用户状态，如果 Token 校验失败，则退出当前用户的登录。
- 跨境支付前置 API 服务：主要是将基于 Dubbo 的用户中台的接口转换为 RESTful API 风格，并暴露给业务产品。
- 统一的 PC 端用户服务页面：PC 端的产品线采用统一的 PC 端用户服务页面将产品接入用户中台。接入流程采用前台模式。前台的 UI 界面、后台的前置 API 服务、用户管理服务及 Token 服务统一使用用户中台的能力，这样能够提高业务产品接入的效率。
- 业务产品线：包括跨境电商、跨境收款、跨境金融和跨境物流等。
- 应用：移动端应用、H5 以及小程序等。业务产品线上的服务提供各种定制化的业务能力，统一采用 RESTful API 和应用完成数据通信。

图 16-3 所示为跨界支付用户的功能架构。

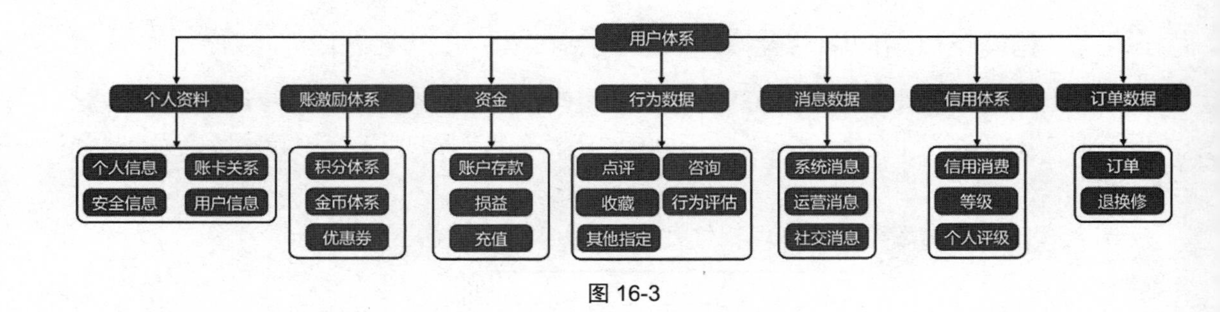

图 16-3

16.2 某娱乐直播公司中台架构

中台战略是一种高度抽象和组织化的架构方法论。企业如果要真正落地这些方法论，需要结合原有的业务产品现状和实际的技术人员的专业技能，做出一些方法论的调整。

某娱乐直播公司在 Spring Cloud Alibaba 的基础上落地中台战略，并采用 Spring Cloud Alibaba 相关的中间件作为实现技术中台和业务中台的底层核心技术。按照实施中台战略的一般步骤，首先落地的就是泛娱乐领域的用户中台。

16.2.1 泛娱乐直播中台架构

在泛娱乐领域，最核心的技术是直播。通过直播可以快速孵化出不同的泛娱乐产品，比如购物类的直播电商产品、游戏类的游戏直播产品、视频表演类的娱乐直播产品，以及社交类的语音和视频类产品。

1. 直播中台架构

图 16-4 所示为一个比较成熟的直播中台架构（业务中台和技术中台）。

直播中台总共分为四层。

（1）技术中台：建设技术中台的目标是标准化底层的基础架构。在直播中台架构中，主要进行了如下标准化改造：

- 统一使用 Spring Cloud Alibaba 作为基础微服务框架，并利用 Spring Cloud Alibaba 将应用接入 Nacos 的注册中心和配置中心。
- 统一使用 Spring Cloud Alibaba RocketMQ Starter 作为客户端，将应用接入 RocketMQ。
- 统一使用 Skywalking 作为全链路监控的基础框架，并利用 Java Agent 技术实现对业务的零侵入。

- 统一使用 Apache Elastic Job 来满足业务中的分布式 Job 的业务场景。
- 统一使用“ELK + Skywalking”的全链路日志平台。

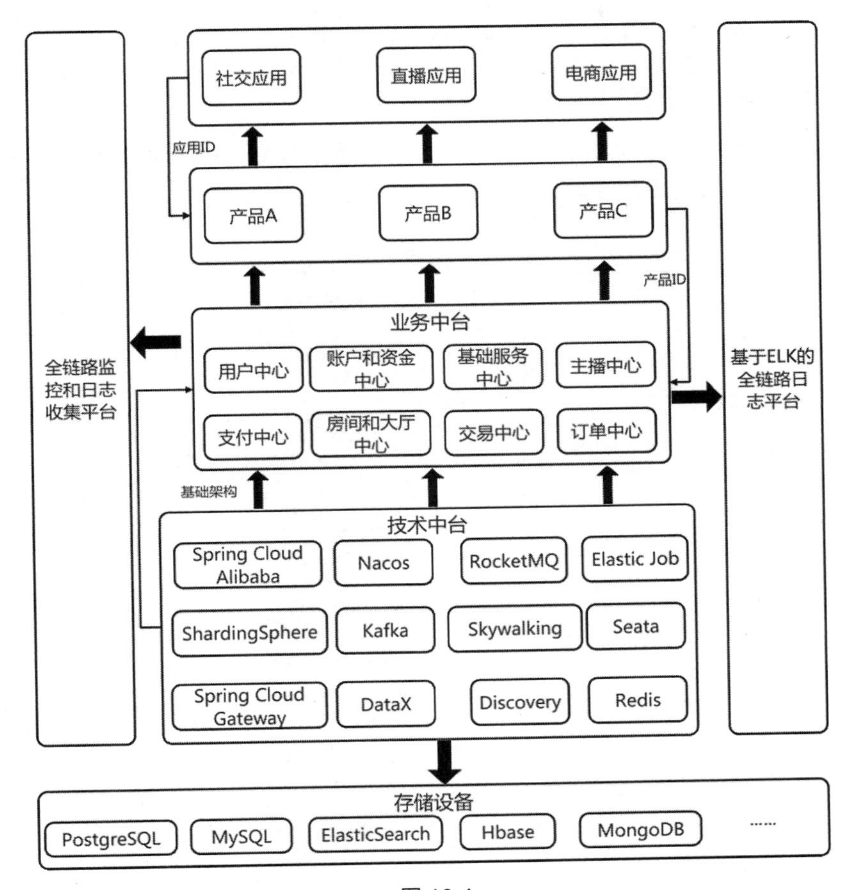

图 16-4

为了标准化基础架构的目标，通常都会采用“基于 Maven 的脚手架框架”维护服务目录结构模板，并利用 Maven 工具一键生成不同类型的服务工程。

（2）业务中台：主要包括用户中心、账户和资金中心、基础服务中心、主播中心、房间和大厅中心、支付中心、交易中心和订单中心。

（3）产品：将基于业务中台形成的前台业务组合起来，包装成一个可以直接交付的业务解决方案。比如，直播平台可以利用业务中台中的通用业务快速搭建一个新的产品。

（4）应用：如果已经有了可以直接交付的产品解决方案，则利用解决方案快速形成一个可视化的 App 应用，并完成业务解决方案的赋能和变现。

在企业中落地技术中台是非常难的。如果说业务长在用户上，并依赖用户生成各种不同的业务场景，那么技术就是实现这些业务场景的工具。

2. 改造之后的服务注册中心和配置中心架构

采用开源的“Spring Cloud Alibaba+Nacos”改造服务注册中心和配置中心。图 16-5 所示为服务改造过程中新旧注册中心和配置中心共存的架构。

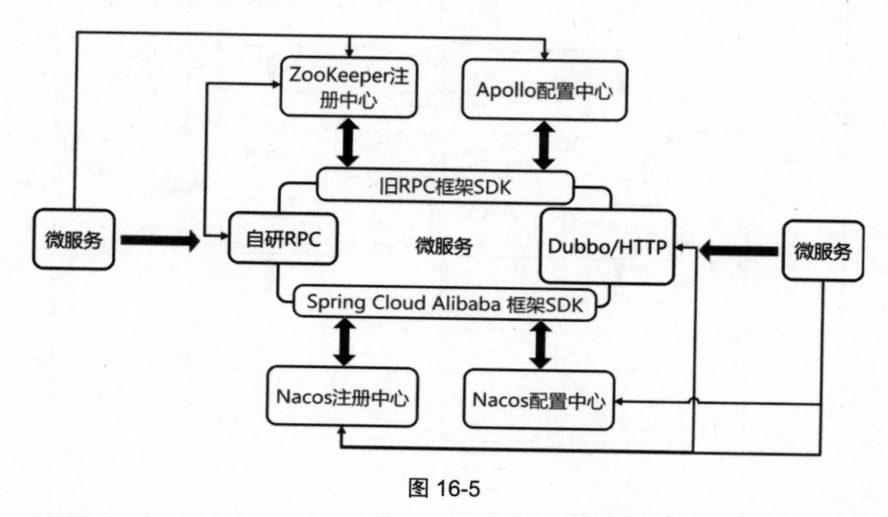

图 16-5

（1）在将微服务的基础框架升级为“Spring Cloud Alibaba + Nacos”的过程中，如果微服务的上下游都存在强依赖关系，则需要微服务同时支持新旧注册中心的 RPC 服务或者 RESTful API 服务。

（2）微服务可以同时使用 Apollo 和 Nacos 作为配置中心，然后逐步将配置信息切换到 Nacos 中。

（3）新项目统一采用“Spring Cloud Alibaba + Nacos”作为基础框架。

（4）业务中台中的服务同时支持 Dubbo 和 HTTP，这样可以通过 HTTP 服务去适配旧的 RPC 框架。

3. 改造之后的 Skywalking 整合产品线的全链路监控架构

使用 Skywalking 整合产品线的全链路监控的最大技术挑战是“需要支持自研 RPC 框架”。Skywalkimg 支持大部分开源的中间件，比如 RocketMQ、Redis 等。如果公司自研 RPC 框架，则在开发人员定制化一个插件之后才能将自研的 RPC 框架接入 Skywalking。

图 16-6 所示为用 Skywalking 整合产品线的全链路监控的架构，主要包括如下。

（1）将产品线上的微服务接入 Skywalking 中，使用 Skywalking 支持的插件去采集分布式链路数据。

（2）运维或者开发人员，按照服务实例维度将 Skywalkimg 客户端插件维护在服务实例的固定目录下。如果服务实例分配给多个服务，则可以将客户端插件维护在指定服务的固定目录下。

（3）将自研的 RPC 框架的插件，选择性地推送到指定服务实例的固定目录下，这样可以比较方便地管理自研的 Skywalking 客户端插件。

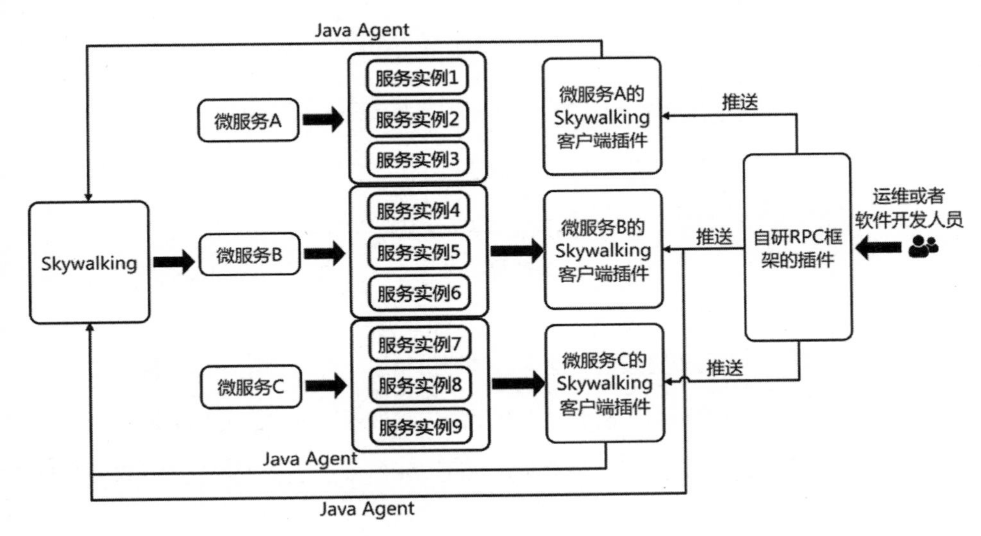

图 16-6

16.2.2 直播用户中台架构

下面介绍直播用户中台的功能架构和拓扑架构。

1. 功能架构

图 16-7 所示为某直播公司用户中台架构。

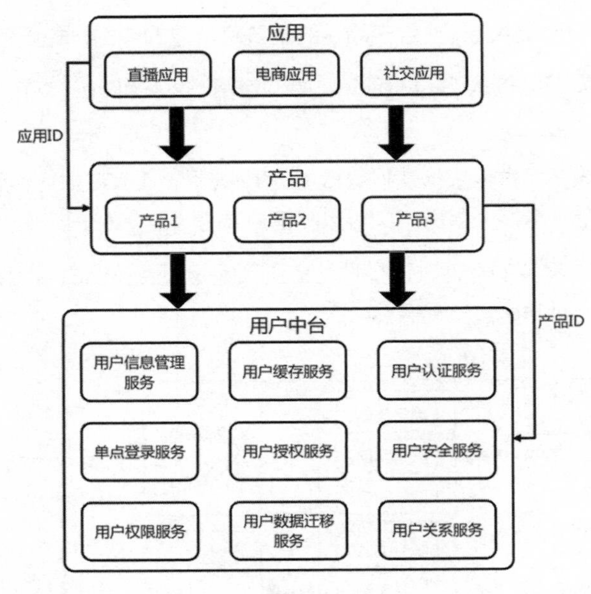

图 16-7

主要功能如下。

（1）用户信息管理服务：用于管理用户信息，并提供登录、注册和用户信息查询等功能。

（2）用户缓存服务：用于缓存用户信息，提升用户中台处理读请求的 QPS 能力。

（3）用户认证服务：用于认证用户，包括实名认证、银行卡认证、人脸识别认证及指纹认证等功能。

（4）单点登录服务：用于管理用户登录的凭证——Token。如果 Token 失效，则重新访问用户缓存服务，生成用户唯一的登录凭证。

（5）用户授权服务：如果接入用户中台的产品有一些开放业务接口能力的场景，则需要使用授权服务统一给第三方应用授权，这样第三方应用才能够调用开放能力。

（6）用户安全服务：用于保护用户信息，防止用户信息泄密，从安全的角度加固用户中台。

（7）用户权限服务：用于管理用户权限、用户角色及用户菜单资源，为后台管理系统提供统一的用户权限相关的功能。

（8）用户数据迁移服务：在用户中台上线的过程中，需要做历史全量数据及实时增量数据的迁移。如果有一些后台系统存在冗余用户信息，则需要利用用户数据迁移服务，实时地从新库同步数据到后台系统的数据库，确保用户数据的一致性。

（9）用户关系服务：用于管理用户之间的关系，比如主播与用户之间的关系。

2. 拓扑架构

图 16-8 所示为直播用户中台的拓扑架构，主要功能如下。

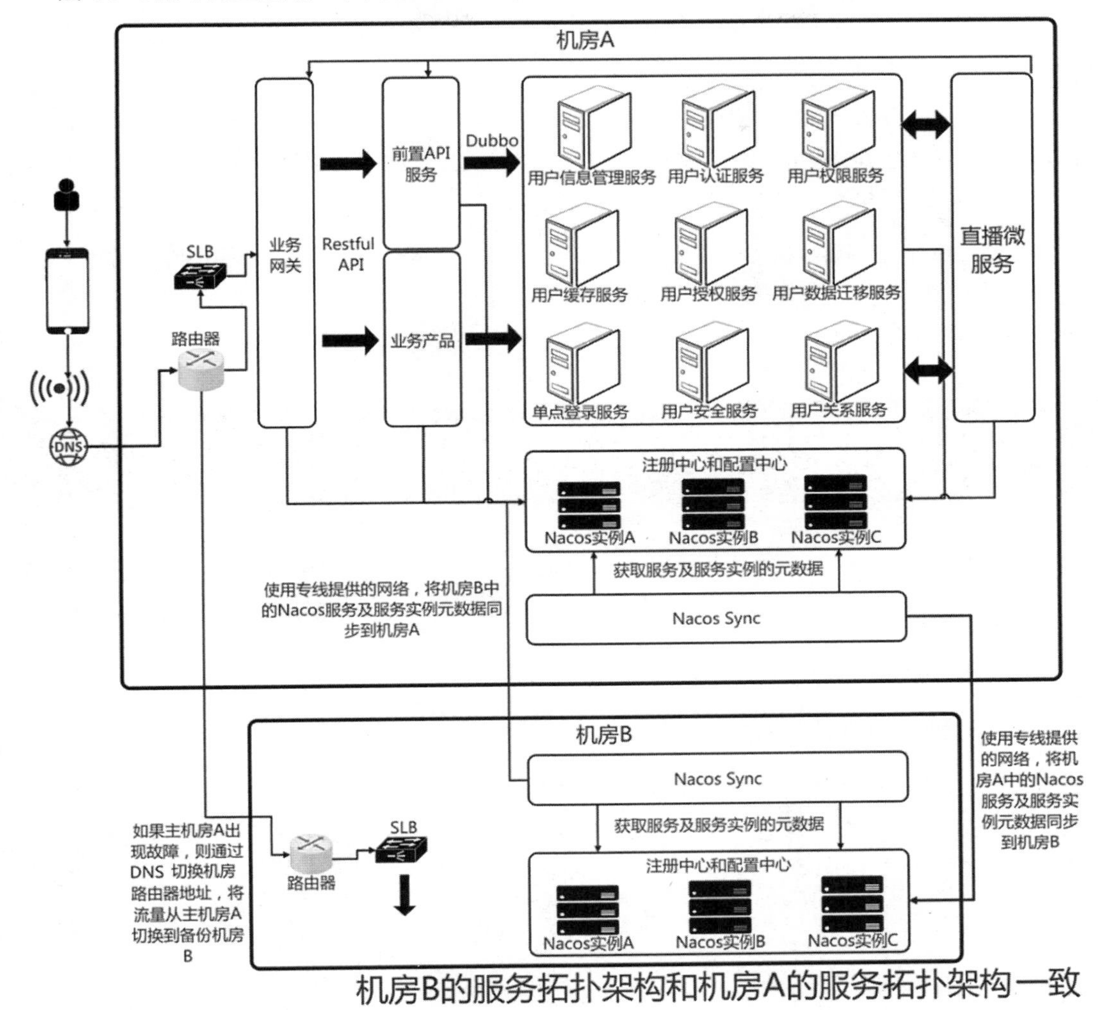

图 16-8

（1）线上采用双机房部署：主机房 A 和备用机房 B 会定期做机房的灾备切换演练，每隔半年将服务从主机房切换到备用机房，并更换机房的角色，这样可以预防由机房软硬件故障导致的服务器不可用问题。

（2）用 Spring Cloud Gateway 作为业务网关，统一治理和管控 API。

（3）用户中台用前置 API 服务暴露 Restful API 给应用。

（4）在业务产品和用户中台之间，采用内部 RPC（比如 Dubbo）作为通信渠道，完成业务产品对用户接口的调用。

（5）在用户中台和直播微服务之间，采用内部 RPC（比如 Dubbo）作为通信渠道，完成直播微服务对用户接口的调用。

（6）主机房 A 和备用机房 B 采用 1：1 对称部署，即备用机房的服务器数量及机器配置与主机房基本一致。如果主机房中的核心业务出现软硬件故障，靠服务自身的集群管理已经无法恢复，则可以快速切换到备用机房，完成核心业务的调用及结果处理。

（7）在主机房 A 和备用机房 B 之间，采用 Nacos Sync 服务将 Nacos 注册中心的服务及服务实例元数据进行双向同步，确保两个机房的 Nacos 注册中心数据的一致性，这样在进行机房切换时能够提高服务的高可用性。

16.3 用“服务双写和灰度发布”来实现中台服务上线过程中的“业务方零停机时间”

如果中台服务改造完成，需要业务方对接，但是业务方要求不能停止业务，（即在业务服务器不能停机的前提完成新服务的升级），则需要服务具备双写和灰度发布的能力。

16.3.1 服务双写架构

图 16-9 所示为服务双写架构。

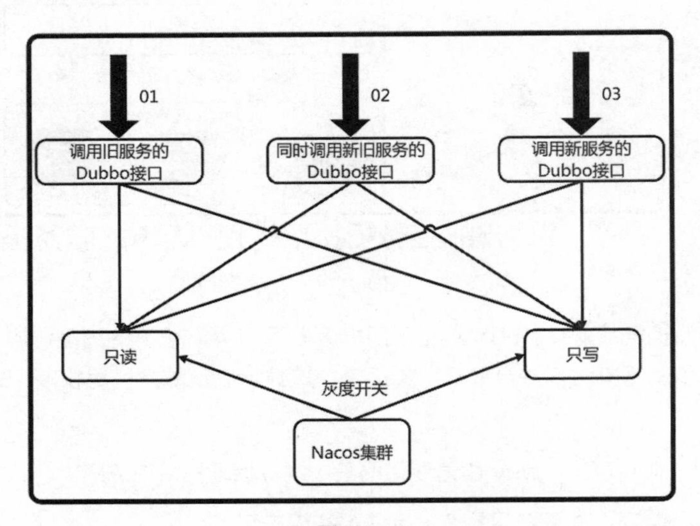

图 16-9

双写架构的主要流程如下。

（1）梳理旧服务中接口中所有方法的类型，类型包括“只读”和“只写”。具体规则是：如果方法中所有的代码逻辑都是读数据，则类型为“只读”；如果方法中所有代码逻辑只要有一处存在写数据，则类型为“只写”。

（2）不改动旧服务接口的请求参数，这样确保服务的订阅者不用更改代码，将对业务的影响降到最低。

（3）在旧服务的接口中增加一个配置项，配置项的 Key 为“remoteMode”，这个配置项有 3 个取值：01、02 和 03。

- 01 代表，直接调用旧服务的接口，完成数据的增加、删减、修改和查询，数据被持久化到旧的数据库中。
- 02 代表，双写，既调用旧服务的接口，又调用新服务的接口，完成数据的增加、删除、修改和查询，数据会同时持久化到新数据库和旧数据库中。
- 03 代表，直接调用新服务的接口，完成数据的增加、删除、修改和查询，数据被持久化到新的数据库中。

16.3.2 服务灰度发布架构

图 16-10 所示为服务灰度发布架构。

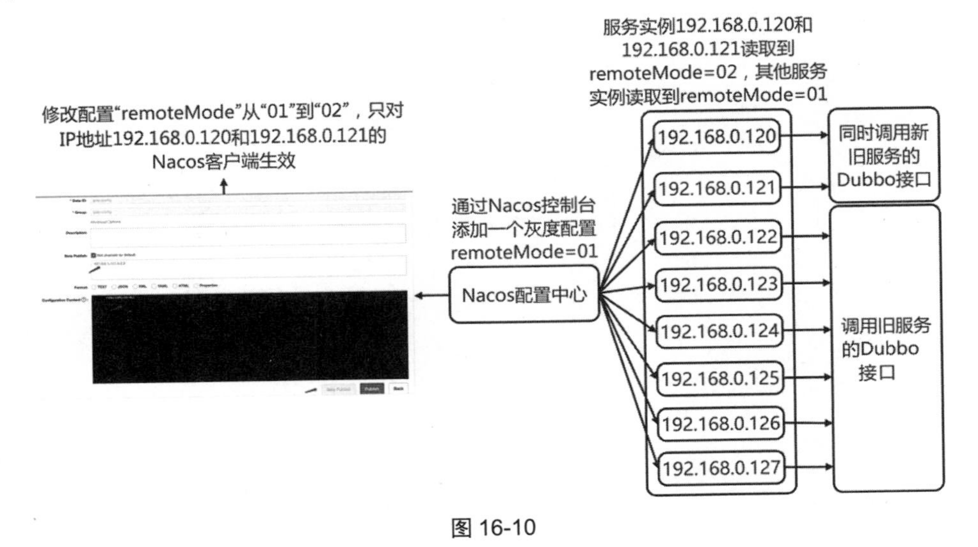

图 16-10

灰度发布的主要流程如下：

（1）在服务启动之前，在 Nacos 配置中心添加一个“remoteMode=01”的配置项，这样在服务启动后，所有的 Nacos 客户端都能读取这个配置项的值。服务可以将配置项作为条件，去控制请求只调用旧服务的 Dubbo 接口。

（2）修改配置项的值为“remoteMode=02”，并指定灰度发布的 IP 地址为“192.168.0.120”和“192.168.0.121”，这样在集群中就只有灰度发布的 IP 地址上的服务实例能够读取最新的值，其他服务实例还是“remoteMode=01”，服务可以将配置项作为条件，去控制请求同时调用新/旧服务的 Dubbo 接口，并完成双写。

（3）修改灰度配置的 IP 地址，将集群中的服务实例的 IP 地址逐个添加到灰度发布的 IP 地址列表中，从而实现“服务双写”的滚动发布。

第 17 章 【项目】异构数据迁移平台——基于 DataX

DataX 是软件开发人员广泛使用的离线数据同步工具/平台，它实现了包括 MySQL、Oracle、SqlServer、Postgre、HDFS、Hive、ADS、HBase 等异构数据源之间的数据高效同步存储。

DataX 作为数据同步框架，将“不同数据源的同步”抽象为从源头数据源读取数据的 Reader 插件，以及向目标端写入数据的 Writer 插件。理论上 DataX 框架可以支持任意数据源类型的数据同步。

DataX 插件体系作为一套生态系统，每接入一套新数据源，则该新加入的数据源即可实现与现有数据源互通。

17.1 搭建环境

MySQL 是软件开发中比较常见的一种数据源，本书的数据源统一采用 MySQL 数据库。

17.1.1 软件环境

DataX 支持以下软件环境。

- 操作系统：支持 Mac 和 Linux。
- JDK：1.8 及以上。
- Python：推荐 2.6.X。

- Apache Maven 3.X：用于编译 DataX 的源码，生成可以部署的 Jar 包。

17.1.2 搭建 MySQL 的异构数据迁移环境

目前 DataX 官方源码暂时不支持 MySQL 的 8.X 版，但是可以通过修改部分源码，从而实现其对 MySQL 8.X 版本的支持。

（1）从官网下载 DataX 源码包，修改源码，具体如下。

① 升级 mysqlwriter 和 mysqlreader 模块中的 Jar 包“mysql-connector-java”的 POM 版本号（从 5.X 到 8.X，本书统一采用 8.0.16）。

② 修改 DataBaseType 类的源码。

修改 MySQL 的驱动程序类，用“com.mysql.cj.jdbc.Driver”替换“com.mysql.jdbc.Driver”，具体修改如下：

```
public enum DataBaseType {
    MySql("mysql", "com.mysql.cj.jdbc.Driver"),
    ...
    OceanBase("oceanbase", "com.alipay.oceanbase.jdbc.Driver");
    ...
}
```

修改 appendJDBCSuffixForWriter() 方法，用“CONVERT_TO_NULL”替换“convertToNull”，具体修改如下：

```
public String appendJDBCSuffixForWriter(String jdbc) {
    switch (this) {
        case MySql:
            suffix = "yearIsDateType=false&zeroDateTimeBehavior=
CONVERT_TO_NULL&rewriteBatchedStatements=true&tinyInt1isBit=false";
        ...
    }
    ...
}
```

（2）使用 Maven 编译 DataX 的源码包，生成压缩包文件“datax.tar.gz”。

（3）使用命令“tar -zxvf datax.tar.gz”解压压缩包，生成文件“datax”。

（4）在文件“datax/job”中新增一个 Job 文件“mysql-sync.json”，具体配置信息如下：

```
{
    "job": {
        "content": [
            {
```

```
            "reader": {
                "name": "mysqlreader",
                "parameter": {
                    "username": "root",
                    "password": "123456huxian",
                    "column": ["id","user_name","sex","age","phone"],
                    "where": "id>0",
                    "connection": [
                        {
                            "table": [
                                "base_user_info"
                            ],
                            "jdbcUrl": [
                                "jdbc:mysql://127.0.0.1:3306/datax_read?
serverTimezone=UTC&useSSL=false"
                            ]
                        }
                    ]
                }
            },
            "writer": {
                "name": "mysqlwriter",
                "parameter": {
                    "username": "root",
                    "password": "123456huxian",
                    "column": ["id","user_name","sex","age","phone"],
                    "connection": [
                        {
                            "table": [
                                "base_user_info"
                            ],
                            "jdbcUrl":"jdbc:mysql://127.0.0.1:3306/
datax_write?serverTimezone=UTC&useSSL=false"
                        }
                    ]
                }
            }
        }
    ],
    "setting": {
        "speed": {
            "channel": 1,
            "byte": 104857600
        },
        "errorLimit": {
```

```
                "record": 10,
                "percentage": 0.05
            }
        }
    }
}
```

（5）修改文件夹“datax/conf”中的配置文件 core.json，具体修改如下：

```
{
    "core": {
        "transport": {
            "channel": {
                "class": "com.alibaba.datax.core.transport.channel.
memory.MemoryChannel",
                "speed": {
                    //①修改 core.transport.channel.speed.byte 为 65536
                    "byte": 65536,
                    //②修改 core.transport.channel.speed.record 为 10000
                    "record": 10000
                }
            }
        }
    }
}
```

（6）创建数据库“datax_read”和“datax_write”，并在两个数据库中创建表“base_user_info”，具体 SQL 语句可以参考本书配套资源中的源码。

（7）在文件夹“datax/bin”中执行命令“python datax.py ../job/mysql-sync.json”，启动同步异构数据的任务。

如图 17-1 所示，任务在执行的过程中，会从数据库“datax_read”的表“base_user_info”中读出数据，并写入数据库“datax_write”的表“base_user_info”中。

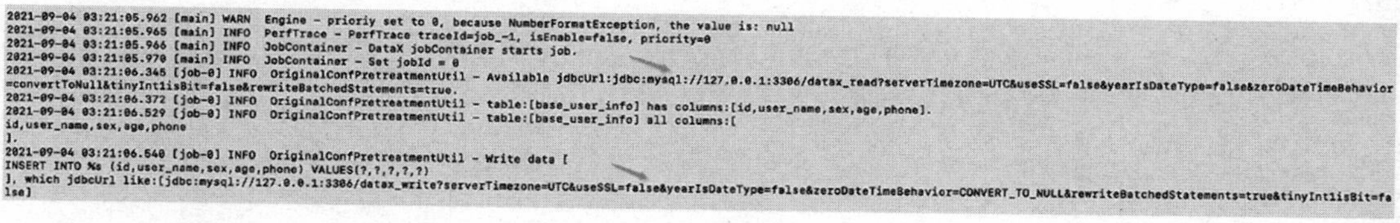

```
2021-09-04 03:21:05.962 [main] WARN  Engine - prioriy set to 0, because NumberFormatException, the value is: null
2021-09-04 03:21:05.965 [main] INFO  PerfTrace - PerfTrace traceId=job_-1, isEnable=false, priority=0
2021-09-04 03:21:05.966 [main] INFO  JobContainer - DataX jobContainer starts job.
2021-09-04 03:21:05.970 [main] INFO  JobContainer - Set jobId = 0
2021-09-04 03:21:06.345 [job-0] INFO  OriginalConfPretreatmentUtil - Available jdbcUrl:jdbc:mysql://127.0.0.1:3306/datax_read?serverTimezone=UTC&useSSL=false&yearIsDateType=false&zeroDateTimeBehavior
=convertToNull&tinyInt1isBit=false&rewriteBatchedStatements=true.
2021-09-04 03:21:06.372 [job-0] INFO  OriginalConfPretreatmentUtil - table:[base_user_info] has columns:[id,user_name,sex,age,phone].
2021-09-04 03:21:06.529 [job-0] INFO  OriginalConfPretreatmentUtil - table:[base_user_info] all columns:[
id,user_name,sex,age,phone
].
2021-09-04 03:21:06.540 [job-0] INFO  OriginalConfPretreatmentUtil - Write data [
INSERT INTO %s (id,user_name,sex,age,phone) VALUES(?,?,?,?,?)
], which jdbcUrl like:[jdbc:mysql://127.0.0.1:3306/datax_write?serverTimezone=UTC&useSSL=false&yearIsDateType=false&zeroDateTimeBehavior=CONVERT_TO_NULL&rewriteBatchedStatements=true&tinyInt1isBit=fa
lse]
```

图 17-1

如图 17-2 所示，datax 会统计本次任务执行的结果。

```
2021-09-04 03:21:16.672 [job-0] INFO  JobContainer -
        [total cpu info] =>
                averageCpu                     | maxDeltaCpu                    | minDeltaCpu
                -1.00%                         | -1.00%                         | -1.00%


        [total gc info] =>
                 NAME                 | totalGCCount       | maxDeltaGCCount    | minDeltaGCCount    | totalGCTime        | maxDeltaGCTime     | minDeltaGCTime
                 PS MarkSweep         | 0                  | 0                  | 0                  | 0.000s             | 0.000s             | 0.000s
                 PS Scavenge          | 0                  | 0                  | 0                  | 0.000s             | 0.000s             | 0.000s

2021-09-04 03:21:16.673 [job-0] INFO  JobContainer - PerfTrace not enable!
2021-09-04 03:21:16.680 [job-0] INFO  StandAloneJobContainerCommunicator - Total 2 records, 49 bytes | Speed 4B/s, 0 records/s | Error 0 records, 0 bytes |  All Task WaitWriterTime 0.000s |  All Task WaitReaderTime 0.000s | Percentage 100.00%
2021-09-04 03:21:16.682 [job-0] INFO  JobContainer -
任务启动时刻                    : 2021-09-04 03:21:05
任务结束时刻                    : 2021-09-04 03:21:16
任务总计耗时                    :                 10s
任务平均流量                    :                4B/s
记录写入速度                    :              0rec/s
读出记录总数                    :                   2
读写失败总数                    :                   0
```

图 17-2

17.2 搭建控制台

本书统一使用开源社区的 datax-web 来搭建 DataX 的可视化控制台，控制台主要包括：后台管理系统 datax-admin 和任务执行器 datax-executor。

本书是基于 datax-web 的源码（从 GitHub 官网下载源码）来搭建可视化控制台，以便读者熟悉 datax-web。

17.2.1 构建部署包

构建部署包的过程如下：

（1）下载 datax-web 的源码压缩包“datax-web-master.zip”，并解压缩生成文件“datax-web-master”。

（2）在文件“datax-web-master”中执行打包命令“mvn clean install-Dmaven.test.skip=true”，生成部署包“datax-web-2.1.2.tar.gz”，并解压缩生成文件“datax-web-2.1.2”。

（3）执行文件“datax-web-2.1.2/bin”中的 Shell 脚本 install.sh，生成 modules 文件，并自动将需要部署的后台管理系统的部署包“datax-admin_2.1.2_1.tar.gz”和任务执行器部署包“datax-executor_2.1.2_1.tar.gz”复制到 modules 文件中，解压缩两个部署包，生成文件夹 datax-admin 和 datax-executor，如图 17-3 所示。

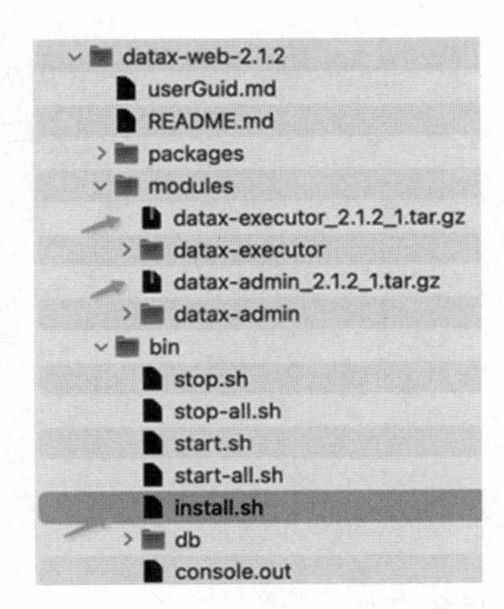

图 17-3

17.2.2 用部署包搭建后台管理系统 datax-admin

datax-web 已经将需要开发人员定制化的配置信息，统一用配置文件“/modules/datax-admin/bin/env.properties”来管理。

（1）在文件夹 datax-admin 中的配置文件 env.properties 中添加定制化的配置信息。

```
#①添加环境变量
WEB_LOG_PATH=${BIN}/../logs
WEB_CONF_PATH=${BIN}/../conf
DATA_PATH=${BIN}/../data
SERVER_PORT=9527
PID_FILE_PATH=${BIN}/dataxadmin.pid
#②添加接收告警的邮箱
MAIL_USERNAME=1234huxian@163.com
MAIL_PASSWORD=123456
#③配置后台管理系统的数据库连接信息
DB_HOST=127.0.0.1
DB_PORT=3306
DB_USERNAME=root
DB_PASSWORD=123456huxian
DB_DATABASE=datax_web
#④添加 debug 端口
REMOTE_DEBUG_SWITCH=true
REMOTE_DEBUG_PORT=7003
```

（2）在后台管理系统的数据库中执行 SQL 语句，SQL 语句路径为“/datax-web-2.1.2/bin/db/datax_web.sql”，目前仅支持 MySQL 数据库。

（3）使用命令“sh datax-admin.sh start”执行文件夹“/modules/datax-admin/bin /”中的脚本“datax-admin.sh”。

（4）访问域名“127.0.0.1:9527/index.html”，并使用用户名和密码（admin/123456）登录 DataX 的后台管理系统。

部署完成后，在台管理系统和任务执行器后会启动两个 Java 进程，其中，任务执行器支持单机部署和集群部署（即可以同时部署多个执行器进程），并使用路由算法进行调度执行器的负载均衡。

17.2.3 用部署包搭建任务执行器 datax-executor

datax-web 已经将需要开发人员定制化的配置信息统一用配置文件“/modules/datax-executor/bin/env.properties”来管理。

（1）在文件夹 datax-executor 中的配置文件 env.properties 中添加定制化的配置信息。

```
#①配置环境变量
SERVICE_LOG_PATH=${BIN}/../logs
SERVICE_CONF_PATH=${BIN}/../conf
DATA_PATH=${BIN}/../data
#②datax json 文件的存放路径
JSON_PATH=${BIN}/../json
#executor_port
EXECUTOR_PORT=9999
#③保持和 datax-admin 端口一致
DATAX_ADMIN_PORT=9527
#④添加 DataX 中 Python 脚本执行的路径
PYTHON_PATH=../../../../datax/bin/datax.py
#⑤datax_Web 服务器端口
SERVER_PORT=9504
#⑥debug 远程调试端口
REMOTE_DEBUG_SWITCH=true
REMOTE_DEBUG_PORT=7004
```

（2）使用命令“sh datax-executor.sh start”执行文件夹“/modules/datax-executor/bin /”中的脚本 datax-executor.sh。

17.2.4 使用可视化控制台执行 MySQL 异构数据迁移

使用可视化控制台，执行 MySQL 异构数据迁移的过程如下。

1. 创建项目

在控制台的“项目管理”中单击“添加”按钮，添加一个项目，如图 17-4 所示。

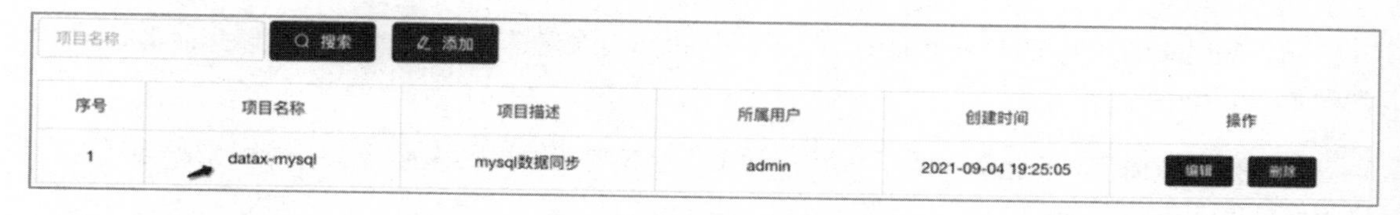

图 17-4

2. 创建 DataX 任务模板

在控制台的“任务管理/DataX 任务模板”中单击“添加”按钮，添加一个 DataX 任务，并关联到项目 datax-mysql 中，如图 17-5 所示。

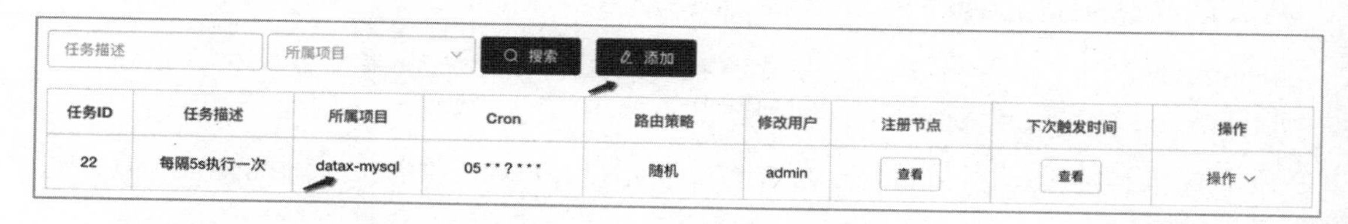

图 17-5

3. 创建 MySQL 数据源

在控制台的“数据源管理”中单击“添加”按钮，添加“datax_read”和“datax_write”数据源，如图 17-6 所示。

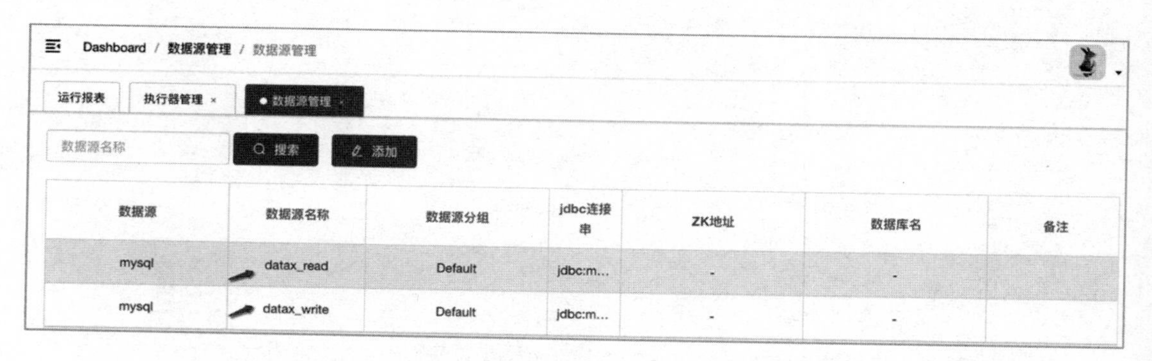

图 17-6

4. 创建任务执行器

在控制台的“执行器管理”中单击“添加”按钮，添加一个任务执行器，如图 17-7 所示。

图 17-7

属性字段描述如下：

① "AppName" 对应任务执行器部署包文件 "/modules/datax-executor/conf/application.yml" 中的属性 "datax.job.executor.appname"，本书配置其值为 "datax-executor"。

② "名称" 为 "datax 执行器"，以提高执行器的可读性。

③ "注册方式" 是指，调度中心获取执行器地址的方式。其中，

- 自动注册：执行器自动进行执行器注册，调度中心通过底层注册表动态发现执行器的机器地址。
- 手动录入：软件开发人员手动填写执行器的地址信息，如有多地址则用逗号分隔，供调度中心使用。

④ "机器地址" 为任务执行器的 IP 地址，"注册方式" 为 "手动录入" 时有效，支持人工维护执行器的地址信息。

5. 构建 DataX 执行的 Json 脚本

构建 DataX 执行的 Json 脚本，总共有四步。

（1）构建 reader 数据源。在控制台的 "任务管理/任务构建" 中单击 "任务构建" 按钮，构建 reader 数据源，如图 17-8 所示。其中，"数据库源" 选中 "datax_read"，"数据库表名" 选中 "base_user_info"，"表所有字段" 勾选 "全选"，单击 "下一步" 按钮。

（2）构建 write 数据源。在控制台中，"数据库源" 选中 "datax_write"，"数据库表名" 选中 "base_user_info"，"表所有字段" 勾选 "全选"，如图 17-9 所示，单击 "下一步" 按钮。

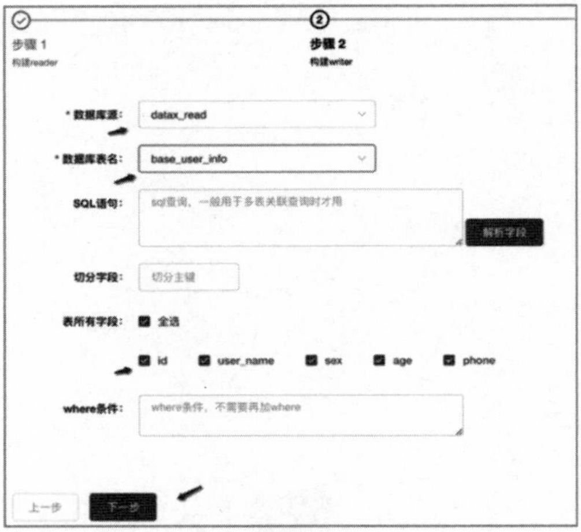

图 17-8

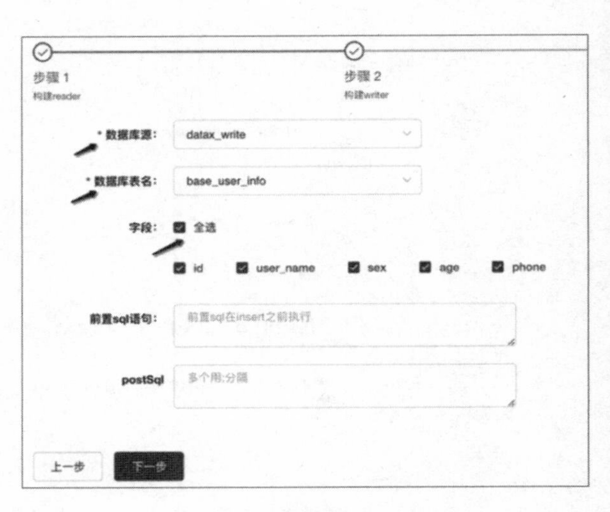

图 17-9

（3）绑定读数据源和写数据源中迁移表“base_user_info”的字段关系，如图 17-10 所示，“源端字段”勾选“全选”，“目标字段”也勾选“全选”（如果不需要全部表字段都签约，则可以勾选部分字段），单击“构建”按钮。

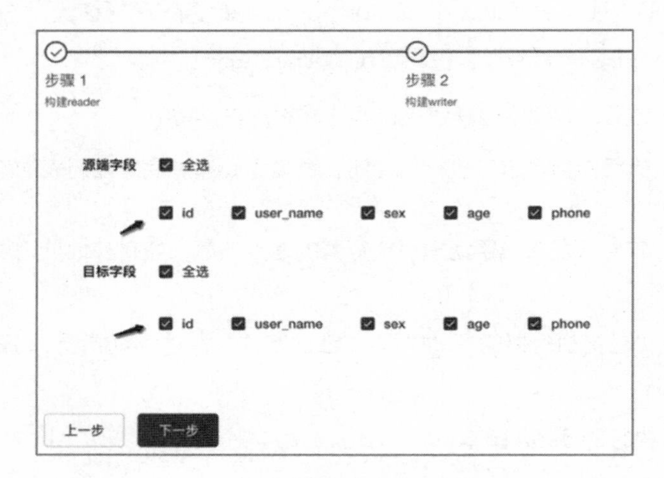

图 17-10

（4）在绑定字段关系之后，单击“构建”按钮构建出 DataX 可以执行的 Json 脚本，如图 17-11 所示。

1.构建　2.选择模板　复制json　(步骤：构建->选择模板->下一步)

```
1  {
2    "job": {
3      "setting": {
4        "speed": {
5          "channel": 3,
6          "byte": 1048576
7        },
8        "errorLimit": {
9          "record": 0,
10         "percentage": 0.02
11       }
12     },
13     "content": [
14       {
15         "reader": {
16           "name": "mysqlreader",
17           "parameter": {
18             "username": "yRjwDFuoPKlqya9h9H2Amg==",
19             "password": "YarYFyOwxUfl+bUJF5dDCA==",
20             "column": [
21               "`id`",
22               "`user_name`",
23               "`sex`",
24               "`age`",
25               "`phone`"
26             ],
27             "splitPk": "",
28             "connection": [
29               {
30                 "table": [
31                   "base_user_info"
32                 ],
33                 "jdbcUrl": [
34                   "jdbc:mysql://127.0.0.1:3306/datax_read?serverTimezone=UTC&useSSL=false"
35                 ]
36               }
37             ]
38           }
39         },
```

图 17-11

单击“选择模板”按钮选中一个 DataX 任务模板，如图 17-12 所示。

选择模板

任务ID	任务描述	所属项目	Cron	路由策略
22	每隔5s执行一次	datax-mysql	0 5 * * ? * * *	随机

共 1 条　10条/页　1　前往 1 页

图 17-12

6. 演示 DataX 任务执行的效果

在任务管理列表中，可以看到一条异构数据迁移的任务，如图 17-13 所示，单击“操作”按钮，选择“查询日志”跳转到日志管理按钮，可以看到任务执行成功的日志，如图 17-14 所示。

ID	任务名称	所属项目	Cron	路由策略	状态	注册节点	下次触发时间	执行状态	操作
27	base_user_info	datax-mysql	0 5 * * ? * * *	随机	停止 启动	查看	查看	无	操作
26	迁移表product_info的"1-2000000"数据到product_detail	datax-mysql	0 5 30 00,03 ? * * *	随机	停止 启动	查看	查看	失败	执行一次 查询日志 编辑

图 17-13

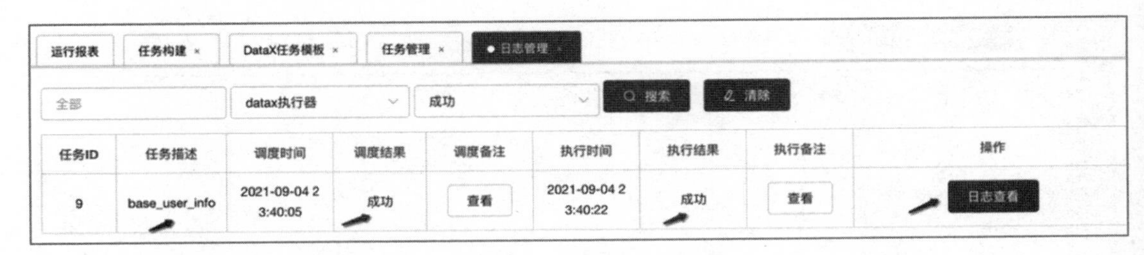

图 17-14

单击“日志查看”按钮，可以查看任务运行的日志，如图 17-15 所示，从数据源“datax_read”的表“base_user_info”成功迁移两条数据到数据源“datax_write”的表“base_user_info”中。

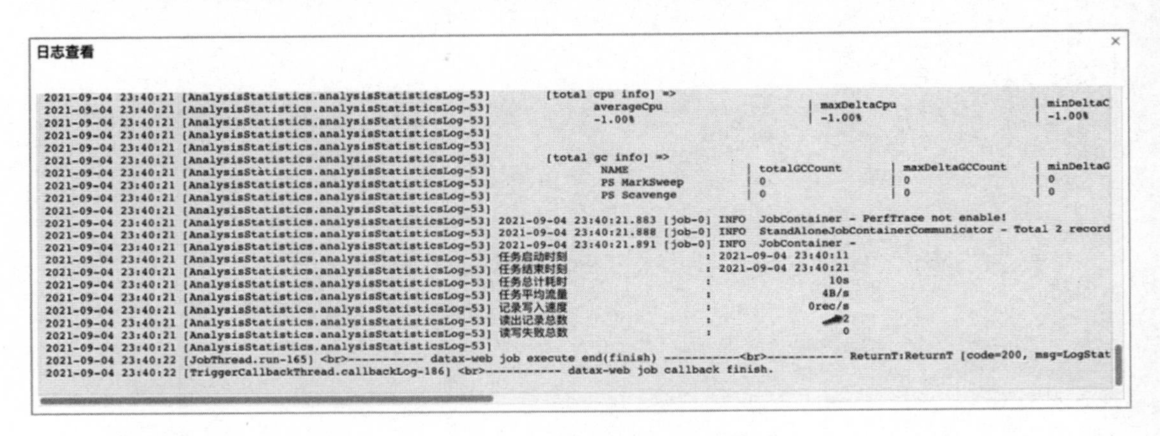

图 17-15

17.3 在 Spring Cloud Alibaba 应用中用 DataX 完成异构数据迁移

本项目的源码在本书配套资源的“chapterseventeen/”目录下。

笔者所在的公司已经在大范围使用 Spring Cloud Alibaba 作为基础框架，去升级原有的 RPC 框架，并启动业务中台战略，梳理和抽取公共的业务模型，从而将公共能力下沉到业务中台中。

在实施业务中台的过程中，需要调研现有业务，并按照“DDD 领域模型”拆分/合并服务，形成不同的能力中心，比如用户中心、商品中心等。在拆分和合并服务的过程中，需要重构代码和数据库表。如果重构了数据库表结构，则在新服务上线时需要优先考虑异构数据源的数据迁移。

本项目演示在数据库表结构重构后，使用 DataX 完成商品数据从旧表向新表的迁移。

1. 基于 DataX 的商品数据迁移的物理架构

图 17-16 所示为基于 DataX 的商品数据迁移的物理架构。

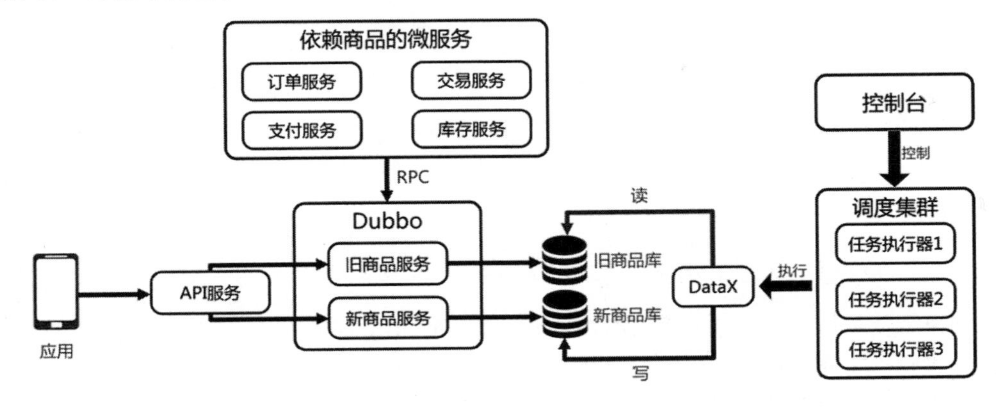

图 17-16

其中包括以下架构组件。

- API 服务：统一采用 RESTful 风格，并依赖基于 Dubbo 的 RPC 服务。其中，商品 API 会依赖新商品服务和旧商品服务。
- Dubbo：Dubbo 服务统一采用 Dubbo 作为基础 RPC 框架，Dubbo 服务底层会封装业务领域中业务功能的实现，比如商品域中的新增/查询商品。
- 依赖商品的微服务：使用基于 RPC 的通信渠道，去调用商品服务的其他业务域的微服务，比如订单服务、支付服务等。
- DataX：从数据源的角度支撑商品数据的迁移。
- 控制台：管理 DataX 迁移任务模板，以及任务调度集群。
- 调度集群：执行数据迁移的机器列表，控制台会采用路由算法（比如轮询算法）完成调度集群的负载均衡。

2. 基于 DataX 的商品数据迁移的功能架构

图 17-17 所示为基于 DataX 的商品数据迁移的功能架构。将旧商品库中的商品信息表“product_info”的数据，迁移到新商品库中基础商品信息表“base_ product_info”和商品详情表“product_detail”中。

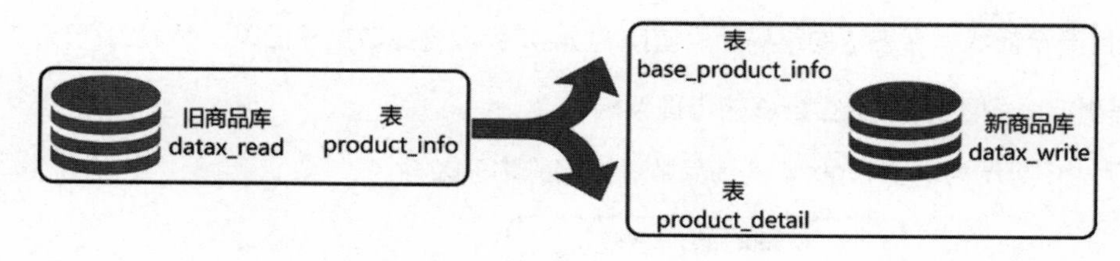

图 17-17

3. 用 Spring Cloud Alibaba 初始化 3 个服务

使用 Spring Cloud Alibaba 作为基础框架快速初始化 3 个服务，具体可以查阅本书配套资源中的服务源码。

如图 17-18 所示，3 个服务的主要功能如下：

- good-old-server 用来模拟旧的商品服务，在旧商品数据库中新增/查询商品数据。
- good-new-server 用来模拟新的商品服务，在新商品数据库中新增/查询商品数据。
- good-api 用来模拟商品 API 服务，通过 RPC 调用新/旧商品服务，并对比从新旧商品库中查询出来的商品数据，模拟“在数据迁移过程中，抽样验证新旧商品库中的商品数据一致性”的业务场景。

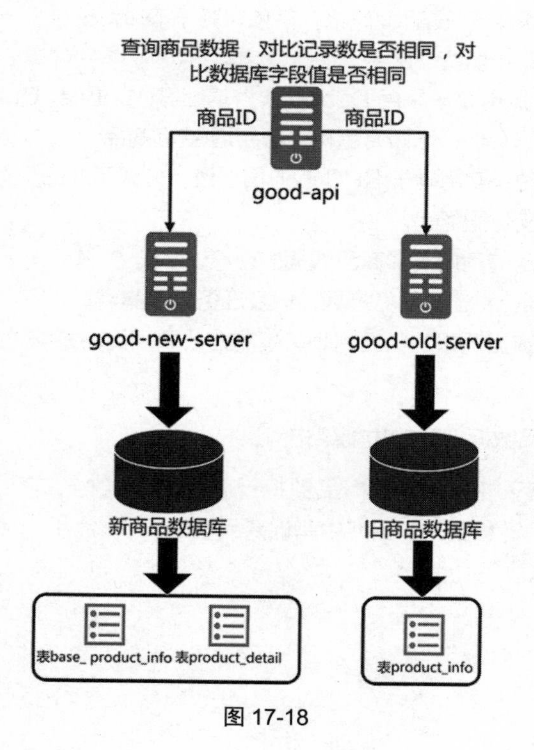

图 17-18

4. 在控制台中配置 DataX 任务，执行数据迁移

本项目在旧商品库“datax_read”的表“product_info”中准备了 1000 万条商品数据（生成商品数据的代码可以参考本书配套资源中的代码）。

（1）将 1000 万条商品数据切割为 5 份，并使用 10 个 DataX 任务迁移数据，如图 17-19 所示。

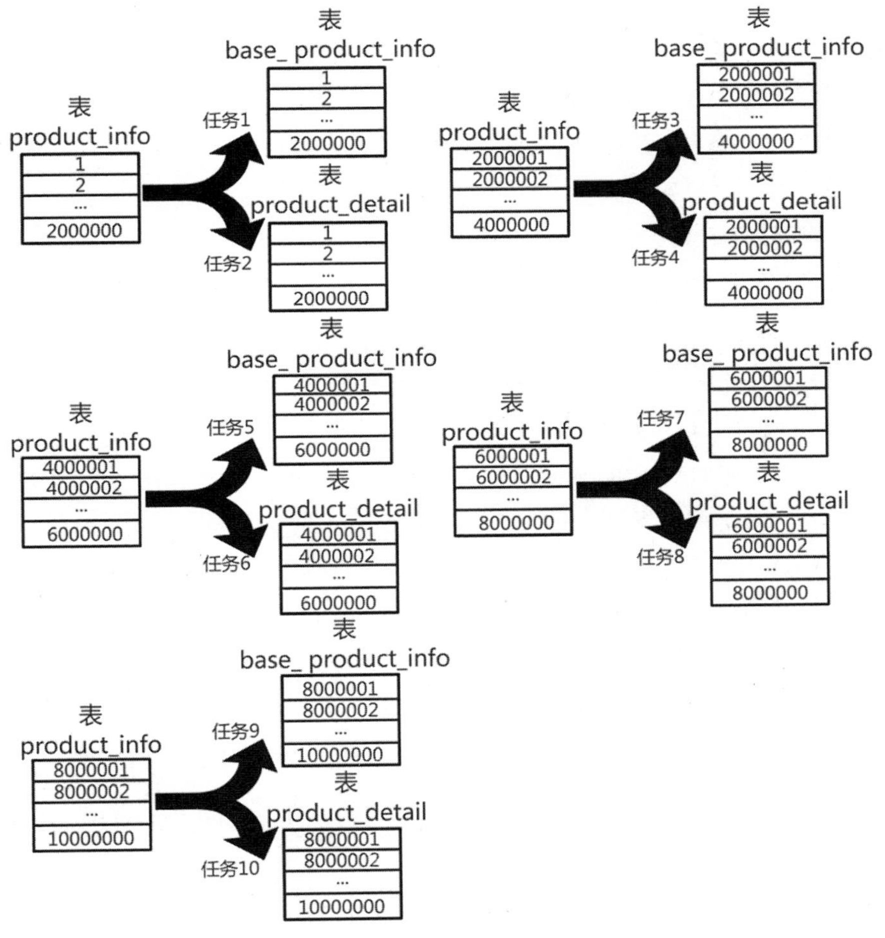

图 17-19

（2）在 DataX 的控制台配置 10 个 DataX 任务，并在迁移的 SQL 语句中添加查询条件，比如“where id<=2000000”。图 17-20 所示为迁移前 200 万条数据的两个 DataX 任务。

13	迁移表product_info的"1-2000000"数据到product_detail	datax-mysql	05 30 00,03 ? * **	随机	停止 启动	查看	查看	无	操作
12	迁移表product_info的"1-2000000"数据到base_product_info	datax-mysql	05 30 00,03 ? * **	随机	停止 启动	查看	查看	无	操作

图 17-20

（3）在控制台中，开启 10 个 DataX 任务，开始进行数据迁移。

（4）在控制台中，查看任务 1 和任务 2 运行的结果，并查看运行日志（其他 8 个任务，读者可以尝试自己运行，并观察结果）。

图 17-21 所示为基础商品信息表“base_ product_info”的迁移结果，任务总耗时 5035s，任务的 TPS 为 397，迁移成功的记录数为 200000，迁移失败的记录数为 0。

```
2021-09-07 23:25:40 [AnalysisStatistics.analysisStatisticsLog-53] 任务启动时刻          : 2021-09-07 22:01:44
2021-09-07 23:25:40 [AnalysisStatistics.analysisStatisticsLog-53] 任务结束时刻          : 2021-09-07 23:25:40
2021-09-07 23:25:40 [AnalysisStatistics.analysisStatisticsLog-53] 任务总计耗时          :               5035s
2021-09-07 23:25:40 [AnalysisStatistics.analysisStatisticsLog-53] 任务平均流量          :           46.50KB/s
2021-09-07 23:25:40 [AnalysisStatistics.analysisStatisticsLog-53] 记录写入速度          :            397rec/s
2021-09-07 23:25:40 [AnalysisStatistics.analysisStatisticsLog-53] 读出记录总数          :             2000000
2021-09-07 23:25:40 [AnalysisStatistics.analysisStatisticsLog-53] 读写失败总数          :                   0
```

图 17-21

图 17-22 所示为商品详情表“product_detail”的迁移结果，任务总耗时 2814s，任务 TPS 为 710，迁移成功的记录数为 200000，迁移失败的记录数为 0。

```
2021-09-13 16:30:35 [AnalysisStatistics.analysisStatisticsLog-53] 2021-09-13 16:30:35.812 [job-0] INFO  JobContainer - PerfTrace not enable!
2021-09-13 16:30:35 [AnalysisStatistics.analysisStatisticsLog-53] 2021-09-13 16:30:35.812 [job-0] INFO  StandAloneJobContainerCommunicator - Total 2000000
2021-09-13 16:30:35 [AnalysisStatistics.analysisStatisticsLog-53] 2021-09-13 16:30:35.815 [job-0] INFO  JobContainer -
2021-09-13 16:30:35 [AnalysisStatistics.analysisStatisticsLog-53] 任务启动时刻          : 2021-09-13 15:43:41
2021-09-13 16:30:35 [AnalysisStatistics.analysisStatisticsLog-53] 任务结束时刻          : 2021-09-13 16:30:35
2021-09-13 16:30:35 [AnalysisStatistics.analysisStatisticsLog-53] 任务总计耗时          :               2814s
2021-09-13 16:30:35 [AnalysisStatistics.analysisStatisticsLog-53] 任务平均流量          :           57.62KB/s
2021-09-13 16:30:35 [AnalysisStatistics.analysisStatisticsLog-53] 记录写入速度          :            710rec/s
2021-09-13 16:30:35 [AnalysisStatistics.analysisStatisticsLog-53] 读出记录总数          :             2000000
2021-09-13 16:30:35 [AnalysisStatistics.analysisStatisticsLog-53] 读写失败总数          :                   0
```

图 17-22

（5）启动服务 good-old-server、good-new-server 和 good-api，用 Postman 发送 HTTP 请求验证新商品数据库和旧商品数据库的数据一致性。

如图 17-23 所示，查询新商品数据库中的基础商品记录为 2000000 条，商品详情记录为 2000000 条。

如图 17-24 所示，查询旧商品数据库中的商品记录为 10000000 条。本项目只迁移了旧商品库中的第一部分商品，总计为 2000000 条，新旧商品数据库读和写的商品数量一致。

如图 17-25 所示，对比新旧商品数据库中商品 ID 为“3428969865897452261”的数据库的结果为“商品信息相同”。

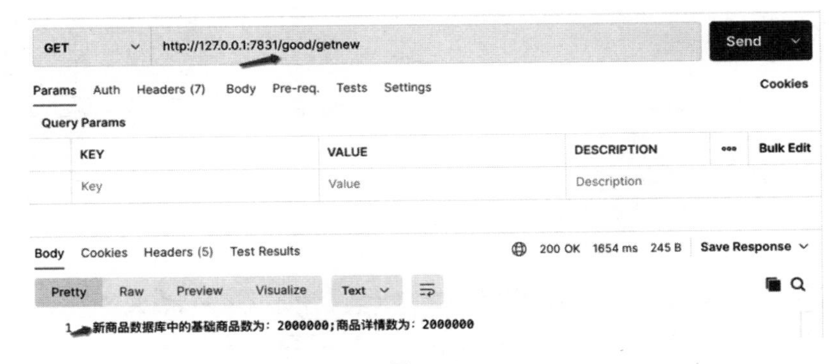

图 17-23

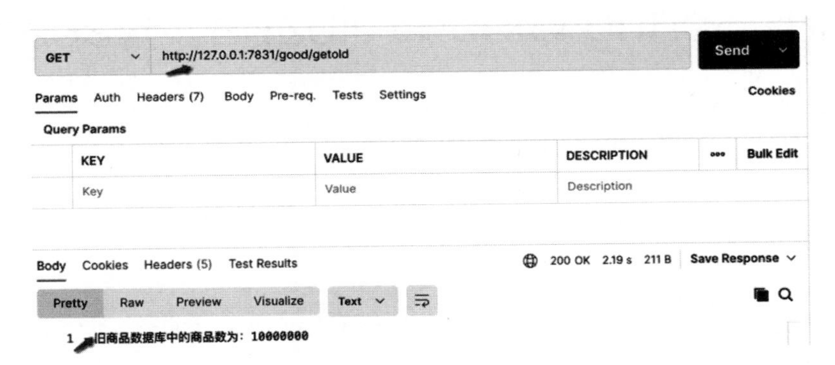

图 17-24

图 17-25

18

第 18 章

【项目】业务链路告警平台——基于 Spring Cloud Alibaba、Nacos 和 Skywalking

在微服务架构中，如果开发人员能够搭建一个可靠的、实时的、精准的业务链路告警平台，并且告警平台能够可视化地将告警推送给开发人员和运维人员，则能够极大地提高微服务架构中微服务系统运行的稳定性。

本项目将利用开源技术（Spring Cloud Alibaba、Nacos、Skywalking、RocketMQ），构建一个基于钉钉的、通用的业务链路告警平台。

18.1 告警平台的整体架构设计

图 18-1 所示为告警平台的整体架构设计。

其中包含如下核心组件。

（1）电商微服务：在告警平台中，电商微服务主要包括交易服务、订单服务、支付服务、商品服务、库存服务，以及物流服务。在本项目中，它们主要用于模拟电商业务场景，并在告警平台中生成业务链路告警。

（2）Nacos 注册中心和配置中心：在告警平台中，电商微服务可以注册到注册中心，并将服务配置属性托管到配置中心，Skywalking 也可以将部分配置信息托管到配置中心。

- A 代表 在Skywalking平台启动时配置告警规则和告警通知回调的Webhook地址
- B 代表 告警服务调用RocketMQ的Open API，查询消息消费进度
- C 代表 Skywalking用Webhook地址访问告警服务的Open API，将服务监控告警通知到告警服务
- D 代表 告警服务调用Skywalking的Graphql API查询服务的错误链路信息
- E 代表 从Nacos的注册中心获取实例的健康状态

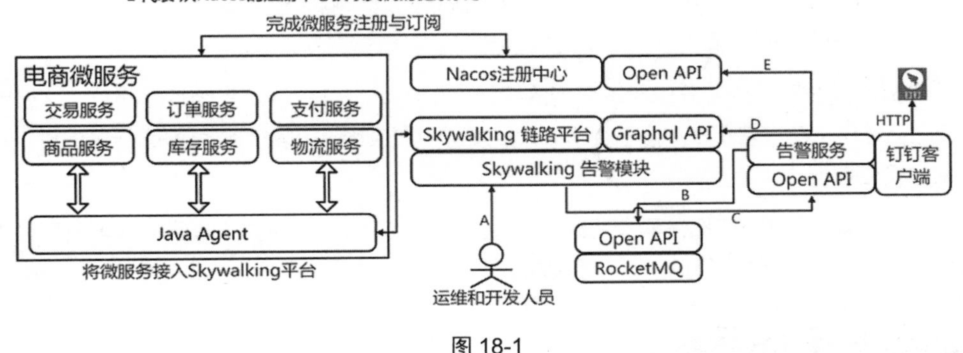

图 18-1

（3）分布式链路追踪平台 Skywalking：在告警平台中，电商微服务可以接入 Skywalking，并完成服务调用的分布式链路追踪。

（4）告警服务：其中包含如下功能。

①利用 Webhook 地址回调告警服务提供的 Open API 接口，并将被 Skywalking 监控的服务的告警消息推送给告警服务。

②主动调用 Skywalking 提供的 Graphql API 接口，查询出现错误的服务链路信息。

③主动调用 Nacos 提供的 Open API 接口，查询注册中心服务健康状态的数据。

④处理 Skywalking 的告警消息、错误的服务链路信息，以及服务的健康状态数据，并将这些处理后的数据通过钉钉客户端推送到指定的钉钉告警群。

⑤主动调用 RocketMQ 的 Open API，查询指定消息主题的消费进度。如果消息堆积量达到告警阈值，则触发告警并通知到指定的钉钉告警群。

告警平台的整体设计流程如下：

（1）搭建 Skywalking 和 Nacos 的集群环境，可以参考本书的第 4 章和第 10 章。

（2）在 IDEA 控制台中启动告警服务。

（3）在 Nacos 控制台中修改 Skywalking 的告警模块中的告警通知的回调地址（Webhook 地址）。

（4）分别启动电商微服务中的交易服务、订单服务等。

（5）模拟故障场景：在整个调用链路中注入一个超时异常，这样 Skywalking 就能捕获链路中的异常信息。在告警平台中会收到链路异常信息，并实时地推送到钉钉告警群中。

（6）模拟 RT（请求响应时间）超过告警规则的阈值：在整个调用链路中，注入一些请求响应的延迟时间，模拟接口的慢响应场景，从而让整个链路的请求响应时间超过告警规则的阈值。在告警规则中会收到慢响应的告警信息，并实时地推送到钉钉告警群中。

（7）模拟电商微服务中部分服务不可用的场景：使用 kill 命令手动关闭几个服务（比如商品服务、库存服务等）。在告警平台中会收到 Nacos 集群中对应服务及实例不可用的告警信息，并实时地推送到钉钉告警群中。

（8）在电商微服务中，模拟在异步发送 RocketMQ 消息的过程中由于消费者不可用而导致消息大量堆积，并实时地推送告警到钉钉告警群中。

18.2 告警服务详细设计

在告警平台中，告警服务需要接收第三方平台（Nacos、Skywalking 等）的监控数据，并二次处理监控数据，且将最终的告警消息推送给钉钉客户端。

告警服务的详细设计如图 18-2 所示。

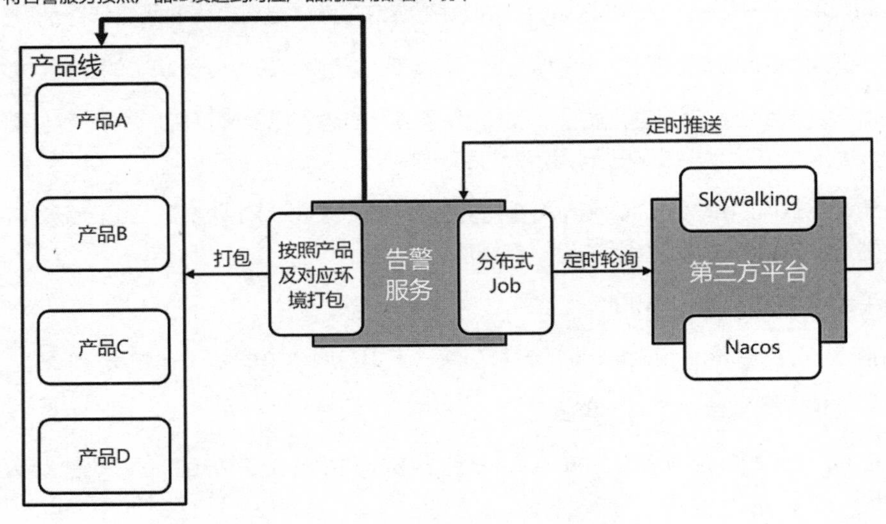

图 18-2

其主要功能如下。

（1）支持产品化部署：一个大型的互联网公司通常都会有多个成熟的产品。告警平台作为基础功能，需要下沉到基础中台中，并作为统一的中台服务，为不同的产品中的微服务提供告警功能。

（2）支持定时拉取告警数据：用分布式 Job 定时地从第三方平台（Skywalking 和 Nacos）拉取告警数据。

（3）支持监控数据的“推”模式：可以接收第三方平台（比如 Skywalking）推送的监控数据。

（4）支持动态的过滤告警消息：在不重启告警服务的前提下，可以在配置中心的控制台中在线修改过滤条件，并过滤指定的告警消息，这样开发人员在告警钉钉群中就接收不到已经过滤的告警消息。

18.2.1 产品化部署设计

图 18-3 所示为告警服务的产品化部署设计。

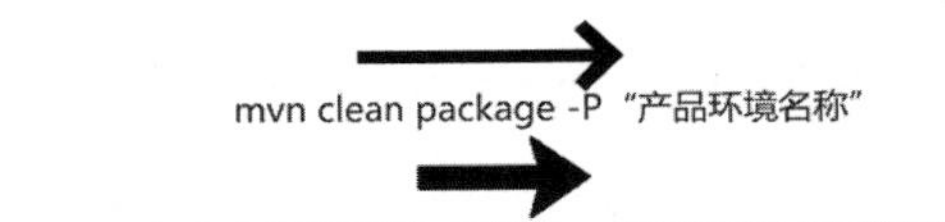

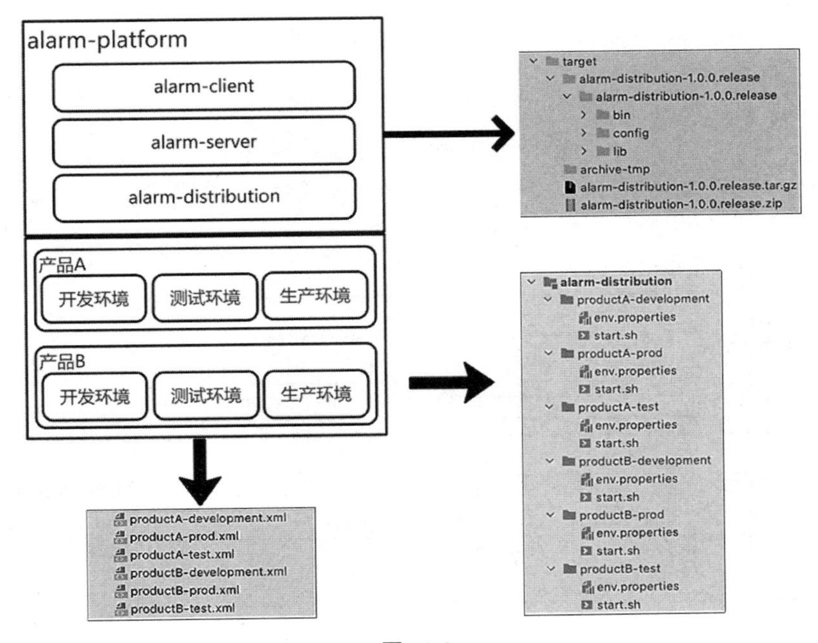

图 18-3

告警服务（alarm–plaform）包含 alarm–client（API）、alarm–server（告警核心代码）和 alarm–distribution（运维部署包），主要设计流程如下。

（1）在 alarm–distribution 中添加产品环境，比如产品 A 的开发环境、测试环境，以及生产环境。在本项目中添加了两个示例产品（产品 A 和产品 B），开发人员在实际的业务项目中可以结合自身的环境来进行配置。

（2）执行 Maven 的打包命令，打包部署的 Jar 包。比如，要给运维人员打包一个产品 A 开发环境的部署包，则可以在 Maven 中执行命令“mvn clean package –PproductA–development”，打包完成后，会在文件“/alarm–distribution/target”中生成部署包“alarm–distribution–1.0.0.release.tar.gz”或者“alarm–distribution–1.0.0.release.zip”。

（3）在线上机器上解压缩部署包后，运维人员可以执行命令“sh start.sh start”来启动告警服务。

（4）如果告警平台需要接入新的产品，则只需要在模块 alarm–distribution 中添加一套个性化的属性文件和启动脚本文件，并用指定的 XML 文件去加载属性和启动脚本。运维人员只需要在自动部署的构建脚本中，用指定产品环境名称（和开发人员约定的，比如产品 A 的开发环境，可以统一约定为 productA–development）将打包命令“mvn clean package –PproductA–development”添加到自动化运维流程中。

18.2.2 Nacos 服务健康告警设计

图 18–4 所示为 Nacos 的服务健康告警设计，具体的流程如下：

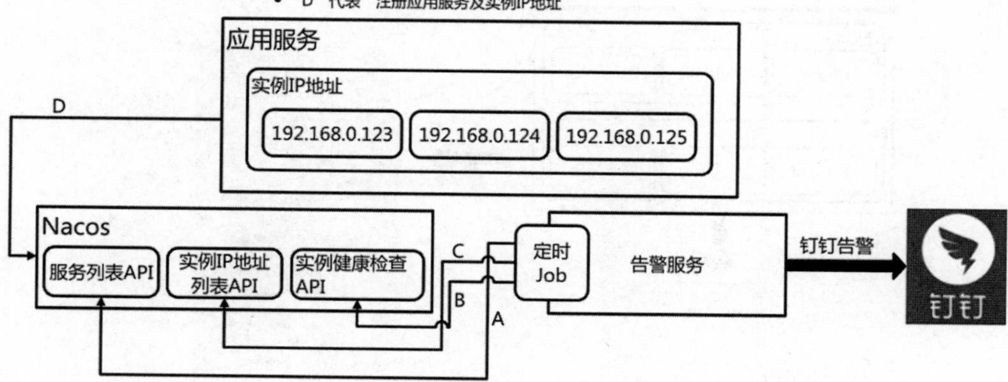

图 18-4

（1）在告警服务中，在服务启动后会启动一个定时 Job 类 NacosAlarmHealthJob（基于 Elastic Job），定时地从注册中心获取服务列表、实例 IP 地址列表，以及实例健康状态数据。

（2）在告警服务中，用缓存来存储从注册中心获取的服务列表，以及实例 IP 地址列表数据，并异步地更新缓存。

（3）在告警服务中，告警服务会异步地对比注册中心和缓存中对应的服务列表，以及实例 IP 地址列表数据的差异性。如果此时应用服务在重启或者应用服务的某一个实例出现不可用，则会准实时地将其从注册中心剔除，这样告警服务就能感知服务和实例的健康状态，从而构造一条钉钉告警消息，通知到指定的钉钉告警群。

（4）在告警服务中，告警服务会调用 Nacos 的实例健康检查的 API，获取指定服务的健康状态。如果应用服务的实例列表中的某一个实例的健康状态被 Nacos 判定为不健康，则会构造一条钉钉告警消息，通知到指定的钉钉告警群。

18.2.3 Skywalking 链路错误告警设计

图 18-5 所示为 Skywalking 链路错误告警设计，具体流程如下：

（1）用 Java Agent 将电商微服务（交易服务、订单服务、支付服务等）接入 Skywalking 平台。

（2）在电商微服务中，模拟服务不可用和网络拥塞的业务场景，这样 Skywalking 平台会追踪到错误的链路。

（3）在告警服务中，通过定时器定时地调用 Skywalking 的 GraphQL API（访问地址为 127.0.0.1:8087/graphql，其中，8087 为 Skywalking 的 UI 控制台的端口号）查询错误的链路信息。

（4）在告警服务中，将“从 Skywalking 中查询到的错误链路信息列表”转换为钉钉告警消息，并实时地推送到钉钉告警群。

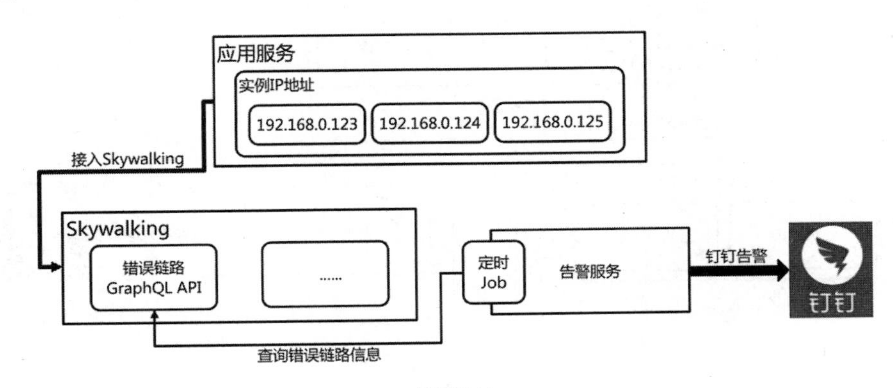

图 18-5

18.2.4 Skywalking 的指标告警设计

图 18-6 所示为 Skywalking 的指标告警设计，具体流程如下：

（1）在告警服务中定义一个 RESTful API 接口（Post 类型）AlarmPushController，在 API 的接口方法 pushAlarmDingDingMessage()中，处理从 Skywalking 平台收集的告警消息（接口路径为“/alarm/alarmPushData”）。

（2）在 Skywalking 中，告警模块会向接口“/alarm/alarmPushData”推送指标告警消息。

（3）在告警服务中，处理 Skywalking 的告警消息，并校验一些白名单规则，将告警消息推送到钉钉告警群。

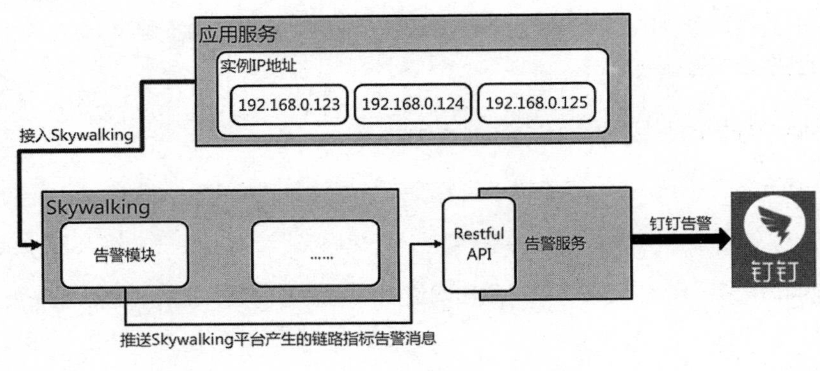

图 18-6

18.2.5 RocketMQ 消息堆积告警设计

图 18-7 所示为 RocketMQ 消息堆积告警设计，具体流程如下。

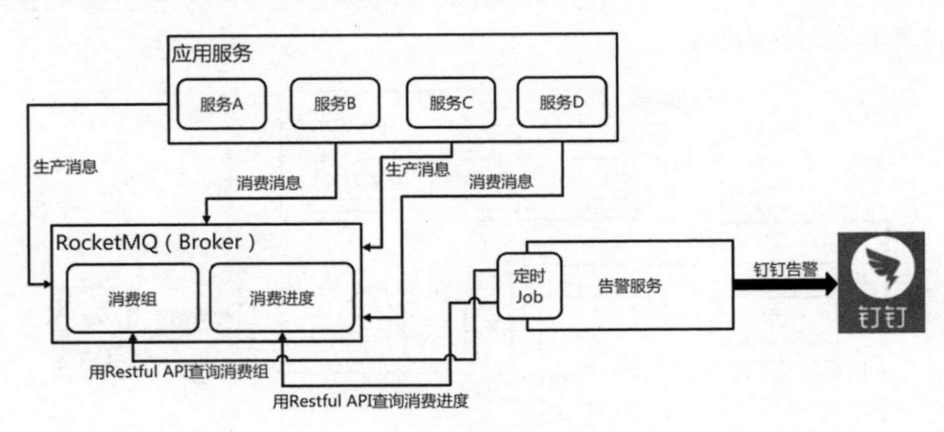

图 18-7

（1）在告警服务中，启动一个定时器类 RocketmqAlarmJob 定时地访问 RocketMQ 的控制台的 HTTP 接口。

- 告警服务访问 RocketMQ 的 RESTful API，获取 RocketMQ 集群中的所有的消费组，访问地址如下所示。

```
127.0.0.1:18000/consumer/groupList.query
```

- 告警服务访问 RocketMQ 的 RESTful API，获取 RocketMQ 集群中的组的消费进度（比如消费组名称为“test”的消费进度），访问地址如下所示。

```
127.0.0.1:18000/consumer/queryTopicByConsumer.query?consumerGroup=test
```

（2）在告警服务中，解析并处理 RocketMQ 的消费进度数据，并生成告警消息通知钉钉告警群。

18.3 分析告警服务的部分源码

本书会提供项目的详细源码，这里就简单地分析一下告警服务的部分源码。

18.3.1 用分布式 Job 类 NacosAlarmHealthJob 实现 Nacos 服务健康告警

在告警服务中，用分布式 Job 类 NacosAlarmHealthJob 去实现 Nacos 服务健康告警，部分代码实现如下所示：

```
    //①从注册中心获取所有的服务名称
    private JSONObject getServiceList(Map<String, String> params) throws
Exception {
        RestResult restResult =
HttpClientUtils.get(nacosAlarmConfig.getNacosIp() +
nacosAlarmConfig.getNacosServiceListApi(), params);
        if (restResult.getCode() != HttpStatus.SC_OK) {
            throw new Exception("HTTP 请求失败" + restResult.getCode() + " message:
" + restResult.getMessage());
        }
        String message = restResult.getMessage();
        return JSON.parseObject(message);
    }
    //②从注册中心获取指定服务的实例 IP 地址列表
    private JSONObject getServiceInstanceList(String serviceName) throws
Exception {
        Map<String, String> params = new HashMap<>(3);
        params.put("namespaceId", nacosAlarmConfig.getNacosNameSpaceId());
        params.put("serviceName", serviceName);
        RestResult restResult = HttpClientUtils
```

```
        .get(nacosAlarmConfig.getNacosIp() +
nacosAlarmConfig.getNacosInstanceListApi(), params);
        if (restResult.getCode() != HttpStatus.SC_OK) {
            throw new Exception("HTTP 请求失败" + restResult.getCode() + " message:
" + restResult.getMessage());
        }
        String message = restResult.getMessage();
        return JSON.parseObject(message);
    }
    //③发送 Nacos 告警消息
    public void sendAlarm(List<NacosAlarmMessage> alarmMessageList) throws
ApiException {
        if (alarmMessageList.isEmpty()) {
            return;
        }
        //④构建钉钉消息请求对象 OapiRobotSendRequest
        OapiRobotSendRequest markDownRequest = new OapiRobotSendRequest();
        markDownRequest.setMsgtype("markdown");
        OapiRobotSendRequest.Markdown markdown = new
OapiRobotSendRequest.Markdown();
        markdown.setTitle("NACOS 告警");
        StringBuffer markDownText = new StringBuffer();
        AtomicInteger successCount = new AtomicInteger();

        alarmMessageList.forEach(info -> {
            //⑤构建告警消息体
            String addText =
                "### NACOS 报警 \n\n" + "### 所属平台 :" +
nacosAlarmConfig.getPlatform() + "\n\n" + "### 命名空间 :" + info
                    .getNameSpace() + "\n\n" + "### 所属服务 :" +
info.getServiceName() + "\n\n" + "### 服务实例 :" + info
                    .getIp() + "\n\n" + "### 告警时间 :" +
DateFormatUtils.format(info.getTime(), "yyyy-MM-dd HH:mm:ss")
                    + "\n\n" + "### 告警内容 :" + info.getMessage() + "\n\n" + "---"
+ "\n\n";
            markDownText.append(addText);
            successCount.getAndIncrement();
        });
        markdown.setText(String.valueOf(markDownText));
        markDownRequest.setMarkdown(markdown);
        String secret = alarmConfig.getSecret();
    String webHook = alarmConfig.getWebhook();
    //⑥获取钉钉客户端对象 DingTalkClient
        DingTalkClient client = DingDingUtils.getClient(secret,webHook);
        OapiRobotSendResponse response = client.execute(markDownRequest);
```

```
        log.info("execute:{}" + response.toString());
        OapiRobotSendRequest request = new OapiRobotSendRequest();
        request.setMsgtype("text");
        OapiRobotSendRequest.Text text = new OapiRobotSendRequest.Text();
        text.setContent("请相关人员尽快处理报警所示异常");
        request.setText(text);
        //⑦添加钉钉消息的责任人，通常是设置手机号码
        OapiRobotSendRequest.At at = new OapiRobotSendRequest.At();
        addResponsible(alarmMessageList, at);
        request.setAt(at);
        client.execute(request);
    }
```

18.3.2 用分布式 Job 类 SkywalkingErrorAlarmJob 实现 Skywalking 链路错误告警

在告警服务中，用分布式 Job 类 SkywalkingErrorAlarmJob 去实现 Skywalking 链路错误告警，部分代码实现如下所示：

```
    //①执行定时 Job
    public void execute(ShardingContext shardingContext) {
        Date date = new Date();
        String nowTime = DateFormatUtils.format(date.getTime(), "yyyy-MM-dd
HHmm");
        String oldTime = DateFormatUtils
                .format(date.getTime() -
Long.parseLong(skywalkingAlarmConfig.getTime()), "yyyy-MM-dd HHmm");
        //②调用 Skywalking 的 GraphQL API 获取错误链路信息
        String paramJson = getParamJson(nowTime, oldTime);
        List<String> errorKeyList;
        try {
            RestResult post =
HttpClientUtils.post(skywalkingAlarmConfig.getReqApi(), paramJson);
            HttpClientUtils.checkHttpResult(post);
            errorKeyList = getErrorKeyList(post.getMessage());
            List<SkywalkingAlarmMessage> skywalkingAlarmMessageList = new
ArrayList<>();
            for (String errorKey : errorKeyList) {
                String message = reqErrorData(errorKey);
                skywalkingErrorAlarmService.
handleMessage(skywalkingAlarmMessageList, message);
            }
            //③过滤掉白名单错误链路告警
            whiteList(skywalkingAlarmMessageList);
            removeDuplicatedAlarm(skywalkingAlarmMessageList);
```

```
            //④用钉钉客户端发送错误链路告警消息
            sendAlarm(skywalkingAlarmMessageList);
        } catch (Exception e) {
            log.error("skywalking 告警失败。", e);
        }
    }
```

18.4 将电商微服务接入告警平台，验证告警平台的实时告警功能

在本项目中，用服务 trade-server、order-server、good-server、pay-server 等去模拟了电商中购买商品的下单和支付的场景。

（1）本项目中购买商品下单的服务调用流程如下：

交易服务（trade-server）→订单服务（order-server）→商品服务（good-server）

（2）本项目中购买商品支付服务流程如下：

交易服务（trade-server）→支付服务（pay-server）→库存服务（inventory-server）→物流服务（logistics-server）

18.4.1 启动告警平台的软件环境

告警平台的软件环境包括：中间件环境、微服务环境及告警服务环境。

（1）告警平台需要依赖一些中间件，比如 Skywalking、Nacos 等：

- 启动 Skywalking。
- 启动 Nacos 的注册中心和配置中心。
- 启动 RocketMQ。
- 启动 ZooKeeper。
- 启动 Redis。
- 启动 Elastic Job。
- 在钉钉上创建钉钉告警群。

（2）启动微服务及告警平台的告警服务。

- 启动微服务，包括 trade- server、order-server、good-server、pay-server 等。
- 启动告警服务 alarm-platform。

在项目中会模拟如下业务故障：

- 通过配置中心，动态地控制在微服务调用过程中植入的调用延迟时间，模拟超时故障，并触发 Skywalking 的告警规则的阈值。
- 使用“kill -9”命令关闭服务进程，模拟服务不可用的技术故障，服务的调用方抛出 RPC 异常，Skywalking 捕获异常链路。
- 使用“kill -9”命令关闭服务进程，Nacos 健康检查会捕获服务不可用的异常。

18.4.2 在购买商品时，在下单过程中验证实时告警功能

具体过程如下。

1. 模拟延迟故障，触发告警平台的 RT（响应时间）告警，并通知到钉钉告警群

模拟延迟故障，触发 Skywalking 的 RT（响应时间）告警，并通知到钉钉告警群，具体过程如下。

（1）在交易服务（trade-server）、订单服务（order-server）和商品服务（good-server）中添加延迟调用的功能，并动态地从配置中心的配置文件 simulated-fault.properties 中读取延迟时间。

3 个服务延迟时间的具体配置信息如下：

```
//订单服务延迟时间为 660ms
order.alarm.delayTime=660
//交易服务延迟时间为 660ms
trade.alarm.delayTime=660
//商品服务延迟时间为 660ms
good.alarm.delayTime=660
```

交易服务延迟调用的具体代码实现如下所示：

```
@PostMapping(value = "/buy")
@ResponseBody
public boolean buy(String goodId){
    boolean tradeResult=tradeService.buy(goodId);
    try {
       //从配置中心读取延迟时间，比如为 500ms，则延迟 500ms 返回执行的结果
       Thread.sleep(Long.valueOf(tradeAlarmConfig.getDelayTime()));
    }catch (InterruptedException e){
        System.out.println(e.getMessage());
    }
    return tradeResult;
}
```

订单服务延迟调用的具体代码实现如下所示：

```
@Override
public Long createOrder(String goodId) {
```

```
    //①模拟创建订单
    Long orderId=distributedService.nextId();
    //②返回订单号
    orderManager.deductionInventory(goodId);
try{
        //③从配置中心读取延迟时间，比如为 500ms，则延迟 500ms 执行的结果
        Thread.sleep(Long.valueOf(orderAlarmConfig.getDelayTime()));
    }catch (InterruptedException e){
        System.out.println(e.getMessage());
    }
    return orderId;
}
```

商品服务延迟调用的具体代码实现如下所示：

```
@Override
public boolean deductionInventory(String goodId) {
    System.out.println("锁住当前商品库存");
    //从配置中心读取延迟时间，比如为 500ms，则延迟 500ms 执行的结果
    try{
        Thread.sleep(Long.valueOf(goodAlarmConfig.getDelayTime()));
    }catch (InterruptedException e){
        System.out.println(e.getMessage());
    }
    return true;
}
```

（2）设置订单服务的订单接口 OrderService 的响应超时时间为 3000ms，具体代码实现如下所示：

```
@DubboService(version = "1.0.0",group = "order-server",timeout = 3000)
public class OrderServiceImpl implements OrderService {
    ...
}
```

（3）设置交易服务中调用订单接口 OrderService 的调用超时时间为 3000ms，具体代码实现如下所示：

```
@DubboReference(version = "1.0.0",group = "order-server",timeout = 3000)
private OrderService orderService;
```

（4）在配置中心的配置文件 alarm.default.alarm-settings 中，配置 Skywalking 告警模块的告警规则，具体配置如下：

```
###响应时间的阈值为 2000ms
rules:
   service_resp_time_rule:
    metrics-name: service_resp_time
```

```
        op: ">"
        threshold: 2000
        period: 2
        count: 1
        silence-period: 1
        message: Response time of service {name} is more than 2000ms in 1 minutes
of last 1minutes.
      endpoint_relation_resp_time_rule:
        metrics-name: endpoint_relation_resp_time
        op: ">"
        threshold: 2000
        period: 2
        count: 1
        silence-period: 1
        message: endpoint_relation_resp_time of endpoint {name} is more than
2000ms in 1 minutes of last 2 minutes
    webhooks:
      - http://127.0.0.1:34769/alarm/alarmPushData
```

触发告警阈值的换算公式为：

实际调用时间 + 延迟时间（比如 1980ms） > 告警阈值时间（比如 2000ms）

如果实际的调用时间大于 2000ms，则触发告警阈值，并生成一条告警消息。

（5）如果使用 swagger-ui 调用交易服务（trade-server），则可以在钉钉告警群中收到一条告警，如图 18-8 所示。

图 18-8

2. 关闭商品服务，触发告警平台的异常链路告警，并通知到钉钉告警群

下线商品服务，触发告警平台的异常链路告警，并通知到钉钉告警群的具体过程如下：

（1）在 Nacos 的控制台中下线商品服务。

（2）用 swagger-ui 调用交易服务（trade-server），可以在钉钉告警群中收到一条商品服务（good-server）不可用的告警，如图 18-9 所示。

图 18-9

18.4.3 在购买商品时，在支付过程中验证实时告警功能

具体过程如下。

1. 关闭物流服务（logistics-server），触发告警平台的服务实例不可用的告警

图 18-10 所示为支付过程中完整的服务请求链路。如果使用“kill -9”命令关闭物流服务（模拟线上服务实例不可用的技术故障场景），则告警平台会生成一条实例不可用的告警，并通知到钉钉告警群。图 18-11 所示为触发的钉钉告警。

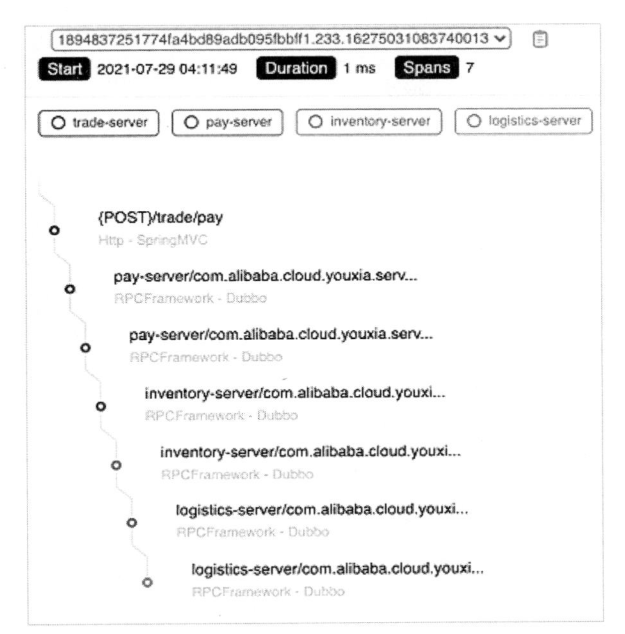

图 18-10

图 18-11

2. 关闭订单服务（order-server），触发告警平台的 RocketMQ 消息堆积的告警

在电商微服务中，在物流服务（logistics-server）完成发货后，需要使用 RocketMQ 通知一条商品已经发货成功的消息给订单服务，消息主题为“sync-order-good-status”。

在使用“kill -9”命令关闭订单服务（order-server）后，消息主题对应的消息会大量堆积，在触发告警阈值（大于消息堆积量 100）后就会产生一条钉钉告警，并通知到钉钉告警群。图 18-12 所示为触发的钉钉告警。

图 18-12